McGraw-Hill

Dictionary of Earth Science

Second Edition

McGraw-Hill

New York Chicago San Francisco Lisbon London Madrid
Mexico City Milan New Delhi San Juan Seoul Singapore
Sydney Toronto

The McGraw-Hill Companies

2 3 4 5 6 7 8 9 0 DOC/DOC 0 9 8 7 6 5 4 3

ISBN 0-07-141045-7

This book is printed on recycled, acid-free paper containing a minimum of 50% recycled, de-inked fiber.

This book was set in Helvetica Bold and Novarese Book by the Clarinda Company, Clarinda, Iowa. It was printed and bound by RR Donnelley, The Lakeside Press.

McGraw-Hill books are available at special quantity discounts to use as premiums and sales promotions, or for use in corporate training programs. For more information, please write to the Director of Special Sales, Professional Publishing, McGraw-Hill, Two Penn Plaza, New York, NY 10121-2298. Or contact your local bookstore.

Library of Congress Cataloging-in-Publication Data

McGraw-Hill dictionary of earth science — 2nd. ed.
p. cm.
"All text in this dictionary was published previously in the McGraw-Hill dictionary of scientific and technical terms, sixth edition, c2003..." — T.p. verso.
ISBN 0-07-141045-7 (alk. paper)
1. Earth sciences—Dictionaries. I. Title: Dictionary of earth science. II. Title: Earth Science. III. McGraw-Hill dictionary of scientific and technical terms. 6th ed.

QE5.M364 2002
550'.3—dc21 2002033188

Contents

Preface v
Staff vi
How to Use the Dictionary vii
Fields and Their Scope ix
Pronunciation Key x
A-Z Terms 1-448
Appendix 449-468
Equivalents of commonly used units for the U.S. Customary System and the metric system 451
Conversion factors for the U.S. Customary System, metric system, and International System 452-455
Geologic column and scale of time 456
Some historical volcanic eruptions 457
Principal regions of a standard earth model 458
Physical properties of some common rocks 458
Approximate concentration of ore elements in earth's crust and in ores 459
Soil orders 459
Elemental composition of earth's crust based on igneous and sedimentary rock 460
World's estimated water supply 460
Cloud classification based on air motion and associated physical characteristics 461
Simplified classification of major igneous rocks on the basis of composition and texture 462
Average chemical composition of igneous rocks (totals reduced to 100%) 463
Dimensions of some major lakes 464
Characteristics of some of the world's major rivers 465
The 100 highest mountain peaks 466

Preface

The *McGraw-Hill Dictionary of Earth Science* provides a compendium of more than 10,000 terms that are central to the broad range of disciplines comprising earth science. The coverage in this Second Edition is focused on the areas of climatology, geochemistry, geodesy, geography, geology, geophysics, hydrology, meteorology, and oceanography, with new terms added and others revised as necessary.

Earth science strives to understand the origins, evolution, and behavior of the earth in a broad context, including the place of the earth in the solar system and the universe. Much of the advances in earth science have resulted from the greatly improved ability to measure and analyze the complex interactions over time of the component parts of the earth, including the atmosphere, the biosphere, the hydrosphere, and the lithosphere. Thus, earth science is highly interdisciplinary, and an understanding of the terminology of the fields covered in this Dictionary is important for an appreciation of its literature and applications.

All of the definitions are drawn from the *McGraw-Hill Dictionary of Scientific and Technical Terms*, Sixth Edition (2003). Each definition is classified according to the field with which it is primarily associated. The pronunciation of each term is provided along with synonyms, acronyms, and abbreviations where appropriate. A guide to the use of the Dictionary on pages vii-viii explains the alphabetical organization of terms, the format of the book, cross referencing, and how synonyms, variant spelling, abbreviations, and similar information are handled. The Pronunciation Key is provided on page x. The Appendix provides conversion tables for commonly used scientific units as well as a revised geologic time scale, periodic table, historical information, and useful listings of data from the varioius disclriplines of earth science.

It is the editors' hope that the Second Edition of the *McGraw-Hill Dictionary of Earth Science* will serve the needs of scientists, engineers, students, teachers, librarians, and writers for high-quality information, and that it will contribute to scientific literacy and communication.

Mark D. Licker
Publisher

Staff

Mark D. Licker, Publisher—Science

Elizabeth Geller, Managing Editor
Jonathan Weil, Senior Staff Editor
David Blumel, Staff Editor
Alyssa Rappaport, Staff Editor
Charles Wagner, Digital Content Manager
Renee Taylor, Editorial Assistant

Roger Kasunic, Vice President—Editing, Design, and Production

Joe Faulk, Editing Manager
Frank Kotowski, Jr., Senior Editing Supervisor

Ron Lane, Art Director

Thomas G. Kowalczyk, Production Manager
Pamela A. Pelton, Senior Production Supervisor

Henry F. Beechhold, Pronunciation Editor
Professor Emeritus of English
Former Chairman, Linguistics Program
The College of New Jersey
Trenton, New Jersey

How to Use the Dictionary

ALPHABETIZATION. The terms in the *McGraw-Hill Dictionary of Earth Science*, Second Edition, are alphabetized on a letter-by-letter basis; word spacing, hyphen, comma, solidus, and apostrophe in a term are ignored in the sequencing. For example, an ordering of terms would be:

aircraft icing	**ARFOR**
air discharge	**Argid**
air-mass analysis	**arid climate**

FORMAT. The basic format for a defining entry provides the term in boldface, the field is small capitals, and the single definition in lightface:

term [FIELD] Definition

A field may be followed by multiple definitions, each introduced by a boldface number:

term [FIELD] **1.** Definition. **2.** Definition. **3.** Definition.

A term may have definitions in two or more fields:

term [CLIMATOL] Definition. [GEOL] Definition.

A simple cross-reference entry appears as:

term *See* another term.

A cross reference may also appear in combination with definitions:

term [CLIMATOL] Definition. [GEOL] *See* another term.

CROSS REFERENCING. A cross-reference entry directs the user to the defining entry. For example, the user looking up "Antarctic vortex" finds:

Antarctic vortex *See* polar vortex.

The user then turns to the "P" terms for the definition. Cross references are also made from variant spellings, acronyms, abbreviations, and symbols.

abs *See* absolute.
bahada *See* bajada.
Ci *See* cirrus cloud.
DDA value *See* depth-duration-area value.

ALSO KNOWN AS . . . , etc. A definition may conclude with a mention of a synonym of the term, a variant spelling, an abbreviation for the term, or other such information, introduced by "Also known as . . . ," "Also spelled . . . ," "Abbreviated . . . ," "Symbolized . . . ," "Derived from . . . ." When a term has more than one definition, the positioning of any of these phrases conveys the extent of applicability. For example:

term [CLIMATOL] **1.** Definition. Also known as synonym. **2.** Definition. Symbolized T.

In the above arrangement, "Also known as . . ." applies only to the first definition; "Symbolized . . ." applies only to the second definition.

term [CLIMATOL] **1.** Definition. **2.** Definition. [GEOL] Definition. Also known as synonym.

In the above arrangement, "Also known as . . ." applies only to the second field.

term [CLIMATOL] Also known as synonym. **1.** Definition. **2.** Definition. [GEOL] Definition.

In the above arrangement, "Also known as . . ." applies to both definitions in the first field.

term Also known as synonym. [CLIMATOL] **1.** Definition. **2.** Definition. [GEOL] Definition.

In the above arrangement, "Also known as . . ." applies to all definitions in both fields.

Fields and Their Scope

[CLIMATOL] **climatology**—That branch of meteorology concerned with the mean physical state of the atmosphere together with its statistical variations in both space and time as reflected in the weather behavior over a period of many years.

[GEOCHEM] **geochemistry**—The field that encompasses the investigation of the chemical composition of the earth, other planets, and the solar system and universe as a whole, as well as the chemical processes that occur within them.

[GEOD] **geodesy**—The subdivision of geophysics which includes determinations of the size and shape of the earth, the earth's gravitational field, and the location of point fixed to the earth's crust in an earth-referred coordinate system.

[GEOGR] **geography**—The science that deals with the description of land, sea, and air and the distribution of plant and animal life, including humans.

[GEOL] **geology**—The study or science of earth, its history, and its life as recorded in the rocks; includes the study of the geologic features of an area, such as the geometry of rock formations, weathering and erosion, and sedimentation.

[GEOPHYS] **geophysics**—The branch of geology in which the principles and practices of physics are used to study the earth and its environment, that is, earth, air, and (by extension) space.

[HYD] **hydrology**—The science dealing with all aspects of the waters on earth, including their occurrence, circulation, and distribution; their chemical and physical properties; and their reaction with the environment, including their relation to living things.

[METEOROL] **meteorology**—The science concerned primarily with the observation of the atmosphere and its phenomena, including temperature, density, winds, clouds, and precipitation.

[OCEANOGR] **oceanography**—The science of the sea, including physical oceanography (the study of the physical properties of seawater and its motion in waves, tides, and currents), marine chemistry, marine geology, and marine biology.

Pronunciation Key

Vowels

a as in b**a**t, th**a**t
ā as in b**ai**t, cr**a**te
ä as in b**o**ther, f**a**ther
e as in b**e**t, n**e**t
ē as in b**ee**t, tr**ea**t
i as in b**i**t, sk**i**t
ī as in b**i**te, l**igh**t
ō as in b**oa**t, n**o**te
ȯ as in b**ough**t, t**au**t
u̇ as in b**oo**k, p**u**ll
ü as in b**oo**t, p**oo**l
ə as in b**u**t, sof**a**
au̇ as in cr**ow**d, p**ow**er
ȯi as in b**oi**l, sp**oi**l
yə as in form**u**la, spectac**u**lar
yü as in f**ue**l, m**u**le

Semivowels/Semiconsonants

w as in **w**ind, t**w**in
y as in **y**et, on**ion**

Stress (Accent)

ˈ precedes syllable with primary stress

ˌ precedes syllable with secondary stress

ˈˌ precedes syllable with variable or indeterminate primary/secondary stress

Consonants

b as in **b**i**b**, dri**bb**le
ch as in **ch**arge, stre**tch**
d as in **d**og, ba**d**
f as in **f**ix, sa**f**e
g as in **g**ood, si**g**nal
h as in **h**and, be**h**ind
j as in **j**oint, di**g**it
k as in **c**ast, bri**ck**
k̲ as in Ba**ch** (used rarely)
l as in **l**oud, be**ll**
m as in **m**ild, su**mm**er
n as in **n**ew, de**n**t
n̲ indicates nasalization of preceding vowel
ŋ as in ri**ng**, si**ng**le
p as in **p**ier, sli**p**
r as in **r**ed, sca**r**
s as in **s**ign, po**s**t
sh as in **su**gar, **sh**oe
t as in **t**imid, ca**t**
th as in **th**in, brea**th**
t̲h̲ as in **th**en, brea**th**e
v as in **v**eil, wea**v**e
z as in **z**oo, crui**s**e
zh as in bei**g**e, trea**s**ure

Syllabication

· Indicates syllable boundary when following syllable is unstressed

aa channel [GEOL] A narrow, sinuous channel in which a lava river moves down and away from a central vent to feed an aa lava flow. { ä'ä 'chan·əl }

aa lava *See* block lava. { ä'ä 'lä·və }

Aalenian [GEOL] Lowermost Middle or uppermost Lower Jurassic geologic time. { ȯ'lēn·ē,ən }

a axis [GEOL] The direction of movement or transport in a tectonite. { 'ā 'ak,sis }

abandoned channel *See* oxbow. { ə'ban·dənd 'chan·əl }

ABC system [GEOD] *See* airborne control system. [GEOPHYS] A procedure in seismic surveying to determine the effect of irregular weathering thickness. { 'ā'bē'sē 'sis·təm }

abioseston [OCEANOGR] A general term for dead organic matter floating in ocean water. { ¦ā,bī·ō'ses·tən }

ablation [GEOL] The wearing away of rocks, as by erosion or weathering. [HYD] The reduction in volume of a glacier due to melting and evaporation. { ə'blā·shən }

ablation area [HYD] The section in a glacier or snowfield where ablation exceeds accumulation. { ə'blā·shən 'er·ē·ə }

ablation cone [HYD] A debris-covered cone of ice, firn, or snow formed by differential ablation. { ə'blā·shən kōn }

ablation factor [HYD] The rate at which a snow or ice surface wastes away. { ə'blā·shən 'fak·tər }

ablation form [HYD] A feature on a snow or ice surface caused by melting or evaporation. { ə'blā·shən fȯrm }

ablation moraine [GEOL] **1.** A layer of rock particles overlying ice in the ablation of a glacier. **2.** Drift deposited from a superglacial position through the melting of underlying stagnant ice. { ə'blā·shən mə'rān }

abnormal anticlinorium [GEOL] An anticlinorium with axial planes of subsidiary folds diverging upward. { ab'nȯr·məl ¦an·tə·kli'nȯ·rē·əm }

abnormal fold [GEOL] An anticlinorium in which there is an upward convergence of the axial surfaces of the subsidiary folds. { ab'nȯr·məl 'fōld }

abnormal magnetic variation [GEOPHYS] The anomalous value in magnetic compass readings made in some local areas containing unknown sources that deflect the compass needle from the magnetic meridian. { ab'nȯr·məl mag'ned·ik ve·rē'ā·shən }

abnormal synclinorium [GEOL] A synclinorium with axial planes of subsidiary folds converging downward. { ab'nȯr·məl ¦sin·kli'nȯ·rē·əm }

a-b plane [GEOL] The surface along which differential movement takes place. { ā¦bē ,plān }

abrade [GEOL] To wear away by abrasion or friction. { ə'brād }

Abraham's tree [METEOROL] The popular name given to a form of cirrus radiatus clouds, consisting of an assemblage of long feathers and plumes of cirrus that seems to radiate from a single point on the horizon. { 'ā·brə,hamz 'trē }

abrasion [GEOL] Wearing away of sedimentary rock chiefly by currents of water laden with sand and other rock debris and by glaciers. { ə'brā·zhən }

abrasion platform [GEOL] An uplifted marine peneplain or plain, according to the smoothness of the surface produced by wave erosion, which is of large area. { ə'brā·zhən 'plat·fȯrm }

abrasive [GEOL] A small, hard, sharp-cornered rock fragment, used by natural agents in abrading rock material or land surfaces. Also known as abrasive ground. { ə'brās·əv }

abrasive ground *See* abrasive. { ə'brās·əv 'graůnd }

abs *See* absolute.

absolute [METEOROL] Referring to the highest or lowest recorded value of a meteorological element, whether at a single station or over an area, during a given period. Abbreviated abs. { ˌab·sə'lüt }

absolute age [GEOL] The geologic age of a fossil, or a geologic event or structure expressed in units of time, usually years. Also known as actual age. { 'ab·səˌlüt 'āj }

absolute drought [METEOROL] In the United Kingdom, a period of at least 15 consecutive days during which no measurable daily precipitation has fallen. { 'ab·səˌlüt ˌdraůt }

absolute geopotential topography *See* geopotential topography. { 'ab·səˌlüt jē·ō·pə'ten·shəl tə'päg·rə·fē }

absolute instability [METEOROL] The state of a column of air in the atmosphere when it has a superadiabatic lapse rate of temperature, that is, greater than the dry-adiabatic lapse rate. Also known as autoconvective instability; mechanical instability. { 'ab·səˌlüt ˌin·stə'bil·ə·dē }

absolute isohypse [METEOROL] A line that has the properties of both constant pressure and constant height above mean sea level. { 'ab·səˌlüt 'ī·sōˌhīps }

absolute linear momentum *See* absolute momentum. { 'ab·səˌlüt 'lin·ē·ər mə'ment·əm }

absolute momentum [METEOROL] The sum of the (vector) momentum of a particle relative to the earth and the (vector) momentum of the particle due to the earth's rotation. Also known as absolute linear momentum. { 'ab·səˌlüt mə'ment·əm }

absolute stability [METEOROL] The state of a column of air in the atmosphere when its lapse rate of temperature is less than the saturation-adiabatic lapse rate. { 'ab·səˌlüt stə'bil·ə·dē }

absolute time [GEOL] Geologic time measured in years, as determined by radioactive decay of elements. { 'ab·səˌlüt 'tīm }

absorption [HYD] Entrance of surface water into the lithosphere. { əb'sȯrp·shən }

abstraction [HYD] **1.** The draining of water from a stream by another having more rapid corroding action. **2.** The part of precipitation that does not become direct runoff. { ab'strak·shən }

abundance [GEOCHEM] The relative amount of a given element among other elements. { ə'bən·dəns }

abyssal [GEOL] *See* plutonic. [OCEANOGR] Pertaining to the abyssal zone. { ə'bis·əl }

abyssal-benthic [OCEANOGR] Pertaining to the bottom of the abyssal zone. { ə'bis·əl 'ben·thik }

abyssal cave *See* submarine fan. { ə'bis·əl 'kāv }

abyssal fan *See* submarine fan. { ə'bis·əl 'fan }

abyssal floor [GEOL] The ocean floor, or bottom of the abyssal zone. { ə'bis·əl 'flȯr }

abyssal gap [GEOL] A gap in a sill, ridge, or rise that lies between two abyssal plains. { ə'bis·əl 'gap }

abyssal hill [GEOL] A hill 2000 to 3000 feet (600 to 900 meters) high and a few miles wide within the deep ocean. { ə'bis·əl 'hil }

abyssal injection [GEOL] The process of driving magmas, originating at considerable depths, up through deep-seated contraction fissures in the earth's crust. { ə'bis·əl in'jek·shən }

abyssal plain [GEOL] A flat, almost level area occupying the deepest parts of many of the ocean basins. { ə'bis·əl 'plān }

abyssal rock [GEOL] Plutonic, or deep-seated, igneous rocks. { ə'bis·əl 'räk }

abyssal theory [GEOL] A theory of the origin of ores involving the separation of ore silicates from the liquid stage during the cooling of the earth. { ə'bis·əl 'thē·ə·rē }

abyssal zone [OCEANOGR] The biogeographic realm of the great depths of the ocean beyond the limits of the continental shelf, generally below 1000 meters. { ə'bis·əl 'zōn }

abyssolith [GEOL] A molten mass of eruptive material passing up without a break from the zone of permanently molten rock within the earth. { ə'bis·ō,lith }

abyssopelagic [OCEANOGR] Pertaining to the open waters of the abyssal zone. { ə'bis·ō·pə'la·jik }

Ac *See* altocumulus cloud.

Acadian orogeny [GEOL] The period of formation accompanied by igneous intrusion that took place during the Middle and Late Devonian in the Appalachian Mountains. { ə'kād·ē·ən ȯr'äj·ə·nē }

accelerated erosion [GEOL] Soil erosion that occurs more rapidly than soil horizons can form from the parent regolith. { ak'sel·ər,ā·dəd i'rō·zhən }

acceptable risk [GEOPHYS] In seismology, that level of earthquake effects which is judged to be of sufficiently low social and economic consequence, and which is useful for determining design requirements in structures or for taking certain actions. { ak¦sep·tə·bəl 'risk }

accessory cloud [METEOROL] A cloud form that is dependent, for its formation and continuation, upon the existence of one of the major cloud genera; may be an appendage of the parent cloud or an immediately adjacent cloudy mass. { ak'ses·ə·rē ,klaůd }

accessory ejecta [GEOL] Pyroclastic material formed from solidified volcanic rocks that are from the same volcano as the ejecta. { ak'ses·ə·rē i'jek·tə }

accessory element *See* trace element. { ak'ses·ə·rē 'el·ə·mənt }

accident [HYD] An interruption in a river that interferes with, or sometimes stops, the normal development of the river system. { 'ak·sə,dent }

accidental ejecta [GEOL] Pyroclastic rock formed from preexisting nonvolcanic rocks or from volcanic rocks unrelated to the erupting volcano. { ¦ak·sə¦den·təl i'jek·tə }

accident block [GEOL] A solid chip of rock broken off from the subvolcanic basement and ejected from a volcano. { 'ak·sə,dent ,bläk }

acclivity [GEOL] A slope that is ascending from a reference point. { ə'kliv·əd·ē }

accordant [GEOL] Pertaining to topographic features that have nearly the same elevation. { ə'kȯrd·ənt }

accordant fold [GEOL] One of several folds that are similarly oriented. { ə'kȯrd·ənt ,fōld }

accordant drainage [HYD] Flow of surface water that follows the dip of the strata over which it flows. Also known as concordant drainage. { ə¦kȯrd·ənt 'drān·ij }

accordant summit level [GEOL] A hypothetical horizontal plane that can be drawn over a broad region connecting mountain summits of similar elevation. { ə'kȯrd·ənt 'səm·ət ,lev·əl }

accretion [GEOL] **1.** Gradual buildup of land on a shore due to wave action, tides, currents, airborne material, or alluvial deposits. **2.** The process whereby stones or other inorganic masses add to their bulk by adding particles to their surfaces. Also known as aggradation. **3.** *See* accretion tectonics. [METEOROL] The growth of a precipitation particle by the collision of a frozen particle (ice crystal or snowflake) with a supercooled liquid droplet which freezes upon contact. { ə'krē·shən }

accretionary lapilli *See* mud ball. { ə'krē·shən,er·ē lə'pi·lē }

accretionary lava ball [GEOL] A rounded ball of lava that occurs on the surface of an aa lava flow. { ə'krē·shən,er·ē 'lä·və ,bȯl }

accretionary ridge [GEOL] A beach ridge located inland from the modern beach, indicating that the coast has been built seaward. { ə'krē·shən,er·ē ,rij }

accretion tectonics [GEOL] The bringing together, or suturing, of terranes; regarded by many geologists as an important mechanism of continental growth. Also known as accretion. { ə'krē·shən tek'tän·iks }

accretion topography [GEOL] Topographic features built by accumulation of sediment. { ə'krē·shən tä'päg·rə·fē }

accretion vein [GEOL] A type of vein formed by the repeated filling of channels followed by their opening because of the development of fractures in the zone undergoing mineralization. { ə'krē·shən ˌvān }

accretion zone [GEOL] Any beach area undergoing accretion. { ə'krē·shən ˌzōn }

accumulated temperature [METEOROL] A value based on the integrated product of the number of degrees that air temperature rises above a given threshold value and the number of days in the period during which this excess is maintained. { ə'kyü·myəˌlād·əd 'tem·prə·chər }

accumulation [HYD] The quantity of snow or other solid form of water added to a glacier or snowfield, such as by precipitation, wind drift, or avalanches. { ə·kyü·myə'lā·shən }

accumulation area [HYD] The portion of a glacier above the firn line, where the accumulation exceeds ablation. Also known as firn field; zone of accumulation. { ə·kyü·myə'lā·shən 'er·ē·ə }

accumulation zone [GEOL] The area where the bulk of the snow contributing to an avalanche was originally deposited. { ə·kyü·myə'lā·shən ˌzōn }

a-c girdle [GEOL] A girdle of points in a petrofabric diagram that have a tread parallel with the plane of the a and c fabric axes. { 'a'sē 'gərd·əl }

achondrite [GEOL] A stony meteorite that contains no chondrules. { ¦ā'känˌdrīt }

acicular ice [HYD] Fresh-water ice composed of many long crystals and layered hollow tubes of varying shape containing air bubbles. Also known as fibrous ice; satin ice. { ə'sik·yə·lər 'īs }

acid clay [GEOL] A type of clay that gives off hydrogen ions when it dissolves in water. { 'as·əd 'klā }

acidic lava [GEOL] Extruded felsic igneous magma which is rich in silica (SiO_2 content exceeds 65). { ə'sid·ik 'lä·və }

acidity coefficient [GEOCHEM] The ratio of the oxygen content of the bases in a rock to the oxygen content in the silica. Also known as oxygen ratio. { ə'sid·ə·tē ˌkō·ə'fish·ənt }

acid precipitation [METEOROL] Rain or snow with a pH of less than 5.6. { 'as·əd prəˌsip·ə'tā·shən }

acid rain [METEOROL] Precipitation in the form of water drops that incorporates anthropogenic acids and acid materials. { ¦as·əd 'rān }

acid soil [GEOL] A soil with pH less than 7; results from presence of exchangeable hydrogen and aluminum ions. { 'as·əd 'sȯil }

acidulous water [HYD] Mineral water either with dissolved carbonic acid or dissolved sulfur compounds such as sulfates. { ə'sij·ə·ləs 'wȯd·ər }

aclinal [GEOL] Without dip; horizontal. { ¦ā'klīn·əl }

aclinic [GEOPHYS] Referring to a situation where a freely suspended magnetic needle remains in a horizontal position. { a'klin·ik }

aclinic line *See* magnetic equator. { a'klin·ik 'līn }

acre-foot [HYD] The volume of water required to cover 1 acre to a depth of 1 foot, hence 43,560 cubic feet; a convenient unit for measuring irrigation water, runoff volume, and reservoir capacity. { 'ā·kər 'fu̇t }

acre-foot per day [HYD] The United States unit of volume rate of water flow. Abbreviated acre-ft/d. { 'ā·kər 'fu̇t pər 'dā }

acre-ft/d *See* acre-foot per day.

acre-in. *See* acre-inch.

acre-inch [HYD] A unit of volume used in the United States for water flow, equal to 3630 cubic feet. Abbreviated acre-in. { 'ā·kər 'inch }

acre-yield [GEOL] The average amount of oil, gas, or water taken from one acre of a reservoir. { 'ā·kər ¦yēld }

acrobatholithic [GEOL] A stage in batholithic erosion where summits of cupolas and stocks are exposed without any exposure of the surface separating the barren interior of the batholith from the mineralized upper part. { ˌak·rə¦bath·ə¦lith·ik }

acromorph [GEOL] A salt dome. { 'ak·rōˌmȯrf }

acrozone *See* range zone. { 'ak·rōˌzōn }

active front [METEOROL] A front, or portion thereof, which produces appreciable cloudiness and, usually, precipitation. { 'ak·tiv frənt }
active glacier [HYD] A glacier in which some of the ice is flowing. { 'ak·tiv 'glā·shər }
active layer [GEOL] That part of the soil which is within the suprapermafrost layer and which usually freezes in winter and thaws in summer. Also known as frost zone. { 'ak·tiv 'lā·ər }
active margin [GEOL] A continental margin that is characterized by earthquakes, volcanic activity, and orogeny resulting from movement of tectonic plates. { 'ak·təv 'mär·jən }
active permafrost [GEOL] Permanently frozen ground (permafrost) which, after thawing by artificial or unusual natural means, reverts to permafrost under normal climatic conditions. { 'ak·tiv 'pər·mə,fròst }
active volcano [GEOL] A volcano capable of venting lava, pyroclastic material, or gases. { 'ak·tiv ,väl'kā·nō }
activity ratio [GEOL] The ratio of plasticity index to percentage of clay-sized minerals in sediment. { ,ak'tiv·əd·ē ,rā·shō }
actual age *See* absolute age. { 'ak·chə·wəl āj }
actual elevation [METEOROL] The vertical distance above mean sea level of the ground at the meteorological station. { 'ak·chə·wəl ,el·ə'vā·shən }
actualism *See* uniformitarianism. { 'ak·chü·ə,liz·əm }
actual pressure [METEOROL] The atmospheric pressure at the level of the barometer (elevation of ivory point), as obtained from the observed reading after applying the necessary corrections for temperature, gravity, and instrumental errors. { 'ak·chə·wəl 'presh·ər }
actual relative movement *See* slip. { 'ak·chə·wəl 'rel·ə·tiv 'müv·mənt }
acute angle block [GEOL] A fault block in which the strike of strata on the down-dip side meets a diagonal fault at an acute angle. { ə'kyüt ¦aŋ·gəl 'bläk }
adakites [GEOL] Rocks formed from lavas that melted from subducting slabs associated with either volcanic arcs or arc/continent collision zones; they were first described from Adak Island in the Aleutians. { 'a·də,kīts }
adalert [GEOPHYS] An advance alert issued by a regional warning center to give prompt warning of a change in solar activity. { 'ad·ə,lərt }
ader wax *See* ozocerite. { 'äd·ər ,waks }
adfreezing [HYD] The process by which one object adheres to another by the binding action of ice; applied to permafrost studies. { ,ad'frēz·iŋ }
adiabat [METEOROL] The relatively constant rate (5.5°F/100 feet or 10°C/kilometer) at which a mass of air cools as it rises. { 'ad·ē·ə,bat }
adiabatic atmosphere [METEOROL] A model atmosphere characterized by a dry-adiabatic lapse rate throughout its vertical extent. { ¦ad·ē·ə¦bad·ik 'at·mə,sfir }
adiabatic chart *See* Stuve chart. { ¦ad·ē·ə¦bad·ik 'chärt }
adiabatic condensation pressure *See* condensation pressure. { ¦ad·ē·ə¦bad·ik ,kän,den 'sā·shən ,presh·ər }
adiabatic condensation temperature *See* condensation. { ¦ad·ē·ə¦bad·ik ,kän,den'sā·shən 'tem·prə·chər }
adiabatic equilibrium [METEOROL] A vertical distribution of temperature and pressure in an atmosphere in hydrostatic equilibrium such that an air parcel displaced adiabatically will continue to possess the same temperature and pressure as its surroundings, so that no restoring force acts on a parcel displaced vertically. Also known as convective equilibrium. { ¦ad·ē·ə¦bad·ik ,ē·kwə'lib·rē·əm }
adiabatic equivalent temperature *See* equivalent temperature. { ¦ad·ē·ə¦bad·ik i'kwiv·ə·lənt 'tem·prə,chər }
adiabatic lapse rate *See* dry adiabatic lapse rate. { ¦ad·ē·ə¦bad·ik 'laps ,rāt }
adiabatic rate *See* dry adiabatic lapse rate. { ¦ad·ē·ə¦bad·ik 'rāt }
adiabatic saturation pressure *See* condensation pressure. { ¦ad·ē·ə¦bad·ik ,sach·ə'rā·shən ,presh·ər }
adiabatic saturation temperature *See* condensation temperature. { ¦ad·ē·ə¦bad·ik ,sach·ə'rā·shən ,tem·prə·chər }

adinole [GEOL] An argillaceous sediment that has undergone albitization at the margin of a basic intrusion. { 'ad·ən,ōl }

adjacent sea [GEOGR] A sea connected with the oceans but semienclosed by land; examples are the Caribbean Sea and North Polar Sea. { ə'jās·ənt 'sē }

adjusted elevation [GEOD] **1.** The elevation resulting from the application of an adjustment correction to an orthometric elevation. **2.** The elevation resulting from the application of both an orthometric correction and an adjustment correction to a preliminary elevation. { ə'jəs·təd ,el·ə'vā·shən }

adjusted stream [HYD] A stream which flows mostly parallel to the strike and as little as necessary in other courses. { ə'jəs·təd 'strēm }

adjustment [GEOD] **1.** The determination and application of corrections to orthometric differences of elevation or to orthometric elevations to make the elevation of all bench marks consistent and independent of the circuit closures. **2.** The placing of detail or control stations in their positions relative to other detail or control stations. { ə'jəst·mənt }

adlittoral [OCEANOGR] Of, pertaining to, or occurring in shallow waters adjacent to a shore. { ,ad'lid·ə·rəl }

admixture [GEOL] One of the lesser or subordinate grades of sediment. { ¦ad¦miks·chər }

adobe [GEOL] Heavy-textured clay soil found in the southwestern United States and in Mexico. { ə'dō·bē }

adobe flats [GEOL] Broad flats that are floored with sandy clay and have been formed from sheet floods. { ə'dō·bē 'flats }

adolescence [GEOL] Stage in the cycle of erosion following youth and preceding maturity. { ,ad·əl'es·əns }

adolescent coast [GEOL] A type of shoreline characterized by low but nearly continuous sea cliffs. { ,ad·əl'es·ənt ,kōst }

adolescent river [HYD] A river with a graded bed and a well-cut channel that reaches base level at its mouth, its waterfalls and lakes of the youthful stage having been destroyed. { ,ad·əl'es·ənt 'riv·ər }

adolescent stream [HYD] A stream characterized by a well-cut, smoothly graded channel that may reach base level at its mouth. { ,ad·əl'es·ənt 'strēm }

adularization [GEOL] Replacement by or introduction of the mineral adularia. { ə,jül·ə·rə'zā·shən }

advance [GEOL] **1.** A continuing movement of a shoreline toward the sea. **2.** A net movement over a specified period of time of a shoreline toward the sea. [HYD] The forward movement of a glacier. { əd'vans }

advection [METEOROL] The process of transport of an atmospheric property solely by the mass motion of the atmosphere. [OCEANOGR] The process of transport of water, or of an aqueous property, solely by the mass motion of the oceans, most typically via horizontal currents. { ,ad'vek·shən }

advectional inversion [METEOROL] An inverted temperature gradient in the air resulting from a horizontal inflow of colder air into an area. { ad'vek·shən·əl in'vər·zhən }

advection fog [METEOROL] A type of fog caused by the horizontal movement of moist air over a cold surface and the consequent cooling of that air to below its dew point. { ,ad'vek·shən ,fäg }

advective hypothesis [METEOROL] The assumption that local temperature changes are the result only of horizontal or isobaric advection. { ,ad'vek·tiv hī'päth·ə·səs }

advective model [METEOROL] A mathematical or dynamic model of fluid flow which is characterized by the advective hypothesis. { ,ad'vek·tiv 'mäd·əl }

advective thunderstorm [METEOROL] A thunderstorm resulting from static instability produced by advection of relatively colder air at high levels or relatively warmer air at low levels or by a combination of both conditions. { ,ad'vek·tiv 'thən·dər,stȯrm }

adventive cone [GEOL] A volcanic cone that is on the flank of and subsidiary to a larger volcano. Also known as lateral cone; parasitic cone. { ad'ven·tiv 'kōn }

adventive crater [GEOL] A crater opened on the flank of a large volcanic cone. { ad'ven·tiv 'krāt·ər }

aeolian *See* eolian. { ē'ōl·ē·ən }

AERO code [METEOROL] An international code used to encode for transmission, in words five numerical digits long, synoptic weather observations of particular interest to aviation operations. { 'e·rō 'kōd }

aerogeography [GEOGR] The geographic study of earth features by means of aerial observations and aerial photography. { ˌe·rō·jē'äg·rə·fē }

aerogeology [GEOL] The geologic study of earth features by means of aerial observations and aerial photography. { ˌe·rō·jē'äl·ə·jē }

aerography [METEOROL] **1.** The study of the air or atmosphere. **2.** The practice of weather observation, map plotting, and maintaining records. **3.** *See* descriptive meteorology. { e'räg·rə·fē }

aerolite *See* stony meteorite. { 'e·rō,līt }

aerological days [METEOROL] Specified days on which additional upper-air observations are made; an outgrowth of the International Polar Year. { ˌe·rə'lä·jə·kəl 'dāz }

aerological diagram [METEOROL] A diagram of atmospheric thermodynamics plotted from upper-atmospheric soundings; usually contains various reference lines such as isobars and isotherms. { ˌe·rə¦lä·jə·kəl 'dī·əˌgram }

aerology [METEOROL] **1.** Synonym for meteorology, according to official usage in the U.S. Navy until 1957. **2.** The study of the free atmosphere throughout its vertical extent, as distinguished from studies confined to the layer of the atmosphere near the earth's surface. { e'rä·lə·jē }

aeromagnetic surveying [GEOPHYS] The mapping of the magnetic field of the earth through the use of electronic magnetometers suspended from aircraft. { ˌe·rō·mag'ned·ik sər'vā·iŋ }

aeronautical climatology [METEOROL] The application of the data and techniques of climatology to aviation meteorological problems. { e·rə'nȯd·ə·kəl ˌklī·mə'täl·ə·je }

aeronautical meteorology [METEOROL] The study of the effects of weather upon aviation. { e·rə'nȯd·ə·kəl ˌmēd·ē·ə'räl·ə·jē }

aeronomy [GEOPHYS] The study of the atmosphere of the earth or other bodies, particularly in relation to composition, properties, relative motion, and radiation from outer space or other bodies. { e'rän·ə·mē }

aeropause [GEOPHYS] A region of indeterminate limits in the upper atmosphere, considered as a transition region between the denser portion of the atmosphere and interplanetary space. { 'e·rəˌpȯz }

aerosiderite [GEOL] A meteorite composed principally of iron. { ˌe·rō'sīd·əˌrīt }

aerosol [METEOROL] A small droplet or particle suspended in the atmosphere and formed from both natural and anthropogenic sources. { 'e·rəˌsȯl }

aerospace environment [GEOPHYS] **1.** The conditions, influences, and forces that are encountered by vehicles, missiles, and so on in the earth's atmosphere or in space. **2.** External conditions which resemble those of atmosphere and space, and in which a piece of equipment, a living organism, or a system operates. { ¦e·rō¦spās in'vī·rən·mənt }

aerothermodynamic border [GEOPHYS] An altitude of about 100 miles (160 kilometers), above which the atmosphere is so rarefied that the skin of an object moving through it at high speeds generates no significant heat. { ¦e·rōˌthər·mō·dī'nam·ik 'bȯrd·ər }

affine deformation [GEOL] A type of deformation in which very thin layers slip against each other so that each moves equally with respect to its neighbors; generally does not result in folding. { ə'fīn ˌdē·fȯr'mā·shən }

affine strain [GEOPHYS] A strain in the earth that does not differ from place to place. { ə'fīn 'strān }

A frame [OCEANOGR] An A-shaped frame used for outboard suspension of oceanographic gear on a research vessel. { 'ā ˌfrām }

Africa [GEOGR] The second largest continent, with an area of 11,700,000 square miles (30,420,000 square kilometers); bisected midway by the Equator, above and below which it shows symmetry of climate and vegetation zones. { 'af·ri·kə }

African superplume [GEOPHYS] A large, discrete, slowly rising plume of heated material in the earth's mantle, beneath southern Africa, believed by some to contribute to the movement of tectonic plates. { ¦af·ri·kən 'sü·pər,plüm }

afterglow [METEOROL] A broad, high arch of radiance or glow seen occasionally in the western sky above the highest clouds in deepening twilight, caused by the scattering effect of very fine particles of dust suspended in the upper atmosphere. { 'af·tər,glō }

aftershock [GEOPHYS] A small earthquake following a larger earthquake and originating at or near the larger earthquake's epicenter. { 'af·tər,shäk }

Aftonian interglacial [GEOL] Post-Nebraska interglacial geologic time. { ,af'ton·ē·ən ,in·tər'glā·shəl }

agalmatolite [GEOL] A soft, waxy, gray, green, yellow, or brown mineral or stone, such as pinite and steatite; used by the Chinese for carving images. Also known as figure stone; lardite; pagodite. { ,a·gəl'mad·əl,īt }

Agassiz orogeny [GEOL] A phase of diastrophism confined to North America Cordillera occurring at the boundary between the Middle and Late Jurassic. { 'ag·ə·sē ȯ'räj·ə·nē }

Agassiz trawl [OCEANOGR] A dredge consisting of a net attached to an iron frame with a hoop at each end that is used to collect organisms, particularly invertebrates, living on the ocean bottom. { 'ag·ə·sē 'trȯl }

Agassiz Valleys [GEOL] Undersea valleys in the Gulf of Mexico between Cuba and Key West. { 'ag·ə·sē 'val·ēz }

agatized wood *See* silicified wood. { 'ag·ə·tīzd 'wu̇d }

age [GEOL] **1.** Any one of the named epochs in the history of the earth marked by specific phases of physical conditions or organic evolution, such as the Age of Mammals. **2.** One of the smaller subdivisions of the epoch as geologic time, corresponding to the stage or the formation, such as the Lockport Age in the Niagara Epoch. { āj }

aged [GEOL] Of a ground configuration, having been reduced to base level. { 'ā·jəd }

age determination [GEOL] Identification of the geologic age of a biological or geological specimen by using the methods of dendrochronology or radiometric dating. { 'āj di,tər·mə'nā·shən }

aged shore [GEOL] A shore long established at a constant level and adjusted to the waves and currents of the sea. { 'ā·jəd 'shȯr }

age of diurnal inequality [GEOPHYS] The time interval between the maximum semimonthly north or south declination of the moon and the time that the maximum effect of the declination upon the range of tide or speed of the tidal current occurs. Also known as age of diurnal tide; diurnal age. { 'āj əv dī'ərn·əl ,in·ē'kwäl·əd·ē }

age of diurnal tide *See* age of diurnal inequality. { 'āj əv dī'ərn·əl ,tīd }

Age of Fishes [GEOL] An informal designation of the Silurian and Devonian periods of geologic time. { 'āj əv 'fish·əz }

Age of Mammals [GEOL] An informal designation of the Cenozoic era of geologic time. { 'āj əv 'mam·əlz }

Age of Man [GEOL] An informal designation of the Quaternary period of geologic time. { 'āj əv 'man }

age of parallax inequality [GEOPHYS] The time interval between the perigee of the moon and the maximum effect of the parallax (distance of the moon) upon the range of tide or speed of tidal current. Also known as parallax age. { 'āj əv 'par·ə,laks ,in·ē'kwäl·əd·ē }

age of phase inequality [GEOPHYS] The time interval between the new or full moon and the maximum effect of these phases upon the range of tide or speed of tidal current. Also known as age of tide; phase age. { 'āj əv 'fāz ,in·ē'kwäl· əd·ē }

age of tide *See* age of phase inequality. { 'āj əv 'tīd }

ageostrophic wind *See* geostrophic departure. { ə'jē·ə¦sträf·ik 'wind }

age ratio [GEOL] The ratio of the amount of daughter to parent isotope in a mineral being dated radiometrically. { 'āj ,rā·shō }

agglomerate [GEOL] A pyroclastic rock composed of angular rock fragments in a matrix of volcanic ash; typically occurs in volcanic vents. { ə'gläm·ə·rət }

agglomeration [METEOROL] The process in which particles grow by collision with and assimilation of cloud particles or other precipitation particles. Also known as coagulation. { ə,gläm·ə'rā·shən }

agglutinate cone *See* spatter cone. { ə'glüt·ən,āt ,kōn }

aggradation [GEOL] *See* accretion. [HYD] A process of shifting equilibrium of stream deposition, with upbuilding approximately at grade. { ,ag·rə'dā·shən }

aggradation recrystallization [GEOL] Recrystallization resulting in the enlargement of crystals. { ,ag·rə'dā·shən rē,kris·tə·lə'zā·shən }

aggraded valley floor [GEOL] The surface of a flat deposit of alluvium which is thicker than the stream channel's depth and is formed where a stream has aggraded its valley. { ə'grād·əd 'val·ē 'flȯr }

aggraded valley plain *See* alluvial plain. { ə'grād·əd 'val·ē 'plān }

aggregate [GEOL] A collection of soil grains or particles gathered into a mass. { 'ag·rə·gət }

aggregate structure [GEOL] A mass composed of separate small crystals, scales, and grains that, under a microscope, extinguish at different intervals during the rotation of the stage. { 'ag·rə·gət 'strək·chər }

aggressive magma [GEOL] A magma that forces itself into place. { ə'gres·iv 'mag·mə }

aggressive water [HYD] Any of the waters which force their way into place. { ə'gres·iv 'wȯd·ər }

agonic line [GEOPHYS] The imaginary line through all points on the earth's surface at which the magnetic declination is zero; that is, the locus of all points at which magnetic north and true north coincide. { ā'gän·ik līn }

agravic [GEOPHYS] Of or pertaining to a condition of no gravitation. { ,ā'grav·ik }

agricere [GEOL] A waxy or resinous organic coating on soil particles. { 'ag·rə,sir }

agricultural geography [GEOGR] A branch of geography that deals with areas of land cultivation and the effect of such cultivation on the physical landscape. { ¦ag·ri¦kəl·chə·rəl jē'ag·rə·fē }

agricultural geology [GEOL] A branch of geology that deals with the nature and distribution of soils, the occurrence of mineral fertilizers, and the behavior of underground water. { ¦ag·rə¦kəl·chə·rəl jē'äl·ə·jē }

Agulhas Current [OCEANOGR] A fast current flowing in a southwestward direction along the southeastern coast of Africa. { ə'gəl·əs 'kər·ənt }

aiguille [GEOL] The needle-top of the summit of certain glaciated mountains, such as near Mont Blanc. { ,ā'gwēl }

aimless drainage [HYD] Drainage without a well-developed system, as in areas of glacial drift or karst topography. { 'ām·ləs 'drān·ij }

airborne control system [GEOD] A survey system for fourth-order horizontal and vertical control surveys involving electromagnetic distance measurements and horizontal and vertical measurements from two or more known positions to a helicopter hovering over the unknown position. Also known as ABC system. { 'er,bȯrn kən'trōl ,sis·təm }

airborne profile [GEOD] Continuous terrain-profile data produced by an absolute altimeter in an aircraft which is making an altimeter-controlled flight along a prescribed course. { 'er,bȯrn 'prō,fīl }

air composition [METEOROL] The kinds and amounts of the constituent substances of air, the amounts being expressed as percentages of the total volume or mass. { 'er ,käm·pə'zish·ən }

aircraft ceiling [METEOROL] After United States weather observing practice, the ceiling classification applied when the reported ceiling value has been determined by a pilot while in flight within 1.5 nautical miles (2.8 kilometers) of any runway of the airport. { 'er,kraft 'sēl·iŋ }

aircraft electrification [METEOROL] **1.** The accumulation of a net electric charge on the surface of an aircraft. **2.** The separation of electric charge into two concentrations of opposite sign on distinct portions of an aircraft surface. { 'er,kraft i,lek·trə·fə'kā·shən }

aircraft icing [METEOROL] The accumulation of ice on the exposed surfaces of aircraft when flying through supercooled water drops (cloud or precipitation). { 'er,kraft 'īs·iŋ }

aircraft report *See* pilot report. { 'er,kraft ri,pȯrt }

aircraft thermometry [METEOROL] The science of temperature measurement from aircraft. { 'er,kraft thər'mäm·ə·trē }

aircraft weather reconnaissance [METEOROL] The making of detailed weather observations or investigations from aircraft in flight. { 'er,kraft 'weth·ər ri,kän·ə,səns }

air current *See* air-earth conduction current. { 'er ,kər·ənt }

air discharge [GEOPHYS] **1.** A form of lightning discharge, intermediate in character between a cloud discharge and a cloud-to-ground discharge, in which the multibranching lightning channel descending from a cloud base does not reach the ground, but succeeds only in neutralizing the space charge distributed in the subcloud layer. **2.** A type of diffuse electrical discharge occasionally reported as occurring in the region above an active thunderstorm. { 'er 'dis,chärj }

air drainage [METEOROL] General term for gravity-induced, downslope flow of relatively cold air. { 'er 'drān·ij }

air-earth conduction current [GEOPHYS] That part of the air-earth current contributed by the electrical conduction of the atmosphere itself; represented as a downward movement of positive space charge in storm-free regions all over the world. Also known as air current. { ¦er ¦ərth kən'dək·shən ,kər·ənt }

air-earth current [GEOPHYS] The transfer of electric charge from the positively charged atmosphere to the negatively charged earth; made up of the air-earth conduction current, a precipitation current, a convection current, and miscellaneous smaller contributions. { ¦er ¦ərth 'kər·ənt }

air gap *See* wind gap. { 'er ,gap }

airglow [GEOPHYS] The quasi-steady radiant emission from the upper atmosphere over middle and low latitudes, as distinguished from the sporadic emission of auroras which occur over high latitudes. Also known as light-of-the-night-sky; night-sky light; night-sky luminescence; permanent aurora. { 'er,glō }

air heave [GEOL] Deformation of plastic sediments on a tidal flat as a result of the growth of air pockets in them; the growth occurs by accretion of smaller air bubbles oozing through the sediment. { 'er ,hēv }

air hoar [HYD] A hoar growing on objects above the ground or snow. { 'er ,hȯr }

airlight [METEOROL] In determinations of visual range, light from sun and sky which is scattered into the eyes of an observer by atmospheric suspensoids (and, to slight extent, by air molecules) lying in the observer's cone of vision. { 'er,līt }

air mass [METEOROL] An extensive body of the atmosphere which approximates horizontal homogeneity in its weather characteristics, particularly with reference to temperature and moisture distribution. { 'er ,mas }

air-mass analysis [METEOROL] In general, the theory and practice of synoptic surface-chart analysis by the so-called Norwegian methods, which involve the concepts of the polar front and of the broad-scale air masses which it separates. { 'er ,mas ə'nal·ə·səs }

air-mass climatology [CLIMATOL] The representation of the climate of a region by the frequency and characteristics of the air masses under which it lies; basically, a type of synoptic climatology. { 'er ,mas klīm·ə'täl·ə·jē }

air-mass precipitation [METEOROL] Any precipitation that can be attributed only to moisture and temperature distribution within an air mass when that air mass is not, at that location, being influenced by a front or by orographic lifting. { 'er ,mas pri,sip·ə'tā·shən }

air-mass shower [METEOROL] A shower that is produced by local convection within an unstable air mass; the most common type of air-mass precipitation. { 'er ,mas 'shau̇·ər }

air-mass source region [METEOROL] An extensive area of the earth's surface over which bodies of air frequently remain for a sufficient time to acquire characteristic temperature and moisture properties imparted by that surface. { 'er ,mas 'sȯrs ,rē·jən }

air parcel [METEOROL] An imaginary body of air to which may be assigned any or all of the basic dynamic and thermodynamic properties of atmospheric air. { 'er ˌpär·səl }

air pocket [METEOROL] An expression used in the early days of aviation for a downdraft; such downdrafts were thought to be pockets in which there was insufficient air to support the plane. { 'er ˌpäk·ət }

airshed [GEOGR] The geographical area associated with a given air supply. [METEOROL] The air supply in a given region. { 'erˌshed }

air shooting [GEOPHYS] In seismic prospecting, the technique of applying a seismic pulse to the earth by detonating a charge or charges in the air. { 'er ˌshüd·iŋ }

air sounding [METEOROL] The act of measuring atmospheric phenomena or determining atmospheric conditions at altitude, especially by means of apparatus carried by balloons or rockets. { 'er ˌsau̇nd·iŋ }

airspace [METEOROL] **1.** Of or pertaining to both the earth's atmosphere and space. Also known as aerospace. **2.** The portion of the atmosphere above a particular land area, especially a nation or other political subdivision. { 'erˌspās }

air temperature [METEOROL] **1.** The temperature of the atmosphere which represents the average kinetic energy of the molecular motion in a small region and is defined in terms of a standard or calibrated thermometer in thermal equilibrium with the air. **2.** The temperature that the air outside of the aircraft is assumed to have as indicated on a cockpit instrument. { 'er ¦tem·prə·chər }

air turbulence [METEOROL] Highly irregular atmospheric motion characterized by rapid changes in wind speed and direction and by the presence, usually, of up and down currents. { 'er ¦tər·byə·ləns }

air volcano [GEOL] An eruptive opening in the earth from which large volumes of gas emanate, in addition to mud and stones; a variety of mud volcano. { ¦er ˌväl¦kā·nō }

airwave [METEOROL] A wavelike oscillation in the pattern of wind flow aloft, usually with reference to the stronger portion of the westerly current. { 'erˌwāv }

airways code *See* United States airways code. { 'erˌwāz ˌkōd }

airways forecast *See* aviation weather forecast. { 'erˌwāz ˌfȯrˌkast }

airways observation *See* aviation weather observation. { 'erˌwāz ˌäb·zər'vā·shən }

Airy isostasy [GEOPHYS] A theory of hydrostatic equilibrium of the earth's surface which contends that mountains are floating on a fluid lava of higher density, and that higher mountains have a greater mass and deeper roots. { ¦er·ē i'säs·tə·sē }

Aitken nuclei [METEOROL] The microscopic particles in the atmosphere which serve as condensation nuclei for droplet growth during the rapid adiabatic expansion produced by an Aitken dust counter. { ¦āt·kən 'nü·klēˌī }

aktological [GEOL] Nearshore shallow-water areas, conditions, sediments, or life. { ˌak·tə'läj·ə·kəl }

Alaska Current [OCEANOGR] A current that flows northwestward and westward along the coasts of Canada and Alaska to the Aleutian Islands. { ə'las·kə 'kər·ənt }

albedo neutrons *See* albedo particles. { al'bēˌdō 'nüˌtränz }

albedo particles [GEOPHYS] Neutrons or other particles, such as electrons or protons, which leave the earth's atmosphere, having been produced by nuclear interactions of energetic particles within the atmosphere. Also known as albedo neutrons. { al'bēˌdō 'pärd·ə·kəlz }

Alberta low [METEOROL] A low centered on the eastern slope of the Canadian Rockies in the province of Alberta, Canada. { al'bərt·ə 'lō }

Albian [GEOL] Uppermost Lower Cretaceous geologic time. { 'al·bē·ən }

albic horizon [GEOL] A soil horizon from which clay and free iron oxides have been removed or in which the iron oxides have been segregated. { 'al·bik hə'rīz·ən }

Albionian [GEOL] Lower Silurian geologic time. { ˌal·bē'ōn·ē·ən }

Alboll [GEOL] A suborder of the soil order Mollisol with distinct horizons, wet for some part of the year; occurs mostly on upland flats and in shallow depressions. { 'alˌbȯl }

alcove [GEOL] A large niche formed by a stream in a face of horizontal strata. { 'alˌkōv }

alcove lands [GEOL] Terrain where the mud rocks or sandy clays and shales that compose the hills (badlands) are interstratified by occasional harder beds; the slopes are terraced. { 'alˌkōv ˌlanz }

alee basin [GEOL] A basin formed in the deep sea by turbidity currents aggrading courses where the currents were deflected around a submarine ridge. { ə'lē ˌbās·ən }

aleishtite [GEOL] A bluish or greenish mixture of dickite and other clay minerals. { ə'lē·ishˌtīt }

Aleutian Current [OCEANOGR] A current setting southwestward along the southern coasts of the Aleutian Islands. { ə'lü·shən ˌkər·ənt }

Aleutian low [METEOROL] The low-pressure center located near the Aleutian Islands on mean charts of sea-level pressure; represents one of the main centers of action in the atmospheric circulation of the Northern Hemisphere. { ə'lü·shən ˌlō }

Alexandrian [GEOL] Lower Silurian geologic time. { ˌal·ig'zan·dre·ən }

Alfisol [GEOL] An order of soils with gray to brown surface horizons, a medium-to-high base supply, and horizons of clay accumulation. { 'al·fəˌsōl }

algal [GEOL] Formed from or by algae. { 'al·gəl }

algal biscuit [GEOL] A disk-shaped or spherical mass, up to 20 centimeters in diameter, made up of carbonate that is probably the result of precipitation by algae. { ¦al·gəl ¦bis·kət }

algal coal [GEOL] Coal formed mainly from algal remains. { 'al·gəl ˌkōl }

algal pit [GEOL] An ablation depression that is small and contains algae. { 'al·gəl ˌpit }

algal reef [GEOL] An organic reef which has been formed largely of algal remains and in which algae are or were the main lime-secreting organisms. { 'al·gəl ˌrēf }

algal ridge [GEOL] Elevated margin of a windward coral reef built by actively growing calcareous algae. { 'al·gəl ˌrij }

algal rim [GEOL] Low rim built by actively growing calcareous algae on the lagoonal side of a leeward reef or on the windward side of a patch reef in a lagoon. { 'al·gəl ˌrim }

algal structure [GEOL] A deposit, most frequently calcareous, with banding, irregular concentric structures, crusts, and pseudo-pisolites or pseudo-concretionary forms resulting from organic, colonial secretion and precipitation. { ¦al·gəl ¦strək·chər }

Algoman orogeny [GEOL] Orogenic episode affecting Archean rocks of Canada about 2.4 billion years ago. Also known as Kenoran orogeny. { al'gōm·ən ȯ'räj·ə·nē }

Algonkian *See* Proterozoic.

alkali emission [GEOPHYS] Light emission from free lithium, potassium, and especially sodium in the upper atmosphere. { 'al·kəˌlī i'mish·ən }

alkali flat [GEOL] A level lakelike plain formed by the evaporation of water in a depression and deposition of its fine sediment and dissolved minerals. { 'al·kəˌlī ˌflat }

alkali lake [HYD] A lake with large quantities of dissolved sodium and potassium carbonates as well as sodium chloride. { 'al·kəˌlī 'lāk }

alkaline soil [GEOL] Soil containing soluble salts of magnesium, sodium, or the like, and having a pH value between 7.3 and 8.5. { 'al·kəˌlīn 'sȯil }

alkali soil [GEOL] A soil, with salts injurious to plant life, having a pH value of 8.5 or higher. { 'al·kəˌlī ˌsȯil }

alkenones [GEOL] Long-chain (37–39 carbon atoms) di-, tri-, and tetraunsaturated methyl and ethyl ketones produced by certain phytoplankton (coccolithophorids), which biosynthetically control the degree of unsaturation (number of carbon-carbon double bonds) in response to the water temperature; the survival of this temperature signal in marine sediment sequences provides a temporal record of sea surface temperatures that reflect past climates. { 'al·kəˌnōnz }

Alleghenian [GEOL] Lower Middle Pennsylvanian geologic time. { ¦al·ə¦gān·ē·ən }

Alleghenian orogeny [GEOL] Pennsylvanian and Early Permian orogenic episode which deformed the rocks of the Appalachian Valley and the Ridge and Plateau provinces. { ¦al·ə¦gān·ē·ən ȯ'räj·ə·nē }

Allende meteorite [GEOL] A meteorite that fell in Mexico in 1969 and contains inclusions that have been radiometrically dated at 4.56×10^9 years, the oldest found so far, presumably indicating the time of formation of the first solid bodies in the solar system. { ai¦yen·de 'mēd·ē·əˌrīt }

Allerod oscillation [CLIMATOL] A temporary increase in temperature during the closing stages of the Pleistocene ice age, dated in Europe about 9850–8850 B.C. { 'al·ə,räd ,äs·ə'lā·shən }

allevardite *See* rectorite. { ,al·ə'vär,dīt }

allochem [GEOL] Sediment formed by chemical or biochemical precipitation within a depositional basin; includes intraclasts, oolites, fossils, and pellets. { 'a·lō,kem }

allochthon [GEOL] A rock that was transported a great distance from its original deposition by some tectonic process, generally related to overthrusting, recumbent folding, or gravity sliding. { ə'läk·thən }

allochthonous coal [GEOL] A type of coal arising from accumulations of plant debris moved from their place of growth and deposited elsewhere. { ə'läk·thə·nəs ,kōl }

allochthonous stream [HYD] A stream flowing in a channel that it did not form. { ə'läk·thə·nəs ,strēm }

allogene [GEOL] A mineral or rock that has been moved to the site of deposition. Also known as allothigene; allothogene. { 'a·lə,jēn }

allogenic *See* allothogenic. { ¦a·lə¦jen·ik }

allophane [GEOL] $Al_2O_3 \cdot SiO_2 \cdot nH_2O$ A clay mineral composed of hydrated aluminosilicate gel of variable composition; P_2O_5 may be present in appreciable quantity. { 'a·lə,fān }

allothigene *See* allogene. { ə'läth·ə,jēn }

allothimorph [GEOL] A metamorphic rock constituent which retains its original crystal outlines in the new rock. { ə'läth·ə,mȯrf }

allothogene *See* allogene. { ə'läth·ə,jēn }

allothogenic [GEOL] Formed from preexisting rocks which have been transported from another location. Also known as allogenic. { ə¦läth·ə¦jen·ik }

alluvial [GEOL] **1.** Of a placer, or its associated valuable mineral, formed by the action of running water. **2.** Pertaining to or consisting of alluvium, or deposited by running water. { ə'lüv·ē·əl }

alluvial cone [GEOL] An alluvial fan with steep slopes formed of loose material washed down the slopes of mountains by ephemeral streams and deposited as a conical mass of low slope at the mouth of a gorge. Also known as cone delta; cone of dejection; cone of detritus; debris cone; dry delta; hemicone; wash. { ə'lüv·ē·əl 'kōn }

alluvial dam [GEOL] A sedimentary deposit which is built by an overloaded stream and dams its channel; especially characteristic of distributaries on alluvial fans. { ə'lüv·ē·əl 'dam }

alluvial deposit *See* alluvium. { ə'lüv·ē·əl di'päz·ət }

alluvial fan [GEOL] A fan-shaped deposit formed by a stream either where it issues from a narrow moutain valley onto a plain or broad valley, or where a tributary stream joins a main stream. { ə'lüv·ē·əl 'fan }

alluvial flat [GEOL] A small alluvial plain having a slope of about 5 to 20 feet per mile (1.5 to 6 meters per 1600 meters) and built of fine sandy clay or adobe deposited during flood. { ə'lüv·ē·əl 'flat }

alluvial ore deposit [GEOL] A deposit in which the valuable mineral particles have been transported and left by a stream. { ə'lüv·ē·əl ¦ȯr di¦päz·ət }

alluvial plain [GEOL] A plain formed from the deposition of alluvium usually adjacent to a river that periodically overflows. Also known as aggraded valley plain; river plain; wash plain; waste plain. { ə'lüv·ē·əl 'plān }

alluvial slope [GEOL] A surface of alluvium which slopes down from mountainsides and merges with the plain or broad valley floor. { ə'lüv·ē·əl 'slōp }

alluvial soil [GEOL] A soil deposit developed on floodplain and delta deposits. { ə'lüv·ē·əl 'sȯil }

alluvial terrace [GEOL] A terraced embankment of loose material adjacent to the sides of a river valley. Also known as built terrace; drift terrace; fill terrace; stream-built terrace; wave-built platform; wave-built terrace. { ə'lüv·ē·əl 'ter·əs }

alluvial valley [GEOL] A valley filled with a stream deposit. { ə'lüv·ē·əl 'val·ē }

alluviation [GEOL] The deposition of sediment by a river. { ə,lüv·ē'ā·shən }

alluvion *See* alluvium. { ə'lüv·ē·ən }

alluvium [GEOL] The detrital materials that are eroded, transported, and deposited by streams; an important constituent of shelf deposits. Also known as alluvial deposit; alluvion. { ə'lüv·ē·əm }

alongshore current *See* littoral current. { ə'lȯŋ,shȯr 'kər·ənt }

alpenglow [METEOROL] A reappearance of sunset colors on a mountain summit after the original mountain colors have faded into shadow; also, a similar phenomenon preceding the regular coloration at sunrise. { 'al·pən,glō }

Alpides [GEOL] Great east-west structural belt including the Alps of Europe and the Himalayas and related mountains of Asia; mostly folded in Tertiary times. { 'al·pə,dēz }

alpine [GEOL] Similar to or characteristic of a lofty mountain or mountain system. { 'al,pīn }

alpine glacier [HYD] A glacier lying on or occupying a depression in mountainous terrain. Also known as mountain glacier. { 'al,pīn 'glā·shər }

Alpine orogeny [GEOL] Jurassic through Tertiary orogeny which affected the Alpides. { 'al,pīn ȯ'räj·ə·nē }

alpinotype tectonics [GEOL] Tectonics of the alpine-type geosynclinal mountain belts characterized by deep-seated plastic folding, plutonism, and lateral thrusting. { al'pē·nō,tīp ,tek'tän·iks }

Altaid orogeny [GEOL] Mountain building in Central Europe and Asia that occurred from the late Carboniferous to the Permian. { ¦al,tād ȯ'räj·ə·nē }

altiplanation [GEOL] A phase of solifluction that may be seen as terracelike forms, flattened summits, and passes that are mainly accumulations of loose rock. { ,al·tə·plā'nā·shən }

altiplanation surface [GEOL] A flat area fronted by scarps a few to hundreds of feet in height; the area ranges from several square rods to hundreds of acres. Also known as altiplanation terrace. { ,al·tə·plā'nā·shən ,sər·fəs }

altiplanation terrace *See* altiplanation surface. { ,al·tə·plā'nā·shən ,ter·əs }

altithermal [GEOPHYS] Period of high temperature, particularly the postglacial thermal optimum. Also known as hypsithermal. { ¦al·tə¦thər·məl }

Altithermal [GEOL] A dry postglacial interval centered about 5500 years ago during which temperatures were warmer than at present. { ¦al·tə¦thər·məl }

altithermal soil [GEOL] Soil recording a period of rising or high temperature. { ¦al·tə¦thər·məl 'sȯil }

altocumulus cloud [METEOROL] A principal cloud type, white or gray or both white and gray in color; occurs as a layer or patch with a waved aspect, the elements of which appear as laminae, rounded masses, or rolls; frequently appears at different levels in a given sky. Abbreviated Ac. { ¦al·tō¦kyüm·yə·ləs 'klau̇d }

altostratus cloud [METEOROL] A principal cloud type in the form of a gray or bluish (never white) sheet or layer of striated, fibrous, or uniform appearance; very often totally covers the sky and may cover an area of several thousand square miles; vertical extent may be from several hundred to thousands of meters. Abbreviated As. { ¦al·tō¦strat·əs 'klau̇d }

alum coal [GEOL] Argillaceous brown coal rich in pyrite in which alum is formed on weathering. { 'al·əm ,kōl }

aluminum ore [GEOL] A natural material from which aluminum may be economically extracted. { ə'lüm·ə·nəm 'ȯr }

alunitization [GEOL] Introduction of or replacement by alunite. { ,al·yə·nə·tə'zā·shən }

alyphite [GEOL] Bitumen that yields a high percentage of open-chain aliphatic hydrocarbons upon distillation. { 'al·ə,fīt }

ambient stress field [GEOPHYS] The distribution and numerical value of the stresses present in a rock environment prior to its disturbance by man. Also known as in-place stress field; primary stress field; residual stress field. { 'am·bē·ənt 'stres ,fēld }

amemolite [GEOL] A stalactite with one or more changes in its axis of growth. { ə'mem·ə,līt }

amictic lake [HYD] A lake that is perennially frozen. { ə'mik·tik 'lāk }

amino acid dating [GEOCHEM] Relative or absolute age determination of materials by measuring the degree of racemization of certain amino acids, which generally increases with geologic age. { ə,mē·nō ¦as·əd ¦dā·diŋ }

Ammanian [GEOL] Middle Upper Cretaceous geologic time. { ,ä'man·ē·ən }

amoeboid fold [GEOL] A fold or structure, such as an anticline, having no prevailing trend or definite shape. { ə'mē,bȯid 'fōld }

amoeboid glacier [HYD] A glacier connected with its snowfield for a portion of the year only. { ə'mē,bȯid 'glā·shər }

amorphous frost [HYD] Hoar frost which possesses no apparent simple crystalline structure; opposite of crystalline frost. { ə'mȯr·fəs 'frȯst }

amorphous peat [GEOL] Peat composed of fine grains of organic matter; it is plastic like wet, heavy soil, with all original plant structures destroyed by decomposition of cellulosic matter. { ə'mȯr·fəs 'pēt }

amorphous sky [METEOROL] A sky characterized by an abundance of fractus clouds, usually accompanied by precipitation falling from a higher, overcast cloud layer. { ə'mȯr·fəs 'skī }

amorphous snow [HYD] A type of snow with irregular crystalline structure. { ə'mȯr·fəs 'snō }

amphidromic [OCEANOGR] Of or pertaining to progression of a tide wave or bulge around a point or center of little or no tide. { ¦am·fə¦dräm·ik }

amphimorphic [GEOL] A rock or mineral formed by two geologic processes. { ,am·fə'mȯr·fik }

amphisapropel [GEOL] Cellulosic ooze containing coarse plant debris. { ,am¦fīz·ə¦prō,pel }

amphitheater [GEOGR] A valley or gulch having an oval or circular floor and formed by glacial action. { 'am·fə,thē·ə·tər }

amphoterite [GEOL] A stony meteorite containing bronzite and olivine with some oligoclase and nickel-rich iron. { am'fäd·ə,rīt }

amygdaloid [GEOL] Lava rock containing amygdules. Also known as amygdaloidal lava. { ə'mig·də,lȯid }

amygdaloidal lava *See* amygdaloid. { ə'mig·də,lȯid·əl'läv·ə }

amygdule [GEOL] **1.** A mineral filling formed in vesicles (cavities) of lava flows; it may be chalcedony, opal, calcite, chlorite, or prehnite. **2.** An agate pebble. { ə'mig,dyül }

anabatic wind [METEOROL] An upslope wind; usually applied only when the wind is blowing up a hill or mountain as the result of a local surface heating, and apart from the effects of the larger-scale circulation. { ¦an·ə¦bad·ik 'wind }

anabranch [HYD] A diverging branch of a stream or river that loses itself in sandy soil or rejoins the main flow downstream. { 'an·ə,branch }

anaclinal [GEOL] Having a downward inclination opposite to that of a stratum. { ¦an·ə¦klīn·əl }

anacoustic zone [GEOPHYS] The zone of silence in space, starting at about 100 miles (160 kilometers) altitude, where the distance between air molecules is greater than the wavelength of sound, and sound waves can no longer be propagated. { ¦an·ə¦kü·stik ,zōn }

anaerobic sediment [GEOL] A highly organic sediment formed in the absence or near absence of oxygen in water that is rich in hydrogen sulfide. { ¦an·ə¦rōb·ik 'sed·ə·mənt }

anafront [METEOROL] A front at which the warm air is ascending the frontal surface up to high altitudes. { 'an·ə,frənt }

analcimization [GEOL] The replacement in igneous rock of feldspars or feldspathoids by analcime. { ə¦nal·sə·mə¦zā·shən }

anallobaric center *See* pressure-rise center. { ə¦nal·ə¦bär·ik 'sen·tər }

analog [METEOROL] A past large-scale synoptic weather pattern which resembles a given (usually current) situation in its essential characteristics. { 'an·əl,äg }

analysis [METEOROL] A detailed study in synoptic meteorology of the state of the atmosphere based on actual observations, usually including a separation of the entity into its component patterns and involving the drawing of families of isopleths for various elements. { ə'nal·ə·səs }

analytical geomorphology *See* dynamic geomorphology. { ˌan·əlˈid·ə·kəl ˌjē·ōˌmörˈfäl·ə·jē }

anamigmatism [GEOL] A process of high-temperature, high-pressure remelting of sediment to yield magma. { ˌan·əˈmig·məˌtiz·əm }

anamorphic zone [GEOL] The zone of rock flow, as indicated by reactions that may involve decarbonation, dehydration, and deoxidation; silicates are built up, and the formation of denser minerals and of compact crystalline structure takes place. { ¦an·ə¦mör·fik ˈzōn }

anamorphism [GEOL] A kind of metamorphism at considerable depth in the earth's crust and under great pressure, resulting in the formation of complex minerals from simple ones. { ˌan·əˈmör·fiz·əm }

anaseism [GEOPHYS] Movement of the earth in a direction away from the focus of an earthquake. { ¦an·ə¦sīz·əm }

anastatic water [HYD] That part of the subterranean water in the capillary fringe between the zone of aeration and the zone of saturation in the soil. { ¦an·ə¦stad·ik ˈwöd·ər }

anatexis [GEOL] A high-temperature process of metamorphosis by which plutonic rock in the lowest levels of the crust is melted and regenerated as a magma. { ˌan·əˈtek·səs }

anathermal [GEOL] A period of time between the age of other strata or units of reference in which the temperature is increasing. { ˌan·əˈthər·məl }

anchieutectic [GEOL] A type of magma which is incapable of undergoing further notable main-stage differentiation because its mineral composition is practically in eutectic proportions. { ¦aŋ·kē·yü¦tek·tik }

anchored dune [GEOL] A sand dune stabilized by growth of vegetation. { ˈaŋ·kərd ˈdün }

anchor ice [HYD] Ice formed beneath the surface of water, as in a lake or stream, and attached to the bottom or to submerged objects. Also known as bottom ice; ground ice. { ˈaŋ·kər ˌīs }

anchor station [OCEANOGR] An anchoring site by a research vessel for the purpose of making a set of scientific observations. { ˈaŋ·kər ˌstā·shən }

anchor stone [GEOL] A rock or pebble that has marine plants attached to it. { ˈaŋ·kər ˌstōn }

Andean-type continental margin [GEOL] A continental margin, as along the Pacific coast of South America, where oceanic lithosphere descends beneath an adjacent continent producing andesitic continental margin volcanism. { ˈan·dē·ən ˌtīp ˌkänt·ənˈent·əl ˈmär·jən }

Andept [GEOL] A suborder of the soil order Inceptisol, formed chiefly in volcanic ash or in regoliths with high components of ash. { ¦an¦dept }

Andes glow *See* Andes lightning. { ˈanˌdēz ˌglō }

andesite line [GEOL] The postulated geographic and petrographic boundary between the andesite-dacite-rhyolite rock association of the margin of the Pacific Ocean and the olivine-basalt-trachyte rock association of the Pacific Ocean basin. { ˈan·dəˌzīt ˌlīn }

andesitic glass [GEOL] A natural glass that is chemically equivalent to andesite. { ˈan·dəˌzīt·ik ˌglas }

Andes lightning [GEOPHYS] Electrical coronal discharges observable often as far as several hundred miles away, generally over any of the mountainous areas of the world when under disturbed electrical conditions. Also known as Andes glow. { ˈanˌdēz ˈlīt·niŋ }

andrite [GEOL] A meteorite composed principally of augite with some olivine and troilite. { ˈanˌdrīt }

anemoclast [GEOL] A clastic rock that was fragmented and rounded by wind. { ¦a·nə·mō¦klast }

anemoclastic [GEOL] Referring to rock that was broken by wind erosion and rounded by wind action. { ¦a·nə·mō¦klas·tik }

anemology [METEOROL] Scientific investigation of winds. { ˌan·əˈmäl·ə·jē }

anemometry [METEOROL] The study of measuring and recording the direction and speed (or force) of the wind, including its vertical component. { ˌan·əˈmäm·ə·trē }
Angara Shield [GEOL] A shield area of crystalline rock in Siberia. { ˌäŋ·gəˈrä ˌshēld }
angle of current [HYD] In stream gaging, the angular difference between 90° and the angle made by the current with a measuring section. { ˈaŋ·gəl əv ˈkər·ənt }
angle of dip *See* dip. { ˈaŋ·gəl əv ˈdip }
angle of shear [GEOL] The angle between the planes of maximum shear which is bisected by the axis of greatest compression. { ˈaŋ·gəl əv ˈshēr }
Angoumian [GEOL] Upper middle Upper Cretaceous (Upper Turonian) geologic time. { ˌänˈgüm·ē·ən }
angrite [GEOL] An achondrite stony meteorite composed principally of augite with a little olivine and troilite. { ˈaŋˌgrīt }
anguclast [GEOL] An angular phenoclast. { ˈaŋ·gyüˌklast }
angular spreading [OCEANOGR] The lateral extension of ocean waves as they move out of the wave-generating area as swell. { ˈaŋ·gyə·lər ˈspred·iŋ }
angular-spreading factor [OCEANOGR] The ratio of the actual wave energy present at a point to that which would have been present in the absence of angular spreading. { ˈaŋ·gyə·lər ˈspred·iŋ ˌfak·tər }
angular unconformity [GEOL] An unconformity in which the older strata dip at a different angle (usually steeper) than the younger strata. { ˈaŋ·gyə·lər ˌən·kənˈfórm·əd·ē }
angular wave number [METEOROL] The number of waves of a given wavelength required to encircle the earth at the latitude of the disturbance. Also known as hemispheric wave number. { ˈaŋ·gyə·lər ¦wāv ˈnəm·bər }
Animikean [GEOL] The middle subdivision of Proterozoic geologic time. Also known as Penokean; Upper Huronian. { ə¦nim·ə¦kē·ən }
animikite [GEOL] An ore of silver, composed of a mixture of sulfides, arsenides, and antimonides, and containing nickel and lead; occurs in white or gray granular masses. { əˈnim·əˌkīt }
Anisian [GEOL] Lower Middle Triassic geologic time. { əˈnis·ē·ən }
annual flood [HYD] The highest flow at a point on a stream during any particular calendar year or water year. { ˈan·yə·wəl ˈfləd }
annual inequality [OCEANOGR] Seasonal variation in water level or tidal current speed, more or less periodic, due chiefly to meteorological causes. { ˈan·yə·wəl ˌin·iˈkwäl·əd·ē }
annual layer [GEOL] **1.** A sedimentary layer deposited, or presumed to have been deposited, during the course of a year; for example, a glacial varve. **2.** A dark layer in a stratified salt deposit containing disseminated anhydrite. { ˈan·yə·wəl ˈlā·ər }
annual magnetic change *See* magnetic annual change. { ˈan·yə·wəl ˌmagˈned·ik ˈchānj }
annual magnetic variation *See* magnetic annual variation. { ˈan·yə·wəl ˌmagˈned·ik ver·ēˈā·shən }
annual storage [HYD] The capacity of a reservoir that can handle a watershed's annual runoff but cannot carry over any portion of the water for longer than the year. { ˈan·yə·wəl ˈstór·ij }
annual variation [GEOPHYS] A component in the change with time in the earth's magnetic field at a specified location that has a period of 1 year. { ˈan·yə·wəl ver·ēˈā·shən }
annular drainage pattern [HYD] A ringlike pattern subsequent in origin and associated with maturely dissected dome or basin structures. { ˈan·yə·lər ˈdrān·ij ˌpad·ərn }
anomalous magma [GEOL] Magma formed or obviously changed by assimilation. { əˈnäm·ə·ləs ˈmag·mə }
anomaly [GEOL] A local deviation from the general geological properties of a region. [METEOROL] The deviation of the value of an element (especially temperature) from its mean value over some specified interval. [OCEANOGR] The difference between conditions actually observed at a serial station and those that would have existed had the water all been of a given arbitrary temperature and salinity. { əˈnäm·ə·lē }
anomaly of geopotential difference *See* dynamic-height anomaly. { əˈnäm·ə·lē əv ˌjē·ō·pə¦ten·shəl ˈdif·rəns }

anorogenic [GEOL] Of a feature, forming during tectonic quiescence between orogenic periods, that is, lacking in tectonic disturbance. { ¦a,nȯ·rō¦jen·ik }

anorogenic time [GEOL] Geologic time when no significant deformation of the crust occurred. { ¦a,nȯ·rō¦jen·ik 'tīm }

anorthositization [GEOL] A process of anorthosite formation by replacement or metasomatism. { ə¦nȯr·thə,sid·ə'zā·shən }

anoxic zone [OCEANOGR] An oxygen-depleted region in a marine environment. { a'nak·sik ,zōn }

Antarctica [GEOGR] A continent roughly centered on the South Pole and surrounded by an ocean consisting of the southern parts of the Atlantic, Pacific, and Indian oceans. { ,ant'ärd·ik·ə }

antarctic air [METEOROL] A type of air whose characteristics are developed in an antarctic region. { ,ant'ärd·ik 'er }

antarctic anticyclone [METEOROL] The glacial anticyclone which has been said to overlie the continent of Antarctica; analogous to the Greenland anticyclone. { ,ant'ärd·ik ,ant·i'sī,klōn }

Antarctic Circle [GEOD] The parallel of latitude approximately 66°32′ south of the Equator. { ,ant'ärd·ik 'sər·kəl }

Antarctic Circumpolar Current [OCEANOGR] The ocean current flowing from west to east through all the oceans around the Antarctic Continent. Also known as West Wind Drift. { ,ant'ärd·ik ,sər·kəm'pōl·ər 'kər·ənt }

Antarctic Convergence [OCEANOGR] The oceanic polar front indicating the boundary between the subantarctic and subtropical waters. Also known as Southern Polar Front. { ,ant'ärd·ik kən'vər·jəns }

antarctic front [METEOROL] The semipermanent, semicontinuous front between the antarctic air of the Antarctic continent and the polar air of the southern oceans; generally comparable to the arctic front of the Northern Hemisphere. { ,ant'ärd·ik 'frənt }

Antarctic Intermediate Water [OCEANOGR] A water mass in the Southern Hemisphere, formed at the surface near the Antarctic Convergence between 45° and 55°S; it can be traced in the North Atlantic to about 25°N. { ,ant'ärd·ik in·tər'mēd·ē·ət 'wȯd·ər }

Antarctic Ocean [GEOGR] A circumpolar ocean belt including those portions of the Atlantic, Pacific, and Indian oceans which reach the Antarctic continent and are bounded on the north by the Subtropical Convergence; not recognized as a separate ocean. { ,ant'ärd·ik ,ō·shən }

Antarctic ozone hole [METEOROL] In the spring, the depletion of stratospheric ozone over the Antarctic region, typically south of 55° latitude, the formation of the hole is explained by the activation of chlorine and the catalytic destruction of O_3, it occurs during September, when the polar regions are sunlit but the air is still cold and isolated from midlatitude air by a strong polar vortex. Also known as ozone hole. { ant¦ärt·ik 'ō,zōn ,hōl }

Antarctic vortex *See* polar vortex. { ant¦ärt·ik 'vȯr,teks }

Antarctic Zone [GEOGR] The region between the Antarctic Circle (66°32′S) and the South Pole. { ,ant'ärd·ik ,zōn }

antecedent platform [GEOL] A submarine platform 165 feet (50 meters) or more below sea level from which barrier reefs and atolls are postulated to grow toward the water's surface. { ,ant·ə'sēd·ənt 'plat,fȯrm }

antecedent precipitation index [METEOROL] A weighted summation of daily precipitation amounts; used as an index of soil moisture. { ,ant·ə'sēd·ənt pri,sip·ə'tā·shən 'in,deks }

antecedent stream [HYD] A stream that has retained its early course in spite of geologic changes since its course was assumed. { ,ant·ə'sēd·ənt 'strēm }

antecedent valley [GEOL] A stream valley that existed before uplift, faulting, or folding occurred and which has maintained itself during and after these events. { ,ant·ə'sēd·ənt 'val·ē }

anteconsequent stream [HYD] A stream consequent to the form assumed by the earth's surface as the result of early movement of the earth but antecedent to later movement. { ˌan·tē'kän·sə·kwənt 'strēm }

antediluvial [GEOL] Formerly referred to time or deposits antedating Noah's flood. { ¦an·tē·də¦lüv·ē·əl }

anthodite [GEOL] Gypsum or aragonite growing in clumps of long needle- or hairlike crystals on the roof or wall of a cave. { 'an·thəˌdīt }

anthracitization [GEOCHEM] The natural process by which bituminous coal is transformed into anthracite coal. { ˌan·thrəˌsīd·ə'zā·shən }

anthracoxene [GEOL] A brownish resin that occurs in brown coal; in ether it dissolves into an insoluble portion, anthrocoxenite, and a soluble portion, schlanite. { ˌan·thrə'käkˌsēn }

anthraxolite [GEOL] Anthracite-like asphaltic material occurring in veins in Precambrian slate of Sudbury District, Ontario. { an'thrak·səˌlīt }

anthraxylon [GEOL] The vitreous-appearing components of coal that are derived from the woody tissues of plants. { an'thrak·səˌlän }

anthropogeography *See* human geography. { ¦an·thrə·pō·jē·'äg·rə·fē }

anticenter [GEOL] The point on the surface of the earth that is diametrically opposite the epicenter of an earthquake. Also known as antiepicenter. { ¦an·tē'sent·ər }

anticlinal [GEOL] Folded as in an anticline. { ¦an·tē¦klīn·əl }

anticlinal axis [GEOL] The median line of a folded structure from which the strata dip on either side. { ¦an·tē¦klīn·əl 'ak·səs }

anticlinal bend [GEOL] An upwardly convex flexure of rock strata in which one limb dips gently toward the apex of the strata and the other dips steeply away from it. { ¦an·tē¦klīn·əl 'bend }

anticlinal mountain [GEOL] Ridges formed by a convex flexure of the strata. { ¦an·tē¦klīn·əl 'maůn·tən }

anticlinal theory [GEOL] A theory relating trapped underground oil accumulation to anticlinal structures. { ¦an·tē¦klīn·əl 'thē·ə·rē }

anticlinal trap [GEOL] A formation in the top of an anticline in which petroleum has accumulated. { ¦ant·i¦klīn·əl 'trap }

anticlinal valley [GEOL] A valley that follows an anticlinal axis. { ¦an·tē¦klīn·əl 'val·ē }

anticline [GEOL] A fold in which layered strata are inclined down and away from the axes. { 'an·tiˌklīn }

anticlinorium [GEOL] A series of anticlines and synclines that form a general arch or anticline. { ˌan·tiˌklī'nor·ē·əm }

anticyclogenesis [METEOROL] The process which creates an anticyclone or intensifies an existing one. { ¦an·tē¦sī·klō¦jen·ə·səs }

anticyclolysis [METEOROL] Any weakening of anticyclonic circulation in the atmosphere. { ¦an·tēˌsī¦kläl·ə·səs }

anticyclone [METEOROL] High-pressure atmospheric closed circulation whose relative direction of rotation is clockwise in the Northern Hemisphere, counterclockwise in the Southern Hemisphere, and undefined at the Equator. Also known as high-pressure area. { ˌan·tē'sīˌklōn }

anticyclonic [METEOROL] Referring to a rotation about the local vertical that is clockwise in the Northern Hemisphere, counterclockwise in the Southern Hemisphere, undefined at the Equator. { ¦an·tēˌsī¦klän·ik }

anticyclonic shear [METEOROL] Horizontal wind shear of such a nature that it tends to produce anticyclonic rotation of the individual air particles along the line of flow. { ¦an·tēˌsī¦klän·ik 'shēr }

anticyclonic winds [METEOROL] The winds associated with a high pressure area and constituting part of an anticyclone. { ¦an·tēˌsī¦klän·ik 'winz }

antidip stream [HYD] A stream that flows in a direction opposite to the general dip of the strata. { ˌant·ē'dip ¦strēm }

antidune [GEOL] A temporary form of ripple on a stream bed analogous to a sand dune but migrating upcurrent. { 'an·tēˌdün }

antiestuarine circulation [OCEANOGR] In an estuary, the inflow of low-salinity surface water over a deeper outflowing (seaward), dense, high-salinity water layer. { ˌan·tē¦es·chə·wəˌrēn ˌser·kyü'lā·shən }

antiform [GEOL] An anticline-like structure whose stratigraphic sequence is not known. { 'an·tēˌfȯrm }

Antilles Current [OCEANOGR] A current formed by part of the North Equatorial Current that flows along the northern side of the Greater Antilles. { an'til·ēz ¦kər·ənt }

antiperthite [GEOL] Natural intergrowth of feldspars formed by separation of sodium feldspar (albite) and potassium feldspar (orthoclase) during slow cooling of molten mixtures; the potassium-rich phase is evolved in a plagioclase host, exactly the inverse of perthite. { ˌan·ti'pərˌthīt }

antipodes [GEOD] Diametrically opposite points on the earth. { an'tip·əˌdēz }

antitrades [METEOROL] A deep layer of westerly winds in the troposphere above the surface trade winds of the tropics. { 'an·tēˌtrādz }

antitriptic wind [METEOROL] A wind for which the pressure force exactly balances the viscous force, in which the vertical transfers of momentum predominate. { ¦an·tē¦trip·tik ¦wind }

antitwilight arch [METEOROL] The pink or purplish band of about 3° vertical angular width which lies just above the antisolar point at twilight; it rises with the antisolar point at sunset and sets with the antisolar point at sunrise. { ¦an·tē¦twīˌlīt 'arch }

Antler orogeny [GEOL] Late Devonian and Early Mississippian orogeny in Nevada, resulting in the structural emplacement of eugeosynclinal rocks over microgeosynclinal rocks. { 'ant·lər ȯ'räj·ə·nē }

anvil *See* incus. { 'an·vəl }

anvil cloud [METEOROL] The popular name given to a cumulonimbus capillatus cloud, a thunderhead whose upper portion spreads in the form of an anvil with a fibrous or smooth aspect; it also refers to such an upper portion alone when it persists beyond the parent cloud. { 'an·vəl ˌklau̇d }

Ao horizon [GEOL] That portion of the A horizon of a soil profile which is composed of pure humus. { ¦ā¦ō hə'rīz·ən }

Aoo horizon [GEOL] Uppermost portion of the A horizon of a soil profile which consists of undecomposed vegetable litter. { ¦ā¦ō¦ō hə'rīz·ən }

apex [GEOL] The part of a mineral vein nearest the surface of the earth. { 'āˌpeks }

aphotic zone [OCEANOGR] The deeper part of the ocean where sunlight is absent. { a'fäd·ik ˌzōn }

apob [METEOROL] An observation of pressure, temperature, and relative humidity taken aloft by means of an aerometeorograph; a type of aircraft sounding. { 'āˌpäb }

apogean tidal currents [OCEANOGR] Tidal currents of decreased speed occurring at the time of apogean tides. { ¦ap·ə¦jē·ən ¦tīd·əl ¦kər·əns }

apogean tides [OCEANOGR] Tides of decreased range occurring when the moon is near apogee. { ¦ap·ə¦jē·ən 'tīdz }

Appalachia [GEOL] Proposed borderland along the southeastern side of North America, seaward of the Appalachian geosyncline in Paleozoic time. { ¦ap·ə¦lā·chə }

Appalachian orogeny [GEOL] An obsolete term referring to Late Paleozoic diastrophism beginning perhaps in the Late Devonian and continuing until the end of the Permian; now replaced by Alleghenian orogeny. { ¦ap·ə¦lā·chən ȯ'räj·ə·nē }

apparent cohesion [GEOL] In soil mechanics, the resistance of particles to being pulled apart due to the surface tension of the moisture film surrounding each particle. Also known as film cohesion. { ə'pa·rənt ˌkō'hē·zhən }

apparent dip [GEOL] Dip of a rock layer as it is exposed in any section not at a right angle to the strike. { ə'pa·rənt 'dip }

apparent movement of faults [GEOL] The apparent motion observed to have occurred in any chance section across a fault. { ə'pa·rənt ¦müv·mənt əv ¦fȯlts }

apparent plunge [GEOL] Inclination of a normal projection of lineation in the plane of a vertical cross section. { ə'pa·rənt 'plənj }

apparent precession *See* apparent wander. { ə'pa·rənt pri'sesh·ən }

apparent vertical [GEOPHYS] The direction of the resultant of gravitational and all other accelerations. Also known as dynamic vertical. { ə'pa·rənt 'verd·ə·kəl }

apparent wander [GEOPHYS] Apparent change in the direction of the axis of rotation of a spinning body, such as a gyroscope, due to rotation of the earth. Also known as apparent precession; wander. { ə'pa·rənt 'wän·dər }

apparent water table *See* perched water table. { ə'pa·rənt 'wȯd·ər ˌtā·bəl }

apple coal [GEOL] Easily mined soft coal that breaks into small pieces the size of apples. { 'ap·əl ˌkōl }

Appleton layer *See* F_2 layer. { 'ap·əl·tən ˌlā·ər }

applied climatology [CLIMATOL] The scientific analysis of climatic data in the light of a useful application for an operational purpose. { ə'plīd ˌklīm·ə'täl·ə·jē }

applied meteorology [METEOROL] The application of current weather data, analyses, or forecasts to specific practical problems. { ə'plīd ˌmēd·ē·ə'räl·ə·jē }

apposition beach [GEOL] One of a series of parallel beaches formed on the seaward side of an older beach. { ˌap·ə'zish·ən ˌbēch }

apron *See* outwash plain; ram. { 'ā·prən }

Aptian [GEOL] Lower Cretaceous geologic time, between Barremian and Albian. Also known as Vectian. { 'ap·tē·ən }

aquagene tuff *See* hyaloclastite. { 'ak·wəˌjēn 'təf }

aqualf [GEOL] A suborder of the soil order Alfisol, seasonally wet and marked by gray or mottled colors; occurs in depressions or on wide flats in local landscapes. { 'ak·wəlf }

Aquent [GEOL] A suborder of the soil order Entisol, bluish gray or greenish gray in color; under water until very recent times; located at the margins of oceans, lakes, or seas. { 'ā·kwənt }

aqueous lava [GEOL] Mud lava produced by the mixing of volcanic ash with condensing volcanic vapor or other water. { 'āk·wē·əs 'läv·ə }

Aquept [GEOL] A suborder of the soil order Inceptisol, wet or drained, which lacks silicate clay accumulation in the soil profiles; surface horizon varies in thickness. { 'ak·wəpt }

aquiclude [GEOL] A porous formation that absorbs water slowly but will not transmit it fast enough to furnish an appreciable supply for a well or spring. { 'ak·wəˌklüd }

aquifer [GEOL] A permeable body of rock capable of yielding quantities of groundwater to wells and springs. [HYD] A subsurface zone that yields economically important amounts of water to wells. { 'ak·wə·fər }

aquifuge [GEOL] An impermeable body of rock which contains no interconnected openings or interstices and therefore neither absorbs nor transmits water. { 'ak·wəˌfyüj }

Aquitanian [GEOL] Lower lower Miocene or uppermost Oligocene geologic time. { ˌak·wə'tān·ē·ən }

aquitard [GEOL] A bed of low permeability adjacent to an aquifer; may serve as a storage unit for groundwater, although it does not yield water readily. { 'ak·wəˌtärd }

Aquod [GEOL] A suborder of the soil order Spodosol, with a black or dark brown horizon just below the surface horizon; seasonally wet, it occupies depressions or wide flats from which water cannot escape easily. { 'ak·wəd }

Aquoll [GEOL] A suborder of the soil order Mollisol, with thick surface horizons; formed under wet conditions, it may be under water at times, but is seasonally rather than continually wet. { 'ak·wȯl }

Aquox [GEOL] A suborder of the soil order Oxisol, seasonally wet, found chiefly in shallow depressions; deeper soil profiles are predominantly gray, sometimes mottled, and contain nodules or sheets of iron and aluminum oxides. { 'ak·wəks }

Aquult [GEOL] A suborder of the soil order Ultisol; seasonally wet, it is saturated with water a significant part of the year unless drained; surface horizon of the soil profile is dark and varies in thickness, grading to gray in the deeper portions; it occurs in depressions or on wide upland flats from which water drains very slowly. { 'ak·wəlt }

Arbuckle orogeny [GEOL] Mid-Pennsylvanian episode of diastrophism in the Wichita and Arbuckle Mountains of Oklahoma. { 'är·bək·əl ȯ'räj·ə·nē }

arc [GEOL] A geologic or topographic feature that is repeated along a curved line on the surface of the earth. { ärk }

Archean [GEOL] A term, meaning ancient, which has been applied to the oldest rocks of the Precambrian; as more physical measurements of geologic time are made, the usage is changing; the term Early Precambrian is preferred. { är'kē·ən }

archeomagnetic dating [GEOPHYS] An absolute dating method based on the earth's shifting magnetic poles. When clays and other rock and soil materials are fired to approximately 1300°F (700°C) and allowed to cool in the earth's magnetic field, they retain a weak magnetism which is aligned with the position of the poles at the time of firing. This allows for dating, for example, of when a fire pit was used, based on the reconstruction of pole position for earlier times. { ¦är·kē·ō,mag¦ned·ik 'dā·diŋ }

Archeozoic [GEOL] **1.** The era during which, or during the latter part of which, the oldest system of rocks was made. **2.** The last of three subdivisions of Archean time, when the lowest forms of life probably existed; as more physical measurements of geologic time are made, the usage is changing; it is now considered part of the Early Precambrian. { ¦är·kē·ə¦zō·ik }

archibenthic zone [OCEANOGR] The biogeographic realm of the ocean extending from a depth of about 665 feet to 2625–3610 feet (200 meters to 800–1100 meters). { ¦är·kē¦ben·thik ,zōn }

archibole *See* positive element. { 'är·kē,bōl }

arching [GEOL] The folding of schists, gneisses, or sediments into anticlines. { 'ärch·iŋ }

archipelagic apron [GEOL] A fan-shaped slope around an oceanic island differing from deep-sea fans in having little, if any, sediment cover. { ¦är·kə·pə¦laj·ik 'ā·prən }

archipelago [GEOGR] **1.** A large group of islands. **2.** A sea that has a large group of islands within it. { ,är·kə'pel·ə,gō }

architectonic [GEOL] Of forces that determine structure. { ¦är·kə,tek¦tän·ik }

arc measurement [GEOD] A survey method used to determine the size of the earth. { ¦ärk ¦mezh·ər·mənt }

arc of parallel [GEOD] A part of an astronomic or a geodetic parallel. { ¦ärk əv 'par·ə,lel }

arctic air [METEOROL] An air mass whose characteristics are developed mostly in winter over arctic surfaces of ice and snow. { ¦ärd·ik *or* 'ärk·tik 'er }

arctic anticyclone *See* arctic high. { ¦ärd·ik ,an·tē'sī,klōn }

Arctic Circle [GEOD] The parallel of latitude 66°32′N (often taken as 661/2°N). { ¦ärd·ik 'sər·kəl }

arctic climate *See* polar climate. { ¦ärd·ik 'klī·mət }

arctic desert *See* polar desert. { ¦ärd·ik 'dez·ərt }

arctic front [METEOROL] The semipermanent, semicontinuous front between the deep, cold arctic air and the shallower, basically less cold polar air of northern latitudes. { ¦ärd·ik 'frənt }

arctic haze [METEOROL] A condition of reduced horizontal and slant visibility (but unimpeded vertical visibility) encountered by aircraft in flight (up to more than 30,000 feet, or 9140 meters) over arctic regions. { ¦ärd·ik 'hāz }

arctic high [METEOROL] A weak high that appears on mean charts of sea-level pressure over the Arctic Basin during late spring, summer, and early autumn. Also known as arctic anticyclone; polar anticyclone; polar high. { 'ärd·ik 'hī }

arctic mist [METEOROL] A mist of ice crystals; a very light ice fog. { 'ärd·ik 'mist }

Arctic Ocean [GEOGR] The north polar ocean lying between North America, Greenland, and Asia. { 'ärd·ik 'ō·shən }

Arctic Oscillation [METEOROL] Atmospheric pressure fluctuations (positive and negative phases) between the polar and middle latitudes (above 45° North) that strengthen and weaken the winds circulating counterclockwise from the surface to the lower stratosphere around the Arctic and, as a result, modulate the severity of the winter weather over most Northern Hemisphere middle and high latitudes. Also known as the Northern Hemisphere annular mode. { ¦ärd·ik ,äs·ə'lā·shən }

arctic sea smoke [METEOROL] Steam fog; but often specifically applied to steam fog rising from small areas of open water within sea ice. { 'ärd·ik ¦sē ,smōk }

Arctic Zone [GEOGR] The area north of the Arctic Circle (66°32′N). { ¦ärd·ik ¦zōn }

arcuate delta [GEOL] A bowed or curved delta with the convex margin facing the body of water. Also known as fan-shaped delta. { 'ärk·yə·wət 'del·tə }

arcuation [GEOL] Production of an arc, as in rock flowage where movement proceeded in a fanlike manner. { ˌärk·yə'wā·shən }

arcus [METEOROL] A dense and horizontal roll-shaped accessory cloud, with more or less tattered edges, situated on the lower front part of the main cloud. { 'ar·kəs }

ARDC model atmosphere *See* standard atmosphere. { ¦ā¦ar¦dē¦sē ˌmäd·əl 'at·məˌsfēr }

Ardennian orogeny [GEOL] A short-lived orogeny during the Ludlovian stage of the Silurian period of geologic time. { är'den·ē·ən ȯ'räj·ə·nē }

area forecast [METEOROL] A weather forecast for a specified geographic area; usually applied to a form of aviation weather forecast. Also known as regional forecast. { 'er·ē·ə 'fȯrˌkast }

areal eruption [GEOL] Volcanic eruption resulting from collapse of the roof of a batholith; the volcanic rocks grade into parent plutonic rocks. { 'er·e·əl i'rəp·shən }

areal geology [GEOL] Distribution and form of rocks or geologic units of any relatively large area of the earth's surface. { 'er·e·əl jē'äl·ə·jē }

arenaceous [GEOL] Of sediment or sedimentary rocks that have been derived from sand or that contain sand. Also known as arenarious; psammitic; sabulous. { ¦a·rə¦nāsh·əs }

arenarious *See* arenaceous. { ¦a·rə¦ner·ē·əs }

arenicolite [GEOL] A hole, groove, or other mark in a sedimentary rock, generally sandstone, interpreted as a burrow made by an arenicolous marine worm or a trail of a mollusk or crustacean. { ˌa·rə'nik·əˌlīt }

Arenigian [GEOL] A European stage including Lower Ordovician geologic time (above Tremadocian, below Llanvirnian). Also known as Skiddavian. { ˌa·rə'nij·ē·ən }

Arent [GEOL] A suborder of the soil order Entisol, consisting of soils formerly of other classifications that have been severely disturbed, completely disrupting the sequence of horizons. { 'a·rənt }

arête [GEOL] Narrow, jagged ridge produced by the merging of glacial cirques. Also known as arris; crib; serrate ridge. { a'rāt }

ARFOR [METEOROL] A code word used internationally to indicate an area forecast; usually applied to an aviation weather forecast. { 'ärˌfȯr }

ARFOT [METEOROL] A code word used internationally to indicate an area forecast with units in the English system; usually applied to an aviation weather forecast. { 'ärˌfōt }

Argid [GEOL] A suborder of the soil order Aridisol, well drained, having a characteristically brown or red color and a silicate accumulation below the surface horizon; occupies older land surfaces in deserts. { 'är·jəd }

argillaceous [GEOL] Of rocks or sediments made of or largely composed of clay-size particles or clay minerals. { ˌär·jə'lā·shəs }

argillation [GEOL] Development of clay minerals by weathering of aluminum silicates. { ˌär·jə'lā·shən }

argillic alteration [GEOL] A rock alteration in which certain minerals are converted to minerals of the clay group. { är'jil·ik ˌȯl·tə'rā·shən }

argilliferous [GEOL] Abounding in or producing clay. { ˌär·jə¦lif·ə·rəs }

Argovian [GEOL] Upper Jurassic (lower Lusitanian), a substage of geologic time in Great Britain. { är'gōv·ē·ən }

arid climate [CLIMATOL] Any extremely dry climate. { 'ar·əd 'klī·mət }

arid erosion [GEOL] Erosion or wearing away of rock that occurs in arid regions, due largely to the wind. { 'ar·əd i'rō·zhən }

Aridisol [GEOL] A soil order characterized by pedogenic horizons; low in organic matter and nitrogen and high in calcium, magnesium, and more soluble elements; usually dry. { a'rid·əˌsȯl }

aridity [CLIMATOL] The degree to which a climate lacks effective, life-promoting moisture. { ə'rid·əd·ē }

aridity coefficient [CLIMATOL] A function of precipitation and temperature designed by W. Gorczynski to represent the relative lack of effective moisture (the aridity) of a place. { ə'rid·əd·ē ˌkō·ə'fish·ənt }

aridity index [CLIMATOL] An index of the degree of water deficiency below water need at any given station; a measure of aridity. { ə'rid·əd·ē ˌinˌdeks }

arid zone *See* equatorial dry zone. { 'ar·əd ˌzōn }

Arikareean [GEOL] Lower Miocene geologic time. { əˌrik·ə'rē·ən }

arm [GEOL] A ridge or a spur that extends from a mountain. [OCEANOGR] A long, narrow inlet of water extending from another body of water. { ärm }

ARMET [METEOROL] An international code word used to indicate an area forecast with units in the metric system. { 'ärˌmet }

armored mud ball [GEOL] A large (0.4–20 inches or 1–50 centimeters in diameter) subspherical mass of silt or clay coated with coarse sand and fine gravel. Also known as pudding ball. { 'är·mərd 'məd ˌbȯl }

Armorican orogeny [GEOL] Little-used term, now replaced by Hercynian or Variscan orogeny. { är'mȯr·ə·kən ȯ'räj·ə·nē }

arrested decay [GEOL] A stage in coal formation where biochemical action ceases. { ə'res·təd di'kā }

arris *See* arête. { 'ar·əs }

arrival time [GEOPHYS] In seismological measurements, the time at which a given wave phase is detected by a seismic recorder. { ə'rī·vəl ˌtīm }

arroyo [GEOL] Small, deep gully produced by flash flooding in arid and semiarid regions of the southwestern United State. { ə'rȯi·ō }

arteritic migmatite [GEOL] Injection gneiss supposedly produced by introduction of pegmatite, granite, or aplite into schist parallel to the foliation. { ¦ard·ə¦rid·ik 'mig·məˌtīt }

artesian aquifer [HYD] An aquifer that is bounded above and below by impermeable beds and that contains artesian water. Also known as confined aquifer. { är'tē·zhən 'ak·wə·fər }

artesian basin [HYD] A geologic structural feature or combination of such features in which water is confined under artesian pressure. { är'tē·zhən 'bās·ən }

artesian leakage [HYD] The slow percolation of water from artesian formations into the confining materials of a less permeable, but not strictly impermeable, character. { är'tē·zhən 'lēk·ij }

artesian spring [HYD] A spring whose water issues under artesian pressure, generally through some fissure or other opening in the confining bed that overlies the aquifer. Also known as fissure spring. { är'tē·zhən 'spriŋ }

artesian water [HYD] Groundwater that is under sufficient pressure to rise above the level at which it encounters a well, but which does not necessarily rise to or above the surface of the ground. { är'tē·zhən 'wȯd·ər }

artesian well [HYD] A well in which the water rises above the top of the water-bearing bed. { är'tē·zhən 'wel }

artificial radiation belt [GEOPHYS] High-energy electrons trapped in the earth's magnetic field as a result of high-altitude nuclear explosions. { ¦ärd·ə¦fish·əl ˌrād·ē'ā·shən ˌbelt }

Artinskian [GEOL] A European stage of geologic time including Lower Permian (above Sakmarian, below Kungurian). { är'tin·skē·ən }

As *See* altostratus cloud.

asar *See* esker. { 'a·sər }

aschistic [GEOL] Pertaining to rocks of minor igneous intrusions that have not been differentiated into light and dark portions but that have essentially the same composition as the larger intrusions with which they are associated. { ā'skis·tik }

aseismic [GEOPHYS] Not subject to the occurrence or destructive effects of earthquakes. { ā'sīz·mik }

ash [GEOL] Volcanic dust and particles less than 4 millimeters in diameter. { ash }

Ashby [GEOL] A North American stage of Middle Ordovician geologic time, forming the upper subdivision of Chazyan, and lying above Marmor and below Porterfield. { 'ash·bē }

ash cone [GEOL] A volcanic cone built primarily of unconsolidated ash and generally shaped somewhat like a saucer, with a rim in the form of a wide circle and a broad central depression often nearly at the same elevation as the surrounding country. { 'ash ˌkōn }

ash fall [GEOL] **1.** A fall of airborne volcanic ash from an eruption cloud; characteristic of Vulcanian eruptions. Also known as ash shower. **2.** Volcanic ash resulting from an ash fall and lying on the ground surface. { 'ash ˌfȯl }

ash field [GEOL] A thick, extensive deposit of volcanic ash. Also known as ash plain. { 'ash ˌfēld }

ash flow [GEOL] **1.** An avalanche of volcanic ash, generally a highly heated mixture of volcanic gases and ash, traveling down the flanks of a volcano or along the surface of the ground. Also known as glowing avalanche; incandescent tuff flow. **2.** A deposit of volcanic ash and other debris resulting from such a flow and lying on the surface of the ground. { 'ash ˌflō }

ash fusibility [GEOL] The gradual softening and melting of coal ash that takes place with increase in temperature as a result of the melting of the constituents and chemical reactions. { 'ash ˌfyüz·ə'bil·əd·ē }

Ashgillian [GEOL] A European stage of geologic time in the Upper Orodovician (above Upper Caradocian, below Llandoverian of Silurian). { ash'gil·yən }

ash plain *See* ash field. { 'ash ˌplān }

ash rock [GEOL] The material of arenaceous texture produced by volcanic explosions. { 'ash ˌräk }

ash shower *See* ash fall. { 'ash ˌshaů·ər }

ash viscosity [GEOL] The ratio of shearing stress to velocity gradient of molten ash; indicates the suitability of a coal ash for use in a slag-tap-type boiler furnace. { 'ash vis'käs·əd·ē }

ashy grit [GEOL] **1.** Pyroclastic material of sand and smaller size. **2.** Mixture of ordinary sand and volcanic ash. { 'ash·ē 'grit }

Asia [GEOGR] The largest continent, comprising the major portion of the broad east-west extent of the Northern Hemisphere land masses. { 'āzh·ə }

asiderite *See* stony meteorite. { ə'sīd·əˌrīt }

Aso lava [GEOL] A type of indurated pyroclastic deposit produced during the explosive eruptions that formed the Aso Caldera of Kyushu, Japan. { 'äs·ō 'läv·ə }

aspect [GEOL] **1.** The general appearance of a specific geologic entity or fossil assemblage as considered more or less apart from relations in time and space. **2.** The direction toward which a valley side or slope faces with respect to the compass or rays of the sun. { 'aˌspekt }

aspect angle [GEOL] The angle between the aspect of a slope and the geographic south (Northern Hemisphere) or the geographic north (Southern Hemisphere). { 'aˌspekt ˌaŋ·gəl }

asperity [GEOL] A type of surface roughness appearing along the interface of two faults. { a'sper·ə·dē }

asphaltic sand [GEOL] Deposits of sand grains cemented together with soft, natural asphalt. { a'sfȯlt·ik 'sand }

asphaltite [GEOL] Any of the dark-colored, solid, naturally occurring bitumens that are insoluble in water, but more or less completely soluble in carbon disulfide, benzol, and so on, with melting points between 250 and 600°F (121 and 316°C); examples are gilsonite and grahamite. { a'sfȯlˌtīt }

asphalt rock [GEOL] Natural asphalt-containing sandstone or dolomite. Also known as asphalt stone; bituminous rock; rock asphalt. { 'aˌsfȯlt 'räk }

asphalt stone *See* asphalt rock. { 'aˌsfȯlt 'stōn }

aspite [GEOL] A cratered volcano with the base wide in relation to the height; for example, Mauna Loa. { 'asˌpīt }

assemblage [GEOL] **1.** A group of fossils that, appearing together, characterize a particular stratum. **2.** A group of minerals that compose a rock. { ə'sem·blij }

assimilation [GEOL] Incorporation of solid or fluid material that was originally in the rock wall into a magma. { əˌsim·ə'lā·shən }

associated corpuscular emission [GEOPHYS] The full complement of secondary charged particles associated with the passage of an x-ray or gamma-ray beam through air. { ə'sō·sē,ād·əd ,kȯr'pəs·kyə·lər i'mish·ən }
Astartian *See* Sequanian. { ə'stär·shən }
asthenolith [GEOL] A body of magma locally melted at any time within any solid portion of the earth. { as'then·ə,lith }
asthenosphere [GEOL] That portion of the upper mantle beneath the rigid lithosphere which is plastic enough for rock flowage to occur; extends from a depth of 30–60 miles (50–100 kilometers) to about 240 miles (400 kilometers) and is seismically equivalent to the low velocity zone. { as'then·ə,sfir }
Astian [GEOL] A European stage of geologic time: upper Pliocene, above Plaisancian, below the Pleistocene stage known as Villafranchian, Calabrian, or Günz. { 'as·tē·ən }
astrobleme [GEOL] A circular-shaped depression on the earth's surface produced by the impact of a cosmic body. { 'as·trō,blēm }
astrogeodetic [GEOD] Pertaining to direct measurements of the earth. { ,as·trō,jē·ə'ded·ik }
astrogeodetic datum orientation [GEOD] Adjustment of the ellipsoid of reference for a particular datum so that the sum of the squares of deflections of the vertical at selected points throughout the geodetic network is made as small as possible. { ,as·trō,jē·ə'ded·ik 'dad·əm ,ȯr·ē·ən'tā·shən }
astrogeodetic deflection [GEOD] The angle at a point between the normal to the geoid and the normal to the ellipsoid of an astrogeodetically oriented datum. Also known as relative deflection. { ,as·trō,jē·ə'ded·ik di'flek·shən }
astrogeodetic leveling [GEOD] A concept whereby the astrogeodetic deflections of the vertical are used to determine the separation of the ellipsoid and the geoid in studying the figure of the earth. Also known as astronomical leveling. { ,as·trō·,jē·ə'ded·ik 'lev·əl·iŋ }
astrogeodetic undulations [GEOD] Separations between the geoid and astrogeodetic ellipsoid. { ,as·trō,jē·ə'ded·ik ,ən·jə'lā·shənz }
astrogravimetric leveling [GEOD] A concept whereby a gravimetric map is used for the interpolation of the astrogeodetic deflections of the vertical to determine the separation of the ellipsoid and the geoid in studying the figure of the earth. { ¦as·trō,grav·ə¦me·trik 'lev·əl·iŋ }
astrogravimetric points [GEOD] Astronomical positions corrected for the deflection of the vertical by gravimetric methods. { ¦as·trō,grav·ə¦me·trik 'pȯins }
astronomical azimuth [GEOD] The angle between the astronomical meridian plane of the observer and the plane containing the observed point and the true normal (vertical) of the observer, measured in the plane of the horizon, preferably clockwise from north. { ,as·trə'näm·ə·kəl 'az·ə·məth }
astronomical equator [GEOD] An imaginary line on the surface of the earth connecting points having 0° astronomical latitude. Also known as terrestrial equator. { ,as·trə'näm·ə·kəl i'kwād·ər }
astronomical latitude [GEOD] Angular distance between the direction of gravity (plumb line) and the plane of the celestial equator; applies only to positions on the earth and is reckoned from the astronomical equator. { ,as·trə'näm·ə·kəl 'lad·ə,tüd }
astronomical leveling *See* astrogeodetic leveling. { ,as·trə'näm·ə·kəl 'lev·əl·iŋ }
astronomical longitude [GEOD] The angle between the plane of the reference meridian and the plane of the local celestial meridian. { ,as·trə'näm·ə·kəl 'län·jə,tüd }
astronomical meridian [GEOD] A line on the surface of the earth connecting points having the same astronomical longitude. Also known as terrestrial meridian. { ,as·trə'näm·ə·kəl mə'rid·ē·ən }
astronomical meridian plane [GEOD] A plane that contains the vertical of the observer and is parallel to the instantaneous rotation axis of the earth. { ,as·trə'näm·ə·kəl mə'rid·ē·ən ,plān }
astronomical parallel [GEOD] A line connecting points having the same astronomical latitude. { ,as·trə'näm·ə·kəl 'par·ə,lel }

astronomical position [GEOD] **1.** A point on the earth whose coordinates have been determined as a result of observation of celestial bodies. Also known as astronomical station. **2.** A point on the earth defined in terms of astronomical latitude and longitude. { ˌas·trəˈnäm·ə·kəl pəˈzish·ən }

astronomical refraction [GEOPHYS] The bending of a ray of celestial radiation as it passes through atmospheric layers of increasing density. { ˌas·trəˈnäm·ə·kəl riˈfrak·shən }

astronomical station *See* astronomical position. { ˌas·trəˈnäm·ə·kəl ˈstā·shən }

astronomical surveying [GEOD] The celestial determination of latitude and longitude; separations are calculated by computing distances corresponding to measured angular displacements along the reference spheroid. { ˌas·trəˈnäm·ə·kəl sərˈvā·iŋ }

astronomical tide [OCEANOGR] An equilibrium tide due to attractions of the sun and moon. { ˌas·trəˈnäm·ə·kəl ˈtīd }

Asturian orogeny [GEOL] Mid-Upper Carboniferous diastrophism. { əˈstu̇r·ē·ən ȯˈräj·ə·nē }

asymmetrical bedding [GEOL] An order in which lithologic types or facies follow one another in a circuitous arrangement so that, for example, the sequence of types 1-2-3-1-2-3-1-2-3 indicates asymmetry (while the sequence 1-2-3-2-1-2-3-2-1 indicates symmetrical bedding). { ¦ā·sə¦me·tri·kəl ˈbed·iŋ }

asymmetrical fold [GEOL] A fold in which one limb dips more steeply than the other. { ¦ā·sə¦me·tri·kəl ˈfōld }

asymmetrical laccolith [GEOL] A laccolith in which the beds dip at conspicuously different angles in different sectors. { ¦ā·sə¦me·tri·kəl ˈlak·əˌlith }

asymmetrical ripple mark [GEOL] The normal form of ripple mark, with short downstream slopes and comparatively long, gentle upstream slopes. { ¦ā·sə¦me·tri·kəl ˈrip·əl ˌmärk }

asymmetrical vein [GEOL] A crustified vein of geologic material with unlike layers on each side. { ¦ā·sə¦me·tri·kəl ˈvān }

asymptote of convergence *See* convergence line. { ˈas·əmˌtōt əv kənˈvər·jəns }

asymptotic cone of acceptance [GEOPHYS] The solid angle in the celestial sphere from which particles have to come in order to contribute significantly to the counting rate of a given neutron monitor on the surface of the earth. { āˌsimˈtäd·ik ¦kōn əv ikˈsep·təns }

asymptotic direction of arrival [GEOPHYS] The direction at infinity of a positively charged particle, with given rigidity, which impinges in a given direction at a given point on the surface of the earth, after passing through the geomagnetic field. { āˌsimˈtäd·ik diˈrek·shən əv əˈrīv·əl }

ataxic [GEOL] Pertaining to unstratified ore deposits. { əˈtak·sik }

ataxite [GEOL] An iron meteorite that lacks the structure of either hexahedrite or octahedrite and contains more than 10% nickel. { əˈtakˌsīt }

atectonic [GEOL] Of an event that occurs when orogeny is not taking place. { ¦ā·tekˈtän·ik }

atectonic pluton [GEOL] A pluton that is emplaced when orogeny is not occurring. { ¦ā·tekˈtän·ik ˈplüˌtän }

Atlantic Ocean [GEOGR] The large body of water separating the continents of North and South America from Europe and Africa and extending from the Arctic Ocean to the continent of Antarctica. { ətˈlan·tik ˈō·shən }

Atlantic-type continental margin [GEOL] A continental margin typified by that of the Atlantic which is aseismic because oceanic and continental lithospheres are coupled. { ətˈlan·tik ˌtīp ˌkänt·ənˈent·əl ˈmär·jən }

atmoclast [GEOL] A fragment of rock broken off in place by atmospheric weathering. { ˈat·məˌklast }

atmogenic [GEOL] Of rocks, minerals, and other deposits derived directly from the atmosphere by condensation, wind action, or deposition from volcanic vapors; for example, snow. { ¦at·mə¦jen·ik }

atmolith [GEOL] A rock precipitated from the atmosphere, that is, an atmogenic rock. { ˈat·məˌlith }

atmophile element [METEOROL] **1.** Any of the most typical elements of the atmosphere (hydrogen, carbon, nitrogen, oxygen, iodine, mercury, and inert gases). **2.** Any of the elements which either occur in the uncombined state or, as volatile compounds, concentrate in the gaseous primordial atmosphere. { 'at·mō,fīl 'el·ə·mənt }

atmosphere [METEOROL] The gaseous envelope surrounding a planet or celestial body. { 'at·mə,sfir }

atmospheric absorption [GEOPHYS] The reduction in energy of microwaves by gases and moisture in the atmosphere. { ¦at·mə¦sfir·ik əb'zȯrp·shən }

atmospheric attenuation [GEOPHYS] A process in which the flux density of a parallel beam of energy decreases with increasing distance from the source as a result of absorption or scattering by the atmosphere. { ¦at·mə¦sfir·ik ə,ten·yə'wā·shən }

atmospheric boundary layer *See* surface boundary layer. { ¦at·mə¦sfir·ik 'baün·drē ,lā·ər }

atmospheric cell [METEOROL] An air parcel that exhibits a specific type of motion within its boundaries, such as the vertical circular motion of the Hadley cell. { ¦at·mə,sfir·ik 'sel }

atmospheric chemistry [METEOROL] The study of the production, transport, modification, and removal of atmospheric constituents in the troposphere and stratosphere. { ¦at·mə¦sfir·ik 'kem·ə·strē }

atmospheric composition [METEOROL] The chemical abundance in the earth's atmosphere of its constituents, including nitrogen, oxygen, argon, carbon dioxide, water vapor, ozone, neon, helium, krypton, methane, hydrogen, and nitrous oxide. { ¦at·mə¦sfir·ik ,käm·pə'zish·ən }

atmospheric condensation [METEOROL] The transformation of water in the air from a vapor phase to dew, fog, or cloud. { ¦at·mə¦sfir·ik ,kän·dən'sā·shən }

atmospheric density [METEOROL] The ratio of the mass of a portion of the atmosphere to the volume it occupies. { ¦at·mə¦sfir·ik 'den·səd·ē }

atmospheric diffusion [METEOROL] The exchange of fluid parcels between regions in the atmosphere in the apparently random motions of a scale too small to be treated by equations of motion. { ¦at·mə¦sfir·ik di'fyü·zhən }

atmospheric disturbance [METEOROL] Any agitation or disruption of the atmospheric steady state. { ¦at·mə¦sfir·ik dis'tər·bəns }

atmospheric duct [GEOPHYS] A stratum of the troposphere within which the refractive index varies so as to confine within the limits of the stratum the propagation of an abnormally large proportion of any radiation of sufficiently high frequency, as in a mirage. { ¦at·mə¦sfir·ik 'dəkt }

atmospheric electric field [GEOPHYS] The atmosphere's electric field strength in volts per meter at any specified point in time and space; near the earth's surface, in fair-weather areas, a typical datum is about 100 and the field is directed vertically in such a way as to drive positive charges downward. { ¦at·mə¦sfir·ik i¦lek·trik 'fēld }

atmospheric electricity [GEOPHYS] The electrical processes occurring in the lower atmosphere, including both the intense local electrification accompanying storms and the much weaker fair-weather electrical activity over the entire globe produced by the electrified storms continuously in progress. { ¦at·mə¦sfir·ik i,lek'tris·əd·ē }

atmospheric evaporation [HYD] The exchange of water between the earth's oceans, lakes, rivers, ice, snow, and soil and the atmosphere. { ¦at·mə¦sfir·ik i,vap·ə'rā·shən }

atmospheric gas [METEOROL] One of the constituents of air, which is a gaseous mixture primarily of nitrogen, oxygen, argon, carbon dioxide, water vapor, ozone, neon, helium, krypton, methane, hydrogen, and nitrous oxide. { ¦at·mə¦sfir·ik 'gas }

atmospheric general circulation [METEOROL] The statistical mean global flow pattern of the atmosphere. { ¦at·mə¦sfir·ik ¦jen·rəl sərk·yə'lā·shən }

atmospheric interference [GEOPHYS] Electromagnetic radiation, caused by natural electrical disturbances in the atmosphere, which interferes with radio systems. Also known as atmospherics; sferics; strays. { ¦at·mə¦sfir·ik ,in·tər'fir·əns }

atmospheric ionization [GEOPHYS] The process by which neutral atmospheric molecules or atoms are rendered electrically charged chiefly by collisions with high-energy particles. { ¦at·mə¦sfir·ik ,ī·ə·nə'zā·shən }

atmospheric lapse rate *See* environmental lapse rate. { ¦at·mə¦sfir·ik 'laps ˌrāt }

atmospheric layer *See* atmospheric shell. { ¦at·mə¦sfir·ik 'lā·ər }

atmospheric physics [GEOPHYS] The study of the physical phenomena of the atmosphere. { ¦at·mə¦sfir·ik 'fiz·iks }

atmospheric radiation [GEOPHYS] Infrared radiation emitted by or being propagated through the atmosphere. { ¦at·mə¦sfir·ik ˌrād·ē'ā·shən }

atmospheric refraction [GEOPHYS] **1.** The angular difference between the apparent zenith distance of a celestial body and its true zenith distance, produced by refraction effects as the light from the body penetrates the atmosphere. **2.** Any refraction caused by the atmosphere's normal decrease in density with height. { ¦at·mə¦sfir·ik ri'frak·shən }

atmospheric region *See* atmospheric shell. { ¦at·mə¦sfir·ik 'rē·jən }

atmospherics *See* atmospheric interference. { ¦at·mə¦sfir·iks }

atmospheric scattering [GEOPHYS] A change in the direction of propagation, frequency, or polarization of electromagnetic radiation caused by interaction with the atoms of the atmosphere. { ¦at·mə¦sfir·ik 'skad·ər·iŋ }

atmospheric shell [METEOROL] Any one of a number of strata or layers of the earth's atmosphere; temperature distribution is the most common criterion used for denoting the various shells. Also known as atmospheric layer; atmospheric region. { ¦at·mə¦sfir·ik 'shel }

atmospheric sounding [METEOROL] A measurement of atmospheric conditions aloft, above the effective range of surface weather observations. { ¦at·mə¦sfir·ik 'sau̇nd·iŋ }

atmospheric structure [METEOROL] Atmospheric characteristics, including wind direction and velocity, altitude, air density, and velocity of sound. { ¦at·mə¦sfir·ik 'strək·chər }

atmospheric suspensoids [METEOROL] Moderately finely divided particles suspended in the atmosphere; dust is an example. { ¦at·mə¦sfir·ik sə'spenˌsoidz }

atmospheric tide [GEOPHYS] Periodic global motions of the earth's atmosphere, produced by gravitational action of the sun and moon; amplitudes are minute except in the upper atmosphere. { ¦at·mə¦sfir·ik 'tīd }

atmospheric turbulence [METEOROL] Apparently random fluctuations of the atmosphere that often constitute major deformations of its state of fluid flow. { ¦at·mə¦sfir·ik 'tər·byə·ləns }

Atokan [GEOL] A North American provincial series in lower Middle Pennsylvanian geologic time, above Morrowan, below Desmoinesian. { ə'tō·kən }

atoll [GEOGR] A ring-shaped coral reef that surrounds a lagoon without projecting land area and that is surrounded by open sea. { 'aˌtȯl }

atoll texture [GEOL] The surrounding of a ring of one mineral with another mineral, or minerals, within and without the ring. Also known as core texture. { 'aˌtȯl ˌteks·chər }

attached dune [GEOL] A dune that has formed around a rock or other geological feature in the path of windblown sand. { ə'tacht 'dün }

attached groundwater [HYD] The portion of subsurface water adhering to pore walls in the soil. { ə'tacht 'grau̇ndˌwȯd·ər }

Atterberg scale [GEOL] A geometric and decimal grade scale for classification of particles in sediments based on the unit value of 2 millimeters and involving a fixed ratio of 10 for each successive grade; subdivisions are geometric means of the limits of each grade. { 'at·ərˌbərg ˌskāl }

Attican orogeny [GEOL] Late Miocene diastrophism. { 'ad·ə·kən ȯ'räj·ə·nē }

attitude [GEOL] The position of a structural surface feature in relation to the horizontal. { 'ad·əˌtüd }

attrital coal [GEOL] A bright coal composed of anthraxylon and of attritus in which the translucent cell-wall degradation matter or translucent humic matter predominates, with the ratio of anthraxylon to attritus being less than 1:3. { ə'trīd·əl 'kōl }

attrition [GEOL] The act of wearing and smoothing of rock surfaces by the flow of water charged with sand and gravel, by the passage of sand drifts, or by the movement of glaciers. { ə'trish·ən }

attritus [GEOL] **1.** Visible-to-ultramicroscopic particles of vegetable matter produced by microscopic and other organisms in vegetable deposits, particularly in swamps and bogs. **2.** The dull gray to nearly black, frequently striped portion of material that makes up the bulk of some coals and alternate bands of bright anthraxylon in well-banded coals. { ə'trīd·əs }

aubrite [GEOL] An enstatite achondrite (meteorite) consisting almost wholly of crystalline-granular enstatite (and clinoenstatite) poor in lime and practically free from ferrous oxide, with accessory oligoclase. Also known as bustite. { 'ō,brīt }

augen kohle *See* eye coal. { 'au̇·gən ,kōl·ə }

aulacogen [GEOL] A major fault-bounded trough considered to be one part of a three-rayed fault system on the domes above mantle hot spots; the other two rays open as proto-ocean basins. { ,au̇'läk·ə·jən }

aureole [GEOL] A ring-shaped contact zone surrounding an igneous intrusion. Also known as contact aureole; contact zone; exomorphic zone; metamorphic aureole; metamorphic zone; thermal aureole. [METEOROL] A poorly developed corona in the atmosphere characterized by a bluish-white disk immediately around the luminous celestial body, as around the sun or moon in the fog. { 'ȯr·ē,ōl }

auriferous [GEOL] Of a substance, especially a mineral deposit, bearing gold. { ȯ'rif·ə·rəs }

aurora [GEOPHYS] The most intense of the several lights emitted by the earth's upper atmosphere, seen most often along the outer realms of the Arctic and Antarctic, where it is called the aurora borealis and aurora australis, respectively; excited by charged particles from space. { ə'rȯr·ə }

aurora australis [GEOPHYS] The aurora of southern latitudes. Also known as southern lights. { ə'rȯr·ə ȯ'strā·ləs }

aurora borealis [GEOPHYS] The aurora of northern latitudes. Also known as northern lights. { ə'rȯr·ə ,bȯr·ē'al·əs }

auroral absorption event [GEOPHYS] A large increase in D-region electron density and associated radio-signal absorption, caused by electron-bombardment of the atmosphere during an aurora or a geomagnetic storm. { ə'rȯr·əl əb'zȯrp·shən i'vent }

auroral caps [GEOPHYS] The regions surrounding the auroral poles, lying between the poles and the auroral zones. { ə'rȯr·əl ,kaps }

auroral electrojet [GEOPHYS] An intense electric current in the magnetosphere, flowing along the auroral zones during a polar substorm. { ə'rȯr·əl i'lek·trə,jet }

auroral forms [GEOPHYS] Auroral display types, of which two are basic: ribbonlike bands and cloudlike surfaces. { ə'rȯr·əl 'fȯrmz }

auroral frequency [GEOPHYS] The percentage of nights on which an aurora is seen at a particular place, or on which one would be seen if clouds did not interfere. { ə'rȯr·əl 'frē·kwən·sē }

auroral isochasm [GEOPHYS] A line connecting places of equal auroral frequency, averaged over a number of years. { ə'rȯr·əl 'ī·sō,kaz·əm }

auroral oval [GEOPHYS] An oval-shaped region centered on the earth's magnetic pole in which auroral emissions occur. { ə'rȯr·əl 'ō·vəl }

auroral poles [GEOPHYS] The points on the earth's surface on which the auroral isochasms are centered; coincide approximately with the magnetic-axis poles of the geomagnetic field. { ə'rȯr·əl 'pōlz }

auroral region [GEOPHYS] The region within 30° geomagnetic latitude of each auroral pole. { ə'rȯr·əl ¦rē·jən }

auroral storm [GEOPHYS] A rapid succession of auroral substorms, occurring in a short period, of the order of a day, during a geomagnetic storm. { ə'rȯr·əl ¦stȯrm }

auroral substorm [GEOPHYS] A characteristic sequence of auroral intensifications and movements occurring around midnight, in which a rapid poleward movement of auroral arcs produces a bulge in the auroral oval. { ə'rȯr·əl 'səb,stȯrm }

auroral zone [GEOPHYS] A roughly circular band around either geomagnetic pole within which there is a maximum of auroral activity; lies about 10–15° geomagnetic latitude from the geomagnetic poles. { ə'rȯr·əl ,zōn }

auroral zone blackout [GEOPHYS] Communication fadeout in the auroral zone most often due to an increase of ionization in the lower atmosphere. { ə'rȯr·əl ¦zōn 'blak,au̇t }

aurora polaris [GEOPHYS] A high-altitude aurora borealis or aurora australis. { ə'rȯr·əl pə'lar·əs }

auster *See* ostria. { 'ȯs·tər }

austral [GEOD] Pertaining to south. { 'ȯs·trəl }

austral axis pole [GEOPHYS] The southern intersection of the geomagnetic axis with the earth's surface. { 'ȯs·trəl ¦ak·səs ,pōl }

Australia [GEOGR] An island continent of 2,941,526 square miles (7,618,517 square kilometers), with low elevation and moderate relief, situated in the southern Pacific. { ȯ'strāl·yə }

australite [GEOL] A tektite found in southern Australia, occurring as glass balls and spheroidal dumbbell forms of green and black, similar to obsidian and probably of cosmic origin. { 'ȯs·trə,līt }

Austrian orogeny [GEOL] A short-lived orogeny during the end of the Early Cretaceous. { 'ȯs·trē·ən ȯ'räj·ə·nē }

authigenic [GEOL] Of constituents that came into existence with or after the formation of the rock of which they constitute a part; for example, the primary and secondary minerals of igneous rocks. { ¦ȯ·thə¦jen·ik }

authigenic sediment [GEOL] Sediment occurring in the place where it was originally formed. { ¦ȯ·thə¦jen·ik 'sed·ə·mənt }

autobarotropy [METEOROL] The state of a fluid which is characterized by both barotropy and piezotropy when the coefficients of barotropy and piezotropy are equal. { ¦ȯd·ō·bə'rä·trə·pē }

autobrecciation [GEOL] The process whereby portions of the first consolidated crust of a lava flow are incorporated into the still-fluid portion. { ¦ȯd·ō,brech·ē'ā·shən }

autochthon [GEOL] A succession of rock beds that have been moved comparatively little from their original site of formation, although they may be folded and faulted extensively. { ȯ'täk·thən }

autochthonous [GEOL] Having been formed or occurring in the place where found. { ȯ'täk·thə·nəs }

autochthonous coal [GEOL] Coal believed to have originated from accumulations of plant debris at the place where the plants grew. Also known as indigenous coal. { ȯ'täk·thə·nas 'kōl }

autochthonous sediment [GEOL] A residual soil deposit formed in place through decomposition. { ȯ'täk·thə·nas 'sed·ə·mənt }

autochthonous stream [HYD] A stream flowing in its original channel. { ȯ'täk·thə·nas 'strēm }

autoclastic [GEOL] Of rock, fragmented in place by folding due to orogenic forces when the rock is not so heavily loaded as to render it plastic. { ¦ȯd·ō¦klas·tik }

autoclastic schist [GEOL] Schist formed in place from massive rocks by crushing and squeezing. { ¦ȯd·ō¦klas·tik 'shist }

autoconsequent falls [HYD] Waterfalls in streams carrying a heavy load of calcium carbonate in solution which develop at particular sites along the stream course where warming, evaporation, and other factors cause part of the solution load to be precipitated. { ¦ȯd·ō'kän·sə·kwənt 'fȯlz }

autoconsequent stream [HYD] A stream in the process of building a fan or an alluvial plain, the course of which is guided by the slopes of the alluvium the stream itself has deposited. { ¦ȯd·ō'kän·sə·kwənt 'strēm }

autoconvection [METEOROL] The phenomenon of the spontaneous initiation of convection in an atmospheric layer in which the lapse rate is equal to or greater than the autoconvective lapse rate. { ¦ȯd·ō·kən'vek·shən }

autoconvection gradient *See* autoconvective lapse rate. { ¦ȯd·ō·kən'vek·shən 'grād·ē·ənt }

autoconvective instability *See* absolute instability. { ¦ȯd·ō·kən'vek·tiv ,in·stə'bil·əd·ē }

autoconvective lapse rate [METEOROL] The largely hypothetical environmental lapse rate of temperature in an atmosphere in which the density is constant with height (homogeneous atmosphere); 3°C per 100 meters in dry air. Also known as autoconvection gradient. { ¦ȯd·ō·kən'vek·tiv 'laps ˌrāt }

autogenetic drainage [HYD] A self-established drainage system developed solely by headwater erosion. { ¦ȯd·ō·jə¦ned·ik 'drān·ij }

autogenetic topography [GEOL] Conformation of land due to the physical action of rain and streams. { ¦ȯd·ō·jə¦ned·ik tə'päg·rə·fē }

autogeosyncline [GEOL] A parageosyncline that subsides as an elliptical basin or trough nearly without associated highlands. Also known as intracratonic basin. { ¦ȯd·ō¦jē·ō'sin‚klīn }

autoinjection *See* autointrusion. { ¦ȯd·ō‚in'jek·shən }

autointrusion [GEOL] A process wherein the residual liquid of a differentiating magma is drawn into rifts formed in the crystal mesh at a late stage by deformation of unspecified origin. Also known as autoinjection. { ¦ȯd·ō·in'trü·zhən }

autolysis [GEOCHEM] Return of a substance to solution, as of phosphate removed from seawater by plankton and returned when these organisms die and decay. { ȯ'täl·ə·səs }

automatic weather station [METEOROL] A weather station at which the services of an observer are not required; usually equipped with telemetric apparatus. { ¦ȯd·ə¦mad·ik 'weth·ər ˌstā·shən }

autonomous underwater vehicle [OCEANOGR] A crewless, untethered submersible which operates independent of direct human control. Abbreviated AUV. { ȯ¦tän·ə·məs ˌən·dər‚wȯd·ər 'vē·ə·kəl }

autophytograph [GEOL] An imprint on a rock surface made by chemical activity of a plant or plant part. { ˌȯd·ə'fīd·ə‚graf }

autopneumatolysis [GEOL] The occurrence of metamorphic changes at the pneumatolytic stage of a cooling magma when temperatures are approximately 400–600°C. { ¦ȯd·ō‚nü·mə'täl·ə·səs }

autumn ice [OCEANOGR] Sea ice in early stage of formation; comparatively salty, and crystalline in appearance. { ¦ȯd·əm 'īs }

Autunian [GEOL] A European stage of Lower Permian geologic time, above the Stephanian of the Carboniferous and below the Saxonian. { ˌō'tün·ē·ən }

AUV *See* autonomous underwater vehicle.

Auversian *See* Ledian. { ˌō'vərzh·ən }

auxiliary fault [GEOL] A branch fault; a minor fault ending against a major one. { ȯg'zil·yə·rē 'fȯlt }

auxiliary plane [GEOL] A plane at right angles to the net slip on a fault plane as determined from analysis of seismic data for an earthquake. { ȯg'zil·yə·rē 'plān }

available moisture [HYD] Moisture in soil that is available for use by plants. { ə'vāl·ə·bəl 'mȯis·chər }

available relief [GEOL] The vertical distance after uplift between the altitude of the original surface and the level at which grade is first attained. { ə'vāl·ə·bəl ri'lēf }

avalanche [HYD] A mass of snow or ice moving rapidly down a mountain slope or cliff. { 'av·ə‚lanch }

avalanche wind [METEOROL] The rush of air produced in front of an avalanche of dry snow or in front of a landslide. { 'av·ə‚lanch ˌwind }

aven *See* pothole. { 'av·ən }

average limit of ice [OCEANOGR] The average seaward extent of ice formation during a normal winter. { 'av·rij 'lim·ət əv 'īs }

aviation weather forecast [METEOROL] A forecast of weather elements of particular interest to aviation, such as the ceiling, visibility, upper winds, icing, turbulence, and types of precipitation or storms. Also known as airways forecast. { ˌā·vē'ā·shən ¦weth·ər ˌfȯr‚kast }

aviation weather observation [METEOROL] An evaluation, according to set procedure, of those weather elements which are most important for aircraft operations. Also known as airways observation. { ˌā·vē'ā·shən ¦weth·ər ˌäb·zər'vā·shən }

Avonian *See* Dinantian. { ə'vōn·ē·ən }

avulsion [HYD] A sudden change in the course of a stream by which a portion of land is cut off, as where a stream cuts across and forms an oxbow. { ə'vəl·shən }

axial compression [GEOL] A compression applied parallel with the cylinder axis in experimental work involving rock cylinders. { 'ak·sē·əl kəm'presh·ən }

axial culmination [GEOL] Distortion of the fold axis upward in a form similar to an anticline. { 'ak·sē·əl ˌkəl·mə'nā·shən }

axial dipole field [GEOPHYS] A postulated magnetic field for the earth, consisting of a dipolar field centered at the earth's center, with its axis coincident with the earth's rotational axis. { 'ak·sē·əl 'diˌpōl ˌfēld }

axial plane [GEOL] A plane that intersects the crest or trough in such a manner that the limbs or sides of the fold are more or less symmetrically arranged with reference to it. Also known as axial surface. { 'ak·sē·əl 'plān }

axial-plane cleavage [GEOL] Rock cleavage essentially parallel to the axial plane of a fold. { 'ak·sē·əl ¦plān ˌklē·vij }

axial-plane foliation [GEOL] Foliation developed in rocks parallel to the axial plane of a fold and perpendicular to the chief deformational pressure. { 'ak·sē·əl ¦plān ˌfō·lē'ā·shən }

axial-plane schistosity [GEOL] Schistosity developed parallel to the axial planes of folds. { 'ak·sē·əl ¦plān ˌshis'täs·əd·ē }

axial-plane separation [GEOL] The distance between axial planes of adjacent anticline and syncline. { 'ak·sē·əl ¦plān sep·ə'rā·shən }

axial stream [HYD] **1.** The chief stream of an intermontane valley, the course of which is along the deepest part of the valley and is parallel to its longer dimension. **2.** A stream whose course is along the axis of an anticlinal or a synclinal fold. { 'ak·sē·əl 'strēm }

axial surface *See* axial plain. { 'ak·sē·əl 'sər·fəs }

axial trace [GEOL] The intersection of the axial plane of a fold with the surface of the earth or any other specified surface; sometimes such a line is loosely and incorrectly called the axis. { 'ak·sē·əl 'trās }

axial trough [GEOL] Distortion of a fold axis downward into a form similar to a syncline. { 'ak·sē·əl 'trȯf }

axinitization [GEOL] The replacement of rocks by axinite, as in the border zones of some granites. { akˌzin·ə·tə'zā·shən }

axis [GEOL] **1.** A line where a folded bed has maximum curvature. **2.** The central portion of a mountain chain. { 'ak·səs }

azimuth [GEOD] Horizontal direction on the earth's surface. { 'az·ə·məth }

azimuth equation [GEOD] A condition equation which expresses the relationship between the fixed azimuths of two lines that are connected by triangulation or traverse. { 'az·ə·məth i'kwā·zhən }

Azoic [GEOL] That portion of the earlier Precambrian time in which there is no trace of life. { ā'zō·ik }

azonal soil [GEOL] Any group of soils without well-developed profile characteristics, owing to their youth, conditions of parent material, or relief that prevents development of normal soil-profile characteristics. Also known as immature soil. { 'āˌzōn·əl 'sȯil }

Azores high [METEOROL] The semipermanent subtropical high over the North Atlantic Ocean, especially when it is located over the eastern part of the ocean; when in the western part of the Atlantic, it becomes the Bermuda high. { 'āˌzȯrz 'hī }

B

back-arc basin [GEOL] The region (small ocean basin) between an island arc and the continental mainland formed during oceanic plate subduction, containing sediment eroded from both. { 'bak,ärk ,bās·ən }

back beach *See* backshore. { 'bak ,bēch }

back-bent occlusion *See* bent-back occlusion. { 'bak ,bent ə'klü·zhən }

backbone [GEOL] **1.** A ridge forming the principal axis of a mountain. **2.** The principal mountain ridge, range, or system of a region. { 'bak,bōn }

backdeep [GEOL] An epieugeosynclinal basin; a nonvolcanic postorogenic geosynclinal basin whose sediments are derived from an uplifted eugeosyncline. { 'bak,dēp }

back-door cold front [METEOROL] A front which leads a cold air mass toward the south and southwest along the Atlantic seaboard of the United States. { 'bak ,dȯr 'kōld ,frənt }

backflooding [HYD] A reversal of flow of water at the water table resulting from changes in precipitation. { 'bak,fləd·iŋ }

backfolding [GEOL] Process in mountain forming in which the folds are overturned toward the interior of an orogenic belt. Also known as backward folding. { 'bak ,fōld·iŋ }

backing [METEOROL] **1.** Internationally, a change in wind direction in a counterclockwise sense (for example, south to east) in either hemisphere of the earth. **2.** In United States usage, a change in wind direction in a counterclockwise sense in the Northern Hemisphere, clockwise in the Southern Hemisphere. { 'bak·iŋ }

backlands [GEOL] A section of a river floodplain lying behind a natural levee. { 'bak,lanz }

backlimb [GEOL] Of the two limbs of an asymmetrical anticline, the one that is more gently dipping. { 'bak,lim }

back radiation *See* counterradiation. { 'bak ,rād·ē'ā·shən }

back reef [GEOGR] The area between a reef and the land. { 'bak ,rēf }

back rush [OCEANOGR] Return of water seaward after the uprush of the waves. { 'bak ,rəsh }

back-set bed [GEOL] Cross bedding that dips in a direction against the flow of a depositing current. { 'bak ,set ,bed }

backshore [GEOL] The upper shore zone that is beyond the advance of the usual waves and tides. Also known as back beach; backshore beach. { 'bak,shȯr }

backshore beach *See* backshore. { 'bak,shȯr ,bēch }

backshore terrace *See* berm. { 'bak,shȯr 'ter·əs }

back slope *See* dip slope. { 'bak ,slōp }

backswamp [GEOL] Swampy depressed area of a floodplain between the natural levees and the edge of the floodplain. { 'bak,swamp }

backthrusting [GEOL] The thrusting in the direction of the interior of an orogenic belt, opposite the general structural trend. { 'bak,thrəst·iŋ }

backward folding *See* backfolding. { 'bak·wərd ¦fōld·iŋ }

backwash [OCEANOGR] **1.** Water or waves thrown back by an obstruction such as a ship or breakwater. **2.** The seaward return of water after a rush of waves onto the beach foreshore. { 'bak,wäsh }

backwash mark [GEOL] A crisscross ridge pattern in beach sand, caused by backwash. { 'bak,wäsh ,märk }

backwash ripple mark [GEOL] Ripple marks that are broad and flat and parallel to the shoreline, with narrow, shallow troughs and crests about 30 centimeters apart; formed by backwash above the maximum wave retreat level. { 'bak,wäsh 'rip·əl ,märk }

backwater [HYD] **1.** A series of connected lagoons, or a creek parallel to a coast, narrowly separated from the sea and connected to it by barred outlets. **2.** Accumulation of water resulting from and held back by an obstruction. **3.** Water reversed in its course by an obstruction. { 'bak,wȯd·ər }

baculite [GEOL] A crystallite that looks like a dark rod. { 'bak·yə,līt }

badlands [GEOGR] An erosive physiographic feature in semiarid regions characterized by sharp-edged, sinuous ridges separated by steep-sided, narrow, winding gullies. { 'bad,lanz }

baffling wind [METEOROL] A wind that is shifting so that nautical movement by sailing vessels is impeded. { ¦baf·liŋ 'wind }

baguio [METEOROL] A tropical cyclone that occurs in the Philippines. { bäg'yō }

bahada *See* bajada. { bə'häd·ə }

bai [METEOROL] A yellow mist prevalent in China and Japan in spring and fall, when the loose surface of the interior of China is churned up by the wind, and clouds of sand rise to a great height and are carried eastward, where they collect moisture and fall as a yellow mist. { bī }

bajada [GEOL] An alluvial plain formed as a result of lateral growth of adjacent alluvial fans until they finally coalesce to form a continuous inclined deposit along a mountain front. Also spelled bahada. { bə'häd·ə }

Bajocian [GEOL] A European stage: the middle Middle or lower Middle Jurassic geologic time; above Toarcian, below Bathonian. { bə'jō·shən }

balanced rock *See* perched block. { 'bal·ənst ,räk }

balance equation [METEOROL] A diagnostic equation expressing a balance between the pressure field and the horizontal field of motion of the atmosphere. { 'bal·əns i'kwā·zhən }

bald [GEOGR] An elevated grassy, treeless area, as on the top of a mountain. { bȯld }

baldheaded anticline [GEOL] An upfold with a crest that has been deeply eroded before later deposition. { 'bȯld,hed·əd 'an·ti,klīn }

Bali wind [METEOROL] A strong east wind at the eastern end of Java. { ¦bäl·ē ¦wind }

ball [GEOL] **1.** A low sand ridge, underwater by high tide, which extends generally parallel with the shoreline; usually separated by an intervening trough from the beach. **2.** A spheroidal mass of sedimentary material. **3.** Common name for a nodule, especially of ironstone. { bȯl }

ball-and-socket joint *See* cup-and-ball joint. { ¦bȯl ən 'säk·ət ,jȯint }

ball coal [GEOL] A variety of coal occurring in spheroidal masses. { 'bȯl ,kōl }

ball ice [OCEANOGR] Numerous floating spheres of sea ice having diameters of 1–2 inches (2.5–5 centimeters), generally in belts similar to slush which forms at the same time. { 'bȯl ,īs }

ball lightning [GEOPHYS] A relatively rare form of lightning, consisting of a reddish, luminous ball, of the order of 1 foot (30 centimeters) in diameter, which may move rapidly along solid objects or remain floating in midair. Also known as globe lightning. { 'bȯl ¦līt·niŋ }

balloon ceiling [METEOROL] The ceiling classification applied when the ceiling height is determined by timing the ascent and disappearance of a ceiling balloon or pilot balloon in United States weather observing practice. { bə'lün ,sēl·iŋ }

balloon drag [METEOROL] A small balloon, loaded with ballast and inflated so that it will explode at a predetermined altitude, which is attached to a larger balloon; frequently used to retard the ascent of a radiosonde during the early part of the flight so that more detailed measurements may be obtained. { bə'lün ,drag }

ballstone [GEOL] **1.** Large mass or concretion of fine, unstratified limestone resulting from growth of coral colonies. **2.** A nodule of rock, especially ironstone, in a stratified unit. { 'bȯl,stōn }

balm [GEOL] A concave cliff or precipice that forms a shelter. { bäm }

Baltic Sea [GEOGR] An intracontinental, Mediterranean-type sea, connected with the North Sea and surrounded by Sweden, Denmark, Germany, Poland, the Baltic States, and Finland. { 'bȯl·tik 'sē }

banco [HYD] A meander or oxbow lake separated from a river by a change in its course. { 'baŋ·kō }

band [GEOD] Any latitudinal strip, designated by accepted units of linear or angular measurement, which circumscribes the earth. [GEOL] A thin layer or stratum of rock that is noticeable because its color is different from the colors of adjacent layers. { band }

banded coal [GEOL] A variety of bituminous and subbituminous coal made up of a sequence of thin lenses of highly lustrous coalified wood or bark interspersed with layers of more or less striated bright or dull coal. { 'ban·dəd 'kōl }

banded iron formation [GEOL] A sedimentary mineral deposit consisting of alternate silica-rich (chert or quartz) and iron-rich layers formed 2.5–3.5 billion years ago; the major source of iron ore. { ¦band·əd 'ī·ərn fȯr,mā·shən }

banded ore [GEOL] Ore made up of layered bands composed either of the same minerals that differ from band to band in color or textures or proportion, or of different minerals. { 'ban·dəd 'ȯr }

banded peat [GEOL] Peat formed of alternate layers of vegetable debris. { 'ban·dəd 'pēt }

banded structure [METEOROL] The appearance of precipitation echoes in the form of long bands as presented on radar plan position indicator (PPI) scopes. { 'ban·dəd 'strək·chər }

banded vein [GEOL] A vein composed of layers of different minerals that lie parallel to the walls. Also known as ribbon vein. { 'ban·dəd 'vān }

band lightning *See* ribbon lightning. { 'band ¦līt·niŋ }

bank [GEOL] **1.** The edge of a waterway. **2.** The rising ground bordering a body of water. **3.** A steep slope or face, generally consisting of unconsolidated material. [OCEANOGR] A relatively flat-topped raised portion of the sea floor occurring at shallow depth and characteristically on the continental shelf or near an island. { baŋk }

bank deposit [GEOL] Mounds, ridges, and terraces of sediment rising above and about the surrounding sea bottom. { ¦baŋk di'päz·ət }

banket [GEOL] A conglomerate containing valuable metal to be exploited. { baŋ'ket }

bankfull stage [HYD] The flow stage of a river in which the stream completely fills its channel and the elevation of the water surface coincides with the bank margins. { ¦baŋk ¦fu̇l ,stāj }

bank gravel *See* bank-run gravel. { 'baŋk ,grav·əl }

bank-inset reef [GEOL] A coral reef situated on island or continental shelves well inside the outer edges. { 'baŋk 'in,set ,rēf }

bank reef [GEOL] A reef which rises at a distance back from the outer margin of rimless shoals. { 'baŋk ,rēf }

bank-run gravel [GEOL] A natural deposit comprising gravel or sand. { 'baŋk ,rən 'grav·əl }

bank sand [GEOL] Deposits occurring in banks or pits and containing a low percentage of clay; used in core making. { 'baŋk ,sand }

bank storage [HYD] Water absorbed in the permeable bed and banks of a lake, reservoir, or stream. { 'baŋk ,stȯr·ij }

banner cloud [METEOROL] A cloud plume often observed to extend downwind from isolated mountain peaks, even on otherwise cloud-free days. Also known as cloud banner. { 'ban·ər ,klau̇d }

bar [GEOL] **1.** Any of the various submerged or partially submerged ridges, banks, or mounds of sand, gravel, or other unconsolidated sediment built up by waves or currents within stream channels, at estuary mouths, and along coasts. **2.** Any band of hard rock, for example, a vein or dike, that extends across a lode. { bär }

baraboo [GEOL] A monadnock buried by a series of strata and then reexposed by the partial erosion of these younger strata. { 'bär·ə,bü }
barat [METEOROL] A heavy northwest squall in Manado Bay on the north coast of the island of Celebes, prevalent from December to February. { bə'rät }
barb [METEOROL] A means of representing wind speed in the plotting of a synoptic chart, being a short, straight line drawn obliquely toward lower pressure from the end of a wind-direction shaft. Also known as feather. { bärb }
Barbados earth [GEOL] A deposit of fossil radiolarians. { bar'bā·dəs ,ərth }
bar beach [GEOL] A straight beach of offshore bars that are separated by shallow bodies of water from the mainland. { 'bär ,bēch }
barbed tributary [HYD] A tributary that enters the main stream in an upstream direction instead of pointing downstream. { 'bärbd 'trib·yə,ter·ē }
barber [METEOROL] A severe storm at sea during which spray and precipitation freeze onto the decks and rigging of ships. { 'bär·bər }
barchan [GEOL] A crescent-shaped dune or drift of windblown sand or snow, the arms of which point downwind; formed by winds of almost constant direction and of moderate speeds. Also known as barchane; barkhan; crescentic dune. { bär'kän }
barchane *See* barchan. { bär'kän }
bar finger sand [GEOL] An elongated lenticular sand body that lies beneath a distributory in a birdfoot delta. { 'bär ¦fiŋ·gər ,sand }
baric topography *See* height pattern. { 'bar·ik tə'päg·rə·fē }
baric wind law *See* Buys-Ballot's law. { 'bar·ik ¦wind ,lò }
barines [METEOROL] Westerly winds in eastern Venezuela. { ba'rēnz }
baring *See* overburden. { 'ba·riŋ }
barkhan *See* barchan. { bär'kän }
baroclinic disturbance [METEOROL] Any migratory cyclone associated with strong baroclinity of the atmosphere, evidenced on synoptic charts by temperature gradients in the constant-pressure surfaces, vertical wind shear, tilt of pressure troughs with height, and concentration of solenoids in the frontal surface near the ground. Also known as baroclinic wave. { ¦bar·ə¦klin·ik di'stər·bəns }
baroclinic field [METEOROL] A distribution of atmospheric pressure and mass such that the specific volume, or density, of air is a function not solely of pressure. { ¦bar·ə¦klin·ik 'fēld }
baroclinic instability [METEOROL] A hydrodynamic instability arising from the existence of a meridional temperature gradient (and hence of a thermal wind) in an atmosphere in quasi-geostrophic equilibrium and possessing static stability. { ¦bar·ə¦klin·ik in·stə'bil·əd·ē }
baroclinic model [METEOROL] A concept of stratification in the atmosphere, involving surfaces of constant pressure intersecting surfaces of constant density. { ¦bar·ə¦klin·ik 'mäd·əl }
baroclinic wave *See* baroclinic disturbance. { ¦bar·ə¦klin·ik 'wāv }
barometer elevation [METEOROL] The vertical distance above mean sea level of the ivory point (zero point) of a weather station's mercurial barometer; frequently the same as station elevation. Also known as elevation of ivory point. { bə'räm·əd·ər el·ə'vā·shən }
barometric tendency *See* pressure tendency. { bar·ə'met·rik 'ten·dən·sē }
barometric tide [GEOPHYS] A daily variation in atmospheric pressure due to the gravitational attraction of the sun and moon. { bar·ə'met·rik 'tīd }
barometric wave [METEOROL] Any wave in the atmospheric pressure field; the term is usually reserved for short-period variations not associated with cyclonic-scale motions or with atmospheric tides. { bar·ə'met·rik 'wāv }
barotropic disturbance [METEOROL] Also known as barotropic wave. **1.** A wave disturbance in a two-dimensional nondivergent flow; the driving mechanism lies in the variation of either vorticity of the basic current or the variation of the vorticity of the earth about the local vertical. **2.** An atmospheric wave of cyclonic scale in which troughs and ridges are approximately vertical. { ,bar·ə'träp·ik dis'tər·bəns }
barotropic field [METEOROL] A distribution of atmospheric pressures and mass such

that the specific volume, or density, of air is a function solely of pressure. { ˌbar·ə'träp·ik 'fēld }

barotropic model [METEOROL] Any of a number of model atmospheres in which some of the following conditions exist throughout the motion: coincidence of pressure and temperature surfaces, absence of vertical wind shear, absence of vertical motions, absence of horizontal velocity divergence, and conservation of the vertical component of absolute vorticity. { ˌbar·ə'träp·ik 'mäd·əl }

barotropic wave *See* barotropic disturbance. { ˌbar·ə'träp·ik ˌwāv }

bar plain [GEOL] A plain formed by a stream without a low-water channel or an alluvial cover. { 'bär ˌplān }

barranca [GEOL] A hole or deep break made by heavy rain; a ravine. { bə'raŋ·kə }

barred basin *See* restricted basin. { ¦bärd ¦bās·ən }

barred beach sequence [GEOL] A sequence comprising longshore bars, barrier beaches, and lagoons that develop when, under low-energy conditions, waves cross a broad continental shelf before impinging on a shoreline where sand-sized sediments are abundant. { ¦bärd 'bēch 'sē·kwəns }

Barremian [GEOL] Lower Cretaceous geologic age, between Hauterivian and Aptian. { bə'rām·ē·ən }

barrens [GEOGR] An area that because of adverse environmental conditions is relatively devoid of vegetation compared with adjacent areas. { 'bar·ənz }

barrier bar [GEOL] Ridges whose crests are parallel to the shore and which are usually made up of water-worn gravel put down by currents in shallow water at some distance from the shore. { 'bar·ē·ər ˌbär }

barrier basin [GEOL] A basin formed by natural damming, for example, by landslides or moraines. { 'bar·ē·ər ˌbās·ən }

barrier beach [GEOL] A single, long, narrow ridge of sand which rises slightly above the level of high tide and lies parallel to the shore, from which it is separated by a lagoon. Also known as offshore beach. { 'bar·ē·ər ˌbēch }

barrier chain [GEOL] A series of barrier spits, barrier islands, and barrier beaches extending along a coastline. { 'bar·ē·ər ˌchān }

barrier flat [GEOL] An area which is relatively flat and frequently occupied by pools of water that separate the seaward edge of the barrier from a lagoon on the landward side. { 'bar·ē·ər ˌflat }

barrier ice *See* shelf ice. { 'bar·ē·ər ˌīs }

barrier island [GEOL] An elongate accumulation of sediment formed in the shallow coastal zone and separated from the mainland by some combination of coastal bays and their associated marshes and tidal flats; barrier islands are typically several times longer than their width and are interrupted by tidal inlets. { 'bar·ē·ər ˌī·lənd }

barrier lagoon [GEOGR] A shallow body of water that separates the shore and a barrier reef. { 'bar·ē·ər lə'gün }

barrier lake [HYD] A small body of water that lies in a basin, retained there by a natural dam or barrier. { 'bar·ē·ər ˌlāk }

barrier reef [GEOL] A coral reef that runs parallel to the coast of an island or continent, from which it is separated by a lagoon. { 'bar·ē·ər ˌrēf }

barrier spit [GEOL] A barrier of sand joined at one of its ends to the mainland. { 'bar·ē·ər ˌspit }

barrier theory of cyclones [METEOROL] A theory of cyclone development, proposed by F.M. Exner, which states that a slow-moving mass of cold air in the path of rapidly eastward-moving warmer air will bring about the formation of low pressure on the lee side of the cold air; analogous to the formation of a dynamic trough on the lee side of an orographic barrier. Also known as drop theory. { 'bar·ē·ər ˌthē·ə·rē əv 'sīˌklōnz }

Barrovian metamorphism [GEOL] A regional metamorphism that can be zoned into facies that are metamorphic. { bə'rōv·ē·ən ˌmed·ə'mȯrˌfiz·əm }

Barstovian [GEOL] Upper Miocene geologic time. { ˌbär'stōv·ē·ən }

bar theory [GEOL] A theory that accounts for thick deposits of salt, gypsum, and other evaporites in terms of increased salinity of a solution in a lagoon caused by evaporation. { 'bär 'thē·ə·rē }

Bartonian [GEOL] A European stage: Eocene geologic time above Auversian, below Ludian. Also known as Marinesian. { bär'tōn·ē·ən }

basal complex *See* basement. { 'bā·səl 'käm,pleks }

basal conglomerate [GEOL] A coarse gravelly sandstone or conglomerate forming the lowest member of a series of related strata which lie unconformably on older rocks; records the encroachment of the seabeach on dry land. { 'bā·səl kən'gläm·ə·rət }

basal groundwater [HYD] A large body of groundwater that floats on and is in hydrodynamic equilibrium with sea water. { 'bā·səl 'grau̇nd,wȯd·ər }

basalt glass *See* tachylite. { bə'sȯlt ,glas }

basaltic dome *See* shield volcano. { bə'sȯl·tik 'dōm }

basaltic magma [GEOL] Mobile rock material of basaltic composition. { bə'sȯl·tik 'mag·mə }

basaltic shell [GEOL] The lower crystal layer of basalt underlying the oceans and beneath the sialic layer of continents. { bə'sȯl·tik 'shel }

basaltiform [GEOL] Similar to basalt in form. { bə'sȯl·tə,fȯrm }

basalt obsidian *See* tachylite. { bə'sȯlt əb'sid·ē·ən }

basal water table [HYD] The water table of basal groundwater. { 'bā·səl 'wȯd·ər ,tā·bəl }

basculating fault *See* wrench fault. { 'ba·skyə,lād·iŋ 'fȯlt }

base exchange [GEOCHEM] Replacement of certain ions by others in clay. { 'bās iks'chānj }

base flow [HYD] The flow of water entering stream channels from groundwater sources in the drainage of large lakes. { 'bās ,flō }

base level [GEOL] That critical plane of erosion and deposition represented by river level on continents and by wave or current base in the sea. { 'bās ,lev·əl }

base-leveled plain [GEOL] Any land surface changed almost to a plain by subaerial erosion. Also known as peneplain. { 'bās ,lev·əld 'plān }

base-leveling epoch *See* gradation period. { 'bās ,lev·əl·iŋ 'ep·ək }

basement [GEOL] **1.** A complex, usually of igneous and metamorphic rocks, that is overlain unconformably by sedimentary strata. Also known as basement rock. **2.** A crustal layer beneath a sedimentary one and above the Mohorovičić discontinuity. **3.** The ancient continental igneous rock base that lies beneath Precambrian rocks. Also known as basal complex; basement complex. { 'bās·mənt }

basement complex *See* basement. { 'bās·mənt 'käm,pleks }

basement rock *See* basement. { 'bās·mənt ,räk }

base station [GEOD] A geographic position whose absolute gravity value is known. { 'bās ,stā·shən }

basic front [GEOL] An advancing zone of granitization enriched in calcium, magnesium, and iron. { 'bā·sik ¦frənt }

basification [GEOL] Development of a more basic rock, usually with more hornblende, biotite, and oligoclase, by contamination of a granitic magma in the assimilation of country rock. { ,bās·ə·fə'kā·shən }

basimesostasis [GEOL] A process of the partial or entire enclosure of plagioclase crystals in a diabase by augite. { ¦bā·zē,mez·ə'stā·səs }

basin [GEOL] **1.** A low-lying area, wholly or largely surrounded by higher land, that varies from a small, nearly enclosed valley to an extensive, mountain-rimmed depression. **2.** An entire area drained by a given stream and its tributaries. **3.** An area in which the rock strata are inclined downward from all sides toward the center. **4.** An area in which sediments accumulate. [OCEANOGR] Deep portion of sea surrounded by shallower regions. { 'bās·ən }

basin accounting *See* hydrologic accounting. { 'bās·ən ə'kau̇nt·iŋ }

basin-and-range structure [GEOL] Regional structure dominated by fault-block mountains separated by basins filled with sediment. { 'bās·ən ən 'ranj ,strək·chər }

basin fold [GEOL] Synclinal and anticlinal folds in structural basins. { 'bās·ən ,fōld }

basining [GEOL] A settlement of earth in the form of basins due to the solution and transportation of underground deposits of salt and gypsum. { 'bās·ən·iŋ }
basin length [GEOL] Length in a straight line from the mouth of a stream to the farthest point on the drainage divide of its basin. { 'bās·ən ˌleŋkth }
basin order [GEOL] A classification of basins according to stream drainage; for example, a first-order basin contains all of the drainage area of a first-order stream. { 'bās·ən ˌȯrd·ər }
basin peat *See* local peat. { 'bās·ən ˌpēt }
basin range [GEOL] A mountain range characteristic of the Great Basin in the western United States and formed by a faulted and tilted block of strata. { 'bās·ən ˌrānj }
basin valley [GEOL] The filled-in depression of large intermountain areas; an example is Salt Lake Valley in Utah. { 'bās·ən ˌval·ē }
basset [GEOL] The outcropping edge of a layer of rock exposed to the surface. { 'bas·ət }
bastion [GEOL] A prominent aggregation of bedrock extending from the mouth of a hanging glacial trough and reaching well into the main glacial valley. { 'bas·chən }
batholite [GEOL] An older massive protrusion of magma that solidifies as coarse crystalline rock in the deep horizons of the earth's crust. { 'bath·əˌlīt }
batholith [GEOL] A body of igneous rock, 40 square miles (100 square kilometers) or more in area, emplaced at great or intermediate depth in the earth's crust. { 'bath·əˌlith }
Bathonian [GEOL] A European stage of geologic time: Middle Jurassic, below Callovian, above Bajocian. Also known as Bathian. { bə'thōn·ē·ən }
bathyal zone [OCEANOGR] The biogeographic realm of the ocean depths between 100 and 1000 fathoms (180 and 1800 meters). { 'bath·ē·əl ˌzōn }
bathymetric biofacies [GEOL] The lateral distribution and character of underwater sedimentary strata. { ¦bath·ə'me·trik ¦bī·ō'fā·shēz }
bathyorographical [GEOD] Concerned with depths of oceans and heights of mountains. { ¦bath·ēˌȯr·ə'graf·ə·kəl }
bathypelagic zone [OCEANOGR] The biogeographic realm of the ocean lying between depths of 500 and 2550 fathoms (900 and 3700 meters). { ¦bath·ə·pə'laj·ik 'zōn }
battery reefs *See* Kimberley reefs. { 'bad·ə·rē ˌrēfs }
batture [GEOL] An elevation of the bed of a river under the surface of the water; sometimes used to signify the same elevation when it has risen above the surface. { ba'tu̇r }
bauxitization [GEOL] Bauxite development from either primary aluminum silicates or secondary clay minerals. { ¦bȯk·sə·də'zā·shən }
bay [GEOGR] **1.** A body of water, smaller than a gulf and larger than a cove in a recess in the shoreline. **2.** A narrow neck of water leading from the sea between two headlands. [GEOPHYS] A simple transient magnetic disturbance, usually an hour in duration, whose appearance on a magnetic record has the shape of a V or a bay of the sea. { bā }
bay bar *See* baymouth bar. { 'bā ˌbär }
bay barrier [GEOL] A narrow shoal or small point of land projecting from the shore across the mouth of a bay and severing the bay's connection with the main body of water. { 'bā ˌbar·ē·ər }
bay delta [GEOL] A usually triangular alluvial deposit formed at the point where the mouth of a stream enters the head of a drowned valley. { 'bā ˌdel·tə }
bay head [GEOL] A swampy region at the head of a bay. { ¦bā ¦hed }
bay-head bar [GEOL] A bar formed a short distance from the shore at the head of a bay. { ¦bā ¦hed ˌbär }
bay-head beach [GEOL] A beach formed around a bay head by storm waves; layers of sediment cover the bay floor and bare rock benches front the headland cliffs. { ¦bā ¦hed ˌbēch }
bay-head delta [GEOL] A delta at the head of an estuary or a bay into which a river discharges because of the margin of the land's late partial submergence. { ¦bā ¦hed ˌdel·tə }

bay ice [OCEANOGR] Sea ice that is young and flat but sufficiently thick to impede navigation. { ¦bā ˌīs }
baymouth bar [GEOL] A bar extending entirely or partially across the mouth of a bay. Also known as bay bar. { 'bā,maůth ˌbär }
bayou [HYD] A small, sluggish secondary stream or lake that exists often in an abandoned channel or a river delta. { 'bī,yü }
bayside beach [GEOL] A beach formed at the side of a bay by materials eroded from nearby headlands and deposited by longshore currents. { 'bā,sīd ˌbēch }
b-c fracture [GEOL] A tension fracture parallel with the fabric plane and normal to the *a* axis. { ¦bē¦sē 'frak·chər }
b-c plane [GEOL] A plane that is perpendicular to the plane of movement and parallel to the *b* direction in that plane. { ¦bē¦sē ˌplān }
beach [GEOL] The zone of unconsolidated material that extends landward from the low-water line to where there is marked change in material or physiographic form or to the line of permanent vegetation. { bēch }
beach cycle [GEOL] Periodic retreat and outbuilding of beaches resulting from waves and tides. { bēch ˌsī·kəl }
beach drift [GEOL] The material transported by drifting of beach. { bēch ˌdrift }
beach face *See* foreshore. { bēch ˌfās }
beach gravel [GEOL] Gravels in which most of the particles cluster about one size. { bēch ˌgrav·əl }
beach plain [GEOL] Embankments of wave-deposited material added to a prograding shoreline. { 'bēch ˌplān }
beach platform *See* wave-cut bench. { 'bēch ˌplat,fȯrm }
beach profile [GEOL] Intersection of a beach's ground surface with a vertical plane perpendicular to the shoreline. { 'bēch 'prō,fīl }
beach ridge [GEOL] A continuous mound of beach material behind the beach that was heaped up by waves or other action. { 'bēch ˌrij }
beach scarp [GEOL] A nearly vertical slope along the beach caused by wave erosion. { 'bēch ˌskärp }
beaded lake *See* paternoster lake. { 'bēd·əd ¦lāk }
bean ore [GEOL] A lenticular, pisolitic aggregate of limonite. { 'bēn ˌȯr }
Beaufort force [METEOROL] A number denoting the speed (or so-called strength) of the wind according to the Beaufort wind scale. Also known as Beaufort number. { 'bō,fərt ˌfȯrs }
Beaufort number *See* Beaufort force. { 'bō,fərt ˌnəm·bar }
Beaufort wind scale [METEOROL] A system of code numbers from 0 to 12 classifying wind speeds into groups from 0–1 mile per hour or 0–1.6 kilometers per hour (Beaufort 0) to those over 75 miles per hour or 121 kilometers per hour (Beaufort 12). { 'bō·fərt 'wind ˌskāl }
bed [GEOL] **1.** The smallest division of a stratified rock series, marked by a well-defined divisional plane from its neighbors above and below. **2.** An ore deposit, parallel to the stratification, constituting a regular member of the series of formations; not an intrusion. [HYD] The bottom of a channel for the passage of water. { bed }
bedded [GEOL] Pertaining to rocks exhibiting depositional layering or bedding formed from consolidated sediments. { 'bed·əd }
bedded vein [GEOL] A lode occupying the position of a bed that is parallel with the enclosing rock stratification. { 'bed·əd ˌvān }
bedding [GEOL] Condition where planes divide sedimentary rocks of the same or different lithology. { 'bed·iŋ }
bedding cleavage [GEOL] Cleavage parallel to the rock bedding. { 'bed·iŋ ˌklēv·ij }
bedding fault [GEOL] A fault whose fault surface is parallel to the bedding plane of the constituent rocks. Also known as bedding-plane fault. { 'bed·iŋ ˌfȯlt }
bedding fissility [GEOL] Primary foliation parallel to the bedding of sedimentary rocks. { 'bed·iŋ fi'sil·əd·ē }
bedding joint [GEOL] A joint parallel to the rock bedding. { 'bed·iŋ ˌjȯint }

bedding plane [GEOL] Any of the division planes which separate the individual strata or beds in sedimentary or stratified rock. { 'bed·iŋ ˌplān }
bedding-plane fault *See* bedding fault. { 'bed·iŋ ˌplān ˌfȯlt }
bedding-plane slip *See* flexural slip. { 'bed·iŋ ˌplān ˌslip }
bedding schistosity [GEOL] Schistosity that is parallel to the rock bedding. { 'bed·iŋ ˌshis'täs·əd·ē }
bedding thrust [GEOL] A thrust fault parallel to bedding. { 'bed·iŋ ˌthrəst }
bedding void [GEOL] A void formed between successive batches of lava that are discharged in a single short activity of a volcano, as well as between flows made a long time apart. { 'bed·iŋ ˌvȯid }
Bedford limestone *See* spergenite. { 'bed·fərd 'līmˌstōn }
bediasite [GEOL] A black to brown tektite found in Texas. { bē'dī·əˌzīt }
bed load [GEOL] Particles of sand, gravel, or soil carried by the natural flow of a stream on or immediately above its bed. Also known as bottom load. { 'bed ˌlōd }
Bedoulian [GEOL] Lower Cretaceous (lower Aptian) geologic time in Switzerland. { bə'dül·ē·ən }
bedrock [GEOL] General term applied to the solid rock underlying soil or any other unconsolidated surficial cover. { 'bedˌräk }
beetle stone *See* septarium. { 'bēd·əl ˌstōn }
beheaded stream [HYD] A water course whose upper portion, through erosion, has been cut off and captured by another water course. { bi'hed·əd ¦strēm }
belat [METEOROL] A strong land wind from the north or northwest which sometimes blows across the southeastern coast of Arabia and is accompanied by a hazy atmosphere due to sand blown from the interior desert. { bā'lät }
belonite [GEOL] A rod- or club-shaped microscopic embryonic crystal in a glassy rock. { 'bel·əˌnīt }
below minimums [METEOROL] Below operational weather limits for aircraft. { bi'lō 'min·əˌməmz }
belt [HYD] A long area or strip of pack ice, with a width of 1 kilometer (0.6 mile) to more than 100 kilometers (60 miles). { belt }
belted plain [GEOL] A plain whose surface has been slowly worn down and sculptured into bands or belts of different levels. { 'bel·təd ¦plān }
belteroporic [GEOL] Of crystals in rocks whose growth was determined by the direction of easiest growth. { ˌbel¦ter·ə'pȯr·ik }
belt of cementation *See* zone of cementation. { ¦belt əv ˌsi·men'tā·shən }
belt of soil moisture *See* belt of soil water. { ¦belt əv 'sȯil ˌmȯis·chər }
belt of soil water [GEOL] The upper subdivision of the zone of aeration limited above by the land surface and below by the intermediate belt; this zone contains plant roots and water available for plant growth. Also known as belt of soil moisture; discrete film zone; soil-water belt; soil-water zone; zone of soil water. { ¦belt əv 'sȯil ˌwȯd·ər }
bench [GEOL] A terrace of level earth or rock that is raised and narrow and that breaks the continuity of a declivity. { bench }
bench gravel [GEOL] Gravel beds found on the sides of valleys above the present stream bottoms, representing parts of the bed of the stream when it was at a higher level. { 'bench ˌgrav·əl }
bench lava [GEOL] Semiconsolidated, crusted basaltic lava forming raised platforms and crags about the edges of lava lakes. Also known as bench magma. { 'bench ˌla·və }
bench magma *See* bench lava. { 'bench ˌmag·mə }
bench placer [GEOL] A placer in ancient stream deposits from 50 to 300 feet (15 to 90 meters) above present streams. { 'bench ˌplās·ər }
bend [GEOL] **1.** A curve or turn occurring in a stream course, bed, or channel which has not yet become a meander. **2.** The land area partly encircled by a bend or meander. { bend }
bending [HYD] Movement of sea ice up or down resulting from lateral pressure exerted

by wind or tide. [OCEANOGR] The first stage in the formation of pressure ice caused by the action of current, wind, tide, or air temperature changes. { 'ben·diŋ }

Benguela Current [OCEANOGR] A strong current flowing northward along the southwestern coast of Africa. { ben'gwel·ə¦kər·ənt }

Benioff zone [GEOPHYS] A zone of earthquake hypocenters distributed on well-defined planes that dips from a shallow depth into the earth's mantle to depths as great as 420 miles (700 kilometers). Also known as Benioff-Wadati zone; Wadati-Benioff zone. { 'ben·ē·ȯf ˌzōn }

Benioff-Wadati zone *See* Benioff zone. { ¦ben·ēˌȯf wə'dä·tē ˌzōn }

bent-back occlusion [METEOROL] An occluded front that has reversed its direction of motion as a result of the development of a new cyclone (usually near the point of occlusion) or, less frequently, as the result of the displacement of the old cyclone along the front. Also known as back-bent occlusion. { 'bent ˌbak ə'klü·zhən }

benthic [OCEANOGR] Of,pertaining to, or living on the bottom or at the greatest depths of a large body of water. Also known as benthonic. { 'ben·thik }

benthonic *See* benthic. { ben'thän·ik }

benthos [OCEANOGR] The floor or deepest part of a sea or ocean. { 'benˌthäs }

bentonite [GEOL] A clay formed from volcanic ash decomposition and largely composed of montmorillonite and beidellite. Also known as taylorite. { 'bent·ənˌīt }

bentu de soli [METEOROL] An east wind on the coast of Sardinia. { 'ben·tü dā 'sōl·ē }

Bergeron-Findeisen theory [METEOROL] The theoretical explanation that precipitation particles form within a mixed cloud (composed of both ice crystals and liquid water drops) because the equilibrium vapor pressure of water vapor with respect to ice is less than that with respect to liquid water at the same temperature. Also known as ice-crystal theory; Wedener-Bergeron process. { 'berzh·əˌrän ¦finˌdīz·ən ˌthē·ə·rē }

bergschrund [HYD] A type of crevice in a glacier; formed when ice and snow break away from a rock face. { 'berk ˌshru̇nt }

berg till *See* floe till. { 'bərg ˌtil }

Bering Sea [GEOGR] A body of water north of the Pacific Ocean, bounded by Siberia, Alaska, and the Aleutian Islands. { 'ber·iŋ 'sē }

berm [GEOL] **1.** A narrow terrace which originates from the interruption of an erosion cycle with rejuvenation of a stream in the mature stage of its development and renewed dissection. **2.** A horizontal portion of a beach or backshore formed by deposit of material as a result of wave action. Also known as backshore terrace; coastal berm. { bərm }

berm crest [GEOL] The seaward limit and usually the highest spot on a coastal berm. Also known as berm edge. { 'bərm ˌkrest }

berm edge *See* berm crest. { 'bərm ˌej }

Bermuda high [METEOROL] The semipermanent subtropical high of the North Atlantic Ocean, especially when it is located in the western part of that ocean area. { bər'myüd·ə ¦hī }

Berriasian [GEOL] Part of or the underlying stage of the Valanginian at the base of the Cretaceous. { ˌber·ē'ā·zhən }

Bessel ellipsoid of 1841 [GEOD] The reference ellipsoid of which the semimajor axis is 6,377,397.2 meters, the semiminor axis is 6,356,079.0 meters, and the flattening or ellipticity equals 1/299.15. Also known as Bessel spheroid of 1841. { 'bes·əl i'lipˌsoid əv ˌā¦tēn¦fȯr·dē'wən }

Bessel spheroid of 1841 *See* Bessel ellipsoid of 1841. { 'bes·əl 'sfirˌȯid əv ˌā¦tēn¦fȯr·dē'wən }

beta plane [GEOPHYS] The model, introduced by C.G. Rossby, of the spherical earth as a plane whose rate of rotation (corresponding to the Coriolis parameter) varies linearly with the north-south direction. { 'bād·ə ˌplān }

betrunked river [GEOL] A river that is shorn of its lower course as a result of submergence of the land margin by the sea. { bē'trəŋkt 'riv·ər }

betwixt mountains *See* median mass. { bə'twikst ˌmau̇nt·ənz }

beveling [GEOL] Planing by erosion of the outcropping edges of strata. { 'bev·ə·liŋ }

B horizon [GEOL] The zone of accumulation in soil below the A horizon (zone of leaching). Also known as illuvial horizon; subsoil; zone of accumulation; zone of illuviation. { 'bē hə'rīz·ən }

bifurcation ratio [HYD] The ratio of number of stream segments of one order to the number of the next higher order. { bī·fər'kā·shən ˌrā·shō }

bight [GEOL] **1.** A long, gradual bend or recess in the coastline which forms a large, open receding bay. **2.** A bend in a river or mountain range. [OCEANOGR] An indentation in shelf ice, fast ice, or a floe. { bīt }

billow cloud [METEOROL] Broad, nearly parallel lines of cloud oriented normal to the wind direction, with cloud bases near an inversion surface. Also known as undulatus. { 'bil·ō ˌklau̇d }

bimaceral [GEOL] A coal microlithotype that consists of a mixture of two macerals. { bī'mas·ə·rəl }

binding coal *See* caking coal. { 'bīn·diŋ ˌkōl }

bing ore [GEOL] The purest lead ore, with the largest crystals of galena. { 'biŋ ˌȯr }

biochemical deposit [GEOL] A precipitated deposit formed directly or indirectly from vital activities of organisms, such as bacterial iron ore and limestone. { ¦bī·ō'kem·ə·kəl di'päz·ət }

biochronology [GEOL] The relative age dating of rock units based on their fossil content. { ˌbī·ō·krə'näl·ə·jē }

bioerosion [OCEANOGR] The process by which animals, through drilling, grazing, and burrowing, erode hard substances such as rocks and coral reefs. { ˌbī·ō·i'rōzh·ən }

biofacies [GEOL] **1.** A rock unit differing in biologic aspect from laterally equivalent biotic groups. **2.** Lateral variation in the biologic aspect of a stratigraphic unit. { 'bī·ōˌfā·shēz }

biofog [METEOROL] A type of steam fog caused by contact between extremely cold air and the warm, moist air surrounding human or animal bodies or generated by human activity. { 'bī·ōˌfäg }

biogenic reef [GEOL] A mass consisting of the hard parts of organisms, or of a biogenically constructed frame enclosing detrital particles, in a body of water; most biogenic reefs are made of corals or associated organisms. { ¦bī·ō¦jen·ik 'rēf }

biogenic sediment [GEOL] A deposit resulting from the physiological activities of organisms. { ¦bī·ō¦jen·ik 'sed·ə·mənt }

biogeochemical cycle [GEOCHEM] The chemical interactions that exist between the atmosphere, hydrosphere, lithosphere, and biosphere. { ˌbī·ōˌjē·ō'kem·ə·kəl 'sīkəl }

biogeochemical prospecting [GEOCHEM] A prospecting technique for subsurface ore deposits based on interpretation of the growth of certain plants which reflect subsoil concentrations of some elements. { ˌbī·ōˌjē·ō'kem·ə·kəl 'präsˌpek·tiŋ }

biogeochemistry [GEOCHEM] A branch of geochemistry that is concerned with biologic materials and their relation to earth chemicals in an area. { ˌbī·ōˌjē·ō'kem·ə·strē }

bioherm [GEOL] A circumscribed mass of rock exclusively or mainly constructed by marine sedimentary organisms such as corals, algae, and stromatoporoids. Also known as organic mound. { 'bī·ōˌhərm }

biolite [GEOL] A concretion formed of concentric layers through the action of living organisms. { 'bī·ōˌlīt }

biological oceanography [OCEANOGR] The study of the flora and fauna of oceans in relation to the marine environment. { ¦bī·ə¦läj·ə·kəl ˌō·shə'näg·rə·fē }

biologic weathering *See* organic weathering. { ¦bī·ə¦läj·ik 'weth·ə·riŋ }

biomarkers [GEOL] Complex organic compounds found in oil, bitumen, rocks, and sediments that are linked with and distinctive of a particular source (such as algae, bacteria, or vascular plants); they are useful dating indicators in stratigraphy and molecular paleontology. Also known as chemical fossils; molecular fossils. { 'bī·ōˌmär·kərz }

biostratigraphic unit [GEOL] A stratum or body of strata that is defined and identified by one or more distinctive fossil species or genera without regard to lithologic or other physical features or relations. { ¦bī·ōˌstrad·ə'graf·ik 'yü·nət }

biostromal limestone [GEOL] Biogenic carbonate accumulations that are laterally uniform in thickness, in contrast to the moundlike nature of bioherms. { ¦bī·ə¦strō·məl 'līm,stōn }
biostrome [GEOL] A bedded structure or layer (bioclastic stratum) composed of calcite and dolomitized calcarenitic fossil fragments distributed over the sea bottom as fine lentils, independent of or in association with bioherms or other areas of organic growth. { 'bī·ə,strōm }
bioturbation [GEOL] The disruption of marine sedimentary structures by the activities of benthic organisms. { ¦bī·o·tər'bā·shən }
bird's-foot delta [GEOL] A delta with long, projecting distributary channels that branch outward like the toes or claws of a bird. { 'bərdz ,fu̇t 'del·tə }
bise [METEOROL] A cold, dry wind which blows from a northerly direction in the winter over the mountainous districts of southern Europe. Also spelled bize. { bēz }
Bishop's ring [METEOROL] A faint, broad, reddish-brown corona occasionally seen in dust clouds, especially those which result from violent volcanic eruptions. { 'bish·əps 'riŋ }
bitter lake [HYD] A lake rich in alkaline carbonates and sulfates. { ¦bid·ər 'lāk }
bitumenite *See* torbanite. { bī'tü·mə,nīt }
bituminization *See* coalification. { bī,tü·mə·nə'zā·shən }
bituminous coal [GEOL] A dark brown to black coal that is high in carbonaceous matter and has 15–50% volatile matter. Also known as soft coal. { bī'tü·mə·nəs 'kōl }
bituminous lignite [GEOL] A brittle, lustrous bituminous coal. Also known as pitch coal. { bī'tü·mə·nəs 'lig,nīt }
bituminous rock *See* asphalt rock. { bī'tü·mə·nəs 'räk }
bituminous sand [GEOL] Sand containing bituminous-like material, such as the tar sands at Athabasca, Canada, from which oil is extracted commercially. { bī'tü·mə·nəs 'sand }
bituminous wood [GEOL] A variety of brown coal having the fibrous structure of wood. Also known as board coal; wood coal; woody lignite; xyloid coal; xyloid lignite. { bī'tü·mə·nəs 'wu̇d }
bize *See* bise. { bēz }
black alkali [GEOL] A deposit of sodium carbonate that has formed on or near the surface in arid to semiarid areas. { ¦blak 'al·kə,lī }
black amber *See* jet coal. { ¦blak 'am·bər }
blackband [GEOL] An earthy carbonate of iron that is present with coal beds. { 'blak,band }
black buran *See* karaburan. { 'blak bü'rän }
black coal *See* natural coke. { ¦blak 'kōl }
black cotton soil *See* regur. { ¦blak ¦kat·ən 'sȯil }
black durain [GEOL] A durain that has high hydrogen content and volatile matter, many microspores, and some vitrain fragments. { ¦blak 'du̇,rān }
black frost [HYD] A dry freeze with respect to its effects upon vegetation, that is, the internal freezing of vegetation unaccompanied by the protective formation of hoarfrost. Also known as hard frost. { ¦blak 'frȯst }
black ice [HYD] A type of ice forming on lake or salt water; compact, and dark in appearance because of its transparency. { 'blak ,īs }
black lignite [GEOL] A lignite with a fixed carbon content of 35–60% and a total carbon content of 73.6–76.2% that contains between 6300 and 8300 Btu per pound; higher in rank than brown lignite. Also known as lignite A. { 'blak 'lig,nīt }
black mud [GEOL] A mud formed where there is poor circulation or weak tides, such as in lagoons, sounds, or bays; the color is due to iron sulfides and organic matter. { ¦blak 'məd }
black sand [GEOL] Heavy, dark, sandlike minerals found on beaches and in stream beds; usually magnetite and ilmenite and sometimes gold, platinum, and monazite are present. { ¦blak 'sand }
Black Sea [GEOGR] A large inland sea, area 163,400 square miles (423,000 square kilometers), bounded on the north and east by the Commonwealth of Independent

States (former U.S.S.R.) on the south and southwest by Turkey, and on the west by Bulgaria and Rumania. { ¦blak 'sē }

black smoker *See* hydrothermal vent. { ¦blak 'smōk·ər }

black snow [HYD] Snow that falls through a particulate-laden atmosphere. { ¦blak 'snō }

black storm *See* karaburan. { ¦blak 'stȯrm }

bladder *See* vesicle. { 'blad·ər }

Blaine formation [GEOL] A Permian red bed formation containing red shale and gypsum beds of marine origin in Oklahoma, Texas, and Kansas. { 'blān fȯr'mā·shən }

Blancan [GEOL] Upper Pliocene or lowermost Pleistocene geologic time. { 'blăŋ·kən }

blanket deposit [GEOL] A flat deposit of ore; its length and width are relatively great compared with its thickness. { 'blaŋ·kət di'päz·ət }

blanket sand [GEOL] A relatively thin body of sand or sandstone covering a large area. Also known as sheet sand. { 'blaŋ·kət ˌsand }

blastic deformation [GEOL] Rock deformation involving recrystallization in which space lattices are destroyed or replaced. { 'blas·tik ˌdēˌfȯr'mā·shən }

blasting [GEOL] Abrasion caused by movement of fine particles against a stationary fragment. { 'blas·tiŋ }

blastopsammite [GEOL] A relict fragment of sandstone that is contained in a metamorphosed conglomerate. { bla'stäp·səˌmīt }

blastopsephitic [GEOL] Descriptive of the structure of metamorphosed conglomerate or breccia. { bla¦stäp·sə¦fid·ik }

bleach spot [GEOL] A green or yellow area in red rocks formed by reduction of ferric oxide around an organic particle. Also known as deoxidation sphere. { 'blēch ˌspät }

blended unconformity [GEOL] An unconformity that is not sharp because the original erosion surface was covered by a thick residual soil that graded downward into the underlying rock. { ¦blen·dəd ˌən·kən'fȯr·məd·ē }

blind [GEOL] Referring to a mineral deposit with no surface outcrop. { blīnd }

blind coal *See* natural coke. { ¦blīnd ¦kōl }

blind drainage *See* closed drainage. { ¦blīnd ¦drā·nij }

blind rollers [OCEANOGR] Long, high swells which have increased in height, almost to the breaking point, as they pass over shoals or run in shoaling water. Also known as blind seas. { ¦blīnd 'rō·lərz }

blind seas *See* blind rollers. { ¦blīnd 'sēz }

blind valley [GEOL] A valley that has been made by a spring from an underground channel which emerged to form a surface stream, and that is enclosed at the head of the stream by steep walls. { 'blīnd 'val·ē }

blink [METEOROL] A brightening of the base of a cloud layer, caused by the reflection of light from a snow- or ice-covered surface. { bliŋk }

blister [GEOL] A domelike protuberance caused by the buckling of the cooling crust of a molten lava before the flowing mass has stopped. { 'blis·tər }

blister hypothesis [GEOL] A theory of the formation of compressional mountains by a process in which radiogenic heat expands and melts a portion of the earth's crust and subcrust, causing a domed regional uplift (blister) on a foundation of molten material that has no permanent strength. { 'blis·tər hī'päth·ə·səs }

blizzard [METEOROL] A severe weather condition characterized by low temperatures and by strong winds bearing a great amount of snow (mostly fine, dry snow picked up from the ground). { 'bliz·ərd }

blob [METEOROL] In radar, oscilloscope evidence of a fairly small-scale temperature and moisture inhomogeneity produced by turbulence within the atmosphere. { bläb }

block clay *See* mélange. { 'bläk ˌklā }

block faulting [GEOL] A type of faulting in which fault blocks are displaced at different orientations and elevations. { 'bläk ˌfȯl·tiŋ }

block glide [GEOL] A translational landslide in which the slide mass moves outward and downward as an intact unit. { 'bläk ˌglīd }

blocking [METEOROL] Large-scale obstruction of the normal west-to-east progress of migratory cyclones and anticyclones. { 'bläk·iŋ }

blocking anticyclone *See* blocking high. { ¦bläk·iŋ ˌan·ti'sīˌklōn }

blocking high [METEOROL] Any high (or anticyclone) that remains nearly stationary or moves slowly compared to the west-to-east motion upstream from its location, so that it effectively blocks the movement of migratory cyclones across its latitudes. Also known as blocking anticyclone. { ¦bläk·iŋ 'hī }

block lava [GEOL] Lava flows which occur as a tumultuous assemblage of angular blocks. Also known as aa lava. { 'bläk ˌläv·ə }

block mountain [GEOL] A mountain formed by the combined processes of uplifting, faulting, and tilting. Also known as fault-block mountain. { 'bläk ˌmau̇n·tən }

blocky iceberg [OCEANOGR] An iceberg with steep, precipitous side and with a horizontal or nearly horizontal upper surface. { ¦bläk·ē 'īsˌbərg }

blood rain [METEOROL] Rain of a reddish color caused by dust particles containing iron oxide that were picked up by the raindrops during descent. { 'bləd ˌrān }

bloom *See* blossom. { blüm }

blossom [GEOL] The oxidized or decomposed outcrop of a vein or coal bed. Also known as bloom. { 'bläs·əm }

blowdown [METEOROL] A wind storm that causes trees or structures to be blown down. { 'blōˌdau̇n }

blowhole [GEOL] A longitudinal tunnel opening in a sea cliff, on the upland side away from shore; columns of sea spray are thrown up through the opening, usually during storms. { 'blōˌhōl }

blowing cave [GEOL] A cave with an alternating air movement. Also known as breathing cave. { ¦blō·iŋ ¦kāv }

blowing dust [METEOROL] Dust picked up locally from the surface of the earth and blown about in clouds or sheets. { ¦blō·iŋ 'dəst }

blowing sand [METEOROL] Sand picked up from the surface of the earth by the wind and blown about in clouds or sheets. { ¦blō·iŋ ¦sand }

blowing snow [METEOROL] Snow lifted from the surface of the earth by the wind to a height of 6 feet (1.8 meters) or more (higher than drifting snow) and blown about in such quantities that horizontal visibility is restricted. { ¦blō·iŋ ¦snō }

blowing spray [METEOROL] Spray lifted from the sea surface by the wind and blown about in such quantities that horizontal visibility is restricted. { ¦blō·iŋ ¦sprā }

blowout [GEOL] Any of the various trough-, saucer-, or cuplike hollows formed by wind erosion on a dune or other sand deposit. [HYD] A bubbling spring which bursts from the ground behind a river levee when water at flood stage is forced under the levee through pervious layers of sand or silt. Also known as sand boil. { 'blōˌau̇t }

blowout dune *See* parabolic dune. { 'blōˌau̇t ˌdün }

blue band [GEOL] **1.** A layer of bubble-free, dense ice found in a glacier. **2.** A bluish clay found as a thin, persistent bed near the base of No. 6 coal everywhere in the Illinois-Indiana coal basin. { ¦blü 'band }

blue ground [GEOL] **1.** The decomposed peridotite or kimberlite that carries the diamonds in the South African mines. **2.** Strata of the coal measures, consisting principally of beds of hard clay or shale. { ¦blü ˌgrau̇nd }

blue ice [HYD] Pure ice in the form of large, single crystals that is blue owing to the scattering of light by the ice molecules; the purer the ice, the deeper the blue. { ¦blü 'īs }

blue magnetism [GEOPHYS] The magnetism displayed by the south-seeking end of a freely suspended magnet; this is the magnetism of the earth's north magnetic pole. { ¦blü 'mag·nəˌtiz·əm }

blue metal [GEOL] The common fine-grained blue-gray mudstone which is part of many of the coal beds of England. { 'blü ˌmed·əl }

blue mud [GEOL] A combination of terrigenous and deep-sea sediments having a bluish gray color due to the presence of organic matter and finely divided iron sulfides. { 'blü ˌməd }

blue-sky scale *See* Linke scale. { ¦blü¦skī 'skāl }

bluff [GEOGR] **1.** A steep, high bank. **2.** A broad-faced cliff. { bləf }
board coal *See* bituminous wood. { 'bȯrd ˌkōl }
bodily tide *See* earth tide. { ¦bäd·əl·ē 'tīd }
body [GEOGR] A separate entity or mass of water, such as an ocean or a lake. [GEOL] An ore body, or pocket of mineral deposit. { 'bäd·ē }
body wave [GEOPHYS] A seismic wave that travels within the earth, as distinguished from one that travels along the surface. { 'bäd·ē ˌwāv }
boehm lamellae [GEOL] Lines or bands with dusty inclusions that are subparallel to the basal plane of quartz. { ¦bāmlə'mel·ē }
bogen structure [GEOL] The structure of vitric tuffs composed largely of shards of glass. { 'bō·gən ˌstrək·chər }
boghead cannel shale [GEOL] A coaly shale that contains much waxy or fatty algae. { 'bägˌhed ¦kan·əl ˌshāl }
boghead coal [GEOL] Bituminous or subbituminous coal containing a large proportion of algal remains and volatile matter; similar to cannel coal in appearance and combustion. { 'bägˌhed ˌkōl }
boiler plate [GEOL] A fairly smooth surface on a cliff, consisting of flush or overlapping slabs of rock, having little or no foothold. [HYD] A crusty, frozen surface of snow. { 'bȯil·ər ˌplāt }
boiling spring [HYD] **1.** A spring which emits water at a high temperature or at boiling point. **2.** A spring located at the head of an interior valley and rising from the bottom of a residual clay basin. **3.** A rapidly flowing spring that develops strong vertical eddies. { 'bȯil·iŋ ˌspriŋ }
bole [GEOL] Any of various red, yellow, or brown earthy clays consisting chiefly of hydrous aluminum silicates. Also known as bolus; terra miraculosa. { bōl }
bolson [GEOL] In the southwestern United States, a basin or valley having no outlet. { bōlˌsän }
bolus *See* bole. { 'bō·ləs }
bomb [GEOL] Any large (greater than 64 millimeters) pyroclast ejected while viscous. { bäm }
bombiccite *See* hartite. { bäm'bēˌchīt }
bomb sag [GEOL] Depressed and deranged laminae mainly found in beds of fine-grained ash or tuff around an included volcanic bomb or block which fell on and became buried in the deposit. { 'bäm ˌsag }
bone bed [GEOL] Several thin strata or layers with many fragments of fossil bones, scales, teeth, and also organic remains. { 'bōn ˌbed }
bone coal [GEOL] Argillaceous coal or carbonaceous shale that is found in coal seams. { 'bōn ˌkōl }
Bononian [GEOL] Upper Jurassic (lower Portlandian) geologic time. { bə'nōn·ē·ən }
book structure [GEOL] A rock structure of numerous parallel sheets of slate alternating with quartz. { 'bu̇k ˌstrək·chər }
boorga *See* burga. { 'bu̇r·gə }
bora [METEOROL] A fall wind whose source is so cold that when the air reaches the lowlands or coast the dynamic warming is insufficient to raise the air temperature to the normal level for the region; hence it appears as a cold wind. { 'bȯr·ə }
bora fog [METEOROL] A dense fog caused when the bora lifts a spray of small drops from the surface of the sea. { 'bȯr·ə ˌfäg }
Boralf [GEOL] A suborder of the soil order Alfisol, dull brown or yellowish brown in color; occurs in cool or cold regions, chiefly at high latitudes or high altitudes. { 'bȯrˌalf }
border facies [GEOL] The outer portion of an igneous intrusion which differs in composition and texture from the main body. { 'bȯrd·ər ˌfā·shēz }
borderland [GEOL] One of the crystalline, continental landmasses postulated to have existed on the exterior (oceanward) side of geosynclines. { 'bȯrd·ərˌland }
borderland slope [GEOL] A declivity which indicates the inner margin of the borderland of a continent. { 'bȯrd·ərˌland 'slōp }
bore [OCEANOGR] **1.** A high, breaking wave of water, advancing rapidly up an estuary.

Also known as eager; mascaret; tidal bore. **2.** A submarine sand ridge, in very shallow water, whose crest may rise to intertidal level. { bȯr }

bornhardt [GEOL] A large dome-shaped granite-gneiss outcrop having the characteristics of an inselberg. { 'bȯrn,härt }

Boroll [GEOL] A suborder of the soil order Mollisol, characterized by a mean annual soil temperature of less than 8°C and by never being dry for 60 consecutive days during the 90-day period following the summer solstice. { 'bȯ,rȯl }

bosporus [GEOGR] A strait connecting two seas or a lake and a sea. { 'bäs·pə·rəs }

boss [GEOL] A large, irregular mass of crystalline igneous rock that formed some distance below the surface but is now exposed by denudation. { bȯs }

botryoid [GEOL] **1.** A mineral formation shaped like a bunch of grapes. **2.** Specifically, such a formation of calcium carbonate occurring in a cave. Also known as clusterite. { 'bä·trē,ȯid }

bottom [GEOL] **1.** The bed of a body of running or still water. **2.** *See* root. { 'bäd·əm }

bottom flow [HYD] A density current that is denser than any section of the surrounding water and that flows along the bottom of the body of water. Also known as underflow. { 'bäd·əm ,flō }

bottom ice *See* anchor ice. { 'bäd·əm ,īs }

bottomland [GEOL] A lowland formed by alluvial deposit about a lake basin or a stream. { 'bäd·əm,land }

bottom load *See* bed load. { 'bäd·əm ,lōd }

bottom moraine *See* ground moraine. { 'bäd·əm mə'rān }

bottomset beds [GEOL] Horizontal or gently inclined layers of finer material carried out and deposited on the bottom of a lake or sea in front of a delta. { 'bäd·əm,set ,bedz }

bottom terrace [GEOL] A landform deposited by streams with moderate or small bottom loads of coarse sand and gravel, and characterized by a broad, sloping surface in the direction of flow and a steep escarpment facing downstream. { 'bäd·əm ,ter·əs }

bottom time [OCEANOGR] In an operation involving diving by personnel, the total time of a dive measured from the moment the diver passes below the water surface until the beginning of the ascent. { 'bäd·əm ,tīm }

bottom water [HYD] Water lying beneath oil or gas in productive formations. [OCEANOGR] The water mass at the deepest part of a water column in the ocean. { 'bäd·əm ,wȯd·ər }

boudin [GEOL] One of a series of sausage-shaped segments found in a boudinage. { bü'dan }

boudinage [GEOL] A structure in which beds set in a softer matrix are divided by cross fractures into segments resembling pillows. { ¦büd·ən¦äzh }

Bouguer correction *See* Bouguer reduction. { bü'ger kə'rek·shən }

Bouguer gravity anomaly [GEOPHYS] A value that corrects the observed gravity for latitude and elevation variations, as in the free-air gravity anomaly, plus the mass of material above some datum (usually sea level) within the earth and topography. { bü'ger 'grav·əd·ē ə'näm·ə·lē }

Bouguer reduction [GEOL] A correction made in gravity work to take account of the station's altitude and the rock between the station and sea level. Also known as Bouguer correction. { bü'ger ri'dək·shən }

Bouguer's halo [METEOROL] A faint, white circular arc of light of about 39° radius around the antisolar point. Also known as Ulloa's ring. { bü'gerz 'hā,lō }

boulder [GEOL] A worn rock with a diameter exceeding 256 millimeters. Also spelled bowlder. { 'bōl·dər }

boulder barricade [GEOL] An accumulation of large boulders that is visible along a coast between low and half tide. { 'bōl·dər ,bar·ə,kād }

boulder belt [GEOL] A long, narrow accumulation of boulders elongately transverse to the direction of glacier movement. { 'bōl·dər ,belt }

boulder clay *See* till. { 'bōl·dər ,klā }

boulder pavement [GEOL] A surface of till with boulders; the till has been abraded to flatness by glacier movement. { 'bōl·dər ˌpāv·mənt }

boulder train [GEOL] Glacial boulders derived from one locality and arranged in a right-angled line or lines leading off in the direction in which the drift agency operated. { 'bōl·dər ˌtrān }

bounce cast [GEOL] A short ridge underneath a stratum fading out gradually in both directions. { 'baủns ˌkast }

bounce dive [OCEANOGR] A dive performed rapidly with an extremely short bottom time in order to keep decompression time to a minimum. { 'baủns ˌdīv }

boundary [GEOL] A line between areas occupied by rocks or formations of different type and age. { 'baủn·drē }

boundary layer [METEOROL] The lower portion of the atmosphere, extending to a height of approximately 1.2 miles (2 kilometers). { 'baủn·drē ˌlā·ər }

boundary wave [GEOPHYS] A seismic wave that propagates along a free surface or an interface between defined layers. { 'baủn·drē ˌwāv }

bourne [HYD] A small intermittent stream in a dry valley. { bủrn }

Bowie formula [GEOPHYS] A correction used for calculation of the local gravity anomaly on earth. { 'bō·ē ˌfȯrm·yə·lə }

bowlder *See* boulder. { 'bōl·dər }

box canyon [GEOGR] A canyon that has steep rock sides and a zigzag course, and is usually closed upstream. { 'bäks ˌkan·yən }

box fold [GEOL] A fold in which the broad, flat top of an anticline or the broad, flat bottom of a syncline is bordered by steeply dipping limbs. { 'bäks ˌfōld }

Box Hole [GEOL] A meteorite crater in central Australia, 575 feet (175 meters) in diameter. { 'bäks ˌhōl }

boxwork [GEOL] Limonite and other minerals which formed at one time as blades or plates along cleavage or fracture planes, after which the intervening material dissolved, leaving the intersecting blades or plates as a network. { 'bäksˌwərk }

brachypinacoid [GEOL] A pinacoid parallel to the vertical and the shorter lateral axis. { ˌbrak·i'pin·əˌkȯid }

brachysyncline [GEOL] A broad, short syncline. { ˌbrak·i'sinˌklīn }

brackish [HYD] **1.** Of water, having salinity values ranging from approximately 0.50 to 17.00 parts per thousand. **2.** Of water, having less salt than sea water, but undrinkable. { 'brak·ish }

Bradfordian [GEOL] Uppermost Devonian geologic time. { ˌbrad'fȯrd·ē·ən }

braided stream [HYD] A stream flowing in several channels that divide and reunite. { 'brād·əd ˌstrēm }

branch [HYD] A small stream that merges into another, generally bigger, stream. { branch }

branching bay *See* estuary. { 'branch·iŋ 'bā }

branchite *See* hartite. { 'branˌchīt }

brass [GEOL] A British term for sulfides of iron (pyrites) in coal. Also known as brasses. { bras }

brasses *See* brass. { 'bras·əz }

brave west winds [METEOROL] A nautical term for the strong and rather persistent westerly winds over the oceans in temperate latitudes, between 40° and 50°S. { ¦brāv ¦west 'winz }

Brazil Current [OCEANOGR] The warm ocean current that flows southward along the Brazilian coast below Natal; the western boundary current in the South Atlantic Ocean. { brə'zil ˌkər·ənt }

breached anticline [GEOL] An anticline that has been more deeply eroded in the center. Also known as scalped anticline. { ¦brēcht 'an·tiˌklīn }

breached cone [GEOL] A cinder cone in which lava has broken through the sides and broken material has been carried away. { 'brēcht ˌkōn }

breadcrust [GEOL] A surficial structure resembling a crust of bread, as the concretions formed by evaporation of salt water. { 'bredˌkrəst }

breadcrust bomb [GEOL] A volcanic bomb with a cracked exterior. { 'bredˌkrəst ˌbäm }

break [GEOGR] A significant variation of topography, such as a deep valley. [GEOL] *See* knickpoint. [METEOROL] **1.** A sudden change in the weather; usually applied to the end of an extended period of unusually hot, cold, wet, or dry weather. **2.** A hole or gap in a layer of clouds. { brāk }

breaker [OCEANOGR] A wave breaking on a shore, over a reef, or other mass in a body of water. { 'brā·kər }

breaker depth [OCEANOGR] The still-water depth measured at the point where a wave breaks. Also known as breaking depth. { 'brā·kər ˌdepth }

breaker terrace [GEOL] A type of shore found in lakes in glacial drift; the terrace is formed from stones deposited by waves. { 'brā·kər ˌter·əs }

breaking depth *See* breaker depth. { 'brāk·iŋ ˌdepth }

breaking-drop theory [GEOPHYS] A theory of thunderstorm charge separation based upon the suggested occurrence of the Lenard effect in thunderclouds, that is, the separation of electric charge due to the breakup of water drops. { ¦brāk·iŋ ¦dräp ˌthē·ə·rē }

breaks in overcast [METEOROL] In United States weather observing practice, a condition wherein the cloud cover is more than 0.9 but less than 1.0. { ¦brāks in 'ō·vərˌkast }

break thrust [GEOL] A thrust fault cutting across one limb of a fold. { 'brāk ˌthrəst }

breakup [HYD] The spring melting of snow, ice, and frozen ground; specifically, the destruction of the ice cover on rivers during the spring thaw. { 'brākˌəp }

breathing cave *See* blowing cave. { 'brēth·iŋ ˌkāv }

breccia dike [GEOL] A dike formed of breccia injected into the country rock. { 'brech·ə ˌdīk }

breeze [METEOROL] **1.** A light, gentle, moderate, fresh wind. **2.** In the Beaufort scale, a wind speed ranging from 4 to 31 miles (6.4 to 49.6 kilometers) per hour. { brēz }

Bretonian orogeny [GEOL] Post-Devonian diastrophism that is found in Nova Scotia. { bre'tōn·ē·ən ȯ'räj·ə·nē }

Bretonian strata [GEOL] Upper Cambrian strata in Cape Breton, Nova Scotia. { bre'tōn·ē·ən 'strad·ə }

brickfielder [METEOROL] A hot, dry, dusty north wind blowing from the interior across the southern coast of Australia. { 'brikˌfēl·dər }

bridled pressure plate [METEOROL] An instrument for measuring air velocity in which the pressure on a plate exposed to the wind is balanced by the force of a spring, and the deflection of the plate is measured by an inductance-type transducer. { ¦brīd·əld 'presh·ər ˌplāt }

briefing *See* pilot briefing. { 'brē·fiŋ }

bright band [METEOROL] The enhanced echo of snow as it melts to rain, as displayed on a range-height indicator scope. { 'brīt ˌband }

bright-banded coal *See* bright coal. { 'brīt ˌban·dəd ¦kōl }

bright coal [GEOL] A jet-black, pitchlike type of banded coal that is more compact than dull coal and breaks with a shell-shaped fracture; microscopic examination shows a consistency of more than 5% anthraxyllon and less than 20% opaque matter. Also known as bright-banded coal; brights. { 'brīt ˌkōl }

brights *See* bright coal. { brīts }

bright segment [GEOPHYS] A faintly glowing band which appears above the horizon after sunset or before sunrise. Also known as crepuscular arch; twilight arch. { 'brīt ˌseg·mənt }

brine [OCEANOGR] Sea water containing a higher concentration of dissolved salt than that of the ordinary ocean. { brīn }

brine spring [HYD] A salt-water spring. { 'brīn ˌspriŋ }

brisa [METEOROL] **1.** A northeast wind which blows on the coast of South America or an east wind which blows on Puerto Rico during the trade wind season. **2.** The northeast monsoon in the Philippines. Also spelled briza. { 'brē·sə }

brisa carabinera *See* carabine. { 'brē·sə ˌkär·ə·bi'ner·ə }

brise carabinée *See* carabine. { ˌbrēz kä·rä·bē'nā }

brisote [METEOROL] The northeast trade wind when it is blowing stronger than usual on Cuba. { brē'sȯ·tā }

briza *See* brisa. { 'brē·zə }

Brocken specter [METEOROL] The illusory appearance of a gigantic figure (actually, the observer's shadow projected on cloud surfaces), observed on the Brocken peak in the Hartz Mountains of Saxony, but visible from other mountaintops under suitable conditions. { 'bräk·ən ˌspek·tər }

broeboe [METEOROL] A strong, dry east wind in the southwestern part of the island of Celebes. { ¦brō·ə¦bō·ə }

broken [METEOROL] Descriptive of a sky cover of from 0.6 to 0.9 (expressed to the nearest tenth). { 'brō·kən }

broken belt [OCEANOGR] The transition zone between open water and consolidated ice. { ¦brō·kən 'belt }

broken stream [HYD] A stream that repeatedly disappears and reappears, such as occurs in an arid region. { ¦brō·kən 'strēm }

broken water [OCEANOGR] Water having a surface covered with ripples or eddies, and usually surrounded by calm water. { ¦brō·kən 'wȯd·ər }

brontides [GEOPHYS] Low, rumbling, thunderlike sounds of short duration, most frequently heard in active seismic regions and believed to be of seismic origin. { 'brän,tīdz }

brown clay *See* red clay. { ¦braün ¦klā }

brown clay ironstone [GEOL] Limonite in the form of concrete masses, often in concretionary nodules. { 'braün ˌklā 'ī·ərn,stōn }

brown coal *See* lignite. { ¦braün ¦kōl }

brown lignite [GEOL] A type of lignite with a fixed carbon content ranging from 30 to 55% and total carbon from 65 to 73.6; contains 6300 Btu per pound (14.65 megajoules per kilogram). Also known as lignite B. { ¦braün 'lig,nīt }

brown snow [METEOROL] Snow intermixed with dust particles. { ¦braün ¦snō }

brown soil [GEOL] Any of a zonal group of soils, with a brown surface horizon which grades into a lighter-colored soil and then into a layer of carbonate accumulation. { ¦braün ¦sȯil }

brown spar [GEOL] Any light-colored crystalline carbonate that contains iron, such as ankerite or dolomite, and is therefore brown. { 'braün ˌspär }

brubru [METEOROL] A squall in Indonesia. { 'brü,brü }

Brückner cycle [CLIMATOL] An alternation of relatively cool-damp and warm-dry periods, forming an apparent cycle of about 35 years. { 'brük·nər ˌsī·kəl }

bruma [METEOROL] A haze that appears in the afternoons on the coast of Chile when sea air is transported inland. { 'brü·mə }

Brunt-Douglas isallobaric wind *See* isallobaric wind. { ¦brənt ¦dəg·ləs ˌī¦sa·lə¦bär·ik 'wind }

brüscha [METEOROL] A northwest wind in the Bergell Valley, Switzerland. { 'brüsh·ə }

Bruxellian [GEOL] Lower middle Eocene geologic time. { brü'sel·yən }

bubble *See* bubble high. { 'bəb·əl }

bubble high [METEOROL] A small high, complete with anticyclonic circulation, of the order of 50 to 300 miles (80 to 480 kilometers) across, often induced by precipitation and vertical currents associated with thunderstorms. Also known as bubble. { 'bəb·əl ˌhī }

bubble pulse [GEOPHYS] An extraneous effect during a seismic survey caused by a bubble formed by a seismic charge, explosion, or spark fired in a body of water. { 'bəb·əl ˌpəls }

bubble train [GEOL] A string or strings of vesicles in lava, indicating the path of rising gas escaping a flow of lava. { 'bəb·əl ˌtrān }

bubble wall fragment [GEOL] A glassy volcanic shard revealing part of a vesicle surface which may be curved or flat. { 'bəb·əl ˌwȯl ˌfrag·mənt }

buckle fold [GEOL] A double flexure of rock beds formed by compression acting in the plane of the folded beds. { 'bək·əl ˌfōld }

buckwheat coal [GEOL] An anthracite coal that passes through 9/16-inch (14-millimeter) holes and over 5/16-inch (8-millimeter) holes in a screen. { 'bək,wēt ,kōl }

budget year [METEOROL] The 1-year period beginning with the start of the accumulation season at the firn line of a glacier or ice cap and extending through the following summer's ablation season. { 'bəj·ət ,yir }

built terrace *See* alluvial terrace. { ¦bilt ¦ter·əs }

bulb glacier [HYD] A glacier formed at the foot of a mountain and out into an open slope; the glacier ends spread out into an ice fan. { 'bəlb ,glā·shər }

bull-eye squall [METEOROL] A squall forming in fair weather, characteristic of the ocean off the coast of South Africa; it is named for the peculiar appearance of the small, isolated cloud marking the top of the invisible vortex of the storm. { 'bəlz ,ī ,skwȯl }

Bunter [GEOL] Lower Triassic geologic time. Also known as Buntsandstein. { 'bu̇n·tər }

Buntsandstein *See* Bunter. { 'bu̇nt·sən,shtīn }

buran [METEOROL] A violent northeast storm of south Russia and central Siberia, similar to the blizzard. { bü'rän }

burden [GEOL] All types of rock or earthy materials overlying bedrock. { 'bərd·ən }

Burdigalian [GEOL] Upper lower Miocene geologic time. { ,bərd·i'gāl·yən }

burga [METEOROL] A storm of wind and sleet in Alaska. Also spelled boorga. { 'bu̇r·gə }

Burgess Shale [GEOL] A fossil deposit in the Canadian Rockies, British Columbia, consisting of a diverse fauna that accumulated in a clay and silt sequence during the Cambrian. { ¦bər·jəs 'shāl }

burial metamorphism [GEOL] A kind of regional metamorphism which affects sediments and interbedded volcanic rocks in a geosyncline without the factors of orogenesis or magmatic intrusions. { 'ber·ē·əl med·ə'mȯr,fiz·əm }

buried hill [GEOL] A hill of resistant older rock over which later sediments are deposited. { 'ber·ēd 'hil }

buried placer [GEOL] Old deposit of a placer which has been buried beneath lava flows or other strata. { 'ber·ēd 'plās·ər }

buried river [GEOL] A river bed which has become buried beneath streams of alluvial drifts or basalt. { 'ber·ēd 'riv·ər }

buried soil *See* paleosol. { 'ber·ēd 'sȯil }

burn off [METEOROL] With reference to fog or low stratus cloud layers, to dissipate by daytime heating from the sun. { ¦bərn 'ȯf }

Busch lemniscate [METEOROL] The locus in the sky, or on a diagrammatic representation thereof, of all points at which the plane of polarization of diffuse sky radiation is inclined 45° to the vertical; a polarization isocline. Also known as neutral line. { ¦bu̇sh lem'nis·kət }

Bushveld Complex [GEOL] In South Africa, an enormous layered intrusion, containing over half the world's platinum, chromium, vanadium, and refractory minerals. { ¦bu̇sh,veld 'käm,pleks }

bustite *See* aubrite. { 'bəs,tīt }

butte [GEOGR] A detached hill or ridge which rises abruptly. { byüt }

buttress sands [GEOL] Sandstone bodies deposited above an unconformity; the upper portion rests upon the surface of the unconformity. { 'bə·trəs ,sanz }

Buys-Ballot's law [METEOROL] A law describing the relationship of the horizontal wind direction in the atmosphere to the pressure distribution: if one stands with one's back to the wind, the pressure to the left is lower than to the right in the Northern Hemisphere; in the Southern Hemisphere the relation is reversed. Also known as baric wind law. { 'bīz bə'läts ,lȯ }

byerite [GEOL] Bituminous coal that does not crack in fire and melts and enlarges upon heating. { 'bī·ə,rīt }

byon [GEOL] Gem-bearing gravel, particularly that with brownish-yellow clay in which corundum, rubies, sapphires, and so forth occur. { 'bī,än }

bysmalith [GEOL] A body of igneous rock that is more or less vertical and cylindrical; it crosscuts adjacent sediments. { 'biz·mə,lith }

C

caballing [OCEANOGR] The mixing of two water masses of identical in situ densities but different in situ temperatures and salinities, such that the resulting mixture is denser than its components and therefore sinks. { kə'bal·iŋ }

cable *See* cable length. { 'kā·bəl }

cable length [OCEANOGR] A unit of distance, originally equal to the length of a ship's anchor cable, now variously considered to be 600 feet (183 meters), 608 feet (185.3 meters, one-tenth of a British nautical mile), or 720 feet or 120 fathoms (219.5 meters). Also known as cable. { 'kā·bəl ˌlengkth }

cacimbo [METEOROL] Local name in Angola for the wet fogs and drizzles noted with onshore winds from the Benguela Current. { kä'simˌbō }

Cainozoic *See* Cenozoic. { ˌkān·ə'zō·ik }

caju rains [METEOROL] In northeastern Brazil, light showers that occur during the month of October. { ¦kä¦zhü ˌrānz }

caking coal [GEOL] A type of coal which agglomerates and softens upon heating; after volatile material has been expelled at high temperature, a hard, gray cellular mass of coke remains. Also known as binding coal. { 'kāk·iŋ ˌkōl }

Calabrian [GEOL] Lower Pleistocene geologic time. { kə'läb·rē·ən }

calcareous crust *See* caliche. { kal'ker·ē·əs 'krəst }

calcareous duricrust *See* caliche. { kal'ker·ē·əs 'dür·iˌkrəst }

calcareous ooze [GEOL] A fine-grained pelagic sediment containing undissolved sand- or silt-sized calcareous skeletal remains of small marine organisms mixed with amorphous clay-sized material. { kal'ker·ē·əs 'üz }

calcareous sinter *See* tufa. { kal'ker·ē·əs 'sin·tər }

calcareous soil [GEOL] A soil containing accumulations of calcium and magnesium carbonate. { kal'ker·ē·əs 'söil }

calcareous tufa *See* tufa. { kal'ker·ē·əs 'tü·fə }

calcification [GEOCHEM] Any process of soil formation in which the soil colloids are saturated to a high degree with exchangeable calcium, thus rendering them relatively immobile and nearly neutral in reaction. { ˌkal·sə·fə'kā·shən }

calcite compensation depth [GEOL] The depth in the ocean (about 5000 meters) below which solution of calcium carbonate occurs at a faster rate than its deposition. Abbreviated CCD. { 'kalˌsīt käm·pən'sā·shən ˌdepth }

calcrete [GEOL] A conglomerate of surficial gravel and sand cemented by calcium carbonate. { ¦kal¦krēt }

calc-silicate [GEOL] Referring to a metamorphic rock consisting mainly of calcite and calcium-bearing silicates. { 'kalk 'sil·ə·kət }

caldera [GEOL] A large collapse depression at a volcano summit that is typically circular to slightly elongate in shape, with dimensions many times greater than any included vent. It ranges from a few miles to 37 miles (60 kilometers) in diameter. It may resemble a volcanic crater in form, but differs in that it is a collapse rather than a constructional feature. { kal'der·ə }

Caledonian orogeny [GEOL] Deformation of the crust of the earth by a series of diastrophic movements beginning perhaps in Early Ordovician and continuing through Silurian, extending from Great Britain through Scandinavia. { ¦kal·ə¦dōn·ē·ən ȯ'räj·ə·nē }

Caledonides [GEOL] A mountain system formed in Late Silurian to Early Devonian time in Scotland, Ireland, and Scandinavia. { ˌkal·əˈdäˌnīdz }

calf *See* calved ice. { kaf }

caliche [GEOL] **1.** Conglomerate of gravel, rock, soil, or alluvium cemented with sodium salts in Chilean and Peruvian nitrate deposits; contains sodium nitrate, potassium nitrate, sodium iodate, sodium chloride, sodium sulfate, and sodium borate. **2.** A thin layer of clayey soil capping auriferous veins (Peruvian usage). **3.** Whitish clay in the selvage of veins (Chilean usage). **4.** A recently discovered mineral vein. A secondary accumulation of opaque, reddish brown to buff or white calcareous material occurring in layers on or near the surface of stony soils in arid and semiarid regions of the southwestern United States; called hardpan, calcareous duricrust, and kanker in different geographic regions. Also known as calcareous crust; croute calcaire; nari; sabach; tepetate. { kəˈlē·chē }

California Current [OCEANOGR] The ocean current flowing southward along the western coast of the United States to northern Baja California. { ¦kal·ə¦fȯr·nyə }

California fog [METEOROL] Fog peculiar to the coast of California and its coastal valleys; off the coast, winds displace warm surface water, causing colder water to rise from beneath, resulting in the formation of fog; in the coastal valleys, fog is formed when moist air blown inland during the afternoon is cooled by radiation during the night. { ¦kal·ə¦fȯr·nyə ˈfäg }

California method [HYD] A form of frequency analysis which employs the return period, a parameter that measures the average time period between the occurrence of a quantity in hydrology and that of an equal or greater quantity, as the plotting position. { ¦kal·ə¦fȯr·nyə ˈmeth·əd }

calina [METEOROL] A haze prevalent in Spain during the summer, when the air becomes filled with dust swept up from the dry ground by strong winds. { kəˈlēn·ə }

Callao painter *See* painter. { kəˈyau̇ ˌpān·tər }

callenia *See* stromatolite. { kəˈlēn·yə }

Callovian [GEOL] A stage in uppermost Middle or lowermost Upper Jurassic which marks a return to clayey sedimentation. { kəˈlōv·ē·ən }

calm [METEOROL] The absence of apparent motion of the air; in the Beaufort wind scale, smoke is observed to rise vertically, or the surface of the sea is smooth and mirrorlike; in U.S. weather observing practice, the wind has a speed under 1 mile per hour or 1 knot (1.6 kilometers per hour). { käm }

calm belt [METEOROL] A belt of latitude in which the winds are generally light and variable; the principal calm belts are the horse latitudes (the calms of Cancer and of Capricorn) and the doldrums. { ˈkäm ˌbelt }

calms of Cancer [METEOROL] One of the two light, variable winds and calms which occur in the centers of the subtropical high-pressure belts over the oceans; their usual position is about latitude 30°N, the horse latitudes. { ¦kämz əv ˈkan·sər }

calms of Capricorn [METEOROL] One of the two light, variable winds and calms which occur in the centers of the subtropical high-pressure belts over the oceans; their usual position is about latitude 30°S, the horse latitudes. { ¦kämz əv ˈkap·riˌkȯrn }

calved ice [OCEANOGR] A piece of ice floating in a body of water after breaking off from a mass of land ice or an iceberg. Also known as calf. { ¦kavd ˈīs }

calving [GEOL] The breaking off of a mass of ice from its parent glacier, iceberg, or ice shelf. Also known as ice calving. { ˈkav·iŋ }

camanchaca *See* garúa. { kä·mänˈchä·kə }

camber [GEOL] **1.** A terminal, convex shoulder of the continental shelf. **2.** A structural feature that is caused by plastic clay beneath a bed flowing toward a valley so that the bed sags downward and seems to be draped over the sides of the valley. { ˈkam·bər }

Cambrian [GEOL] The lowest geologic system that contains abundant fossils of animals, and the first (earliest) geologic period of the Paleozoic era from 570 to 500 million years ago. { ˈkam·brē·ən }

Campanian [GEOL] European stage of Upper Cretaceous. { kamˈpan·ē·ən }

Canadian Shield *See* Laurentian Shield. { kəˈnād·ē·ən ˈshēld }

canal [GEOGR] A long, narrow arm of the sea extending far inland, between islands, or between islands and the mainland. { kə'nal }

Canary Current [OCEANOGR] The prevailing southward flow of water along the northwestern coast of Africa. { kə'ner·ē ˌkər·ənt }

Canastotan [GEOL] Lower Upper Silurian geologic time. { kə'nas·tə·tən }

Candlemas crack *See* Candlemas Eve winds. { 'kan·dəl·məs ˌkrak }

Candlemas Eve winds [METEOROL] Heavy winds often occurring in Great Britain in February or March (Candlemas is February 2). Also known as Candlemas crack. { 'kan·dəl·məs ˌēv 'winz }

cannel coal [GEOL] A fine-textured, highly volatile bituminous coal distinguished by a greasy luster and blocky, conchoidal fracture; burns with a steady luminous flame. Also known as cannelite. { 'kan·əl ˌkōl }

cannelite *See* cannel coal. { 'kan·əlˌīt }

canneloid [GEOL] **1.** Coal that resembles cannel coal. **2.** Coal intermediate between bituminous and cannel. **3.** Durain laminae in banded coal. **4.** Cannel coal of anthracite or semianthracite rank. { 'kan·əlˌȯid }

cannel shale [GEOL] A black shale formed by the accumulation of an aquatic ooze rich in bituminous organic matter in association with inorganic materials such as silt and clay. { 'kan·əl ˌshāl }

Canterbury northwester [METEOROL] A strong northwest foehn descending the New Zealand Alps onto the Canterbury plains of South Island, New Zealand. { ˌkan·təˌber·ē ˌnȯrth'wes·tər }

canyon [GEOGR] A chasm, gorge, or ravine cut in the surface of the earth by running water; the sides are steep and form cliffs. { 'kan·yən }

canyon bench [GEOL] A steplike level of hard strata in the walls of deep valleys in regions of horizontal strata. { 'kan·yən ˌbench }

canyon fill [GEOL] Loose, unconsolidated material which fills a canyon to a depth of 50 feet (15 meters) or more during periods between great floods. { 'kan·yən ˌfil }

canyon wind [METEOROL] **1.** Also known as gorge wind. **2.** The mountain wind of a canyon; that is, the nighttime down-canyon flow of air caused by cooling at the canyon walls. **3.** Any wind modified by being forced to flow through a canyon or gorge; its speed may be increased as a jet-effect wind, and its direction is rigidly controlled. { 'kan·yən ¦wind }

capacity of the wind [GEOL] The total weight of airborne particles (soil and rock) of given size, shape, and specific gravity, which can be carried in 1 cubic mile (4.17 cubic kilometers) of wind blowing at a given speed. { kə'pas·əd·ē əv thə ¦wind }

cap cloud [METEOROL] An approximately stationary cloud, or standing cloud, on or hovering above an isolated mountain peak; formed by the cooling and condensation of humid air forced up over the peak. Also known as cloud cap. { 'kap ˌklau̇d }

cape [GEOGR] A prominent point of land jutting into a body of water. Also known as head; headland; mull; naze; ness; point; promontory. { kāp }

cape doctor [METEOROL] The strong southeast wind which blows on the South African coast. { 'kāp ˌdäk·tər }

Cape Horn Current [OCEANOGR] That part of the west wind drift flowing eastward in the immediate vicinity of Cape Horn, and then curving northeastward to continue as the Falkland Current. { ˌkāp ˌhȯrn 'kər·ənt }

capillary [GEOL] A fissure or a crack in a formation which provides a route for flow of water or hydrocarbons. { 'kap·əˌler·ē }

capillary ejecta *See* Pele's hair. { 'kap·əˌler·ē i'jek·tə }

capillary fringe [HYD] The lower subdivision of the zone of aeration that overlies the zone of saturation and in which the pressure of water in the interstices is lower than atmospheric. { 'kap·əˌler·ē ¦frinj }

capillary migration [HYD] Movement of water produced by the force of molecular attraction between rock material and the water. { 'kap·əˌler·ē mī'grā·shən }

capillary water [HYD] Soil water held by capillarity as a continuous film around soil particles and in interstices between particles above the phreactic line. { 'kap·əˌler·ē ˌwȯd·ər }

capped column [HYD] A form of ice crystal consisting of a hexagonal column with plate or stellar crystals (so-called caps) at its ends and sometimes at intermediate positions; the caps are perpendicular to the column. { 'kapt 'käl·əm }

capping [GEOL] **1.** Consolidated barren rock overlying a mineral or ore deposit. **2.** *See* gossan. { 'kap·iŋ }

cap rock [GEOL] **1.** An overlying, generally impervious layer or stratum of rock that overlies an oil- or gas-bearing rock. **2.** Barren vein matter, or a pinch in a vein, supposed to overlie ore. **3.** A hard layer of rock, usually sandstone, a short distance above a coal seam. **4.** An impervious body of anhydrite and gypsum in a salt dome. { 'kap ˌräk }

capture [GEOCHEM] In a crystal structure, the substitution of a trace element for a lower-valence common element. [HYD] The natural diversion of the headwaters of one stream into the channel of another stream having greater erosional activity and flowing at a lower level. Also known as piracy; river capture; river piracy; robbery; stream capture; stream piracy; stream robbery. { 'kap·chər }

carabine [METEOROL] In France and Spain, a sudden and violent wind. Also known as brisa carabinera; brise carabinée. { kär·ə·bēn *or* kär·ə·bē·nā }

Caradocian [GEOL] Lower Upper Ordovician geologic time. { kar·ə'dō·shən }

carapace [GEOL] The upper normal limb of a fold having an almost horizontal axial plane. { 'kar·əˌpās }

carbohumin *See* ulmin. { ˌkär·bō'hyü·mən }

carbonaceous chondrite [GEOL] A chondritic meteorite that contains a relatively large amount of carbon and has a resulting dark color. Also known as carbonaceous meteorite. { ˌkär·bə'nā·shəs 'känˌdrīt }

carbonaceous meteorite *See* carbonaceous chondrite. { kär·bə'nā·shəs 'mēd·ē·əˌrīt }

carbonaceous shale [GEOL] Shale rich in carbon. { kär·bə'nā·shəs 'shāl }

carbonate cycle [GEOCHEM] The biogeochemical carbonate pathways, involving the conversion of carbonate to CO_2 and HCO_3, the solution and deposition of carbonate, and the metabolism and regeneration of it in biological systems. { 'kär·bə·nət ˌsī·kəl }

carbonate reservoir [GEOL] An underground oil or gas trap formed in reefs, clastic limestones, chemical limestones, or dolomite. { 'kär·bə·nət 'rez·əvˌwär }

carbonate spring [HYD] A type of spring containing dissolved carbon dioxide gas. { 'kär·bə·nət 'spriŋ }

carbon cycle [GEOCHEM] The cycle of carbon in the biosphere, in which plants convert carbon dioxide to organic compounds that are consumed by plants and animals, and the carbon is returned to the biosphere in the form of inorganic compounds by processes of respiration and decay. { 'kär·bən ˌsī·kəl }

carbonation [GEOCHEM] A process of chemical weathering whereby minerals that contain soda, lime, potash, or basic oxides are changed to carbonates by the carbonic acid in air or water. { ˌkär·bə'nā·shən }

Carboniferous [GEOL] A division of late Paleozoic rocks and geologic time including the Mississippian and Pennsylvanian periods. { ˌkär·bə'nif·ə·rəs }

carbonification *See* coalification. { kärˌbän·ə·fə'kā·shən }

carbon isotope ratio [GEOL] Ratio of carbon-12 to either of the less common isotopes, carbon-13 or carbon-14, or the reciprocal of one of these ratios; if not specified, the ratio refers to carbon-12/carbon-13. Also known as carbon ratio. { ¦kar·bən 'is·əˌtōp ˌrā·shō }

carbonite *See* natural coke. { 'kär·bəˌnīt }

carbonization [GEOCHEM] **1.** In the coalification process, the accumulation of residual carbon by changes in organic material and their decomposition products. **2.** Deposition of a thin film of carbon by slow decay of organic matter underwater. **3.** A process of converting a carbonaceous material to carbon by removal of other components. { ˌkär·bə·nə'zā·shən }

carbon-nitrogen-phosphorus ratio [OCEANOGR] The relatively constant relationship between the concentrations of carbon (C), nitrogen (N), and phosphorus (P) in plankton, and N and P in sea water, owing to removal of the elements by the organisms in the same proportions in which the elements occur and their return upon decomposition of the dead organisms. { ¦kär·bən ¦nī·trə·jən ¦fäs·fə·rəs ˌrā·shō }

carbon pool [GEOCHEM] A reservoir with the capacity to store and release carbon, such as soil, terrestrial vegetation, the ocean, and the atmosphere. { 'kär·bən ˌpül }

carbon ratio [GEOL] **1.** The ratio of fixed carbon to fixed carbon plus volatile hydrocarbons in a coal. **2.** *See* carbon isotope ratio. { 'kär·bən ˌrā·shō }

carbon-ratio theory [GEOL] The theory that the gravity of oil in any area is inversely proportional to the carbon ratio of the coal. { 'kär·bən ˌrā·shō ˌthē·ə·rē }

carbon sequestration [GEOCHEM] The uptake and storage of atmospheric carbon in, for example, soil and vegetation. { ˌkär·bən ˌsē·kwes'trā·shən }

carbon sink [GEOCHEM] A reservoir that absorbs or takes up atmospheric carbon; for example, a forest or an ocean. { 'kär·bən ˌsiŋk }

carcenet [METEOROL] A very cold and violent gorge wind in the eastern Pyrenees (upper Aude valley). { kärs'nā }

cardinal point [GEOD] Any of the four principal directions: north, east, south, or west of a compass. { 'kärd·nəl ¦pȯint }

cardinal winds [METEOROL] Winds from the four cardinal points of the compass, that is, north, east, south, and west winds. { 'kärd·nəl ˌwinz }

Caribbean Current [OCEANOGR] A water current flowing westward through the Caribbean Sea. { kar·ə'bē·ən 'kər·ənt }

Caribbean Sea [GEOGR] One of the largest and deepest enclosed basins in the world, surrounded by Central and South America and the West Indian island chains. { kar·ə'bē·ən 'sē }

Carnian [GEOL] Lower Upper Triassic geologic time. Also spelled Karnian. { 'kärn·ē·ən }

Carolina Bays [GEOGR] Shallow, marshy, often ovate depressions on the coastal plain of the mideastern and southeastern United States of unknown origin. { ˌkar·ə'lī·nə 'bāz }

carry-over [HYD] The portion of the stream flow during any month or year derived from precipitation in previous months or years. { 'kar·ē ˌō·vər }

cascade [GEOL] A landform structure formed by gravity collapse, consisting of a bed that buckles into a series of folds as it slides down the flanks of an anticline. [HYD] A small waterfall or series of falls descending over rocks. { ka'skād }

Cascadian orogeny [GEOL] Post-Tertiary deformation of the crust of the earth in western North America. { ka'skād·ē·ən ȯ'räj·ə·nē }

cascading glacier [HYD] A glacier broken by numerous crevasses because of passing over a steep irregular bed, giving the appearance of a cascading stream. { ka'skād·iŋ 'glā·shər }

case hardening [GEOL] Formation of a mineral coating on the surface of porous rock by evaporation of a mineral-bearing solution. { 'kās ˌhärd·ən·iŋ }

Cassadagan [GEOL] Middle Upper Devonian geologic time, above Chemungian. { kə'sad·ə·gən }

Casselian *See* Chattian. { ka'sel·yən }

Cassiar orogeny [GEOL] Orogenic episode in the Canadian Cordillera during late Paleozoic time. { 'kas·ē·ər ȯ'räj·ə·nē }

castellanus [METEOROL] A cloud species with at least a fraction of its upper part presenting some vertically developed cumuliform protuberances (some of which are more tall than wide) which give the cloud a crenellated or turreted appearance. Previously known as castellatus. { ˌkas·tə'län·əs }

castellatus *See* castellanus. { ˌkas·təˌläd·əs }

castings *See* fecal pellets. { 'kast·iŋz }

CAT *See* clear-air turbulence.

catachosis [GEOL] Fracturing or crushing of rock during metamorphism. { ˌkad·ə'kō·səs }

cataclasis [GEOL] Deformation of rock by fracture and rotation of aggregates or mineral grains. { ˌkad·ə'klā·səs }

cataract [HYD] A waterfall of considerable volume with the vertical fall concentrated in one sheer drop. { 'kad·ə,rakt }
catastrophism [GEOL] The theory that most features in the earth were produced by the occurrence of sudden, short-lived, worldwide events. { kə'tas·trə,fiz·əm }
catazone [GEOL] The deepest zone of rock metamorphism where high temperatures and pressures prevail. { 'kad·ə,zōn }
catchment area [GEOGR] The rural-urban outskirts of a particular city. [HYD] *See* drainage basin. { 'kach·mənt ,er·ē·ə }
catchment glacier *See* snowdrift glacier. { 'kach·mənt ,glā·shər }
catena [GEOL] A group of soils derived from uniform or similar parent material which nonetheless show variations in type because of differences in topography or drainage. { kə'tē·nə }
cat's paw [METEOROL] A puff of wind; a light breeze affecting a small area, as one that causes patches of ripples on the surface of water. { 'kats ,pȯ }
cauldron subsidence [GEOL] **1.** A structure formed by the lowering along a steep ring fracture of a more or less cylindrical block, usually 1 to 10 miles (1.6 to 16 kilometers) in diameter, into a magma chamber. **2.** The process of forming such a structure. { 'kȯl·drən səb'sī·dəns }
caustobiolith [GEOL] Combustible organic rock formed by direct accumulation of plant materials; includes coal peat. { ¦kȯ,stō'bī·ə,lith }
cavaburd [METEOROL] Shetland Islands term for a thick fall of snow. Also spelled kavaburd. { 'ka·və·bərd }
cavaliers [METEOROL] The local term, in the vicinity of Montpelier, France, for the days near the end of March or the beginning of April when the mistral is usually strongest. { kä'väl·yā }
cave [GEOL] A natural, hollow chamber or series of chambers and galleries beneath the earth's surface, or in the side of a mountain or hill, with an opening to the surface. { kāv }
cave breccia [GEOL] Sharp fragments of limestone debris deposited on the floor of a cave. { ¦kāv 'brech·ə }
cave formation *See* speleothem. { 'kāv fȯr'mā·shən }
cave pearl [GEOL] A small, smooth, rounded concretion of calcite or aragonite, formed by concentric precipitation about a nucleus and usually found in limestone caves. { 'kāv ,pərl }
caver [METEOROL] A gentle breeze in the Hebrides, west of Scotland. Also spelled kaver. { 'kāv·ər }
cavern [GEOL] An underground chamber or series of chambers of indefinite extent carved out by rock springs in limestone. { 'kav·ərn }
cavernous [GEOL] **1.** Having many caverns or cavities. **2.** Producing caverns. **3.** Of or pertaining to a cavern, that is, suggesting vastness. { 'kav·ər·nəs }
c axis [GEOL] The reference axis perpendicular to the plane of movement of rock or mineral strata. { 'sē ,ak·səs }
cay [GEOL] **1.** A flat coral island. **2.** A flat mound of sand built up on a reef slightly above high tide. **3.** A small, low coastal islet or emergent reef composed largely of sand or coral. { kā }
cay sandstone [GEOL] Firmly cemented or friable coral sand formed near the base of coral reef cays. { ¦kā 'san,stōn }
Cayugan [GEOL] Upper Silurian geologic time. { kī'yü·gən }
Cazenovian [GEOL] Lower Middle Devonian geologic time. { kaz·ə'nōv·ē·ən }
Cc *See* cirrocumulus cloud.
CCD *See* calcite compensation depth.
ceiling [METEOROL] In the United States, the height ascribed to the lowest layer of clouds or of obscuring phenomena when it is reported as broken, overcast, or obscuration and not classified as thin or partial. { 'sē·liŋ }
ceiling classification [METEOROL] In aviation weather observations, a description or explanation of the manner in which the height of the ceiling is determined. { 'sē·liŋ ,klas·ə·fə'kā·shən }

celestial geodesy [GEOD] The branch of geodesy which utilizes observations of near celestial bodies and earth satellites to determine the size and shape of the earth. { sə'les·chəl jē'äd·ə·sē }

cellular convection [METEOROL] An organized, convective, fluid motion characterized by the presence of distinct convection cells or convective units, usually with upward motion (away from the heat source) in the central portions of the cell, and sinking or downward flow in the cell's outer regions. { 'sel·yə·lər kən'vek·shən }

cellular soil *See* polygonal ground. { 'sel·yə·lər 'sȯil }

cement [GEOL] Any chemically precipitated material, such as carbonates, gypsum, and barite, occurring in the interstices of clastic rocks. { si'ment }

cementation [GEOL] The precipitation of a binding material around minerals or grains in rocks. { ˌsēˌmen'tā·shən }

cement gravel [GEOL] Gravel consolidated by clay, silica, calcite, or other binding material. { si'ment ˌgrav·əl }

Cenomanian [GEOL] Lower Upper Cretaceous geologic time. { ¦sen·ə¦mān·ē·ən }

cenote *See* pothole. { sə'nōd·ē }

Cenozoic [GEOL] The youngest of the eras, or major subdivisions of geologic time, extending from the end of the Mesozoic Era to the present, or Recent. Also spelled as Cainozoic. { ¦sen·ə¦zō·ik }

center jump [METEOROL] The formation of a second low-pressure center within an already well-developed low-pressure center; the latter diminishes in magnitude as the center of activity shifts or appears to jump to the new center. { 'sen·tər ¦jəmp }

center of action [METEOROL] A semipermanent high or low atmospheric pressure system at the surface of the earth; fluctuations in the intensity, position, orientation, shape, or size of such a center are associated with widespread weather changes. { 'sen·tər əv 'ak·shən }

center of falls *See* pressure-fall center. { 'sen·tər əv 'fȯlz }

center of rises *See* pressure-rise center. { 'sen·tər əv 'rīz·əz }

central pressure [METEOROL] At any given instant, the atmospheric pressure at the center of a high or low; the highest pressure in a high, the lowest pressure in a low. { 'sen·trəl 'presh·ər }

central valley *See* rift valley. { 'sen·trəl 'val·ē }

central water [OCEANOGR] Upper water mass associated with the central region of oceanic gyre. { 'sen·trəl 'wȯd·ər }

centrifugal drainage pattern *See* radial drainage pattern. { ˌsen'trif·i·gəl 'drān·ij ˌpad·ərn }

centroclinal [GEOL] Referring to geologic strata dipping toward a common center, as in a structural basin. { ¦sen·trō¦klīn·əl }

centrosphere [GEOL] The central core of the earth. Also known as the barysphere. { 'sen·trəˌsfir }

cers [METEOROL] A term for the mistral in Catalonia, Narbonne, and parts of Provence (southern France and northeastern Spain). { sərz }

C figure *See* C index. { 'sē ˌfig·yər }

chain [GEOL] A series of interconnected or related natural features, such as lakes, islands, or seamounts, arranged in a longitudinal sequence. { chān }

chain lightning [GEOPHYS] A rare form of lightning in a long zigzag or apparently broken line. { 'chān ˌlīt·niŋ }

chalazoidite *See* mud ball. { 'kal·əˌzȯiˌdīt }

chalcophile [GEOL] Having an affinity for sulfur and therefore massing in greatest concentration in the sulfide phase of a molten mass. { 'kal·kəˌfīl }

challiho [METEOROL] Strong southerly winds which blow for about 40 days in spring in some parts of India; sometimes the winds are violent, causing blinding dust storms. { 'chäl·əˌhō }

Champlainian [GEOL] Middle Ordovician geologic time. { ˌsham'plān·ē·ən }

Chandler motion *See* polar wandering. { 'chand·lər ˌmō·shən }

Chandler period [GEOPHYS] The period of the Chandler wobble. { 'chand·lər ˌpir·ē·əd }

Chandler wobble [GEOPHYS] A movement in the earth's axis of rotation, the period of motion being about 14 months. Also known as Eulerian nutation. { 'chand·lər ˌwäb·əl }
chandui *See* chanduy. { 'chän·dwē }
chanduy [METEOROL] A cool, descending wind at Guayaquil, Ecuador, which blows during the dry season (July to November). Also spelled chandui. { 'chän·dwē }
change chart [METEOROL] A chart indicating the amount and direction of change of some meteorological element during a specified time interval; for example, a height-change chart or pressure-change chart. Also known as tendency chart. { 'chānj ˌchärt }
change of tide [OCEANOGR] A reversal of the direction of motion (rising or falling) of a tide, or in the set of a tidal current. Also known as turn of the tide. { 'chānj əv ˌtīd }
channel [HYD] The deeper portion of a waterway carrying the main current. { 'chan·əl }
channel control [HYD] A condition whereby the stage of a stream is controlled only by discharge and the general configuration of the stream channel, that is, the contours of its bed, banks, and floodplains. { 'chan·əl kən'trōl }
channel fill [GEOL] Accumulations of sand and detritus in a stream channel where the transporting capacity of the water is insufficient to remove the material as rapidly as it is delivered. { 'chan·əl ˌfil }
channel frequency *See* stream frequency. { 'chan·əl ˌfrē·kwən·sē }
channel-lag deposit [GEOL] Coarse residual material left as accumulations in the channel in the normal processes of the stream. { 'chan·əl ˌlag diˌpäz·ət }
channel morphology *See* river morphology. { 'chan·əl ˌmȯr'fäl·ə·jē }
channel-mouth bar [GEOL] A bar formed where moving water enters a body of still water, due to decreased velocity. { 'chan·əl ˌmau̇th ˌbär }
channel net [HYD] Stream channel pattern within a drainage basin. { 'chan·əl ˌnet }
channel order *See* stream order. { 'chan·əl ˌȯrd·ər }
channel pattern [HYD] The configuration of a limited reach of a river channel as seen in plan view from an airplane. { 'chan·əl ˌpad·ərn }
channel roughness [GEOL] A measure of the resistivity offered by the material constituting stream channel margins to the flow of water. { 'chan·əl ˌrəf·nəs }
channel sand [GEOL] A sandstone or sand deposited in a stream bed or other channel eroded into the underlying bed. { 'chan·əl ˌsand }
channel segment *See* stream segment. { 'chan·əl ˌseg·mənt }
channel splay *See* floodplain splay. { 'chan·əl ˌsplā }
channel width [GEOL] The distance across a stream or channel as measured from bank to bank near bankful stage. { 'chan·əl ˌwidth }
Chapman equation [GEOPHYS] A theoretical relation describing the distribution of electron density with height in the upper atmosphere. { 'chap·mən iˌkwā·zhən }
Chapman region [GEOPHYS] A hypothetical region in the upper atmosphere in which the distribution of electron density with height can be described by Chapman's theoretical equation. { 'chap·mən ˌrē·jən }
character [GEOPHYS] A distinctive aspect of a seismic event, for example, the waveform. { 'kar·ik·tər }
Charmouthian [GEOL] Middle Lower Jurassic geologic time. { chär'mau̇th·ē·ən }
charnockite series [GEOL] A series of plutonic rocks compositionally similar to the granitic rock series but characterized by the presence of orthopyroxene. { 'chär·nəˌkīt ˌsir·ēz }
charted depth [OCEANOGR] The vertical distance from the tidal datum to the bottom. { 'char·təd 'depth }
Charybdis *See* Galofaro. { kə'rib·dəs }
chassignite [GEOL] An achondritic stony meteorite composed chiefly of olivine (95); resembles dunite. { 'shas·ənˌyīt }
chatter mark [GEOL] A scar on the surface of bedrock made by the abrasive action of drift carried at the base of a glacier. { 'chad·ər ˌmärk }

Chattian [GEOL] Upper Oligocene geologic time. Also known as Casselian. { 'chad·ē·ən }

Chautauquan [GEOL] Upper Devonian geologic time, below Bradfordian. { shə'täk·wən }

Chazyan [GEOL] Middle Ordovician geologic time. { 'chaz·ē·ən }

check observation [METEOROL] An aviation weather observation taken primarily for aviation radio broadcast purposes; usually abbreviated to include just those elements of a record observation that have an important affect on aircraft operations. { 'chek ˌäb·zər'vā·shən }

chemical denudation [GEOL] Wasting of the land surface by water transport of soluble materials into the sea. { 'kem·i·kəl ˌdē·nü'dā·shən }

chemical fossils *See* biomarkers. { ¦kem·i·kəl 'fäs·əlz }

chemical precipitates [GEOL] A sediment formed from precipitated materials as distinguished from detrital particles that have been transported and deposited. { 'kem·i·kəl pri'sip·əˌtāts }

chemical remanent magnetization [GEOPHYS] Permanent magnetization of rocks acquired when a magnetic material, such as hematite, is grown at low temperature through the oxidation of some other iron mineral, such as magnetite or goethite; the growing mineral becomes magnetized in the direction of any field which is present. Abbreviated CRM. { 'kem·i·kəl 'rem·ə·nənt ˌmag·nət·ə'zā·shən }

chemical reservoir [GEOL] An underground oil or gas trap formed in limestones or dolomites deposited in quiescent geologic environments. { 'kem·i·kəl 'rez·əvˌwär }

chemical weathering [GEOCHEM] A weathering process whereby rocks and minerals are transformed into new, fairly stable chemical combinations by such chemical reactions as hydrolysis, oxidation, ion exchange, and solution. Also known as decay; decomposition. { 'kem·i·kəl 'weth·ə·riŋ }

chemocline [HYD] The transition in a meromictic lake between the mixolimnion layer (at the top) and the monimolimnion layer (at the bottom). { 'kē·məˌklīn }

chemosphere [METEOROL] The vaguely defined region of the upper atmosphere in which photochemical reactions take place; generally considered to include the stratosphere (or the top thereof) and the mesosphere, and sometimes the lower part of the thermosphere. { 'kē·mōˌsfir }

chemostratigraphy [GEOCHEM] The correlation and dating of marine sediments and sedimentary rocks through the use of trace-element concentrations, molecular fossils, and certain isotopic ratios that can be measured on components of the rocks. { ˌkē·mō·strə'tig·rə·fē }

Chemungian [GEOL] Middle Upper Devonian geologic time, below Cassodagan. { ke'mən·jē·ən }

chenier [GEOL] A continuous ridge of beach material built upon swampy deposits; often supports trees, such as pines or evergreen oaks. { 'shen·yā }

chergui [METEOROL] An eastern or southeastern desert wind in Morocco (North Africa), especially in the north; it is persistent, very dry and dusty, hot in summer, cold in winter. { 'chər·gwē }

Chernozem [GEOL] One of the major groups of zonal soils, developed typically in temperate to cool, subhumid climate; the Chernozem soils in modern classification include Borolls, Ustolls, Udolls, and Xerolls. Also spelled Tchernozem. { ¦chər·nəz¦yȯm }

chertification [GEOL] A process of replacement by silica in limestone in the form of fine-grained quartz or chalcedony. { ˌchərd·ə·fə'ka·shən }

Chesterian [GEOL] Upper Mississippian geologic time. { che'stir·ē·ən }

chestnut coal [GEOL] Anthracite coal small enough to pass through a round mesh of $1^5/_8$ inches (3.1 centimeters) but too large to pass through a round mesh of $1^{13}/_{16}$ inches (1.7 centimeters). { 'chesˌnət ˌkōl }

Chestnut soil [GEOL] One of the major groups of zonal soils, developed typically in temperate to cool, subhumid to semiarid climate; the Chestnut soils in modern classification include Ustolls, Borolls, and Xerolls. { 'chesˌnət ¦sȯil }

chevron fold [GEOL] An accordionlike fold with limbs of equal length. { 'shev·rən ˌfōld }
chibli *See* ghibli. { 'chib·lē }
chichili *See* chili. { chi'chil·ē }
Chideruan [GEOL] Uppermost Permian geologic time. { chi'der·ə·wən }
chili [METEOROL] A warm, dry, descending wind in Tunisia, resembling the sirocco; in southern Algeria it is called chichili. { 'chil·ē }
chill wind factor [METEOROL] An arbitrary index, developed by the Canadian Army, to correlate the performance of equipment and personnel in an Arctic winter; it is equal to the sum of the wind speed in miles per hour and the negative of the Fahrenheit temperature; the term is not to be confused with wind chill. { ¦chil ¦wind ˌfak·tər }
chimney *See* pipe; spouting horn. { 'chimˌnē }
chimney cloud [METEOROL] A cumulus cloud in the tropics that has much greater vertical than horizontal extent. { 'chimˌnē ˌklaůd }
chimney rock [GEOL] **1.** A chimney-shaped remnant of a rock cliff whose sides have been cut into and carried away by waves and the gravel beach. **2.** A rock column rising above its surroundings. { 'chimˌnē ˌräk }
chinook [METEOROL] The foehn on the eastern side of the Rocky Mountains. { shə'nůk }
chinook arch [METEOROL] A foehn cloud formation appearing as a bank of clouds over the Rocky Mountains, generally a flat layer of altostratus, heralding the approach of a chinook. { shə'nůk 'ärch }
chlorinity [OCEANOGR] A measure of the chloride and other halogen content, by mass, of sea water. { klə'rin·əd·ē }
chlorosity [OCEANOGR] The chlorine and bromide content of one liter of sea water; equals the chlorinity of the sample times its density at 20°C. { klə'räs·əd·ē }
chocolate gale *See* chocolatero. { 'chäk·lət 'gāl }
chocolatero [METEOROL] A moderate norther in the Gulf region of Mexico. Also known as chocolate gale. { ˌchō·kō·lä'ter·ō }
chondrite [GEOL] A stony meteorite containing chondrules. { 'känˌdrīt }
chondrule [GEOL] A spherically shaped body consisting chiefly of pyroxene or olivine minerals embedded in the matrix of certain stony meteorites. { 'känˌdrül }
choppy sea [OCEANOGR] In popular usage, short, rough, irregular wave motion on a sea surface. { ¦chäp·ē ¦sē }
C horizon [GEOL] The portion of the parent material in soils which has been penetrated with roots. { 'sē hə'rīz·ən }
chromel [METEOROL] An alloy containing 90% nickel and 10% chromium, used to form the chromel-alumel thermocouple. { 'krōˌmel }
chromocratic *See* melanocratic. { ˌkrō·mə'krad·ik }
chron [GEOL] **1.** The time unit equivalent to the stratigraphic unit, subseries, and geologic name of a division of geologic time. **2.** The geochronological equivalent of chronozone. { krän }
chronolith *See* time-stratigraphic unit. { 'krän·əˌlith }
chronolithologic unit *See* time-stratigraphic unit. { ¦krän·ə¦lith·ə'läj·ik 'yü·nət }
chronostratic unit *See* time-stratigraphic unit. { ¦krän·ə¦strad·ik 'yü·nət }
chronostratigraphic unit *See* time-stratigraphic unit. { ¦krän·ə¦strad·ə'graf·ik 'yü·nət }
chronostratigraphic zone *See* chronozone. { ˌkrän·əˌstrad·ə'graf·ik 'zōn }
chronostratigraphy [GEOL] A division of stratigraphy that uses age determination and time sequence of rock strata to develop an interpretation of the earth's geologic history. { ˌkrän·ə·strə'tig·rə·fē }
chronozone [GEOL] **1.** A formal time-stratigraphic unit used to specify strata equivalent in time span to a zone in another type of classification, for example, a biostratigraphic zone. Also known as chronostratigraphic zone. **2.** The smallest subdivision of chronostratigraphic units, below stage, composed of rocks formed during a chron of geologic time. { krän·əˌzōn }
chubasco [METEOROL] A severe thunderstorm with vivid lightning and violent squalls

coming from the land on the west coast of Nicaragua and Costa Rica in Central America. { chü'bä,skō }

Chubb [GEOL] A meteorite crater in Ungava, Quebec, Canada. { chəb }

churada [METEOROL] A severe rain squall in the Mariana Islands (western Pacific Ocean) during the northeast monsoon; these squalls occur from November to April or May, but especially from January through March. { chü'rä·də }

churn hole *See* pothole. { 'chərn ,hōl }

chute [HYD] A short channel across a narrow land area which bypasses a bend in a river; formed by the river's breaking through the land. { shüt }

Ci *See* cirrus cloud.

CI *See* temperature-humidity index.

Cincinnatian [GEOL] Upper Ordovician geologic time. { sin·sə'nad·ē·ən }

cinder [GEOL] Fine-grained pyroclastic material ranging in diameter from 0.16 to 1.28 inch (4 to 32 millimeters). { 'sin·dər }

cinder coal *See* natural coke. { 'sin·dər ,kōl }

cinder cone [GEOL] A conical elevation formed by the accumulation of volcanic debris around a vent. { 'sin·dər ,kōn }

C index [GEOPHYS] A subjectively obtained daily index of geomagnetic activity, in which each day's record is evaluated on the basis of 0 for quiet, 1 for moderately disturbed, and 2 for very disturbed. Also known as C figure; magnetic character figure. { 'sē ,in,deks }

circle of equal altitude [GEOD] A circle on the surface of the earth, on every point of which the altitude of a given celestial body is the same at a given instant; the pole of this circle is the geographical position of the body, and the great-circle distance from this pole to the circle is the zenith distance of the body. { 'sər·kəl əv 'ē·kwəl 'al·tə,tüd }

circle of illumination [GEOL] The edge of the sunlit hemisphere, which forms a circular boundary separating the earth into a light half and a dark half. { 'sər·kəl əv ə,lü·mə'nā·shən }

circle of inertia *See* inertial circle. { 'sər·kəl əv i'nər·shə }

circle of latitude [GEOD] A meridian of the terrestrial sphere along which latitude is measured. { 'sər·kəl əv 'lad·ə,tüd }

circle of longitude *See* parallel. { 'sər·kəl əv 'län·jə,tüd }

circular coal *See* eye coal. { 'sər·kyə·lər ¦kōl }

circular vortex [METEOROL] An atmospheric flow in parallel planes in which streamlines and other isopleths are concentric circles about a common axis; an atmospheric model of easterly and westerly winds is a circular vortex about the earth's polar axis. { 'sər·kyə·lər 'vȯr,teks }

circulation [METEOROL] For an air mass, the line integral of the tangential component of the velocity field about a closed curve. [OCEANOGR] A water current flow occurring within a large area, usually in a closed circular pattern. { ,sər·kyə·'lā·shən }

circulation flux [METEOROL] Flux due to mean atmospheric motion as opposed to eddy flux; the dominant flux in low latitudes. { ,sər·kyə·'lā·shən ,fləks }

circulation index [METEOROL] A measure of the magnitude of one of several aspects of large-scale atmospheric circulation patterns; indices most frequently measured represent the strength of the zonal (east-west) or meridional (north-south) components of the wind, at the surface or at upper levels, usually averaged spatially and often averaged in time. { ,sər·kyə·'lā·shən 'in,deks }

circulation pattern [METEOROL] The general geometric configuration of atmospheric circulation usually applied, in synoptic meteorology, to the large-scale features of synoptic charts and mean charts. { ,sər·kyə·'lā·shən ,pad·ərn }

circumpolar [GEOGR] Located around one of the polar regions of earth. { ¦sər·kəm'pō·lər }

circumpolar westerlies *See* westerlies. { ¦sər·kəm'pō·lər 'wes·tər,lēz }

circumpolar whirl *See* polar vortex. { ¦sər·kəm'pō·lər 'wərl }

cirque [GEOL] A steep elliptic to elongated enclave high on mountains in calcareous districts, usually forming the blunt end of a valley. Also known as corrie; cwm. { sərk }

cirque lake [HYD] A small body of water occupying a cirque. { 'sərk ˌlāk }

cirriform [METEOROL] Descriptive of clouds composed of small particles, mostly ice crystals, which are fairly widely dispersed, usually resulting in relative transparency and whiteness and often producing halo phenomena not observed with other cloud forms. { 'sir·əˌfȯrm }

cirrocumulus cloud [METEOROL] A principal cloud type, appearing as a thin, white path of cloud without shadows, composed of very small elements in the form of grains, ripples, and so on. Abbreviated Cc. { ¦sir·ō'kyü·myə·ləs ¦klaủd }

cirrostratus cloud [METEOROL] A principal cloud type, appearing as a whitish veil, usually fibrous but sometimes smooth, which may totally cover the sky and often produces halo phenomena, either partial or complete. Abbreviated Cs. { ¦sir·ō'strad·əs ¦klaủd }

cirrus cloud [METEOROL] A principal cloud type composed of detached cirriform elements in the form of white, delicate filaments, of white (or mostly white) patches, or narrow bands. Abbreviated Ci. { 'sir·əs ¦klaủd }

cistern [GEOL] A hollow that holds water. { 'sis·tərn }

Claibornian [GEOL] Middle Eocene geologic time. { ˌkler'bȯrn·ē·ən }

Clairaut's formula [GEOD] An approximate formula for gravity at the earth's surface, assuming that the earth is an ellipsoid; states that the gravity is equal to $g_e\,[1+(^5/_2\,m' - f)\sin^2\theta]$, where θ is the latitude, g_e is the gravity at the equator, m' is the ratio of centrifugal acceleration to gravity at the equatorial surface, and f is the earth's flattening, equal to $(a - b)/a$, where a is the semimajor axis and b is the semiminor axis. { 'kler·ȯz ˌfȯr·myə·lə }

clarain [GEOL] A coal lithotype appearing as stratifications parallel to the bedding plane and usually having a silky luster and scattered or diffuse reflection. Also known as clarite. { 'klaˌrān }

Clarendonian [GEOL] Lower Pliocene or upper Miocene geologic time. { ˌkla·rən'dōn·ē·ən }

clarite *See* clarain. { 'klaˌrīt }

clarke [GEOCHEM] A unit of the average abundance of an element in the earth's crust, expressed as a percentage. Also known as crustal abundance. { klärk }

Clarke ellipsoid of 1866 [GEOD] The reference ellipsoid adopted by the U.S. Coast and Geodetic Survey of 1880 for charting North America. { 'klärk ə'lipˌsȯid əv ā'tēn 'sik·stē'siks }

clarodurain [GEOL] A transitional lithotype of coal composed of vitrinite and other macerals, principally micrinite and exinite. { ¦kla·rō'dủˌrān }

clarofusain [GEOL] A transitional lithotype of coal composed of fusinite and vitrinite and other macerals. { ¦kla·rō'fyüˌzān }

clarovitrain [GEOL] A transitional lithotype of coal rock composed primarily of the maceral vitrinite, with lesser amounts of other macerals. { ¦kla·rō'viˌtrān }

clast [GEOL] An individual grain, fragment, or constituent of detrital sediment or sedimentary rock produced by physical breakdown of a larger mass. { klast }

clastation *See* weathering. { kla'stā·shən }

clastic [GEOL] Rock or sediment composed of clasts which have been transported from their place of origin, as sandstone and shale. { 'klas·tik }

clastic dike [GEOL] A tabular-shaped sedimentary dike composed of clastic material and transecting the bedding of a sedimentary formation; represents invasion by extraneous material along a crack of the containing formation. { 'klas·tik 'dīk }

clastic pipe [GEOL] A cylindrical body of clastic material having an irregular columnar or pillarlike shape, standing approximately vertically through enclosing formations (usually limestone), and measuring a few centimeters to 50 meters (165 feet) in diameter and 1 to 60 meters (3 to 200 feet) in height. { 'klas·tik 'pīp }

clastic ratio [GEOL] The ratio of the percentage of clastic rocks to that of nonclastic rocks in a geologic section. Also known as detrital ratio. { 'klas·tik 'rā·shō }

clastic reservoir [GEOL] An underground oil or gas trap formed in clastic limestone. { 'klas·tik 'rez·əv,wär }
clastic sediment [GEOL] Deposits of clastic materials transported by mechanical agents. Also known as mechanical sediment. { 'klas·tik 'sed·ə·mənt }
clastic wedge [GEOL] The sediments of the exogeosyncline, derived from the tectonic landmasses of the adjoining orthogeosyncline. { 'klas·tik 'wej }
clathrate *See* gas hydrate. { 'klath,rāt }
clathrate hydrate *See* gas hydrate. { ¦klath,rāt 'hī,drāt }
clay [GEOL] **1.** A natural, earthy, fine-grained material which develops plasticity when mixed with a limited amount of water; composed primarily of silica, alumina, and water, often with iron, alkalies, and alkaline earths. **2.** The fraction of an earthy material containing the smallest particles, that is, finer than 3 micrometers. { klā }
Clay Belt [GEOL] A lowland area bordering on the western and southern portions of Hudson and James bays in Canada, composed of clays and silts recently deposited in large glacial lakes during the withdrawal of the continental glaciers. { 'klā ,belt }
clay gall [GEOL] A dry, curled clay shaving derived from dried, cracked mud and embedded and flattened in a sand stratum. { 'klā ,gȯl }
clay loam [GEOL] Soil containing 27–40% clay, 20–45% sand, and the remaining portion silt. { ¦klā 'lōm }
clay marl [GEOL] A chalky clay, whitish with a smooth texture. { ¦klā 'märl }
claypan [GEOL] A stratum of compact, stiff, relatively impervious noncemented clay; can be worked into a soft, plastic mass if immersed in water. { 'klā,pan }
clay plug [GEOL] Sediment, with a great deal of organic muck, deposited in a cutoff river meander. { ¦klā ¦pləg }
clay shale [GEOL] **1.** Shale composed wholly or chiefly of clayey material which becomes clay again on weathering. **2.** Consolidated sediment composed of up to 10% sand and having a silt to clay ratio of less than 1:2. { ¦klā ¦shāl }
clay soil [GEOL] A fine-grained inorganic soil which forms hard lumps when dry and becomes sticky when wet. { ¦klā ¦sȯil }
claystone [GEOL] Indurated clay, consisting predominantly of fine material of which a major proportion is clay mineral. { 'klā,stōn }
clay vein [GEOL] A body of clay which is similar to an ore vein in form and fills a crevice in a coal seam. Also known as dirt slip. { 'klā ,vān }
clear [METEOROL] **1.** After United States weather observing practice, the state of the sky when it is cloudless or when the sky cover is less than 0.1 (to the nearest tenth). **2.** To change from a stormy or cloudy weather condition to one of no precipitation and decreased cloudiness. { klir }
clear-air turbulence [METEOROL] A meteorological phenomenon occurring in the upper troposphere and lower stratosphere, in which high-speed aircraft are subject to violent updrafts and downdrafts. Abbreviated CAT. { ,klir 'er 'tərb·yə·ləns }
clear ice [HYD] Generally, a layer or mass of ice which is relatively transparent because of its homogeneous structure and small number and size of air pockets. { ¦klir ¦īs }
cleat [GEOL] Vertical breakage planes found in coal. Also spelled cleet. { klēt }
cleavage [GEOL] Splitting, or the tendency to split, along parallel, closely positioned planes in rock. { 'klēv·ij }
cleavage banding [GEOL] A compositional banding, usually formed from incompetent material such as argillaceous rocks, that is parallel to the cleavage rather than the bedding. { 'klēv·ij ,band·iŋ }
cliff [GEOGR] A high, steep, perpendicular or overhanging face of a rock; a precipice. { klif }
cliff of displacement *See* fault scarp. { 'klif əv dis'plā·smənt }
Cliftonian [GEOL] Middle Middle Silurian geologic time. { klif'tän·ē·ən }
climagram *See* climatic diagram. { 'klī·mə,gram }
climagraph *See* climatic diagram. { 'klī·mə,graf }
climate [CLIMATOL] The long-term manifestations of weather. { 'klī·mət }
climate change [METEOROL] Any change in global temperatures and precipitation over time due to natural variability or to human activity. { 'klī·mət ,chānj }

climate control [CLIMATOL] Schemes for artificially altering or controlling the climate of a region. { 'klī·mət kən'trōl }

climate model [CLIMATOL] A mathematical representation of the earth's climate system capable of simulating its behavior under present and altered conditions. { 'klī·mət ˌmäd·əl }

climatic change [CLIMATOL] The long-term fluctuation in rainfall, temperature, and other aspects of the earth's climate. { klī'mad·ik 'chānj }

climatic classification [CLIMATOL] The division of the earth's climates into a system of contiguous regions, each one of which is defined by relative homogeneity of the climate elements. { klī'mad·ik ˌklas·ə·fə'kā·shən }

climatic controls [CLIMATOL] The relatively permanent factors which govern the general nature of the climate of a portion of the earth, including solar radiation, distribution of land and water masses, elevation and large-scale topography, and ocean currents. { klī'mad·ik kən'trōlz }

climatic cycle [CLIMATOL] A long-period oscillation of climate which recurs with some regularity, but which is not strictly periodic. Also known as climatic oscillation. { klī'mad·ik 'sī·kəl }

climatic diagram [CLIMATOL] A graphic presentation of climatic data; generally limited to a plot of the simultaneous variations of two climatic elements, usually through an annual cycle. Also known as climagram; climagraph; climatograph; climogram; climograph. { klī'mad·ik 'dī·əˌgram }

climatic divide [CLIMATOL] A boundary between regions having different types of climate. { klī'mad·ik də'vīd }

climatic factor [CLIMATOL] Climatic control, but regarded as including more local influences; thus city smoke and the extent of the builtup metropolitan area are climatic factors, but not climatic controls. { klī'mad·ik 'fak·tər }

climatic forecast [CLIMATOL] A forecast of the future climate of a region; that is, a forecast of general weather conditions to be expected over a period of years. { klī'mad·ik 'fórˌkast }

climatic optimum [CLIMATOL] The period in history (about 5000–2500 B.C.) during which temperatures were warmer than at present in nearly all parts of the world. { klī'mad·ik 'äp·tə·məm }

climatic oscillation *See* climatic cycle. { klī'mad·ik ˌäs·ə'lā·shən }

climatic prediction [METEOROL] The description of the future state of the climate, that is, the average or expected atmospheric and earth-surface conditions, for example, temperature, precipitation, humidity, winds, and their range of variability. Seasonal and interannual climate predictions, made many months in advance, provide useful information for planners and policy makers. { klī'mad·ik prə'dik·shən }

climatic province [CLIMATOL] A region of the earth's surface characterized by an essentially homogeneous climate. { klī'mad·ik 'prä·vəns }

climatic snow line [METEOROL] The altitude above which a flat surface (fully exposed to sun, wind, and precipitation) would experience a net accumulation of snow over an extended period of time; below this altitude, ablation would predominate. { klī'mad·ik 'snō ˌlīn }

climatic zone [CLIMATOL] A belt of the earth's surface within which the climate is generally homogeneous in some respect; an elemental region of a simple climatic classification. { klī'mad·ik ˌzōn }

climatochronology [GEOL] The absolute age dating of recent geologic events by using the oxygen isotope ratios in ice, shells, and so on. { klīˌmad·ō·krə'näl·ə·jē }

climatograph *See* climatic diagram. { klī'mad·əˌgraf }

climatography [CLIMATOL] A quantitative description of climate, particularly with reference to the tables and charts which show the characteristic values of climatic elements at a station or over an area. { ˌklī·mə'täg·rə·fē }

climatological forecast [METEOROL] A weather forecast based upon the climate of a region instead of upon the dynamic implications of current weather, with consideration given to such synoptic weather features as cyclones and anticyclones, fronts, and the jet stream. { ˌklī·məd·əl'äj·ə·kəl 'fórˌkast }

climatological station elevation [CLIMATOL] The elevation above mean sea level chosen as the reference datum level for all climatological records of atmospheric pressure in a given locality. { ˌklī·məd·əl'äj·ə·kəl 'stā·shən ˌel·ə'vā·shən }

climatological station pressure [CLIMATOL] The atmospheric pressure computed for the level of the climatological station elevation, used to give all climatic records a common reference; it may or may not be the same as station pressure. { ˌklī·məd·əl'äj·ə·kəl 'stā·shən ˌpresh·ər }

climatological substation [CLIMATOL] A weather-observing station operated (by an unpaid volunteer) for the purpose of recording climatological observations. { ˌklī·məd·əl'äj·ə·kəl 'səbˌstā·shən }

climatology [METEOROL] That branch of meteorology concerned with the mean physical state of the atmosphere together with its statistical variations in both space and time as reflected in the weather behavior over a period of many years. { ˌklī·mə'täl·ə·jē }

climbing dune [GEOL] A dune that develops on the windward side of mountains or hills. { 'klīm·iŋ 'dün }

climogram *See* climatic diagram. { 'klī·məˌgram }

climograph *See* climatic diagram. { 'klī·məˌgraf }

clinker [GEOL] Burnt or vitrified stony material, as ejected by a volcano or formed in a furnace. { 'kliŋ·kər }

clinoform [GEOL] A subaqueous landform, such as the continental slope of the ocean or the foreset bed of a delta. { 'klī·nəˌfórm }

clint [GEOL] A hard or flinty rock, such as a projecting rock or ledge. { klint }

Clintonian [GEOL] Lower Middle Silurian geologic time. { klin'tōn·ē·ən }

clog snow [HYD] A skiing term for wet, sticky, new snow. { 'kläg ˌsnō }

close [METEOROL] Colloquially, descriptive of oppressively still, warm, moist air, frequently applied to indoor conditions. { klōs }

closed drainage [HYD] Drainage in which the surface flow of water collects in sinks or lakes having no surface outlet. Also known as blind drainage. { ¦klōzd 'drā·nij }

closed fold [GEOL] A fold whose limbs have been compressed until they are parallel, and whose structure contour lines form a closed loop. Also known as tight fold. { ¦klōzd ¦fōld }

closed high [METEOROL] A high that may be completely encircled by an isobar or contour line. { ¦klōzd 'hī }

closed lake [HYD] A lake that does not have a surface effluent and that loses water by evaporation or by seepage. { ¦klōzd 'lāk }

closed low [METEOROL] A low that may be completely encircled by an isobar or contour line, that is, an isobar or contour line of any value, not necessarily restricted to those arbitrarily chosen for the analysis of the chart. { ¦klōzd 'lō }

closed sea [OCEANOGR] **1.** That part of the ocean enclosed by headlands, within narrow straits, or within other landforms. **2.** That part of the ocean within the territorial jurisdiction of a country. { ¦klōzd 'sē }

close-joints cleavage *See* slip cleavage. { ¦klōs ¦jóins 'klē·vij }

close sand *See* tight sand. { ¦klōs ¦sand }

closure [GEOL] The vertical distance between the highest and lowest point on an anticline which is enclosed by contour lines. { 'klō·zhər }

cloud [METEOROL] Suspensions of minute water droplets or of ice crystals produced by the condensation of water vapor. { klaůd }

cloud absorption [GEOPHYS] The absorption of electromagnetic radiation by the waterdrops and water vapor within a cloud. { 'klaůd əb'sórp·shən }

cloudage *See* cloud cover. { 'klaů·dij }

cloud albedo [GEOPHYS] The fraction of the incident solar radiation reflected to space by clouds, which depends on the drop size, liquid water content, water vapor content, and thickness of the cloud, and the sun's elevation. The smaller the drops and the greater the liquid water content, the greater the cloud albedo. { ¦klaůd al¦bēd·ō }

cloud band [METEOROL] A broad band of clouds, about 10 to 100 or more miles (16 to 160 kilometers) wide, and varying in length from a few tens of miles to hundreds of miles. { 'klaůd ,band }

cloud bank [METEOROL] A fairly well-defined mass of cloud observed at a distance; covers an appreciable portion of the horizon sky, but does not extend overhead. { 'klaůd ,baŋk }

cloud banner *See* banner cloud. { 'klaůd ,ban·ər }

cloud bar [METEOROL] **1.** A heavy bank of clouds that appears on the horizon with the approach of an intense tropical cyclone (hurricane or typhoon); it is the outer edge of the central cloud mass of the storm. **2.** Any long, narrow, unbroken line of cloud, such as a crest cloud or an element of billow cloud. { 'klaůd ,bär }

cloud base [METEOROL] For a given cloud or cloud layer, that lowest level in the atmosphere at which the air contains a perceptible quantity of cloud particles. { 'klaůd ,bās }

cloudburst [METEOROL] In popular terminology, any sudden and heavy fall of rain, usually of the shower type, and with a fall rate equal to or greater than 100 millimeters (3.94 inches) per hour. Also known as rain gush; rain gust. { 'klaůd,bərst }

cloud cap *See* cap cloud. { 'klaůd ,kap }

cloud classification [METEOROL] **1.** A scheme of distinguishing and grouping clouds according to their appearance and, where possible, to their process of formation. **2.** A scheme of classifying clouds according to their altitudes: high, middle, or low clouds. **3.** A scheme of classifying clouds according to their particulate composition: water clouds, ice-crystal clouds, or mixed clouds. { 'klaůd ,klas·ə·fə'kā·shən }

cloud cover [METEOROL] That portion of the sky cover which is attributed to clouds, usually measured in tenths of sky covered. Also known as cloudage; cloudiness. { 'klaůd ,kəv·ər }

cloud crest *See* crest cloud. { 'klaůd ,krest }

cloud deck [METEOROL] The upper surface of a cloud. { 'klaůd ,dek }

cloud discharge [GEOPHYS] A lightning discharge occurring between a positive charge center and a negative charge center, both of which lie in the same cloud. Also known as cloud flash; intracloud discharge. { 'klaůd 'dis,chärj }

cloud droplet [METEOROL] A particle of liquid water from a few micrometers to tens of micrometers in diameter, formed by condensation of atmospheric water vapor and suspended in the atmosphere with other drops to form a cloud. { 'klaůd ,dräp·lət }

cloud echo [METEOROL] The radar target signal returned from clouds alone, as detected by cloud detection radars or other very-short-wavelength equipment. { 'klaůd ,ek·ō }

cloud flash *See* cloud discharge. { 'klaůd ,flash }

cloud formation [METEOROL] **1.** The process by which various types of clouds are formed, generally involving adiabatic cooling of ascending moist air. **2.** A particular arrangement of clouds in the sky, or a striking development of a particular cloud. { 'klaůd fȯr'mā·shən }

cloud height [METEOROL] The absolute altitude of the base of a cloud. { 'klaůd ,hīt }

cloudiness *See* cloud cover. { 'klaůd·ē·nəs }

cloud layer [METEOROL] An array of clouds, not necessarily all of the same type, whose bases are at approximately the same level; may be either continuous or composed of detached elements. { 'klaůd ,lā·ər }

cloud level [METEOROL] **1.** A layer in the atmosphere in which are found certain cloud genera; three levels are usually defined: high, middle, and low. **2.** At a particular time, the layer in the atmosphere bounded by the limits of the bases and tops of an existing cloud form. { 'klaůd ,lev·əl }

cloud modification [METEOROL] Any process by which the natural course of development of a cloud is altered by artificial means. { 'klaůd ,mäd·ə·fə'kā·shən }

cloud particle [METEOROL] A particle of water, either a drop of liquid water or an ice crystal, comprising a cloud. { 'klaůd ,pärd·ə·kəl }

cloud-phase chart [METEOROL] A chart designed to indicate and distinguish supercooled water clouds from ice-crystal clouds. { 'klaůd ,fāz ,chärt }

cloud physics [METEOROL] The study of the physical and dynamical processes governing the structure and development of clouds and the release from them of snow, rain, and hail. { 'klau̇d ˌfiz·iks }

cloud seeding [METEOROL] Any technique carried out with the intent of adding to a cloud certain particles that will alter its natural development. { 'klau̇d ˌsēd·iŋ }

cloud shield [METEOROL] The principal cloud structure of a typical wave cyclone, that is, the cloud forms found on the cold-air side of the frontal system. { 'klau̇d ˌshēld }

cloud street [METEOROL] A line of cumuliform clouds frequently one cumulus element wide, but ranging upward in width so that it is sometimes difficult to differentiate between streets and bands. { 'klau̇d ˌstrēt }

cloud symbol [METEOROL] One of a set of specified ideograms that represent the various cloud types of greatest significance or those most commonly observed, and entered on a weather map as part of a station model. { 'klau̇d ˌsim·bəl }

cloud system [METEOROL] An array of clouds and precipitation associated with a cyclonic-scale feature of atmospheric circulation, and displaying typical patterns and continuity. Also known as nephsystem. { 'klau̇d ˌsis·təm }

cloud-to-cloud discharge [GEOPHYS] A lightning discharge occurring between a positive charge center of one cloud and a negative charge center of a second cloud. Also known as intercloud discharge. { ¦klau̇d tə ¦klau̇d 'disˌchärj }

cloud-to-ground discharge [GEOPHYS] A lightning discharge occurring between a charge center (usually negative) in the cloud and a center of opposite charge at the ground. Also known as ground discharge. { ¦klau̇d tə ¦grau̇nd 'disˌchärj }

cloud top [METEOROL] The highest level in the atmosphere at which the air contains a perceptible quantity of cloud particles for a given cloud or cloud layer. { 'klau̇d ˌtäp }

cloudy [METEOROL] The character of a day's weather when the average cloudiness, as determined from frequent observations, is more than 0.7 for the 24-hour period. { 'klau̇d·ē }

clough [GEOGR] A cleft in a hill; a ravine or narrow valley. { kləf }

cluse [GEOL] A narrow gorge, trench, or water gap with steep sides that cuts transversely through an otherwise continuous ridge. { klüz }

clusterite *See* botryoid. { 'klə·stəˌrīt }

coagulation *See* agglomeration. { kōˌag·yə'lā·shən }

Coahuilan [GEOL] A North American provincial series in Lower Cretaceous geologic time, above the Upper Jurassic and below the Comanchean. { kō·ə'wēl·ən }

coal [GEOL] The natural, rocklike, brown to black derivative of forest-type plant material, usually accumulated in peat beds and progressively compressed and indurated until it is finally altered into graphite or graphite-like material. { kōl }

coal ball [GEOL] A subspherical mass containing mineral matter embedded with plant material, found in coal seams and overlying beds of the late Paleozoic. { 'kōl ˌbōl }

coal bed [GEOL] A seam or stratum of coal parallel to the rock stratification. Also known as coal rake; coal seam. { 'kōl ˌbed }

coal breccia [GEOL] Angular fragments of coal within a coal bed. { 'kōl ˌbrech·ə }

coal clay *See* underclay. { 'kōl ˌklā }

coalescence [METEOROL] In cloud physics, merging of two or more water drops into a single larger drop. { ˌkō·ə'les·əns }

coalescence efficiency [METEOROL] The fraction of all collisions which occur between waterdrops of a specified size and which result in actual merging of two drops into a single larger drop. { ˌkō·ə'les·əns i'fish·ən·sē }

coalescence process [METEOROL] The growth of raindrops by the collision and coalescence of cloud drops or small precipitation particles. { ˌkō·ə'les·əns ˌpräs·əs }

coalification [GEOL] Formation of coal from plant material by the processes of diagenesis and metamorphism. Also known as bituminization; carbonification; incarbonization; incoalation. { ˌkōl·ə·fə'kā·shən }

Coal Measures [GEOL] The sequence of rocks typically containing coal of the Upper Carboniferous. { 'kōl ˌmezh·ərz }

coal pebbles [GEOL] Rounded masses of coal occurring in sedimentary rock. { 'kōl ˌpeb·əlz }

coal petrology [GEOL] The science that deals with the origin, history, occurrence, structure, chemical composition, and classification of coal. { 'kōl pə'träl·ə·jē }
coal rake *See* coal bed. { 'kōl ˌrāk }
coal seam *See* coal bed. { 'kōl ˌsēm }
coarse fragment [GEOL] A rock or mineral fragment in the soil with an equivalent diameter greater than 0.08 inch (2 millimeters). { 'kȯrs 'frag·mənt }
coast [GEOGR] The general region of indefinite width that extends from the sea inland to the first major change in terrain features. { kōst }
coastal berm *See* berm. { 'kōs·təl 'bərm }
coastal current [OCEANOGR] An offshore current flowing generally parallel to the shoreline with a relatively uniform velocity. { 'kōs·təl 'kər·ənt }
coastal dune [GEOL] A mobile mound of windblown material found along many sea and lake shores. { 'kōs·təl 'dün }
coastal ice *See* fast ice. { 'kōs·təl 'īs }
coastal landform [GEOGR] The characteristic features and patterns of land in a coastal zone subject to marine and subaerial processes of erosion and deposition. { 'kōs·təl 'landˌfȯrm }
coastal plain [GEOL] An extensive, low-relief area that is bounded by the sea on one side and by a high-relief province on the landward side. Its geologic province actually extends beyond the shoreline across the continental shelf; it is linked to the stable part of a continent on the trailing edge of a plate. Typically, it has strata that dip gently and uniformly toward the sea. { 'kōs·təl 'plān }
coastal sediment [GEOL] The mineral and organic deposits of deltas, lagoons, and bays, barrier islands and beaches, and the surf zone. { 'kōs·təl 'sed·ə·mənt }
coast ice *See* fast ice. { 'kōst ˌīs }
coastline [GEOGR] **1.** The line that forms the boundary between the shore and the coast. **2.** The line that forms the boundary between the water and the land. { 'kōstˌlīn }
coast shelf *See* submerged coastal plain. { 'kōst ˌshelf }
cobble [GEOL] A rock fragment larger than a pebble and smaller than a boulder, having a diameter in the range of 64–256 millimeters (2.5–10.1 inches), somewhat rounded or otherwise modified by abrasion in the course of transport. { 'käb·əl }
cobble beach *See* shingle beach. { 'käb·əl ˌbēch }
Coblentzian [GEOL] Upper Lower Devonian geologic time. { kō'blens·ē·ən }
coccolith ooze [GEOL] A fine-grained pelagic sediment containing undissolved sand- or silt-sized particles of coccoliths mixed with amorphous clay-sized material. { 'käk·əˌlith ˌüz }
cockeyed bob [METEOROL] A thunder squall occurring during the summer, on the northwest coast of Australia. { 'käkˌīd 'bäb }
cockpit karst *See* cone karst. { 'käkˌpit 'karst }
cocurrent line [OCEANOGR] A line through places having the same tidal current hour. { kō'kər·ənt 'līn }
cognate [GEOL] Pertaining to contemporaneous fractures in a system with regard to time of origin and deformational type. { ˌkägˌnāt }
cognate ejecta [GEOL] Essential or accessory pyroclasts derived from the magmatic materials of a current volcanic eruption. { ˌkägˌnāt ē'jek·tə }
coherent deposit [GEOL] A consolidated sedimentary deposit that is not easily shattered. { kōˈhir·ənt di'päz·ət }
cohesionless [GEOL] Referring to a soil having low shear strength when dry, and low cohesion when wet. Also known as frictional; noncohesive. { kō'hē·zhən·ləs }
cohesiveness [GEOL] Property of unconsolidated fine-grained sediments by which the particles stick together by surface forces. { kō'hē·siv·nəs }
cohesive soil [GEOL] A sticky soil, such as clay or silt; its shear strength equals about half its unconfined compressive strength. { kō'hē·siv 'sȯil }
coke coal *See* natural coke. { 'kōk ˌkōl }
cokeite [GEOL] Naturally occurring coke formed by the action of magma on coal or by natural combustion of coal. { 'kōˌkīt }
coking coal [GEOL] A very soft bituminous coal suitable for coking. { 'kok·iŋ ˌkōl }

col [GEOL] A high, sharp-edged pass occurring in a mountain ridge, usually produced by the headward erosion of opposing cirques. [METEOROL] The point of intersection of a trough and a ridge in the pressure pattern of a weather map; it is the point of relatively lowest pressure between two highs and the point of relatively highest pressure between two lows. Also known as neutral point; saddle point. { käl }

colatitude [GEOD] Ninety degrees minus the latitude. { kō'lad·ə,tüd }

cold-air drop *See* cold pool. { 'kōld ¦er ,dräp }

cold-air outbreak *See* polar outbreak. { 'kōld ¦er 'au̇t,brāk }

cold anticyclone *See* cold high. { 'kōld ,an·tē'sī,klōn }

cold-core cyclone *See* cold low. { ¦kōld ¦kȯr 'sī,klōn }

cold-core high *See* cold high. { ¦kōld ¦kȯr 'hī }

cold-core low *See* cold low. { ¦kōld ¦kȯr 'lō }

cold dome [METEOROL] A cold air mass, considered as a three-dimensional entity. { 'kōld ,dōm }

cold drop *See* cold pool. { 'kōld ,dräp }

cold front [METEOROL] Any nonoccluded front, or portion thereof, that moves so that the colder air replaces the warmer air; the leading edge of a relatively cold air mass. { 'kōld ,frənt }

cold-front-like sea breeze [METEOROL] Sea breeze that forms over the ocean, moves slowly toward the land, and then moves inland quite suddenly. Also known as sea breeze of the second kind. { 'kōld ,frənt ,līk 'sē ,brēz }

cold-front thunderstorm [METEOROL] A thunderstorm attending a cold front. { 'kōld ,frənt 'thən·dər,stȯrm }

cold glacier [GEOL] A glacier whose base is at a temperature much below 32°F (0°C) and frozen to the bedrock, resulting in insignificant movement and almost no erosion. { ¦kōld 'glā·shər }

cold high [METEOROL] At a given level in the atmosphere, any high that is generally characterized by colder air near its center than around its periphery. Also known as cold anticyclone; cold-core high. { 'kōld 'hī }

cold low [METEOROL] At a given level in the atmosphere, any low that is generally characterized by colder air near its center than around its periphery. Also known as cold-core cyclone; cold-core low. { 'kōld 'lō }

cold pole [CLIMATOL] The location which has the lowest mean annual temperature in its hemisphere. { 'kōld ,pōl }

cold pool [METEOROL] A region of relatively cold air surrounded by warmer air; the term is usually applied to cold air of appreciable vertical extent that has been isolated in lower latitudes as part of the formation of a cutoff low. Also known as cold-air drop; cold drop. { 'kōld ,pül }

cold tongue [METEOROL] In synoptic meteorology, a pronounced equatorward extension or protrusion of cold air. { 'kōld ,təŋ }

cold wall [OCEANOGR] The line or surface along which two water masses of significantly different temperature are in contact. { 'kōld ,wȯl }

cold-water desert [GEOGR] An arid, often foggy region characterized by sparse precipitation because incoming airstreams are cooled over an offshore coastal current and deposit rain over the sea. { ¦kōld ,wȯd·ər 'dez·ərt }

cold-water sphere [OCEANOGR] Those portions of the ocean water having a temperature below 8°C. Also known as oceanic stratosphere. { 'kōld ,wȯd·ər ,sfir }

cold wave [METEOROL] A rapid fall in temperature within 24 hours to a level requiring substantially increased protection to agriculture, industry, commerce, and social activities. { 'kōld ,wāv }

colk *See* pothole. { kōk }

colla [METEOROL] In the Philippines, a fresh or strong (less than 39–46 miles per hour, or 63–74 kilometers per hour) south to southwest wind, accompanied by heavy rain and severe squall. Also known as colla tempestade. { 'kōl·yə }

collada [METEOROL] A strong wind (35–50 miles per hour, or 56–80 kilometers per hour) in the Gulf of California, blowing from the north or northwest in the upper part, and from the northeast in the lower part of the gulf. { kə'yäd·ə }

collapse breccia [GEOL] Angular rock fragments derived from the collapse of rock overlying a hollow space. { kə'laps ˌbrech·ə }

collapse caldera [GEOL] A caldera formed primarily as a result of collapse due to withdrawal of magmatic support. { kə'laps kal'dir·ə }

collapse sink [GEOL] A sinkhole resulting from local collapse of a cavern that has been enlarged by solution and erosion. { kə'laps ˌsiŋk }

collapse structure [GEOL] A structure resulting from rock slides under the influence of gravity. Also known as gravity-collapse structure. { kə'laps ˌstrək·chər }

colla tempestade *See* colla. { 'kä·lə ˌtem·pə'städ·ə }

collective [METEOROL] In aviation weather observations, a group of observations transmitted in prescribed order by stations on the same long-line teletypewriter circuit. Also known as sequence. { kə'lek·tiv }

collinite [GEOL] The maceral, of collain consistency, of jellified plant material precipitated from solution and hardened; a variety of euvitrinite. { 'käl·əˌnīt }

collision cross section *See* cross section. { kə'lizh·ən 'krȯs ˌsek·shən }

collision efficiency [METEOROL] The fraction of all water-drops which, initially moving on a collision course with respect to other drops, actually collide (make surface contact) with the other drops. { kə'lizh·ən i'fish·ən·sē }

colloform [GEOL] Pertaining to the rounded, globular texture of mineral formed by colloidal precipitation. { 'käl·əˌfȯrm }

colloidal instability [METEOROL] A property attributed to clouds, by which the particles of the cloud tend to aggregate into masses large enough to precipitate. { kə'lȯid·əl in·stəˌbil·əd·ē }

colluvium [GEOL] Loose, incoherent deposits at the foot of a slope or cliff, brought there principally by gravity. { kə'lü·vē·əm }

Coloradoan [GEOL] Middle Upper Cretaceous geologic time. { ˌkäl·ə'rad·ə·wən }

Colorado low [METEOROL] A low which makes its first appearance as a definite center in the vicinity of Colorado on the eastern slopes of the Rocky Mountains; analogous to the Alberta low. { ˌkal·ə'rad·ō ˌlō }

columnar jointing [GEOL] Parallel, prismatic columns that are formed as a result of contraction during cooling in basaltic flow and other extrusive and intrusive rocks. Also known as columnar structure; prismatic jointing; prismatic structure. { kə'ləm·nər 'jȯint·iŋ }

column *See* geologic column; stalacto-stalagmite. { 'käl·əm }

columnar resistance [GEOPHYS] The electrical resistance of a column of air 1 centimeter square, extending from the earth's surface to some specified altitude. { kə'ləm·nər ri'zis·təns }

columnar section [GEOL] A vertical strip or scale drawing of the strip taken from a given area or locality showing the sequence of the rock units and their stratigraphic relationship, and indicating the thickness, lithology, age, classification, and fossil content of the rock units. Also known as section. { kə'ləm·nər 'sek·shən }

columnar structure *See* columnar jointing. { kə'ləm·nər ˌstrək·chər }

comagmatic province *See* petrographic province. { ¦kō·mag'mad·ik 'prä·vəns }

Comanchean [GEOL] A North American provincial series in Lower and Upper Cretaceous geologic time, above Coahuilan and below Gulfian. { kə'man·chē·ən }

comber [OCEANOGR] A deep-water wave of long, curling character with a high, breaking crest pushed forward by a strong wind. { 'kōm·ər }

combination coefficient [GEOPHYS] A measure of the specific rate of disappearance of small ions in the atmosphere due to either union with neutral Aitken nuclei to form new large ions, or union with large ions of opposite sign to form neutral Aitken nuclei. { ˌkäm·bə'nā·shən ¦kō·i'fish·ənt }

combination trap [GEOL] Underground reservoir structure closure, deformation, or fault where reservoir rock covers only part of the structure. { ˌkäm·bə'nā·shən ¦trap }

combined water [GEOCHEM] Water attached to soil minerals by means of chemical bonds. { kəm'bīnd 'wȯd·ər }

combustible shale *See* tasmanite. { kəm'bəs·tə·bəl ˌshāl }

combustion nucleus [METEOROL] A condensation nucleus formed as a result of industrial or natural combustion processes. { kəm'bəs·chən ˌnü·klē·əs }

comendite [GEOL] A white, sodic rhyolite containing alkalic amphibole or pyroxene. { kə'menˌdīt }

comfort index *See* temperature-humidity index. { 'kəm·fərt ˌinˌdeks }

Comleyan [GEOL] Lower Cambrian geologic time. { 'käm·lā·ən }

common establishment *See* high-water full and change. { ¦käm·ən ə'stab·lish·mənt }

compaction [GEOL] Process by which soil and sediment mass loses pore space in response to the increasing weight of overlying material. { kəm'pak·shən }

compass points [GEOD] The 32 divisions of a compass at intervals of $11^1/_4$, with each division further divided into quarter points. Also known as points of the compass. { 'käm·pəs ˌpȯins }

compensation depth [OCEANOGR] The depth at which the light intensity is just sufficient to bring about a balance between the oxygen produced and that consumed by algae. { ˌkäm·pən'sā·shən ˌdepth }

competence [GEOL] The ability of the wind to transport solid particles either by rolling, suspension, or saltation (intermittent rolling and suspension); usually expressed in terms of the weight of a single particle. [HYD] The ability of a stream, flowing at a given velocity, to move the largest particles. { 'käm·pəd·əns }

competent beds [GEOL] Beds or strata capable of withstanding the pressures of folding without flowing or changing in original thickness. { ¦käm·pəd·ənt ¦bedz }

complementary rocks [GEOL] Rocks which are differentiated from the same magma, and whose average composition is the same as the parent magma. { ˌkäm·plə'men·trē 'räks }

complex [GEOL] An assemblage of rocks that has been folded together, intricately mixed, involved, or otherwise complicated. { 'kämˌpleks }

complex climatology [CLIMATOL] Analysis of the climate of a single space, or comparison of the climates of two or more places, by the relative frequencies of various weather types or groups of such types; a type is defined by the simultaneous occurrence within specified narrow limits of each of several weather elements. { 'käm ˌpleks ˌklī·mə'täl·ə·jē }

complex dune [GEOL] A dune of varying forms, often very large, and produced by variable, shifting winds and the merging of various dune types. { 'kämˌpleks 'dün }

complex fold [GEOL] A fold whose axial line is also folded. { 'kämˌpleks 'fōld }

complex low [METEOROL] An area of low atmospheric pressure within which more than one low-pressure center is found. { 'kämˌpleks 'lō }

complex tombolo [GEOL] A system resulting when several islands and the mainland are interconnected by a complex series of tombolos. Also known as tombolo cluster; tombolo series. { 'kämˌpleks 'täm·bəˌlō }

composite cone [GEOL] A large volcanic cone constructed of lava and pyroclastic material in alternating layers. { kəm'päz·ət 'kōn }

composite dike [GEOL] A dike consisting of several intrusions differing in chemical and mineralogical composition. { kəm'päz·ət 'dīk }

composite flash [GEOPHYS] A lightning discharge which is made up of a series of distinct lightning strokes with all strokes following the same or nearly the same channel, and with successive strokes occurring at intervals of about 0.05 second. Also known as multiple discharge. { kəm'päz·ət 'flash }

composite fold [GEOL] A fold having smaller folds on its limbs. { kəm'päz·ət 'fōld }

composite grain [GEOL] A sedimentary clast formed of two or more original particles. { kəm'päz·ət 'grān }

composite sequence [GEOL] An ideal sequence of cyclic sediments containing all the lithological types in their proper order. { kəm'päz·ət 'sē·kwəns }

composite sill [GEOL] A sill consisting of several intrusions differing in chemical and mineralogical compositions. { kəm'päz·ət 'sil }

composite topography [GEOL] A topography whose features have developed in two or more erosion cycles. { kəm'päz·ət tə'päg·rə·fē }

composite unconformity [GEOL] An unconformity that has resulted from more than one episode of nondeposition and possible erosion. { kəm'päz·ət ,ən·kən'fȯr·məd·ē }

composite vein [GEOL] A large fracture zone composed of parallel ore-filled fissures and converging diagonals, whose walls and intervening country rock have been replaced to a certain degree. { kəm'päz·ət 'vān }

composite volcano *See* stratovolcano. { kəm'päz·ət väl'kā·nō }

compositional maturity [GEOL] Concept of a type of maturity in sedimentary rocks in which a sediment approaches the compositional end product to which formative processes drive it. { ,käm·pə'zish·ən·əl mə'chůr·əd·ē }

compound alluvial fan [GEOL] Structure formed by the lateral growth and merger of fans made by neighboring streams. { 'käm,pau̇nd ə¦lü·vē·əl ¦fan }

compound fault [GEOL] A zone or series of essentially parallel faults, closely spaced. { 'käm,pau̇nd 'fȯlt }

compound ripple marks [GEOL] Complex ripple marks of great diversity which originate by simultaneous interference of wave oscillation with current action. { 'käm ,pau̇nd 'rip·əl ,märks }

compound valley glacier [HYD] A glacier composed of several ice streams emanating from different tributary valleys. { 'käm,pau̇nd 'val·ē ,glā·shər }

compound volcano [GEOL] **1.** A volcano consisting of a complex of two or more cones. **2.** A volcano with an associated volcanic dome. { 'käm,pau̇nd väl'kā·nō }

compression [GEOD] *See* flattening. [GEOL] A system of forces which tend to decrease the volume or shorten rocks. { kəm'presh·ən }

concentration [HYD] The ratio of the area of the sea covered by ice to the total area of sea surface. { ,kän·sən'trā·shən }

concentration time [HYD] The time required for water to travel from the most remote portion of a river basin to the basin outlet; it varies with the quantity of flow and channel conditions. { ,kän·sən'trā·shən ,tīm }

concentric faults [GEOL] Faults that are arranged concentrically. { kən'sen·trik 'fȯlts }

concentric fold [GEOL] A fold in which the original thickness of the strata is unchanged during deformation. Also known as parallel fold. { kən'sen·trik 'fōld }

concentric fractures [GEOL] A system of fractures concentrically arranged about a center. { kən'sen·trik 'frak·chərz }

concentric weathering *See* spheroidal weathering. { kən'sen·trik 'weth·ə·riŋ }

conchoidal [GEOL] Having a smoothly curved surface; used especially to describe the fracture surface of a mineral or rock. { käŋ'kȯid·əl }

concordant body [GEOL] An intrusive igneous body whose contacts are parallel to the bedding of the country rock. Also known as concordant injection; concordant pluton. { kən 'kȯrd·ənt ¦bäd·ē }

concordant coastline [GEOL] A coastline parallel to the land structures which form the margin of an ocean basin. { kən'kȯrd·ənt 'kōst,līn }

concordant drainage *See* accordant drainage. { kən'kȯrd·ənt 'drān·ij }

concordant injection *See* concordant body. { kən'kȯrd·ənt in'jek·shən }

concordant pluton *See* concordant body. { kən'kȯrd·ənt 'plü,tän }

concretion [GEOL] A hard, compact mass of mineral matter in the pores of sedimentary or fragmental volcanic rock; represents a concentration of a minor constituent of the enclosing rock or of cementing material. { kän'krē·shən }

concretionary [GEOL] Tending to grow together, forming concretions. { kən'krē·shə,ner·ē }

concretioning [GEOL] The process of forming concretions. { kən'krē·shən·iŋ }

concussion fracture [GEOL] Radiating system of fractures in a shock-metamorphosed rock. { kən'kəsh·ən ,frak·chər }

condensate field [GEOL] A petroleum field developed in predominantly gas-bearing reservoir rocks, but within which condensation of gas to oil commonly occurs with decreases in field pressure. { 'kän·dən,sāt ,fēld }

condensation [METEOROL] The process by which water vapor becomes a liquid such as dew, fog, or cloud or a solid like snow; condensation in the atmosphere is brought about by either of two processes: cooling of air to its dew point, or addition of enough water vapor to bring the mixture to the point of saturation (that is, the relative humidity is raised to 100). { ˌkän·dən'sā·shən }

condensation cloud [METEOROL] A mist or fog of minute water droplets that temporarily surrounds the fireball following an atomic detonation in a comparatively humid atmosphere. { ˌkän·dən'sā·shən ˌklau̇d }

condensation nucleus [METEOROL] A particle, either liquid or solid, upon which condensation of water vapor begins in the atmosphere. { ˌkän·dən'sā·shən 'nü·klē·əs }

condensation pressure [METEOROL] The pressure at which a parcel of moist air expanded dry adiabatically reaches saturation. Also called adiabatic condensation pressure; adiabatic saturation pressure. { ˌkän·dən'sā·shən ˌpresh·ər }

condensation temperature [METEOROL] The temperature at which a parcel of moist air expanded dry adiabatically reaches saturation. Also known as adiabatic condensation temperature; adiabatic saturation temperature. { ˌkän·dən'sā·shən 'tem·prə·chər }

condensation trail [METEOROL] A visible trail of condensed water vapor or ice particles left behind an aircraft, an airfoil, or such, in motion through the air. Also known as contrail; vapor trail. { ˌkän·dən'sā·shən ˌtrāl }

conditional instability [METEOROL] The state of a column of air in the atmosphere when its lapse rate of temperature is less than the dry adiabatic lapse rate but greater than the saturation adiabatic lapse rate. { kən'dish·ən·əl ˌin·stə'bil·əd·ē }

conductive equilibrium *See* isothermal equilibrium. { kən'dək·tiv ˌē·kwə'lib·rē·əm }

conductivity *See* permeability. { ˌkänˌdək'tiv·əd·ē }

conductivity current *See* air-earth conduction current. { ˌkänˌdək'tiv·əd·ē ˌkər·ənt }

conduit [GEOL] A water-filled underground passage that is always under hydrostatic pressure. { 'kän·də·wət }

cone [GEOL] A mountain, hill, or other landform having relatively steep slopes and a pointed top. { kōn }

cone delta *See* alluvial cone. { 'kōn ˌdel·tə }

cone dike *See* cone sheet. { 'kōn ˌdīk }

cone-in-cone structure [GEOL] The structure of a concretion characterized by the development of a succession of cones one within another. { 'kōn in 'kōn 'strək·chər }

cone karst [GEOL] A type of karst, typical of tropical regions, characterized by a pattern of steep, convex sides and slightly concave floors. Also known as cockpit karst; Kegel karst. { 'kōn ˌkärst }

Conemaughian [GEOL] Upper Middle Pennsylvanian geologic time. { ˌkän·ə'mȯg·ē·ən }

cone of dejection *See* alluvial cone. { 'kōn əv di'jek·shən }

cone of depression [HYD] The depression in the water table around a well defining the area of influence of the well. Also known as cone of influence. { 'kōn əv di'presh·ən }

cone of detritus *See* alluvial cone. { 'kōn əv di'trīd·əs }

cone of escape [GEOPHYS] A hypothetical cone in the exosphere, directed vertically upward, through which an atom or molecule would theoretically be able to pass to outer space without a collision. { 'kōn əv ə'skāp }

cone of influence *See* cone of depression. { 'kōn əv 'in·flü·əns }

cone sheet [GEOL] An accurate dike forming part of a concentric set that dips inward toward the center of the arc. Also known as cone dike. { 'kōn ˌshēt }

Conewangoan [GEOL] Upper Upper Devonian geologic time. { ˌkän·ə'waŋ·gə·wən }

confined aquifer *See* artesian aquifer. { kən'fīnd 'ak·wə·fər }

confining bed [GEOL] An impermeable bed adjacent to an aquifer. { kən'fīn·iŋ ˌbed }

confining pressure [GEOL] An equal, all-sided pressure, such as lithostatic pressure produced by overlying rocks in the crust of the earth. { kən'fīn·iŋ ˌpresh·ər }

confluence [HYD] **1.** A stream formed from the flowing together of two or more streams. **2.** The place where such streams join. { 'känˌflü·əns }

conformable [GEOL] **1.** Pertaining to the contact of an intrusive body when it is aligned with the internal structures of the intrusion. **2.** Referring to strata in which layers are formed above one another in an unbroken, parallel order. { kən'fȯr·mə·bəl }

conformity [GEOL] The shared and undisturbed correspondence between adjacent sedimentary strata that have been deposited in orderly sequence with little or no indication of time lapses. { kən'fōr·məd·ē }

confused sea [OCEANOGR] A highly disturbed water surface without a single, well-defined direction of wave travel. { kən'fyüzd 'sē }

congelifluction *See* gelifluction. { kən,jel·ə'flək·shən }

congelifraction [GEOL] The splitting or disintegration of rocks as the result of the freezing of the water contained. Also known as frost bursting; frost riving; frost shattering; frost splitting; frost weathering; frost wedging; gelifraction; gelivation. { kən¦jel·ə¦frak·shən }

congeliturbate [GEOL] Soil or unconsolidated earth which has been moved or disturbed by frost action. { kən,jel·ə'tər·bət }

congeliturbation [GEOL] The churning and stirring of soil as a result of repeated cycles of freezing and thawing; includes frost heaving and surface subsidence during thaws. Also known as cryoturbation; frost churning; frost stirring; geliturbation. { kən,jel·ə·tər'bā·shən }

conglomerate [GEOL] Cemented, rounded fragments of water-worn rock or pebbles, bound by a siliceous or argillaceous substance. { kən'gläm·ə·rət }

conglomeratic mudstone *See* paraglomerate. { kən¦gläm·ə¦rad·ik 'məd,stōn }

congruent melting [GEOL] Melting of a solid substance to a liquid identical in composition. { kən'grü·ənt 'melt·iŋ }

Coniacian [GEOL] Lower Senonian geologic time. { ,kän·ē'ā·shən }

conjugate [GEOL] **1.** Pertaining to fractures in which both sets of veins or joints show the same strike but opposite dip. **2.** Pertaining to any two sets of veins or joints lying perpendicular. { 'kän·jə·gət }

conjugate joint system [GEOL] Two joint sets with a symmetrical pattern arranged about another structural feature or an inferred stress axis. { 'kän·jə·gət ¦jöint 'sis·təm }

connate [GEOL] Referring to materials involved in sedimentary processes that are contemporaneous with surrounding materials. { kə'nāt }

connate water [HYD] Water entrapped in the interstices of igneous rocks when the rocks were formed; usually highly mineralized. { kə'nāt 'wȯd·ər }

connecting bar *See* tombolo. { kə'nekt·iŋ ,bär }

Conrad discontinuity [GEOPHYS] A relatively abrupt discontinuity in the velocity of elastic waves in the earth, increasing from 6.1 to 6.4–6.7 kilometers per second; occurs at various depths and marks contact of granitic and basaltic layers. { 'kän,rad dis,känt·ən'ü·əd·ē }

consanguineous [GEOL] Of a natural group of sediments or sedimentary rocks, having common or related origin. { ¦kän·saŋ¦gwin·ē·əs }

consequent [GEOL] Of, pertaining to, or characterizing movements of the earth resulting from the external transfer of material in the process of gradation. { 'kän·sə·kwənt }

consequent stream [GEOL] A stream whose course is determined by the slope of the land. Also known as superposed stream. { 'kän·sə·kwənt ,strēm }

consequent valley [GEOL] **1.** A valley whose direction depends on corrugation. **2.** A valley formed by the widening of a trench cut by a consequent stream. { 'kän·sə·kwənt ,val·ē }

conservative concentrations [OCEANOGR] Concentrations such as heat content or salinity occurring in bodies of water that are altered locally, except at the boundaries, by processes of diffusion and advection only. { kən'sər·və·tiv ,kän·sən'trā·shənz }

consolidated ice [OCEANOGR] Ice which has been compacted into a solid mass by wind and ocean currents and covers an area of the ocean. { kən'säl·ə,dād·əd 'īs }

consolidation [GEOL] **1.** Processes by which loose, soft, or liquid earth become coherent and firm. **2.** Adjustment of a saturated soil in response to increased load; involves squeezing of water from the pores and a decrease in void ratio. { kən,säl·ə'dā·shən }

constant-height chart [METEOROL] A synoptic chart for any surface of constant geometric altitude above mean sea level (a constant-height surface), usually containing plotted data and analyses of the distribution of such variables as pressure, wind, temperature, and humidity at that altitude. Also known as constant-level chart; fixed-level chart; isohypsic chart. { ¦kän·stənt ¦hīt ˌchärt }
constant-height surface [METEOROL] A surface of constant geometric or geopotential altitude measured with respect to mean sea level. Also known as constant-level surface; isohypsic surface. { ¦kän·stənt ¦hīt ˌsər·fəs }
constant-level chart *See* constant-height chart. { ¦kän·stənt ¦lev·əl 'chärt }
constant-level surface *See* constant-height surface. { ¦kän·stənt ¦lev·əl 'sər·fəs }
constant-pressure chart [METEOROL] The synoptic chart for any constant-pressure surface, usually containing plotted data and analyses of the distribution of height of the surface, wind temperature, humidity, and so on. Also known as isobaric chart; isobaric contour chart. { ¦kän·stənt ¦presh·ər ˌchärt }
constant-pressure surface *See* isobaric surface. { ¦kän·stənt ¦presh·ər ˌsər·fəs }
constituent number [OCEANOGR] One of the harmonic elements in a mathematical expression for the tide-producing force, and in corresponding formulas for the tide or tidal current. { kən'stich·ə·wənt 'nəm·bər }
consumptive use [HYD] The total annual land water loss in an area, due to evaporation and plant use. { kən'səm·div 'yüs }
contact [GEOL] The surface between two different kinds of rocks. { 'känˌtakt }
contact aureole *See* aureole. { 'känˌtakt 'ȯr·ēˌōl }
contact metasomatism [GEOL] One of the main local processes of thermal metamorphism that is related to intrusion of magmas; takes place in rocks or near their contact with a body of igneous rock. { 'känˌtakt ˌmed·ə'sō·məˌtiz·əm }
contact vein [GEOL] **1.** A variety of fissure vein formed by deposition of minerals in a fault fissure at a rock contact. **2.** A replacement vein formed by mineralized solutions percolating along the more permeable surface areas of the contact. { 'känˌtakt ˌvān }
contact zone *See* aureole. { 'känˌtakt ˌzōn }
contamination [GEOL] A process in which the chemical composition of a magma changes due to the assimilation of country rocks. [HYD] The addition to water of any substance or property that prevents its use without further treatment. { kənˌtam·ə'nā·shən }
contemporaneous [GEOL] **1.** Formed, existing, or originating at the same time. **2.** Of a rock, developing during formation of the enclosing rock. { kənˌtem·pə'rā·nē·əs }
continent [GEOGR] A protuberance of the earth's crustal shell, with an area of several million square miles and sufficient elevation so that much of it is above sea level. { 'känt·ən·ənt }
continental accretion [GEOL] The theory that continents have grown by the addition of new continental material around an original nucleus, mainly through the processes of geosynclinal sedimentation and orogeny. { ¦känt·ən¦ent·əl ə'krē·shən }
continental air [METEOROL] A type of air whose characteristics are developed over a large land area and which therefore has relatively low moisture content. { ¦känt·ən¦ent·əl 'er }
continental anticyclone *See* continental high. { ¦känt·ən¦ent·əl ˌan·tē'sīˌklōn }
continental borderland [GEOL] The area of the continental margin between the shoreline and the continental slope. { ¦känt·ən¦ent·əl 'bor·dərˌland }
continental climate [CLIMATOL] Climate characteristic of the interior of a landmass of continental size, marked by large annual, daily, and day-to-day temperature ranges, low relative humidity, and a moderate or small irregular rainfall; annual extremes of temperature occur soon after the solstices. { ¦känt·ən¦ent·əl 'klī·mət }
continental crust [GEOL] The basement complex of rock, that is, metamorphosed sedimentary and volcanic rock with associated igneous rocks mainly granitic, that underlies the continents and the continental shelves. { ¦känt·ən¦ent·əl 'krəst }
continental deposits [GEOL] Sedimentary deposits laid down within a general land area. { ¦känt·ən¦ent·əl di'päz·əts }
continental displacement *See* continental drift. { ¦känt·ən¦ent·əl di'splās·mənt }

continental divide [GEOL] A drainage divide of a continent, separating streams that flow in opposite directions; for example, the divide in North America that separates watersheds of the Pacific Ocean from those of the Atlantic Ocean. { ¦känt·ən¦ent·əl di'vīd }

continental drift [GEOL] The concept of continent formation by the fragmentation and movement of land masses on the surface of the earth. Also known as continental displacement. { ¦känt·ən¦ent·əl 'drift }

continental geosyncline [GEOL] A geosyncline filled with nonmarine sediments. { ¦känt·ən¦ent·əl ¦jē·ō'sin,klīn }

continental glacier [HYD] A sheet of ice covering a large tract of land, such as the ice caps of Greenland and the Antarctic. { ¦känt·ən¦ent·əl 'glā·shər }

continental growth [GEOL] The processes contributing to growth of continents at the expense of ocean basins. { ¦känt·ən¦ent·əl 'grōth }

continental heat flow [GEOPHYS] The amount of thermal energy escaping from the earth through the continental crust per unit area and unit time. { ¦känt·ən¦ent·əl 'hēt ,flō }

continental high [METEOROL] A general area of high atmospheric pressure which on mean charts of sea-level pressure is seen to overlie a continent during the winter. Also known as continental anticyclone. { ¦känt·ən¦ent·əl 'hī }

continentality [CLIMATOL] The degree to which a point on the earth's surface is in all respects subject to the influence of a land mass. { ,känt·ən·en'tal·əd·ē }

continental margin [GEOL] Those provinces between the shoreline and the deep-sea bottom; generally consists of the continental borderland, shelf, slope, and rise. { ¦känt·ən¦ent·əl 'mär·jən }

continental mass [GEOGR] The continental land rising more or less abruptly from the ocean floor and also the shallow submerged areas surrounding this land. { ¦känt·ən¦ent·əl 'mas }

continental nucleus [GEOL] A large area of basement rock consisting of basaltic and more mafic oceanic crust and periodotitic mantle from which it is postulated that continents have grown. Also known as continental shield; cratogene; shield. { ¦känt·ən¦ent·əl 'nü·klē·əs }

continental plate [GEOL] Thick continental crust. { ¦känt·ən¦ent·əl 'plāt }

continental plateau *See* tableland. { ¦känt·ən¦ent·əl plə'tō }

continental platform *See* continental shelf. { ¦känt·ən¦ent·əl 'plat,fȯrm }

continental polar air [METEOROL] Polar air having low surface temperature, low moisture content, and (especially in its source regions) great stability in the lower layers. { ¦känt·ən¦ent·əl ¦pō·lər 'er }

continental rise [GEOL] A transitional part of the continental margin; a gentle slope with a generally smooth surface, built up by the shedding of sediments from the continental block, and located between the continental slope and the abyssal plain. { ¦känt·ən¦ent·əl 'rīz }

continental shelf [GEOL] The zone around a continent, that part of the continental margin extending from the shoreline and the continental slope; composes with the continental slope the continental terrace. Also known as continental platform; shelf. { ¦känt·ən¦ent·əl 'shelf }

continental shield *See* shield. { ¦känt·ən¦ent·əl 'shēld }

continental slope [GEOL] The part of the continental margin consisting of the declivity from the edge of the continental shelf extending down to the continental rise. { ¦känt·ən¦ent·əl 'slōp }

continental terrace [GEOL] The continental shelf and slope together. { ¦känt·ən¦ent·əl 'ter·əs }

continental tropical air [METEOROL] A type of tropical air produced over subtropical arid regions; it is hot and very dry. { ¦känt·ən¦ent·əl ¦träp·ə·kəl 'er }

continent formation [GEOL] A series of six or seven major episodes, resulting from the buildup of radioactive heat and then the melting or partial melting of the earth's interior; the molten rock melt rises to the surface, differentiating into less primitive lavas; the continent then nucleates, differentiates, and grows from oceanic crust and mantle. { ¦känt·ən¦ent·əl fər'mā·shən }

continuity chart [METEOROL] A chart maintained for weather analysis and forecasting upon which are entered the positions of significant features (pressure centers, fronts, instability lines, through lines, ridge lines) of the regular synoptic charts at regular intervals in the past. { ,känt·ən'ü·əd·ē ,chärt }

continuous leader *See* dart leader. { kən¦tin·yə·wəs 'lēd·ər }

continuous permafrost zone [GEOL] Regional zone predominantly underlain by permanently frozen subsoil that is not interrupted by pockets of unfrozen ground. { kən¦tin·yə·wəs 'pər·mə,fròst ,zōn }

continuous profiling [GEOL] A method of shooting in seismic exploration in which uniformly placed seismometer stations along a line are shot from holes spaced along the same line so that each hole records seismic ray paths geometrically identical with those from adjacent holes. { kən¦tin·yə·wəs 'prō,fīl·iŋ }

contour-change line *See* height-change line. { 'kän,tür ,chānj ,līn }

contour code [METEOROL] A code in which data on the topography of constant-pressure surfaces are transmitted; a modification of the international analysis code. { 'kän,tür ,kōd }

contourite [OCEANOGR] A marine sediment deposited by swift ocean-bottom currents that generally flow along contours. { 'kän,tü,rīt }

contour line [METEOROL] A line on a weather map connecting points of equal atmospheric pressure, temperature, or such. { 'kän,tür ,līn }

contour microclimate [CLIMATOL] That portion of the microclimate which is directly attributable to the small-scale variations of ground level. { 'kän,tür ¦mī·krō¦klī·mət }

contraction hypothesis [GEOL] Theory that shrinking of the earth is the cause of compression folding and thrusting. { kən'trak·shən hī'päth·ə·səs }

contrail *See* condensation trail. { 'kän,trāl }

contrail-formation graph [METEOROL] A graph containing the parameters pressure, temperature, and relative humidity for critical values at which condensation trails (contrails) form; used as an aid in forecasting the formation of condensation trails. { 'kän,trāl fòr'mā·shən ,graf }

contra solem [METEOROL] Characterizing air motion that is counterclockwise in the Northern Hemisphere and clockwise in the Southern Hemisphere; literally, against the sun. { 'kän·trə 'sō,lem }

contrastes [METEOROL] Winds a short distance apart blowing from opposite quadrants, frequent in the spring and fall in the western Mediterranean. { kòn'tras,tēz }

contributory *See* tributary. { kən'trib·yə,tòr·ē }

control day [METEOROL] One of several days on which the weather is supposed (according to folklore) to provide the key for the weather of a subsequent period. Also known as key day. { kən'trōl ,dā }

control-tower visibility [METEOROL] The visibility that is observed from an airport control tower. { kən'trōl ¦taù·ər ,viz·ə'bil·əd·ē }

convection [METEOROL] Atmospheric motions that are predominantly vertical, resulting in vertical transport and mixing of atmospheric properties. [OCEANOGR] Movement and mixing of ocean water masses. { kən'vek·shən }

convectional stability *See* static stability. { kən'vek·shən·əl stə'bil·əd·ē }

convection cell [GEOPHYS] A concept in plate tectonics that accounts for the lateral or the upward and downward movement of subcrustal mantle material as due to heat variation in the earth. [METEOROL] An atmospheric unit in which organized convective fluid motion occurs. { kən'vek·shən ,sel }

convection current [GEOPHYS] Mass movement of subcrustal or mantle material as the result of temperature variations. [METEOROL] Any current of air involved in convection; usually, the upward-moving portion of a convection circulation, such as a thermal or the updraft in cumulus clouds. Also known as convective current. { kən'vek·shən ,kər·ənt }

convection stability *See* static stability. { kən'vek·shən stə'bil·əd·ē }

convection theory of cyclones [METEOROL] A theory of cyclone development proposing that the upward convection of air (particularly of moist air) due to surface heating can be of sufficient magnitude and duration that the surface inflow of air will attain appreciable cyclonic rotation. { kən'vek·shən ,thē·ə·rē əv 'sī,klōnz }

convective activity [METEOROL] Generally, manifestations of convection in the atmosphere, alluding particularly to the development of convective clouds and resulting weather phenomena, such as showers, thunderstorms, squalls, hail, and tornadoes. { kən'vek·div ak'tiv·əd·ē }

convective cloud [METEOROL] A cloud which owes its vertical development, and possibly its origin, to convection. { kən'vek·div 'klaud }

convective-cloud-height diagram [METEOROL] A graph used as an aid in estimating the altitude of the base of convective clouds; since its basis is the same as that for the dew-point formula, only the surface temperature and dew point need be known to use the diagram. { kən¦vek·div 'klaud ,hīt ,dī·ə,gram }

convective condensation level [METEOROL] On a thermodynamic diagram, the point of intersection of a sounding curve (representing the vertical distribution of temperature in an atmospheric column) with the saturation mixing-ratio line corresponding to the average mixing ratio in the surface layer (that is, approximately the lowest 1500 feet, or 450 meters). { kən'vek·div ,kän·den'sā,shən ,lev·əl }

convective equilibrium *See* adiabatic equilibrium. { kən'vek·div ,ē·kwə'lib·rē·əm }

convective instability [METEOROL] The state of an unsaturated layer or column of air in the atmosphere whose wet-bulb potential temperature (or equivalent potential temperature) decreases with elevation. Also known as potential instability. { kən'vek·div in·stə'bil·əd·ē }

convective overturn *See* overturn. { kən'vek·div 'ō·vər,tərn }

convective precipitation [METEOROL] Precipitation from convective clouds, generally considered to be synonymous with showers. { kən'vek·div prə,sip·ə'tā·shən }

convective region [METEOROL] An area particularly favorable for the formation of convection in the lower atmosphere, or one characterized by convective activity at a given time. { kən'vek·div ,rē·jən }

convergence [GEOL] Diminution of the interval between geologic horizons. [HYD] The line of demarcation between turbid river water and clear lake water. [METEOROL] The increase in wind setup observed beyond that which would take place in an equivalent rectangular basin of uniform depth, caused by changes in platform or depth. [OCEANOGR] A condition in the ocean in which currents or water masses having different densities, temperatures, or salinities meet; results in the sinking of the colder or more saline water. { kən'vər·jəns }

convergence line [METEOROL] Any horizontal line along which horizontal convergence of the airflow is occurring. Also known as asymptote of convergence. { kən'vər·jəns ,līn }

convergent precipitation [METEOROL] A synoptic type of precipitation caused by local updrafts of moist air. { kən'vər·jənt prə,sip·ə'tā·shən }

convergent zone paths [OCEANOGR] The velocity structure of permanent deep sound channels that produces focusing regions at distant intervals from a shallow source. { kən'vər·jənt 'zōn ,pathz }

convolute bedding [GEOL] The extremely contorted laminae usually confined to a single layer of sediment, resulting from subaqueous slumping. { 'kän·və,lüt ,bed·iŋ }

convolution [GEOL] **1.** The process of developing convolute bedding. **2.** A structure resulting from a convolution process, such as a small-scale but intricate fold. { ,kän·və'lü·shən }

cooking snow *See* water snow. { 'kuk·iŋ ,snō }

cooperative observer [METEOROL] An unpaid observer who maintains a meteorological station for the U.S. National Weather Service. { kō'äp·rəd·iv əb'zər·vər }

coorongite [GEOL] A boghead coal in the peat stage. { kō'ä·rən,jīt }

Copenhagen water *See* normal water. { ¦kō·pən¦häg·ən ,wȯd·ər }

copper ore [GEOL] Rock containing copper minerals. { 'käp·ər ,ȯr }

coprolite [GEOL] Petrified excrement. { 'käp·rə,līt }

coral head [GEOL] A small reef patch of coralline material. Also known as coral knoll. { 'kä·rəl ¦hed }
coral knoll *See* coral head. { 'kä·rəl ¦nōl }
coral mud [GEOL] Fine-grade deposits of coral fragments formed around coral islands and coasts bordered by coral reefs. { 'kär·əl ˌməd }
coral pinnacle [GEOL] A sharply upward-projecting growth of coral rising from the floor of an atoll lagoon. { 'kär·əl 'pin·ə·kəl }
coral reef [GEOL] A ridge or mass of limestone built up of detrital material deposited around a framework of skeletal remains of mollusks, colonial coral, and massive calcareous algae. { 'kär·əl ˌrēf }
coral-reef lagoon [GEOGR] The central, shallow body of water of an atoll or the water separating a barrier reef from the shore. { 'kär·əl ˌrēf lə'gün }
coral-reef shoreline [GEOL] A shoreline formed by reefs composed of coral polyps. Also known as coral shoreline. { 'kär·əl ˌrēf 'shȯrˌlīn }
coral sand [GEOL] Coarse-grade deposits of coral fragments formed around coral islands and coasts bordered by coral reefs. { 'kär·əl ˌsand }
coral shoreline *See* coral-reef shoreline. { 'kär·əl 'shȯrˌlīn }
cordillera [GEOGR] A mountain range or group of ranges, including valleys, plains, rivers, lakes, and so on, forming the main mountain axis of a continent. { ˌkȯrd·əl'er·ə }
cordilleran geosyncline [GEOL] The Devonian geosynclinal region of western North America. { ˌkȯrd·əl'er·ən ˌjē·ō'sinˌklīn }
cordonazo [METEOROL] A southerly wind of hurricane force generated along the western coast of Mexico when a tropical cyclone passes offshore in a northerly direction. { ˌkȯrd·ən'ä·sō }
core [GEOL] **1.** Center of the earth, beginning at a depth of 2900 kilometers. Also known as earth core. **2.** A vertical, cylindrical boring of the earth from which composition and stratification may be determined; in oil or gas well exploration the presence of hydrocarbons or water are items of interest. [OCEANOGR] That area within a layer of ocean water where parameters such as temperature, salinity, or velocity reach extreme values. { kȯr }
core analysis [GEOL] The use of core samples taken from the borehole during drilling to give information on strata age, composition, and porosity, and the presence of hydrocarbons or water along the length of the borehole. { 'kȯr ə'nal·ə·səs }
core intersection [GEOL] **1.** The point in a borehole where an ore vein or body is encountered as shown by the core. **2.** The width or thickness of the ore body, as shown by the core. Also known as core interval. { 'kȯr ¦in·tərˌsek·shən }
core interval *See* core intersection. { 'kȯr ¦in·tər·vəl }
core logging [GEOL] The analysis of the strata through which a borehole passes by the taking of core samples at predetermined depth intervals as the well is drilled. { 'kȯr ˌläg·iŋ }
core sample [GEOL] A sample of rock, soil, snow, or ice obtained by driving a hollow tube into the undisturbed medium and withdrawing it with its contained sample or core. { 'kȯr ˌsam·pəl }
corestone [GEOL] A rounded or broadly rectangular joint block of granite formed as a result of subsurface weathering in a manner similar to a tor but entirely separated from the bedrock. { 'kȯrˌstōn }
Coriolis parameter [GEOPHYS] Twice the component of the earth's angular velocity about the local vertical $2\Omega \sin \phi$, where Ω is the angular speed of the earth and ϕ is the latitude; the magnitude of the Coriolis force per unit mass on a horizontally moving fluid parcel is equal to the product of the Coriolis parameter and the speed of the parcel. { kȯr·ē'ō·ləs pə'ram·əd·ər }
corneite [GEOL] A biotite-hornfels formed during deformation of shale by folding. { 'kȯr·nēˌīt }
corn snow *See* spring snow. { 'kȯrn ˌsnō }
coromell [METEOROL] A land breeze from the south at La Paz, Mexico, near the mouth

of the Gulf of California, prevailing from November to May; it sets in at night and usually persists until 8 or 10 a.m. { kȯr·ə'mel }

corona [GEOL] A mineral zone that is usually radial about another mineral or at the area between two minerals. Also known as kelyphite. [METEOROL] A set of one or more prismatically colored rings of small radii, concentrically surrounding the disk of the sun, moon, or other luminary when veiled by a thin cloud; due to diffraction by numerous waterdrops. { kə'rō·nə }

corona method [GEOPHYS] A method of estimating drop sizes in clouds by utilizing measurements of the angular radii of the rings of a corona. { kə'rō·nə 'meth·əd }

corrasion [GEOL] Mechanical wearing away of rock and soil by the action of solid materials moved along by wind, waves, running water, glaciers, or gravity. Also known as mechanical erosion. { kə'rā·zhən }

corrected altitude [METEOROL] The indicated altitude corrected for temperature deviation from the standard atmosphere. Also known as true altitude. { kə'rek·təd 'al·tə,tüd }

corrected establishment *See* mean high-water lunitidal interval. { kə'rek·təd i'stab·lish·mənt }

correlation [GEOL] **1.** The determination of the equivalence or contemporaneity of geologic events in separated areas. **2.** As a step in seismic study, the selecting of corresponding phases, taken from two or more separated seismometer spreads, of seismic events seemingly developing at the same geologic formation boundary. { ,kär·ə'lā·shən }

corrie *See* cirque. { kȯr·ē }

corrosion [GEOCHEM] Chemical erosion by motionless or moving agents. { kə'rō·zhən }

cosmic sediment [GEOL] Particles of extraterrestrial origin which are observed as black magnetic spherules in deep-sea sediments. { 'käz·mik 'sed·ə·mənt }

cosmic spherules [GEOCHEM] Solidified, millimeter-sized to microscopic, rounded particles of extraterrestrial materials that melted either during high-velocity entry into the atmosphere or during hypervelocity impact of large meteoroids onto the earth's surface. { ¦käz·mik 'sfe·rülz }

cotton-belt climate [CLIMATOL] A type of warm climate characterized by dry winters and rainy summers; that is, a monsoon climate, in contrast to a Mediterranean climate. { 'kät·ən ,belt ,klī·mət }

coulee [GEOL] **1.** A thick, solidified sheet or stream of lava. **2.** A steep-sided valley or ravine, sometimes with a stream at the bottom. { kü'lā }

counterradiation [GEOPHYS] The downward flux of atmospheric radiation passing through a given level surface, usually taken as the earth's surface. Also known as back radiation. { ¦kau̇nt·ər,rād·ē'ā·shən }

country rock [GEOL] **1.** Rock that surrounds and is penetrated by mineral veins. **2.** Rock that surrounds and is invaded by an igneous intrusion. { ¦kən·trē 'räk }

Couvinian [GEOL] Lower Middle Devonian geologic time. { kü'vin·ē·ən }

cove [GEOGR] **1.** A small, narrow, sheltered bay, inlet, or creek on a coast. **2.** A deep recess or hollow occurring in a cliff or steep mountainside. { kōv }

cowshee *See* kaus. { 'kau̇·shē }

crachin [METEOROL] A period of light rain accompanied by low stratus clouds and poor visibility which frequently occurs in the China Sea between January and April. { krä·chin }

crag [GEOL] A steep, rugged point or eminence of rock, as one projecting from the side of a mountain. { krag }

crater [GEOL] **1.** A large, bowl-shaped topographic depression with steep sides. **2.** A rimmed structure at the summit of a volcanic cone; the floor is equal to the vent diameter. { 'krād·ər }

crater cone [GEOL] A cone built around a volcanic vent by lava extruded from the vent. { 'krād·ər ,kōn }

crater lake [HYD] A fresh-water lake formed by the accumulation of rain and groundwater in a caldera or crater. { 'krād·ər ,lāk }

craton [GEOL] A large, stable portion of the continental crust. Cratons are the broad heartlands of continents with subdued topography, encompassing the largest areas of most continents. { 'krā,tän }

cream ice *See* sludge. { 'krēm ,is }

creek [HYD] A natural stream of water, smaller than a river but larger than a brook. { krēk }

creep [GEOL] A slow, imperceptible downward movement of slope-forming rock or soil under sheer stress. { krēp }

crenitic [GEOL] Relating to or resulting from the raising of subterranean minerals by the action of spring water. { krə'nid·ik }

crenulation cleavage *See* slip cleavage. { ,kren·yə'lā·shən ,klēv·ij }

crepuscular arch *See* bright segment. { krə'pəs·kyə·lər 'ärch }

crescent beach [GEOL] A crescent-shaped beach at the head of a bay or the mouth of a stream entering the bay, with the concave side facing the sea. { 'kres·ənt ,bēch }

crescentic dune *See* barchan. { krə'sen·tik 'dün }

crescentic lake *See* oxbow lake. { krə'sen·tik 'lāk }

crestal plane [GEOL] The plane formed by joining the crests of all beds of an anticline. { 'krest·əl ,plān }

crest cloud [METEOROL] A type of standing cloud which forms along a mountain ridge, either on the ridge, or slightly above and leeward of it, and remains in the same position relative to the ridge. Also known as cloud crest. { 'krest ,klaủd }

crest length [OCEANOGR] The length of a wave measured along its crest. Also known as crest width. { 'krest ,leŋkth }

crest line [GEOL] The line connecting the highest points on the same bed of an anticline in an infinite number of cross sections. { 'krest ,līn }

crest stage [HYD] The highest stage reached at a point along a stream culminating a rise by waters of that stream. { 'krest ,stāj }

crest width *See* crest length. { 'krest ,width }

Cretaceous [GEOL] In geological time, the last period of the Mesozoic Era, preceded by the Jurassic Period and followed by the Tertiary Period; it extended from 144 million years to 65 million years before present. { kri'tā·shəs }

crevasse [GEOL] An open, nearly vertical fissure in a glacier or other mass of land ice or the earth, especially after earthquakes. { krə'vas }

crevasse deposit [GEOL] Kame deposited in a crevasse. { krə'vas di'päz·ət }

crevasse hoar [HYD] Ice crystals which form and grow in glacial crevasses and in other cavities where a large cooled space is formed and in which water vapor can accumulate under calm, still conditions. { krə'vas ,hȯr }

criador [METEOROL] The rain-bringing west wind of northern Spain. { krē·ə'dȯr }

crib *See* arête. { krib }

critical bottom slope [GEOL] The depth distribution in which depth d of an ocean increases with latitude ϕ according to an equation of the form $d = d_0 \sin \phi +$ constant. { 'krid·ə·kəl 'bäd·əm ,slōp }

critical density [GEOL] That degree of density of a saturated, granular material below which, as it is rapidly deformed, it will decrease in strength and above which it will increase in strength. { 'krid·ə·kəl 'den·səd·ē }

critical depth [HYD] In a water channel, that depth at which the flow is at its minimum energy with respect to the bottom of the channel. { 'krid·ə·kəl 'depth }

critical frequency [GEOPHYS] The minimum frequency of a vertically directed radio wave which will penetrate a particular layer in the ionosphere; for example, all vertical radio waves with frequencies greater than the E-layer critical frequency will pass through the E layer. Also known as penetration frequency. { 'krid·ə·kəl 'frē·kwən·sē }

critical level of escape [GEOPHYS] **1.** That level, in the atmosphere, at which a particle moving rapidly upward will have a probability of $1/e$ (e is base of natural logarithm) of colliding with another particle on its way out of the atmosphere. **2.** The level at which the horizontal mean free path of an atmospheric particle equals the scale height of the atmosphere. { 'krid·ə·kəl 'lev·əl əv ə'skāp }

critical slope [HYD] The channel slope or grade that is equal to the loss of head per foot resulting from flow at a depth which will provide uniform flow at critical depth. { 'krid·ə·kəl 'slōp }

crivetz [METEOROL] A wind blowing from the northeast quadrant in Rumania and southern Russia, especially a cold boralike wind from the north-northeast, characteristic of the climate of Rumania. { krə'vets }

CRM *See* chemical remanent magnetization.

Croixian [GEOL] Upper Cambrian geologic time. { 'krȯi·ən }

Cromwell Current [OCEANOGR] An eastward-setting subsurface current that extends about $1^1/_2°$ north and south of the Equator, and from about 150°E to 92°W. { 'kräm ˌwel ˌkər·ənt }

crop out *See* outcrop. { 'kräp ˌau̇t }

cross-bedding [GEOL] The condition of having laminae lying transverse to the main stratification planes of the strata; occurs only in granular sediments. Also known as cross-lamination; cross-stratification. { ¦krȯs 'bed·iŋ }

crosscutting relationships [GEOL] Relationships which may occur between two adjacent rock bodies, where the relative age may be determined by observing which rock "cuts" the other, for example, a granitic dike cutting across a sedimentary unit. { 'krȯsˌkəd·iŋ ri'lā·shənˌships }

cross fault [GEOL] **1.** A fault whose strike is perpendicular to the general trend of the regional structure. **2.** A minor fault that intersects a major fault. { 'krȯs ˌfȯlt }

cross fold [GEOL] A secondary fold whose axis is perpendicular or oblique to the axis of another fold. Also known as subsequent fold; superimposed fold; transverse fold. { 'krȯs ˌfōld }

cross joint [GEOL] A fracture in igneous rock perpendicular to the lineation caused by flow magma. Also known as transverse joint. { 'krȯs ˌjȯint }

cross-lamination *See* cross-bedding. { ¦krȯs lam·ə'nā·shən }

cross sea [OCEANOGR] A series of waves or swell crossing another wave system at an angle. { ¦krȯs ¦sē }

cross section [GEOL] **1.** A diagram or drawing that shows the downward projection of surficial geology along a vertical plane, for example, a portion of a stream bed drawn at right angles to the mean direction of the flow of the stream. **2.** An actual exposure or cut which reveals geological features. { 'krȯs ˌsek·shən }

cross-stratification *See* cross-bedding. { 'krȯs ˌstrad·ə·fə'kā·shən }

cross valley *See* transverse valley. { 'krȯs ˌval·ē }

crosswind [METEOROL] A wind which has a component directed perpendicularly to the course (or heading) of an exposed, moving object. { 'krȯsˌwind }

croute calcaire *See* caliche. { ˌkrüt kal'ker }

crude oil [GEOL] A comparatively volatile liquid bitumen composed principally of hydrocarbon, with traces of sulfur, nitrogen, or oxygen compounds; can be removed from the earth in a liquid state. { ¦krüd 'ȯil }

crumb structure [GEOL] A soil condition in which the particles are crumblike aggregates; suitable for agriculture. { 'krəm ˌstrək·chər }

crush breccia [GEOL] A breccia formed in place by mechanical fragmentation of rock during movements of the earth's crust. { 'krəsh ˌbrech·ə }

crush conglomerate [GEOL] Beds similar to a fault breccia, except that the fragments are rounded by attrition. Also known as tectonic conglomerate. { 'krəsh kən'gläm·ə·rət }

crush fold [GEOL] A fold of large dimensions that may involve considerable minor folding and faulting such as would produce a mountain chain or an oceanic deep. { 'krəsh ˌfōld }

crush zone [GEOL] A zone of fault breccia on fault gouge. { 'krəsh ˌzōn }

crust [GEOL] The outermost solid layer of the earth, mostly consisting of crystalline rock and extending no more than a few miles from the surface to the Mohorovičić discontinuity. Also known as earth crust. [HYD] A hard layer of snow lying on top of a soft layer. { krəst }

crustal motion [GEOL] Movement of the earth's crust. { ¦krəst·əl 'mō·shən }

crustal plate *See* tectonic plate. { 'krəst·əl ˌplāt }
cryoconite [GEOL] A dark, powdery dust transported by wind and deposited on the surface of snow or ice; found, however, mainly in cryoconite holes. { krī'äk·əˌnīt }
cryoconite hole [GEOL] A cylindrical dust well filled with cryoconite; absorbs solar radiation, causing melting of glacier ice around and below it. { krī'äk·əˌnīt ˌhōl }
cryogenic period [GEOL] A time period in geologic history during which large bodies of ice appeared at or near the poles and climate favored the formation of continental glaciers. { ˌkrī·ə'jen·ik ¦pir·ē·əd }
cryolaccolith *See* hydrolaccolith. { ¦krī·ō'lak·əˌlith }
cryology [HYD] The study of ice and snow. { krī'äl·ə·jē }
cryomorphology [GEOL] The branch of geomorphology that treats the processes and topographic features of regions where the ground is permanently frozen. { ¦krī·ō·mȯr'fäl·ə·jē }
cryopedology [GEOL] A branch of geology that deals with the study of intensive frost action and permanently frozen ground. { ¦krī·ō·pə'däl·ə·jē }
cryoplanation [GEOL] Land erosion at high latitudes or elevations due to processes of intensive frost action. { ¦krī·ō·plə'nā·shən }
cryosphere [GEOL] That region of the earth in which the surface is perennially frozen. { 'krī·əˌsfir }
cryostatic pressure [GEOL] Hydrostatic pressure exerted on soil and rocks when soil water freezes. { 'krī·əˌstad·ik 'presh·ər }
cryoturbation *See* congeliturbation. { ¦krī·ō·tər'bā·shən }
cryptoclastic [GEOL] Composed of extremely fine, almost submicroscopic, broken or fragmental particles. { ¦krip·tə¦klas·tik }
cryptoclimatology [CLIMATOL] The science of climates of confined spaces (cryptoclimates); basically, a form of microclimatology. Also spelled kryptoclimatology. { ¦krip·tōˌklī·mə'täl·ə·jē }
cryptocrystalline [GEOL] Having a crystalline structure but of such a fine grain that individual components are not visible with a magnifying lens. { ¦krip·tō'krist·əl·ən }
cryptovolcanic [GEOL] A small, nearly circular area of highly disturbed strata in which there is no evidence of volcanic materials to confirm the origin as being volcanic. { ¦krip·tō·väl'kan·ik }
crystalline frost [HYD] Hoarfrost that exhibits a relatively simple macroscopic crystalline structure. { 'kris·tə·lən 'frȯst }
crystalline porosity [GEOL] Porosity in crystalline limestone and dolomite, making possible underground oil reservoirs. { 'kris·tə·lən pə'räs·əd·ē }
crystallite [GEOL] A small, rudimentary form of crystal which is of unknown mineralogic composition and which does not polarize light. { 'kris·təˌlīt }
crystalloblastic series [GEOL] A series of metamorphic minerals ordered according to decreasing formation energy, so crystals of a listed mineral have a tendency to form idioblastic outlines at surfaces of contact with simultaneously developed crystals of all minerals in lower positions. { 'kris·tə·lə'blas·tik 'sirˌēz }
crystalloblastic texture [GEOL] A crystalline texture resulting from metamorphic recrystallization under conditions of high viscosity and directed pressure. { 'kris·tə·lə'blas·tik 'teks·chər }
crystal sandstone [GEOL] Siliceous sandstone in which deposited silica is precipitated upon the quartz grains in crystalline position. { ¦krist·əl 'sandˌstōn }
crystal settling [GEOL] Sinking of crystals in magma from the liquid in which they formed, by the action of gravity. { ¦krist·əl ¦set·liŋ }
crystal tuff [GEOL] Consolidated volcanic ash in which crystals and crystal fragments predominate. { ¦krist·əl 'təf }
crystal-vitric tuff [GEOL] Consolidated volcanic ash composed of 50–75% crystal fragments and 25–50% glass fragments. { ¦krist·əl ¦vi·trik 'təf }
crystosphene [HYD] A buried sheet or mass of ice, as in the tundra of northern America, formed by the freezing of rising and spreading springwater beneath alluvial deposits. { ¦kris·tə¦sfēn }
Cs *See* cirrostratus cloud.

cuesta [GEOGR] A gently sloping plain which terminates in a steep slope on one side. { 'kwes·tə }
culmination [GEOL] A high point on the axis of a fold. { kəl·mə'nā·shən }
cum sole [GEOPHYS] With the sun; hence anticyclonic or clockwise in the Northern Hemisphere. { ˌku̇m 'sōl·ə }
cumuliform cloud [METEOROL] A fundamental cloud type, showing vertical development in the form of rising mounds, domes, or towers. { 'kyü·myə·ləˌfȯrm ˌklau̇d }
cumulonimbus calvus cloud [METEOROL] A species of cumulonimbus cloud evolving from cumulus congestus: the protuberances of the upper portion have begun to lose the cumuliform outline; they loom and usually flatten, then transform into a whitish mass with a more or less diffuse outline and vertical striation; cirriform cloud is not present, but the transformation into ice crystals often proceeds with great rapidity. { ¦kyü·myə·lō'nim·bəs 'kal·vəs ˌklau̇d }
cumulonimbus capillatus cloud [METEOROL] A species of cumulonimbus cloud characterized by the presence of distinct cirriform parts, frequently in the form of an anvil, a plume, or a vast and more or less disorderly mass of hair, and usually accompanied by a thunderstorm. { ¦kyü·myə·lō'nim·bəs kap·ə'lad·əs ˌklau̇d }
cumulonimbus cloud [METEOROL] A principal cloud type, exceptionally dense and vertically developed, occurring either as isolated clouds or as a line or wall of clouds with separated upper portions. { ¦kyü·myə·lō'nim·bəs ˌklau̇d }
cumulus [GEOCHEM] The accumulation of minerals which have precipitated from a liquid without having been modified by later crystallization. { 'kyü·myə·ləs }
cumulus cloud [METEOROL] A principal type of cloud in the form of individual, detached elements which are generally dense and possess sharp nonfibrous outlines; these elements develop vertically, appearing as rising mounds, domes, or towers, the upper parts of which often resemble a cauliflower. { 'kyü·myə·ləs ˌklau̇d }
cumulus congestus cloud [METEOROL] A strongly sprouting cumulus species with generally sharp outline and sometimes a great vertical development, and with cauliflower or tower aspect. { 'kyü·myə·ləs kən'jes·təs ˌklau̇d }
cumulus humilis cloud [METEOROL] A species of cumulus cloud characterized by small vertical development and a generally flattened appearance, vertical growth is usually restricted by the existence of a temperature inversion in the atmosphere, which in turn explains the unusually uniform height of the cloud. Also known as fair-weather cumulus. { 'kyü·myə·ləs 'hyü·mə·ləs ˌklau̇d }
cumulus mediocris cloud [METEOROL] A cloud species unique to the species cumulus, of moderate vertical development, the upper protuberances or sproutings being not very marked; there may be a small cauliflower aspect; while this species does not give any precipitation, it frequently develops into cumulus congestus and cumulonimbus. { 'kyü·myə·ləs mē·dē'ō·krəs ˌklau̇d }
cup-and-ball joint [GEOL] A dish-shaped transverse fracture which divides a basalt column into segments. Also known as ball-and-socket joint. { ˌkəp ən 'bȯl ˌjȯint }
cup crystal [HYD] A crystal of ice in the form of a hollow hexagonal cup; a common form of depth hoar. { 'kəp ˌkrist·əl }
cupola [GEOL] An isolated, upward-projecting body of plutonic rock that lies near a larger body; both bodies are presumed to unite at depth. { 'kyü·pə·lə }
cupped pebble [GEOL] A pebble fragment that has become hollow after being subjected to solution. { 'kəpt ¦peb·əl }
current-bedding [GEOL] Cross-bedding resulting from water or air currents. { 'kər·ənt ˌbed·iŋ }
current constants [OCEANOGR] Tidal current relations that remain practically constant for any particular locality. { 'kər·ənt 'kän·stəns }
current curve [OCEANOGR] In marine operations, a graphic representation of the flow of a current, consisting of a rectangular-coordinate graph on which speed is represented by the ordinates and time by the abscissas. { 'kər·ənt ˌkərv }
current cycle [OCEANOGR] A complete set of tidal current conditions, as those occurring during a tidal day, lunar month, or Metonic cycle. { 'kər·ənt ˌsī·kəl }

current diagram [OCEANOGR] A graph showing the average speeds of flood and ebb currents throughout the current cycle for a considerable part of a tidal waterway. { 'kər·ənt 'dī·ə,gram }

current difference [OCEANOGR] In marine operations, the difference between the time of slack water or strength of current at a subordinate station and its reference station. { 'kər·ənt ,dif·rəns }

current drift [HYD] A broad, shallow, slow-moving ocean or lake current. { 'kər·ənt ,drift }

current ellipse [OCEANOGR] In marine operations, a graphic representation of a rotary current, in which the speed and direction of the current at various hours of the current cycle are represented by radius vectors; a line connecting the ends of the radius vectors approximates an ellipse. { 'kər·ənt ə,lips }

current hour [OCEANOGR] The average time interval between the moon's transit over the meridian of Greenwich and the time of the following strength of flood current modified by the times of slack water and strength of ebb. { 'kər·ənt ,aù·ər }

current lineation *See* parting lineation. { 'kər·ənt lin·ē'ā·shən }

current mark [GEOL] Any structure formed by direct or indirect action of a water current on a sedimentary surface. { 'kər·ənt ,märk }

current ripple [GEOL] A type of ripple mark having a long, gentle slope toward the direction from which the current flows, and a shorter, steeper slope on the lee side. { 'kər·ənt ,rip·əl }

current rips [OCEANOGR] Small waves formed on the surface of water by the meeting of opposing ocean currents; vertical oscillation, rather than progressive waves, is characteristic of current rips. { 'kər·ənt ,rips }

current tables [OCEANOGR] Tables listing predictions of the time and speeds of tidal currents at various places. { 'kər·ənt ,tā·bəlz }

curtain [GEOL] **1.** A thin sheet of dripstone that hangs or projects from a cave wall. **2.** A rock formation connecting two adjacent bastions. { 'kərt·ən }

curvature correction [GEOD] The correction applied in some geodetic work to take account of the divergence of the surface of the earth (spheroid) from a plane. { 'kər·və·chər kə'rek·shən }

cusp [GEOL] One of a series of low, crescent-shaped mounds of beach material separated by smoothly curved, shallow troughs spaced at more or less regular intervals along and generally perpendicular to the beach face. Also known as beach cusp. [GEOPHYS] Any of the funnel-shaped regions in the magnetosphere extending from the front magnetopause to the polar ionosphere, and filled with solar wind plasma. { kəsp }

cuspate bar [GEOL] A crescentic bar joining with the shore at each end. { 'kə,spāt ,bär }

cuspate ripple mark *See* linguoid ripple mark. { 'kə,spāt 'rip·əl ,märk }

custard winds [METEOROL] Cold easterly winds on the northeastern coast of England. { 'kəs·tərd ,winz }

cut and fill [GEOL] **1.** Lateral corrosion of one side of a meander accompanied by deposition on the other. **2.** A sedimentary structure consisting of a small filled-in channel. { ¦kət ən 'fil }

cutbank [GEOL] The concave bank of a winding stream that is maintained as a steep or even overhanging cliff by the action of water at its base. { 'kət,baŋk }

cutinite [GEOL] A variety of exinite consisting of plant cuticles. { 'kyüt·ən,īt }

cutoff [GEOL] A new, relatively short channel formed when a stream cuts through the neck of an oxbow or horseshoe bend. { 'kət,ȯf }

cutoff high [METEOROL] A warm high which has become displaced out of the basic westerly current, and lies to the north of this current. { 'kət,ȯf ,hī }

cutoff lake *See* oxbow lake. { 'kət,ȯf ,lāk }

cutoff low [METEOROL] A cold low which has become displaced out of the basic westerly current, and lies to the south of this current. { 'kət,ȯf ,lō }

cutout *See* horseback. { 'kət,aůt }

cut platform *See* wave-cut platform. { 'kət ,plat,fȯrm }

cutting-off process [METEOROL] A sequence of events by which a warm high or cold low, originally within the westerlies, becomes displaced either poleward (cutoff high) or equatorward (cutoff low) out of the westerly current; this process is evident at very high levels in the atmosphere, and it frequently produces, or is part of the production of, a blocking situation. { 'kəd·iŋ ˌöf ˌpräs·əs }

cwm *See* cirque. { küm }

cycle of erosion *See* geomorphic cycle. { 'sī·kəl əv i'rō·zhən }

cycle of sedimentation [GEOL] **1.** Also known as sedimentary cycle. **2.** A series of related processes and conditions appearing repeatedly in the same sequence in a sedimentary deposit. **3.** The sediments deposited from the beginning of one cycle to the beginning of a second cycle of the spread of the sea over a land area, consisting of the original land sediments, followed by those deposited by shallow water, then deep water, and then the reverse process of the receding water. *See* cyclothem. { 'sī·kəl əv ˌsed·ə·mən'tā·shən }

cyclic salt [OCEANOGR] Salt removed from the sea as spray, blown inland, and returned to its source by land drainage. { 'sīk·lik 'sölt }

cyclic sedimentation [GEOL] Deposition of various kinds of sediment in a repeated regular sequence. { 'sīk·lik ˌsed·ə·mən'tā·shən }

cyclogenesis [METEOROL] Any development or strengthening of cyclonic circulation in the atmosphere. { ¦sī·klō'jen·ə·səs }

cyclolysis [METEOROL] The weakening or decay of cyclonic circulation in the atmosphere. { sī'kläl·ə·səs }

cyclone [METEOROL] A low-pressure region of the earth's atmosphere with roundish to elongated-oval ground plan, in-moving air currents, centrally upward air movement, and generally outward movement at various higher elevations in the troposphere. { 'sīˌklōn }

cyclone family [METEOROL] A series of wave cyclones occurring in the interval between two successive major outbreaks of polar air, and traveling along the polar front, usually eastward and poleward. { 'sīˌklōn ˌfam·lē }

cyclone wave [METEOROL] **1.** A disturbance in the lower troposphere, of wavelength 1000–2500 kilometers; cyclone waves are recognized on synoptic charts as migratory high- and low-pressure systems. **2.** A frontal wave at the crest of which there is a center of cyclonic circulation, that is, the frontal wave of a wave cyclone. { 'sīˌklōn ˌwāv }

cyclonic [GEOPHYS] Having a sense of rotation about the local vertical that is the same as that of the earth's rotation: as viewed from above, counterclockwise in the Northern Hemisphere, clockwise in the Southern Hemisphere, undefined at the Equator. { sī'klän·ik }

cyclonic scale [METEOROL] The scale of the migratory high-and low-pressure systems (or cyclone waves) of the lower troposphere, with wavelengths of 1000–2500 kilometers. Also known as synoptic scale. { sī'klän·ik 'skāl }

cyclonic shear [METEOROL] Horizontal wind shear of such a nature that it contributes to the cyclonic vorticity of the flow; that is, it tends to produce cyclonic rotation of the individual air particles along the line of flow. { sī'klän·ik 'shir }

cyclopean stairs [GEOL] The landscape that results in a glacial trough after the ice has melted away, and that consists of an irregular series of rock steps, with steep cliffs on the down-valley side and small lakes in the shallow excavated depressions of the rock steps. { ˌsī·klə'pē·ən 'sterz }

cyclostrophic wind [METEOROL] The horizontal wind velocity for which the centripetal acceleration exactly balances the horizontal pressure force. { ¦sī·klō¦strä·fik 'wind }

cyclothem [GEOL] A rock stratigraphic unit associated with unstable shelf of interior basin conditions, in which the sea has repeatedly covered the land. { 'si·kləˌthem }

D

Dacian [GEOL] Lower upper Pliocene geologic time. { 'dā·shən }

dacite [GEOL] Very fine crystalline or glassy rock of volcanic origin, composed chiefly of sodic plagioclase and free silica with subordinate dark-colored minerals. { 'dā,sīt }

dacite glass [GEOL] A natural glass formed by rapid cooling of dacite lava. { 'dā,sīt ,glas }

dactylitic [GEOL] Of a rock texture, characterized by fingerlike projections of a mineral that penetrate another mineral. { dak·tə'lid·ik }

dadur [METEOROL] In India, a wind blowing down the Ganges Valley from the Siwalik hills at Hardwar. { dä'důr }

daily forecast [METEOROL] A forecast for periods of from 12 to 48 hours in advance. { ¦dā·lē 'fȯr,kast }

daily mean [METEOROL] The average value of a meteorological element over a period of 24 hours. { ¦dā·lē ,mēn }

daily retardation [OCEANOGR] The amount of time by which corresponding tidal phases grow later day by day; averages approximately 50 minutes. { ¦dā·lē ,re,tär'dā·shən }

daily variation [GEOPHYS] Oscillation occurring in the earth's magnetic field in a 1-day period. { ¦dā·lē ,ver·ē'ā·shən }

Dakotan [GEOL] Lower Upper Cretaceous geologic time. { də'kot·ən }

damp air [METEOROL] Air that has a high relative humidity. { ¦damp ¦er }

damp haze [METEOROL] Small water droplets or very hygroscopic particles in the air, reducing the horizontal visibility somewhat, but to not less than 11/4 miles (2 kilometers); similar to a very thin fog, but the droplets or particles are more scattered than in light fog and presumably smaller. { ¦damp 'hāz }

dancing dervish *See* dust whirl. { ¦dan·siŋ 'dər·vish }

dancing devil *See* dust whirl. { ¦dan·siŋ ,dev·əl }

dangerous semicircle [METEOROL] The half of the circular area of a tropical cyclone having the strongest winds and heaviest seas, where a ship tends to be drawn into the path of the storm. { 'dān·jə·rəs 'sem·i,sər·kəl }

Danian [GEOL] Lowermost Paleocene or uppermost Cretaceous geologic time. { 'dān·ē·ən }

dark segment [METEOROL] A bluish-gray band appearing along the horizon opposite the rising or setting sun and lying just below the antitwilight arch. Also known as earth's shadow. { 'därk ,seg·mənt }

Darling shower [METEOROL] A dust storm caused by cyclonic winds in the vicinity of the River Darling in Australia. { 'där·liŋ ,shaů·ər }

dart leader [GEOPHYS] The leader which, after the first stroke, initiates each succeeding stroke of a composite flash of lightning. Also known as continuous leader. { 'därt ,lēd·ər }

Darwin-Doodson system [GEOPHYS] A method for predicting tides by expressing them as sums of harmonic functions of time. { ¦där·wən ¦düd·sən ,sis·təm }

Darwin glass [GEOL] A highly siliceous, vesicular glass shaped in smooth blobs or twisted shreds, found in the Mount Darwin range in western Tasmania. Also known as queenstownite. { 'där·wən ,glas }

datum [GEOD] The latitude and longitude of an initial point; the azimuth of a line from this point. [GEOL] The top or bottom of a bed of rock on which structure contours are drawn. { 'dad·əm, 'dād·əm, *or* 'däd·əm }

Davian [GEOL] A subdivision of the Upper Cretaceous in Europe; a limestone formation with abundant hydrocorals, bryozoans, and mollusks in Denmark; marine limestone and nonmarine rocks in southeastern France; and continental formations in the Davian of Spain and Portugal. { 'dä·vē·ən }

Davidson Current [OCEANOGR] A coastal countercurrent of the Pacific Ocean running north, inshore of the California Current, along the western coast of the United States (from northern California to Washington to at least latitude 48°N) during the winter months. { 'dā·vəd·sən ˌkər·ənt }

DDA value *See* depth-duration-area value. { ¦dē¦dē'ā ˌval·yü }

dead [GEOL] In economic geology, designating a region with no economic value. { ded }

dead cave [GEOL] A cave where there is no moisture or no growth of mineral deposits associated with moisture. { ¦ded 'kāv }

dead sea [HYD] A body of water that has undergone precipitation of its rock salt, gypsum, or other evaporites. { ¦ded 'sē }

Dead Sea [GEOGR] A salt lake between Jordan and Israel. { ¦ded 'sē }

dead water [OCEANOGR] The mass of eddying water associated with formation of internal waves near the keel of a ship; forms under a ship of low propulsive power when it negotiates water which has a thin layer of fresher water over a deeper layer of more saline water. { 'ded ˌwȯd·ər }

deaister *See* doister. { dē'ās·ter }

debris [GEOL] Large fragments arising from disintegration of rocks and strata. { də'brē }

debris avalanche [GEOL] The sudden and rapid downward movement of incoherent mixtures of rock and soil on deep slopes. { də'brē 'av·əˌlanch }

debris cone [GEOL] **1.** A mound of fine-grained debris piled atop certain boulders moved by a landslide. **2.** A mound of ice or snow on a glacier covered with a thin layer of debris. { də'brē ˌkōn }

debris fall [GEOL] A relatively free downward or forward falling of unconsolidated or poorly consolidated earth or rocky debris from a cliff, cave, or arch. { də'brē ˌfȯl }

debris flow [GEOL] A variety of rapid mass movement involving the downslope movement of high-density coarse clast-bearing mudflows, usually on alluvial fans. { də'brē ˌflō }

debris glacier [HYD] A glacier formed from ice fragments that have fallen from a larger and taller glacier. { də'brē ˌglā·shər }

debris line *See* swash mark. { də'brē ˌlīn }

debris slide [GEOL] A type of landslide involving a rapid downward sliding and forward rolling of comparatively dry, unconsolidated earth and rocky debris. { də'brē ˌslīd }

debris slope *See* talus slope. { də'brē ˌslōp }

decay [GEOCHEM] *See* chemical weathering. [OCEANOGR] In ocean-wave studies, the loss of energy from wind-generated ocean waves after they have ceased to be acted on by the wind; this process is accompanied by an increase in length and a decrease in height of the wave. { di'kā }

decay area [OCEANOGR] The area into which ocean waves travel (as swell) after leaving the generating area. { di'kā ˌer·ē·ə }

decay distance [OCEANOGR] The distance through which ocean waves pass after leaving the generating area. { di'kā ˌdis·təns }

decay of waves [OCEANOGR] The decrease in height and increase in length of waves after leaving a generating area and passing through a calm, or region of lighter winds. { di'kā əv 'wavz }

Deccan basalt [GEOL] Fine-grained, nonporphyritic, tholeiitic basaltic lava consisting essentially of labradorite, clinopyroxene, and iron ore; found in the Deccan region of southeastern India. Also known as Deccan trap. { 'dek·ən bə'sȯlt }

Deccan trap *See* Deccan basalt. { 'dek·ən 'trap }

declination [GEOPHYS] The angle between the magnetic and geographical meridians, expressed in degrees and minutes east or west to indicate the direction of magnetic north from true north. Also known as magnetic declination; variation. { ˌdek·lə'nā·shən }

declivity [GEOL] **1.** A slope descending downward from a point of reference. **2.** A downward deviation from the horizontal. { də'kliv·əd·ē }

décollement [GEOL] Folding or faulting of sedimentary beds by sliding over the underlying rock. { dā'käl·mənt }

decomposition *See* chemical weathering. { dēˌkäm·pə'zish·ən }

decrement *See* groundwater discharge. { 'dek·rə·mənt }

decrepitation [GEOPHYS] Breaking up of mineral substances when exposed to heat; usually accompanied by a crackling noise. { diˌkrep·ə'tā·shən }

decussate structure [GEOL] A crisscross microstructure of certain minerals; most noticeable in rocks composed predominantly of minerals with a columnar habit. { 'dek·əˌsāt ˌstrək·chər }

dedolomitization [GEOL] Destruction of dolomite to form calcite and periclase, usually by contact metamorphism at low pressures. { dēˌdō·ləˌmīd·ə'zā·shən }

deep [OCEANOGR] An area of great depth in the ocean, representing a depression in the ocean floor. { dēp }

deep-casting [OCEANOGR] Sampling ocean water at great depths by lowering a number of self-sealing bottles, usually made of brass or bronze, on a cable. { 'dēp ˌkast·iŋ }

deep easterlies *See* equatorial easterlies. { ¦dēp 'ē·stər·lēz }

deepening [METEOROL] A decrease in the central pressure of a pressure system on a constant-height chart, or an analogous decrease in height on a constant-pressure chart. { 'dēp·ə·niŋ }

deep inland sea [GEOGR] A sea adjacent to but in restricted communication with the sea; depth exceeds 660 feet (200 meters). { 'dēp 'in·lənd 'sē }

deep-marine sediments [GEOL] Sedimentary environments occurring in water deeper than 200 meters (660 feet), seaward of the continental shelf break, on the continental slope and the basin. { ˌdēp·mə¦rēn 'sed·ə·mins }

deep scattering layer [OCEANOGR] The stratified populations of organisms which scatter sound in most oceanic waters. { ¦dēp 'skad·ə·riŋ ˌlā·ər }

deep-sea basin [GEOL] A depression of the sea floor more or less equidimensional in form and of variable extent. { ¦dēp ¦sē 'bās·ən }

deep-sea channel [GEOL] A trough-shaped valley of low relief beyond the continental rise on the deep-sea floor. Also known as mid-ocean canyon. { ¦dēp ¦sē 'chan·əl }

deep-sea plain [GEOL] A broad, almost level area forming the predominant portion of the ocean floor. { ¦dēp ¦sē 'plān }

deep-seated *See* plutonic. { ¦dēp 'sēd·əd }

deep-sea trench [GEOL] A long, narrow depression of the deep-sea floor having steep sides and containing the greatest ocean depths; formed by depression, to several kilometers' depth, of the high-velocity crustal layer and the mantle. { ¦dēp ¦sē 'trench }

deep trades *See* equatorial easterlies. { ¦dēp 'trādz }

deep water [OCEANOGR] An ocean area where depth of the water layer is greater than one-half the wave length. { 'dēp 'wȯd·ər }

deep-water wave [OCEANOGR] A surface wave whose length is less than twice the depth of the water. Also known as short wave. { 'dēp ˌwȯd·ər ˌwāv }

Deerparkian [GEOL] A North American stage of geologic time in the Lower Devonian, above Helderbergian and below Onesquethawan. { dir'pärk·ē·ən }

deflation [GEOL] The sweeping erosive action of the wind over the ground. { di'flā·shən }

deflation basin [GEOL] A topographic depression formed by deflation. { di'flā·shən ˌbās·ən }

deflation lake [HYD] A lake in a basin that was formed primarily by wind erosion, especially in arid or semiarid regions. { di'flā·shən ˌlāk }

deflection angle [GEOD] The angle at a point on the earth between the direction of a plumb line (the vertical) and the perpendicular (the normal) to the reference spheroid; this difference seldom exceeds 30 seconds of arc. { di'flek·shən ˌaŋ·gəl }

deflection of the vertical [GEOD] The angle between the direction of gravity, defining astronomical latitude and longitude, and the normal to the reference ellipsoid defining geodetic latitude and longitude. { di'flek·shən əv thə 'vərd·ə·kəl }

deformation fabric [GEOL] The space orientation of rock elements produced by external stress on the rock. { ˌdef·ərˌmā·shən ˌfab·rik }

deformation lamella [GEOL] A type of slipband in the crystalline grains of a material (particularly quartz) produced by intracrystalline slip during tectonic deformation. { ˌdef·ərˌmā·shən ləˌmel·ə }

degenerative recrystallization *See* degradation recrystallization. { di'jen·ə·rəd·iv rēˌkrist·əl·ə'zā·shən }

deglaciation [HYD] Exposure of an area from beneath a glacier or ice sheet as a result of shrinkage of the ice by melting. { dēˌglās·ē'ā·shən }

degradation [GEOL] The wearing down of the land surface by processes of erosion and weathering. [HYD] **1.** Lowering of a stream bed. **2.** Shrinkage or disappearance of permafrost. { ˌdeg·rə'dā·shən }

degradation recrystallization [GEOL] Recrystallization resulting in a decrease in the size of crystals. Also known as degenerative recrystallization; grain diminution. { ˌdeg·rə'dā·shən rēˌkrist·əl·ə'zā·shən }

degrading stream [HYD] A stream actively deepening its channel or valley and capable of transporting more load than is presently provided. { də'grād·iŋ ˌstrēm }

degrees of frost [METEOROL] In England, the number of degrees Fahrenheit that the temperature falls below the freezing point; thus a day with a minimum temperature of 27°F may be designated as a day of five degrees of frost. { di'grēz əv 'frȯst }

Deister phase [GEOL] A subdivision of the late Ammerian phase of the Jurassic period between the Kimmeridgian and lower Portlandian. { 'dī·stər ˌfāz }

dell [GEOGR] A small, secluded valley or vale. { del }

Delmontian [GEOL] Upper Miocene or lower Pliocene geologic time. { del'män·chən }

delta [GEOL] An alluvial deposit, usually triangular in shape, at the mouth of a river, stream, or tidal inlet. { 'del·tə }

delta geosyncline *See* exogeosyncline. { 'del·tə ˌjē·ō'sinˌklīn }

deltaic deposits [GEOL] Sedimentary deposits in a delta. { del'tā·ik di'päz·əts }

delta moraine *See* ice-contact delta. { 'del·tə mə'rān }

delta plain [GEOL] A plain formed by deposition of silt at the mouth of a stream or by overflow along the lower stream courses. { 'del·tə ˌplān }

delta region [METEOROL] A region in the atmosphere characterized by difluence. { 'del·tə ˌrējən }

demorphism *See* weathering. { dē'mȯr·fiz· əm }

dendritic drainage [HYD] Irregular stream branching, with tributaries joining the main stream at all angles. { den'drid·ik 'drān·ij }

dendritic valleys [GEOL] Treelike extensions of the valleys in a region lying upon horizontally bedded rock. { den'drid·ik 'val·ēz }

dendrochronology [GEOL] The science of measuring time intervals and dating events and environmental changes by reading and dating growth layers of trees as demarcated by the annual rings. { ¦den·drō·krə'näl·ə·jē }

dendroclimatology [METEOROL] The study of the tree-ring record to reconstruct climate history, based on the fact that temperature, precipitation, and other climatic variables affect tree growth. { ˌden·drōˌkī·mə'täl·ə·jē }

dendrogeomorphology [GEOGR] The use of tree-ring data to study earth surface processes. Scientists can date when trees were killed (by dating the outer ring of the tree) or bent (by analyzing when dramatic changes in tree growth occurred) by mass movements, such as landslides and snow avalanches. { ˌden·drōˌgē·ō·mȯr'fäl·ə·jē }

dendrohydrology [HYD] The science of determining hydrologic occurrences by the comparison of tree ring thickness with streamflow or precipitation. Also known as tree-ring hydrology. { ˌden·dro·hī'dräl·ə·jē }

density altitude [METEOROL] The altitude, in the standard atmosphere, at which a given density occurs. { 'den·səd·ē 'al·tə,tüd }

density channel [METEOROL] A channel used to investigate a density current; for example, in experiments relating to the behavior of cold masses of air in the atmosphere and related frontal structures. { 'den·səd·ē ,chan·əl }

density current [METEOROL] Intrusion of a dense air mass beneath a lighter air mass; the usage applies to cold fronts. [OCEANOGR] *See* turbidity current. { 'den·səd·ē ,kər·ənt }

density ratio [METEOROL] The ratio of the density of the air at a given altitude to the air density at the same altitude in a standard atmosphere. { 'den·səd·ē ,rā·shō }

denudation [GEOL] General wearing away of the land; laying bare of subjacent lands. { ,dē·nü'dā·shən }

deoxidation sphere *See* bleach spot. { dē,äk·sə'dā·shən ,sfir }

departure [METEOROL] The amount by which the value of a meteorological element differs from the normal value. { di'pär·chər }

depegram [METEOROL] On a diagram having entropy and temperature as coordinates, a curve representing the distribution of the dew point as a function of pressure for a given sounding of the atmosphere. { 'dep·ə,gram }

depéq [METEOROL] Strong winds over Loet Tawar (Sumatra, East Indies) during the southwest monsoon. { də'pek }

depergelation [HYD] The act or process of thawing permafrost. { dē,pər·jə'lā·shən }

depocenter [GEOL] A site of maximum deposition. { 'dep·ə,sen·tər }

deposit [GEOL] Consolidated or unconsolidated material that has accumulated by a natural process or agent. { də'päz·ət }

deposition [GEOL] The laying, placing, or throwing down of any material; specifically, the constructive process of accumulation into beds, veins, or irregular masses of any kind of loose, solid rock material by any kind of natural agent. { ,dep·ə'zish·ən }

depositional dip *See* primary dip. { ,dep·ə'zish·ən·əl 'dip }

depositional sequence [GEOL] A major but informal assemblage of formations or groups and supergroups, bounded by regionally extensive unconformities at both their base and top and extending over broad areas of continental cratons. { ,dep·ə'zish·ən·əl 'sē·kwəns }

depositional strike [GEOL] Sedimentary deposits that are continuous laterally on a gently sloping surface. { ,dep·ə'zish·ən·əl 'strīk }

depression [GEOL] **1.** A hollow of any size on a plain surface having no natural outlet for surface drainage. **2.** A structurally low area in the crust of the earth. [METEOROL] An area of low pressure; usually applied to a certain stage in the development of a tropical cyclone, to migratory lows and troughs, and to upper-level lows and troughs that are only weakly developed. Also known as low. { di'presh·ən }

depression spring [HYD] A type of gravity spring that flows onto the land surface because the surface slopes down to the water table. { di'presh·ən 'spriŋ }

depression storage [HYD] Water retained in puddles, ditches, and other depressions in the surface of the ground. { di'presh·ən ,stȯr·ij }

depth [OCEANOGR] The vertical distance from a specified sea level to the sea floor. { depth }

depth contour *See* isobath. { 'depth ,kän,tu̇r }

depth curve *See* isobath. { 'depth ,kərv }

depth-duration-area value [METEOROL] The average depth of precipitation that has occurred within a specified time interval over an area of given size. Abbreviated DDA value. { ¦depth də¦rā·shən 'er·ē·ə ,val·yü }

depth hoar [HYD] A layer of ice crystals formed between the ground and snow cover by sublimation. Also known as sugar snow. { 'depth ,hȯr }

depth of compensation [GEOPHYS] That depth at which density differences occurring in the earth's crust are compensated isostatically; calculated to be between 62 and 70–73 miles (100 and 113–117 kilometers). [HYD] The depth in a body of water at which illuminance has diminished to the extent that oxygen production through photosynthesis and oxygen consumption through respiration by plants are equal; it is the lower boundary of the euphotic zone. { 'depth əv ˌkäm·pən'sā·shən }

depth zone [GEOL] A zone within the earth giving rise to different metamorphic assemblages. [OCEANOGR] Any one of four oceanic environments: the littoral, neritic, bathyal, and abyssal zones. { 'depth ˌzōn }

derecho *See* plow wind. { dā'rā·chō }

derived gust velocity [METEOROL] The maximum velocity of a sharp-edged gust that would produce a given acceleration on a particular airplane flown in level flight at the design cruising speed of the aircraft at a given air density. { də'rīvd 'gəst vəˌläs·əd·ē }

descendant [GEOL] A topographic feature that is formed from the mass beneath an older topographic form, now removed. { di'sen·dənt }

descriptive climatology [CLIMATOL] Climatology as presented by graphic and verbal description, without going into causes and theory. { di'skrip·tiv klī·mə'täl·ə·jē }

descriptive meteorology [METEOROL] A branch of meteorology which deals with the description of the atmosphere as a whole and its various phenomena, without going into theory. Also known as aerography. { di'skrip·tiv mēd·ē·ə'räl·ə·jē }

desert [GEOGR] **1.** A wide, open, comparatively barren tract of land with few forms of life and little rainfall. **2.** Any waste, uninhabited tract, such as the vast expanse of ice in Greenland. { 'dez·ərt }

desert climate [CLIMATOL] A climate type which is characterized by insufficient moisture to support appreciable plant life; that is, a climate of extreme aridity. { ¦dez·ərt ¦klī·mət }

desert crust *See* desert pavement. { ¦dez·ərt ¦krəst }

desert devil *See* dust whirl. { 'dez·ərt ˌdev·əl }

desert pavement [GEOL] A mosaic of pebbles and large stones which accumulate as the finer dust and sand particles are blown away by the wind. Also known as desert crust. { ¦dez·ərt 'pāv·mənt }

desert peneplain *See* pediplain. { ¦dez·ərt 'pen·əˌplān }

desert plain *See* pediplain. { ¦dez·ərt 'plān }

desert polish [GEOL] A smooth, shining surface imparted to rocks and other hard substances by the action of windblown sand and dust of desert regions. { ¦dez·ərt 'päl·ish }

desert soil [GEOL] In early United States classification systems, a group of zonal soils that have a light-colored surface soil underlain by calcareous material and a hardpan. { ¦dez·ərt 'sȯil }

desert varnish *See* rock varnish. { ¦dez·ərt 'vär·nish }

desert wind [METEOROL] A wind blowing off the desert, which is very dry and usually dusty, hot in summer but cold in winter, and with a large diurnal range of temperature. { ¦dez·ərt 'wind }

desiccation [HYD] The permanent decrease or disappearance of water from a region, caused by a decrease of rainfall, a failure to maintain irrigation, or deforestation or overcropping. { ˌdes·ə'kā·shən }

desiccation breccia [GEOL] Fragments of a mud-cracked layer of sediment deposited with other sediments. { ˌdes·ə'kā·shən ˌbrech·ə }

desiccation crack *See* mud crack. { ˌdes·ə'kā·shən ˌkrak }

design climatology [CLIMATOL] The scientific analysis of climatic data for the purpose of improving the design of equipment and structures intended to operate in or withstand extremes of climate. { di'zīn klī·mə'täl·ə·jē }

design water depth [OCEANOGR] **1.** A value based on the sum of the vertical distance from the nominal water level to the ocean bottom and the height of the tides, both astronomical and storm. **2.** The greatest water depth in which an offshore drilling well is able to maintain its operations. { di'zīn 'wȯd·ər ˌdepth }

desilication [GEOCHEM] Removal of silica, as from rock or a magma. { dēˌsil·ə'kā·shən }

Des Moinesian [GEOL] Lower Middle Pennsylvanian geologic time. { də'mȯin·ē·ən }

detached core [GEOL] The inner bed or beds of a fold that may become separated or pinched off from the main body of the strata due to extreme folding and compression. { di'tacht 'kȯr }

detrainment [METEOROL] The transfer of air from an organized air current to the surrounding atmosphere. { dē'trān·mənt }

detrital fan *See* alluvial fan. { də'trīd·əl 'fan }

detrital ratio *See* clastic ratio. { də'trīd·əl 'rā·shō }

detrital remanent magnetization [GEOPHYS] Magnetization acquired by magnetic grains during formation of a sedimentary rock. Abbreviated DRM. { də'trīd·əl 'rem·ə·nənt 'mag·nəd·ə'zā·shən }

detrital reservoir [GEOL] A clastic or detrital-granular reservoir, classified by rock type and other factors such as sediments (quartzose-type, graywacke, or arkose sediments). { də'trīd·əl 'rez·əv,wär }

detrital sediment [GEOL] Accumulations of the organic and inorganic fragmental products of the weathering and erosion of land transported to the place of deposition. { də'trīd·əl 'sed·ə·mənt }

detritus [GEOL] Any loose material removed directly from rocks and minerals by mechanical means, such as disintegration or abrasion. { də'trīd·əs }

deuteric [GEOL] Of or pertaining to alterations in igneous rock during the later stages and as a direct result of consolidation of magma or lava. Also known as epimagmatic; paulopost. { dü'tir·ik }

development [GEOL] The progression of changes in fossil groups which have succeeded one another during deposition of the strata of the earth. [METEOROL] The process of intensification of an atmospheric disturbance, most commonly applied to cyclones and anticyclones. { də'vel·əp·mənt }

development index [METEOROL] An index used as an aid in forecasting cyclogenesis; the development index I is defined most frequently as the difference in divergence between two well-separated, tropospheric, constant-pressure surfaces. Also known as relative divergence. { də'vel·əp·mənt ,in,deks }

deviatoric stress [GEOL] A condition in which the stress components operating at a point in a body are not the same in every direction. Also known as differential stress. { ¦dēv·ē·ə¦tȯr·ik 'stres }

Devonian [GEOL] The fourth period of the Paleozoic Era, covering the geological time span between about 412 and 354 $\times$ 10^6 years before present. { di'vō·nē·ən }

De Vries effect [GEOCHEM] A relatively short-term oscillation, on the order of 100 years, in the radiocarbon content of the atmosphere, and the resulting variation in the apparent radiocarbon age of samples. { də'vrēz i'fekt }

dew [HYD] Water condensed onto grass and other objects near the ground, the temperatures of which have fallen below the dew point of the surface air because of radiational cooling during the night but are still above freezing. { dü }

de Witte relation [GEOPHYS] Graphical plot of the relation between electrical conductivity and distance over which the conductivity is measured through reservoir rock with clay minerals, (the effect is similar to two parallel electrical circuits), the current passing through the conducting clay minerals and the water-filled pores. { də'wit rē'lā·shən }

dew point [METEOROL] The temperature at which air becomes saturated when cooled without addition of moisture or change of pressure; any further cooling causes condensation. Also known as dew-point temperature. { 'dü ,pȯint }

dew-point depression [METEOROL] The number of degrees the dew point is found to be lower than the temperature. { 'dü ,pȯint di'presh·ən }

dew-point formula [METEOROL] A formula for the calculation of the approximate height of the lifting condensation level; employed to estimate the height of the base of convective clouds, under suitable atmospheric and topographic conditions. { 'dü ,pȯint ,fȯr·myə·lə }

dew-point spread [METEOROL] The difference in degrees between the air temperature and the dew point. { 'dü ,pȯint ,spred }

dew-point temperature *See* dew point. { 'dü ,pȯint 'tem·prə·chər }

dextral drag fold [GEOL] A drag fold in which the trace of a given surface bed is displaced to the right. { 'dek·strəl 'drag ˌfōld }

dextral fault [GEOL] A strike-slip fault in which an observer approaching the fault sees the opposite block as having moved to the right. Also known as right-lateral fault; right-lateral slip fault; right-slip fault. { 'dek·strəl 'fȯlt }

dextral fold [GEOL] An asymmetric fold in which the long limb appears to be offset to the right to an observer looking along the long limb. { 'dek·strəl 'fōld }

D horizon [GEOL] A soil horizon sometimes occurring below a B or C horizon, consisting of unweathered rock. { 'dē hə'rīz·ən }

DI *See* temperature-humidity index.

diachronous [GEOL] Of a rock unit, varying in age in different areas or cutting across time planes or biostratigraphic zones. Also known as time-transgressive. { dī'ak·rə·nəs }

diaclinal [GEOL] Pertaining to a stream crossing a fold, perpendicular to the strike of the underlying strata it traverses. { ¦dī·ə¦klīn·əl }

diagenesis [GEOL] Chemical and physical changes occurring in sediments during and after their deposition but before consolidation. { ˌdī·ə'jen·ə·səs }

diagnostic equation [METEOROL] Any equation governing a system which contains no time derivative and therefore specifies a balance of quantities in space at a moment of time; examples are a hydrostatic equation or a balance equation. { ˌdī·əg'näs·tik i'kwā·zhən }

diagonal fault [GEOL] A fault whose strike is diagonal or oblique to the strike of the adjacent strata. Also known as oblique fault. { dī'ag·ən·əl 'fȯlt }

diagonal joint [GEOL] A joint having its strike oblique to the strike of the strata of the sedimentary rock, or to the cleavage plane of the metamorphic rock in which it occurs. Also known as oblique joint. { dī'ag·ən·əl 'jȯint }

diamond matrix [GEOL] The rock material in which diamonds are formed. { 'dī·mənd 'mā·triks }

diapir [GEOL] A dome or anticlinal fold in which a mobile plastic core has ruptured the more brittle overlying rock. Also known as diapiric fold; piercement dome; piercing fold. { 'dī·əˌpir }

diapiric fold *See* diapir. { ¦dī·ə¦pir·ik 'fōld }

diastem [GEOL] A temporal break between adjacent geologic strata that represents nondeposition or local erosion but not a change in the general regimen of deposition. { 'dī·əˌstem }

diastrophism [GEOL] **1.** The general process or combination of processes by which the earth's crust is deformed. **2.** The results of this deforming action. { dī'as·trəˌfiz·əm }

diatomaceous earth [GEOL] A yellow, white, or light-gray, siliceous, porous deposit made of the opaline shells of diatoms; used as a filter aid, paint filler, adsorbent, abrasive, and thermal insulator. Also known as kieselguhr; tripolite. { ¦dī·ə·tə¦mā·shəs 'ərth }

diatomaceous ooze [GEOL] A pelagic, siliceous sediment composed of more than 30% diatom tests, up to 40% calcium carbonate, and up to 25% mineral grains. { ¦dī·ə·tə¦mās·shəs 'üz }

diatomite [GEOL] Dense, chert-like, consolidated diatomaceous earth. { dī'ad·əˌmīt }

diatreme [GEOL] A circular volcanic vent produced by the explosive energy of gas-charged magmas. { 'dī·əˌtrēm }

dictyonema bed [GEOL] A thin shale bed rich in remains of graptolites of the genus *Dictyonema*. { ˌdik·tē·ə'nē·mə ˌbed }

difference of latitude [GEOD] The shorter arc of any meridian between the parallels of two places, expressed in angular measure. { 'dif·rəns əv 'lad·əˌtüd }

difference of longitude [GEOD] The smaller angle at the pole or the shorter arc of a parallel between the meridians of two places, expressed in angular measure. { 'dif·rəns əv 'län·jəˌtüd }

differential analysis [METEOROL] Synoptic analysis of change charts or of vertical differential charts (such as thickness charts) obtained by the graphical or numerical subtraction of the patterns of some meteorological variable at two times or two levels. { 'dif·ə'ren·chəl ə'nal·ə·səs }

differential chart [METEOROL] A chart showing the amount and direction of change of a meteorological quantity in time or space. { ˌdif·ə'ren·chəl 'chärt }

differential compaction [GEOL] Compression in sediments, such as sand or limestone, as the weight of overburden causes reduction in pore space and forcing out of water. { ˌdif·ə'ren·chəl kəm'pak·shən }

differential erosion [GEOL] Rapid erosion of one area of the earth's surface relative to another. { ˌdif·ə'ren·chəl i'rō·zhən }

differential fault *See* scissors fault. { ˌdif·ə'ren·chəl 'fȯlt }

differential stress *See* deviatoric stress. { ˌdif·ə'ren·chəl 'stres }

diffuse aurora [GEOPHYS] A widespread and relatively uniform type of aurora which is easily overlooked from the ground but is prominent in satellite pictures. { də'fyüs ə'rȯr·ə }

diffuse front [METEOROL] A front across which the characteristics of wind shift and temperature change are weakly defined. { də'fyüs 'frənt }

diffusion [METEOROL] The exchange of fluid parcels (and hence the transport of conservative properties) between regions in space, in the apparently random motions of the parcels on a scale too small to be treated by the equations of motion; the diffusion of momentum (viscosity), vorticity, water vapor, heat (conduction), and gaseous components of the atmospheric mixture have been studied extensively. { də'fyü·zhən }

diffusion diagram [METEOROL] A diagram for displaying the comparative properties of various diffusion processes, with coordinates of the mean free path or mixing length and mean molecular speed or diffusion velocity, for molecular or eddy diffusion, respectively; each point of the diagram determines diffusivity. { də'fyü·zhən ˌdī·əˌgram }

diffusive equilibrium [METEOROL] The steady state resulting from the diffusion process, primarily of interest when external forces and sources and sinks exist within the field; in such a state the constituent gases of the atmosphere would be distributed independently of each other, the heavier decreasing more rapidly with height than the lighter; but the presence of turbulent mixing precludes establishment of complete diffusive equilibrium. { də'fyü·ziv ˌē·kwə'lib·rē·əm }

digitation [GEOL] A secondary recumbent anticline emanating from a larger recumbent anticline. { ˌdij·ə'tā·shən }

dike [GEOL] A tabular body of igneous rock that cuts across adjacent rocks or cuts massive rocks. { dīk }

dike ridge [GEOL] Any small wall-like ridge created by differential erosion. { 'dīk ˌrij }

dike set [GEOL] A small group of dikes arranged linearly or parallel to each other. { 'dīk ˌset }

dike swarm [GEOL] A large group of parallel, linear, or radially oriented dikes. { 'dīk ˌswȯrm }

dilatancy [GEOL] Expansion of deformed masses of granular material, such as sand, due to rearrangement of the component grains. { dī'lāt·ən·sē }

dimictic lake [HYD] A lake which circulates twice a year. { dī'mik·tik 'lāk }

dimmerfoehn [METEOROL] A rare form of foehn where, during a very strong upper wind from the south, a pressure difference of 12 millibars or more exists between the south and north sides of the Alps; a stormy foehn wind then overleaps the upper valleys in the northern slopes, reaches the ground in the lower parts of the valleys, and enters the foreground as a very strong wind; the foehn wall and the precipitation area extend beyond the crest across the almost calm surface area in the upper valleys. { 'dim·ərˌfān }

Dinantian [GEOL] Lower Carboniferous geologic time. Also known as Avonian. { di'nan·chən }

Dinarides [GEOGR] A mountain system, east of the Adriatic Sea, in Yugoslavia. { di'nar·əˌdēz }

dip [GEOL] **1.** The angle that a stratum or fault plane makes with the horizontal. Also known as angle of dip; formation dip; true dip. **2.** A pronounced depression in the land surface. { dip }

dip fault [GEOL] A type of fault that strikes parallel with the dip of the strata involved. { 'dip ˌfȯlt }

dip joint [GEOL] A joint that strikes approximately at right angles to the cleavage or bedding of the constituent rock. { 'dip ˌjȯint }

dip log [GEOL] A log of the dips of formations traversed by boreholes. { 'dip ˌläg }

dipmeter log [GEOL] A dip log produced by reading of the direction and angle of formation dip as analyzed from impulses from a dipmeter consisting of three electrodes 120° apart in a plane perpendicular to the borehole. { 'dipˌmēd·ər ˌläg }

dip reversal *See* reversal of dip. { 'dip ri'vər·səl }

dip slip [GEOL] The component of a fault parallel to the dip of the fault. Also known as normal displacement. { 'dip ˌslip }

dip slope [GEOL] A slope of the surface of the land determined by and conforming approximately to the dip of the underlying rocks. Also known as back slope; outface. { 'dip ˌslōp }

dip stream [HYD] A consequent stream that flows in the direction of the dip of the strata it traverses. { 'dip ˌstrēm }

dip-strike symbol [GEOL] A geologic symbol used on maps to show the strike and dip of a planar feature. { 'dip ˌstrīk ˌsim·bəl }

direct cell [METEOROL] A closed thermal circulation in a vertical plane in which the rising motion occurs at higher potential temperature than the sinking motion. { də¦rekt 'sel }

direction *See* trend. { də'rek·shən }

directional structure [GEOL] Any sedimentary structure having directional significance; examples are cross-bedding and ripple marks. Also known as vectorial structure. { də'rek·shən·əl 'strək·chər }

direct stratification *See* primary stratification. { də'rekt ˌstrad·ə·fə'kā·shən }

direct tide [GEOPHYS] A gravitational solar or lunar tide in the ocean or atmosphere which is in phase with the apparent motions of the attracting body, and consequently has its local maxima directly under the tide-producing body, and on the opposite side of the earth. { də¦rekt 'tīd }

dirt band [GEOL] A dark layer in a glacier representing a former surface, usually a summer surface, where silt and debris accumulated. { 'dərt ˌband }

dirt bed [GEOL] A buried soil containing partially decayed organic material; sometimes occurs in glacial drift. { 'dərt ˌbed }

dirt slip *See* clay vein. { 'dərt ˌslip }

discomfort index *See* temperature-humidity index. { dis'kəm·fərt ˌinˌdeks }

disconformity [GEOL] Unconformity between parallel beds or strata. { ˌdis·kən'fȯr·məd·ē }

discontinuity [GEOL] **1.** An interruption in sedimentation. **2.** A surface that separates unrelated groups of rocks. [GEOPHYS] A boundary at which the velocity of seismic waves changes abruptly. { disˌkänt·ən'ü·əd·ē }

discontinuous reaction series [GEOL] The branch of Bowen's reaction series that include olivine, pyroxene, amphibole, and biotite; each change in the series represents an abrupt change in phase. { ˌdis·kən'tin·yə·wəs rē'ak·shən ˌsir·ēz }

discordance [GEOL] An unconformity characterized by lack of parallelism between strata which touch without fusion. { di'skȯrd·əns }

discordant pluton [GEOL] An intrusive igneous body that cuts across the bedding or foliation of the intruded formations. { di'skȯrd·ənt 'plüˌtän }

discrete-film zone *See* belt of soil water. { di'skrēt ˌfilm ˌzōn }

disharmonic fold [GEOL] A fold in which changes in form or magnitude occur with depth. { ˌdis·här'män·ik 'fōld }

dishpan experiment [METEOROL] A model experiment carried out by differential heating of fluid in a flat, rotating pan; it establishes similarity with the atmosphere and is used to reproduce many important features of the general circulation and, on a smaller scale, atmospheric motion. { 'dishˌpan ikˌsper·ə·mənt }

dislocation [GEOL] Relative movement of rock on opposite sides of a fault. Also known as displacement. { ˌdis·lō'kā·shən }

dislocation breccia *See* fault breccia. { ˌdis·lō'kā·shən 'brech·ə }

dismicrite [GEOL] Fine-grained limestone of obscure origin, resembling micrite but containing sparry calcite bodies. { diz'mīˌkrīt }

dispersal pattern [GEOCHEM] Distribution pattern of metals in soil, rock, water, or vegetation. { də'spər·səl ˌpad·ərn }

dispersed elements [GEOCHEM] Elements which form few or no independent minerals but are present as minor ingredients in minerals of abundant elements. { də'spərst 'el·ə·mənts }

displaced ore body [GEOL] An ore body which has been subjected to displacement or disruption after its initial deposition. { dis'plāst 'ȯr ˌbäd·ē }

displacement *See* dislocation. { dis'plās·mənt }

display loss *See* visibility factor. { di'splā ˌlȯs }

dissected topography [GEOGR] Physical features marked by erosive cutting. { də'sek·təd tə'päg·rə·fē }

dissection [GEOL] Destruction of the continuity of the land surface by erosive cutting of valleys or ravines into a relatively even surface. { də'sek·shən }

dissipation constant [GEOPHYS] In atmospheric electricity, a measure of the rate at which a given electrically charged object loses its charge to the surrounding air. { ˌdis·ə'pā·shən ˌkän·stənt }

dissolved load [HYD] Material carried in solution by a stream or river. { dīˈzälvd 'lōd }

distorted water [METEOROL] A multimolecular layer of water, at the boundary between a mass of liquid water and the surrounding vapor, whose structure is not identical with that of bulk water. { di'stȯrd·əd 'wȯd·ər }

distortional wave *See* S wave. { di'stȯr·shən·əl 'wāv }

distributary [HYD] An irregular branch flowing out from a main stream and not returning to it, as in a delta. Also known as distributary channel. { də'strib·yəˌter·ē }

distributary channel *See* distributary. { də'strib·yəˌter·ē ˌchan·əl }

distributed fault *See* fault zone. { di'strib·yəd·əd ˌfȯlt }

distribution graph [HYD] A statistically derived hydrograph for a storm of specified duration, graphically representing the percent of total direct runoff passing a point on a stream, as a function of time; usually presented as a histogram or table of percent runoff within each of successive short time intervals. { ˌdis·trə'byü·shən ˌgraf }

distributive fault *See* step fault. { di'strib·yəd·iv 'fȯlt }

district forecast [METEOROL] In U.S. Weather Bureau usage, a general weather forecast for conditions over an established geographical "forecast district." { 'di·strikt 'fȯrˌkast }

disturbance [GEOL] Folding or faulting of rock or a stratum from its original position. [METEOROL] **1.** Any low or cyclone, but usually one that is relatively small in size and effect. **2.** An area where weather, wind, pressure, and so on show signs of the development of cyclonic circulation. **3.** Any deviation in flow or pressure that is associated with a disturbed state of the weather, such as cloudiness and precipitation. **4.** Any individual circulatory system within the primary circulation of the atmosphere. { də'stər·bəns }

diurnal age *See* age of diurnal inequality. { dī'ərn·əl 'āj }

diurnal inequality [OCEANOGR] The difference between the heights of the two high waters or the two low waters of a lunar day. { dī'ərn·əl ˌin·ə'kwäl·əd·ē }

diurnal range *See* great diurnal range. { dī'ərn·əl 'rānj }

diurnal tide [OCEANOGR] A tide in which there is only one high water and one low water each lunar day. { dī'ərn·əl 'tīd }

diurnal variation [GEOPHYS] Daily variations of the earth's magnetic field at a given point on the surface, with both solar and lunar periods having their source in the horizontal movements of air in the ionosphere. { dī'ərn·əl ˌver·ē'ā·shən }

divagation [HYD] Lateral shifting of the course of a stream caused by extensive deposition of alluvium in its bed and frequently accompanied by the development of meanders. { ˌdiv·əˈgā·shən }

divergence [METEOROL] The two-dimensional horizontal divergence of the velocity field. [OCEANOGR] A horizontal flow of water, in different directions, from a common center or zone. { dəˈvər·jəns }

divergence loss [GEOPHYS] During geophysical prospecting, the portion of the power lost in transmitting signals that is caused by the spreading of seismic or sound rays by the geometry of the geologic features. { dəˈvər·jəns ˌlȯs }

Divesian *See* Oxfordian. { dəˈvēzh·ən }

divide [GEOGR] A ridge or section of high ground between drainage systems. { dəˈvīd }

Djulfian [GEOL] Upper upper Permian geologic time. { ˈju̇l·fē·ən }

D layer [GEOL] The lower mantle of the earth, between a depth of 600 and 1800 miles (1000 and 2900 kilometers). [GEOPHYS] The lowest layer of ionized air above the earth, occurring in the D region only in the daytime hemisphere; reflects frequencies below about 50 kilohertz and partially absorbs higher-frequency waves. { ˈdē ˌlā·ər }

Dobson unit [METEOROL] The unit of measure for atmospheric ozone; one Dobson unit is equal to 2.7×10^{16} ozone molecules per square centimeter, which would be equivalent to a layer of ozone 0.001 centimeter thick, at 1 atmosphere and 0°C. { ˈdäb·sən ˌyü·nət }

doctor [METEOROL] A cooling sea breeze in the tropics. { ˈdäk·tər }

dog days [CLIMATOL] The period of greatest heat in summer. { ˈdȯg ˌdāz }

dogger [GEOL] Concretionary masses of calcareous sandstone or ironstone. { ˈdȯg·ər }

doister [METEOROL] In Scotland, a severe storm from the sea. Also known as deaister; dyster. { ˈdȯis·tər }

doldrums [METEOROL] A nautical term for the equatorial trough, with special reference to the light and variable nature of the winds. Also known as equatorial calms. { ˈdōlˌdrəmz }

doline [GEOL] A general term for a closed depression in an area of karst topography that is formed either by solution of the surficial limestone or by collapse of underlying caves. { dəˈlēn }

dolocast [GEOL] The cast or impression of a dolomite crystal. { ˈdō·ləˌkast }

dolomitization [GEOL] Conversion of limestone to dolomite rock by replacing a portion of the calcium carbonate with magnesium carbonate. { ˌdō·lə·məd·əˈzā·shən }

dome [GEOL] **1.** A circular or elliptical, almost symmetrical upfold or anticlinal type of structural deformation. **2.** A large igneous intrusion whose surface is convex upward. { dōm }

Domerian [GEOL] Upper Charmouthian geologic time. { dōˈmer·ə·ən }

Donau glaciation [GEOL] A Pleistocene glacial time unit in the Alps region in Europe. { ˈdōˌnau̇ glā·sēˈā·shən }

doodlebug [GEOL] Also known as douser. **1.** Any unscientific device or apparatus, such as a divining rod, used to locate subsurface water, minerals, gas, or oil. **2.** A scientific instrument used for locating minerals. { ˈdüd·əlˌbəg }

dopplerite [GEOL] A naturally occurring gel of humic acids found in peat bags or where an aqueous extract from a low-rank coal can collect. { ˈdäp·ləˌrīt }

double ebb [OCEANOGR] An ebb current comprising two maxima of velocity that are separated by a smaller ebb velocity. { ¦dəb·əl ˈeb }

double tide [OCEANOGR] A high tide comprising two maxima of nearly identical height separated by a relatively small depression, or low tide comprising two minima separated by a relatively small elevation. { ¦dəb·əl ˈtīd }

doubly plunging fold [GEOL] A fold that plunges in opposite directions, either away from or toward a central point. { ¦dəb·lē ˌplənj·iŋ ˈfōld }

douser *See* doodlebug. { ˈdau̇s·ər }

down [GEOL] **1.** Hillock of sand thrown up along the coast by the sea or the wind. **2.** A flat eminence on the top of a hill or mountain. { dau̇n }

downcutting [GEOL] Stream erosion in which the cutting is directed in a downward direction. { 'dau̇n,kəd·iŋ }

downdip [GEOL] Pertaining to a position parallel to or in the direction of the dip of a stratum or bed. { 'dau̇n,dip }

downrush [METEOROL] A term sometimes applied to the strong downward-flowing air current that marks the dissipating stages of a thunderstorm. { 'dau̇n,rəsh }

downstream [HYD] In the direction of flow, as a current or waterway. { 'dau̇n,strēm }

downthrow [GEOL] The side of a fault whose relative movement appears to have been downward. { 'dau̇n,thrō }

downwarp [GEOL] A segment of the earth's crust that is broadly bent downward. { 'dau̇n,wȯrp }

downwelling *See* sinking. { 'dau̇n,wel·iŋ }

Dowtonian [GEOL] Uppermost Silurian or lowermost Devonian geologic time. { dau̇ 'tōn·ē·ən }

drag fold [GEOL] A minor fold formed in an incompetent bed by movement of a competent bed so as to subject it to couple; the axis is at right angles to the direction in which the beds slip. { 'drag ,fōld }

drag mark [GEOL] Long, even mark usually having longitudinal striations produced by current drag of an object across a sedimentary surface. { 'drag ,märk }

drainage [HYD] The pattern followed by the waters of an area as they pass or flow off in surface or subsurface streams. { 'drān·ij }

drainage area *See* drainage basin. { 'drān·ij ,er·ē·ə }

drainage basin [HYD] An area in which surface runoff collects and from which it is carried by a drainage system, as a river and its tributaries. Also known as catchment area; drainage area; feeding ground; gathering ground; hydrographic basin. { 'drān·ij ,bā·sən }

drainage density [HYD] Ratio of the total length of all channels in a drainage basin to the basin area. { 'drān·ij ,den·səd·ē }

drainage divide [GEOL] **1.** The border of a drainage basin. **2.** The boundary separating adjacent drainage basins. { 'drān·ij də,vīd }

drainage lake [HYD] An open lake which loses water via a surface outlet or whose level is essentially controlled by effluent discharge. { 'drān·ij ,lāk }

drainage pattern [HYD] The configuration of a natural or artificial drainage system; stream patterns reflect the topography and rock patterns of the area. { 'drān·ij ,pad·ərn }

drainage ratio [HYD] The ratio expressing runoff compared with precipitation in a specific area for a given time period. { 'drān·ij ,rā·shō }

drainage system [HYD] A surface stream or a body of impounded surface water, together with all other such streams and bodies that are tributary, by which a geographical area is drained. { 'drān·ij ,sis·təm }

drainage wind *See* gravity wind. { 'drān·ij ,wind }

draping [GEOL] Structural concordance of the strata overlying a limestone reef or other hard core to the surface of the reef or core. { 'drāp·iŋ }

drawdown [HYD] The magnitude of the change in water surface level in a well, reservoir, or natural body of water resulting from the withdrawal of water. { 'drȯ,dau̇n }

D region [GEOPHYS] The region of ionosphere up to about 60 miles (97 kilometers) above the earth, below the E and F regions, in which the D layer forms. { 'dē ,rē·jən }

dreikanter [GEOL] A pebble with three facets shaped by sandblasting. { 'drī,kän·tər }

Dresbachian [GEOL] Lower Croixan geologic time. { drez'bäk·ē·ən }

drewite [GEOL] Calcareous ooze composed of impalpable calcareous material. { 'drü,īt }

drift [GEOL] **1.** Rock material picked up and transported by a glacier and deposited elsewhere. **2.** Detrital material moved and deposited on a beach by waves and currents. [OCEANOGR] *See* drift current. { drift }

drift bottle [OCEANOGR] A bottle which is released into the sea for studying currents; contains a card, identifying the date and place of release, to be returned by the finder with date and place of recovery. Also known as floater. { 'drift ,bäd·əl }

drift card [OCEANOGR] A card, such as is used in a drift bottle, encased in a buoyant, waterproof envelope and released in the same manner as a drift bottle. { 'drift ˌkärd }

drift current [OCEANOGR] **1.** A wide, slow-moving ocean current principally caused by winds. Also known as drift; wind drift; wind-driven current. **2.** Current determined from the differences between dead reckoning and a navigational fix. { 'drift ˌkə·rənt }

drift dam [GEOL] A dam formed by glacial drift in a stream valley. { 'drift ˌdam }

drift glacier *See* snowdrift glacier. { 'drift ˌglā·shər }

drift ice [OCEANOGR] Sea ice that has drifted from its place of formation. { 'drift ˌīs }

drift ice foot *See* ramp. { 'drift ˌīs ˌfu̇t }

drifting snow [METEOROL] Wind-driven snow raised from the surface of the earth to a height of less than 6 feet (1.8 meters). { 'drif·tiŋ 'snō }

drift station [OCEANOGR] **1.** A scientific station established on the ice of the Arctic Ocean, generally based on an ice flow. **2.** A set of observations made over a period of time from a drifting vessel. { 'drift ˌstā·shən }

drift terrace *See* alluvial terrace. { 'drift ˌter·əs }

drip [HYD] Condensed or otherwise collected moisture falling from leaves, twigs, and so forth. { drip }

dripping drop atomization [HYD] A type of natural gravitational atomization process in which there is periodic emission of drops from the bottom side of a surface to which a liquid is fed continuously, as in dripping of water from leaves. { ¦drip·iŋ ¦dräp ˌad·ə·mə'zā·shən }

dripstone [GEOL] A cave feature, such as a stalagmite, which is formed by precipitation of calcium carbonate or another mineral from dripping water. { 'drip,stōn }

driven snow [METEOROL] Snow which has been moved by wind and collected into snowdrifts. { ¦driv·ən 'snō }

drizzle [METEOROL] Very small, numerous, and uniformly dispersed water drops that may appear to float while following air currents; unlike fog droplets, drizzle falls to the ground; it usually falls from low stratus clouds and is frequently accompanied by low visibility and fog. { 'driz·əl }

drizzle drop [METEOROL] A drop of water of diameter 0.2 to 0.5 millimeter falling through the atmosphere; however, all water drops of diameter greater than 0.2 millimeter are frequently termed raindrops, as opposed to cloud drops. { 'driz·əl ˌdräp }

DRM *See* detrital remanent magnetization.

drop [HYD] The difference in water-surface elevations that is measured up- and downstream from a narrowing in the stream. { dräp }

droplet [METEOROL] A water droplet in the atmosphere; there is no defined size limit separating droplets from drops of water, but sometimes a maximum diameter of 0.2 millimeter is the limit for droplets. { 'dräp·lət }

drop-size distribution [METEOROL] The frequency distribution of drop sizes (diameters, volumes) that is characteristic of a given cloud or rainfall. { 'dräp ˌsīz ˌdis·trə'byü·shən }

dropsonde observation [METEOROL] An evaluation of the significant radio signals received from a descending dropsonde, and usually presented in terms of height, temperature, and dew point at the mandatory and significant pressure levels; it is comparable to a radiosonde observation. { 'dräp,sänd ˌäb·sər'vā·shən }

dropstone [GEOL] A rock that was carried by a glacier or iceberg, and deposited as the ice melted. { 'dräp,stōn }

drop theory *See* barrier theory of cyclones. { 'dräp ˌthē·ə·rē }

drought [CLIMATOL] A period of abnormally dry weather sufficiently prolonged so that the lack of water causes a serious hydrologic imbalance (such as crop damage, water supply shortage, and so on) in the affected area; in general, the term should be reserved for relatively extensive time periods and areas. { drau̇t }

drowned atoll [GEOL] An atoll which has not reached the water surface. { ¦drau̇nd 'a,tȯl }

drowned coast [GEOL] A shoreline transformed from a hilly land surface to an archipelago of small islands by inundation by the sea. { ¦drau̇nd 'kōst }

drowned river mouth *See* estuary. { ¦draủnd 'riv·ər ˌmaủth }
drowned stream [HYD] A stream that has been flooded over by the ocean. Also known as flooded stream. { ¦draủnd 'strēm }
drowned valley [GEOL] A valley whose lower part has been inundated by the sea due to submergence of the land margin. { ¦draủnd 'val·ē }
droxtal [HYD] An ice particle measuring 10–20 micrometers in diameter, formed by direct freezing of supercooled water droplets at temperatures below −30°C. { 'dräk·stəl }
drumlin [GEOL] A hill of glacial drift or bedrock having a half-ellipsoidal streamline form like the inverted bowl of a spoon, with its long axis paralleling the direction of movement of the glacier that fashioned it. { 'drəm·lən }
drumlinoid *See* rock drumlin. { 'drəm·ləˌnȯid }
druse [GEOL] A small cavity in a rock or vein encrusted with aggregates of crystals of the same minerals which commonly constitute the enclosing rock. { drüz }
drusy [GEOL] Of or pertaining to rocks containing numerous druses. { 'drüz·ē }
dry adiabat [METEOROL] A line of constant potential temperature on a thermodynamic diagram. { ¦drī 'ad·ē·əˌbat }
dry adiabatic lapse rate [METEOROL] A special process lapse rate of temperature, defined as the rate of decrease of temperature with height of a parcel of dry air lifted adiabatically through an atmosphere in hydrostatic equilibrium. Also known as adiabatic lapse rate; adiabatic rate. { ¦drī ˌad·ē·ə¦bad·ik 'laps ˌrāt }
dry adiabatic process [METEOROL] An adiabatic process in a system of dry air. { ¦drī ˌad·ē·ə¦bad·ik 'präs·əs }
dry air [METEOROL] Air that contains no water vapor. { ¦drī 'er }
dry climate [CLIMATOL] **1.** In W. Köppen's climatic classification, the major category which includes steppe climate and desert climate, defined strictly by the amount of annual precipitation as a function of seasonal distribution and of annual temperature. **2.** In C. W. Thornwaite's climatic classification, any climate type in which the seasonal water surplus does not counteract seasonal water deficiency, and having a moisture index of less than zero; included are the dry subhumid, semiarid, and arid climates. { ¦drī 'klī·mət }
dry delta *See* alluvial fan. { ¦drī 'del·tə }
dry-dock iceberg *See* valley iceberg. { 'drī ˌdäk 'īsˌbərg }
dry firn *See* polar firn. { ¦drī 'fərn }
dry fog [METEOROL] A fog that does not moisten exposed surfaces. { ¦drī 'fäg }
dry freeze [HYD] The freezing of the soil and terrestrial objects caused by a reduction of temperature when the adjacent air does not contain sufficient moisture for the formation of hoarfrost on exposed surfaces. { ¦drī 'frēz }
dry haze [METEOROL] Fine dust or salt particles in the air, too small to be individually apparent but in sufficient number to reduce horizontal visibility, and to give the atmosphere a characteristic hazy appearance. { ¦drī 'hāz }
dry-hot-rock geothermal system [GEOL] A water-deficient hydrothermal reservoir dominated by the presence of rocks at depths in which large quantities of heat are stored. { ¦drī 'hät ˌräk jē·ō¦thər·məl 'sis·təm }
dryline [METEOROL] The boundary separating warm dry air from warm moist air along which thunderstorms and tornadoes may develop. { 'drīˌlīn }
dry permafrost [GEOL] A loose and crumbly permafrost which contains little or no ice. { ¦drī 'pər·məˌfrȯst }
dry quicksand [GEOL] An accumulation of alternate layers of firmly compacted sand and loose sand that cannot support heavy loads. { ¦drī 'kwikˌsand }
dry sand [GEOL] **1.** A formation, underlying the production sand, into which oil has leaked due to careless drilling practices. **2.** A nonproductive oil sand. { ¦drī ¦sand }
dry season [CLIMATOL] In certain types of climate, an annually recurring period of one or more months during which precipitation is at a minimum for the region. { 'drī ˌsēz·ən }
dry spell [CLIMATOL] A period of abnormally dry weather, generally reserved for a less extensive, and therefore less severe, condition than a drought; in the United States,

describes a period lasting not less than 2 weeks, during which no measurable precipitation was recorded. { 'drī ˌspel }

drystone [GEOL] A stalagmite or stalactite formed by dropping water. { 'drīˌstōn }

dry tongue [METEOROL] In synoptic meteorology, a pronounced protrusion of relatively dry air into a region of higher moisture content. { ¦drī ¦təŋ }

dry valley [GEOL] A valley, usually in a chalk or karst type of topography, that has no permanent water course along the valley floor. { ¦drī 'val·ē }

dry wash [GEOL] A wash, arroyo, or coulee whose bed lacks water. { 'drī ˌwäsh }

Dst [GEOPHYS] The "storm-time" component of variation of the terrestrial magnetic field, that is, the component which correlates with the interval of time since the onset of a magnetic storm; used as an index of intensity of the ring current.

duct [GEOPHYS] The space between two air layers, or between an air layer and the earth's surface, in which microwave beams are trapped in ducting. Also known as radio duct; tropospheric duct. { dəkt }

ducting [GEOPHYS] An atmospheric condition in the troposphere in which temperature inversions cause microwave beams to refract up and down between two air layers, so that microwave signals travel 10 or more times farther than the normal line-of-sight limit. Also known as superrefraction; tropospheric ducting. { 'dək·tiŋ }

dull coal [GEOL] A component of banded coal with a grayish color and dull appearance, consisting of small anthraxylon constituents in addition to cuticles and barklike constituents embedded in the attritus. { ¦dəl ¦kōl }

dune [GEOL] A mound or ridge of unconsolidated granular material, usually of sand size and of durable composition (such as quartz), capable of movement by transfer of individual grains entrained by a moving fluid. { dün }

dune complex [GEOGR] The totality of topographic forms, especially dunes, which comprise the moving landscape. { 'dün ˌkämˌpleks }

duplicatus [METEOROL] A cloud variety composed of superposed layers, sheets, or patches, at slightly different levels and sometimes partly merged. { ˌdü·plə'käd·əs }

durain [GEOL] A hard, granular ingredient of banded coal which occurs in lenticels and shows a close, firm texture. Also known as durite. { 'düˌrān }

Durargid [GEOL] A great soil group constituting a subdivision of the Argids, indicating those soils with a hardpan cemented by silica and called a duripan. { dür'är·jəd }

duration [OCEANOGR] The interval of time of the rising or falling tide, or the length of time of flood or ebb tidal currents. { də'rā·shən }

duricrust [GEOL] The case-hardened soil crust formed in semiarid climates by precipitation of salts; contains aluminous, ferruginous, siliceous, and calcareous material. { 'dür·əˌkrəst }

durinite [GEOL] The principal maceral of durain; a heterogeneous material, semiopaque in section (including all parts of plants); micrinite, exinite, cutinite, resinite, collinite, xylinite, suberinite, and fusinite may be present. { 'dür·əˌnīt }

duripan [GEOL] A horizon in mineral soil characterized by cementation by silica. { 'dür·əˌpan }

durite *See* durain. { 'dü·rīt }

düsenwind [METEOROL] The mountain-gap wind of the Dardanelles; a strong east-northeast wind which blows out of the Dardanelles into the Aegean Sea, penetrating as far as the island of Lemnos, and caused by a ridge of high pressure over the Black Sea. { 'dēz·ənˌvint }

dust [GEOL] Dry solid matter of silt and clay size (less than 1/16 millimeter). { dəst }

dust avalanche [GEOL] An avalanche of dry, loose snow. { 'dəst ˌav·əˌlanch }

dust bowl [CLIMATOL] A name given, early in 1935, to the region in the south-central United States afflicted by drought and dust storms, including parts of Colorado, Kansas, New Mexico, Texas, and Oklahoma, and resulting from a long period of deficient rainfall combined with loosening of the soil by destruction of the natural vegetation; dust bowl describes similar regions in other parts of the world. { 'dəst ˌbōl }

dust devil [METEOROL] A small but vigorous whirlwind, usually of short duration, rendered visible by dust, sand, and debris picked up from the ground; diameters range from about 10 to 100 feet (3 to 30 meters), and average height is about 600 feet (180 meters). { 'dəst ˌdev·əl }

dust-devil effect [GEOPHYS] In atmospheric electricity, rather sudden and short-lived change (positive or negative) of the vertical component of the atmospheric electric field that accompanies passage of a dust devil near an instrument sensitive to the vertical gradient. { 'dəst ˌdev·əl iˌfekt }

dust horizon [METEOROL] The top of a dust layer which is confined by a low-level temperature inversion and has the appearance of the horizon when viewed from above, against the sky; the true horizon is usually obscured by the dust layer. { 'dəst həˌrīz·ən }

dust storm [METEOROL] A strong, turbulent wind carrying large clouds of dust. { 'dəst ˌstȯrm }

dust well [HYD] A pit in an ice surface produced when small, dark particles on the ice are heated by sunshine and sink down into the ice. { 'dəst ˌwel }

dust whirl [METEOROL] A rapidly rotating column of air over a dry and dusty or sandy area, carrying dust, leaves, and other light material picked up from the ground; when well developed, it is known as a dust devil. Also known as dancing dervish; dancing devil; desert devil; sand auger; sand devil. { 'dəst ˌwərl }

duty of water [HYD] The total volume of irrigation water required to mature a particular type of crop, including consumptive use, evaporation and seepage from ditches and canals, and the water eventually returned to streams by percolation and surface runoff. { 'düd·ē əv 'wȯd·ər }

dwey *See* dwigh. { dwā }

dwigh [METEOROL] In Newfoundland, a sudden shower or snow storm. Also known as dwey; dwoy. { dwī }

dwoy *See* dwigh. { dwȯi }

Dwyka tillite [GEOL] A glacial Permian deposit that is widespread in South Africa. { də¦vīk·ə 'tiˌlīt }

dynamic climatology [CLIMATOL] The climatology of atmospheric dynamics and thermodynamics, that is, a climatological approach to the study and explanation of atmospheric circulation. { dī¦nam·ik ˌklī·mə'täl·ə·jē }

dynamic forecasting *See* numerical forecasting. { dī¦nam·ik 'fȯrˌkast·iŋ }

dynamic geomorphology [GEOL] The quantitative analysis of steady-state, self-regulatory geomorphic processes. Also known as analytical geomorphology. { dī¦nam·ik ˌjē·ō·mȯr'fäl·ə·jē }

dynamic height [GEOPHYS] As measured from sea level, the distance above the geoid of points on the same equipotential surface, in terms of linear units measured along a plumb line at a given latitude, generally 45°. { dī¦nam·ik 'hīt }

dynamic-height anomaly [OCEANOGR] The excess of the actual geopotential difference, between two given isobaric surfaces, over the geopotential difference in a homogeneous water column of salinity 35 per mille and temperature 0°C. Also known as anomaly of geopotential difference. { dī¦nam·ik ¦hīt ə'näm·ə·lē }

dynamic metamorphism [GEOL] Metamorphism resulting exclusively or largely from rock deformation, principally faulting and folding. Also known as dynamometamorphism. { dī¦nam·ik ˌmed·ə'mȯrˌfiz·əm }

dynamic meteorology [METEOROL] The study of atmospheric motions as solutions of the fundamental equations of hydrodynamics or other systems of equations appropriate to special situations, as in the statistical theory of turbulence. { dī¦nam·ik mēd·ē·ə'räl·ə·jē }

dynamic roughness [OCEANOGR] A quantity, designated z_0, dependent on the shape and distribution of the roughness elements of the sea surface, and used in calculations of wind at the surface. Also known as roughness length. { dī¦nam·ik 'rəf·nəs }

dynamic thickness [OCEANOGR] The vertical separation between two isobaric surfaces in the ocean. { dī¦nam·ik 'thik·nəs }

dynamic trough [METEOROL] A pressure trough formed on the lee side of a mountain range across which the wind is blowing almost at right angles. Also known as lee trough. { dī¦nam·ik 'trȯf }

dynamic vertical *See* apparent vertical. { dī¦nam·ik 'vərd·ə·kəl }

dynamo effect [GEOPHYS] A process in the ionosphere in which winds and the resultant movement of ionization in the geomagnetic field give rise to induced current. { 'dī·nə,mō i,fekt }

dynamometamorphism *See* dynamic metamorphism. { ¦dī·nə,mō,med·ə'mȯr,fiz·əm }

dynamo theory [GEOPHYS] The hypothesis which explains the regular daily variations in the earth's magnetic field in terms of electrical currents in the lower ionosphere, generated by tidal motions of the ionized air across the earth's magnetic field. { 'dī·nə,mō ,thē·ə·rē }

dyster *See* doister. { 'dī·stər }

E

earth [GEOL] **1.** Solid component of the globe, distinct from air and water. **2.** Soil; loose material composed of disintegrated solid matter. { ərth }

earth crust *See* crust. { 'ərth ˌkrəst }

earth current [GEOPHYS] A current flowing through the ground and due to natural causes, such as the earth's magnetic field or auroral activity. Also known as telluric current. { 'ərth ˌkə·rənt }

earth-current storm [GEOPHYS] Irregular fluctuations in an earth current in the earth's crust, often associated with electric field strengths as large as several volts per kilometer, and superimposed on the normal diurnal variation of the earth currents. { 'ərth ˌkə·rənt ˌstȯrm }

earth figure [GEOD] The shape of the earth. { 'ərth ˌfig·yər }

earthflow [GEOL] A variety of mass movement involving the downslope slippage of soil and weathered rock in a series of subparallel sheets. { 'ərthˌflō }

earth hummock [GEOL] A small, dome-shaped uplift of soil caused by the pressure of groundwater. Also known as earth mound. { 'ərth ˌhəm·ək }

earth interior [GEOL] The portion of the earth beneath the crust. { ¦ərth in¦tir·ē·ər }

earth-layer propagation [GEOPHYS] **1.** Propagation of electromagnetic waves through layers of the earth's atmosphere. **2.** Electromagnetic wave propagation through layers below the earth's surface. { 'ərth ˌlā·ər ˌpräp·əˌgā·shən }

earth mound *See* earth hummock. { 'ərth ˌmau̇nd }

earth movements [GEOPHYS] Movements of the earth, comprising revolution about the sun, rotation on the axis, precession of equinoxes, and motion of the surface of the earth relative to the core and mantle. { 'ərth ¦müv·məns }

earth oscillations [GEOPHYS] Any rhythmic deformations of the earth as an elastic body; for example, the gravitational attraction of the moon and sun excite the oscillations known as earth tides. { ¦ərth ˌäs·ə'lā·shənz }

earth pillar [GEOL] A tall, conical column of earth materials, such as clay or landslide debris, that has been sheltered from erosion by a cap of hard rock. { 'ərth ¦pil·ər }

earthquake [GEOPHYS] A sudden movement of the earth caused by the abrupt release of accumulated strain along a fault in the interior. The released energy passes through the earth as seismic waves (low-frequency sound waves), which cause the shaking. { 'ərthˌkwāk }

earthquake tremor *See* tremor. { 'ərthˌkwāk ˌtrem·ər }

earthquake zone [GEOL] An area of the earth's crust in which movements, sometimes with associated volcanism, occur. Also known as seismic area. { 'ərthˌkwāk ˌzōn }

earth radiation *See* terrestrial radiation. { ¦ərth ˌrād·ē'ā·shən }

Earth Radiation Budget Experiment [METEOROL] A satellite observational program to study the earth's radiation budget. Abbreviated ERBE. { ˌərth ˌrād·ē¦ā·shən ¦bəj·ət ikˌsper·ə·mənt }

earth shadow [METEOROL] Any shadow projecting into a hazy atmosphere from mountain peaks at times of sunrise or sunset. { 'ərth ˌshad·ō }

earth's shadow *See* dark segment. { ¦ərths 'shad·ō }

earth system [GEOPHYS] The atmosphere, oceans, biosphere, cryosphere, and geosphere, together. { 'ərth ˌ sis·təm }

earth tide [GEOPHYS] The periodic movement of the earth's crust caused by forces of the moon and sun. Also known as bodily tide. { 'ərth ˌtīd }
earth tremor *See* tremor. { 'ərth ˌtrem·ər }
earth wax *See* ozocerite. { 'ərth,waks }
east [GEOD] The direction 90° to the right of north. { ēst }
East Africa Coast Current [OCEANOGR] A current that is influenced by the monsoon drifts of the Indian Ocean, flowing southwestward along the Somalia coast in the Northern Hemisphere winter and northeastward in the Northern Hemisphere summer. Also known as Somali Current. { ¦ēst ¦af·rə·kə ¦kōst ˌkə·rənt }
East Australia Current [OCEANOGR] The current which is formed by part of the South Equatorial Current and flows southward along the eastern coast of Australia. { ¦ēst ȯ'strāl·yə ˌkə·rənt }
easterly wave [METEOROL] A long, weak migratory low-pressure trough occurring in the tropics. { 'ēs·tər·lē 'wāv }
Eastern Hemisphere [GEOGR] The half of the earth lying mostly to the east of the Atlantic Ocean, including Europe, Africa, and Asia. { ¦ē·stərn 'hem·ə,sfir }
East Greenland Current [OCEANOGR] A current setting south along the eastern coast of Greenland and carrying water of low salinity and low temperature. { ¦ēst 'grēn·lənd ˌkə·rənt }
east point [GEOD] That intersection of the prime vertical with the horizon which lies to the right of the observer when facing north. { 'ēst ˌpȯint }
ebb-and-flow structure [GEOL] Rock strata with alternating horizontal and cross-bedded layers, believed to have been produced by ebb and flow of tides. { ¦eb ən 'flō ˌstrək·chər }
ebb current [OCEANOGR] The tidal current associated with the decrease in the height of a tide. { 'eb ˌkə·rənt }
ebb tide [OCEANOGR] The portion of the tide cycle between high water and the following low water. Also known as falling tide. { 'eb ˌtīd }
echelon faults [GEOL] Separate, parallel faults having steplike trends. { 'esh·ə,län ˌfȯls }
echosonde [GEOPHYS] An acoustic sounding instrument used to study meteorological disturbances in the lower atmosphere such as wind velocity and turbulence. { 'ek·ō,sänd }
ecnephias [METEOROL] A squall or thunderstorm in the Mediterranean. { ek·nə'fē·əs }
economic geography [GEOGR] A branch of geography concerned with the relations of physical environment and economic conditions to the manufacture and distribution of commodities. { ˌek·ə'näm·ik jē'äg·rə·fē }
economic geology [GEOL] **1.** Application of geologic knowledge to materials usage and principles of engineering. **2.** The study of metallic ore deposits. { ˌek·ə'näm·ik jē'äl·ə·jē }
ectohumus [GEOL] An accumulation of organic matter on the soil surface with little or no mixing with mineral material. Also known as mor; raw humus. { ¦ek·tō'hyü·məs }
eddy correlation [METEOROL] A method of studying the effects of sea surface on the air above it by measuring simultaneous fluctuations of the horizontal and vertical components of the airflow from the mean. { 'ed·ē 'kä·rə,lā·shən }
eddy mill *See* pothole. { 'ed·ē ˌmil }
Edenian [GEOL] Lower Cincinnatian geologic stage in North America, above the Mohawkian and below Maysvillian. { ˌē'dēn·ē·ən }
edge water [GEOL] In reservoir structures, the subsurface water that surrounds the gas or oil. { 'ej ˌwȯd·ər }
edge wave [OCEANOGR] An ocean wave moving parallel to the coast, with crests normal to the coastline; maximum amplitude is at shore, with amplitude falling off exponentially farther from shore. { 'ej ˌwāv }
effective atmosphere [GEOPHYS] **1.** That part of the atmosphere which effectively influences a particular process or motion, its outer limits varying according to the terms

of the process or motion considered. **2.** *See* optically effective atmosphere. { ə¦fek·tiv 'at·mə,sfir }

effective gust velocity [METEOROL] The vertical component of the velocity of a sharp-edged gust that would produce a given acceleration on a particular airplane flown in level flight at the design cruising speed of the aircraft and at a given air density. { ə¦fek·tiv 'gəst və,läs·əd·ē }

effective porosity [GEOL] A property of earth containing interconnecting interstices, expressed as a percent of bulk volume occupied by the interstices. { ə¦fek·tiv pə'räs·əd·ē }

effective precipitable water [METEOROL] That part of the precipitable water which, in theory, can actually fall as precipitation. { ə¦fek·tiv prə'sip·əd·ə·bəl 'wȯd·ər }

effective precipitation [HYD] **1.** The part of precipitation that reaches stream channels as runoff. Also known as effective rainfall. **2.** In irrigation, the portion of the precipitation which remains in the soil and is available for consumptive use. { ə¦fek·tiv prə,sip·ə'tā·shən }

effective pressure *See* effective stress. { ə¦fek·tiv 'presh·ər }

effective radiation *See* effective terrestrial radiation. { ə¦fek·tiv ,rād·ē'ā·shən }

effective rainfall *See* effective precipitation. { ə¦fek·tiv 'rān,fȯl }

effective snowmelt [HYD] The part of snowmelt that reaches stream channels as runoff. { ə¦fek·tiv 'snō,melt }

effective stress [GEOL] The average normal force per unit area transmitted directly from particle to particle of a rock or soil mass. Also known as effective pressure; intergranular pressure. { ə¦fek·tiv 'stres }

effective temperature [METEOROL] The temperature at which motionless, saturated air would induce, in a sedentary worker wearing ordinary indoor clothing, the same sensation of comfort as that induced by the actual conditions of temperature, humidity, and air movement. { ə¦fek·tiv 'tem·prə·chər }

effective terrestrial radiation [GEOPHYS] The amount by which outgoing infrared terrestrial radiation of the earth's surface exceeds downcoming infrared counterradiation from the sky. Also known as effective radiation; nocturnal radiation. { ə¦fek·tiv tə¦res·trē·əl ,rā·dē'ā·shən }

effluent [HYD] **1.** Flowing outward or away from. **2.** Liquid which flows away from a containing space or a main waterway. { ə'flü·ənt }

effluent stream [HYD] A stream that is fed by seeping groundwater. { ə'flü·ənt 'strēm }

effusive stage [GEOL] The second cooling stage for volcanic rocks. { e'fyü·siv ,stāj }

e-folding height *See* scale height. { 'ē,fōl·diŋ ,hīt }

Egnell's law [METEOROL] The rule stating that above any fixed place the velocity of straight or nearly straight winds in the upper half of the troposphere increases with height at roughly the same rate that the density of the air decreases. { 'eg,nelz ,lȯ }

Egyptian asphalt [GEOL] A glance pitch (bituminous mixture similar to asphalt) found in the Arabian Desert. { i'jip·shən 'as,fȯlt }

8D technique [METEOROL] A technique for using the radiosonde observation to determine the presence of liquid water-droplets in supercooled clouds in saturated or nearly saturated layers of air; for each reported level in the sounding, the negative value of eight times the dew-point spread (−8D) is plotted on the pseudoadiabatic chart (or equivalent chart); where the temperature sounding lies to the left of the −8D curve, liquid droplet clouds are considered to be present, and icing is possible on aircraft flying in the cloud layer. Also known as frost-point technique. { ¦āt 'dē tek,nēk }

einkanter [GEOL] A stone shaped by windblown sand only upon one facet. { 'īn,kän·tər }

ejecta [GEOL] Material which is discharged by a volcano. { ē'jek·tə }

Ekman convergence [OCEANOGR] A zone of convergence of warm surface water caused by Ekman transport, creating a marked depression of the ocean's thermocline in the affected area. { 'ek·mən kən'vər·jəns }

Ekman layer [METEOROL] The layer of transition between the surface boundary layer of the atmosphere, where the shearing stress is constant, and the free atmosphere, which is treated as an ideal fluid in approximate geostrophic equilibrium. Also known as spiral layer. { 'ek·mən ˌlā·ər }

Ekman spiral [METEOROL] A theoretical representation that a wind blowing steadily over an ocean of unlimited depth and extent and uniform viscosity would cause, in the Northern Hemisphere, the immediate surface water to drift at an angle of 45° to the right of the wind direction, and the water beneath to drift further to the right, and with slower and slower speeds, as one goes to greater depths. { 'ek·mən ˌspī·rəl }

Ekman transport [OCEANOGR] The movement of ocean water caused by wind blowing steadily over the surface; occurs at right angles to the wind direction. { 'ek·mən ˌtransˌpȯrt }

elastic bitumen *See* elaterite. { i'las·tik bī'tü·mən }

elastic rebound theory [GEOL] A theory which attributes faulting to stresses (in the form of potential energy) which are being built up in the earth and which, at discrete intervals, are suddenly released as elastic energy; at the time of rupture the rocks on either side of the fault spring back to a position of little or no strain. { i'las·tik 'rēˌbau̇nd ˌthē·ə·rē }

elaterite [GEOL] A light-brown to black asphaltic pyrobitumen that is moderately soft and elastic. Also known as elastic bitumen; mineral caoutchouc. { i'lad·əˌrīt }

E layer [GEOPHYS] A layer of ionized air occurring at altitudes between 60 and 72 miles (100 and 120 kilometers) in the E region of the ionosphere, capable of bending radio waves back to earth. Also known as Heaviside layer; Kennelly-Heaviside layer. { 'ē ˌlā·ər }

elbow [GEOGR] A sharp change in direction of a coast line, channel, bank, or so on. { 'elˌbō }

ELDORA *See* Electra Doppler Radar. { el'dȯr·ə }

Electra Doppler Radar [METEOROL] An airborne Doppler radar used for detecting and measuring weather phenomena, as well as meteorological research. Abbreviated ELDORA. { i¦lek·trə ¦däp·lər 'rāˌdär }

electrical storm [METEOROL] A popular term for a thunderstorm. { i'lek·trə·kəl 'stȯrm }

electrical thickness [OCEANOGR] The vertical measure between the surface of an ocean current and an isokinetic point having a value of about one-tenth the surface speed. { i'lek·trə·kəl 'thik·nəs }

electrification ice nucleus [METEOROL] An ice nucleus that is formed by the fragmentation of dendritic crystals exposed to an electric field strength of several hundred volts per centimeter; it is a type of fragmentation nucleus. { iˌlek·trə·fə¦kā·shən 'īs ˌnü·klē·əs }

electrodynamic drift [GEOPHYS] Motion of charged particles in the upper atmosphere due to the combined effect of electric and magnetic fields; in the ionospheric F region and above, the drift velocity is perpendicular to both the electric and magnetic fields. { iˌlek·trō·dī'nam·ik 'drift }

electrofiltration [GEOL] Counterprocess during electrical logging of well boreholes, in which mud filtrate forced through the mud cake produces an emf in the mud cake opposite a permeable bed, positive in the direction of filtrate flow. { iˌlek·trō·fil'trā·shən }

electrogram [METEOROL] A record, usually automatically produced, which shows the time variations of the atmospheric electric field at a given point. { i'lek·trəˌgram }

electrojet [GEOPHYS] A stream of intense electric current moving in the upper atmosphere around the equator and in polar regions. { i'lek·trəˌjet }

electrostatic coalescence [METEOROL] **1.** The coalescence of cloud drops induced by electrostatic attractions between drops of opposite charges. **2.** The coalescence of two cloud or rain drops induced by polarization effects resulting from an external electric field. { iˌlek·trə'stad·ik kō·ə'les·əns }

elephant *See* elephanta. { 'el·ə·fənt }

elephanta [METEOROL] A strong southeasterly wind on the Malabar coast of southwest

India in September and October, at the end of the southwest monsoon, bringing thundersqualls and heavy rain. Also known as elephant; elephanter. { ˌel·əˈfan·tə }

elephanter *See* elephanta. { ˌel·əˈfan·tər }

elephant-hide pahoehoe [GEOL] A type of pahoehoe on whose surface are innumerable tummuli, broad swells, and pressure ridges which impart the appearance of elephant hide. { ˈel·ə·fənt ˌhīd paˈhō·ēˌhō·ē }

elerwind [METEOROL] A wind of Sun Valley north of Kufstein, in the Tyrol. { ˈel·ərˌvint }

elevation of ivory point *See* barometer elevation. { ˌel·əˈvā·shən əv ˈīv·rē ˌpoint }

ellipsoidal lava *See* pillow lava. { əˌlipˈsȯid·əl ˈläv·ə }

El Niño [METEOROL] A warming of the tropical Pacific Ocean that occurs roughly every 4–7 years. { el ˈnēn·yō }

El Niño Southern Oscillation [OCEANOGR] **1.** The irregular cyclic swing in atmospheric pressure in the tropical Pacific. **2.** The irregular cyclic swing of warm and cold phases in the tropical Pacific. Abbreviated ENSO. { el ¦nēn·yō ¦sə<u>th</u>·ərn ˌäs·əˈlā·shən }

Elsasser's radiation chart [METEOROL] A radiation chart developed by W. M. Elsasser for the graphical solution of the radiative transfer problems of importance in meteorology: given a radiosonde record of the vertical variation of temperature and water vapor content, one can find with this chart such quantities as the effective terrestrial radiation, net flux of infrared radiation at a cloud base or a cloud top, and radiative cooling rates. { ˈel·zə·sərz rād·ēˈā·shən ˌchärt }

elutriation [GEOL] The washing away of the lighter or finer particles in a soil, especially by the action of raindrops. { ēˌlü·trēˈā·shən }

eluvial [GEOL] Of, composed of, or relating to eluvium. { ēˈlüv·ē·əl }

eluvial placer [GEOL] A placer deposit that is concentrated near the decomposed outcrop of the source. { ēˈlüv·ē·əl ˈplā·sər }

eluviation [HYD] The process of transporting dissolved or suspended materials in the soil by lateral or downward water flow when rainfall exceeds evaporation. { ēˌlü·veˈā·shən }

eluvium [GEOL] Disintegrated rock material formed and accumulated in situ or moved by the wind alone. { ēˈlü·vē·əm }

elve [METEOROL] A transient luminous event that occurs over a thunderstorm, constituting a broad disk of illumination typically at an altitude of 85–90 kilometers (51–54 miles) with a thickness of about 6 kilometers (4 miles). { elv }

elvegust [METEOROL] A cold descending squall in the upper parts of Norwegian fjords. Also known as sno. { ˈel·vəˌgəst }

embacle [HYD] The piling up of ice in a stream after a refreeze, and the pile so formed. { emˈbak·əl }

embata [METEOROL] A local onshore southwest wind caused by the reversal of the northeast trade winds in the lee of the Canary Islands. { emˈbä·tä }

embatholithic [GEOL] Pertaining to ore deposits associated with a batholith where exposure of the batholith and country rock is about equal. { emˌbath·əˈlith·ik }

embayed [GEOGR] Formed into a bay. { emˈbād }

embayed coastal plain [GEOL] A coastal plain that has been partly sunk beneath the sea, thereby forming a bay. { emˈbād ¦kōst·əl ˈplān }

embayed mountain [GEOL] A mountain that has been depressed enough for sea water to enter the bordering valleys. { emˈbād ˈmau̇nˌtən }

embayment [GEOGR] Indentation in a shoreline forming a bay. [GEOL] **1.** Act or process of forming a bay. **2.** A reentrant of sedimentary rock into a crystalline massif. { emˈbā·mənt }

embouchure [GEOL] **1.** The mouth of a river. **2.** A river valley widened into a plain. { ¦äm·bə¦shu̇r }

emerged shoreline *See* shoreline of emergence. { ə¦mərjd ˈshȯrˌlīn }

emergence [GEOL] **1.** Dry land which was part of the ocean floor. **2.** The act or process of becoming an emergent land mass. [HYD] *See* resurgence. { əˈmər·jəns }

emissary sky [METEOROL] A sky of cirrus clouds which are either isolated or in small, separated groups; so called because this formation often is one of the first indications of the approach of a cyclonic storm. { ˈem·əˌser·ē ˌskī }

emission [METEOROL] A natural or anthropogenic discharge of particulate, gaseous, or soluble waste material or pollution into the air. { i'mish·ən }

emission control [METEOROL] A strategy for reducing or preventing atmospheric pollution, such as a catalytic converter used for pollutant removal from automotive exhaust. { i'mish·ən kən,trōl }

emplacement [GEOL] Intrusion of igneous rock or development of an ore body in older rocks. { em'plās·mənt }

encrinal limestone [GEOL] A limestone consisting of more than 10% but less than 50% of fossil crinoidal fragments. { en'krīn·əl 'līm·stōn }

end moraine [GEOL] An accumulation of drift in the form of a ridge along the border of a valley glacier or ice sheet. { 'end mə,rān }

endobatholithic [GEOL] Pertaining to ore deposits along projecting portions of a batholith. { ¦en·dō·bath·ə'lid·ik }

endocast *See* steinkern. { 'en·dō,kast }

endogenetic *See* endogenic. { ¦en·dō·jə'ned·ik }

endogenic [GEOL] Of or pertaining to a geologic process, or its resulting feature such as a rock, that originated within the earth. Also known as endogenetic; endogenous. { ¦en·dō¦jen·ik }

endogenous *See* endogenic. { en'däj·ə·nəs }

endometamorphism [GEOL] A phase of contact metamorphism involving changes in an igneous rock due to assimilation of portions of the rocks invaded by its magma. { ¦en·dō,med·ə'mȯr,fiz·əm }

endoreism *See* endorheism. { ,en·dō'rē,iz·əm }

endorheism [HYD] A drainage pattern of a basin or region in which little or none of the surface drainage reaches the ocean. Also spelled endoreism. { ,en·dō'rē,iz·əm }

en echelon [GEOL] Referring to an overlapped or staggered arrangement of geologic features. { 'en ,esh·ə,län }

en echelon fault blocks [GEOL] A belt in which the individual fault blocks trend approximately 45° to the trend of the entire fault belt. { 'en ,esh·ə,län 'fȯlt ,bläks }

energy budget [CLIMATOL] The energy pools, the directions of energy flow, and the rates of energy transformations quantified within a physical or ecological system. { 'en·ər·jē ,bəj·ət }

energy coefficient [OCEANOGR] The ratio between the energy transmitted forward in a wave per unit crest length at a point in shallow water, and the energy transmitted forward in a wave per unit crest length in deep water. { 'en·ər·jē ,kō·i'fish·ənt }

energy level [GEOL] The kinetic energy supplied by waves or current action in an aqueous sedimentary environment either at the interface of deposition or several meters above. { 'en·ər·jē ,lev·əl }

energy transfer [METEOROL] The transfer of energy of a given form among different scales of motion; for example, kinetic energy may be transferred between the zonal and meridional components of the wind, or between the mean and eddy components of the wind. { 'en·ər·jē ,trans·fər }

englacial [HYD] Of or pertaining to the inside of a glacier. { en'glā·shəl }

engysseismology [GEOPHYS] Seismology dealing with earthquake records made close to the disturbance. { ¦en·jə·sīz'mäl·ə·jē }

ensialic geosyncline [GEOL] A geosyncline whose geosynclinal prism accumulates on a sialic crust and contains clastics. { en·sē'al·ik ,jē·ō'sin,klīn }

ensimatic geosyncline [GEOL] A geosyncline whose geosynclinal prism accumulates on a simatic crust and is composed largely of volcanic rock or sediments of volcanic debris. { en·sə'mad·ik ,jē·ō'sin,klīn }

ENSO *See* El Niño Southern Oscillation. { 'en,sō }

ensonification field [OCEANOGR] The area of the sea floor that is acoustically imaged in the course of a sonar survey. { en,sän·ə·fə'kā·shən ,fēld }

enstatite chondrite [GEOL] A type of chrondritic meteorite consisting almost entirely of enstatite, with metal inclusions that may be abundant and are usually low in nickel. { 'en·stə,tīt 'kän,drīt }

enterolithic [GEOL] Of or pertaining to structures, such as small folds, formed in evaporites due to flowage or hydration. { ˌent·ə·rə'lith·ik }
Entisol [GEOL] An order of soil having few or faint horizons. { 'ent·əˌsȯl }
entrail pahoehoe [GEOL] A type of pahoehoe having a surface that resembles an intertwined mass of entrails. { 'enˌtrāl pə'hō·ēˌhō·ē }
entrainment [HYD] The pickup and movement of sediment as bed load or in suspension by current flow. [METEOROL] The mixing of environmental air into a preexisting organized air current so that the environmental air becomes part of the current. [OCEANOGR] The transfer of fluid by friction from one water mass to another, usually occurring between currents moving in respect to each other. { en'trān·mənt }
entrance region [METEOROL] The region of confluence at the upwind extremity of a jet stream. { 'en·trəns ˌrē·jən }
entrapment [GEOL] The underground trapping of oil or gas reserves by folds, faults, domes, asphaltic seals, unconformities, and such. { en'trap·mənt }
entrenched meander [HYD] A deepened meander of a river which is carried downward further below the valley surface in which the meander originally formed. Also known as inherited meander. { en'trencht mē'an·dər }
entrenched stream [HYD] A stream that flows in a valley or narrow trench cut into a plain or relatively level upland. Also spelled intrenched stream. { en'trencht 'strēm }
envelope orography [METEOROL] A method for developing a numerical model for weather forecasting in which it is assumed that mountain passes and valleys are filled mostly with stagnant air, thus increasing the average height of the model mountains and enhancing the blocking effect. { 'en·vəˌlōp ȯ'räg·rə·fē }
environmental lapse rate [METEOROL] The rate of decrease of temperature with elevation in the atmosphere. Also known as atmospheric lapse rate. { in¦vī·ərn¦mənt·əl 'laps ˌrāt }
environment of sedimentation [GEOL] A more or less destructive geomorphologic setting in which sediments are deposited as beach environment. { in¦vī·ərn¦mənt əv ˌsed·ə·men'tā·shən }
Eocambrian [GEOL] Pertaining to the thick sequences of strata conformably underlying Lower Cambrian fossils. Also known as Infracambrian. { ˌē·ō'kam·brē·ən }
Eocene [GEOL] The next to the oldest of the five major epochs of the Tertiary period (in the Cenozoic era). { 'ē·əˌsēn }
Eogene *See* Paleogene. { 'ē·əˌjēn }
eolation [GEOL] Any action of wind on the land. { ˌē·ə'lā·shən }
eolian [METEOROL] Pertaining to the action or the effect of the wind, as in eolian sounds or eolian deposits (of dust). Also spelled aeolian. { ē'ōl·yən }
eolian dune [GEOL] A dune resulting from entrainment of grains by the flow of moving air. { ē'ōl·yən 'dün }
eolian erosion [GEOL] Erosion due to the action of wind. { ē'ōl·yən ə'rō·zhən }
eolianite [GEOL] A sedimentary rock consisting of clastic material which has been deposited by wind. { ē'ōl·yəˌnīt }
eolian ripple mark [GEOL] A mark made in sand by the wind. { ē'ōl·yən 'rip·əl ˌmärk }
eolian sand [GEOL] Deposits of sand arranged by the wind. { ē'ōl·yən 'sand }
eolian soil [GEOL] A type of soil ranging from sand dunes to loess deposits whose particles are predominantly of silt size. { ē'ōl·yən 'sȯil }
eonothem [GEOL] A chronostratigraphic unit, above erathem, composed of rocks formed during an eon of geologic time. { 'ēn·əˌthem }
eötvös [GEOPHYS] A unit of horizontal gradient of gravitational acceleration, equal to a change in gravitational acceleration of 10^{-9} galileo over a horizontal distance of 1 centimeter. { 'ət·vəsh }
epeiric sea *See* epicontinental sea. { ə'pīr·ik 'sē }
epeirogeny [GEOL] Movements which affect large tracts of the earth's crust. { ˌeˌpī'räj·ə·nē }
ephemeral gully [GEOL] A channel that forms in a cultivated field when precipitation exceeds the rate of soil infiltration. { ə¦fem·ə·rəl ¦gəl·ē }

ephemeral stream [HYD] A stream channel which carries water only during and immediately after periods of rainfall or snowmelt. { ə'fem·ə·rəl 'strēm }

epicenter [GEOL] A point on the surface of the earth which is directly above the seismic focus of an earthquake and where the earthquake vibrations reach first. { 'ep·ə,sen·tər }

epiclastic [GEOL] Pertaining to the texture of mechanically deposited sediments consisting of detrital material from preexistent rocks. { ¦ep·ə'klas·tik }

epicontinental [GEOL] Located upon a continental plateau or platform. { ¦ep·ə,kant·ən'ent·əl }

epicontinental sea [OCEANOGR] That portion of the sea lying upon the continental shelf, and the portions which extend into the interior of the continent with similar shallow depths. Also known as epeiric sea; inland sea. { ¦ep·ə,kant·ən'ent·əl 'sē }

epidotization [GEOL] The introduction of epidote into, or the formation of epidote from, rocks. { ,ep·ə,dōd·ə'zā·shən }

epieugeosyncline [GEOL] Deep troughs formed by subsidence which have limited volcanic power and overlie a eugeosyncline. { ¦ep·ē,yü,jē·ō'sin,klīn }

epigene [GEOL] **1.** A geologic process originating at or near the earth's surface. **2.** A structure formed at or near the earth's surface. { 'ep·ə,jēn }

epigenesis [GEOL] Alteration of the mineral content of rock due to outside influences. { ,ep·ə'jen·ə·səs }

epigenetic [GEOL] Produced or formed at or near the surface of the earth. { ¦ep·ə·jə¦ned·ik }

epilimnion [HYD] A fresh-water zone of relatively warm water in which mixing occurs as a result of wind action and convection currents. { ,ep·ə'lim·nē,än }

epimagma [GEOL] A gas-free, vesicular to semisolid magmatic residue of pasty consistency formed by cooling and loss of gas from liquid lava in a lava lake. { ,ep·ə'mag·ma }

epimagmatic *See* deuteric. { ,ep·ə·mag'mad·ik }

epipelagic [OCEANOGR] Of or pertaining to the portion of oceanic zone into which enough light penetrates to allow photosynthesis. { ¦ep·ə·pə'laj·ik }

epipelagic zone [OCEANOGR] The region of an ocean extending from the surface to a depth of about 600 feet (200 meters); light penetrates this zone, allowing photosynthesis. { ¦ep·ə·pə'laj·ik 'zōn }

episode [GEOL] A distinctive event or series of events in the geologic history of a region or feature. { 'ep·ə,sōd }

epithermal [GEOL] Pertaining to mineral veins and ore deposits formed from warm waters at shallow depth, at temperatures ranging from 50–200°C, and generally at some distance from the magmatic source. { ¦ep·ə'thər·məl }

epithermal deposit [GEOL] Ore deposit formed in and along openings in rocks by deposition at shallow depths from ascending hot solutions. { ¦ep·ə'thər·məldə'päz·ət }

epizone [GEOL] **1.** The zone of metamorphism characterized by moderate temperature, low hydrostatic pressure, and powerful stress. **2.** The outer depth zone of metamorphic rocks. { 'ep·ə,zōn }

epoch [GEOL] A major subdivision of a period of geologic time. { 'ep·ək }

equator [GEOD] The great circle around the earth, equally distant from the North and South poles, which divides the earth into the Northern and Southern hemispheres; the line from which latitudes are reckoned. { ē'kwād·ər }

equatorial air [METEOROL] The air of the doldrums or the equatorial trough; distinguished somewhat vaguely from the tropical air of the trade-wind zones. { ,e·kwə'tör·ē·əl 'er }

equatorial axis [GEOD] The diameter of the earth described between two points on the equator. { ,e·kwə'tör·ē·əl 'ak·səs }

equatorial bulge [GEOD] The excess of the earth's equatorial diameter over the polar diameter. { ,e·kwə'tör·ē·əl 'bəlj }

equatorial calms *See* doldrums. { ,e·kwə'tör·ē·əl 'kämz }

equatorial convergence zone *See* intertropical convergence zone. { ,e·kwə'tȯr·ē·əl kən'vər·jəns ,zōn }

Equatorial Countercurrent [OCEANOGR] An ocean current flowing eastward (counter to and between the westward-flowing North Equatorial Current and South Equatorial Current) through all the oceans. { ,e·kwə'tȯr·ē·əl 'kau̇nt·ər,kər·ənt }

Equatorial Current *See* North Equatorial Current; . *See* South Equatorial Current. { ,e·kwə'tȯr·ē·əl 'kə·rənt }

equatorial dry zone [CLIMATOL] An arid region existing in the equatorial trough; the most famous dry zone is situated a little south of the equator in the central Pacific. Also known as arid zone. { ,e·kwə'tȯr·ē·əl 'drī ,zōn }

equatorial easterlies [METEOROL] The trade winds in the summer hemisphere when they are very deep, extending at least 5 to 6 miles (8 to 10 kilometers) in altitude, and generally not topped by upper westerlies; if upper westerlies are present, they are too weak and shallow to influence the weather. Also known as deep easterlies; deep trades. { ,e·kwə'tȯr·ē·əl 'ēs·tər·lēz }

equatorial electrojet [GEOPHYS] A concentration of electric current in the atmosphere found in the magnetic equator. { ,e·kwə'tȯr·ē·əl ə'lek·trə,jet }

equatorial front *See* intertropical front. { ,e·kwə'tȯr·ē·əl 'frənt }

equatorial radius [GEOD] The radius assigned to the great circle making up the terrestrial equator; approximately 6,378,139 meters (20,925,653 feet). { ,e·kwə'tȯr·ē·əl 'rād·ē·əs }

equatorial tide [OCEANOGR] **1.** A lunar fortnightly tide. **2.** A tidal component with a period of 328 hours. { ,e·kwə'tȯr·ē·əl 'tīd }

equatorial trough [METEOROL] The quasicontinuous belt of low pressure lying between the subtropical high-pressure belts of the Northern and Southern hemispheres. Also known as meteorological equator. { ,e·kwə'tȯr·ē·əl 'trȯf }

Equatorial Undercurrent [OCEANOGR] **1.** A subsurface current flowing from west to east in the Indian Ocean near the 450-foot (150-meter) depth at the equator during the time of the Northeast Monsoon. **2.** A permanent subsurface current in the equatorial region of the Atlantic and Pacific oceans. { ,e·kwə'tȯr·ē·əl 'ən·dər,kə·rənt }

equatorial vortex [METEOROL] A closed cyclonic circulation with the equatorial trough. { ,e·kwə'tȯr·ē·əl 'vȯr,teks }

equatorial wave [METEOROL] A wavelike disturbance of the equatorial easterlies that extends across the equatorial trough. { ,e·kwə'tȯr·ē·əl 'wāv }

equatorial westerlies [METEOROL] The westerly winds occasionally found in the equatorial trough and separated from the mid-latitude westerlies by the broad belt of easterly trade winds. { ,e·kwə'tȯr·ē·əl 'wes·tər·lēz }

equigeopotential surface *See* geopotential surface. { ¦ē·kwə·jē·ō·pə'ten·chəl 'sər·fəs }

equilibrium line [HYD] The level on a glacier where the net balance equals zero and accumulation equals ablation. { ,ē·kwə'lib·rē·əm ,līn }

equilibrium profile *See* profile of equilibrium. { ,ē·kwə'lib·rē·əm 'prō,fīl }

equilibrium solar tide [GEOPHYS] The form of the atmosphere which is determined solely by gravitational forces in the absence of any rotation of the earth relative to the sun. { ,ē·kwə'lib·rē·əm ¦sō·lər 'tīd }

equilibrium spheroid [GEOPHYS] The shape that the earth would attain if it were entirely covered by a tideless ocean of constant depth. { ,ē·kwə'lib·rē·əm 'sfir,ȯid }

equilibrium theory [OCEANOGR] An ocean water model which assumes instantaneous response of water bodies to the tide-producing forces of the moon and sun to form an equilibrium surface, and disregards the effects due to friction, inertia, and irregular distribution of land masses. { ,ē·kwə'lib·rē·əm ,thē·ə·rē }

equilibrium tide [OCEANOGR] The hypothetical tide due to the tide-producing forces of celestial bodies, particularly the sun and moon. { ,ē·kwə'lib·rē·əm ,tīd }

equinoctial rains [METEOROL] Rainy seasons which occur regularly at or shortly after the equinoxes in many places within a few degrees of the equator. { ,ē·kwə'näk·shəl 'rānz }

equinoctial storm [METEOROL] In semipopular belief, a violent storm of wind and rain which is supposed, both in the United States and in Britain, to occur at or near the time of the equinox. Also known as line gale; line storm. { ˌē·kwə'näk·shəl 'stȯrm }

equinoctial tide [OCEANOGR] A tide occurring near an equinox. { ˌē·kwə'näk·shəl 'tīd }

equiparte [METEOROL] In Mexico, heavy cold rains during October to January, which last for several days. Also known as equipatos. { ¦e·kwē¦pär·tā }

equipatos *See* equiparte. { ¦e·kwē¦pä·tōs }

equiphase zone [GEOPHYS] That region in space where the difference in phase of two radio signals is indistinguishable. { 'e·kwəˌfāz ˌzōn }

equipotential surface [GEOPHYS] A surface characterized by the potential being constant everywhere on it for the attractive forces concerned. { ¦e·kwə·pə'ten·chəl 'sər·fəs }

equivalent barotropic model [METEOROL] A model atmosphere characterized by frictionless and adiabatic flow and by hydrostatic quasigeostrophic equilibrium, and in which the vertical shear of the horizontal wind is assumed to be proportional to the horizontal wind itself. { i'kwiv·ə·lənt bar·ə'träp·ik 'mäd·əl }

equivalent diameter *See* nominal diameter. { i'kwiv·ə·lənt dī'am·əd·ər }

equivalent height *See* virtual height. { i'kwiv·ə·lənt 'hīt }

equivalent potential temperature [METEOROL] The potential temperature corresponding to the adiabatic equivalent temperature. { i'kwiv·ə·lənt pə¦ten·chəl 'tem·prə·chər }

equivalent temperature [METEOROL] **1.** The temperature that an air parcel would have if all water vapor were condensed out at constant pressure, the latent heat released being used to heat the air. Also known as isobaric equivalent temperature. **2.** The temperature that an air parcel would have after undergoing the following theoretical process: dry-adiabatic expansion until saturated, pseudoadiabatic expansion until all moisture is precipitated out, and dry adiabatic compression to the initial pressure; this is the equivalent temperature as read from a thermodynamic chart and is always greater than the isobaric equivalent temperature. Also known as adiabatic equivalent temperature; pseudoequivalent temperature. { i'kwiv·ə·lənt 'tem·prə·chər }

equivoluminal wave *See* S wave. { ¦e·kwə·və¦lüm·ə·nəl 'wāv }

era [GEOL] A unit of geologic time constituting a subdivision of an eon and comprising one or more periods. { 'ir·ə }

eradiation *See* terrestrial radiation. { iˌrād·ē'ā·shən }

erathem [GEOL] A chronostratigraphic unit, below eonothem and above system, composed of rocks formed during an era of geologic time. { 'er·əˌthem }

ERBE *See* Earth Radiation Budget Experiment. { 'ərˌbē }

erg [GEOGR] A large expanse of the earth's surface that is covered with sand, generally blown by wind into dune formations. { ərg }

Erian [GEOL] Middle Devonian geologic time; a North American provincial series. { 'i·rē·ən }

Erian orogeny [GEOL] One of the orogenies during Phanerozoic geologic time, at the end of the Silurian; the last part of the Caledonian orogenic era. Also known as Hibernian orogeny. { 'i·rē·ən ȯ'räj·ə·nē }

eroding velocity [GEOL] The minimum average velocity required for eroding homogeneous material of a given particle size. { ə'rōd·iŋ və'läs·əd·ē }

erosion [GEOL] **1.** The loosening and transportation of rock debris at the earth's surface. **2.** The wearing away of the land, chiefly by rain and running water. { ə'rō·zhən }

erosional unconformity [GEOL] The surface that separates older, eroded rocks from younger, overlying sediments. { ə'rō·zhən·əl ˌən·kən'fȯr·məd·ē }

erosion cycle [GEOL] A postulated sequence of conditions through which a new landmass proceeds as it wears down, classically the concept of youth, maturity, and old age, as stated by W.M. Davis; an original landmass is uplifted above base level, cut by canyons, gradually converted into steep hills and wide valleys, and is finally reduced to a flat lowland at or near base level. { ə'rō·zhən ˌsī·kəl }

erosion pavement [GEOL] A layer of pebbles and small rocks that prevents the soil underneath from eroding. { ə'rō·zhən ˌpāv·mənt }

erosion platform *See* wave-cut platform. { ə'rō·zhən ˌplatˌfȯrm }

erosion ridge [HYD] One of a group of ridges on the surface of snow; formed by the corrosive action of wind-blown snow. { ə'rō·zhən ˌrij }

erosion surface [GEOL] A land surface shaped by agents of erosion. { ə'rō·zhən ˌsər·fəs }

erratic [GEOL] A rock fragment that has been transported a great distance, generally by glacier ice or floating ice, and differs from the bedrock on which it rests. { ə'rad·ik }

ertor [METEOROL] The effective (radiational) temperature of the ozone layer (region). { 'ərˌtȯr }

eruption [GEOL] The ejection of solid, liquid, or gaseous material from a volcano. { i'rəp·shən }

Erzgebirgian orogeny [GEOL] Diastrophism of the early Late Carboniferous. { 'erts·gəˌbər·jən ȯ'räj·ə·nē }

escar *See* esker. { 'es·kər }

escarpment [GEOL] A cliff or steep slope of some extent, generally separating two level or gently sloping areas, and produced by erosion or by faulting. Also known as scarp. { ə'skärp·mənt }

eschar *See* esker. { 'es·kər }

eskar *See* esker. { 'es·kər }

esker [GEOL] A sinuous ridge of constructional form, consisting of stratified accumulations, glacial sand, and gravel. Also known as asar; eschar; eskar; osar; serpent kame. { 'es·kər }

espalier drainage *See* trellis drainage. { e'spal·yər ˌdrān·ij }

establishment [OCEANOGR] The interval of time between the transit (upper or lower) of the moon and the next high water at a place. { i'stab·lish·mənt }

estuarine circulation [OCEANOGR] In an estuary, the outflow (seaward) of low-salinity surface water over a deeper inflowing layer of dense, high-salinity water. { 'es·chə·wəˌrēn ˌsər·kyə¦lā·shən }

estuarine deposit [GEOL] A sediment deposited at the heads and floors of estuaries. { 'es·chə·wəˌrēn də'päz·ət }

estuarine environment [OCEANOGR] The physical conditions and influences of an estuary. { 'es·chə·wəˌrēn en'vī·rən·mənt }

estuarine oceanography [OCEANOGR] The study of the chemical, physical, biological, and geological properties of estuaries. { 'es·chə·wəˌrēn ˌō·shə'näg·rə·fē }

estuary [GEOGR] A semienclosed coastal body of water which has a free connection with the open sea and within which sea water is measurably diluted with fresh water. Also known as branching bay; drowned river mouth; firth. { 'es·chə ˌwer·ē }

etesian climate *See* Mediterranean climate. { ə'tē·zhən 'klī·mət }

etesians [METEOROL] The prevailing northerly winds in summer in the eastern Mediterranean, and especially the Aegean Sea; basically similar to the monsoon and equivalent to the maestro of the Adriatic Sea. { ə'tē·zhənz }

ethmolith [GEOL] A downward tapering, funnel-shaped, discordant intrusion of igneous rocks. { 'eth·məˌlith }

eugeosyncline [GEOL] The internal volcanic belt of an orthogeosyncline. { yüˌjē·ō'sinˌklīn }

Eulerian nutation *See* Chandler wobble. { ȯi'ler·ē·ən nyü'tā·shən }

Eulerian wind [METEOROL] A wind motion only in response to the pressure force; the cyclostrophic wind is a special case of the Eulerian wind, which is limited in its meteorological applicability to those situations in which the Coriolis effect is negligible. { ȯi'ler·ē·ən 'wind }

eulittoral [OCEANOGR] A subdivision of the benthic division of the littoral zone of the marine environment, extending from high-tide level to about 200 feet (60 meters), the lower limit for abundant growth of attached plants. { yü'lid·ə·rəl }

eupelagic *See* pelagic. { yü·pə'laj·ik }

euphotic [OCEANOGR] Of or constituting the upper levels of the marine environment down to the limits of effective light penetration for photosynthesis. { yü'fäd·ik }

Euramerica [GEOL] The continent that was composed of Europe and North America during most of the Mesozoic Era. { ˌyür·ə'mer·ə·kə }

Europe [GEOGR] A great western peninsula of the Eurasian landmass, usually called a continent; its eastern limits are arbitrary and are conventionally drawn along the water divide of the Ural Mountains, the Ural River, the Caspian Sea, and the Caucasus watershed to the Black Sea. { 'yu̇r·əp }

eustacy [OCEANOGR] Worldwide fluctuations of sea level due to changing capacity of the ocean basins or the volume of ocean water. { 'yü·stə·sē }

eutrophic [HYD] Pertaining to a lake containing a high concentration of dissolved nutrients; often shallow, with periods of oxygen deficiency. { yü'träf·ik }

euxinic [HYD] Of or pertaining to an environment of restricted circulation and stagnant or anaerobic conditions. { yük'sin·ik }

evaporation capacity *See* evaporative power. { iˌvap·ə'rā·shən kəˌpas·əd·ē }

evaporation current [OCEANOGR] An ocean current resulting from the accumulation of water through precipitation and river runoff at one point, and loss by evaporation at another point. { iˌvap·ə'rā·shən ˌkə·rənt }

evaporation power *See* evaporative power. { iˌvap·ə'rā·shən ˌpau̇·ər }

evaporative capacity *See* evaporative power. { i'vap·əˌrād·iv kə'pas·əd·ē }

evaporative power [METEOROL] A measure of the degree to which the weather or climate of a region is favorable to the process of evaporation; it is usually considered to be the rate of evaporation, under existing atmospheric conditions, from a surface of water which is chemically pure and has the temperature of the lowest layer of the atmosphere. Also known as evaporation capacity; evaporation power; evaporative capacity; evaporativity; potential evaporation. { i'vap·əˌrād·iv 'pau̇·ər }

evaporativity *See* evaporative power. { iˌvap·ə·rə'tiv·əd·ē }

evaporite [GEOL] Deposits of mineral salts from sea water or salt lakes due to evaporation of the water. { i'vap·əˌrīt }

evapotranspiration [HYD] Discharge of water from the earth's surface to the atmosphere by evaporation from lakes, streams, and soil surfaces and by transpiration from plants. Also known as fly-off; total evaporation; water loss. { iˌvap·ōˌtranz·pə'rā·shən }

event [GEOL] An incident which is of probable tectonic significance but whose full implications are unknown. { i'vent }

evorsion [GEOL] The process of pothole formation in riverbeds; plays an important role in denudation. { ē'vȯr·shən }

excessive precipitation [METEOROL] Precipitation (generally in the form of rain) of an unusually high rate of fall; although often used qualitatively, several meteorological services have adopted quantitative limits. { ek'ses·iv prəˌsip·ə'tā·shən }

exchange capacity [GEOL] The ability of a soil material to participate in ion exchange as measured by the quantity of exchangeable ions in a given unit of the material. { iks'chānj kəˌpas·əd·ē }

exfoliation *See* sheeting. { eksˌfō·lē'ā·shən }

exfoliation dome [GEOL] A large rounded dome-shaped structure produced in massive homogeneous coarse-grained rocks (usually igneous) by exfoliation. { eksˌfō·lē'ā·shən ˌdōm }

exfoliation joint *See* sheeting structure. { eksˌfō·lē'ā·shən ˌjȯint }

exhalation [GEOPHYS] The process by which radioactive gases escape from the surface layers of soil or loose rock, where they are formed by decay of radioactive salts. { ˌeks·ə'lā·shən }

exhaust trail [METEOROL] A visible condensation trail (contrail) that forms when the water vapor of an aircraft exhaust is mixed with and saturates (or slightly supersaturates) the air in the wake of the aircraft. { ig'zȯst ˌtrāl }

exhumation [GEOL] The uncovering or exposure through erosion of a former surface, landscape, or feature that had been buried by subsequent deposition. { ˌeks·yü'mā·shən }

exhumed *See* resurrected. { ig'zyümd }

exinite [GEOL] A hydrogen-rich maceral group consisting of spore exines, cuticular matter, resins, and waxes; includes sporinite, cutinite, alginite, and resinite. Also known as liptinite. { 'ek·sə,nīt }

exit region [METEOROL] The region of difluence at the downwind extremity of a jet stream. { 'eg·zət ,rē·jən }

exocline [GEOL] An inverted anticline or syncline. { 'ek·sə,klīn }

exogeosyncline [GEOL] A parageosyncline that lies along the cratonal border and obtains its clastic sediments from erosion of the adjacent orthogeosynclinal belt outside the craton. Also known as delta geosyncline; foredeep; transverse basin. { ¦ek·sō,jē·ō'sin,klīn }

exomorphic zone *See* aureole. { ¦ek·sə¦mór·fik ,zōn }

exorheic [GEOL] Referring to a basin or region characterized by external drainage. { ek·sə'rē·ik }

exosphere [METEOROL] An outermost region of the atmosphere, estimated at 300–600 miles (500–1000 kilometers), where the density is so low that the mean free path of particles depends upon their direction with respect to the local vertical, being greatest for upward-traveling particles. Also known as region of escape. { 'ek·sō,sfir }

exotic stream [HYD] A stream that crosses a desert as it flows to the sea, or any stream which derives most of its water from the drainage system of another region. { ig'zäd·ik 'strēm }

expanded foot [HYD] A broad, bulblike or fan-shaped ice mass formed where a valley glacier flows beyond its confining walls and extends onto an adjacent lowland at the bottom of a mountain slope. { ik'spand·əd 'fút }

expansion fissures [GEOL] A system of fissures which radiate randomly and pass through feldspars and other minerals adjacent to olivine crystals that have been replaced by serpentine. { ik'span·shən ,fish·ərz }

expansion joint *See* sheeting structure. { ik'span·shən ,jóint }

explosion crater [GEOL] A volcanic crater formed by explosion and commonly developed along rift zones on the flanks of large volcanoes. { ik'splō·zhən ,krād·ər }

explosion tuff [GEOL] A tuff whose constituent ash particles are in the place they fell after being ejected from a volcanic vent. { ik'splō·zhən ,təf }

explosive index [GEOL] The percentage of pyroclastics in the material from a volcanic eruption. { ik'splō·siv 'in,deks }

exponential atmosphere *See* isothermal atmosphere. { ,ek·spə'nen·chəl 'at·mə,sfir }

exposure [METEOROL] The general surroundings of a site, with special reference to its openness to winds and sunshine. { ik'spō·zhər }

exsolution [GEOL] A phenomenon during which molten rock solutions separate when cooled. { ¦ek·sə'lü·shən }

exsolution lamellae [GEOL] Layers of sedimentary rock that solidify from solution by either precipitation or secretion. { ¦ek·sə'lü·shən lə'mel·ē }

exsurgence *See* resurgence. { ek'sər·jəns }

extended forecast [METEOROL] In general, a forecast of weather conditions for a period extending beyond 2 days from the day of issue. Also known as long-range forecast. { ik'stend·əd 'fór,kast }

extended-range forecast *See* medium-range forecast. { ik¦stend·əd ¦rānj 'fór,kast }

extended stream [HYD] A stream lengthened by the extension of its downstream course; the course is through a newly emerged land such as a coastal plain. { ik ¦stend·əd 'strēm }

extended valley [GEOL] **1.** A valley that is lengthened downstream either by a regression of the sea or by uplift of the coastal region. **2.** A valley formed by or containing an extended stream. { ik¦stend·əd 'val·ē }

extending flow [HYD] A glacial flow pattern in which velocity increases as the distance downstream becomes greater. { ik'stend·iŋ ,flō }

extensional fault *See* tension fault. { ik'sten·chən·əl 'fólt }

extension fracture [GEOL] A fracture that develops perpendicular to the direction of greatest stress and parallel to the direction of compression. { ik'sten·chən ,frak·chər }

extension joints [GEOL] Fractures that form parallel to a compressive force. { ik'sten·chən ˌjȯins }
external forcing [CLIMATOL] The influence on the earth system by solar radiation. { ikˌstərn·əl 'fȯrs·iŋ }
extinction [HYD] The drying up of lake by either water loss or destruction of the lake basin. { ek'stiŋk·shən }
extraordinary component *See* extraordinary wave. { ik'strȯr·dənˌer·ē kəm'pō·nənt }
extraordinary wave [GEOPHYS] Magneto-ionic wave component which, when viewed below the ionosphere in the direction of propagation, has clockwise or counterclockwise elliptical polarization respectively, accordingly as the earth's magnetic field has a positive or negative component in the same direction. Also known as X wave. { ik'strȯr·dənˌer·ē 'wāv }
extratropical cyclone [METEOROL] Any cyclone-scale storm that is not a tropical cyclone. Also known as extratropical low; extratropical storm. { ¦ek·strə¦träp·i·kəl 'sīˌklōn }
extratropical low *See* extratropical cyclone. { ¦ek·strə¦träp·i·kəl 'lō }
extratropical storm *See* extratropical cyclone. { ¦ek·strə¦träp·i·kəl 'stȯrm }
extravasation [GEOL] The eruption of lava from a vent in the earth. { ikˌstrav·ə'sā·shən }
extreme [CLIMATOL] The highest, and in some cases the lowest, value of a climatic element observed during a given period or during a given month or season of that period; if this is the whole period for which observations are available, it is the absolute extreme. { ek'strēm }
extrusion [GEOL] Emission of magma or magmatic materials at the surface of the earth. { ek'strü·zhən }
extrusive rock *See* volcanic rock. { ik'strü·siv 'räk }
exudation vein *See* segregated vein. { ˌek·syə'dā·shənˌvān }
eye coal [GEOL] Coal characterized by small, circular or elliptic structural disks that reflect light and are arranged in parallel planes either in or normal to the bedding. Also known as augen kohle; circular coal. { 'ī ˌkōl }
eye of the storm [METEOROL] The center of a tropical cyclone, marked by relatively light winds, confused seas, rising temperature, lowered relative humidity, and often by clear skies. { 'ī əv thə 'stōrm }
eye of the wind [METEOROL] The point or direction from which the wind is blowing. { 'ī əv thə 'wind }
eye wall [METEOROL] A zone at the periphery of the eye of the storm where winds reach their highest speed. { 'ī ˌwȯl }

F

fabric [GEOL] The spatial orientation of the elements of a sedimentary rock. { 'fab·rik }

face [GEOL] **1.** The main surface of a landform. **2.** The original surface of a layer of rock. { fās }

facet [GEOGR] Any part of an intersecting surface that constitutes a unit of geographic study, for example, a flat or a slope. { 'fas·ət }

faceted pebble [GEOL] A pebble with three or more faces naturally worn flat and meeting at sharp angles. { 'fas·əd·əd 'peb·əl }

faceted spur [GEOL] A spur or ridge with an inverted-V face resulting from faulting or from the trimming, beveling, or truncating motion of streams, waves, or glaciers. { 'fas·əd·əd 'spər }

facies [GEOL] Any observable attribute or attributes of a rock or stratigraphic unit, such as overall appearance or composition, of one part of the rock or unit as contrasted with other parts of the same rock or unit. { 'fā·shēz }

facies map [GEOL] A stratigraphic map indicating distribution of sedimentary facies within a specific geologic unit. { 'fā· shēz ,map }

facsimile chart [METEOROL] Any graphic form of weather information, usually a type of synoptic chart, which has been reproduced by facsimile equipment. Also known as fax chart; fax map. { fak'sim·ə·lē ,chärt }

fahlband [GEOL] A stratum containing metal sulfides; occurs in crystalline rock. { 'fäl,bänt }

fair [METEOROL] Generally descriptive of pleasant weather conditions, with regard for location and time of year; it is subject to popular misinterpretation, for it is a purely subjective description; when this term is used in forecasts of the U.S. Weather Bureau, it is meant to imply no precipitation, less than 0.4 sky cover of low clouds, and no other extreme conditions of cloudiness or windiness. { fer }

fair-weather cumulus *See* cumulus humilis cloud. { 'fer ,weth·ər 'kyü·myə·ləs }

Falkland Current [OCEANOGR] An ocean current flowing northward along the Argentine coast. { 'fȯk·lənd 'kə·rənt }

fallback [GEOL] Fragmented ejecta from an impact or explosion crater during formation which partly refills the true crater almost immediately. { 'fȯl,bak }

falling tide *See* ebb tide. { 'fȯl·iŋ 'tīd }

fall line [GEOL] **1.** The zone or boundary between resistant rocks of older land and weaker strata of plains. **2.** The line indicated by the edge over which a waterway suddenly descends, as in waterfalls. { 'fȯl ,līn }

fallout winds [METEOROL] Tropospheric winds that carry the radioactive fallout materials, observed by standard winds-aloft observation techniques. { 'fȯl,au̇t ,winz }

fall-streak hole [METEOROL] A hole occurring in a cloud layer of supercooled water droplets; produced by the local freezing of some of the droplets and their coversion into fallout, frequently in a streak form. { 'fȯl ,strēk ,hōl }

fall streaks *See* virga. { 'fȯl ,strēks }

Fallstreifen *See* virga. { 'fäl,strīf·ən }

fall wind [METEOROL] A strong, cold, downslope wind, differing from a foehn in that the initially cold air remains relatively cold despite adiabatic warming upon descent, and from the gravity wind in that it is a larger-scale phenomenon prerequiring an accumulation of cold air at high elevations. { 'fȯl ,wind }

false bedding [GEOL] An inclined bedding produced by currents. { ¦fȯls 'bed·iŋ }
false cirrus cloud [METEOROL] Cirrus composed of the debris of the upper frozen parts of a cumulonimbus cloud. { ¦fȯls 'sir·əs ˌklaủd }
false cleavage [GEOL] **1.** A weak cleavage at an angle to the slaty cleavage. **2.** Spaced surfaces about a millimeter apart along which a rock splits. { ¦fȯls 'klēv·ij }
false drumlin *See* rock drumlin. { ¦fȯls 'drəm·lən }
false ice foot [OCEANOGR] Ice that forms along a beach terrace and attaches to it just above the high-water mark; derived from water coming from melting snow above the terrace. { ¦fȯls 'īs ˌfủt }
false oolith *See* pseudo-oolith. { ¦fȯls 'ō,ō,līth }
false warm sector [METEOROL] The sector, in a horizontal plane, between the occluded front and a secondary cold front of an occluded cyclone. { ¦fȯls 'wȯrm ˌsek·tər }
fan [GEOL] A gently sloping, fan-shaped feature usually found near the lower termination of a canyon. { fan }
fan fold [GEOL] A fold of strata in which both limbs are overturned, forming a syncline or anticline. { 'fan ˌfōld }
fanglomerate [GEOL] Coarse material in an alluvial fan, with the rock fragments being only slightly worn. { fan'gläm·ə·rət }
fan-shaped delta *See* arcuate delta. { 'fan ˌshāpt 'del·tə }
farinaceous [GEOL] Of a rock or sediment, having a texture that is mealy, soft, and friable, for example, a limestone or a pelagic ooze. { ¦far·ə¦nā·shəs }
farmer's year [CLIMATOL] In the United Kingdom, the 12-month period starting with the Sunday nearest March 1. { 'fär·mərz 'yir }
fassaite [GEOCHEM] $Ca(Mg,Ti,Al)(Al,Si)_2O_6$ A mineral found in the millimeter-sized rocklets or refractory inclusions of carbonaceous chondrite meteorites. { 'fas·əˌyīt }
fastest mile [METEOROL] Over a specified period (usually the 24-hour observational day), the fastest speed, in miles per hour, of any mile of wind, with its accompanying direction. { ¦fas·təst 'mīl }
fast ice [HYD] Any type of sea, river, or lake ice attached to the shore (ice foot, ice shelf), beached (shore ice), stranded in shallow water, or frozen to the bottom of shallow waters (anchor ice). Also known as landfast ice. [OCEANOGR] Sea ice generally remaining in the position where originally formed and sometimes attaining a considerable thickness; it is attached to the shore or over shoals where it may be held in position by islands, grounded icebergs, or polar ice. Also known as coastal ice; coast ice. { ¦fast ¦īs }
fast ion *See* small ion. { ¦fast 'īˌän }
fathom [OCEANOGR] The common unit of depth in the ocean, equal to 6 feet (1.8288 meters). { 'fath·əm }
fathom curve *See* isobath. { 'fath·əm ˌkərv }
fault [GEOL] A fracture in rock along which the adjacent rock surfaces are differentially displaced. { fȯlt }
fault basin [GEOL] A region depressed in relation to surrounding regions and separated from them by faults. { 'fȯlt ˌbās·ən }
fault block [GEOL] A rock mass that is bounded by faults; the faults may be elevated or depressed and not necessarily the same on all sides. { 'fȯlt ˌbläk }
fault-block mountain *See* block mountain. { 'fȯlt ˌbläk ˌmaủnt·ən }
fault breccia [GEOL] The assembly of angular fragments found frequently along faults. Also known as dislocation breccia. { 'fȯlt ˌbrech·ə }
fault cliff *See* fault scarp. { 'fȯlt ˌklif }
fault escarpment *See* fault scarp. { 'fȯlt eˌskärp·mənt }
faulting [GEOL] The fracturing and displacement processes which produce a fault. { 'fȯl·tiŋ }
fault ledge *See* fault scarp. { 'fȯlt ˌlej }
fault line [GEOL] Intersection of the fault surface with the surface of the earth or any other horizontal surface of reference. Also known as fault trace. { 'fȯltˌlīn }
fault-line scarp [GEOL] A cliff produced when a soft rock erodes against hard rock at a fault. { 'fȯltˌlīn ˌskärp }

fault plane [GEOL] A planar fault surface. { 'fölt ˌplān }
fault rock [GEOL] A rock often found along a fault plane and made up of fragments formed by the crushing and grinding which accompany a dislocation. { 'fölt ˌräk }
fault scarp [GEOL] A steep cliff formed by movement along one side of a fault. Also known as cliff of displacement; fault cliff; fault escarpment; fault ledge. { 'fölt ˌskärp }
fault separation [GEOL] Apparent displacement of a fault measured on the basis of disrupted linear features. { 'fölt ˌsep·əˌrā·shən }
fault strike [GEOL] The angular direction, with respect to north, of the intersection of the fault surface with a horizontal plane. { 'fölt ˌstrīk }
fault system [GEOL] Two or more fault sets which interconnect. { 'fölt ˌsis·təm }
fault terrace [GEOL] A step on a slope, produced by displacement of two parallel faults. { 'fölt ˌter·əs }
fault throw [GEOL] The amount of vertical displacement of rocks due to faulting. { 'fölt ˌthrō }
fault trace *See* fault line. { 'fölt ˌtrās }
fault trap [GEOL] Oil or gas reservoir formed by a structural trap limited in one or more directions by subterranean geological faulting. { 'fölt ˌtrap }
fault-trough lake *See* sag pond. { 'fölt ˌtröf ˌlāk }
fault vein [GEOL] A mineral vein deposited in a fault fissure. { 'fölt ˌvān }
fault wall [GEOL] The mass of rock on a particular side of a fault. { 'fölt ˌwöl }
fault zone [GEOL] A fault expressed as an area of numerous small fractures. Also known as distributed fault. { 'fölt ˌzōn }
faunizone [GEOL] A bed characterized by fossils of a particular assemblage of fauna. { 'fön·əˌzōn }
fax chart *See* facsimile chart. { 'faks ˌchärt }
fax map *See* facsimile chart. { 'faks ˌmap }
feather *See* barb. { 'feth·ər }
feather joint [GEOL] One of a series of joints in a fault zone formed by shear and tension. Also known as pinnate joint. { 'feth·ər ˌjöint }
fecal pellets [GEOL] Mainly the excreta of invertebrates occurring in marine deposits and as fossils in sedimentary rocks. Also known as castings. { 'fē·kəl 'pel·əts }
feeder [GEOL] A small ore-bearing vein which merges with a larger one. [HYD] *See* tributary. { 'fēd·ər }
feeder beach [GEOL] A beach that is artificially widened and nourishes downdrift beaches by natural littoral currents or forces. { 'fēd·ər ˌbēch }
feeder current [OCEANOGR] A current which flows parallel to the shore before converging with other such currents and forming the neck of a rip current. { 'fēd·ər ˌkə·rənt }
feeding ground *See* drainage basin. { 'fēd·iŋ ˌgraünd }
feldspathization [GEOL] Formation of feldspar in a rock usually as a result of metamorphism leading toward granitization. { ˌfelˌspa·thə'zā·shən }
feldspathoid [GEOL] Aluminosilicates of sodium, potassium, or calcium that are similar in composition to feldspars but contain less silica than the corresponding feldspar. { 'felˌspaˌthöid }
felsenmeer [GEOL] A flat or gently sloping veneer of angular rock fragments occurring on moderate mountain slopes above the timber line. { 'felz·ənˌmer }
felty [GEOL] Referring to a pilotaxitic texture in which the microlites are randomly oriented. { 'fel·tē }
fen [GEOGR] Peat land covered by water, especially in the upper regions of old estuaries and around lakes, that can be drained only artificially. { fen }
fen peat *See* low-moor peat. { 'fen ˌpēt }
ferricrete [GEOL] A conglomerate of surficial sand and gravel held together by iron oxide resulting from percolating solutions of iron salts. { 'fer·əˌkrēt }
ferriferous [GEOL] Of a sedimentary rock, iron-rich. { fə'rif·ə·rəs }
Ferrod [GEOL] A suborder of the soil order Spodosol that is well drained and contains an iron accumulation with little organic matter. { 'feˌräd }
fetch [OCEANOGR] **1.** The distance traversed by waves without obstruction. **2.** An area of the sea surface over which seas are generated by a wind having a constant speed

and direction. **3.** The length of the fetch area, measured in the direction of the wind in which the seas are generated. Also known as generating area. { fech }

fiard *See* fjard. { fē'ärd }

fibratus [METEOROL] A cloud species characterized by a fine hairlike or striated composition, the filaments of which are usually distinctly separated from each other; the extremities of these filaments are always thin and never terminated by tufts or hooks. Also known as filosus. { fi'bräd·əs }

Fibrist [GEOL] A suborder of the soil order Histosol, consisting mainly of recognizable plant residues or sphagnum moss and saturated with water most of the year. { 'fī·brəst }

fibrous ice *See* acicular ice. { 'fī·brəs 'īs }

FIDO [METEOROL] A system for artificially dissipating fog, in which gasoline or other fuel is burned at intervals along an airstrip to be cleared. Derived from fog investigation dispersal operations. { 'fī·dō }

fiducial temperature [METEOROL] That temperature at which, in a specified latitude, the reading of a particular barometer does not require temperature or latitude correction. { fə'dü·shəl 'tem·prə·chər }

field [GEOL] A region or area with a particular mineral resource, for example, a gold field. [GEOPHYS] That area or space in which a particular geophysical effect, such as gravity or magnetism, occurs and can be measured. { fēld }

field capacity [HYD] The maximum amount of water that a soil can retain after gravitational water has drained away. { 'fēld kə,pas·əd·ē }

field changes [METEOROL] With regard to thunderstorm electricity, the rapid variations in the vertical component of the electric field strength at the earth's surface. { 'fēld ,chānj·əz }

field focus [GEOPHYS] The total area or volume occupied by an earthquake source. { 'fēld ,fō·kəs }

field geology [GEOL] The study of rocks and rock materials in their environment and in their natural relations to one another. { 'fēld jē,äl·ə·jē }

field moisture [HYD] Water in the ground above the water table. { 'fēld ,mȯis·chər }

field pressure [GEOL] The pressure of natural gas in the underground formations from which it is produced. { 'fēld ,presh·ər }

figure of the earth [GEOD] A precise geometric shape of the earth. { 'fig·yər əv thē 'ərth }

figure stone *See* agalmatolite. { 'fig·yər ,stōn }

filiform lapilli *See* Pele's hair. { 'fil·ə,fȯrm lə'pil·ē }

fillet lightning *See* ribbon lightning. { 'fil·ət ,līt·niŋ }

filling [METEOROL] An increase in the central pressure of a pressure system on a constant-height chart, or an analogous increase in height on a constant-pressure chart; the term is commonly applied to a low rather than to a high. { 'fil·iŋ }

fill terrace *See* alluvial terrace. { 'fil ,ter·əs }

filosus *See* fibratus. { fī'lō·səs }

fine admixture [GEOL] The smaller size grades of a sediment of mixed size grades. { ¦fīn 'ad,miks·chər }

fine earth [GEOL] A soil which can be passed through a 2-millimeter sieve without grinding its primary particles. { ¦fīn 'ərth }

fine gravel [GEOL] Gravel consisting of particles with a diameter range of 1 to 2 millimeters. { ¦fīn 'grav·əl }

fine sand [GEOL] Sand grains between 0.25 and 0.125 millimeter in diameter. { ¦fīn 'sand }

finger [GEOL] The tendency for gas which is displacing liquid hydrocarbons in a heterogeneous reservoir rock system to move forward irregularly (in fingers), rather than on a uniform front. { 'fiŋ·gər }

finger coal *See* natural coke. { 'fiŋ·gər ,kōl }

finger lake [HYD] A long, comparatively narrow lake, generally glacial in origin; may occupy a rock basin in the floor of a glacial trough or be confined by a morainal dam across the lower end of the valley. { 'fiŋ·gər ,lāk }

finite closed aquifer [HYD] The part of a subterranean reservoir containing water (aquifer) in which the aquifer is limited (finite), with no water flow across the exterior reservoir boundary. { ¦fī,nīt ¦klōzd 'ak·wə·fər }
fiord *See* fjord. { fyȯrd }
fireclay [GEOL] **1.** A clay that can resist high temperatures without becoming glassy. **2.** Soft, embedded, white or gray clay rich in hydrated aluminum silicates or silica and deficient in alkalies and iron. { 'fīr ,klā }
fire fountain *See* lava fountain. { 'fīr ,fau̇nt·ən }
fire weather [METEOROL] The state of the weather with respect to its effect upon the kindling and spreading of forest fires. { 'fīr ,weth·ər }
firn [HYD] Material transitional between snow and glacier ice; it is formed from snow after existing through one summer melt season and becomes glacier ice when its permeability to liquid water drops to zero. Also known as firn snow. { fərn }
firn basin *See* firn field. { 'fərn ,bās·ən }
firn field [HYD] The accumulation area or upper region of a glacier where snow accumulates and firn is secreted. Also known as firn basin. { 'fərn ,fēld }
firn ice *See* iced firn. { 'fərn ,īs }
firnification [HYD] The process of firn formation from snow and of transformation of firn into glacier ice. { ,fər·nə·fə'kā·shən }
firn limit *See* firn line. { 'fərn ,lim·ət }
firn line [GEOL] **1.** The regional snow line on a glacier. **2.** The line that divides the ablation area of a glacier from the accumulation area. Also known as firn limit. { 'fərn ,līn }
firn snow *See* firn; old snow. { 'fərn ,snō }
first bottom [GEOL] The floodplain of a river, below the first terrace. { ¦fərst 'bäd·əm }
first gust [METEOROL] The sharp increase in wind speed often associated with the early mature stage of a thunderstorm cell; it occurs with the passage of the discontinuity zone which is the boundary of the cold-air downdraft. { ¦fərst 'gəst }
first-order climatological station [METEOROL] A meteorological station at which autographic records or hourly readings of atmospheric pressure, temperature, humidity, wind, sunshine, and precipitation are made, together with observations at fixed hours of the amount and form of clouds and notes on the weather. { ¦fərst ,ȯrd·ər klī·mə·tə¦läj·ə·kəl 'stā·shən }
first-order relief [GEOGR] Relief features on the largest scale, consisting of continental platforms and ocean basins. { 'fərst ¦ȯr·dər 'ri·lēf }
first-order station [METEOROL] After U.S. National Weather Service practice, any meteorological station that is staffed in whole or in part by National Weather Service (Civil Service) personnel, regardless of the type or extent of work required of that station. { ¦fərst ,ȯrd·ər 'stā·shən }
firth *See* estuary. { fərth }
Fischer ellipsoid of 1960 [GEOD] The reference ellipsoid of which the semimajor axis is 6,378,166.000 meters, the semiminor axis is 6,356,784.298 meters, and the flattening or ellipticity is 1/298.3. Also known as Fischer spheroid of 1960. { 'fish·ər ə'lip,sȯid əv ¦nīn,tēn 'siks·tē }
Fischer spheroid of 1960 *See* Fischer ellipsoid of 1960. { 'fish·ər 'sfir,ȯid əv ¦nīn,tēn 'siks·tē }
fissile [GEOL] Capable of being split along the line of the grain or cleavage plane. { 'fis·əl }
fission-track dating [GEOL] A method of dating geological specimens by counting the radiation-damage tracks produced by spontaneous fission of uranium impurities in minerals and glasses. { 'fish·ən ,trak ,dād·iŋ }
fissure [GEOL] **1.** A high, narrow cave passageway. **2.** An extensive crack in a rock. { 'fish·ər }
fissure spring *See* artesian spring. { 'fish·ər ,spriŋ }
fissure system [GEOL] A group of fissures having the same age and generally parallel strike and dip. { 'fish·ər ,sis·təm }

fissure vein [GEOL] A mineral deposit in a cleft or crack in the rock material of the earth's crust. { 'fish·ər ˌvān }

fitness figure [METEOROL] In the United Kingdom, a measure of the "fitness" of the weather at an airport for the safe landing of aircraft; the figure F is computed on the basis of corrected values of visibility and cloud height; observed visibility is adjusted according to intensity of precipitation, and cloud height is corrected for height of nearby obstructions and cloud amount; further corrections are applied for the cross-runway component of the wind. Also known as fitness number. { 'fit·nəs ˌfig·yər }

fitness number *See* fitness figure. { 'fit·nəs ˌnəm·bər }

five-and-ten system [METEOROL] The most common system for representing wind speed, to the nearest 5 knots, in symbolic form on synoptic charts, consisting of drawing the appropriate number of half-barbs, barbs, and pennants from the end of the wind-direction shaft; in this system, a half-barb represents 5 knots, a barb 10 knots, a pennant 50 knots. { ¦fīv ən 'ten ˌsis·təm }

five-day forecast [METEOROL] A forecast of the average weather conditions and large-scale synoptic features in a 5-day period; a type of extended forecast. { 'fīv ˌdā 'fȯrˌkast }

fixed-level chart *See* constant-height chart. { ¦fikst ˌlev·əl 'chärt }

fjard [GEOGR] A small, narrow, and irregular inlet of the sea with low banks on either side. Also spelled fiard. { fē'ärd }

fjord [GEOGR] A narrow, deep inlet of the sea between high cliffs or steep slopes. Also spelled fiord. { fyȯrd }

fjord valley [GEOGR] A deep, narrow channel occupied by the sea and extending inland about 50–100 miles (80–160 kilometers). { 'fyȯrd ˌval·ē }

flaggy [GEOL] **1.** Of bedding, consisting of strata 4–40 inches (10–100 centimeters) in thickness. **2.** Of rock, tending to split into layers of suitable thickness (0.4–2 inches or 1–5 centimeters) for use as flagstones. { 'flag·ē }

flagstone [GEOL] **1.** A hard, thin-bedded sandstone, firm shale, or other rock that splits easily along bedding planes or joints into flat slabs. **2.** A piece of flagstone used for making pavement or covering the side of a house. { 'flagˌstōn }

flamboyant structure [GEOL] The optical continuity of crystals or grains as disturbed by a structure that is divergent. { flam'bȯi·ənt 'strək·chər }

flan [METEOROL] In Scotland, a sudden gust or squall of wind from land. { flan }

Flanders storm [METEOROL] In England, a heavy fall of snow coming with the south wind. { 'flan·dərz ˌstȯrm }

Flandrian transgression [OCEANOGR] The rapid rise of the North Sea between 8000 and 3000 B.C. from about 180 feet (55 meters) below to about 20 feet (6 meters) below its present level. { 'flan·drē·ən tranz'gresh·ən }

flank *See* limb. { flaŋk }

flaser [GEOL] Streaky layer of parallel, scaly aggregates that surrounds the lenticular bodies of granular material in flaser structure; caused by pressure and shearing during metamorphism. { 'flā·zər }

flaser gabbro [GEOL] A cataclastic gabbro that contains augen of feldspar or quartz surrounded by flakes of mica or chlorite. { 'flā·zər 'gaˌbrō }

flaser structure [GEOL] **1.** A metamorphic structure in which small lenses and layers of granular material are surrounded by a matrix of sheared, crushed material, resembling a crude flow structure. Also known as pachoidal structure. **2.** A primary sedimentary structure consisting of fine-sand or silt lenticles that are aligned and cross-bedded. { 'flā·zər ˌstrək·chər }

flash flood [HYD] A sudden local flood of short duration and great volume; usually caused by heavy rainfall in the immediate vicinity. { ¦flash ¦fləd }

flat [GEOGR] A level tract of land. [GEOL] *See* mud flat. { flat }

flat-lying [GEOL] Of mineral deposits and coal seams, having a relatively flat dip, up to 5°. { 'flat ˌlī·iŋ }

flattening [GEOD] The ratio of the difference between the equatorial and polar radii of the earth; the flattening of the earth is the ellipticity of the spheroid; the magnitude of the flattening is sometimes expressed as the numerical value of the reciprocal of the flattening. Also known as compression. { 'flat·ən·iŋ }

flaw [METEOROL] An English nautical term for a sudden gust or squall of wind. [OCEANOGR] **1.** The seaward edge of fast ice. **2.** A shore lead just outside fast ice. { flȯ }

flaxseed ore [GEOL] Iron ore composed of disk-shaped oauolites that have been partially flattened parallel to the bedding plane. { 'flak,sēd ,ȯr }

F layer [GEOPHYS] An ionized layer in the F region of the ionosphere which consists of the F_1 and F_2 layers in the day hemisphere, and the F_2 layer alone in the night hemisphere; it is capable of reflecting radio waves to earth at frequencies up to about 50 megahertz. { 'ef ,lā·ər }

F_1 layer [GEOPHYS] The ionosphere layer beneath the F_2 layer during the day, at a virtual height of 120–180 miles (200–300 kilometers), being closest to earth around noon; characterized by a distinct maximum of free-electron density, except at high latitudes during winter, when the layer is not detectable. { ¦ef 'wən ,lā·ər }

F_2 layer [GEOPHYS] The highest constantly observable ionosphere layer, characterized by a distinct maximum of free-electron density at a virtual height from about 135 miles (225 kilometers) in the polar winter to more than 240 miles (400 kilometers) in daytime near the magnetic equator. Also known as Appleton layer. { ¦ef 'tü ,lā·ər }

flexible sandstone [GEOL] A variety of itacolumite that consists of fine grains and occurs in thin layers. { ,flek·sə·bəl 'san,stōn }

flexural slip [GEOL] The slipping of sedimentary strata along bedding planes during folding, producing disharmonic folding and, when extreme, découllement. Also known as bedding-plane slip. { 'flek·shə·rəl 'slip }

flexure [GEOL] **1.** A broad, domed structure. **2.** A fold. { 'flek·shər }

flight briefing *See* pilot briefing. { 'flīt ,brēf·iŋ }

flight forecast [METEOROL] An aviation weather forecast for a specific flight. { 'flīt ,fȯr,kast }

flight visibility [METEOROL] Average visibility in a forward direction from an aircraft in flight. { 'flīt ,viz·ə,bil·əd·ē }

flight-weather briefing *See* pilot briefing. { 'flīt ,weth·ər ,brēf·iŋ }

flint clay [GEOL] A hard, smooth, flintlike fireclay; when it is ground, it develops no plasticity, and it breaks with conchoidal fracture. { 'flint ,klā }

flist [METEOROL] In Scotland, a keen blast or shower accompanied by a squall. { flist }

float [GEOL] An isolated, displaced rock or ore fragment. { flōt }

float coal [GEOL] Small, irregularly shaped, isolated deposits of coal embedded in sandstone or in siltstone. Also known as raft. { 'flōt ,kōl }

floater *See* drift bottle. { 'flōd·ər }

floating ice [OCEANOGR] Any form of ice floating in water, including grounded ice and drifting land ice. { ¦flōd·iŋ 'īs }

float mineral [GEOL] Small ore fragments carried from the ore bed by the action of water or by gravity; a float mineral often leads to discovery of mines. { 'flōt ,min·rəl }

floccus [METEOROL] A cloud species in which each element is a small tuft with a rounded top and a ragged bottom. { 'fläk·əs }

floe [OCEANOGR] A piece of floating sea ice other than fast ice or glacier ice; may consist of a single fragment or of many consolidated fragments, but is larger than an ice cake and smaller than an ice field. Also known as ice floe. { flō }

floeberg [OCEANOGR] A mass of hummocked ice formed by the piling up of many ice floes by lateral pressure; an extreme form of pressure ice; may be more than 50 feet (15 meters) high and resemble an iceberg. { 'flō,bərg }

floe till [GEOL] **1.** A glacial till resulting from the intact deposition of a grounded iceberg in a lake bordering an ice sheet. **2.** A lacustrine clay with boulders, stones, and other glacial matter dropped into it by melting icebergs. Also known as berg till. { 'flō ,til }

flood [HYD] The condition that occurs when water overflows the natural or artificial

confines of a stream or other body of water, or accumulates by drainage over low-lying areas. [OCEANOGR] The highest point of a tide. { fləd }
flood basalt *See* plateau basalt. { 'fləd bə,sȯlt }
flood basin [GEOL] **1.** The tract of land actually submerged during the highest known flood in a specific region. **2.** The flat, wide area lying between a low, sloping plain and the natural levee of a river. { 'fləd ,bās·ən }
flood current [OCEANOGR] The tidal current associated with the increase in the height of a tide. { 'fləd ,kə·rənt }
flooded stream *See* drowned stream. { ¦fləd·əd 'strēm }
flood flow [HYD] Stream discharge during a flood. { 'fləd ,flō }
flood fringe *See* pondage land. { 'fləd ,frinj }
flood icing *See* icing. { 'fləd ,īs·iŋ }
flooding ice *See* icing. { 'fləd·iŋ ,īs }
floodplain [GEOL] The relatively smooth valley floors adjacent to and formed by alluviating rivers which are subject to overflow. { 'fləd,plān }
floodplain splay [GEOL] A small alluvial fan or other outspread deposit formed where an overloaded stream breaks through a levee (artificial or natural) and deposits its material (often coarse-grained) on the floodplain. Also known as channel splay. { 'fləd,plān ,splā }
flood plane [HYD] The position of a stream's water surface during a particular flood. { 'fləd ,plān }
flood routing [HYD] The process of computing the progressive time and shape of a flood wave at successive points along a river. Also known as storage routing; streamflow routing. { 'fləd ,rüd·iŋ }
flood stage [HYD] The stage, on a fixed river gage, at which overflow of the natural banks of the stream begins to cause damage in any portion of the reach for which the gage is used as an index. { 'fləd ,stāj }
flood tide [OCEANOGR] **1.** That period of tide between low water and the next high water. **2.** A tide at its highest point. { 'fləd ,tīd }
floor [GEOL] **1.** The rock underlying a stratified or nearly horizontal deposit, corresponding to the footwall of more steeply dipping deposits. **2.** A horizontal, flat ore body. { flȯr }
Florida Current [OCEANOGR] A fast current that sets through the Straits of Florida to a point north of Grand Bahama Island, where it joins the Antilles Current to form the Gulf Stream. { 'flär·ə·də ,kə·rənt }
flowage [GEOL] *See* flow. [HYD] Flooding of water onto adjacent land. { 'flō·ij }
flow [GEOL] Any rock deformation that is not instantly recoverable without permanent loss of cohesion. Also known as flowage; rock flowage. { flō }
flowage line [GEOL] A contour line at the edge of a body of water, such as a reservoir, representing a given water level. { 'flō·ij ,līn }
flow banding [GEOL] An igneous rock structure resulting from flowing of magmas or lavas and characterized by alternation of mineralogically unlike layers. { 'flō ,band·iŋ }
flow breccia [GEOL] A breccia formed with the movement of lava flow while the flow is still in motion. { 'flō ,brech·ə }
flow cleavage [GEOL] Rock cleavage in which solid flow of rock accompanies recrystallization. Also known as slaty cleavage. { 'flō ,klē·vij }
flow earth *See* solifluction mantle. { 'flō ,ərth }
flow fold [GEOL] Folding in beds, composed of relatively plastic rock, that assume any shape impressed upon them by the more rigid surrounding rocks or by the general stress pattern of the deformed zone; there are no apparent surfaces of slip. { 'flō ,fōld }
flow line [HYD] A contour of the water level around a body of water. { 'flō ,līn }
flow regime [HYD] A range of streamflows having similar bed forms, flow resistance, and means of transporting sediment. { 'flō rə,zhēm }
flow slide [GEOL] A slide of waterlogged material in which the slip surface is not well defined. { 'flō ,slīd }

flowstone [GEOL] Deposits of calcium carbonate that accumulated against the walls of a cave where water flowed on the rock. { 'flō,stōn }

flow structure [GEOL] A primary sedimentary structure due to underwater slump or flow. { 'flō ,strək·chər }

flow velocity [GEOL] In soil, a vector point function used to indicate rate and direction of movement of water through soil per unit of time, perpendicular to the direction of flow. { 'flō və'läs·əd·ē }

fluctuation [OCEANOGR] **1.** Wavelike motion of water. **2.** The variations of water-level height from mean sea level that are not due to tide-producing forces. { ,flək·chə'wā·shən }

fluid geometry [GEOL] Fluid distribution in reservoir strata controlled by rock effective pore-size distribution, rock wettability characteristics in relation to the fluids present, method of producing saturation, and rock heterogeneity. { ¦flü·əd jē'äm·ə·trē }

fluid saturation [GEOL] Measure of the gross void space in a reservoir rock that is occupied by a fluid. { ¦flü·əd ,sach·ə'rā·shən }

flume [GEOL] A ravine with a stream flowing through it. { flüm }

fluolite *See* pitchstone. { 'flü·ə,līt }

fluoridation [GEOCHEM] Formation in rocks of fluorine-containing minerals such as fluorite or topaz. { flür·ə'dā·shən }

flurry [METEOROL] A brief shower of snow accompanied by a gust of wind, or a sudden, brief wind squall. { 'flər·ē }

flushing period [HYD] The interval of time required for a quantity of water equal to the volume of a lake to pass through the lake outlet; computed by dividing lake volume by mean flow rate of the outlet. { 'fləsh·iŋ ,pir·ē·əd }

flute [GEOL] **1.** A natural groove running vertically down the face of a rock. **2.** A groove in a sedimentary structure formed by the scouring action of a turbulent, sediment-laden water current, and having a steep upcurrent end. { flüt }

flute cast [GEOL] A raised, oblong, or subconical welt on the bottom surface of a siltstone or sandstone bed formed by the filling of a flute. { 'flüt ,kast }

Fluvent [GEOL] A suborder of the soil order Entisol that is well-drained with visible marks of sedimentation and no identifiable horizons; occurs in recently deposited alluvium along streams or in fans. { 'flü·vənt }

fluvial [HYD] **1.** Pertaining to or produced by the action of a stream or river. **2.** Existing, growing, or living in or near a river or stream. { 'flü·vē·əl }

fluvial cycle of erosion *See* normal cycle. { 'flü·vē·əl 'sī·kəl əv ə'rō·zhən }

fluvial deposit [GEOL] A sedimentary deposit of material transported by or suspended in a river. { ¦flü·vē·əl di'päz·ət }

fluvial sand [GEOL] Sand laid down by a river or stream. { ¦flü·vē·əl 'sand }

fluvial soil [GEOL] Soil laid down by a river or stream. { ¦flü·vē·əl 'sȯil }

fluviatile [GEOL] Resulting from river action. { 'flü·vē·ə,tīl }

fluviology [HYD] The science of rivers. { flü·vē'äl·ə·jē }

fluviomorphology *See* river morphology. { ¦flü·vē·ō·mȯr'fäl·ə·jē }

flying veins [GEOL] A series of mineral-deposit veins which overlap or intersect in a branchlike pattern. { ¦flī·iŋ 'vānz }

fly-off *See* evapotranspiration. { 'flī,ȯf }

flysch [GEOL] Deposits of dark, fine-grained, thinly bedded sandstone shales and of clay, thought to be deposited by turbidity currents and originally defined as rock formations on the northern and southern borders of the Alps. { flīsh }

foam *See* pumice. { fōm }

foam crust [HYD] A snow surface feature that looks like small overlapping waves, like sea foam on a beach, occurring during the ablation of the snow surface and may further develop into a more pronounced wedge-shaped form, known as plowshares. { 'fōm ,krəst }

foam line [OCEANOGR] The front of a wave as it moves toward the shore, after the wave has broken. { 'fōm ,līn }

foam mark [GEOL] A surface sedimentary structure comprising a pattern of barely visible ridges and hollows formed where wind-driven sea foam passes over a surface of wet sand. { 'fōm ˌmärk }

focus [GEOPHYS] The center of an earthquake and the origin of its elastic waves within the earth. { 'fō·kəs }

foehn [METEOROL] A warm, dry wind on the lee side of a mountain range, the warmth and dryness being due to adiabatic compression as the air descends the mountain slopes. Also spelled föhn. { fān }

foehn air [METEOROL] The warm, dry air associated with foehn winds. { 'fān ˌer }

foehn cloud [METEOROL] Any cloud form associated with a foehn, but usually signifying only those clouds of the lenticularis species formed in the lee wave parallel to the mountain ridge. { 'fān ˌklau̇d }

foehn cyclone [METEOROL] A cyclone formed (or at least enhanced) as a result of the foehn process on the lee side of a mountain range. { 'fān 'sīˌklōn }

foehn island [METEOROL] An isolated area where the foehn has reached the ground, in contrast to the surrounding area where foehn air has not replaced colder surface air. { 'fān 'ī·lənd }

foehn nose [METEOROL] As seen on a synoptic surface chart, a typical deformation of the isobars in connection with a well-developed foehn situation; a ridge of high pressure is produced on the windward slopes of the mountain range, while a foehn trough forms on the lee side; the isobars "bulge" correspondingly, giving a noselike configuration. { 'fān ˌnōz }

foehn pause [METEOROL] **1.** A temporary cessation of the foehn at the ground, due to the formation or intrusion of a cold air layer which lifts the foehn above the valley floor. **2.** The boundary between foehn air and its surroundings. { 'fān ˌpȯz }

foehn period [METEOROL] The duration of continuous foehn conditions at a given location. { 'fān ˌpir·ē·əd }

foehn phase [METEOROL] One of three stages to describe the development of the foehn in the Alps: the preliminary phase, when cold air at the surface is separated from warm dry air aloft by a subsidence inversion; the anticyclonic phase, when the warm air reaches a station as the result of the cold air flowing out from the plain; and the stationary phase or cyclonic phase, when the foehn wall forms and the downslope wind becomes appreciable. { 'fān ˌfāz }

foehn storm [METEOROL] A type of destructive storm which frequently occurs in October in the Bavarian Alps. { 'fān ˌstȯrm }

foehn trough [METEOROL] The dynamic trough formed in connection with the foehn. { 'fān ˌtrȯf }

foehn wall [METEOROL] The steep leeward boundary of flat, cumuliform clouds formed on the peaks and upper windward sides of mountains during foehn conditions. { 'fān ˌwȯl }

fog [METEOROL] Water droplets or, rarely, ice crystals suspended in the air in sufficient concentration to reduce visibility appreciably. { fäg }

fogbank [METEOROL] A fairly well-defined mass of fog observed in the distance, most commonly at sea. { 'fägˌbaŋk }

fog deposit [HYD] The deposit of an ice coating on exposed surfaces by a freezing fog. { 'fäg dīˌpäz·ət }

fog dispersal [METEOROL] Artificial dissipation of a fog by means such as seeding or heating. { 'fäg diˌspərs·əl }

fog drip [HYD] Water dripping to the ground from trees or other objects which have collected the moisture from drifting fog; the dripping can be as heavy as light rain, as sometimes occurs among the redwood trees along the coast of northern California. { 'fäg ˌdrip }

fog drop [METEOROL] An elementary particle of fog, physically the same as a cloud drop. Also known as fog droplet. { 'fäg ˌdräp }

fog droplet *See* fog drop. { 'fäg ˌdräp·lət }

fog horizon [METEOROL] The top of a fog layer which is confined by a low-level temperature inversion so as to give the appearance of the horizon when viewed from above against the sky; the true horizon is usually obscured by the fog in such instances. { 'fäg hə,rīz·ən }

fog scale [METEOROL] A classification of fog intensity based on its effectiveness in decreasing horizontal visibility; such practice is not current in United States weather observing procedures. { 'fäg ,skāl }

fog wind [METEOROL] Humid east wind which crosses the divide of the Andes east of Lake Titicaca and descends on the west in violent squalls; probably the same as puelche. { 'fäg ,wind }

föhn *See* foehn. { fān }

fold [GEOL] A bend in rock strata or other planar structure, usually produced by deformation; folds are recognized where layered rocks have been distorted into wavelike form. { fōld }

fold belt *See* orogenic belt. { 'fōld ,belt }

folding [GEOL] Compression of planar structure in the formation of fold structures. { 'fōld·iŋ }

fold system [GEOL] A group of folds with common trends and characteristics. { 'fōld ,sis·təm }

foliaceous [GEOL] Having a leaflike or platelike structure composed of thin layers of minerals. { ,fō·lē'ā·shəs }

foliated ice [HYD] Large masses of ice which grow in thermal contraction cracks in permafrost. Also known as ice wedge. { 'fō·lē,ād·əd 'īs }

foliation [GEOL] A laminated structure formed by segregation of different minerals into layers that are parallel to the schistosity. { ,fō·lē'ā·shən }

Folist [GEOL] A suborder of the soil order Histosol, consisting of wet forest litter resting on rock or rubble. { 'fäl·əst }

following wind [METEOROL] **1.** A wind blowing in the direction of ocean-wave advance. **2.** *See* tailwind. { ¦fäl·ə·wiŋ 'wind }

foothills [GEOGR] A region of relatively low, rounded hills at the base of, or on the periphery of, a mountain range. { 'fu̇t,hilz }

footwall [GEOL] The mass of rock that lies beneath a fault, an ore body, or a mine working. Also known as heading side; heading wall; lower plate. { 'fu̇t,wȯl }

ford [HYD] A shallow and usually narrow part of a stream, estuary, or other body of water that may be crossed; for example, by wading or by a wheeled land vehicle. { fȯrd }

forearc [GEOL] The area between the trench and the volcanic arc of a subduction zone. { 'fȯr,ärk }

forebulge [GEOL] An uplift at the edge of a glacier caused by tilting of the lithosphere. { 'fȯr,bəlj }

forecast [METEOROL] A statement of expected future meteorological occurrences. { 'fȯr,kast }

forecasting [METEOROL] Procedures for extrapolation of the future characteristics of weather on the basis of present and past conditions. { 'fȯr,kast·iŋ }

forecast period [METEOROL] The time interval for which a forecast is made. { 'fōr,kast ,pir·ē·əd }

forecast-reversal test [METEOROL] A test used to evaluate the adequacy of a given method of forecast verification; the same verification method is applied, simultaneously, to a given forecast and to a fabricated forecast of opposite conditions; comparison of the verification scores gives an indication of the value of the verification system. { ¦fȯr,kast ri'vər·səl ,test }

forecast verification [METEOROL] Any process for determining the accuracy of a weather forecast by comparing the predicted weather with the observed weather of the forecast period; used to test forecasting skills and methods. { ¦fȯr,kast ,ver·ə·fə'kā·shən }

foredeep [GEOL] **1.** A long, narrow depression that borders an orogenic belt, such as an island arc, on the convex side. **2.** *See* exogeosyncline. { 'fȯr,dēp }

foredune [GEOL] A coastal dune or ridge that is parallel to the shoreline of a large lake or ocean and is stabilized by vegetation. { 'fȯr,dün }

foreland [GEOGR] An extensive area of land jutting out into the sea. [GEOL] **1.** A lowland area onto which piedmont glaciers have moved from adjacent mountains. **2.** A stable part of a continent bordering an orogenic or mobile belt. { 'fȯr·lənd }

foreland facies *See* shelf facies. { 'fȯr·lənd ˌfā·shēz }

Forel scale [OCEANOGR] A scale of yellows, greens, and blues for recording the color of sea water as seen against the white background of a Secchi disk. { fȯ'rel ˌskāl }

forerunner [OCEANOGR] Low, long-period ocean swell which commonly precedes the main swell from a distant storm, especially a tropical cyclone. { 'fȯrˌrən·ər }

foreset bed [GEOL] One of a series of inclined symmetrically arranged layers of a cross-bedding unit formed by deposition of sediments that rolled down a steep frontal slope of a delta or dune. { 'fȯrˌset ˌbed }

foreshock [GEOPHYS] A tremor which precedes a larger earthquake or main shock. { 'fȯrˌshäk }

foreshore [GEOL] The zone that lies between the ordinary high- and low-watermarks and is daily traversed by the rise and fall of the tide. Also known as beach face. { 'fȯrˌshȯr }

forest climate *See* humid climate. { 'fär·əst ¦klī·mət }

forest wind [METEOROL] A light breeze which blows from forests toward open country on calm clear nights. { ¦fär·əst 'wind }

forked lightning [GEOPHYS] A common form of lightning, in a cloud-to-ground discharge, which exhibits downward-directed branches from the main lightning channel. { ¦fȯrkt 'līt·niŋ }

formation [GEOL] Any assemblage of rocks which have some common character and are mappable as a unit. { fȯr'mā·shən }

formation factor [GEOCHEM] The ratio between the conductivity of an electrolyte and that of a rock saturated with the same electrolyte. Also known as resistivity factor. [GEOL] A function of the porosity and internal geometry of a reservoir rock system, expressed as $F = \phi^{-m}$, where ϕ is the fractional porosity of the rock, and m is the cementation factor (pore-opening reduction). { fȯr'mā·shən ˌfak·tər }

formation pressure *See* reservoir pressure. { fȯr'mā·shən ˌpresh·ər }

formation resistivity [GEOPHYS] Electrical resistivity of reservoir formations measured by electrical log sondes; used for clues to formation lithography and fluid content. { fȯr'mā·shən riˌzis'tiv·əd·ē }

formation water [HYD] Water present with petroleum or gas in reservoirs. Also known as oil-reservoir water. { fȯr'mā·shən ˌwȯd·ər }

Forrel cell [METEOROL] A type of atmospheric circulation in which air moves away from the thermal equator at low latitude levels and in the opposite direction in higher latitudes. { fə'rel ˌsel }

fortnightly tide [OCEANOGR] A tide occurring at intervals of one-half the period of oscillation of the moon, approximately 2 weeks. { ˌfȯrt¦nīt·lē 'tīd }

Forty Saints' storm [METEOROL] A southerly gale in Greece, occurring a little before the equinox in March. { ¦fȯrd·ē ¦sāns ˌstȯrm }

forward scatter [GEOPHYS] The scattering of radiant energy into the hemisphere of space bounded by a plane normal to the direction of the incident radiation and lying on the side toward which the incident radiation was advancing. { ¦fȯr·wərd 'skad·ər }

fossil dune [GEOL] An ancient desert dune. { ¦fäs·əl 'dün }

fossil fuel [GEOL] Any hydrocarbon deposit that may be used for fuel; examples are petroleum, coal, and natural gas. { ¦fäs·əl 'fyül }

fossil ice [HYD] **1.** Relatively old ground ice found in regions of permafrost. **2.** Underground ice in regions where present-day temperatures are not low enough to have formed it. { ¦fäs·əl 'īs }

fossil permafrost *See* passive permafrost. { ¦fäs·əl 'pər·məˌfrȯst }

fossil reef [GEOL] An ancient reef. { ¦fäs·əl 'rēf }

fossil resin [GEOL] A natural resin in geologic deposits which is an exudate of long-buried plant life; for example, amber, retinite, and copal. { ¦fäs·əl 'rez·ən }

fossil soil *See* paleosol. { ¦fäs·əl 'sȯil }

fossil turbulence [METEOROL] Inhomogeneities of temperature and humidity remaining in the air after the motion which produced them has subsided and the density has become uniform; causes scattering of radio waves, and lumpy clouds when air is rising. { ¦fäs·əl 'tər·byə·ləns }

fossil wax *See* ozocerite. { ¦fäs·əl 'waks }

foundation coefficient [GEOPHYS] A coefficient which expresses how much stronger the effect of an earthquake is on a given rock than it would be on an undisturbed crystalline rock under the same conditions. { faủn'dā·shən ,kō·i,fish·ənt }

founder [GEOL] To sink under water either by depression of the land or by rise of sea level, especially in reference to large crustal masses, islands, or significant portions of continents. { 'faủn·dər }

4-D chart [METEOROL] A chart showing the field of D values (deviations of the actual altitudes along a constant-pressure surface from the standard atmosphere altitude of that surface) in terms of the three dimensions of space and one of time; it is a form of a four-dimensional display of pressure altitude; the space dimensions are represented by D-value contours, and the time dimension is provided by tau-value lines. { ¦fȯr ¦dē ,chärt }

four-way dip [GEOPHYS] In seismic prospecting, dip determined by an array of geophones which are set up at points in four directions from a shot point; three of the locations are essential and the fourth serves as a control point. { 'fȯr ,wā 'dip }

fowan [METEOROL] A dry, scorching wind of the United Kingdom and the Isle of Man. { faủ·ən }

fractoconformity [GEOL] The relation between conformable strata, where faulting of the older beds occurs at the same time as deposition of the newer beds. { ¦frak·tō·kən'fȯr·məd·ē }

fracture [GEOL] A crack, joint, or fault in a rock due to mechanical failure by stress. Also known as rupture. { 'frak·shər }

fracture cleavage [GEOL] Cleavage that occurs in deformed but only slightly metamorphosed rocks along closely spaced, parallel joints and fractures. { 'frak·shər ,klēv·ij }

fracture-plane inclination [GEOL] Gradient or inclination of the plane of fracture formed in a reservoir formation. { 'frak·shər ,plān ,in·klə'nā·shən }

fracture system [GEOL] A stress-related group of contemporaneous fractures. { 'frak·shər ,sis·təm }

fracture zone [GEOL] An elongate zone on the deep-sea floor that is of irregular topography and often separates regions of different depths; frequently crosses and displaces the mid-oceanic ridge by faulting. { 'frak·shər ,zōn }

fractus [METEOROL] A cloud species in which the cloud elements are irregular but generally small in size, and which presents a ragged, shredded appearance, as if torn; these characteristics change ceaselessly and often rapidly. { 'frak·təs }

fragipan [GEOL] A dense, natural subsurface layer of hard soil with relatively slow permeability to water, mostly because of its extreme density or compactness rather than its high clay content or cementation. { 'fraj·ə,pan }

fragmentation nucleus [METEOROL] A tiny ice particle broken from a large ice crystal, serving as an ice nucleus; that is, a growth center for a new ice crystal. { ,frag·mən'tā·shən ,nü·klē·əs }

framboid [GEOL] A microscopic aggregate of pyrite grains, often occurring in spheroidal clusters. { 'fram,bȯid }

framework [GEOL] **1.** In a sediment or sedimentary rock, the rigid arrangement created by particles that support one another at contact points. **2.** A fixed calcareous structure impervious to waves, built by sedentary organisms (for example, sponges, corals, and bryozoans) in a high-energy environment. { 'frām,wərk }

Franconian [GEOL] A North American stage of geologic time; the middle Upper Cambrian. { fraŋ'kō·nē·ən }

frazil [HYD] Ice crystals which form in supercooled water that is too turbulent to permit coagulation of the crystals into sheet ice. { 'fra·zəl }

frazil ice [HYD] A spongy or slushy accumulation of frazil in a body of water. Also known as needle ice. { 'fra·zəl ,īs }

free air *See* free atmosphere. { ¦frē 'er }

free-air anomaly *See* free-air gravity anomaly. { ˈfrē ˌer əˈnäm·ə·lē }

free-air gravity anomaly [GEOPHYS] A measure of the mass excesses and deficiencies within the earth; calculated as the difference between the measured gravity and the theoretical gravity at sea level and a free-air coefficient determined by the elevation of the measuring station. Also known as free-air anomaly. { ˈfrē ˌer ˈgrav·əd·ē əˌnäm·ə·lē }

free-air temperature [METEOROL] Temperature of the atmosphere, obtained by a thermometer located so as to avoid as completely as practicable the effects of extraneous heating. { ˈfrē ˌer ˈtem·prə·chər }

free atmosphere [GEOPHYS] That portion of the earth's atmosphere, above the planetary boundary layer, in which the effect of the earth's surface friction on the air motion is negligible and in which the air is usually treated (dynamically) as an ideal fluid. Also known as free air. { ¦frē ˈat·məˌsfir }

free-burning coal *See* noncaking coal. { ¦frē ˌbərn·iŋ ˈkōl }

free end *See* free face. { ¦frē ˈend }

free face [GEOL] A vertical or steeply inclined layer of rock from which weathered material falls to form talus at its base. { ¦frē ˈfās }

free foehn *See* high foehn. { ¦frē ¦fān }

free meander [HYD] A stream meander that displaces itself very easily by lateral corrasion. { ¦frē mēˈan·dər }

freestone [GEOL] Stone, particularly a thick-bedded, even-textured, fine-grained sandstone, that breaks freely and is able to be cut and dressed with equal facility in any direction without tending to split. { ˈfrēˌstōn }

free-traveling wave *See* progressive wave. { ˈfrē ˌtrav·ə·liŋ ˈwāv }

free-water content *See* water content. { ˈfrē ˌwȯd·ər ˌkän·tent }

free-water elevation *See* water table. { ˈfrē ˌwȯd·ər el·əˌvā·shən }

free-water surface *See* watertable. { ˈfrē ˌwȯd·ər ¦sər·fəs }

freeze-out lake [HYD] A shallow lake which may be deeply frozen over for long periods of time. { ˈfrēz ˌau̇t ˌlāk }

freeze-up [HYD] The formation of a continuous ice cover on a body of water. { ˈfrēzˌəp }

freezing drizzle [METEOROL] Drizzle that falls in liquid form but freezes upon impact with the ground to form a coating of glaze. { ¦frēz·iŋ ¦driz·əl }

freezing level [METEOROL] The lowest altitude in the atmosphere over a given location, at which the air temperature is 0°C; the height of the 0°C constant-temperature surface. { ¦frēz·iŋ ¦lev·əl }

freezing-level chart [METEOROL] A synoptic chart showing the height of the 0°C constant-temperature surface by means of contour lines. { ˈfrēz·iŋ ˌlev·əl ˌchärt }

freezing nucleus [METEOROL] Any particle which, when present within a mass of supercooled water, will initiate growth of an ice crystal about itself. { ˈfrēz·iŋ ˌnü·klē·əs }

freezing precipitation [METEOROL] Any form of liquid precipitation that freezes upon impact with the ground or exposed objects; that is, freezing rain or freezing drizzle. { ¦frēz·iŋ prəˌsip·əˈtā·shən }

freezing rain [METEOROL] Rain that falls in liquid form but freezes upon impact to form a coating of glaze upon the ground and on exposed objects. { ¦frēz·iŋ ˈrān }

F region [GEOPHYS] The general region of the ionosphere in which the F_1 and F_2 layers tend to form. { ˈef ˌrē·jən }

fresh [GEOL] Unweathered in reference to a rock or rock surface. [METEOROL] Pertaining to air which is stimulating and refreshing. { fresh }

fresh breeze [METEOROL] In the Beaufort wind scale, a wind whose speed is 17 to 21 knots (19 to 24 miles per hour, or 31 to 39 kilometers per hour). { ¦fresh ˈbrēz }

freshet [HYD] **1.** The annual spring rise of streams in cold climates as a result of melting snow. **2.** A flood resulting from either rain or melting snow; usually applied only to small streams and to floods of minor severity. **3.** A small fresh-water stream. { ˈfresh·ət }

fresh gale [METEOROL] In the Beaufort wind scale, a wind whose speed is from 34 to 40 knots (39 to 46 miles per hour, or 63 to 74 kilometers per hour). { ¦fresh ˈgāl }

fresh ice *See* newly formed ice. { ¦fresh 'īs }
fresh water [HYD] Water containing no significant amounts of salts, such as in rivers and lakes. { ¦fresh 'wȯd·ər }
Fresnian [GEOL] A North American stage of upper Eocene geologic time, above Narizian and below Refugian. { 'frez·nē·ən }
friagem [METEOROL] A period of cold weather in the middle and upper parts of the Amazon Valley and in eastern Bolivia, occurring during the dry season in the Southern Hemisphere winter. Also known as vriajem. { 'frī·ə,jem }
frictional *See* cohesionless. { 'frik·shən·əl }
friction crack [GEOL] A short, crescent-shaped crack in glaciated rock produced by a localized increase in friction between rock and ice, oriented transverse to the direction of ice flow. { 'frik·shən ,krak }
friction depth [OCEANOGR] The depth at which the velocity of wind-driven current becomes negligible compared to the surface velocity; sometimes referred to as the depth of the Ekman layer. { 'frik·shən ,depth }
friction layer *See* surface boundary layer. { 'frik·shən ,lā·ər }
friction velocity [METEOROL] A reference wind velocity defined by the relation

$$u \sqrt{|\tau/\rho|}$$

where τ is the Reynolds stress, ρ the density, and u the friction velocity. { 'frik·shən və'läs·əd·ē }
fringe joint [GEOL] A small-scale joint peripheral to, and usually at a 5–25° angle from the face of, the master joint. { 'frinj ,jȯint }
fringe ore [GEOL] Ore located on the outer boundary of a mineralization pattern or halo. Also known as halo ore. { 'frinj ,ȯr }
fringe region [METEOROL] The upper portion of the exosphere, where the cone of escape equals or exceeds 180°; in this region the individual atoms have so little chance of collision that they essentially travel in free orbits, subject to the earth's gravitation, at speeds imparted by the last collision. Also known as spray region. { 'frinj ,rē·jən }
fringing reef [GEOL] A coral reef attached directly to or bordering the shore of an island or continental landmass. { ¦frin·jiŋ 'rēf }
frog storm [METEOROL] The first bad weather in spring after a warm period. Also known as whippoorwill storm. { 'fräg ,stȯrm }
front [METEOROL] A sloping surface of discontinuity in the troposphere, separating air masses of different density or temperature. { frənt }
front abutment pressure [GEOPHYS] The release of energy in the superincumbent strata above the seam induced by the extraction of the seam. { ¦frənt ə'bət·mənt ,presh·ər }
frontal apron *See* outwash plain. { ¦frənt·əl 'ā·prən }
frontal contour [METEOROL] The line of intersection of a front (frontal surface) with a specified surface in the atmosphere, usually a constant-pressure surface; with respect to only one surface, this line is usually called the front. { ¦frənt·əl 'kän·tu̇r }
frontal cyclone [METEOROL] Any cyclone associated with a front; often used synonymously with wave cyclone or with extratropical cyclone (as opposed to tropical cyclones, which are nonfrontal). { ¦frənt·əl 'sī,klōn }
frontal fog [METEOROL] Fog associated with frontal zones and frontal passages. { ¦frənt·əl 'fäg }
frontal inversion [METEOROL] A temperature inversion in the atmosphere, encountered upon vertical ascent through a sloping front (or frontal zone). { ¦frənt·əl in'vər·zhən }
frontal lifting [METEOROL] The forced ascent of the warmer, less-dense air at and near a front, occurring whenever the relative velocities of the two air masses are such that they converge at the front. { ¦frənt·əl 'lift·iŋ }
frontal occlusion *See* occluded front. { ¦frənt·əl ə'klü·zhən }
frontal passage [METEOROL] The passage of a front over a point on the earth's surface. { ¦frənt·əl 'pas·ij }
frontal plain *See* outwash plain. { ¦frənt·əl ¦plān }
frontal precipitation [METEOROL] Any precipitation attributable to the action of a front;

used mainly to distinguish this type from air-mass precipitation and orographic precipitation. { ¦frənt·əl prə,sip·ə'tā·shən }

frontal profile [METEOROL] The outline of a front as seen on a vertical cross section oriented normal to the frontal surface. { ¦frənt·əl 'prō,fīl }

frontal strip [METEOROL] The presentation of a front, on a synoptic chart, as a frontal zone; that is, two lines, rather than a single line, are drawn to represent the boundaries of the zone; a rare usage. { 'frənt·əl ,strip }

frontal system [METEOROL] A system of fronts as they appear on a synoptic chart. { 'frənt·əl ,sis·təm }

frontal thunderstorm [METEOROL] A thunderstorm associated with a front; limited to thunderstorms resulting from the convection induced by frontal lifting. { 'frənt·əl 'thən·dər,stȯrm }

frontal wave [METEOROL] A horizontal, wavelike deformation of a front in the lower levels, commonly associated with a maximum of cyclonic circulations in the adjacent flow; it may develop into a wave cyclone. { 'frənt·əl ,wāv }

frontal zone [METEOROL] The three-dimensional zone or layer of large horizontal density gradient, bounded by frontal surfaces and surface front. { 'frənt·əl ,zōn }

frontogenesis [METEOROL] **1.** The initial formation of a frontal zone or front. **2.** The increase in the horizontal gradient of an air mass property, mainly density, and the formation of the accompanying features of the wind field that typify a front. { ¦frən·tō¦jen·ə·səs }

frontogenetic function [METEOROL] A kinematic measure of the tendency of the flow in an air mass to increase the horizontal gradient of a conservative property. { ¦frən·tō·jə¦ned·ik 'fəŋk·shən }

frontolysis [METEOROL] **1.** The dissipation of a front or frontal zone. **2.** In general, a decrease in the horizontal gradient of an air mass property, principally density, and the dissipation of the accompanying features of the wind field. { ,frən'täl·ə·səs }

front slope *See* scarp slope. { 'frənt ,slōp }

frost [HYD] A covering of ice in one of its several forms, produced by the sublimation of water vapor on objects colder than 32°F (0°C). { frȯst }

frost action [GEOL] **1.** The weathering process caused by cycles of freezing and thawing of water in surface pores, cracks, and other openings. **2.** Alternate or repeated cycles of freezing and thawing of water contained in materials; the term is especially applied to disruptive effects of this action. { 'frȯst ,ak·shən }

frost boil [GEOL] **1.** An accumulation of water and mud released from ground ice by accelerated spring thawing. **2.** A low mound formed by local differential frost heaving at a location most favorable for the formation of segregated ice and accompanied by the absence of an insulating cover of vegetation. { 'frȯst ,bȯil }

frost bursting *See* congelifraction. { 'frȯst ,bərst·iŋ }

frost churning *See* congeliturbation. { 'frȯst ,chərn·iŋ }

frost climate [CLIMATOL] The coldest temperature province in C. W. Thornthwaite's climatic classification: the climate of the ice cap regions of the earth, that is, those regions perennially covered with snow and ice. { 'frȯst ,klī·mət }

frost day [METEOROL] An observational day on which frost occurs. { 'frȯst ,dā }

frost feathers *See* ice feathers. { 'frȯst ,feth·ərz }

frost flakes *See* ice fog. { 'frȯst ,flāks }

frost flowers *See* ice flowers. { 'frȯst ,flau̇·ərz }

frost fog *See* ice fog. { 'frȯst ,fäg }

frost hazard [METEOROL] The risk of damage by frost, expressed as the probability or frequency of killing frost on different dates during the growing season, or as the distribution of dates of the last killing frost of spring or the first of autumn. { 'frȯst ,haz·ərd }

frost heaving [GEOL] The lifting and distortion of a surface due to internal action of frost resulting from subsurface ice formation; affects soil, rock, pavement, and other structures. { 'frȯst ,hēv·iŋ }

frost hollow [METEOROL] A small, low-lying zone which experiences frequent and severe frosts owing to the accumulation of cold night air; often severe where hills block the afternoon sunshine. { 'fròst ˌhäl·ō }

frostless zone [METEOROL] The warmest part of a slope above a valley floor, lying between the layer of cold air which forms over the valley floor on calm clear nights and the cold hill tops or plateaus; the air flowing down the slopes is warmed by mixing with the air above ground level, and to some extent also by adiabatic compression. Also known as green belt; verdant zone. { 'fròst·ləs ˌzōn }

frost line [GEOL] **1.** The maximum depth of frozen ground during the winter. **2.** The lower limit of the permafrost. { 'fròst ˌlīn }

frost mound [GEOL] A hill and knoll associated with frozen ground in a permafrost region, containing a core of ice. Also known as soffosian knob; soil blister. { 'fròst ˌmau̇nd }

frost pocket [METEOROL] A parcel of cold air in a hollow or at a valley floor, occurring when nighttime terrestrial radiation is greatest on valley slopes. { 'fròst ˌpäk·ət }

frost point [METEOROL] The temperature to which atmospheric moisture must be cooled to reach the point of saturation with respect to ice. { 'fròst ˌpòint }

frost-point technique *See* 8D technique. { 'fròst ˌpòint tekˌnēk }

frost riving *See* congelifraction. { 'fròst ˌrīv·iŋ }

frost shattering *See* congelifraction. { 'fròst ˌshad·ə·riŋ }

frost smoke [METEOROL] **1.** A rare type of fog formed in the same manner as a steam fog, but at colder temperatures so that it is composed of ice particles instead of water droplets. **2.** *See* steam fog. { 'fròst ˌsmōk }

frost splitting *See* congelifraction. { 'fròst ˌsplid·iŋ }

frost stirring *See* congelifraction. { 'fròst ˌstər·iŋ }

frost table [GEOL] An irregular surface in the ground which, at any given time, represents the penetration of thawing into seasonally frozen ground. { 'fròst ˌtā·bəl }

frost thrusting [GEOL] Lateral dislocation of soil and rock materials by the action of freezing and resulting expansion of soil water. { 'fròst ˌthrəst·iŋ }

frost weathering *See* congelifraction. { 'fròst ˌweth·ə·riŋ }

frost wedging *See* congelifraction. { 'fròst ˌwej·iŋ }

frost zone *See* seasonally frozen ground. { 'fròst ˌzōn }

frozen fog *See* ice fog. { ¦frōz·ən 'fäg }

frozen ground [GEOL] Soil having a temperature below freezing, generally containing water in the form of ice. Also known as gelisol; merzlota; taele; tjaele. { ¦frōz·ən 'grau̇nd }

frozen precipitation [METEOROL] Any form of precipitation that reaches the ground in frozen form; that is, snow, snow pellets, snow grains, ice crystals, ice pellets, and hail. { ¦frōz·ən prəˌsip·ə'tā·shən }

fucoid [GEOL] A tunnellike marking on a sedimentary structure identified as a trace fossil but not referred to a described genus. { 'fyüˌkòid }

fulgurite [GEOL] A glassy, rootlike tube formed when a lightning stroke terminates in dry sandy soil; the intense heating of the current passing down into the soil along an irregular path fuses the sand. { 'fu̇l·gəˌrīt }

fuller's earth [GEOL] A natural, fine-grained earthy material, such as a clay, with high adsorptive power; consists principally of hydrated aluminum silicates; used as an adsorbent in refining and decolorizing oils, as a catalyst, and as a bleaching agent. { ¦fu̇l·ərz ¦ərth }

fully arisen sea *See* fully developed sea. { 'fu̇l·ē əˌriz·ən 'sē }

fully developed sea [OCEANOGR] The maximum ocean waves or sea state that can be produced by a given wind force blowing over sufficient fetch, regardless of duration. Also known as fully arisen sea. { 'fu̇l·ē diˌvel·əpt 'sē }

fumarole [GEOL] A hole, usually found in volcanic areas, from which vapors or gases escape. { 'fyü·məˌrōl }

fumulus [METEOROL] A very thin cloud veil at any level, so delicate that it may be almost invisible. { 'fyü·myə·ləs }

fundamental circle *See* primary great circle. { ¦fən·də¦ment·əl 'sər·kəl }

fundamental complex [GEOL] An agglomeration of metamorphic rocks underlying sedimentary or unmetamorphosed rocks; specifically, an agglomeration of Archean rocks supporting a geological column. { ¦fən·də¦ment·əl 'käm,pleks }
fundamental jelly *See* ulmin. { ¦fən·də¦ment·əl 'jel·ē }
fundamental strength [GEOPHYS] The maximum stress that a geological structure can withstand without creep under certain conditions but without reference to time. { ¦fən·də¦ment·əl 'streŋkth }
fundamental substance *See* ulmin. { ¦fən·də¦ment·əl 'səb·stəns }
funnel cloud [METEOROL] The popular term for the tornado cloud, often shaped like a funnel with the small end nearest the ground. { 'fən·əl ,klau̇d }
furiani [METEOROL] A southwest wind that blows in the vicinity of the Po River, Italy, and is vehement and short-lived, followed by a gale from the south or southeast. { fu̇·rē'ä·nē }
furnace cupola *See* cupola. { 'fər·nəs ,kyü·pə·lə }
fusain [GEOL] The local lithotype strands or patches, characterized by silky luster, fibrous structure, friability, and black color. Also known as mineral charcoal; mother-of-coal. { 'fyü,zān }
fusinite [GEOL] The micropetrological constituent of fusain which consists of carbonized woody tissue. { 'fyüz·ən,īt }
fusinization [GEOL] The process of formation of fusain in coal. { ,fyüz·ən·ə'zā·shən }
fusion crust [GEOL] A thin, glassy coating, usually black and rerely more than 1 millimeter thick, which is formed by ablation on the surface of a meteorite. { 'fyü·zhən ,krəst }

G

gaign [METEOROL] A cross-mountain wind that causes clouds to form on the crests of mountains in Italy. { gān }

gale [METEOROL] **1.** An unusually strong wind. **2.** In storm-warning terminology, a wind of 28–47 knots (52–87 kilometers per hour). **3.** In the Beaufort wind scale, a wind whose speed is 28–55 knots (52–102 kilometers per hour). { gāl }

galerna *See* galerne. { gə'lər·nə }

galerne [METEOROL] A squally northwesterly wind that is cold, humid, and showery, occurring in the rear of a low-pressure area over the English Channel and off the Atlantic coast of France and northern Spain. Also known as galerna; galerno; giboulee. { gə'lərn }

galerno *See* galerne. { gə'lər·nō }

gale warning [METEOROL] A storm warning for marine interests of impending winds from 28 to 47 knots (52–87 kilometers per hour), signaled by two triangular red pennants by day, and a white lantern over a red lantern by night. { 'gāl ˌwȯrn·iŋ }

gallego [METEOROL] A cold, piercing, northerly wind in Spain and Portugal. { gə'yā·gō }

gallery [GEOL] **1.** A horizontal, or nearly horizontal, underground passage. **2.** A subsidiary passage in a cave at a higher level than the main passage. { 'gal·rē }

Galofaro [OCEANOGR] A whirlpool in the Strait of Messina, between Sicily and Italy; formerly called Charybdis. { ˌgäl·ə'fä·rō }

gangue [GEOL] The valueless rock or aggregates of minerals in an ore. { gaŋ }

gap [GEOGR] Any sharp, deep notch in a mountain ridge or between hills. { gap }

garbin [METEOROL] A sea breeze; in southwest Frange it refers to a southwesterly sea breeze which sets in about 9 a.m., reaches its maximum toward 2 p.m., and ceases about 5 p.m. { gär'ba }

garúa [METEOROL] A dense fog or drizzle from low stratus clouds on the west coast of South America, creating a raw, cold atmosphere that may last for weeks in winter, and supplying a limited amount of moisture to the area. Also known as camanchaca. { gä'rü·ə }

gas cap [GEOPHYS] The gas immediately in front of a meteoroid as it travels through the atmosphere. { 'gas ˌkap }

gas clathrate *See* gas hydrate. { ¦gas 'klathˌrāt }

gas column [GEOL] The difference in elevation between the highest and lowest parts of the various producing zones of a gas-producing formation. { 'gas ˌka·ləm }

gas-condensate reservoir [GEOL] Hydrocarbon reservoir in which conditions of temperature and pressure have resulted in the condensation of the heavier hydrocarbon constituents from the reservoir gas. { ¦gas 'känd·ənˌsāt ˌrez·əvˌwär }

gas-filled porosity [GEOL] A reservoir formation in which the pore space is filled by gas instead of liquid hydrocarbons. { 'gas ˌfild pə'räs·əd·ē }

gas floor [GEOL] In a sedimentary basin, the depth below which there is no economic accumulation of gaseous hydrocarbons. { 'gas ˌflȯr }

gash fracture [GEOL] Open gashes that are formed diagonally to a fault or fault zone. { 'gash ˌfrak·chər }

gas hydrate [GEOCHEM] A naturally occurring solid composed of crystallized water (ice) molecules, forming a rigid lattice of cages (a clathrate) with most of the cages containing a molecule of natural gas, mainly methane. Also known as clathrate hydrate, gas clathrate. { ¦gas 'hī,drāt }

gash vein [GEOL] A mineralized fissure that extends a short distance vertically. { 'gash ,vān }

gas pocket [GEOL] A gas-filled cavity in rocks, especially above an oil pocket. { 'gas ,päk·ət }

gas reservoir [GEOL] An accumulation of natural gas found with or near accumulations of crude oil in the earth's crust. { ¦gas ¦rez·əv,wär }

gas sand [GEOL] A stratum of sand or porous sandstone from which natural gas may be extracted. { 'gas ,sand }

gas spurt [GEOL] An accumulation of organic matter on certain strata caused by escaping gas. { 'gas ,spərt }

gas zone [GEOL] A rock formation containing gas under a pressure large enough to force the gas out if tapped from the surface. { 'gas ,zōn }

gathering ground *See* drainage basin. { 'gath·ə·riŋ ,graùnd }

gaufrage *See* plaiting. { gō'fräzh }

Gauss point *See* cardinal point. { 'gaùs ,pòint }

geanticline [GEOL] A broad land uplift; refers to the land mass from which sediments in a geosyncline are derived. { ,jē'ant·i,klīn }

gebli *See* ghibli. { 'geb·lē }

geg [METEOROL] A desert dust whirl of China and Tibet. { geg }

geking [OCEANOGR] Obtaining measurements of ocean movements with a geomagnetic electrokinetograph (GEK). { 'jē·kiŋ }

gelifluction [GEOL] The slow, continuous downslope movement of rock debris and water-saturated soil that occurs above frozen ground, as in most polar regions and in many high mountain ranges. Also known as congelifluction; gelisolifluction. { ¦jel·ə¦flək·shən }

gelifraction *See* congeliturbation. { ¦jel·ə¦frak·shən }

gelisol *See* frozen ground. { 'jel·ə,sòl }

gelisolifluction *See* gelifluction. { jə,las·ə'fiək·shən }

geliturbation *See* congeliturbation. { ,jel·ə,ter'bāsh·ən }

gelivation *See* congelifraction. { ¦jel·ə¦vā·shən }

gelose *See* ulmin. { 'je,lōs }

gemstone [GEOL] A mineral or petrified organic matter suitable for use in jewelry. { 'jem,stōn }

gending [METEOROL] A local dry wind in the northern plains of Java that resembles the foehn, caused by a wind crossing the mountains near the south coast and pushing between the volcanoes. { 'jen·diŋ }

general circulation [METEOROL] The complete statistical description of atmospheric motions over the earth. Also known as planetary circulation. { ¦jen·rəl ,sər·kyə'lā·shən }

generalized hydrostatic equation [GEOPHYS] The vertical component of the vector equation of motion in natural coordinates when the acceleration of gravity is replaced by the virtual gravity; for most purposes it is identical to the hydrostatic equation. { 'jen·rə,līzd ,hī·drə¦stad·ik i'kwā·zhən }

generalized transmission function [GEOPHYS] In atmospheric-radiation theory, a set of values, variable with wavelength, each one of which represents an average transmission coefficient for a small wavelength interval and for a specified optical path through the absorbing gas in question. { 'jen·rə,līzd tranz¦mish·ən ,fəŋk·shən }

genesis rocks [GEOL] Rocks that have retained their character from nearly 4.6×10^9 years ago, when planets were still occulting out of the cloud of dust and gas referred to as the solar nebula; examples are meteorites and asteroids. { 'jen·ə·səs ,räks }

genetic facies [GEOL] An ancient deposit of rocks which have been formed by similar sedimentary processes. { jə¦ned·ik 'fā·shēz }

Genoa cyclone [METEOROL] A cyclone, or low, which appears to have formed or developed in the vicinity of the Gulf of Genoa. Also known as Genoa low. { 'jen·ə·wə 'sī,klōn }

Genoa low *See* Genoa cyclone. { 'jen·ə·wə 'lō }

gentle breeze [METEOROL] In the Beaufort wind scale, a wind whose speed is from 7 to 10 knots (13–19 kilometers per hour). { ¦jent·əl 'brēz }

geo [GEOGR] A narrow coastal inlet bordered by steep cliffs. Also spelled gio. { 'gyō }

geobotanical prospecting [GEOL] The use of the distribution, appearance, and growth anomalies of plants in locating ore deposits. { ¦jē·ō·bə¦tan·ə·kəl 'präs·pek·tiŋ }

geocentric latitude [GEOD] Of a position on the earth's surface, the angle between a line to the center of the earth and the plane of the equator. { ¦jē·ō¦sen·trik 'lad·ə,tüd }

geocentric longitude [GEOD] At a position on the earth's surface, the angle between the plane of the reference meridian and a plane through the polar axis and a line from the position in question to the center of mass of the earth. { ¦jē·ō¦sen·trik 'län·jə,tüd }

geocentric vertical [GEOD] The direction of the radius vector drawn from the center of the earth through the location of the observer. Also known as geometric vertical. { ¦jē·ō¦sen·trik 'verd·ə·kəl }

geochemical anomaly [GEOCHEM] Above-average concentration of a chemical element in a sample of rock, soil, vegetation, stream, or sediment; indicative of nearby mineral deposit. { ¦jē·ō¦kem·ə·kəl ə'näm·ə·lē }

geochemical balance [GEOCHEM] The proportional distribution, and the migration rate, in the global fractionation of elements, minerals, or compounds; for example, the distribution of quartz in igneous rocks, its liberation by weathering, and its redistribution into sediments and, in solution, into lakes, rivers, and oceans. { ¦jē·ō¦kem·ə·kəl 'bal·əns }

geochemical cycle [GEOCHEM] During geologic changes, the sequence of stages in the migration of elements between the lithosphere, hydrosphere, and atmosphere. { ¦jē·ō¦kem·ə·kəl 'sī·kəl }

geochemical evolution [GEOCHEM] **1.** A change in any constituent of a rock beyond that amount present in the parent rock. **2.** A change in chemical composition of a major segment of the earth during geologic time, as the oceans. { ¦jē·ō¦kem·ə·kəl ,ev·ə'lü·shən }

geochemistry [GEOL] The study of the chemical composition of the various phases of the earth and the physical and chemical processes which have produced the observed distribution of the elements and nuclides in these phases. { ¦jē·ō¦kem·ə·strē }

geochron *See* isochron. { 'jē·ə,krän }

geochronology [GEOL] **1.** The dating of the events in the earth's history. **2.** A system of dating developed for the purposes of study of the earth's history. { ¦jē·ō·krə 'näl·ə·jē }

geochronometry [GEOL] The study of the absolute age of the rocks of the earth based on the radioactive decay of isotopes, such as ^{238}U, ^{235}U, ^{232}Th, ^{87}Rb, ^{40}K, and ^{14}C, present in minerals and rocks. { ¦jē·ō·krə'näm·ə·trē }

geocorona [GEOPHYS] The outermost part of the earth's atmosphere, consisting of extremely attenuated hydrogen extending to perhaps 15 earth radii, that emits Lyman-alpha radiation under the action of sunlight. { ¦jē·ō·kə'rō·nə }

geocosmogony [GEOL] The study of the origin of the earth. { ¦jē·ō,käz'mäj·ə·nē }

geode [GEOL] A roughly spheroidal, hollow body lined inside with inward-projecting, small crystals; found frequently in limestone beds but may occur in shale. { 'jē,ōd }

geodesy [GEOPHYS] A subdivision of geophysics which includes determination of the size and shape of the earth, the earth's gravitational field, and the location of points fixed to the earth's crust in an earth-referred coordinate system. { jē'äd·ə·sē }

geodetic astronomy [GEOD] The branch of geodesy which utilizes astronomic observations to extract geodetic information. { ¦jē·ə¦ded·ik ə'strän·ə·mē }

geodetic coordinates [GEOD] The quantities latitude, longitude, and elevation which define the position of a point on the surface of the earth with respect to the reference spheroid. { ¦jē·ə¦ded·ik kō'ȯrd·ən,əts }

geodetic datum [GEOD] A datum consisting of five quantities: the latitude and longitude of an initial point, the azimuth of a line from this point, and two constants necessary to define the terrestrial spheroid. { ¦jē·ə¦ded·ik 'dad·əm }

geodetic equator [GEOD] The great circle midway between the poles of revolution of the earth, connecting points of 0° geodetic latitude. { ¦jē·ə¦ded·ik i'kwād·ər }

geodetic gravimetry [GEOD] Worldwide relative measurements of gravitational acceleration used in geodetic studies of the earth. { ¦jē·ə¦ded·ik grə'vim·ə·trē }

geodetic latitude [GEOD] Angular distance between the plane of the equator and a normal to the spheroid; a geodetic latitude differs from the corresponding astronomical latitude by the amount of the meridional component of station error. Also known as geographic latitude; topographical latitude. { ¦jē·ə¦ded·ik 'lad·ə,tüd }

geodetic line [GEOD] The shortest line between any two points on the surface of the spheroid. { ¦jē·ə¦ded·ik 'līn }

geodetic longitude [GEOD] The angle between the plane of the reference meridian and the plane through the polar axis and the normal to the spheroid; a geodetic longitude differs from the corresponding astronomical longitude by the amount of the prime-vertical component of station error divided by the cosine of the latitude. Also known as geographic longitude. { ¦jē·ə¦ded·ik 'lān·jə·tüd }

geodetic meridian [GEOD] A line on a spheroid connecting points of equal geodetic longitude. Also known as geographic meridian. { ¦jē·ə¦ded·ik mə'rid·ē·ən }

geodetic parallel [GEOD] A line connecting points of equal geodetic latitude. Also known as geographic parallel. { ¦jē·ə¦ded·ik 'par·ə,lel }

geodetic position [GEOD] **1.** A point on the earth, the coordinates of which have been determined by triangulation from an initial station, whose location has been established as a result of astronomical observations, the coordinates depending upon the reference spheroid used. **2.** A point on the earth, defined in terms of geodetic latitude and longitude. { ¦jē·ə¦ded·ik pə'zish·ən }

geodynamic height *See* dynamic height. { ¦jē·ō·dī¦nam·ik 'hīt }

geodynamics [GEOPHYS] The branch of geophysics concerned with measuring, modeling, and interpreting the configuration and motion of the crust, mantle, and core of the earth. { ¦jē·ō·dī¦nam·iks }

geodynamo [GEOPHYS] The self-sustaining process responsible for maintaining the earth's magnetic field in which the kinetic energy of convective motion of the earth's liquid core is converted into magnetic energy. { ,jē·ō'dī·nə·mō }

geoeconomy [GEOGR] The study of economic conditions that are influenced by geographic factors. { ¦jē·ō·i¦kän·ə·mē }

geoelectricity *See* terrestrial electricity. { ¦jē·ō·i,lek'tris·əd·ē }

geoflex *See* orocline. { 'jē·ə,fleks }

geognosy [GEOL] The science dealing with the solid body of the earth as a whole, occurrences of minerals and rocks, and the origin of these and their relations. { jē'äg·nə·sē }

geographical coordinates [GEOGR] Spherical coordinates, designating both astronomical and geodetic coordinates, defining a point on the surface of the earth, usually latitude and longitude. Also known as terrestrial coordinates. { ¦jē·ə¦graf·ə·kəl kō'ȯrd·ən·əts }

geographical cycle *See* geomorphic cycle. { ¦jē·ə¦graf·ə·kəl 'sī·kəl }

geographical position [GEOGR] Any position on the earth defined by means of its geographical coordinates, either astronomical or geodetic. { ¦jē·ə¦graf·ə·kəl pə'zish·ən }

Geographic Information Systems [GEOGR] Computer-based technologies for the storage, manipulation, and analysis of geographically referenced information. { ,jē·ō¦graf·ik ,in·fər'mā·shən ,sis·təmz }

geographic latitude *See* geodetic latitude. { ¦jē·ə¦graf·ik 'lad·ə,tüd }

geographic longitude *See* geodetic longitude. { ¦jē·ə¦graf·ik 'län·jə,tüd }

geographic meridian *See* geodetic meridian. { ¦jē·ə¦graf·ik mə'rid·ē·ən }

geographic parallel *See* geodetic parallel. { ¦jē·ə¦graf·ik 'par·ə,lel }

geographic position [GEOGR] The position of a point on the surface of the earth expressed in terms of geographical coordinates either geodetic or astronomical. { ¦jē·ə¦graf·ik pə'zish·ən }

geographic range [GEOD] The extreme distance at which an object or light can be seen when limited only by the curvature of the earth and the heights of the object and the observer. { ¦jē·ə¦graf·ik 'rānj }

geographic vertical [GEOD] A line perpendicular to the surface of the geoid; it is the direction in which the force of gravity acts. { ¦jē·ə¦graf·ik 'vərd·ə·kəl }

geohydrology [HYD] The science dealing with underground water, often referred to as hydrogeology. { ¦jē·ō,hī'dräl·ə·jē }

geoid [GEOD] The figure of the earth considered as a sea-level surface extended continuously over the entire earth's surface. { 'jē,ȯid }

geoisotherm [GEOPHYS] The locus of points of equal temperature in the interior of the earth; a line in two dimensions or a surface in three dimensions. Also known as geotherm; isogeotherm. { ¦jē·ō'ī·sə,thərm }

geolith *See* rock-stratigraphic unit. { 'jē·ə,lith }

geologic age [GEOL] **1.** Any great time period in the earth's history marked by special phases of physical conditions or organic development. **2.** A formal geologic unit of time that corresponds to a stage. **3.** An informal geologic time unit that corresponds to any stratigraphic unit. { ¦jē·ə¦läj·ik 'āj }

geological oceanography [GEOL] The study of the floors and margins of the oceans, including descriptions of topography, composition of bottom materials, interaction of sediments and rocks with air and sea water, the effects of movements in the mantle on the sea floor, and action of wave energy in the submarine crust of the earth. Also known as marine geology; submarine geology. { ¦jē·ə¦läj·ə·kəl ,ō·shə 'näg·rə·fē }

geological survey [GEOL] **1.** An organization making geological surveys and studies. **2.** A systematic geologic mapping of a terrain. { ¦jē·ə¦läj·ə·kəl 'sər,vā }

geological transportation [GEOL] Shifting of material by the action of moving water, ice, or air. { ¦jē·ə¦läj·ə·kəl ,tranz·pər'tā·shən }

geologic climate *See* paleoclimate. { ¦jē·ə¦läj·ik 'klī·mət }

geologic column [GEOL] **1.** The vertical sequence of strata of various ages found in an area or region. Also known as column. **2.** The geologic time scale as represented by rocks. { ¦jē·ə¦läj·ik 'käl·əm }

geologic erosion *See* normal erosion. { ¦jē·ə¦läj·ik ə'rō·zhən }

geologic log [GEOL] A graphic presentation of the lithologic or stratigraphic units or both traversed by a borehole; used in petroleum and mining engineering as well as geological surveys. { ¦jē·ə¦läj·ik 'läg }

geologic map [GEOL] A representation of the geologic surface or subsurface features by means of signs and symbols and with an indicated means of orientation; includes nature and distribution of rock units, and the occurrence of structural features, mineral deposits, and fossil localities. { ¦jē·ə¦läj·ik 'map }

geologic noise [GEOPHYS] Disturbances in observed data caused by random inhomogeneities in surface and near-surface material. { ¦jē·ə¦läj·ik'nȯiz }

geologic province [GEOL] An area in which geologic history has been the same. { ¦jē·ə¦läj·ik 'präv·əns }

geologic section [GEOL] Any succession of rock units found at the surface or below ground in an area. Also known as section. { ¦jē·ə¦läj·ik 'sek·shən }

geologic structure [GEOL] The total structural features in an area. { ¦jē·ə¦läj·ik ,strək·chər }

geologic thermometry *See* geothermometry. { ¦jē·ə¦läj·ik thər'mäm·ə·trē }

geologic time [GEOL] The period of time covered by historical geology, from the end of the formation of the earth as a separate planet to the beginning of written history. { ¦jē·ə¦läj·ik 'tīm }

geologic time scale [GEOL] The relative age of various geologic periods and the absolute time intervals. { ¦jē·ə¦läj·ik 'tīm ,skāl }

geologist [GEOL] An individual who specializes in the geological sciences. { jē'äl·ə·jəst }

geomagnetic coordinates [GEOPHYS] A system of spherical coordinates based on the best fit of a centered dipole to the actual magnetic field of the earth. { ¦jē·ō·mag¦ned·ik kō'ȯrd·ən·əts }

geomagnetic cutoff [GEOPHYS] The minimum energy of a cosmic-ray particle able to reach the top of the atmosphere at a particular geomagnetic latitude. { ¦jē·ō·mag¦ned·ik 'kə,dȯf }

geomagnetic dipole [GEOPHYS] The magnetic dipole caused by the earth's magnetic field. { ¦jē·ō·mag¦ned·ik 'dī,pōl }

geomagnetic equator [GEOPHYS] That terrestrial great circle which is 90° from the geomagnetic poles. { ¦jē·ō·mag¦ned·ik i'kwād·ər }

geomagnetic field [GEOPHYS] The earth's magnetic field. { ¦jē·ō·mag¦ned·ik 'fēld }

geomagnetic field reversal [GEOPHYS] Reversed magnetization in sedimentary and igneous rock, that is, polarized opposite to the mean geomagnetic field. { ¦jē·ō·mag¦ned·ik 'fēld ,ri,vər·səl }

geomagnetic latitude [GEOPHYS] The magnetic latitude that a location would have if the field of the earth were to be replaced by a dipole field closely approximating it. { ¦jē·ō·mag¦ned·ik 'lad·ə,tüd }

geomagnetic longitude [GEOPHYS] Longitude that is determined around the geomagnetic axis instead of around the rotation axis of the earth. { ¦jē·ō·mag¦ned·ik 'län·jə,tüd }

geomagnetic meridian [GEOPHYS] A semicircle connecting the geomagnetic poles. { ¦jē·ō·mag¦ned·ik mə'rid·ē·ən }

geomagnetic noise [GEOPHYS] Unwanted frequencies caused by fluctuations in the geomagnetic field of the earth. { ¦jē·ō·mag¦ned·ik 'nȯiz }

geomagnetic pole [GEOPHYS] Either of two antipodal points marking the intersection of the earth's surface with the extended axis of a powerful bar magnet assumed to be located at the center of the earth and having a field approximating the actual magnetic field of the earth. { ¦jē·ō·mag¦ned·ik 'pōl }

geomagnetic reversal [GEOPHYS] Reversed magnetization of the earth's magnetic dipole. { ¦jē·ō·mag¦ned·ik ri'vər·səl }

geomagnetic secular variation *See* secular variation. { ¦jē·ō·mag¦ned·ik ¦sek·yə·lər ver·ē'ā·shən }

geomagnetic storm *See* magnetic storm. { ¦jē·ō·mag¦ned·ik 'stȯrm }

geomagnetic variation [GEOPHYS] Temporal changes in the geomagnetic field, both long-term (secular) and short-term (transient). { ¦jē·ō·mag¦ned·ik ver·ē'ā·shən }

geomagnetism [GEOPHYS] **1.** The magnetism of the earth. Also known as terrestrial magnetism. **2.** The branch of science that deals with the earth's magnetism. { ¦jē·ō'mag·nə,tiz·əm }

geometrical dip [GEOD] The vertical angle, at the eye of an observer, between the horizontal and a straight line tangent to the surface of the earth. { ¦jē·ə¦me·trə·kəl 'dip }

geometrical horizon [GEOD] The intersection of the celestial sphere and an infinite number of straight lines radiating from the eye of the observer and tangent to the earth's surface. { ¦jē·ə¦me·trə·kəl hə'rīz·ən }

geometric latitude *See* reduced latitude. { ¦jē·ə¦me·trik 'lad·ə,tüd }

geometric vertical *See* geocentric vertical. { ¦jē·ə¦me·trik 'vərd·ə·kəl }

geomorphic cycle [GEOL] The cycle of change in the surface configuration of the earth. Also known as cycle of erosion; geographical cycle. { ¦jē·ō¦mȯr·fik 'sī·kəl }

geomorphology [GEOL] The study of the origin of secondary topographic features which are carved by erosion in the primary elements and built up of the erosional debris. { ¦jē·ō·mȯr'fäl·ə·jē }

geophysical fluid dynamics [GEOPHYS] The study of the naturally occurring, large-scale flows in the atmosphere and oceans, such as in weather patterns, atmospheric fronts, ocean currents, coastal upwelling, and the El Niño phenomenon. { ,jē·ō¦fiz·ə·kəl ¦flü·əd dī'nam·iks }

geophysicist [GEOPHYS] An individual who specializes in geophysics. { ¦jē·ə'fiz·ə·sist }

geophysics [GEOL] The physics of the earth and its environment, that is, earth, air, and (by extension) space. { ¦jē·ə'fiz·iks }

geopotential height [GEOPHYS] The height of a given point in the atmosphere in units proportional to the potential energy of unit mass (geopotential) at this height, relative to sea level. { ¦jē·ō·pə'ten·chəl 'hīt }

geopotential number [GEOPHYS] The numerical value C that is assigned to a given geopotential surface when expressed in geopotential units (1 gpu = 1 meter × 1 kilogal). { ¦jē·ō·pə'ten·chəl 'nəm·bər }

geopotential surface [GEOPHYS] A surface of constant geopotential, that is, a surface along which a parcel of air could move without undergoing any changes in its potential energy. Also known as equigeopotential surface; level surface. { ¦jē·ō·pə'ten·chəl 'sər·fəs }

geopotential thickness [GEOPHYS] The difference in the geopotential height of two constant-pressure surfaces in the atmosphere, proportional to the appropriately defined mean air temperature between the two surfaces. { ¦jē·ō·pə'ten·chəl 'thik·nəs }

geopotential topography [GEOPHYS] The topography of any surface as represented by lines of equal geopotential; these lines are the contours of intersection between the actual surface and the level surfaces (which everywhere are normal to the direction of the force of gravity), and are spaced at equal intervals of dynamic height. Also known as absolute geopotential topography. { ¦jē·ō·pə'ten·chəl tə'päg·rə·fē }

geopotential unit [GEOPHYS] A unit of gravitational potential used in describing the earth's gravitational field; it is equal to the difference in gravitational potential of two points separated by a distance of 1 meter when the gravitational field has a strength of 10 meters per second squared and is directed along the line joining the points. Abbreviated gpu. { ¦jē·ō·pə'ten·chəl 'yü·nət }

geopressure [GEOPHYS] An unusually high pressure exerted by a subsurface formation. { 'jē·ō,presh·ər }

geopressurized geothermal system [GEOL] A geothermal system dominated by the presence of hot fluids under high pressure (brine plus methane) and having higher-than-normal temperatures because of their low thermal conductivity, the presence of interbedded shale layers, or the existence of local, exothermic chemical reactions. { ¦jē·ō'presh·ə,rīzd ¦jē·ō¦thər·məl 'sis·təm }

Georges Banks [GEOL] An elevation beneath the sea east of Cape Cod, Massachusetts. { ¦jór·jəz 'baŋks }

georgiaite [GEOL] Any of a group of North American tektites, 134 million years of age, found in Georgia. { 'jór·jə,īt }

geosere [GEOL] A series of ecological climax communities following each other in geologic time and changing in response to changing climate and physical conditions. { 'jē·ō,sir }

geosol [GEOL] A body of sediment or rock composed of one or more soil horizons. { 'jē·ə,sōl }

geosophy [GEOGR] The study of the nature and expression of geographical knowledge, both past and present. { je'äs·ə·fē }

geosphere [GEOL] **1.** The solid mass of earth, as distinct from the atmosphere and hydrosphere. **2.** The lithosphere, hydrosphere, and atmosphere combined. { 'jē·ō,sfir }

geostatic pressure *See* ground pressure. { ¦jē·ō¦stad·ik 'presh·ər }

geostatistics [GEOL] A branch of applied statistics that focuses on mathematical description and analysis of geological observations. { ,jē·ō·stə'tis·tiks }

geostrophic [GEOPHYS] Pertaining to deflecting force resulting from the earth's rotation. { ¦jē·ō¦sträf·ik }

geostrophic approximation [GEOPHYS] The assumption that the geostrophic current can represent the actual horizontal current. Also known as geostrophic assumption. { ¦jē·ō¦sträf·ik ə¦präk·sə'mā·shən }

geostrophic assumption *See* geostrophic approximation. { ¦jē·ō¦sträf·ik ə'səm·shən }

geostrophic current [GEOPHYS] A current defined by assuming the existence of an exact balance between the horizontal pressure gradient force and the Coriolis force. { ¦jē·ō¦sträf·ik 'kə·rənt }

geostrophic departure [METEOROL] A vector representing the difference between the real wind and the geostrophic wind. Also known as ageostrophic wind; geostrophic deviation. { ¦jē·ō¦sträf·ik di'pär·chər }

geostrophic deviation *See* geostrophic departure. { ¦jē·ō¦sträf·ik ˌdē·vē'ā·shən }

geostrophic distance [METEOROL] The distance (in degrees latitude) along a constant-pressure surface over which the change in height (in feet) is equal to the geostrophic wind speed (in knots). { ¦jē·ō¦sträf·ik 'dis·təns }

geostrophic equation [GEOPHYS] An equation, used to compute geostrophic current speed, which represents a balance between the horizontal pressure gradient force and the Coriolis force. { ¦jē·ō¦sträf·ik i'kwā·shən }

geostrophic equilibrium [GEOPHYS] A state of motion of a nonviscous fluid in which the horizontal Coriolis force exactly balances the horizontal pressure force at all points of the field so described. { ¦jē·ō¦sträf·ik ˌē·kwə'lib·rē·əm }

geostrophic flow [GEOPHYS] A form of gradient flow where the Coriolis force exactly balances the horizontal pressure force. { ¦jē·ō¦sträf·ik 'flō }

geostrophic flux [METEOROL] The transport of an atmospheric property by means of the geostrophic wind. { ¦jē·ō¦sträf·ik 'fləks }

geostrophic vorticity [METEOROL] The vorticity of the geostrophic wind. { ¦jē·ō¦sträf·ik vȯr'tis·əd·ē }

geostrophic wind [METEOROL] That horizontal wind velocity for which the Coriolis acceleration exactly balances the horizontal pressure force. { ¦jē·ō¦sträf·ik 'wind }

geostrophic-wind level [METEOROL] The lowest level at which the wind becomes geostrophic in the theory of the Ekman spiral. Also known as gradient-wind level. { ¦jē·ō¦sträf·ik 'wind ˌlev·əl }

geostrophic-wind scale [METEOROL] A graphical device used for the determination of the speed of the geostrophic wind from the isobar or contour-line spacing on a synoptic chart; it is a nomogram representing solutions of the geostrophic-wind equation. { ¦jē·ō¦sträf·ik 'wind ˌskāl }

geosynclinal couple *See* orthogeosyncline. { ¦jē·ōˌsin'klīn·əl 'kəp·əl }

geosynclinal cycle *See* tectonic cycle. { ¦jē·ōˌsin'klīn·əl 'sī·kəl }

geotectonics *See* tectonics. { ¦jē·ō¦tek'tän·iks }

geosynclinal facies [GEOL] A sedimentary facies marked by great thickness, a generally argillaceous character, and few carbonate rocks. { ¦jē·ōˌsin'klīn·əl 'fā·shēz }

geosyncline [GEOL] A linear part of the earth's crust, hundreds of kilometers long and tens of kilometers wide, that subsided during millions of years as it received thousands of meters of sedimentary and volcanic accumulations. { ¦jē·ō'sinˌklīn }

geotectogene *See* tectogene. { ¦jē·ō¦tek·təˌjēn }

geotectonic cycle *See* orogenic cycle. { ¦jē·ō·tek'tän·ik 'sī·kəl }

geotectonics *See* tectonics. { ¦jē·ō¦tek'tän·iks }

geotherm *See* geoisotherm. { 'jē·ōˌthərm }

geothermal [GEOPHYS] Pertaining to heat within the earth. { ¦jē·ō¦thər·məl }

geothermal energy [GEOPHYS] Thermal energy contained in the earth; can be used directly to supply heat or can be converted to mechanical or electrical energy. { ¦jē·ōˌthərm·əl 'en·ər·jē }

geothermal gradient [GEOPHYS] The change in temperature with depth of the earth. { ¦jē·ō¦thər·məl 'grād·ē·ənt }

geothermal system [GEOL] Any regionally localized geological setting where naturally occurring portions of the earth's internal heat flow are transported close enough to the earth's surface by circulating steam or hot water to be readily harnessed for use; examples are the Geysers Region of northern California and the hot brine fields in the Imperial Valley of southern California. { ¦jē·ō¦thər·məl 'sis·təm }

geothermometer [GEOL] A mineral that yields information about the temperature range within which it was formed. Also known as geologic thermometer. { ¦jē·ō·thər'mäm·əd·ər }

geothermometry [GEOL] Measurement of the temperatures at which geologic processes occur or occurred. Also known as geologic thermometry. { ¦jē·ō·thər'mäm·ə·trē }

gestalt [METEOROL] A complex of weather elements occurring in a familiar form, and though not necessarily referring to basic hydrodynamical or thermodynamical quantities, may persist for an appreciable length of time and is often considered to be an entity in itself. { ge'shtält }

geyser [HYD] A natural spring or fountain which discharges a column of water or steam into the air at more or less regular intervals. { 'gī·zər }

gharbi [METEOROL] A fresh westerly wind of oceanic origin in Morocco. { 'gär·bē }

gharra [METEOROL] Hard squalls from the northeast in Libya and Africa that are sudden and frequent, and are accompanied by heavy rain and thunder. { 'gä·rä }

ghibli [METEOROL] A hot, dust-bearing, desert wind in North Africa, similar to the foehn. Also known as chibli; gebli; gibleh; gibli; kibli. { 'gib·lē }

giant's cauldron *See* giant's kettle. { ¦jī·əns 'kȯl·drən }

giant's kettle [GEOL] A cylindrical hole bored in bedrock beneath a glacier by water falling through a deep moulin or by boulders rotating in the bed of a meltwater stream. Also known as giant's cauldron; moulin pothole; potash kettle. { ¦jī·əns 'ked·əl }

gibleh *See* ghibli. { 'gib·lə }

gibli *See* ghibli. { 'gib·lē }

giboulee *See* galerne. { jə'bü·lē }

gio *See* geo. { gyō }

glacial [GEOL] Pertaining to an interval of geologic time which was marked by an equatorward advance of ice during an ice age; the opposite of interglacial; these intervals are variously called glacial periods, glacial epochs, glacial stages, and so on. [HYD] Pertaining to ice, especially in great masses such as sheets of land ice or glaciers. { 'glā·shəl }

glacial abrasion [GEOL] Alteration of portions of the earth's surface as a result of glacial flow. { ¦glā·shəl ə'brā·zhən }

glacial accretion [GEOL] Deposition of material as a result of glacial flow. { ¦glā·shəl ə'krē·shən }

glacial advance [GEOL] **1.** Increase in the thickness and area of a glacier. **2.** A time period equal to that increase. { ¦glā·shəl əd'vans }

glacial anticyclone [METEOROL] A type of semipermanent anticyclone which overlies the ice caps of Greenland and Antarctica. Also known as glacial high. { ¦glā·shəl ˌan·ti'sīˌklōn }

glacial boulder [GEOL] A boulder moved to a point distant from its original site by a glacier. { ¦glā·shəl 'bōl·dər }

glacial deposit [GEOL] Material carried to a point beyond its original location by a glacier. { ¦glā·shəl di'päz·ət }

glacial drift [GEOL] All rock material in transport by glacial ice, and all deposits predominantly of glacial origin made in the sea or in bodies of glacial meltwater, including rocks rafted by icebergs. { ¦glā·shəl 'drift }

glacial epoch [GEOL] **1.** Any of the geologic epochs characterized by an ice age; thus, the Pleistocene epoch may be termed a glacial epoch. **2.** Generally, an interval of geologic time which was marked by a major equatorward advance of ice; the term has been applied to an entire ice age or (rarely) to the individual glacial stages which make up an ice age. { ¦glā·shəl 'ep·ək }

glacial erosion [GEOL] Movement of soil or rock from one point to another by the action of the moving ice of a glacier. Also known as ice erosion. { ¦glā·shəl ə'rō·zhən }

glacial flour *See* rock flour. { ¦glā·shəl 'flau̇·ər }

glacial flow *See* glacier flow. { ¦glā·shəl 'flō }

glacial geology [GEOL] The study of land features resulting from glaciation. { ¦glā·shəl jē'äl·ə·jē }

glacial high *See* glacial anticyclone. { ¦glā·shəl 'hī }

glacial ice [HYD] Ice that is flowing or that exhibits evidence of having flowed. { ¦glā·shəl 'īs }
glacial lake [GEOL] A lake that exists because of the effects of the glacial period. { ¦glā·shəl 'lāk }
glacial lobe [HYD] A tonguelike projection from a continental glacier's main mass. { ¦glā·shəl 'lōb }
glacial maximum [GEOL] The time or position of the greatest extent of any glaciation; most frequently applied to the greatest equatorward advance of Pleistocene glaciation. { ¦glā·shəl 'mak·sə·məm }
glacial mill *See* moulin. { ¦glā·shəl 'mil }
glacial outwash *See* outwash. { ¦glā·shəl 'au̇t,wäsh }
glacial period [GEOL] **1.** Any of the geologic periods which embraced an ice age; for example, the Quaternary period may be called a glacial period. **2.** Generally, an interval of geologic time which was marked by a major equatorward advance of ice. { ¦glā·shəl 'pir·ē·əd }
glacial plucking *See* plucking. { ¦glā·shəl 'plək·iŋ }
glacial retreat [GEOL] A condition occurring when backward melting at the front of a glacier takes place at a rate exceeding forward motion. { ¦glā·shəl ri'trēt }
glacial scour [GEOL] Erosion resulting from glacial action, whereby the surface material is removed and the rock fragments carried by the glacier abrade, scratch, and polish the bedrock. Also known as scouring. { ¦glā·shəl 'skau̇r }
glacial striae [GEOL] Scratches, commonly parallel, on smooth rock surfaces due to glacial abrasion. { ¦glā·shəl 'strī,ī }
glacial till *See* till. { ¦glā·shəl 'til }
glacial trough [GEOL] A deep U-shaped valley with steep sides that leads down from a cirque and was excavated by a glacier. { ¦glā·shəl 'trȯf }
glacial varve *See* varve. { ¦glā·shəl 'värv }
glaciated terrain [GEOL] A region that once bore great masses of glacial ice; a distinguishing feature is marks of glaciation. { 'glā·shē,ād·əd tə'rān }
glaciation [GEOL] Alteration of any part of the earth's surface by passage of a glacier, chiefly by glacial erosion or deposition. [METEOROL] The transformation of cloud particles from waterdrops to ice crystals, as in the upper portion of a cumulonimbus cloud. { ,glā·shē'ā·shən }
glaciation limit [GEOPHYS] For a given locality, the lowest altitude at which glaciers can develop. { ,glā·shē'ā·shən ,lim·ət }
glacier [HYD] A mass of land ice, formed by the further recrystallization of firn, flowing slowly (at present or in the past) from an accumulation area to an area of ablation. { 'glā·shər }
glacieret *See* snowdrift ice. { ¦glā·shə¦ret }
glacier flow [HYD] The motion that exists within a glacier's body. Also known as glacial flow. { 'glā·shər ,flō }
glacier front [HYD] The leading edge of a glacier. { 'glā·shər ,frənt }
glacier ice [HYD] Any ice that is or was once a part of a glacier, consolidated from firn by further melting and refreezing and by static pressure; for example, an iceberg. { 'glā·shər ,īs }
glacier mill *See* moulin. { 'glā·shər ,mil }
glacier pothole *See* moulin. { 'glā·shər 'pät,hōl }
glacier table [GEOL] A stone block supported by an ice pedestal above the surface of a glacier. { 'glā·shər ,tā·bəl }
glacier well *See* moulin. { 'glā·shər ,wel }
glacier wind [METEOROL] A shallow gravity wind along the icy surface of a glacier, caused by the temperature difference between the air in contact with the glacier and free air at the same altitude. { 'glā·shər ,wind }
glacioeustasy [GEOL] Changes in sea level due to storage or release of water from glacier ice. { ¦glās·ē·ō'yü·stə·sē }
glaciofluvial [GEOL] Pertaining to streams fed by melting glaciers, or to the deposits and landforms produced by such streams. { ¦glā·shē·ō¦flü·vē·əl }

glacioisostasy [GEOL] Lithospheric depression or rebound due to the weight or melting of glacier ice. { ˌglā·sē·ō·ī'sās·tə·sē }

glaciolacustrine [GEOL] Pertaining to lakes fed by melting glaciers, or to the deposits forming therein. { ¦glā·shē·ō·lə'kəs·trən }

glaciology [GEOL] A broad field encompassing all aspects of the study of ice: glaciers, the largest ice masses on earth; ice that forms on rivers, lakes, and the sea; ice in the ground, including both permafrost and seasonal ice such as that which disrupts roads; ice that crystallizes directly from the air on structures such as airplanes and antennas, and all forms of snow research, including hydrological and avalanche forecasting. { ˌglā·shē'äl·ə·jē }

glaçon [OCEANOGR] A piece of sea ice which is smaller than a medium-sized floe. { gla'sōn }

glance pitch [GEOL] A variety of asphaltite having brilliant conchoidal fracture, and resembling gilsonite but having higher specific gravity and percentage of fixed carbon. { 'glans ˌpich }

glare ice [HYD] Ice with a smooth, shiny surface. { 'gler ˌīs }

glass [METEOROL] In nautical terminology, a contraction for "weather glass" (a mercury barometer). { glas }

glave *See* glaves. { 'glä·və }

glaves [METEOROL] A foehnlike wind of the Faroe Islands. Also known as glave; glavis. { 'glä·vəs }

glavis *See* glaves. { 'glä·vəs }

glaze [HYD] A coating of ice, generally clear and smooth but usually containing some air pockets, formed on exposed objects by the freezing of a film of supercooled water deposited by rain, drizzle, or fog, or possibly condensed from supercooled water vapor. Also known as glaze ice; glazed frost; verglas. { glāz }

glessite [GEOL] Fossil resin similar to amber. { 'gleˌsīt }

gley [GEOL] A sticky subsurface layer of clay in some waterlogged soils. { glā }

glide fold *See* shear fold. { 'glīd ˌfōld }

glime [HYD] An ice coating with a consistency intermediate between glaze and rime. { glīm }

glimmer ice [HYD] Ice newly formed within the cracks or holes of old ice, or on the puddles on old ice. { 'glim·ər ˌīs }

global climate change [CLIMATOL] The periodic fluctuations in global temperatures and precipitation, such as the glacial (cold) and interglacial (warm) cycles. { ¦glō·bəl 'klī·mət ˌchānj }

global radiation [GEOPHYS] The total of direct solar radiation and diffuse sky radiation received by a horizontal surface of unit area. { 'glō·bəl ˌrād·ē'ā·shən }

global sea [OCEANOGR] All the seawater of the earth considered as a single ocean constantly intermixing. { 'glō·bəl 'sē }

global warming potential [METEOROL] The ratio of global warming or radiative forcing from 1 kilogram of a greenhouse gas to 1 kilogram of carbon dioxide over 100 years, expressed per mole or per kilogram; it provides a way to calculate the contribution of each greenhouse gas to the annual increase in radiative forcing. { ¦glō·bəl 'wȯrm·iŋ pəˌten·chəl }

globe lightning *See* ball lightning. { 'glōb ˌlīt·niŋ }

globigerina ooze [GEOL] A pelagic sediment consisting of than 30% calcium carbonate in the form of foraminiferal tests of which *Globigerina* is the dominant genus. { glōˌbij·ə'rī·nə ˌüz }

globulite [GEOL] A small, isotropic, globular of spherulelike crystallite; usually dark in color and found in glassy extrusive rocks. { 'gläb·yəˌlīt }

gloom [METEOROL] The condition existing when daylight is very much reduced by dense cloud or smoke accumulation above the surface, the surface visibility not being materially reduced. { glüm }

gloup [GEOL] An opening in the roof of a sea cave. { glüp }

glowing avalanche *See* ash flow. { ¦glō·iŋ 'av·əˌlanch }

glowing cloud *See* nuée ardente. { ¦glō·iŋ 'klau̇d }

glyptolith *See* ventifact. { 'glip·tə,lith }
gobi [GEOL] Sedimentary deposits in a synclinal basin. { 'gō·bē }
Goldschmidt's mineralogical phase rule [GEOL] The rule that the probability of finding a system with degrees of freedom less than two is small under natural rock-forming conditions. { 'gōl,shmits ,min·ə·rə'läj·ə·kəl ¦fāz ,rül }
golfada [METEOROL] A heavy gale of the Mediterranean. { gȯl'fäd·ə }
Gondwana [GEOL] The ancient continent that is supposed to have fragmented and drifted apart during the Triassic to form eventually the present continents. Also known as Gondwanaland. { gänd'wä·nə }
Gondwanaland *See* Gondwana. { gän'dwän·ə,land }
gorge [GEOGR] A narrow passage between mountains or the walls of a canyon, especially one with steep, rocky walls. [OCEANOGR] A collection of solid matter obstructing a channel or a river, as an ice gorge. { gȯrj }
gorge wind *See* canyon wind. { 'gȯrj ,wind }
gosling blast [METEOROL] A sudden squall of rain or sleet in England. Also known as gosling storm. { 'gäz·liŋ ,blast }
gosling storm *See* gosling blast. { 'gäz·liŋ ,stȯrm }
gossan [GEOL] A rusty, ferruginous deposit filling the upper regions of mineral veins and overlying a sulfide deposit; formed by oxidation of pyrites. Also known as capping; gozzan; iron hat. { 'gas·ən }
Gotlandian [GEOL] A geologic time period recognized in Europe to include the Ordovician; it appears before the Devonian. { gät'lan·dē·ən }
gouge [GEOL] Soft, pulverized mixture of rock and mineral material found along shear (fault) zones and produced by the differential movement across the plane of slippage. { gaủj }
gowk storm [METEOROL] In England, a storm or gale occurring at about the end of April or the beginning of May. { 'gəủk ,stȯrm }
gozzan *See* gossan. { 'gäz·ən }
gpu *See* geopotential unit.
graben [GEOL] A block of the earth's crust, generally with a length much greater than its width, that has dropped relative to the blocks on either side. { 'grä·bən }
gradation [GEOL] **1.** The leveling of the land, or the bringing of a land surface or area to a uniform or nearly uniform grade or slope through erosion, transportation, and deposition. **2.** Specifically, the bringing of a stream bed to a slope at which the water is just able to transport the material delivered to it. { grā'dā·shən }
gradation period [GEOL] The time during which the base level of the sea remains in one position. Also known as base-leveling epoch. { grā'dā·shən ,pir·ē·əd }
grade [GEOL] The slope of the bed of a stream, or of a surface over which water flows, upon which the current can just transport its load without either eroding or depositing. { grād }
grade correction *See* slope correction. { 'grād kə,rek·shən }
graded [GEOL] Brought to or established at grade. { 'grād·əd }
graded bedding [GEOL] A stratification in which each stratum displays a gradation in the size of grains from coarse below to fine above. { ¦grād·əd 'bed·iŋ }
graded profile *See* profile of equilibrium. { ¦grād·əd 'prō,fīl }
graded stream [HYD] A stream in which, over a period of years, slope is adjusted to yield the velocity required for transportation of the load supplied from the drainage basin. { ¦grād·əd 'strēm }
grade scale [GEOL] A continuous scale of particle sizes divided into a series of size classes. { 'grād ,skāl }
gradient [GEOL] The rate of descent or ascent (steepness of slope) of any topographic feature, such as streams or hillsides. { 'grād·ē·ənt }
gradient current [OCEANOGR] A current defined by assuming that the horizontal pressure gradient in the sea is balanced by the sum of the Coriolis and bottom frictional forces; at some distance from the bottom the effect of friction becomes negligible, and above this the gradient and geostrophic currents are equivalent. Also known as slope current. { 'grād·ē·ənt ,kə·rənt }

gradient flow [METEOROL] Horizontal frictionless flow in which isobars and streamlines coincide, or equivalently, in which the tangential acceleration is everywhere zero; the balance of normal forces (pressure force, Coriolis force, centrifugal force) is then given by the gradient wind equation. { 'grād·ē·ənt ,flō }

gradient wind [METEOROL] A wind for which Coriolis acceleration and the centripetal acceleration exactly balance the horizontal pressure force. { 'grād·ē·ənt ,wind }

gradient-wind level *See* geostrophic-wind level. { 'grād·ē·ənt ,wind ,lev·əl }

grading [GEOL] The gradual reduction of the land to a level surface; for example, erosion of land to base level by streams. { 'grād·iŋ }

grahamite *See* mesosiderite. { 'grā·ə,mīt }

grain [GEOL] The particles or discrete crystals that make up a sediment or rock. [HYD] The particles which make up settled snow, firn, and glacier ice. { grān }

grain diminution *See* degradation recrystallization. { 'grān dim·yə'nish·ən }

grain size [GEOL] Average size of mineral particles composing a rock or sediment. { 'grān ,sīz }

Grand Banks [GEOGR] Banks off southeastern Newfoundland, important for cod fishing. { 'grand ¦baŋks }

granite series [GEOL] A sequence of products that evolve continuously during crustal fusion; earlier products tend to be deep-seated, syntectonic, and granodioritic, and later products tend to be shallower, late syntectonic, or postsyntectonic, and more potassic. { 'gran·ət ,sir·ēz }

granite wash [GEOL] Material eroded from granites and redeposited, forming a rock with the same major mineral constituents as the original rock. { 'gran·ət ,wäsh }

granitic batholith [GEOL] A granitic shield mass intruded as the fusion of older formations. { grə'nid·ik 'bath·ə,lith }

granular ice [HYD] Ice composed of many tiny, opaque, white or milky pellets or grains frozen together and presenting a rough surface; this is the type of ice deposited as rime and compacted as névé. { 'gran·yə·lər 'īs }

granular snow *See* snow grains. { 'gran·yə·lər 'snō }

granule [GEOL] A somewhat rounded rock fragment ranging in diameter from 2 to 4 millimeters; larger than a coarse sand grain and smaller than a pebble. { 'gran·yül }

grapestone [GEOL] A cluster of sand-size grains, such as calcareous pellets, held together by incipient cementation shortly after deposition; the outer surface is lumpy, resembling a bunch of grapes. { 'grāp,stōn }

grapevine drainage *See* trellis drainage. { 'grāp,vīn ,drān·ij }

graphic texture [GEOL] A pattern of rocks that is similar to cuneiform characters. { 'graf·ik ,teks·chər }

graptolite shale [GEOL] Shale containing an abundance of extinct colonial marine organisms known as graptolites. { 'grap·tə,līt 'shāl }

grassland climate *See* subhumid climate. { 'gras,land ,klī·mət }

grass minimum [METEOROL] The minimum temperature shown by a minimum thermometer exposed in an open situation with its bulb on the level of the tops of the grass blades of short turf. { 'gras 'min·ə·məm }

grass temperature [METEOROL] The temperature registered by a thermometer with its bulb at the level of the tops of the blades of grass in short turf. { 'gras ,tem·prə·chər }

graupel *See* snow pellets. { 'grau̇·pəl }

gravel [GEOL] A loose or unconsolidated deposit of rounded pebbles, cobbles, or boulders. { 'grav·əl }

gravel bank [GEOL] A natural mound or exposed face of gravel, particularly such a place from which gravel is dug. { 'grav·əl ,baŋk }

gravel desert *See* reg. { 'grav·əl ¦dez·ərt }

gravimetric geodesy [GEOD] The science that utilizes measurements and characteristics of the earth's gravity field, as well as theories regarding this field, to deduce the shape of the earth and, in combination with arc measurements, the earth's size. { grav·ə'me·trik jē'äd·ə·sē }

gravitational convection *See* thermal convection. { ,grav·ə'tā·shən·əl kən'vek·shən }

gravitational settling [GEOL] A movement of sediment resulting from gravitational forces. { ˌgrav·əˈtā·shən·əl ˈset·liŋ }
gravitational sliding [GEOL] Extensive sliding of strata down a slope of an uplifted area. Also known as sliding. { ˌgrav·əˈtā·shən·əl ˈslīd·iŋ }
gravitational tide [OCEANOGR] An atmospheric tide due to gravitational attraction of the sun and moon. { ˌgrav·əˈtā·shən·əl ˈtīd }
gravitational water [HYD] Soil water of a temporary character that results from prolonged infiltration from above and which moves downward to the groundwater zone in response to gravity. { ˌgrav·əˈtā·shən·əl ˈwȯd·ər }
gravity anomaly [GEOPHYS] The difference between the observed gravity and the theoretical or predicted gravity. { ˈgrav·əd·ē əˌnäm·ə·lē }
gravity-collapse structure *See* collapse structure. { ˈgrav·əd·ē kə¦laps ˌstrək·chər }
gravity drainage [HYD] Withdrawal of water from strata as a result of gravitational forces. { ˈgrav·əd·ē ˈdrān·ij }
gravity drainage reservoir [GEOL] A reservoir in which production is significantly affected by gas, oil, and water separating under the influence of gravity while production takes place. { ˈgrav·əd·ē ˈdrān·ij ˌrez·əvˌwär }
gravity erosion *See* mass erosion. { ˈgrav·əd·ē iˌrō·zhən }
gravity fault *See* normal fault. { ˈgrav·əd·ē ˌfȯlt }
gravity flow [HYD] A form of glacier movement in which the flow of the ice results from the downslope gravitational component in an ice mass resting on a sloping floor. { ˈgrav·əd·ē ˌflō }
gravity map [GEOPHYS] A map of gravitational variations in an area displaying gravitational highs and lows. { ˈgrav·əd·ē ˌmap }
gravity slope [GEOL] The relatively steep slope on a hillside above the wash slope; usually situated at the angle of repose of the material eroded from it. { ˈgrav·əd·ē ˌslōp }
gravity spring [HYD] A spring that issues under the influence of gravity, not internal pressure. { ˈgrav·əd·ē ˌspriŋ }
gravity tide [GEOPHYS] Cyclic motion of the earth's surface caused by interaction of gravitational forces of the moon, sun, and earth. { ˈgrav·əd·ē ˌtīd }
gravity wind [METEOROL] A wind (or component thereof) directed down the slope of an incline and caused by greater air density near the slope than at the same levels some distance horizontally from the slope. Also known as drainage wind; katabatic wind. { ˈgrav·əd·ē ˌwind }
grease ice [HYD] A kind of slush with a greasy appearance, formed from the congelation of ice crystals in the early stages of freezing. Also known as ice fat; lard ice. { ˈgrēs ˌīs }
Great Basin high [METEOROL] A high-pressure system centered over the Great Basin of the western United States; it is a frequent feature of the surface chart in the winter season. { ¦grāt ¦bas·ən ˈhī }
great circle [GEOD] A circle, or near circle, described on the earth's surface by a plane passing through the center of the earth. { ˈgrāt ¦sər·kəl }
great-circle direction [GEOD] Horizontal direction of a great circle, expressed as angular distance from a reference direction. { ˈgrāt ¦sər·kəl diˈrek·shən }
great-circle distance [GEOD] The length of the shorter arc of the great circle joining two points. { ˈgrāt ¦sər·kəl ˈdis·təns }
great diurnal range [OCEANOGR] The difference in height between mean higher high water and mean lower low water. Also known as diurnal range. { ˈgrāt dīˌərn·əl ˈrānj }
greater ebb [OCEANOGR] The stronger of two ebb currents occurring during a tidal day. { ˈgrād·ər ¦eb }
greater flood [OCEANOGR] The stronger of two flood currents occurring during a tidal day. { ˈgrād·ər ¦fləd }
Great Ice Age [GEOL] The Pleistocene epoch. { ¦grāt ˈīs ˌāj }
great soil group [GEOL] A group of soils having common internal soil characteristics; a subdivision of a soil order. { ¦grāt ˈsȯil ˌgrüp }

great tropic range [OCEANOGR] The difference in height between tropic higher high water and tropic lower low water. { ¦grāt 'träp·ik ˌrānj }

greco [METEOROL] An Italian name for the northeast wind. { 'grek·ō }

green belt *See* frostless zone. { 'grēn ˌbelt }

greenhouse effect [METEOROL] The effect created by the earth's atmosphere in trapping heat from the sun; the atmosphere acts like a greenhouse. { 'grēnˌhaủs iˌfekt }

greenhouse gases [METEOROL] Gases whose concentration is small and varies, mostly due to anthropogenic factors; they absorb heat from incoming solar radiation but do not allow long-wave radiation to reflect back into space. They include carbon dioxide, methane, and nitrous oxide, as well as, water vapor, carbon monoxide, nitrogen oxide, nitrogen dioxide, and ozone. { 'grēnˌhaủs ˌgas·əz }

Greenland anticyclone [METEOROL] The glacial anticyclone which is supposed to overlie Greenland; analogous to the Antarctic anticyclone. { 'grēn·lənd ˌant·i'sīˌklōn }

green mud [GEOL] **1.** A fine-grained, greenish terrigenous mud or oceanic ooze found near the edge of a continental shelf at depths of 300–7500 feet (90–2300 meters). **2.** A deep-sea terrigenous deposit characterized by the presence of a considerable proportion of glauconite and calcium carbonate. { ¦grēn 'məd }

greensand [GEOL] A greenish sand consisting principally of grains of glauconite and found between the low-water mark and the inner mud line. { 'grēnˌsand }

green sky [METEOROL] A greenish tinge to part of the sky, supposed by seamen to herald wind or rain, or in some cases, a tropical cyclone. { 'grēn ¦skī }

green snow [HYD] A snow surface that has attained a greenish tint as a result of the growth within it of certain microscopic algae. { 'grēn ¦snō }

greenstone belts [GEOL] Oceanic and island arclike sequences that are similar to, and run to the south and north of, the Swaziland System. { 'grēnˌstōn ˌbelts }

Greenwich meridian [GEOD] The meridian passing through Greenwich, England, and serving as the reference for Greenwich time; it also serves as the origin of measurement of longitude. { 'gren·ich mə'rid·ē·ən }

gregale [METEOROL] The Maltese and best-known variant of a term for a strong northeast wind in the central and western Mediterranean and adjacent European land areas; it occurs either with high pressure over central Europe or the Balkans and low pressure over Libya, when it may continue for up to 5 days, or with the passage of a depression to the south or southeast, when it lasts only 1 or 2 days; it is most frequent in winter. { grā'gä·lā }

Grenville orogeny [GEOL] A Precambrian mountain-forming epoch. { 'gren·vəl ȯ'räj·ə·nē }

grike [GEOL] A vertical fissure developed along a joint in limestone by dissolution of some of the rock. Also spelled gryke. { grīk }

grit [GEOL] **1.** A hard, sharp granule, as of sand. **2.** A coarse sand. { grit }

groove [GEOL] Glaciated marks of large size on rock. { grüv }

groove casts [GEOL] Rounded or sharp, crested, rectilinear ridges that are a few millimeters high and a few centimeters long; found on the undersurfaces of sandstone layers lying on mudstone. { 'grüv ˌkasts }

gross-austausch [METEOROL] The exchange of air mass properties and the associated momentum and energy transports produced on a worldwide scale by the migratory large-scale disturbances of middle latitudes. { 'grōs 'aủsˌtaủsh }

ground [GEOL] **1.** Any rock or rock material. **2.** A mIneralized deposit. **3.** Rock in which a mineral deposit occurs. { graủnd }

ground discharge *See* cloud-to-ground discharge. { 'graủnd ¦disˌchärj }

grounded ice *See* stranded ice. { 'graủnd·əd ¦īs }

ground fog [METEOROL] A fog that hides less than 0.6 of the sky and does not extend to the base of any clouds that may lie above it. { 'graủnd ˌfäg }

ground frost [METEOROL] In British usage, a freezing condition injurious to vegetation, which is considered to have occurred when a minimum thermometer exposed to the sky at a point just above a grass surface records a temperature (grass temperature) of 30.4°F (-0.9°C) or below. { 'graủnd ˌfrȯst }

ground ice [HYD] **1.** A body of clear ice in frozen ground, most commonly found in more or less permanently frozen ground (permafrost), and may be of sufficient age to be termed fossil ice. Also known as stone ice; subsoil ice; subterranean ice; underground ice. **2.** *See* anchor ice. { 'graůnd ,īs }

ground ice mound [GEOL] A frost mound containing bodies of ice. Also known as ice mound. { 'graůnd ,īs ,maůnd }

ground inversion *See* surface inversion. { 'graůnd in,vər·zhən }

ground layer *See* surface boundary layer. { 'graůnd ,lā·ər }

ground moraine [GEOL] Rock material carried and deposited in the base of a glacier. Also known as bottom moraine; subglacial moraine. { 'graůnd mə,rān }

ground noise [GEOPHYS] In seismic exploration, disturbance of the ground due to some cause other than the shot. { 'graůnd ,nȯiz }

ground pressure [GEOPHYS] The pressure to which a rock formation is subjected by the weight of the superimposed rock and rock material or by diastrophic forces created by movements in the rocks forming the earth's crust. Also known as geostatic pressure; lithostatic pressure; rock pressure. { 'graůnd ,presh·ər }

ground streamer [METEOROL] An upward advancing column of high-ion density which rises from a point on the surface of the earth toward which a stepped leader descends at the start of a lightning discharge. { 'graůnd ,strē·mər }

ground swell [OCEANOGR] A swell passing through shallow water, characterized by a marked increase in height in water shallower than one-tenth wavelength. { 'graůnd ,swel }

ground-to-cloud discharge [GEOPHYS] A lightning discharge in which the original streamer processes start upward from an object located on the ground. { 'graůnd tə ¦klaůd 'dis,chärj }

ground truth measurements [GEOPHYS] Measurements of various properties, such as temperature and land utilization, which are conducted on the ground to calibrate observations made from satellites or aircraft. { 'graůnd ,trüth ,mezh·ər·məns }

ground visibility [METEOROL] In aviation terminology, the horizontal visibility observed at the ground, that is, surface visibility or control-tower visibility. { 'graůnd ,viz·ə'bil· əd·ē }

groundwater [HYD] All subsurface water, especially that part that is in the zone of saturation. { 'graůnd,wȯd·ər }

groundwater decrement *See* groundwater discharge. { 'graůnd,wȯd·ər ¦dek·rə·mənt }

groundwater depletion curve [HYD] A recession curve of streamflow, so adjusted that the slope of the curve represents the runoff (depletion rate) of the groundwater; it is formed by the observed hydrograph during prolonged periods of no precipitation. Also known as groundwater recession. { 'graůnd,wȯd·ər di'plē·shən ,kərv }

groundwater discharge [HYD] **1.** Water released from the zone of saturation. **2.** Release of such water. Also known as decrement; groundwater decrement; phreatic-water discharge. { 'graůnd,wȯd·ər 'dis,chärj }

groundwater flow [HYD] That portion of the precipitation that has been absorbed by the ground and has become part of the groundwater. { 'graůnd,wȯd·ər ,flō }

groundwater hydrology [HYD] The study of the occurrence, circulation, distribution, and properties of any liquid water residing beneath the surface of the earth. { ,graůnd ,wȯd·ər hī'dräl·ə·jē }

groundwater increment *See* recharge. { 'graůnd,wȯd·ər ¦iŋ·krə·mənt }

groundwater level [HYD] The level below which the rocks and subsoil are full of water. { 'graůnd,wȯd·ər ,lev·əl }

groundwater recession *See* groundwater depletion curve. { 'graůnd,wȯd·ər ri'sesh·ən }

groundwater recharge *See* recharge. { 'graůnd,wȯd·ər 'rē,chärj }

groundwater replenishment *See* recharge. { 'graůnd,wȯdər ri'plen·ish·mənt }

group [GEOL] A lithostratigraphic material unit comprising several formations. { grüp }

group number [OCEANOGR] The first two numbers in the argument number in A. T. Doodson's scheme for predicting tides. { 'grüp ,nəm·bər }

grower's year [CLIMATOL] In the United Kingdom, the 12-month period starting November 6, and referring to the cycle of seasonal change of weather. { 'grō·ərz ¦yir }

growl [GEOPHYS] Noise heard when strata are subjected to great pressure. { graúl }

growler [OCEANOGR] A small piece of floating sea ice, usually a fragment of an iceberg or floeberg; it floats low in the water, and its surface often is heavily pitted; it often appears greenish in color. Also known as bergy-bit. { 'graúl·ər }

growth lattice [GEOL] The rigid, reef-building, inplace framework of an organic reef, consisting of skeletons of sessile organisms and excluding reef-flank and other associated fragmental deposits. Also known as organic lattice. { 'grōth ˌlad·əs }

grus *See* gruss. { grüs }

gruss [GEOL] A loose accumulation of fragmental products formed from the weathering of granite. Also spelled grus. { grüs }

gryke *See* grike. { grīk }

Guadalupian [GEOL] A North American provincial series in the Lower and Upper Permian, above the Leonardian and below the Ochoan. { ¦gwäd·əl¦ü·pē·ən }

guba [METEOROL] In New Guinea, a rain squall on the sea. { 'gü·bä }

guest element *See* trace element. { 'gest ¦el·ə·mənt }

Guiana Current [OCEANOGR] A current flowing northwestward along the northeastern coast of South America. { gī'an·ə ¦kə·rənt }

Guinea Current [OCEANOGR] A current flowing eastward along the southern coast of northwestern Africa into the Gulf of Guinea. { 'gin·ē ¦kə·rənt }

gulch [GEOGR] A gulley, sometimes occupied by a torrential stream. { gəlch }

gulf [GEOGR] **1.** An abyss or chasm. **2.** A large extension of the sea partially enclosed by land. { gəlf }

Gulfian [GEOL] A North American provincial series in Upper Cretaceous geologic time, above the Comanchean and below the Paleocene of the Tertiary. { 'gəlf·ē·ən }

Gulf Stream [OCEANOGR] A relatively warm, well-defined, swift, relatively narrow, northward-flowing ocean current which originates north of Grand Bahama Island where the Florida Current and the Antilles Current meet, and which eventually becomes the eastward-flowing North Atlantic Current. { 'gəlf ¦strēm }

Gulf Stream Countercurrent [OCEANOGR] **1.** A surface current opposite to the Gulf Stream, one current component on the Sargasso Sea side and the other component much weaker, on the inshore side. **2.** A predicted, but as yet unobserved, large current deep under the Gulf Stream but opposite to it. { 'gəlf ¦strēm 'kaúnt·ərˌkə·rənt }

Gulf Stream eddy [OCEANOGR] A cutoff meander of the Gulf Stream. { 'gəlf ¦strēm 'ed·ē }

Gulf Stream front [OCEANOGR] The pronounced horizontal temperature gradient that defines a cross section of the Gulf Stream. { 'gəlf ¦strēm 'frənt }

Gulf Stream meander [OCEANOGR] One of the changeable, winding bends in the Gulf Stream; such bends intensify as the Gulf Stream merges into North Atlantic Drift and break up into detached eddies at times, at about 40°N. { 'gəlf ¦strēm mē'and·ər }

Gulf Stream system [OCEANOGR] The Florida Current, Gulf Stream, and North Atlantic Current, collectively. { 'gəlf ¦strēm 'sis·təm }

gully [GEOGR] A narrow ravine. { 'gəl·ē }

gully erosion [GEOL] Erosion of soil by running water. { 'gəl·ē i¦rō·zhən }

gully-squall [METEOROL] A nautical term for a violent squall of wind from mountain ravines on the Pacific side of Central America. { 'gəl·ē ˌskwól }

gumbo [GEOL] A soil that forms a sticky mud when wet. { 'gəm·bō }

gumbotil [GEOL] Deoxidized, leached clay that contains siliceous stones. { 'gəm·bōˌtil }

Günz [GEOL] A European stage of geologic time, in the Pleistocene (above Astian of Pliocene, below Mindel); it is the first stage of glaciation of the Pleistocene in the Alps. { gints }

Günz-Mindel [GEOL] The first interglacial stage of the Pleistocene in the Alps, between Günz and Mindel glacial stages. { 'gints 'mind·əl }

gust [METEOROL] A sudden, brief increase in the speed of the wind; it is of a more transient character than a squall and is followed by a lull or slackening in the wind speed. { gəst }

gustiness [METEOROL] A quality of airflow characterized by gusts. { 'gəs·tē·nəs }
gustiness components [METEOROL] **1.** The ratios, to the mean wind speed, of the average magnitudes of the component fluctuations of the wind along three mutually perpendicular axes. **2.** The ratios of the root-mean-squares of the eddy velocities to the mean wind speed. Also known as intensity of turbulence. { 'gəs·tē·nəs kəm,pō·nəns }
gustiness factor [METEOROL] A measure of the intensity of wind gusts; it is the ratio of the total range of wind speeds between gusts and the intermediate periods of lighter wind to the mean wind speed, averaged over both gusts and lulls. { 'gəs·tē·nəs ,fak·tər }
gut [GEOL] **1.** A narrow water passage such as a strait. **2.** A channel deeper than the surrounding water; generally formed by water in motion. { gət }
guti weather [METEOROL] In Rhodesia, a dense stratocumulus overcast, frequently with drizzle, occurring mainly in early summer, and associated with easterly winds that invade the interior, bringing in cool and stable maritime air when an anticyclone moves eastward south of Africa. { 'güd·ē ,weth·ər }
guttra [METEOROL] In Iran, sudden squalls in May. { 'gü·trə }
guxen [METEOROL] A cold wind of the Alps in Switzerland. { 'gu̇k·sən }
guyot [GEOL] A seamount, usually deeper than 100 fathoms (180 meters), having a smooth platform top. Also known as tablemount. { gē'ō }
guzzle [METEOROL] In the Shetland Islands, an angry blast of wind, dry and parching. { 'gəz·əl }
gypcrete [GEOL] A type of duricrust composed of hydrous calcium sulfate. { 'jip,krēt }
gypsite [GEOL] A variety of gypsum consisting of dirt and sand; found as an efflorescent deposit in arid regions, overlying gypsum. Also known as gypsum earth. { 'jip,sīt }
gypsum earth *See* gypsite. { 'jip·səm ¦ərth }
gyre [OCEANOGR] A closed circulatory system that is larger than a whirlpool or eddy. { jīr }
gyttja [GEOL] A fresh-water anaerobic mud containing an abundance of organic matter; capable of supporting aerobic life. { 'yi,chä }

Haanel depth rule [GEOPHYS] A rule for estimating the depth of a magnetic body, provided the body may be considered magnetically equivalent to a single pole; the depth of the pole is then equal to the horizontal distance from the point of maximum vertical magnetic intensity to the points where the intensity is one-third of the maximum value. { 'hän·əl 'depth ˌrül }

haar [METEOROL] A wet sea fog or very fine drizzle which drifts in from the sea in coastal districts of eastern Scotland and northeastern England; it occurs most frequently in summer. { här }

haboob [METEOROL] A strong wind and sandstorm or duststorm in the northern and central Sudan, especially around Khartum, where the average number is about 24 haboobs a year. { hə'büb }

hadal [OCEANOGR] Pertaining to the environment of the ocean trenches, over 4 miles (6.5 kilometers) in depth. { 'hād·əl }

hade [GEOL] **1.** The angle of inclination of a fault as measured from the vertical. **2.** The inclination angle of a vein or lode. { hād }

Hadean [GEOL] The period (more than 3800 million years ago) extending for several hundred millions of years from the end of the accretion of the earth to the formation of the oldest recognized rocks. { 'hā·dē·Pen }

Hadley cell [METEOROL] A direct, thermally driven, and zonally symmetric circulation first proposed by George Hadley as an explanation for the trade winds; it consists of the equatorward movement of the trade winds between about latitude 30° and the equator in each hemisphere, with rising wind components near the equator, poleward flow aloft, and finally descending components at about latitude 30° again. { 'had·lē ˌsel }

haff [GEOGR] A freshwater lagoon separated from the sea by a sandbar. { haf }

hail [METEOROL] Precipitation composed of lumps of ice formed in strong updrafts in cumulonimbus clouds, having a diameter of at least 0.2 inch (5 millimeters), most hailstones are spherical or oblong, some are conical, and some are bumpy and irregular. { hāl }

hail stage [METEOROL] Thermodynamic process of freezing of suspended water drops in adiabatically rising air with temperature below the freezing point, under the assumption that release of latent heat of fusion maintains constant temperature until all water is frozen. { 'hāl ˌstāj }

hailstone [METEOROL] A single unit of hail, ranging in size from that of a pea to that of a grapefruit, or from less than 1/4 inch (6 millimeters) to more than 5 inches (13 centimeters) diameter; may be spheroidal, conical, or generally irregular in shape. { 'hāl,stōn }

hairstone [GEOL] Quartz embedded with hairlike crystals of rutile, actinolite, or other mineral. { 'her,stōn }

halcyon days [METEOROL] A period of fine weather. { 'hal·sē·ən ¦dāz }

haldenhang *See* wash slope. { 'hal·dən,haŋ }

half-arc angle [METEOROL] The elevation angle of that point which a given observer regards as the bisector of the arc from his zenith to his horizon; a measure of the apparent degree of flattening of the dome of the sky. { 'haf ˌärk 'aŋ·gəl }

half tide [OCEANOGR] The condition when the tide is at the level between any given high tide and the following or preceding low tide. Also known as mean tide. { 'haf ˌtīd }
half-tide level [OCEANOGR] The level midway between mean high water and mean low water. { 'haf ˌtīd ˌlev·əl }
Hallian [GEOL] A North American stage of Pleistocene geologic time, above the Wheelerian and below the Recent. { 'hȯl·ē·ən }
halmeic [GEOL] Referring to minerals or sediments derived directly from sea water. Also known as halmyrogenic; halogenic. { hal'mē·ik }
halmyrogenic *See* halmeic. { halˌmī·rə'jen·ik }
halmyrolysis [GEOCHEM] Postdepositional chemical changes that occur while sediment is on the sea floor. { ˌhal·mə'räl·ə·səs }
halo [GEOL] A ring or crescent surrounding an area of opposite sign; it is a diffusion of a high concentration of the sought mineral into surrounding ground or rock; it is encountered in mineral prospecting and in magnetic and geochemical surveys. [METEOROL] Any one of a large class of atmospheric optical phenomena which appear as colored or whitish rings and arcs about the sun or moon when seen through an ice crystal cloud or in a sky filled with falling ice crystals. { 'hā·lō }
halocline [OCEANOGR] A well-defined vertical gradient of salinity in the oceans and seas. { 'hal·əˌklīn }
halogenic *See* halmeic. { ¦hal·ə¦jen·ik }
halokinesis *See* salt tectonics. { ¦hal·ə·kə'nē·səs }
halomorphic [GEOCHEM] Referring to an intrazonal soil whose features have been strongly affected by either neutral or alkali salts, or both. { ¦hal·ə¦mȯr·fik }
halo ore *See* fringe ore. { 'hā·lō ˌȯr }
hamada [GEOL] A barren desert surface composed of consolidated material usually consisting of exposed bedrock, but sometimes of consolidated sedimentary material. { hə'mä·də }
hanger *See* hanging wall. { 'haŋ·ər }
hanging *See* hanging wall. { 'haŋ·iŋ }
hanging glacier [HYD] A glacier lying above a cliff or steep mountainside; as the glacier advances, calving can cause ice avalanches. { 'haŋ·iŋ ¦glā·shər }
hanging side *See* hanging wall. { 'haŋ·iŋ ¦sīd }
hanging valley [GEOL] A valley whose floor is higher than the level of the shore or other valley to which it leads. { 'haŋ·iŋ ¦val·ē }
hanging wall [GEOL] The rock mass above a fault plane, vein, lode, ore body, or other structure. Also known as hanger; hanging; hanging side. { 'haŋ·iŋ ¦wȯl }
harbor [GEOGR] Any body of water of sufficient depth for ships to enter and find shelter from storms or other natural phenomena. Also known as port. { 'här·bər }
harbor reach [GEOGR] The stretch of a river or estuary which leads directly to the harbor. { 'här·bər ˌrēch }
hard freeze [HYD] A freeze in which seasonal vegetation is destroyed, the ground surface is frozen solid underfoot, and heavy ice is formed on small water surfaces such as puddles and water containers. { 'härd ¦frēz }
hard frost *See* black frost. { 'härd ¦frȯst }
hardpan *See* caliche. { 'härdˌpan }
hard rime [HYD] Opaque, granular masses of rime deposited chiefly on vertical surfaces by a dense super-cooled fog; it is more compact and amorphous than soft rime, and may build out into the wind as glazed cones or feathers. { 'härd ¦rīm }
hard rock [GEOL] Rock which needs drilling and blasting for removal. { 'härd ¦räk }
Harlechian [GEOL] A European stage of geologic time: Lower Cambrian. { här'lek·ē·ən }
harmatan *See* harmattan. { ˌhär·mə'tan }
harmattan [METEOROL] A dry, dust-bearing wind from the northeast or east which blows in West Africa especially from late November until mid-March; it originates in the Sahara as a desert wind and extends southward to about 5°N in January and 18°N in July. Also spelled harmatan; harmetan; hermitan. { ˌhär·mə'tan }
harmetan *See* harmattan. { ˌhär·mə'tan }

harmonic folding [GEOL] Folding in the earth's surface, with no sharp changes with depth in the form of the folds. { här'män·ik 'fōld·iŋ }

harmonic prediction [OCEANOGR] A method used in predicting the tides and tidal currents by combining the harmonic constituents into a single tide curve. { här'män·ik prə'dik·shən }

harmonic tide plane *See* Indian spring low water. { här'män·ik 'tīd ˌplān }

hartite [GEOL] A white, crystalline, fossil resin that is found in lignites. Also known as bombiccite; branchite; hofmannite; josen. { 'härˌtīt }

haster [METEOROL] In England, a violent rain storm. { 'has·tər }

haud [METEOROL] In Scotland, a squall. { hȯd }

Hauterivian [GEOL] A European stage of geologic time, in the Lower Cretaceous, above Valanginian and below Barremian. { ō·trə'vē·ən }

havgul *See* havgull. { 'hav·gəl }

havgula *See* havgull. { 'hav·gəl·ə }

havgull [METEOROL] Cold, damp wind blowing from the sea during summer in Scotland and Norway. Also known as havgul; havgula. { 'hav·gəl }

haycock [HYD] An isolated ice cone created on land ice or shelf ice because of pressure or ice movement. { 'hāˌkäk }

haze [METEOROL] Fine dust or salt particles dispersed through a portion of the atmosphere; the particles are so small that they cannot be felt, or individually seen with the naked eye, but they diminish horizontal visibility and give the atmosphere a characteristic opalescent appearance that subdues all colors. { hāz }

haze droplet [METEOROL] Any small liquid droplet contributing to an atmospheric haze condition. { 'hāz ˌdräp·lət }

haze factor [METEOROL] The ratio of the luminance of a mist or fog through which an object is viewed to the luminance of the object. { 'hāz ˌfak·tər }

haze horizon [METEOROL] The top of a haze layer which is confined by a low-level temperature inversion and has the appearance of the horizon when viewed from above against the sky. { 'hāz həˌriz·ən }

haze layer [METEOROL] A layer of haze in the atmosphere, usually bounded at the top by a temperature inversion and frequently extending downward to the ground. { 'hāz lā·ər }

haze level *See* haze line. { 'hāz ˌlev·əl }

haze line [METEOROL] The boundary surface in the atmosphere between a haze layer and the relatively clean, transparent air above the top of a haze layer. Also known as haze level. { 'hāz ˌlīn }

hazel sandstone [GEOL] An arkosic, iron-bearing redbed sandstone from the Precambrian found in western Texas. { 'ha·zəl 'sanˌstōn }

head *See* headland. { hed }

head erosion *See* headward erosion. { 'hed iˌrō·zhən }

heading side *See* footwall. { 'hed·iŋ ˌsīd }

heading wall *See* footwall. { 'hed·iŋ ˌwȯl }

headland [GEOGR] **1.** A high, steep-faced promontory extending into the sea. Also known as head; mull. **2.** High ground surrounding a body of water. { 'hed·lənd }

headwall [GEOL] The steep cliff at the back of a cirque. { 'hed,wȯl }

headward erosion [GEOL] Erosion caused by water flowing at the head of a valley. Also known as head erosion; headwater erosion. { 'hed·wərd i'rō·zhən }

headwater erosion *See* headward erosion. { 'hedˌwȯd·ər i'rō·zhən }

headwaters [HYD] The source and upstream waters of a stream. { 'hedˌwȯd·ərz }

heat balance [GEOPHYS] The equilibrium which exists on the average between the radiation received by the earth and atmosphere from the sun and that emitted by the earth and atmosphere. { 'hēt ˌbal·əns }

heat budget [GEOPHYS] Amount of heat needed to raise a lake's water from the winter temperature to the maximum summer temperature. { 'hēt ˌbəj·ət }

heat equator [METEOROL] **1.** The line which circumscribes the earth and connects all points of highest mean annual temperature for their longitudes. **2.** The parallel of latitude of 10°N, which has the highest mean temperature of any latitude. Also known as thermal equator. { 'hēt i¦kwād·ər }

heat flow province [GEOPHYS] A geographic area in which the heat flow and heat production are linearly related. { 'hēt ¦flō ˌpräv·əns }

heating degree-day [METEOROL] A form of degree-day used as an indication of fuel consumption; in United States usage, one heating degree-day is given for each degree that the daily mean temperature departs below the base of 65°F (where the Celsius scale is used, the base is usually 19°C). { 'hēd·iŋ di'grē ˌdā }

heat island effect [METEOROL] In urban areas with tall buildings, an atmospheric condition in which heat and pollutants create a haze dome that prevents warm air from rising and being cooled at a normal rate, especially in the absence of strong winds. { 'hēt ˌī·lənd iˌfekt }

heat lightning [GEOPHYS] Nontechnically, the luminosity observed from ordinary lightning too far away for its thunder to be heard. { 'hēt ˌlīt·niŋ }

heat low *See* thermal low. { 'hēt ¦lō }

heat storage [OCEANOGR] The tendency of the ocean to act as a heat reservoir; results in smaller daily and annual variations in temperature over the sea. { 'hēt ˌstȯr·ij }

heat thunderstorm [METEOROL] In popular terminology, a thunderstorm of the air mass type which develops near the end of a hot, humid summer day. { 'hēt ¦thən·dərˌstȯrm }

heat wave [METEOROL] A period of abnormally and uncomfortably hot and usually humid weather; the condition must prevail at least 1 day to be a heat wave, but conventionally the term is reserved for periods of several days to several weeks. Also known as hot wave; warm wave. { 'hēt ˌwāv }

heave [GEOL] The horizontal component of the slip, measured at right angles to the strike of the fault. [OCEANOGR] The motion imparted to a floating body by wave action. { hēv }

Heaviside layer *See* E layer. { 'hev·ēˌsīd ˌlā·ər }

heavy floe [OCEANOGR] A mass of sea ice that is more than 10 feet (3 meters) thick. Also known as heavy ice. { 'hev·ē 'flō }

heavy ice *See* heavy floe. { 'hev·ē 'īs }

heavy ion *See* large ion. { 'hev·ē 'īˌän }

hedreocraton [GEOL] A craton that influenced later continental development. { ˌhed·rē·ō'krāˌtän }

height-change chart [METEOROL] A chart indicating the change in height of a constant-pressure surface over a specified previous time interval; comparable to a pressure-change chart. { 'hīt ˌchānj ˌchärt }

height-change line [METEOROL] A line of equal change in height of a constant-pressure surface over a specified previous interval of time; the lines drawn on a height-change chart. Also known as contour-change line; isallohypse. { 'hīt ˌchānj ˌlīn }

height of tide [OCEANOGR] Vertical distance from the chart datum to the level of the water at any time; it is positive if the water level is higher than the chart datum. { 'hīt əv 'tīd }

height pattern [METEOROL] The general geometric characteristics of the distribution of height of a constant-pressure surface as shown by contour lines on a constant-pressure chart. Also known as baric topography; isobaric topography; pressure topography. { 'hīt ˌpad·ərn }

Helderbergian [GEOL] A North American stage of geologic time, in the lower Lower Devonian. { ˌhel·dər'bərg·ē·ən }

helictite [GEOL] A speleothem whose origin is similar to that of a stalactite or stalagmite but that angles or twists in an irregular fashion. { 'hē·likˌtīt }

heliosphere [GEOPHYS] The region in the ionosphere where helium ions are predominant (sometimes there may be no region in which helium ions dominate). { 'hē·lē·əˌsfir }

heliotropic wind [METEOROL] A subtle, diurnal component of the wind velocity leading to a diurnal shift of the wind or turning of the wind with the sun, produced by the east-to-west progression of daytime surface heating. { ˌhē·lē·ə'träp·ik 'wind }

Helmert's formula [GEOPHYS] A formula for the acceleration due to gravity in terms of the latitude and the altitude above sea level. { 'hel·mərts ˌfȯr·myə·lə }
helm wind [METEOROL] A strong, cold northeasterly wind blowing down into the Eden valley from the western slope of the Crossfell Range in northern England. { 'helm ¦wind }
hemicone *See* alluvial cone. { 'he·mēˌkōn }
hemipelagic region [OCEANOGR] The region of the ocean extending from the edge of a shelf to the pelagic environment; roughly corresponds to the bathyal zone, in which the bottom is 660 to 3300 feet (200 to 1000 meters) below the surface. { ¦he·mē·pə'laj·ik 'rē·jən }
hemipelagic sediment [GEOL] Deposits containing terrestrial material and the remains of pelagic organisms, found in the ocean depths. { ¦he·mē·pə'laj·ik 'sed·ə·mənt }
hemipelagite [OCEANOGR] Deep-sea mud deposits in which more than 25% of the fraction of particles coarser than 5 micrometers is of terrigenous, volcanogenic, or neritic origin. { ˌhem·ē'pel·əˌjīt }
hemisphere [GEOGR] A half of the earth divided into north and south sections by the equator, or into an east section containing Europe, Asia, and Africa, and a west section containing the Americas. { 'he·mēˌsfir }
hemispheric wave number *See* angular wave number. { ˌhe·mē'sfir·ik ¦wāv ˌnəm·bər }
Hemist [GEOL] A suborder of the soil order Histosol, consisting of partially decayed plant residues and saturated with water most of the time. { 'he·mist }
Hercynian geosyncline [GEOL] A principal area of geosynclinal sediment accumulation in Devonian time; found in south-central and southern Europe and northern Africa. { hər'sin·ē·ən ˌjē·ō'sinˌklīn }
Hercynian orogeny *See* Variscan orogeny. { hər'sin·ē·ən ȯ'räj·ə·nē }
hervidero *See* mud volcano. { ¦hər·və¦der·ō }
heterochronism [GEOL] A phenomenon in which two similar geologic deposits may not be of the same age even though they underwent like processes of formation. { ˌhed·ə'räk·rəˌniz·əm }
heterogeneous reservoir [GEOL] Formation with two or more noncommunicating sand members, each possibly with different specific- and relative-permeability characteristics. { ˌhed·ə·rə'jē·nē·əs 'rez·əvˌwär }
heterosphere [METEOROL] The upper portion of a two-part division of the atmosphere (the lower portion is the homosphere) according to the general homogeneity of atmospheric composition; characterized by variation in composition, and in mean molecular weight of constituent gases; starts at 50–62 miles (80–100 kilometers) above the earth and therefore closely coincides with the ionosphere and the thermosphere. { 'hed·ə·rəˌsfir }
Hettangian [GEOL] A stage of Lower Jurassic geologic time. { he'tan·jē·ən }
hexagonal column [METEOROL] One of the many forms in which ice crystals are found in the atmosphere; this crystal habit is characterized by hexagonal cross-section in a plane perpendicular to the long direction (principal axis, optic axis, or *c* axis) of the columns; it differs from that found in hexagonal platelets only in that environmental conditions have favored growth along the principal axis rather than perpendicular to that axis. { hek'sag·ə·nəl 'käl·əm }
hexagonal platelet [METEOROL] A small ice crystal of the hexagonal tabular form; the distance across the crystal from one side of the hexagon to the opposite side may be as large as about 1 millimeter, and the thickness perpendicular to this dimension is of the order of one-tenth as great; this crystal form is usually formed at temperatures of −10 to −20°C by sublimation; at higher temperatures the apices of the hexagon grow out and develop dendritic forms. { hek'sag·ə·nəl 'plāt·lət }
hexahedrite [GEOL] An iron meteorite composed of single crystals or aggregates of kamacite, usually containing 4–6% nickel in the metal phase. { ˌhek·sə'heˌdrīt }
hiatus [GEOL] A gap in a rock sequence due to a lack of deposition of a bed or to erosion of beds. { hī'ād·əs }
hibernal [METEOROL] Of or pertaining to winter. { hī'bərn·əl }
Hibernian orogeny *See* Erian orogeny. { hī'bər·nē·ən ȯ'räj·ə·nē }

hiemal climate [CLIMATOL] Climate pertaining to winter. { 'hī·ə·məl ¦klī·mət }

hieroglyph [GEOL] Any sort of sedimentary mark or structure occurring on a bedding plane. { 'hī·rə,glif }

high [METEOROL] An area of high pressure, referring to a maximum of atmospheric pressure in two dimensions (closed isobars) in the synoptic surface chart, or a maximum of height (closed contours) in the constant-pressure chart; since a high is, on the synoptic chart, always associated with anticyclonic circulation, the term is used interchangeably with anticyclone. { hī }

high aloft *See* upper-level anticyclone. { ¦hī ə'lȯft }

high-altitude station [METEOROL] A weather observing station at a sufficiently high elevation to be nonrepresentative of conditions near sea level; 6500 feet (about 2000 meters) has been given as a reasonable lower limit. { 'hī ¦al·tə,tüd 'stā·shən }

high-angle fault [GEOL] A fault with a dip greater than 45°. { ¦hī ,aŋ·gəl 'fȯlt }

high clouds [METEOROL] Types of clouds whose mean lower level is above 20,000 feet (6100 meters); principal clouds in this group are cirrus, cirrocumulus, and cirrostratus. { 'hī ¦klau̇dz }

high-energy environment [GEOL] An aqueous sedimentary environment which features a high energy level and turbulent motion, created by waves, currents, or surf, which prevents the settling and piling up of fine-grained sediment. { 'hī ,en·ər·jē in'vī·ərn·mənt }

higher high water [OCEANOGR] The higher of two high tides occurring during a tidal day. { 'hī·ər ¦hī ,wȯd·ər }

higher low water [OCEANOGR] The higher of two low tides occurring during a tidal day. { 'hi·ər ¦lō ,wȯd·ər }

high foehn [METEOROL] The occurrence of warm, dry air above the level of the general surface, accompanied by clear skies, resembling foehn conditions; it is due to subsiding air in an anticyclone, above a cold surface layer; in such circumstances the mountain peaks may be warmer than the lowlands. Also known as free foehn. { ¦hī 'fān }

high fog [METEOROL] The frequent fog on the slopes of the coastal mountains of California, especially applied when the fog overtops the range and extends as stratus over the leeward valleys. { ¦hī 'fäg }

high index [METEOROL] A relatively high value of the zonal index which, in middle latitudes, indicates a relatively strong westerly component of wind flow and the characteristic weather features attending such motion; a synoptic circulation pattern of this type is commonly called a high-index situation. { 'hī ¦in,deks }

highland [GEOGR] **1.** Any relatively large area of elevated or mountainous land standing prominently above adjacent low areas. **2.** The higher land of a region. [GEOL] **1.** A lofty headland, cliff, or other high platform. **2.** A dissected mountain region composed of old folded rocks. { 'hī·lənd }

highland climate *See* mountain climate. { 'hīlənd ¦klī·mət }

highland glacier [HYD] A semicontinuous ice cap or glacier that covers the highest or central portion of a mountainous area and partly reflects irregularities of the land surface lying beneath it. Also known as highland ice. { 'hī·lənd ¦glā·shər }

highland ice *See* highland glacier. { 'hī·lənd ¦īs }

high-level anticyclone *See* upper-level anticyclone. { 'hī ,lev·əl an·tē'sī,klōn }

high-level cyclone *See* upper-level cyclone. { 'hī ,lev·əl 'sī,klōn }

high-level ridge *See* upper-level ridge. { 'hī ,lev·əl 'rij }

high-level thunderstorm [METEOROL] Generally, a thunderstorm based at a comparatively high altitude in the atmosphere, roughly 8000 feet (2400 meters) or higher. { 'hī ,lev·əl 'thən·dər,stȯrm }

high-level trough *See* upper-level trough. { 'hī ,lev·əl 'trȯf }

high plain [GEOGR] A large area of level land situated above sea level. { 'hī ¦plān }

high-pressure area *See* anticyclone. { 'hī ¦presh·ər 'er·ē·ə }

high-rank coal [GEOL] Coal consisting of less than 4% moisture when air-dried, or more than 84% carbon. { 'hī ,raŋk 'kōl }

high tide [OCEANOGR] The maximum height reached by a rising tide. Also known as high water. { 'hī ¦tīd }

high-volatile bituminous coal [GEOL] A bituminous coal composed of more than 31% volatile matter. { 'hī ¦väl·əd·əl bə¦tü·mə·nəs 'kōl }

high water *See* high tide. { 'hī ¦wȯd·ər }

high-water full and change [GEOPHYS] The average interval of time between the transit (upper or lower) of the full or new moon and the next high water at a place. Also known as common establishment; vulgar establishment. { 'hī ¦wȯd·ər 'fu̇l· ən 'chānj }

high-water inequality [OCEANOGR] The difference between the heights of the two high tides during a tidal day. { 'hī ¦wȯd·ər ,in·ə'kwäl·əd·ē }

high-water line [OCEANOGR] The intersection of the plane of mean high water with the shore. { 'hī ¦wȯd·ər ,līn }

high-water lunitidal interval [GEOPHYS] The interval of time between the transit (upper or lower) of the moon and the next high water at a place. { 'hī ¦wȯd·ər ¦lün·ə¦tīd·əl 'in·tər·vəl }

high-water platform *See* wave-cut bench. { 'hī ¦wȯd·ər 'plat,fȯrm }

high-water quadrature [OCEANOGR] The average high-water interval when the moon is at quadrature. { 'hī ¦wȯd·ər'kwäd·rə·chər }

high-water springs *See* mean high-water springs. { 'hī ¦wȯd·ər 'spriŋz }

high-water stand [OCEANOGR] The condition at high tide when there is no change in the height of the water. { 'hī ¦wȯd·ər 'stand }

hill [GEOGR] A land surface feature characterized by strong relief; it is a prominence smaller than a mountain. { hil }

hill creep [GEOL] Slow gravity movement of rock and soil waste down a steep hillside. Also known as hillside creep. { 'hil ,krēp }

hillock [GEOL] A small, low hill. { 'hil·ək }

hillside creep *See* hill creep. { 'hil,sīd ,krēp }

Hilt's law [GEOL] The law that in a small area the deeper coals are of higher rank than those above them. { 'hilts ,lȯ }

hinge fault [GEOL] A fault whose movement is an angular or rotational one on a side of an axis that is normal to the fault plane. { 'hinj ,fȯlt }

hinge line [GEOL] **1.** The line separating the region in which a beach has been thrust upward from that in which it is horizontal. **2.** A line in the plane of a hinge fault separating the part of a fault along which thrust or reverse movement occurred from that having normal movement. { 'hinj ,līn }

hinterland [GEOL] **1.** The region behind the coastal district. **2.** The terrain on the back of a folded mountain chain. **3.** The moving block which forces geosynclinal sediments toward the foreland. { 'hin·tər,land }

historical climate [CLIMATOL] A climate of the historical period (the past 7000 years). { hi'stär·ə·kəl 'klī·mət }

historical geology [GEOL] A branch of geology concerned with the systematic study of bedded rocks and their relations in time and the study of fossils and their locations in a sequence of bedded rocks. { hi'stär·ə·kəl jē'äl·ə·jē }

Histosol [GEOL] An order of wet soils consisting mostly of organic matter, popularly called peats and mucks. { 'his·tə,sȯl }

hitch [GEOL] **1.** A fault of strata common in coal measures, accompanied by displacement. **2.** A minor dislocation of a vein or stratum not exceeding in extent the thickness of the vein or stratum. { hich }

hoar crystal [HYD] An individual ice crystal in a deposit of hoarfrost; always grows by sublimation. { 'hȯr ,krist·əl }

hoarfrost [HYD] A deposit of interlocking ice crystals formed by direct sublimation on objects. Also known as white frost. { 'hȯr,frȯst }

hofmannite *See* hartite. { 'häf·mə,nīt }

hogback [GEOL] Alternate ridges and ravines in certain areas of mountains, caused by erosive action of mountain torrents. { 'häg,bak }

Holocene [GEOL] An epoch of the Quaternary Period from the end of the Pleistocene, around 10,000 years ago, to the present. Also known as Postglacial; Recent. { 'hō·lə,sēn }

holomictic lake [HYD] A lake whose water circulates completely from top to bottom. { ¦häl·ō¦mik·tik 'lāk }

holostratotype [GEOL] The originally defined stratotype. { ¦häl·ō'strad·ə,tīp }

homobront *See* isobront. { ¦häm·ə¦bränt }

homocline [GEOL] Any rock unit in which the strata exhibit the same dip. { 'hä·mə,klīn }

homogeneous atmosphere [METEOROL] A hypothetical atmosphere in which the density is constant with height. { ,hä·mə'jē·nē·əs 'at·mə,sfir }

homologous [GEOL] **1.** Referring to strata, in separated areas, that are correlatable (contemporaneous) and are of the same general character or facies, or occupy analogous structural positions along the strike. **2.** Pertaining to faults, in separated areas, that have the same relative position or structure. { hə'mäl·ə·gəs }

homopause [GEOPHYS] The level of transition between the homosphere and the heterosphere; it lies about 50 to 56 miles (80 to 90 kilometers) above the earth. Also known as turbopause. { 'hä·mə,pöz }

homosphere [METEOROL] The lower portion of a two-part division of the atmosphere (the upper portion is the heterosphere) according to the general homogeneity of atmospheric composition; the region in which there is no gross change in atmospheric composition, that is, all of the atmosphere from the earth's surface to about 50 to 62 miles (80–100 kilometers). { 'hä·mə,sfir }

honeycomb formation [GEOL] A rock stratum containing large cavities or caverns. { 'hən·ē,kōm fȯr,mā·shən }

hook [GEOGR] The end of a spit of land that is turned toward shore. Also known as hooked spit; recurved spit. { hu̇k }

horizon [GEOL] **1.** The surface separating two beds. **2.** One of the layers, each of which is a few inches to a foot thick, that make up a soil. { hə'rīz·ən }

horizon distance [GEOD] The distance, at any given azimuth, to the point on the earth's surface constituting the horizon for some specified observer. { hə'rīz·ən ¦dis·təns }

horizontal displacement *See* strike slip. { ,här·ə'zänt·əl dis'plās·mənt }

horizontal fold *See* nonplunging fold. { ,här·ə'zänt·əl 'fōld }

horizontal intensity [GEOPHYS] The strength of the horizontal component of the earth's magnetic field. { ,här·ə'zänt·əl in'ten·səd·ē }

horizontal pressure force [GEOPHYS] The horizontal pressure gradient per unit mass, $-\alpha\nabla_H p$, where α is the specific volume, p the pressure, and ∇_H the horizontal component of the del operator; this force acts normal to the horizontal isobars toward lower pressure; it is one of the three important forces appearing in the horizontal equations of motion, the others being the Coriolis force and friction. { ,här·ə'zänt·əl 'presh·ər ,fȯrs }

horizontal separation *See* strike slip. { ,här·ə'zänt·əl ,sep·ə'rā·shən }

horn [GEOL] A topographically high, sharp, pyramid-shaped mountain peak produced by the headward erosion of mountain glaciers; the Matterhorn is the classic example. { hȯrn }

horse [GEOL] A large rock caught along a fault. { hȯrs }

horseback [GEOL] A low and sharp ridge of sand, gravel, or rock. { 'hȯrs,bak }

horse latitudes [METEOROL] The belt of latitudes over the oceans at approximately 30–35°N and S where winds are predominantly calm or very light and weather is hot and dry. { 'hȯrs ¦lad·ə,tüdz }

horseshoe bend *See* oxbow. { 'hȯr,shü ,bend }

horseshoe lake *See* oxbow lake. { 'hȯr,shü ,lāk }

horsetail ore [GEOL] An ore occurring in fractures which diverge from a larger fracture. { 'hȯrs,tāl ,ȯr }

horst [GEOL] **1.** A block of the earth's crust uplifted along faults relative to the rocks on either side. **2.** A mass of the earth's crust limited by faults and standing in relief. **3.** One of the older mountain masses limiting the Alps on the west and north. **4.** A knobby ledge of limestone beneath a thin soil mantle. { hȯrst }

Horton number [HYD] A dimensionless number that is formed by the product of runoff intensity and erosion proportionality factor; expresses the relative intensity of erosion on the slopes of a drainage basin. { 'hȯrt·ən ˌnəm·bər }
host rock [GEOL] Rock which serves as a host for other rocks or for mineral deposits. { 'hōst ˌräk }
hot belt [CLIMATOL] The belt around the earth within which the annual mean temperature exceeds 20°C. { 'hät ˌbelt }
hot spring [HYD] A thermal spring whose water temperature is above 98°F (37°C). { 'hät ¦spriŋ }
hot wave *See* heat wave. { 'hät ˌwāv }
hot wind [METEOROL] General term for winds characterized by intense heat and low relative humidity, such as summertime desert winds or an extreme foehn. { 'hät ¦wind }
hourly observation *See* record observation. { 'au̇·ər·lē ˌäb·sər'vā·shən }
howardite [GEOL] An achondritic stony meteorite composed chiefly of calcic plagioclase and orthopyroxene. { 'hau̇·ərˌdīt }
huangho deposit [GEOL] A coastal-plain deposit comprising alluvium spread over a level surface (such as a floodplain) but extending into marine beds of equivalent age. { ¦hwäŋ¦hō diˌpäz·ət }
Hubbard Glacier [GEOGR] A valley glacier which reaches tidewater from a source area of Mount St. Elias of Alaska and the Yukon. { 'həb·ərd 'glā·shər }
human geography [GEOGR] The study of the characteristics and phenomena of the earth's surface that relate directly to or are due to human activities. Also known as anthropogeography. { ¦yü·mən jē¦äg·rə·fē }
Humboldt Current *See* Peru Current. { 'həmˌbōlt ¦kə·rənt }
Humboldt Glacier [HYD] The largest Arctic iceberg, at latitude 79°, with a seaward front extending 65 miles (105 kilometers). { 'həmˌbōlt ¦glā·shər }
humic [GEOL] Pertaining to or derived from humus. { 'hyü·mik }
humic-cannel coal *See* pseudocannel coal. { 'hyü·mik ¦kan·əl 'kōl }
humic coal [GEOL] A coal whose attritus is composed mainly of transparent humic degradation material. { 'hyü·mik 'kōl }
humid climate [CLIMATOL] A climate whose typical vegetation is forest. Also known as forest climate. { 'hyü·məd ¦klī·mət }
humidity [METEOROL] Atmospheric water vapor content, expressed in any of several measures, such as relative humidity. { hyü'mid·əd·ē }
humidity coefficient [METEOROL] A measure of the precipitation effectiveness of a region; it recognizes the exponential relationship of temperature versus plant growth and is expressed as humidity coefficient = $P/(1.07)^t$, where P is the precipitation in centimeters, and t is the mean temperature in degrees Celsius for the period in question; the denominator approximately doubles with each 10°C rise in temperature. { hyü'mid·əd·ē ˌkō·i'fish·ənt }
humidity index [CLIMATOL] An index of the degree of water surplus over water need at any given station; it is calculated as humidity index = $100s/n$, where s (the water surplus) is the sum of the monthly differences between precipitation and potential evapotranspiration for those months when the normal precipitation exceeds the latter, and where n (the water need) is the sum of monthly potential evapotranspiration for those months of surplus. { hyü'mid·əd·ē ˌinˌdeks }
humidity mixing ratio [METEOROL] The amount of water vapor mixed with one unit mass of dry air, usually expressed as grams of water vapor per kilogram of air. { hyü'mid·əd·ē 'mik·siŋ ˌrā·shō }
humidity province [CLIMATOL] A region in which the precipitation effectiveness of its climate produces a definite type of biological consequence, in particular the climatic climax formations of vegetation (rain forest, tundra, and the like). { hyü'mid·əd·ē ˌpräv·əns }
humification Formation of humus. { ˌhyü·mə·fə'kā·shən }
humin *See* ulmin. { 'hyü·mən }
hummock [GEOL] A rounded or conical knoll, mound, hillock, or other small elevation,

generally of equal dimensions and not ridgelike. Also known as hammock. [HYD] A mound, hillock, or pile of broken floating ice, either fresh or weathered, that has been forced upward by pressure, as in an ice field or ice floe. { 'həm·ək }

hummocked ice [OCEANOGR] Pressure ice, characterized by haphazardly arranged mounds or hillocks; it has less definite form, and show the effects of greater pressure, than either rafted ice or tented ice, but in fact may develop from either of those. { 'həm·əkt 'īs }

hummocky [GEOL] Any topographic surface characterized by rounded or conical mounds. { 'həm·ə·kē }

Humod [GEOL] A suborder of the soil order Spodosol having an accumulation of humus, and of aluminum but not iron. { 'hyü,mäd }

humodurite *See* translucent attritus. { ¦hyü·mə¦dů,rīt }

humogelite *See* ulmin. { hyü'mäj·əl,īt }

Humox [GEOL] A suborder of the soil order Oxisol that is high in organic matter, well drained but moist all or nearly all year, and restricted to relatively cool climates and high altitudes for Oxisols. { 'hyü,mäks }

Humult [GEOL] A suborder of the soil order Ultisol, well drained with a moderately thick surface horizon; formed under conditions of high rainfall distributed evenly over the year; common in southeastern Brazil. { 'hyü·məlt }

humus [GEOL] The amorphous, ordinarily dark-colored, colloidal matter in soil; a complex of the fractions of organic matter of plant, animal, and microbial origin that are most resistant to decomposition. { 'hyü·məs }

Huronian [GEOL] The lower system of the restricted Proterozoic. { hyü'rō·nē·ən }

hurricane [METEOROL] A tropical cyclone of great intensity; any wind reaching a speed of more than 73 miles per hour (117 kilometers per hour) is said to have hurricane force. { 'hər·ə,kān }

hurricane band *See* spiral band. { 'hər·ə,kān ,band }

hurricane-force wind [METEOROL] In the Beaufort wind scale, a wind whose speed is 64 knots (117 kilometers per hour) or higher. { 'hər·ə,kān ,fórs ,wind }

hurricane monitoring buoy [METEOROL] A free-floating automatic weather station designed as an expendable instrument in connection with hurricane and typhoon monitoring and forecasting services. { 'hər·ə,kān 'män·ə·triŋ ,bói }

hurricane radar band *See* spiral band. { 'hər·ə,kān 'rā,där ,band }

hurricane surge *See* hurricane wave. { 'hər·ə,kān ¦sərj }

hurricane tide *See* hurricane wave. { 'hər·ə,kān ¦tīd }

hurricane warning [METEOROL] A warning of impending winds of hurricane force; for maritime interests, the storm warning signals for this condition are two square red flags with black centers by day, and a white lantern between two red lanterns by night. { 'hər·ə,kān ,wórn·iŋ }

hurricane watch [METEOROL] An announcement for a specific area that hurricane conditions pose a threat; residents are cautioned to take stock of their preparedness needs but, otherwise, are advised to continue normal activities. { 'hər·ə,kān ,wäch }

hurricane wave [OCEANOGR] As experienced on islands and along a shore, a sudden rise in the level of the sea associated with a hurricane. Also known as hurricane surge; hurricane tide. { 'hər·ə,kān ¦wāv }

hurricane wind [METEOROL] In general, the severe wind of an intense tropical cyclone (hurricane or typhoon); the term has no further technical connotation, but is easily confused with the strictly defined hurricane-force wind. Also known as typhoon wind. { 'hər·ə,kān ¦wind }

hyaline [GEOL] Transparent and resembling glass. { 'hī·ə·lən }

hyalobasalt *See* tachylite. { ¦hī·ə·lō·bə'sólt }

hyaloclastite [GEOL] A tufflike deposit formed by the flowing of basalt under water and ice and its consequent fragmentation. Also known as aquagene tuff. { ¦hī·ə·lō'kla,stīt }

hyalopsite *See* obsidian. { ,hī·ə'läp,sīt }

hydatogenesis [GEOL] Crystallization and deposition of minerals from aqueous solutions. { ¦hīd·ə·tō'jen·ə·səs }

hydraulic current [OCEANOGR] A current in a channel, due to a difference in the water level at the two ends. { hī'drȯ·lik 'kə·rənt }

hydraulic discharge [HYD] The direct discharge of groundwater from the zone of saturation upon the land or into a body of surface water. { hī'drȯ·lik 'dis,chärj }

hydraulic gradient [HYD] The slope of the hydraulic grade line of a stream. { hī'drȯ·lik 'grād·ē·ənt }

hydraulic profile [HYD] A vertical section of an aquifer's potentiometric surface. { hi'drȯl·lik 'prō,fīl }

hydraulic ratio [GEOL] The weight of a heavy mineral multiplied by 100 and divided by the weight of a hydraulically equivalent light mineral. { hi'drȯ·lik 'rā·shō }

hydrocast [OCEANOGR] A series of water samplers on a single hydrographic wire which obtain samples simultaneously. { 'hī·drə,kast }

hydrogeochemistry [GEOCHEM] The study of the chemical characteristics of ground and surface waters as related to areal and regional geology. { ¦hī·drō,jē·ō'kem·ə·strē }

hydrogeology [HYD] The science dealing with the occurrence of surface and ground water, its utilization, and its functions in modifying the earth, primarily by erosion and deposition. { ¦hī·drō·jē'äl·ə·jē }

hydrograph [HYD] A graphical representation of stage, flow, velocity, or other characteristics of water at a given point as a function of time. { 'hī·drə,graf }

hydrographic basin *See* drainage basin. { 'hī·drə'graf·ik 'bās·ən }

hydrographic cruise [OCEANOGR] Exploration of a body of water for hydrographic surveys. { 'hī·drə'graf·ik 'krüz }

hydrographic survey [OCEANOGR] Survey of a water area with particular reference to tidal currents, submarine relief, and any adjacent land. { 'hī·drə'graf·ik 'sər,vā }

hydrographic table [OCEANOGR] Tabular arrangement of data relating sea-water density to salinity, temperature, and pressure. { 'hī·drə'graf·ik 'tā·bəl }

hydrography [GEOGR] Science which deals with the measurement and description of the physical features of the oceans, lakes, rivers, and their adjoining coastal areas, with particular reference to their control and utilization. { hī'dräg·rə·fē }

hydrolaccolith [GEOL] A frost mound, 0.3–20 feet (0.1–6 meters) in height, having a core of ice and resembling a laccolith in section. Also known as cryolaccolith. { ,hī·drə'lak·ə,lith }

hydrologic accounting [HYD] A systematic summary of the terms (inflow, outflow, and storage) of the storage equation as applied to the computation of soil-moisture changes, groundwater changes, and so forth; an evaluation of the hydrologic balance of an area. Also known as basin accounting; water budget. { ¦hī·drə¦läj·ik ə'kaůnt·iŋ }

hydrologic cycle [HYD] The complete cycle through which water passes, from the oceans, through the atmosphere, to the land, and back to the ocean. Also known as water cycle. { ¦hī·drə¦läj·ik 'sī·kəl }

hydrologic sequence [GEOL] A series of soil sections from differentiated parent material that shows increasing lack of drainage downslope. { ¦hī·drə¦läj·ik 'sē·kwəns }

hydrologist [HYD] An individual who specializes in hydrology. { hī'dräl·ə·jəst }

hydrolyzate [GEOL] A sediment characterized by elements such as aluminum, potassium, or sodium which are readily hydrolyzed. { hī'dräl·ə,zāt }

hydrometamorphism [GEOL] Alteration of rocks by material carried in solution by water without the influence of high temperature or pressure. { ,hī·drə,med·ə'mȯr·fiz·əm }

hydrometeor [HYD] **1.** Any product of condensation or sublimation of atmospheric water vapor, whether formed in the free atmosphere or at the earth's surface. **2.** Any water particles blown by the wind from the earth's surface. { ¦hī·drō'mēd·ē·ər }

hydrometeorology [METEOROL] That part of meteorology of direct concern to hydrologic problems, particularly to flood control, hydroelectric power, irrigation, and similar fields of engineering and water resources. { ¦hī·drō,mēd·ē·ə'räl·ə·jē }

hydromica [GEOL] Any of several varieties of muscovite, especially illite, which are less elastic than mica, have a pearly luster, and sometimes contain less potash and more water than muscovite. Also known as hydrous mica. { ¦hī·drō'mī·kə }

hydromorphic [GEOL] Referring to an intrazonal soil with characteristics that were developed in the presence of excess water all or part of the time. { ¦hī·drə'mȯr·fik }

hydrosphere [HYD] The water portion of the earth as distinguished from the solid part (lithosphere) and from the gaseous outer envelope (atmosphere). { 'hī·drə,sfir }

hydrostatic approximation [METEOROL] The assumption that the atmosphere is in hydrostatic equilibrium. { ,hī·drə'stad·ik ə,präk·sə'mā·shən }

hydrostatic assumption [GEOPHYS] **1.** The assumption that the pressure of seawater increases by 1 atmosphere (101,325 pascals) over approximately 33 feet (10 meters) of depth, the exact value depending on the water density and the local acceleration of gravity. **2.** Specifically, the assumption that fluid is not undergoing vertical accelerations, hence the vertical component of the passive gradient force per unit mass is equal to g, the local acceleration due to gravity. { ,hī·drə'stad·ik ə'səm·shən }

hydrostatic stability *See* static stability. { ,hī·drə'stad·ik stə'bil·əd·ē }

hydrothermal [GEOL] Of or pertaining to heated water, to its action, or to the products of such action. { ,hī·drə'thər·məl }

hydrothermal alteration [GEOL] Rock or mineral phase changes that are caused by the interaction of hydrothermal liquids and wall rock. { ,hī·drə'thər·məl ,ȯl·tə'rā·shən }

hydrothermal deposit [GEOL] A mineral deposit precipitated from a hot, aqueous solution. { ,hī·drə'thər·məl di'päz·ət }

hydrothermal solution [GEOL] Hot, residual watery fluids derived from magmas during the later stages of their crystallization and commonly containing large amounts of dissolved metals which are deposited as ore veins in fissures along which the solutions often move. { ,hī·drə'thər·məl sə'lü·shən }

hydrothermal synthesis [GEOL] Mineral synthesis in the presence of heated water. { ,hī·drə'thər·məl 'sin·thə·səs }

hydrothermal vent [OCEANOGR] A hot spring on the ocean floor, found mostly along mid-oceanic ridges, where heated fluids exit from cracks in the earth's crust. Iron, sulfur, and other materials precipitate from these waters to form dark clouds. Also known as black smoker. { ,hī·drə¦thərm·əl 'vent }

hydrous mica *See* hydromica. { 'hī·drəs 'mī·kə }

hyetal coefficient *See* pluviometric coefficient. { 'hī·əd·əl ,kō·i'fish·ənt }

hyetal equator [CLIMATOL] A line (or transition zone) which encircles the earth (north of the geographical equator) and lies between two belts that typify the annual time distribution of rainfall in the lower latitudes of each hemisphere; a form of meteorological equator. { 'hī·əd·əl i'kwād·ər }

hyetal region [CLIMATOL] A region in which the amount and seasonal variation of rainfall are of a given type. { 'hī·əd·əl ,rē·jən }

hyetograph [CLIMATOL] A map or chart displaying temporal or areal distribution of precipitation. { hī'ed·ə,graf }

hyetography [CLIMATOL] The study of the annual variation and geographic distribution of precipitation. { ,hī·ə'täg·rə·fē }

hyetology [METEOROL] The science which treats of the origin, structure, and various other features of all the forms of precipitation. { ,hī·ə'täl·ə·jē }

hygrokinematics [METEOROL] The descriptive study of the motion of water substances in the atmosphere. { ¦hī·grə,kin·ə'mad·iks }

hygrology [METEOROL] The study which deals with the water vapor content (humidity) of the atmosphere. { hī'gräl·ə·jē }

hygroscopic coefficient [HYD] The percentage of water that a soil will absorb and hold in equilibrium in a saturated atmosphere. { ¦hī·grə¦skäp·ik ,kō·i'fish·ənt }

hygroscopic water [HYD] The component of soil water that is held adsorbed on the surface of soil particles and is not available to vegetation. { ¦hī·grə¦skäp·ik 'wȯd·ər }

hyperacoustic zone [GEOPHYS] The region in the upper atmosphere, between 62 and 100 miles (100 and 160 kilometers), where the distance between the rarefied air molecules roughly equals the wavelength of sound, so that sound is transmitted with less volume than at lower levels. { ¦hī·pər·ə¦kü·stik ,zōn }

hyperpycnal inflow [HYD] A denser inflow that occurs when a sediment-laden fluid flows down the side of a basin and along the bottom as a turbidity current. { ¦hī·pər¦pik·nəl 'in,flō }

hypersaline [GEOL] Geologic material with high salinity. { ¦hī·pər'sā,lēn }

hypocenter [GEOPHYS] The point along a fault where an earthquake is initiated. { 'hī·pə,sent·ər }

hypogene [GEOL] **1.** Of minerals or ores, formed by ascending waters. **2.** Of geologic processes, originating within or below the crust of the earth. { 'hī·pə,jēn }

hypolimnion [HYD] The lower level of water in a stratified lake, characterized by a uniform temperature that is generally cooler than that of other strata in the lake. { ¦hī·pō'lim·nē,än }

hypomagma [GEOL] Relatively immobile, viscous lava that forms at depth beneath a shield volcano, is undersaturated with gases, and initiates volcanic activity. { ¦hī·pō¦mag·mə }

hypopycnal inflow [HYD] Flowing water of lower density than the body of water into which it flows. { ,hī·pō'pik·nəl 'in,flō }

hypothermal [GEOL] Referring to the high-temperature (300–500°C) environment of hypothermal deposits. { ¦hī·pō'thər·məl }

Hypsithermal *See* Altithermal. { ,hip·sə'thərm·əl }

hypsography [GEOGR] The science of measuring or describing elevations of the earth's surface with reference to a given datum, usually sea level. { hip'säg·rə·fē }

hypsometric formula [GEOPHYS] A formula, based on the hydrostatic equation, for either determining the geopotential difference or thickness between any two pressure levels, or for reducing the pressure observed at a given level to that at some other level. { ,hip·sə'me·trik 'fȯr·myə·lə }

I

IAC *See* international analysis code.

Ibe wind [METEOROL] A local strong wind which blows through the Dzungarian Gate (western China), a gap in the mountain ridge separating the depression of Lakes Balkash and Ala Kul from that of Lake Ebi Nor; the wind resembles the foehn and brings a sudden rise of temperature, in winter from about −15 to about 30°F (−20 to −1°C). { ˌē·be ˌwind }

ice accretion [HYD] The process by which a layer of ice builds up on solid objects which are exposed to freezing precipitation or to supercooled fog or cloud droplets. { 'īs əˌkrē·shən }

ice age [GEOL] A major interval of geologic time during which extensive ice sheets (continental glaciers) formed over many parts of the world. { 'īs ˌāj }

Ice Age *See* Pleistocene. { 'īs ˌāj }

ice apron [HYD] **1.** The snow and ice attached to the walls of a cirque. **2.** The ice that is flowing from an ice sheet over the edge of a plateau. **3.** A piedmont glacier's lobe. **4.** Ice that adheres to a wall of a valley below a hanging glacier. { 'īs ˌā·prən }

ice band [HYD] A layer of ice in firn or snow. { 'īs ˌband }

ice barrier [HYD] The periphery of the Antarctic ice sheet; or used generally for any ice dam. { 'īs ˌbar·ē·ər }

ice bay [OCEANOGR] A baylike recess in the edge of a large ice floe or ice shelf. Also known as ice bight. { 'īs ˌbā }

ice belt [OCEANOGR] A band of fragments of sea ice in otherwise open water. Also known as ice strip. { 'īs ˌbelt }

iceberg [OCEANOGR] A large mass of glacial ice broken off and drifted from parent glaciers or ice shelves along polar seas; it is distinguished from polar pack ice, which is sea ice, and from frozen seawater, whose rafted or hummocked fragments may resemble small icebergs. { 'īsˌbərg }

ice bight *See* ice bay. { 'īs ˌbīt }

ice blink [METEOROL] A relatively bright, usually yellowish-white glare on the underside of a cloud layer, produced by light reflected from an ice-covered surface such as pack ice; used in polar regions with reference to the sky map; ice blink is not as bright as snow blink, but much brighter than water sky or land sky. { 'īsˌbliŋk }

ice boundary [HYD] At any given time, the boundary between fast ice and pack ice or between areas of different concentrations of pack ice. { 'īs ˌbaủn·drē }

ice bridge [OCEANOGR] Surface river ice of sufficient thickness to impede or prevent navigation. { 'īs ˌbrij }

ice cake [HYD] A single, usually relatively flat piece of ice of any size in a body of water. { 'īs ˌkāk }

ice calving *See* calving. { 'īs ˌkav·iŋ }

ice canopy *See* pack ice. { 'īs ˌkam·ə·pē }

ice cap [HYD] **1.** A perennial cover of ice and snow in the shape of a dome or plate on the summit area of a mountain through which the mountain peaks emerge. **2.** A perennial cover of ice and snow on a flat land mass such as an Arctic island. { 'īs ˌkap }

ice-cap climate *See* perpetual frost climate. { 'īs ˌkap ˌklīm·ət }

ice cascade *See* icefall. { 'īs ka'skād }

ice cave [GEOL] A cave that is cool enough to hold ice through all or most of the warm season. [HYD] A cave in ice such as a glacier formed by a stream of melted water. { 'īs ˌkāv }

ice clearing *See* polyn'ya. { 'īs ˌklir·iŋ }

ice concentration [OCEANOGR] In sea ice reporting, the ratio of the areal extent of ice present and the total areal extent of ice and water. Also known as ice cover. { 'īs ˌkäns·ən'trā·shən }

ice-contact delta [GEOL] A delta formed by a stream flowing between a valley slope and the margin of glacial ice. Also known as delta moraine; morainal delta. { 'īs ¦kän,tak ˌdel·tə }

ice cover *See* ice concentration. { 'īs ˌkəv·ər }

ice crust [HYD] A type of snow crust; a layer of ice, thicker than a film crust, upon a snow surface, formed by the freezing of meltwater or rainwater which has flowed onto it. { 'īs ˌkrəst }

ice-crystal cloud [METEOROL] A cloud consisting entirely of ice crystals, such as cirrus (in this sense distinguished from water clouds and mixed clouds), and having a diffuse and fibrous appearance quite different from that typical of water droplet clouds. { 'īs ˌkrist·əl ˌklau̇d }

ice-crystal fog *See* ice fog. { 'īs ˌkrist·əl ˌfäg }

ice-crystal haze [METEOROL] A type of very light ice fog composed only of ice crystals and at times observable to altitudes as great as 20,000 feet (6100 meters), and usually associated with precipitation of ice crystals. { 'īs ˌkrist·əl ˌhāz }

ice-crystal theory *See* Bergeron-Findeisen theory. { 'īs ˌkrist·əl ˌthē·ə·rē }

ice day [CLIMATOL] A day on which the maximum air temperature in a thermometer shelter does not rise above 32°F (0°C), and ice on the surface of water does not thaw. { 'īs ˌdā }

ice desert [CLIMATOL] Any polar area permanently covered by ice and snow, with no vegetation other than occasional red snow or green snow. { 'īs ˌdez·ərt }

iced firn [HYD] A mixture of glacier ice and firn; firn permeated with meltwater and then refrozen. Also known as firn ice. { īst ¦fərn }

ice erosion [GEOL] **1.** Erosion due to freezing of water in rock fractures. **2.** *See* glacial erosion. { 'īs i'rō·zhən }

icefall [HYD] That portion of a glacier where a sudden steepening of descent causes a chaotic breaking up of the ice. Also known as ice cascade. { 'īs ˌfȯl }

ice fat *See* grease ice. { 'īs ¦fat }

ice feathers [HYD] A type of hoarfrost formed on the windward side of terrestrial objects and on aircraft flying from cold to warm air layers. Also known as frost feathers. { 'īs ˌfeth·ərz }

ice field [HYD] A mass of land ice resting on a mountain region and covering all but the highest peaks. [OCEANOGR] A flat sheet of sea ice that is more than 5 miles (8 kilometers) across. { 'īs ˌfēld }

ice floe *See* floe. { 'īs ˌflō }

ice flowers [HYD] **1.** Formations of ice crystals on the surface of a quiet, slowly freezing body of water. **2.** Delicate tufts of hoarfrost that occasionally form in great abundance on an ice or snow surface. Also known as frost flowers. **3.** Frost crystals resembling a flower, formed on salt nuclei on the surface of sea ice as a result of rapid freezing of sea water. Also known as salt flowers. { 'īs ˌflau̇·ərz }

ice fog [METEOROL] A type of fog composed of suspended particles of ice, partly ice crystals 20–100 micrometers in diameter but chiefly, especially when dense, droxtals 12–20 micrometers in diameter; occurs at very low temperatures and usually in clear, calm weather in high latitudes. Also known as frost flakes; frost fog; frozen fog; ice-crystal fog; pogonip; rime fog. { 'īs ˌfäg }

ice foot [OCEANOGR] Sea ice firmly frozen to a polar coast at the high-tide line and unaffected by tide; this fast ice is formed by the freezing of seawater during ebb tide, and of spray, and it is separated from the floating sea ice by a tide crack. { 'īs ¦fu̇t }

ice-free [HYD] **1.** Referring to a harbor, river, estuary, and so on, when there is not sufficient ice present to interfere with navigation. **2.** Descriptive of a water surface completely free of ice. { 'īs ¦frē }

ice fringe [HYD] An ice deposit on plant surfaces, not of hoarfrost from atmospheric water vapor, but of moisture exuded from the stems of plants and appearing as frosted fringes or ribbons. Also known as ice ribbon. [OCEANOGR] A belt of sea ice extending a short distance from the shore. { 'īs ˌfrinj }

ice front [HYD] The floating vertical cliff forming the seaward face or edge of an ice shelf or other glacier that enters water. { 'īs ˌfrənt }

ice gland [HYD] A column of ice in the granular snow at the top of a glacier. { 'īs ˌgland }

ice gruel [HYD] A type of slush formed by the irregular freezing together of ice crystals. { 'īs ˌgrül }

ice island [OCEANOGR] A large tabular fragment of shelf ice found in the Arctic Ocean and having an irregular surface, thickness of 15–50 meters (50–165 feet), and an area between a few thousand square meters and 500 square kilometers (200 square miles) or more. { 'īs ¦ī·lənd }

ice-island iceberg [OCEANOGR] An iceberg having a conical or dome-shaped summit, often mistaken by mariners for ice-covered islands. { 'īs ¦ī·lənd 'īsˌbərg }

ice jam [HYD] **1.** An accumulation of broken river ice caught in a narrow channel, frequently producing local floods during a spring breakup. **2.** Fields of lake or sea ice thawed loose from the shores in early spring, and blown against the shore, sometimes exerting great pressure. { 'īs ˌjam }

ice-laid drift *See* till. { 'īs ¦lād ˌdrift }

Iceland agate *See* obsidian. { 'īs·lənd 'ag·ət }

Icelandic low [METEOROL] **1.** The low-pressure center located near Iceland (mainly between Iceland and southern Greenland) on mean charts of sea-level pressure. **2.** On a synoptic chart, any low centered near Iceland. { 'īs'land·ik 'lō }

ice layer [HYD] An ice crust covered with new snow; when exposed at a glacier front or in crevasses, the ice layers viewed in cross section are termed ice bands. { 'īs ˌlā·ər }

ice mantle *See* ice sheet. { 'īs ˌmant·əl }

ice mound *See* ground ice mound. { 'īs ˌmau̇nd }

ice nucleus [METEOROL] Any particle which may act as a nucleus in formation of ice crystals in the atmosphere. { 'īs ˌnü·klē·əs }

ice pack *See* pack ice. { 'īs ˌpak }

ice pellets [METEOROL] A type of precipitation consisting of transparent or translucent pellets of ice 0.2 inch (5 millimeters) or less in diameter; may be spherical, irregular, or (rarely) conical in shape. { 'īs ˌpel·əts }

ice period [CLIMATOL] The interval between the first appearance and the final dissipation of ice during any year in a given locale. { 'īs ˌpir·ē·əd }

ice pillar [HYD] A column of glacial ice covered with stones or debris which tend to protect the ice from melting. { 'īs ˌpil·ər }

ice pole [GEOGR] The approximate center of the most consolidated portion of the arctic pack ice, near 83 or 84°N and 160°W. Also known as pole of inaccessibility. { 'īs ¦pōl }

ice push [GEOL] Lateral pressure that is caused by expansion of shoreward-moving ice on a lake or a bay of the sea and that follows a rise in temperature. Also known as ice shove; ice thrust. { 'īs ˌpu̇sh }

icequake [HYD] The crash or concussion that accompanies the breakup of ice masses, frequently owing to contraction from the extreme cold. { 'īsˌkwāk }

ice-rafting [GEOL] The transporting of rock and other minerals, of all sizes, on or within icebergs, ice floes, river drift, or other forms of floating ice. { 'īs ˌraf·tiŋ }

ice ribbon *See* ice fringe. { 'īs ˌrib·ən }

ice rind [HYD] A thin but hard layer of sea ice, river ice, or lake ice, which is either a new encrustation upon old ice or a single layer of ice usually found in bays and fiords, where fresh water freezes on top of slightly colder sea water. { 'īs ˌrīnd }

ice run [HYD] The initial stage in the spring or summer breakup of river ice, being an exceedingly rapid process, seldom taking more than 1 day. { 'īs ˌrən }
ice sheet [HYD] A thick glacier, more than 19,300 square miles (50,000 square kilometers) in area, forming a cover of ice and snow that is continuous over a land surface and moving outward in all directions. Also known as ice mantle. { 'īs ˌshēt }
ice shelf [OCEANOGR] A thick sheet of ice with a fairly level or undulating surface, formed along a polar coast and in shallow bays and inlets, fastened to the shore along one side but mostly afloat and nourished by annual accumulation of snow and by the seaward extension of land glaciers. { 'īs ˌshelf }
ice shove *See* ice push. { 'īs ˌshəv }
ice storm [METEOROL] A storm characterized by a fall of freezing precipitation, forming a glaze on terrestrial objects that creates many hazards. Also known as silver storm. { 'īs ˌstȯrm }
ice stream [HYD] A current of ice flowing in an ice sheet or ice cap; usually moves toward an ocean or to an ice shelf. { 'īs ˌstrēm }
ice strip *See* ice belt. { 'īs ˌstrip }
ice thrust *See* ice push. { 'īs ˌthrəst }
ice tongue [HYD] Any narrow extension of a glacier or ice shelf, such as a projection floating in the sea or an outlet glacier of an ice cap. { 'īs ˌtəŋ }
ice wall [HYD] A cliff of ice forming the seaward margin of a glacier that is not afloat. { 'īs ˌwȯl }
ice wedge *See* foliated ice. { 'īs ˌwej }
ichnofacies [GEOL] A recurrent assemblage of ichnofossils that represent certain environmental conditions. { ¦ik·nō¦fā,shēz }
ichnofossil *See* trace fossil. { 'ik·nə,fäs·əl }
ichor [GEOL] A fluid rich in mineralizers. { 'ī,kȯr }
icicle [HYD] Ice shaped like a narrow cone, hanging point downward from a roof, fence, or other sheltered or heated source from which water flows and freezes in below-freezing air. { ī,sik·əl }
icing [HYD] **1.** Any deposit or coating of ice on an object, caused by the impingement and freezing of liquid (usually supercooled) hydrometeors. **2.** A mass or sheet of ice formed on the ground surface during the winter by successive freezing of sheets of water that may seep from the ground, from a river, or from a spring. Also known as flood icing; flooding ice. { 'ī·siŋ }
icing level [METEOROL] The lowest level in the atmosphere at which an aircraft in flight does, or could, encounter aircraft icing conditions over a given locality. { 'ī·siŋ ˌlev·əl }
ICL *See* lifting condensation level.
ideogenous *See* syngenetic. { ˌid·ē'äj·ə·nəs }
idioblast [GEOL] A mineral constituent of a metamorphic rock formed by recrystallization which is bounded by its own crystal faces. { 'id·ē·ō,blast }
IFR terminal minimums [METEOROL] The operational weather limits concerned with minimum conditions of ceiling and visibility at an airport under which aircraft may legally approach and land under instrument flight rules; these minimums frequently are in the form of a sliding scale, and also vary with aircraft type, pilot experience, and from airport to airport. { ¦ī¦ef'är ¦tər·mən·əl 'min·ə·məmz }
IFR weather *See* instrument weather. { ¦ī¦ef'är 'weth·ər }
igneous meteor [GEOPHYS] A visible electric discharge in the atmosphere; lightning is the most common and important type, but types of corona discharge are also included. { 'ig·nē·əs ¦mēd·ē·ər }
igneous province *See* petrographic province. { 'ig·nē·əs präv·əns }
IGY *See* International Geophysical Year.
Illinoian [GEOL] The third glaciation of the Pleistocene in North America, between the Yarmouth and Sangamon interglacial stages. { ¦il·ə¦nȯi·ən }
illumination climate [METEOROL] Also known as light climate. **1.** The worldwide distribution of natural light from the sun and sky (direct solar radiation plus diffuse sky

radiation) as received on a horizontal surface. **2.** The character of total illumination at any given place. { ə,lü·mə,nā·shən ,klī·mət }

illuvial [GEOL] Pertaining to a region or material characterized by the accumulation of soil by the illuviation of another zone or material. { i'lü·vē·əl }

illuvial horizon *See* B horizon. { i'lü·vē·əl hə'rīz·ən }

illuviation [GEOL] The deposition of colloids, soluble salts, and small mineral particles in an underlying layer of soil. { i,lü·vē'ā·shən }

illuvium [GEOL] Material leached by chemical or other processes from one soil horizon and deposited in another. { i'lü·vē·əm }

imbricate structure [GEOL] **1.** A sedimentary structure characterized by shingling of pebbles all inclined in the same direction with the upper edge of each leaning downstream or toward the sea. Also known as shingle structure. **2.** Tabular masses that overlap one another and are inclined in the same direction. Also known as schuppen structure; shingle-block structure. { 'im·brə·kət ,strək·chər }

imbrication [GEOL] Formation of an imbricate structure. Also known as shingling. { ,im·brə'kā·shən }

immature soil *See* azonal soil. { ¦im·ə'chu̇r 'sȯil }

impact cast *See* prod cast. { 'im,pakt ,kast }

impact crater [GEOL] A crater formed on a planetary surface by the impact of a projectile. { 'im,pakt ,krād·ər }

impactite [GEOL] Glassy fused rock or meteor fragments resulting from heat of impact of a meteor on the earth. { 'im,pak,tīt }

impact mark *See* prod mark. { 'im,pakt ,märk }

impression [GEOL] A form left on a soft soil surface by plant parts; the soil hardens and usually the imprint is a concave feature. { im'presh·ən }

imprint *See* overprint. { 'im,print }

impsonite [GEOL] A black, asphaltic pyrobitumen with a high fixed-carbon content derived from the metamorphosis of petroleum. { 'im·sə,nīt }

inactive front [METEOROL] A front, or portion thereof, that produces very little cloudiness and no precipitation, as opposed to an active front. Also known as passive front. { in'ak·tiv 'frənt }

incandescent tuff flow *See* ash flow. { ,in·kən'des·ənt 'təf ,flō }

incarbonization *See* coalification. { in,kär·bə·nə'zā·shən }

Inceptisol [GEOL] A soil order characterized by soils that are usually moist, with pedogenic horizons of alteration of parent materials but not of illuviation. { in'sep·tə,sȯl }

incised meander [GEOL] A deep, tortuous valley cut by a meandering stream that was rejuvenated. { in'sīzd mē'an·dər }

inclination [GEOL] The angle at which a geological body or surface deviates from the horizontal or vertical; often used synonymously with dip. [GEOPHYS] In magnetic inclination, the dip angle of the earth's magnetic field. Also known as magnetic dip. { ,iŋ·klə'nā·shən }

inclination of the wind [METEOROL] The angle between the direction of the wind and the isobars. { ,iŋ·klə'nā·shən əv the 'wind }

inclined bedding [GEOL] A type of bedding in which the strata dip in the direction of current flow. { in'klīnd 'bed·iŋ }

inclined contact [GEOL] A contact plane of gas or oil with water underlying, in which the plane slopes or is inclined. { in'klīnd 'kän,takt }

incoalation *See* coalification. { ,in·kō'lā·shən }

incoherent [GEOL] Pertaining to a rock or deposit that is loose or unconsolidated, or that is unable to hold together firmly or solidly. { ,in·kō'hir·ənt }

incompetent bed [GEOL] A bed not combining sufficient firmness and flexibility to transmit a thrust and to lift a load by bending. { in'käm·pəd·ənt 'bed }

incongruous [GEOL] Of a drag fold, having an axis and axial surface not parallel to the axis and axial surface of the main fold to which it is related. { in'käŋ·grü·əs }

increment *See* recharge. { 'iŋ·krə·mənt }

incumbent [GEOL] Lying above, said of a stratum that is superimposed or overlies another stratum. { in'kəm·bənt }

incus [METEOROL] A supplementary cloud feature peculiar to cumulonimbus capillatus; the spreading of the upper portion of cumulonimbus when this part takes the form of an anvil with a fibrous or smooth aspect. Also known as anvil; thunderhead. { 'iŋ·kəs }

indefinite ceiling [METEOROL] After United States weather observing practice, the ceiling classification applied when the reported ceiling value represents the vertical visibility upward into surface-based, atmospheric phenomena (except precipitation), such as fog, blowing snow, and all of the lithometeors. Formerly known as ragged ceiling. { in'def·ə·nət 'sēl·iŋ }

index bed *See* key bed. { 'in,deks ,bed }

index cycle [METEOROL] A roughly cyclic variation in the zonal index. { 'in,deks ,sī·kəl }

index of aridity [CLIMATOL] A measure of the precipitation effectiveness or aridity of a region, given by the following relationship: index of aridity = P/(T + 10), where P is the annual precipitation in centimeters, and T the annual mean temperature in degrees Celsius. { 'in,deks əv ə'rid·əd·ē }

index plane [GEOL] A surface used as a reference point in determining geological structure. { 'in,deks ,plān }

Indiana limestone *See* spergenite. { ,in·dē'a·nə 'līm,stōn }

Indian Ocean [GEOGR] The smallest and geologically the most youthful of the three oceans, whose surface area is 29,300,000 square miles (75,900,000 square kilometers); it is bounded on the north by India, Pakistan, and Iran; on the east by the Malay Peninsula; on the south by Antarctica; and on the west by the Arabian peninsula and Africa. { 'in·dē·ən 'ō·shən }

Indian spring low water [OCEANOGR] An arbitrary tidal datum approximating the level of the mean of the lower low waters at spring time, first used in waters surrounding India. Also known as harmonic tide plane; Indian tide plane. { 'in·dē·ən ¦spriŋ ¦lō 'wȯd·ər }

Indian summer [CLIMATOL] A period, in mid-or late autumn, of abnormally warm weather, generally clear skies, sunny but hazy days, and cool nights; in New England, at least one killing frost and preferably a substantial period of normally cool weather must precede this warm spell in order for it to be considered a true Indian summer; it does not occur every year, and in some years there may be two or three Indian summers; the term is most often heard in the northeastern United States, but its usage extends throughout English-speaking countries. { 'in·dē·ən 'səm·ər }

Indian tide plane *See* Indian spring low water. { 'in·dē·ən 'tīd ,plān }

indicated air temperature [METEOROL] The uncorrected reading from the free air temperature gage. Also known as outside air temperature. { 'in·də,kād·əd 'er ,tem·prə·chər }

indigenous coal *See* autochthonous coal. { in'dij·ə·nəs kōl }

indirect cell [METEOROL] A closed circulation in a vertical plane in which the rising motion occurs at lower potential temperature than the descending motion, thus forming an energy sink. { ,in·də'rekt 'sel }

indirect stratification *See* secondary stratification. { ,in·də'rekt ,strad·ə·fə'kā·shən }

induced magnetization [GEOPHYS] That component of a rock's magnetization which is proportional to, and has the same direction as, the ambient magnetic field. { in'düst ,mag·nə·tə'zā·shən }

induction method [GEOPHYS] In studies of the radioactivity of the atmosphere, a technique for estimating the concentration of the radioactive gases by exposing a negatively charged wire to the air and then using an ionization chamber to count the activity of the radioactive deposit formed on the wire. { in'dək·shən ,meth·əd }

induration [GEOL] **1.** The hardening of a rock material by the application of heat or pressure or by the introduction of a cementing material. **2.** A hardened mass formed by such processes. **3.** The hardening of a soil horizon by chemical action to form a hardpan. { ,in·də'rā·shən }

inertial flow [GEOPHYS] Frictionless flow in a geopotential surface in which there is no pressure gradient; the centrifugal and Coriolis accelerations must therefore be equal and opposite, and the constant inertial wind speed V_i is given by $V_i = fR$, where f is the Coriolis parameter and R the radius of curvature of the path. { i'nər·shəl 'flō }

industrial climatology [CLIMATOL] A type of applied climatology which studies the effect of climate and weather on industry's operations; the goal is to provide industry with a sound statistical basis for all administrative and operational decisions which involve a weather factor. { in'dəs·trē·əl ˌklī·mə'täl·ə·jē }

industrial geography [GEOGR] A branch of geography that deals with location, raw materials, products, and distribution, as influenced by geography. { in'dəs·trē·əl jē'äg·rə·fē }

industrial meteorology [METEOROL] The application of meteorological information and techniques to industrial problems. { in'dəs·trē·əl ˌmē·dē·ə'räl·ə·jē }

inertia currents [OCEANOGR] Currents resulting after the cessation of wind in a generating area or after the water movement has left the generating area; circular currents with a period of one-half pendulum day. { i'nər·shə ˌkə·rəns }

inertial circle [METEOROL] A loop in the path of an air parcel in inertial flow, which is approximately circular if the latitudinal displacement is small. Also known as circle of inertia. [OCEANOGR] The circle described by inertial motion in a body of ocean water and having a radius $R = C/f$, where C is the particle velocity in a given direction and f is the Coriolis parameter. { i'nər·shəl ˌsər·kəl }

inertial theory [OCEANOGR] The theory associated with the motion of an ocean current under the influences of inertia and the Coriolis force, which cause it to take a circular path. { i'nər·shəl 'thē·ə·rē }

inertia period [OCEANOGR] The time required for a given particle to complete an inertia circle. { i'nər·shə ˌpir·ē·əd }

inertinite [GEOL] A carbon-rich maceral group, which includes micrinite, sclerotinite, fusinite, and semifusinite. { i'nərt·ənˌīt }

inface *See* scarp slope. { 'inˌfās }

infancy [GEOL] The initial (youthful) or very early stage of the cycle of erosion characterized by smooth, nearly level erosional surfaces dissected by narrow stream gorges, numerous depressions filled by marshy lakes and ponds, and shallow streams. Also known as topographic infancy. { 'in·fən·sē }

infiltration [GEOL] Deposition of mineral matter among the pores or grains of a rock by permeation of water carrying the matter in solution. [HYD] Movement of water through the soil surface into the ground. { ˌin·fil'trā·shən }

infiltration capacity [HYD] The maximum rate at which water enters the soil or other porous material in a given condition. { ˌin·fil'trā·shən kə'pas·əd·ē }

infiltration vein [GEOL] Vein deposited in rock by percolating water. { ˌin·fil'trā·shən vān }

infinite aquifer [HYD] The portion of a formation that contains water, and for which the exterior boundary is at an effectively infinite distance from the oil reservoir. { 'in·fə·nət 'ak·wə·fər }

influent stream [HYD] A stream that contributes water to the zone of saturation of groundwater and develops bank storage. Also known as losing stream. { 'inˌflü·ənt 'strēm }

Infracambrian *See* Eocambrian. { ¦in·frə'kam·brē·ən }

infralateral tangent arcs [METEOROL] Two oblique, colored arcs, convex toward the sun and tangent to the halo of 46° at points below the altitude of the sun, produced by refraction (90° effective prism angle) in hexagonal columnar ice crystals whose principal axes are horizontal but randomly directed in azimuth; if the sun's elevation exceeds about 68°, the arcs cannot appear. { ¦in·frə¦lad·ə·rəl 'tan·jənt ˌärks }

infrared window [GEOPHYS] A frequency region in the infrared where there is good transmission of electromagnetic radiation through the atmosphere. { ¦in·frə¦red 'win·dō }

infusorial earth [GEOL] Formerly, and incorrectly, a soft rock or an earthy substance composed of siliceous remains of diatoms. { ˌin·fyə'sȯr·ē·əl 'ərth }

ingrown meander [GEOL] A meander of a stream with an undercut bank on one side and a gentle slope on the other. { 'inˌgrōn mē'an·dər }

initial condition [METEOROL] A prescription of the state of a dynamical system at some specified time; for all subsequent times the equations of motion and boundary conditions determine the state of the system; the appropriate synoptic weather charts, for example, constitute a (discrete) set of initial conditions for a forecast; in many contexts, initial conditions are considered as boundary conditions in the dimension of time. { i'nish·əl kən'dishðən }

initial detention *See* surface storage. { i'nish·əl di'ten·chən }

initial dip *See* primary dip. { i'nish·əl 'dip }

initial landform [GEOL] A landform that is produced directly by epeirogenic, orogenic, or volcanic activity, and whose original features are only slightly modified by erosion. { i'nish·əl 'land,fōrm }

injection [GEOL] Also known as intrusion; sedimentary injection. **1.** A process by which sedimentary material is forced under abnormal pressure into a preexisting rock or deposit. **2.** A structure formed by an injection process. { in'jek·shən }

injection well *See* recharge well. { in'jek·shən ,wel }

inland [GEOGR] Interior land, not bordered by the sea. { 'in·lənd }

inland ice [HYD] Ice composing the inner portion of a continental glacier or large ice sheet; applied particularly to Greenland ice. { 'in·lənd 'īs }

inland sea *See* epicontinental sea. { 'in·lənd 'sē }

inland water [GEOGR] **1.** A lake, river, or other body of water wholly within the boundaries of a state. **2.** An interior body of water not bordered by the sea. { 'in·lənd 'wȯd·ər }

inlet [GEOGR] **1.** A short, narrow waterway connecting a bay or lagoon with the sea. **2.** A recess or bay in the shore of a body of water. **3.** A waterway flowing into a larger body of water. { 'in,let }

inlier [GEOL] A circular or elliptical area of older rocks surrounded by strata that are younger. { 'in,lī·ər }

inner core [GEOL] The central part of the earth's core, extending from a depth of 3160 miles (5100 kilometers) to the center of the earth. Also known as siderosphere. { ¦in·ər 'kȯr }

inner harbor [GEOGR] The part of a harbor more remote from the sea, as contrasted with the outer harbor; this expression is normally used only in a harbor that is clearly divided into parts, by a narrow passageway or artificial structure; the inner harbor generally has additional protection and is often the principal berthing area. { ¦in·ər 'här·bər }

inner mantle *See* lower mantle. { ¦in·ər 'mant·əl }

inosilicate [GEOL] A class or structural type of silicate in which the SiO_4 tetrahedrons are linked together by the sharing of oxygens to form linear chains of indefinite length. { ¦in·ō'sil·ə,kāt }

in-place stress field *See* ambient stress field. { ,in ,plās 'stres ,fēld }

inselberg [GEOL] A large, steep-sided residual hill, knob, or mountain, generally rocky and bare, rising abruptly from an extensive, nearly level lowland erosion surface in arid or semiarid regions. Also known as island mountain. { 'in·səl,bərg }

insequent stream [HYD] A stream that has developed on the present surface, but not consequent upon it, and seemingly not controlled or adjusted by the rock structure and surface features. { in'sē·kwənt 'strēm }

inshore [GEOGR] **1.** Located near the shore. **2.** Indicating a shoreward position. { 'in'shȯr }

inshore current [OCEANOGR] The horizontal movement of water inside the surf zone, including longshore and rip currents. { 'in,shȯr 'kə·rənt }

inshore zone [GEOL] The zone of variable width extending from the shoreline at low tide through the breaker zone. { 'in,shȯr 'zōn }

insoluble residue [GEOL] Material remaining after a geological specimen is dissolved in hydrochloric or acetic acid. { in'säl·yə·bəl 'rez·ə,dü }

inspissation [GEOCHEM] Thickening of an oil deposit by evaporation or oxidation, resulting, for example, after long exposure in pitch or gum formation. { ,in·spi'sā·shən }

instability line [METEOROL] Any nonfrontal line or band of convective activity in the atmosphere; this is the general term and includes the developing, mature, and dissipating stages; however, when the mature stage consists of a line of active thunderstorms, it is properly termed a squall line; therefore, in practice, instability line often refers only to the less active phases. { ˌin·stə'bil·əd·ē ˌlīn }

instrumented buoy [OCEANOGR] An uncrewed floating structure for the mounting, operation, data collection, and transmission of meteorological and oceanographic parameter-measuring systems. { ¦in·strə¦men·təd 'bȯi }

instrument weather [METEOROL] Route or terminal weather conditions of sufficiently low visibility to require the operation of aircraft under instrument flight rules (IFR). Also known as IFR weather. { 'in·strə·mənt ˌweth·ər }

intake *See* recharge. { 'inˌtāk }

intake area *See* recharge area. { 'inˌtāk ˌer·ē·ə }

integrated drainage [HYD] Drainage resulting after folding and faulting of a surface under arid conditions; the streams by working headward have joined basins across intervening mountains or ridges. { 'in·təˌgrād·əd 'drān·ij }

intensity of turbulence *See* gustiness components. { in'ten·səd·ē əv 'tər·byə·ləns }

interbedded [GEOL] Having beds lying between other beds with different characteristics. { ¦in·tər¦bed·əd }

intercalation [GEOL] A layer located between layers of different character. { inˌtər·kə'lā·shən }

intercardinal point [GEOD] Any of the four directions midway between the cardinal points, that is, northeast, southeast, southwest, and northwest. Also known as quadrantal point. { ¦in·tər'kärd·ən·əl ˌpȯint }

interception [HYD] **1.** The process by which precipitation is caught and retained on vegetation or structures and subsequently evaporated without reaching the ground. **2.** That part of the precipitation intercepted by vegetation. [METEOROL] **1.** The loss of sunshine, a part of which may be intercepted by hills, trees, or tall buildings. **2.** The depletion of part of the solar spectrum by atmospheric gases and suspensoids; this commonly refers to the absorption of ultraviolet radiation by ozone and dust. { ˌin·tər'sep·shən }

intercloud discharge *See* cloud-to-cloud discharge. { 'in·tərˌklau̇d 'disˌchärj }

intercontinental sea [GEOGR] A large body of salt water extending between two continents. { in·tərˌkant·ən'ent·əl 'sē }

interface *See* seismic discontinuity. { 'in·tərˌfās }

interference ripple mark [GEOL] A pattern resulting from two sets of symmetrical ripples formed by waves crossing at right angles. { ˌin·tər'fir·əns 'rip·əl ˌmärk }

interflow [HYD] The water, derived from precipitation, that infiltrates the soil surface and then moves laterally through the upper layers of soil above the water table until it reaches a stream channel or returns to the surface at some point downslope from its point of infiltration. { 'in·tərˌflō }

interfluve [GEOL] The area of land between two rivers, usually an upland or ridge between two adjacent valleys that contain streams flowing in approximately the same direction. { 'in·tərˌflüv }

intergelisol *See* pereletok. { ¦in·tər'jel·əˌsȯl }

interglacial [GEOL] Pertaining to or formed during a period of geologic time between two successive glacial epochs or between two glacial stages. { ¦in·tər'glā·shəl }

intergranular pressure *See* effective stress. { ¦in·tər'gran·yə·lər 'presh·ər }

interlobate moraine *See* intermediate moraine. { ¦in·tər'lōˌbāt mə'rān }

intermediate ion [METEOROL] An atmospheric ion of size and mobility intermediate between the small ion and the large ion. { ˌin·tər'mēd·ē·ət 'īˌän }

intermediate moraine [GEOL] A type of lateral moraine formed at the junction of two adjacent glacial lobes. Also known as interlobate moraine. { ˌin·tər'mēd·ē·ət mə'rān }

intermittent current [OCEANOGR] A unidirectional current interrupted at intervals. { ¦in·tər¦mit·ənt 'kə·rənt }

intermittent spring [HYD] A spring that ceases flow after a long dry spell but flows again after heavy rains. { ¦in·tər¦mit·ənt 'spriŋ }

intermittent stream [HYD] A stream which carries water a considerable portion of the time, but which ceases to flow occasionally or seasonally because bed seepage and evapotranspiration exceed the available water supply. { ¦in·tər¦mit·ənt 'strēm }

intermontane [GEOL] Located between or surrounded by mountains. { ¦in·tər ¦män,tān }

intermontane glacier [GEOL] A glacier that is formed by the confluence of several valley glaciers and occupies a trough between separate ranges of mountains. { ¦in·tər¦män,tān 'glā·shər }

intermontane trough [GEOL] **1.** A subsiding area in an island arc of the ocean, lying between the stable elements of a region. **2.** A basinlike area between mountains. { ¦in·tər¦män,tān 'trȯf }

internal cast *See* steinkern. { in'tərn·əl 'kast }

internal drift current [OCEANOGR] Motion in an underlying layer of water caused by shearing stresses and friction created by current in a top layer that has different density. { in'tərn·əl 'drift ,kə·rənt }

internal erosion [GEOL] Erosion effected within a compacting sediment by movement of water through the larger pores. { in'tərn·əl i'rō·zhən }

internal sedimentation [GEOL] Accumulation of clastic or chemical sediments derived from the surface of, or within, a more or less consolidated carbonate sediment (mud or silt); deposited in secondary cavities formed in the host rock (after its deposition) by bending of laminae or by internal erosion or solution. { in'tərn·əl ,sed·ə·mən'tā·shən }

international analysis code [METEOROL] An internationally recognized code for communicating details of synoptic chart analyses. Abbreviated IAC. { ¦in·tər¦nash·ən·əl ə'nal·ə·səs ,kōd }

international ellipsoid of reference [GEOD] The reference ellipsoid, based upon the Hayford spheroid, the semimajor axis of which is 6,378,388 meters; the flattening or ellipticity equals 1/297; by computation the semiminor axis is 6,356,911.946 meters. Also known as international spheroid. { ¦in·tər¦nash·ən·əl ə'lip,sȯid əv 'ref·rəns }

International Geophysical Year [GEOPHYS] An internationally accepted period, extending from July 1957 through December 1958, for concentrated and coordinated geophysical exploration, primarily of the solar and terrestrial atmospheres. Abbreviated IGY. { ¦in·tər¦nash·ən·əl ,jē·ō'fiz·ə·kəl ,yir }

international gravity formula [GEOD] A formula for the acceleration of gravity at the earth's surface, stating that the acceleration of gravity is equal to $9.780327[1 + 5.3024 \times 10^{-3} \sin^2 \phi - 5.8 \times 10^{-6} \sin^2 2\phi]$ m/s^2, where ϕ is the latitude. { ¦in·tər¦nash·ən·əl 'grav·əd·ē ,fȯr·myə·lə }

International Ice Patrol [OCEANOGR] An organization established in 1914 to protect shipping by providing iceberg warnings. { ¦in·tər¦nash·ən·əl 'īs pə,trōl }

international index numbers [METEOROL] A system of designating meteorological observing stations by number, established and administered by the World Meteorological Organization; under this scheme, specified areas of the world are divided into blocks, each bearing a two-number designator; stations within each block have an additional unique three-number designator, the numbers generally increasing from east to west and from south to north. { ¦in·tər¦nash·ən·əl 'in,deks ,nəm·bərz }

International Polar Year [METEOROL] The years 1882 and 1932, during which participating nations undertook increased observations of geophysical phenomena in polar (mostly arctic) regions; the observations were largely meteorological, but included such as auroral and magnetic studies. { ¦in·tər¦nash·ən·əl 'pō·lər ,yir }

International Quiet Sun Year [GEOPHYS] An international cooperative effort, similar to the International Geophysical Year and extending through 1964 and 1965, to study the sun and its terrestrial and planetary effects during the minimum of the 11-year cycle of solar activity. Abbreviated IQSY. Also known as the International Year of the Quiet Sun. { ¦in·tər¦nash·ən·əl ¦kwī·ət 'sən ,yir }

international spheroid *See* international ellipsoid of reference. { ¦in·tər¦nash·ən·əl 'sfir,ȯid }

international synoptic code [METEOROL] A synoptic code approved by the World Meteorological Organization in which the observable meteorological elements are encoded and transmitted in words of five numerical digits length. { ¦in·tər¦nash·ən·əl sə¦näp·tik ,kōd }

International Year of the Quiet Sun *See* International Quiet Sun Year. { ¦in·tər¦nash·ən·əl 'yir əv thə ¦kwī·ət 'sən }

internides [GEOL] The internal part of an orogenic belt, farthest away from the craton, which is commonly the site of a eugeosyncline during its early phases and is later subjected to plastic folding and plutonism. Also known as primary arc. { in'tər·nə,dēz }

interpluvial [GEOL] Pertaining to an episode or period of geologic time that was dryer than the pluvial period occurring before or after it. { ,in·tər'plü·vē·əl }

interstadial [GEOL] Pertaining to a period during a glacial stage in which the ice retreated temporarily. { ,in·tər'städ·ē·əl }

interstice [GEOL] A pore space within a rock or soil. { in'tərs·təs }

interstitial water [HYD] Subsurface water contained in pore spaces between the grains of rock and sediments. { ¦in·tər¦stish·əl 'wȯd·ər }

interstitial water saturation [HYD] The water content of a subterranean reservoir formation. { ¦in·ter¦stish·əl ¦wȯd·ər 'sach·ə'rā·shən }

intertidal zone [OCEANOGR] The part of the littoral zone above low-tide mark. { ¦in·tər'tīd·əl ,zōn }

intertropical convergence zone [METEOROL] The axis, or a portion thereof, of the broad trade-wind current of the tropics; this axis is the dividing line between the southeast trades and the northeast trades (of the Southern and Northern hemispheres, respectively). Also known as equatorial convergence zone; meteorological equator. { ¦in·tər'träp·ə·kəl kən'vər·jəns ,zōn }

intertropical front [METEOROL] The interface or transition zone occurring within the equatorial trough between the Northern and Southern hemispheres. Also known as equatorial front; tropical front. { ¦in·tər'träp·ə·kəl 'frənt }

interurban [GEOGR] Connecting or extending between urban areas. { ¦in·tər'ər·bən }

intraclast [GEOL] A fragment of limestone formed by erosion within a basin of deposition and redeposited there to form a new sediment. { 'in·trə,klast }

intracloud discharge *See* cloud discharge. { 'in·trə,klaủd 'dis,chärj }

intracratonic basin *See* autogeosyncline. { ¦in·trə·krə'tän·ik 'bā·sən }

intraformational conglomerate [GEOL] **1.** A conglomerate in which clasts and the matrix are contemporaneous in origin. **2.** A conglomerate formed in the midst of a geologic formation. { ¦in·trə,fȯr'māsh·ən·əl kən'gläm·ə·rət }

intraformational fold [GEOL] A minor fold confined to a sedimentary layer lying between undeformed beds. { in·trə,fȯr'māsh·ən·əl 'fōld }

intrastratal solution [GEOCHEM] A chemical attrition of the constituents of a rock after deposition. { ¦in·trə'strad·əl sə'lü·shən }

intratelluric [GEOL] **1.** Pertaining to a phenocryst that is formed earlier than its matrix. **2.** Pertaining to a period in which igneous rocks crystallized prior to their eruption. **3.** Located, formed, or originating at great depths within the earth. { ¦in·trə·tə'lyủr·ik }

intrazonal soil [GEOL] A group of soils with well-developed characteristics that reflect the dominant influence of some local factor of relief, parent material, or age over the usual effect of vegetation and climate. { ,in·trə'zōn·əl 'sȯil }

intrusion [GEOL] **1.** The process of emplacement of magma in preexisting rock. Also known as injection; invasion; irruption. **2.** A large-scale sedimentary injection. Also known as sedimentary intrusion. **3.** Any rock mass formed by an intrusive process. { in'trü·zhən }

inundation [HYD] Flooding, by the rise and spread of water, of a land surface that is not normally submerged. { ,i·nən'dā·shən }

invasion [GEOL] **1.** The movement of one material into a porous reservoir area that has been occupied by another material. **2.** *See* intrusion; transgression. { in'vā·zhən }

inverna [METEOROL] A southeast wind of Lake Maggiore, Italy. { in'vər·nä }

inversion [GEOL] **1.** Development of inverted relief through which anticlines are transformed into valleys and synclines are changed into mountains. **2.** The occupancy by a lava flow of a ravine or valley that occurred in the side of a volcano. **3.** A diagenetic process in which unstable minerals are converted to a more stable form without a change in chemical composition. [METEOROL] A departure from the usual decrease or increase with altitude of the value of an atmospheric property, most commonly temperature. { in'vər·zhən }

inversion layer [METEOROL] The atmosphere layer through which an inversion occurs. { in'vər·zhən ˌlā·ər }

inverted *See* overturned. { in'vərd·əd }

inverted plunge [GEOL] A plunge of a fold whose inclination has been carried past the vertical, so that the plunge is less than 90° in the direction opposite from the original attitude; younger rocks plunge beneath the older rocks. { in'vərd·əd 'plənj }

inverted relief [GEOL] A topographic configuration that is opposite to that of the geologic structure, for example, where a valley occupies the site of an anticline. { in'vərd·əd ri'lēf }

ion atmosphere *See* ion cloud. { 'īˌän 'at·məˌsfir }

ion cloud [GEOPHYS] An inhomogeneity or patch of unusually great ion density in one of the regular regions of the ionosphere; such patches occur quite often in the E region. { 'īˌän ˌklau̇d }

ion column [GEOPHYS] The trail of ionized gases in the trajectory of a meteoroid entering the upper atmosphere; a part of the composite phenomenon known as a meteor. Also known as meteor trail. { 'īˌän ˌkäl·əm }

ionic ratio [OCEANOGR] The ratio by weight of a major constituent of seawater to the chloride ion content; for example, SO_4/Cl = 0.1396, Ca/Cl = 0.02150, Mg/Cl = 0.06694. { ī'än·ik 'rā·shō }

ionized layers [GEOPHYS] Layers of increased ionization within the ionosphere produced by cosmic radiation; responsible for absorption and reflection of radio waves and important in connection with communications and tracking of satellites and other space vehicles. { 'ī·əˌnīzd 'lā·ərz }

ionosphere [GEOPHYS] That part of the earth's upper atmosphere which is sufficiently ionized by solar ultraviolet radiation so that the concentration of free electrons affects the propagation of radio waves; its base is at about 40 or 50 miles (70 or 80 kilometers) and it extends to an indefinite height. { ī'an·əˌsfir }

ionospheric disturbance [GEOPHYS] A temporal variation in electron concentration in the ionosphere that is caused by solar activity and that makes the heights of the ionosphere layers go beyond the normal limits for a location, date, and time of day. { ˌīˌän·ə'sfir·ik di'stər·bəns }

ionospheric D scatter meteor burst [GEOPHYS] Phenomenon affecting ionospheric scatter communications resulting from the penetration of meteors through the D region of the ionospheric layer. { ˌīˌän·ə'sfir·ik 'dē ˌskad·ər 'mēd·ē·ər ˌbərst }

ionospheric storm [GEOPHYS] A turbulence in the F region of the ionosphere, usually due to a sudden burst of radiation from the sun; it is accompanied by a decrease in the density of ionization and an increase in the virtual height of the region. { ˌīˌän·ə'sfir·ik 'stȯrm }

Iowan glaciation [GEOL] The earliest substage of the Wisconsin glacial stage; occurred more than 30,000 years ago. { 'ī·ə·wən ˌglā·sē'ā·shən }

ipsonite [GEOL] The final stage of weathered asphalt; a black, infusible substance, only slightly soluble in carbon disulfide, containing 50–80% fixed carbon and very little oxygen. { 'ip·səˌnīt }

IQSY *See* International Quiet Sun Year.

iridescent cloud [METEOROL] An ice-crystal cloud which exhibits brilliant spots or borders of colors, usually red and green, observed up to about 30° from the sun. { ˌir·ə'des·ənt 'klau̇d }

irisation [METEOROL] The coloration exhibited by iridescent clouds and at times along the borders of lenticular clouds. { ī·rə'sā·shən }

Irish Sea [GEOGR] A marginal sea of the Atlantic Ocean between Ireland and England, approximately 53°N latitude and 5°W longitude. { 'ī·rish 'sē }

IRM *See* isothermal remanent magnetization.

Irminger Current [OCEANOGR] An ocean current that is one of the terminal branches of the Gulf Stream system, flowing west off the southern coast of Iceland. { 'ər·miŋ·ər 'kə·rənt }

iron formation [GEOL] Sedimentary, low-grade iron ore bodies consisting mainly of chert or fine-grained quartz and ferric oxide segregated in bands or sheets irregularly mingled. { 'ī·ərn fȯr'mā·shən }

iron hat *See* gossan. { 'ī·ərn 'hat }

iron ore [GEOL] Rocks or deposits containing compounds from which iron can be extracted. { 'ī·ərn 'ȯr }

iron-stony meteorite *See* stony-iron meteorite. { 'ī·ərn ˌstō·nē 'mēd·ē·əˌrīt }

iron winds [METEOROL] Northeasterly winds of Central America, prevalent during February and March, and blowing steadily for several days at a time. { 'ī·ərn 'winz }

irregular crystal [METEOROL] A snow particle, sometimes covered by a coating of rime, composed of small crystals randomly grown together; generally, component crystals are so small that the crystalline form of the particle can be seen only through a magnifying glass or microscope. { i'reg·yə·lər 'krist·əl }

irregular iceberg *See* pinnacled iceberg. { i'reg·yə·lər 'īsˌbərg }

irrotational strain [GEOL] Strain in which the orientation of the axes of strain does not change. Also known as nonrotational strain. { ¦ir·ə'tā·shən·əl 'strān }

irruption *See* intrusion. { i'rəp·shən }

Irvingtonian [GEOL] A stage of geologic time in southern California, in the lower Pleistocene, below the Rancholabrean. { ˌər·viŋ'tō·nē·ən }

isallobar [METEOROL] A line of equal change in atmospheric pressure during a specified time interval; an isopleth of pressure tendency; a common form is drawn for the three-hourly local pressure tendencies on a synoptic surface chart. { īs'al·əˌbär }

isallobaric [METEOROL] Of equal or constant pressure change; this may refer either to the distribution of equal pressure tendency in space or to the constancy of pressure tendency with time. { ī¦sal·ə¦bar·ik }

isallobaric high *See* pressure-rise center. { ī¦sal·ə¦bar·ik 'hī }

isallobaric low *See* pressure-fall center. { ī¦sal·ə¦bar·ik 'lō }

isallobaric maximum *See* pressure-rise center. { ī¦sal·ə¦bar·ik 'mak·sə·məm }

isallobaric minimum *See* pressure-fall center. { ī¦sal·ə¦bar·ik 'min·ə·məm }

isallobaric wind [METEOROL] The wind velocity whose Coriolis force exactly balances a locally accelerating geostrophic wind. Also known as Brunt-Douglas isallobaric wind. { ī¦sal·ə¦bar·ik 'wind }

isallohypse *See* height-change line. { ī'sal·əˌhips }

isallohypsic wind [METEOROL] An isallobaric wind, using height tendency in a constant-pressure surface instead of pressure tendency in a constant-height surface. { ī¦sal·ə¦hip·sik 'wind }

isallotherm [METEOROL] A line connecting points of equal change in temperature within a given time period. { ī'sal·əˌthərm }

isanabat [METEOROL] A line drawn through points of equal vertical component of wind velocity; positive values indicate upward motion, negative values indicate downward motion. { ī'san·əˌbat }

isanakatabar [METEOROL] A line on a chart of equal atmospheric-pressure range during a specified time interval. { ī¦san·ə¦kad·əˌbär }

isanomal *See* isanomalous line. { ˌī·sə'näm·əl }

isanomalous line [METEOROL] A line drawn through geographical points having equal anomaly of some meteorological quantity. Also known as isanomal. { ˌī·sə'näm·ə·ləs ¦līn }

isaurore *See* isochasm. { 'ī·səˌrȯr }

isentropic chart [METEOROL] A constant-entropy chart; a synoptic chart presenting the distribution of meteorological elements in the atmosphere on a surface of constant potential temperature (equivalent to an isentropic surface); it usually contains the plotted data and analysis of such elements as pressure (or height), wind, temperature, and moisture at that surface. { ¦īs·ən¦träp·ik 'chärt }

isentropic condensation level *See* lifting condensation level. { ¦īs·ən¦träp·ik ˌkänd·ən'sā·shən ˌlev·əl }

isentropic map [GEOL] A map indicating constant entropy function for facies. { ¦īs·ən'träp·ik 'map }

isentropic mixing [METEOROL] Any atmospheric mixing process which occurs within an isentropic surface; the fact that many atmospheric motions are reversible adiabatic processes renders this type of mixing important, and exchange coefficients have been computed therefor. { ¦īs·ən'träp·ik 'mik·siŋ }

isentropic surface [METEOROL] A surface in space in which potential temperature is everywhere equal. { ¦īs·ən'träp·ik 'sər·fəs }

isentropic thickness chart [METEOROL] A thickness chart of an atmospheric layer bounded by two selected isentropic surfaces (surfaces of constant potential temperature); the thickness of such a layer is directly proportional to the static instability of that layer; hence, these charts have been called instability charts. Also known as thick-thin chart. { ¦īs·ən'träp·ik 'thik·nəs ˌchärt }

isentropic weight chart [METEOROL] A chart of atmospheric pressure difference between two selected isentropic surfaces (surfaces of constant potential temperature); the greater the pressure difference the greater the weight of the air column separating the two surfaces. { ¦īs·ən'träp·ik 'wāt ˌchärt }

island [GEOGR] A tract of land smaller than a continent and surrounded by water; normally in an ocean, sea, lake, or stream. { 'ī·lənd }

island arc [GEOGR] A group of volcanic islands, usually situated in a curving arch-like pattern that is convex toward the open ocean, having a deep trench or trough on the convex side and usually enclosing a deep basin on the concave side; formed by volcanic activity associated with oceanic plate subduction at convergent plate margins. Also known as volcanic arc. { 'ī·lənd ¦ärk }

island mountain *See* inselberg. { 'ī·lənd ¦maůnt·ən }

isobar [METEOROL] A line drawn through all points of equal atmospheric pressure along a given reference surface, such as a constant-height surface (notably mean sea level on surface charts), an isentropic surface, or the vertical plan of a synoptic cross section. { 'ī·səˌbär }

isobaric chart *See* constant-pressure chart. { ¦i·sə¦bär·ik 'chärt }

isobaric contour chart *See* constant-pressure chart. { ¦i·sə¦bär·ik 'kän·tůr ˌchärt }

isobaric divergence [METEOROL] The horizontal divergence in a constant-pressure surface; expressed in a system of coordinates with pressure as an independent variable. { ¦i·sə¦bär·ik də'vər·jəns }

isobaric equivalent temperature *See* equivalent temperature. { ¦i·sə¦bär·ik i¦kwiv·ə·lənt 'tem·prə·chər }

isobaric map [METEOROL] A map depicting points in the atmosphere of equal barometric pressure. { ¦i·sə¦bär·ik 'map }

isobaric surface [METEOROL] A surface on which the pressure is uniform. Also known as constant-pressure surface. { ¦ī·sə¦bär·ik 'sər·fəs }

isobaric topography *See* height pattern. { ¦i·sə¦bär·ik tə'päg·rə·fē }

isobaric vorticity [METEOROL] Relative vorticity in a constant-pressure surface, that is, expressed in a system of coordinates with pressure as an independent variable. { ¦i·sə¦bär·ik vȯr'tis·əd·ē }

isobath [OCEANOGR] A contour line connecting points of equal water depths on a chart. Also known as depth contour; depth curve; fathom curve. { 'i·səˌbath }

isobathytherm [OCEANOGR] A line or surface showing the depth in oceans or lakes at which points have the same temperatures. { ˌī·sə'bath·əˌthərm }

isobront [METEOROL] A line drawn through geographical points at which a given phase of thunderstorm activity occurred simultaneously. Also known as homobront. { 'ī·səˌbränt }

isocarb [GEOCHEM] A line on a map that connects points of equal content of fixed carbon in coal. { 'ī·sə,kärb }

isoceraunic [METEOROL] Indicating or having equal frequency or intensity of thunderstorm activity. Also spelled isokeraunic. { ¦ī·sō·sə'rȯn·ik }

isoceraunic line [METEOROL] A line drawn through geographical points at which some phenomenon connected with thunderstorms has the same frequency or intensity; used for lines of equal frequency of lightning discharges. { ¦ī·sō·sə'rȯn·ik 'līn }

isochasm [GEOPHYS] A line connecting points on the earth's surface at which the aurora is observed with equal frequency. Also known as isaurore. { 'ī·sə,kaz·əm }

isochron [GEOCHEM] A line on a graph defined by data for rocks of the same age with the same initial lead isotopic composition, the slope of which is proportional to the age. Also known as geochron. { 'ī·sə,krän }

isoclinal *See* isoclinic line. { ¦ī·sə'klīn·əl }

isoclinal chart [GEOPHYS] A chart showing isoclinic lines. Also known as isoclinic chart. { ¦ī·sə'klīn·əl 'chärt }

isocline [GEOL] A fold of strata so tightly compressed that parts on each side dip in the same direction. { 'ī·sə,klīn }

isoclinic chart *See* isoclinal chart. { ¦ī·sə¦klin·ik 'chärt }

isoclinic line [GEOPHYS] A line connecting points on the earth's surface which have the same magnetic dip. Also known as isoclinal. { ¦ī·sə¦klin·ik 'līn }

isodrosotherm [METEOROL] An isogram of dew-point temperature. { ,ī·sə'dräs·ə,thərm }

isodynamic line [GEOPHYS] One of the lines on a map of a magnetic field that connect points having equal strengths of the earth's field. { ¦ī·sō·dī'nam·ik 'līn }

isofacies map [GEOL] A stratigraphic map showing the distribution of one or more facies within a particular stratigraphic unit. { ¦ī·sə¦fā·shēz ,map }

isofronts-preiso code [METEOROL] A code in which data on isobars and fronts at sea level (or earth's surface) are encoded and transmitted; a modified form of the international analysis code. { ¦ī·sə,frəns ¦prē¦ī·sō ,kōd }

isogal [GEOPHYS] A contour line on a map connecting points of equal gravity values on the earth's surface. { 'ī·sə,gal }

isogam [GEOPHYS] A line joining points on the earth's surface having the same value of the acceleration of gravity. { 'ī·sə,gam }

isogeotherm *See* geoisotherm. { ¦ī·sō'jē·ə,thərm }

isogonic line [GEOPHYS] **1.** Any of the lines on a chart or map showing the same direction of the wind vector. **2.** Any of the lines on a chart or map connecting points of equal magnetic variation. { 'ī·sə¦gän·ik ,līn }

isograd [GEOL] A line on a map joining those rocks comprising the same metamorphic grade. { 'ī·sə,grad }

isogradient [METEOROL] A line connecting points having the same horizontal gradient of atmospheric pressure, temperature, and so on. { ¦ī·sə'grād·ē·ənt }

isohaline [OCEANOGR] **1.** Of equal or constant salinity. **2.** A line on a chart connecting all points of equal salinity. { ,ī·sō'hā,lēn }

isohel [METEOROL] A line drawn through geographical points having the same duration of sunshine (or other function of solar radiation) during any specified time period. { 'ī·sō,hel }

isohume [GEOL] A line of a map or chart connecting points of equal moisture content in a coal bed. [METEOROL] A line drawn through points of equal humidity on a given surface; an isopleth of humidity; the humidity measures used may be the relative humidity or the actual moisture content (specific humidity or mixing ratio). { 'ī·sə,hyüm }

isohyet [METEOROL] A line drawn through geographic points recording equal amounts of precipitation for a specified period or for a particular storm. { ¦ī·sə¦hī·ət }

isohypsic chart *See* constant-height chart. { ,ī·sə'hip·sik 'chärt }

isohypsic surface *See* constant-height surface. { ,ī·sə'hip·sik 'sər·fəs }

isokeraunic *See* isoceraunic. { ¦ī·sō·kə'rȯn·ik }

isolith [GEOL] A line on a contour-type map that denotes the aggregate thickness of a single lithology in a stratigraphic succession composed of one or more lithologies. { 'ī·sə,lith }

isolith map [GEOL] A contour-line map depicting the thickness of an exclusive lithology. { 'ī·sə,lith ,map }

isomagnetic [GEOPHYS] Of or pertaining to lines connecting points of equality in some magnetic element. { ¦ī·sō·mag¦ned·ik }

isoneph [METEOROL] A line drawn through all points on a map having the same amount of cloudiness. { 'ī·sə,nef }

isopach map [GEOL] Map of the areal extent and thickness variation of a stratigraphic unit; used in geological exploration for oil and for underground structural analysis. { 'ī·sə,pak ,map }

isopachous line [GEOL] One of the lines drawn on a map to indicate equal thickness. { ¦ī·sō¦pak·əs ,līn }

isopectic [CLIMATOL] A line on a map connecting points at which ice begins to form at the same time of winter. { ,i·sə'pek·tik }

isopiestic line [HYD] A line indicating on a map the piezometric surface of an aquifer. { ¦ī·sə'pī·es·tik ,līn }

isopleth [METEOROL] **1.** A line of equal or constant value of a given quantity with respect to either space or time. Also known as isogram. **2.** More specifically, a line drawn through points on a graph at which a given quantity has the same numerical value (or occurs with the same frequency) as a function of the two coordinate variables. Also known as isarithm. { 'ī·sə,pleth }

isopluvial [METEOROL] A line on a map drawn through geographical points having the same pluvial index. { ¦ī·sō'plü·vē·əl }

isopor [GEOPHYS] An imaginary line connecting points on the earth's surface having the same annual change in a magnetic element. { 'ī·sə,pȯr }

isopotential level *See* potentiometric surface. { ¦ī·sō·pə'ten·chəl 'lev·əl }

isopycnic [METEOROL] A line on a chart connecting all points of equal or constant density. { ¦ī·sō¦pik·nik }

isopycnic level [METEOROL] Specifically, a level surface in the atmosphere, at about 5 miles (8 kilometers) altitude, where the air density is approximately constant in space and time; this level corresponds to the maximum upper-tropospheric interdiurnal pressure variation. { ¦ī·sō¦pik·nik ,lev·əl }

isoseismal [GEOPHYS] Pertaining to points having equal intensity of earthquake shock, or to a line on a map of the earth's surface connecting such points. { ¦ī·sə'sīz·məl }

isoshear [METEOROL] A line on a chart of equal magnitude of vertical wind shear. { 'ī·sə,shir }

isostasy [GEOPHYS] A theory of the condition of approximate equilibrium in the outer part of the earth, such that the gravitational effect of masses extending above the surface of the geoid in continental areas is approximately counterbalanced by a deficiency of density in the material beneath those masses, while deficiency of density in ocean waters is counterbalanced by an excess in density of the material under the oceans. { ī'säs·tə·sē }

isostatic adjustment *See* isostatic compensation. { ¦ī·sə'stad·ik ə'jəs·mənt }

isostatic anomaly [GEOPHYS] A gravity anomaly based on a generalized hypothesis that the gravitational effect of masses above sea level is approximately compensated by a density deficiency of the subsurface materials. { ¦ī·sə'stad·ik ə'näm·ə·lē }

isostatic compensation [GEOL] The process in which lateral transport at the surface of the earth by erosion or deposition is compensated by lateral movements in a subcrustal layer. Also known as isostatic adjustment; isostatic correction. { ¦ī·sə'stad·ik ,käm·pən'sā·shən }

isostatic correction *See* isostatic compensation. { ¦ī·sə'stad·ik kə'rek·shən }

isotach [METEOROL] A line in a given surface connecting points with equal wind speed. Also known as isokinetic; isovel. { 'ī·sə,tak }

isotach chart [METEOROL] A synoptic chart showing the distribution of wind by means of isotachs. { 'ī·sə,tak ,chärt }

isothere [CLIMATOL] A line on a map connecting points having the same mean summer temperature. { 'ī·sə,thir }

isotherm [GEOPHYS] A line on a chart connecting all points of equal or constant temperature. { 'ī·sə,thərm }

isothermal atmosphere [METEOROL] An atmosphere in hydrostatic equilibrium, in which the temperature is constant with height and the pressure decreases exponentially upward. Also known as exponential atmosphere. { ¦ī·sə¦thər·məl 'at·mə,sfir }

isothermal chart [GEOPHYS] A map showing the distribution of air temperature (or sometimes sea-surface or soil temperature) over a portion of the earth or at some level in the atmosphere; places of equal temperature are connected by lines called isotherms. { ¦ī·sə¦thər·məl 'chärt }

isothermal equilibrium [METEOROL] The state of an atmosphere at rest, uninfluenced by any external agency, in which the conduction of heat from one part to another has produced, after a sufficient length of time, a uniform temperature throughout its entire mass. Also known as conductive equilibrium. { ¦ī·sə¦thər·məl ,ē·kwə'lib·rē·əm }

isothermal layer [METEOROL] The approximately isothermal region of the atmosphere immediately above the tropopause. { ¦ī·sə¦thər·məl 'lā·ər }

isothermal remanent magnetization [GEOPHYS] A spurious magnetization induced by lightning strikes that produce large surface electrical currents. Abbreviated IRM. { ¦ī·sə¦thər·məl ¦rem·ə·nənt ,mag·nə·tə'zā·shən }

isothermobath [OCEANOGR] A line connecting points having the same temperature in a diagram of a vertical section of the ocean. { ,ī·sə'thər·mə,bath }

isotherm ribbon [METEOROL] A zone of crowded isotherms on a synoptic upper-level chart; the temperature gradient is many times greater than normally encountered in the atmosphere. { 'ī·sə,thərm ,rib·ən }

isotimic [METEOROL] Pertaining to a quantity which has equal value in space at a particular time. { ¦ī·sə¦tim·ik }

isotimic line [METEOROL] On a given reference surface in space, a line connecting points of equal value of some quantity; most of the lines drawn in the analysis of synoptic charts are isotimic lines. { ¦ī·sə¦tim·ik 'līn }

isotimic surface [METEOROL] A surface in space on which the value of a given quantity is everywhere equal; isotimic surfaces are the common reference surfaces for synoptic charts, principally constant-pressure surfaces and constant-height surfaces. { ¦ī·sə¦tim·ik 'sər·fəs }

isovel *See* isotach. { 'ī·sə,vel }

isthmus [GEOGR] A narrow strip of land having water on both sides and connecting two large land masses. { 'is·məs }

itabirite [GEOL] A laminated, metamorphosed, oxide-facies iron formation in which the original chert or jasper bands have been recrystallized into megascopically distinguished grains of quartz and in which the iron is present as thin layers of hematite, magnetite, or martite. { ,ēd·ə'bi,rīt }

I-type magma [GEOL] Magma formed from igneous source materials. { 'ī ,tīp ¦mag·mə }

J

Jacobshavn Glacier [HYD] A glacier on the west coast of Greenland at latitude 68°N; it is the most productive glacier in the Northern Hemisphere, calving about 1400 icebergs yearly. { 'yä·kəps,häf·ən 'glā·shər }

January thaw [CLIMATOL] A period of mild weather popularly supposed to recur each year in late January in New England and other parts of the northeastern United States. { 'jan·yə,wer·ē 'thȯ }

Japan Current *See* Kuroshio Current. { jə'pan 'kə·rənt }

jaspoid *See* tachylite. { 'jas,pȯid }

jauch *See* jauk. { yau̇k }

jauk [METEOROL] A local name for the foehn in the Klagenfurt basin of Austria; it may come from the south, but is developed as a north foehn. Also spelled jauch. { yau̇k }

jaw [GEOL] The side of a narrow passage such as a gorge. { jȯ }

jelly *See* ulmin. { 'jel·ē }

jet coal [GEOL] A hard, lustrous, pure black variety of lignite, occurring in isolated masses in bituminous shale; thought to be derived from waterlogged driftwood. Also known as black amber. { 'jet ¦kōl }

jet-effect wind [METEOROL] A wind which is increased in speed through the channeling of air by some mountainous configuration, such as a narrow mountain pass or canyon. { ¦jet i¦fekt ,wind }

jet streak [METEOROL] A region within the jet stream exhibiting wind speeds higher than the jet stream itself. { 'jet ,strēk }

jet stream [METEOROL] A relatively narrow, fast-moving wind current flanked by more slowly moving currents; observed principally in the zone of prevailing westerlies above the lower troposphere, and in most cases reaching maximum intensity with regard to speed and concentration near the troposphere. { 'jet ,strēm }

Jevons effect [METEOROL] The effect upon the measurement of rainfall caused by the presence of the rain gage; in 1861 W.S. Jevons pointed out that the rain gage causes a disturbance in airflow past it, and this carries part of the rain past the gage which would normally be captured. { 'jev·ənz i,fekt }

J function [GEOPHYS] A dimensionless mathematical relationship to correlate capillary pressure data of similar geologic formations. { 'jā ,fəŋk·shən }

jochwinde [METEOROL] The mountain-gap wind of the Tauern Pass in the Alps. { 'yōk ,vin·də }

joint [GEOL] A fracture that traverses a rock and does not show any discernible displacement of one side of the fracture relative to the other. { jȯint }

joint block [GEOL] A body of rock that is bounded by joints. { 'jȯint ,bläk }

joint drag *See* kink band. { 'jȯint ,drag }

jointing [GEOL] A condition of rock characterized by joints. { 'jȯint·iŋ }

joint plane [GEOL] The surface of fracturing or potential fracture of a joint. { 'jȯint ,plān }

joint set [GEOL] A group of parallel joints in a geologic formation. { 'jȯint ,set }

joint system [GEOL] Two or more joint sets. { 'jȯint ,sis·təm }

joint vein [GEOL] A small vein in a joint. { 'jȯint ,vān }

joran *See* juran. { 'jȯr·ən }

josen *See* hartite. { 'jō·sən }
junction streamer [GEOPHYS] The streamer process by which negative charge centers at successively higher altitudes in a thundercloud are believed to be "tapped" for discharge by lightning. { 'jəŋk·shən ˌstrēm·ər }
junk wind [METEOROL] A south or southeast monsoon wind, favorable for the sailing of junks; the wind is known in Thailand, China, and Japan. { 'jəŋkˌwind }
junta [METEOROL] A wind blowing through Andes Mountain passes, sometimes reaching hurricane force. { 'hůn·tə }
Jura *See* Jurassic. { 'jůr·ə }
juran [METEOROL] A wind blowing from the Jura Mountains in Switzerland from the northwest toward Lake Geneva; it is a cold and snowy wind and may be very turbulent, especially in spring. Also spelled joran. { 'jůr·ən }
Jurassic [GEOL] Also known as Jura. **1.** The second period of the Mesozoic era of geologic time. **2.** The corresponding system of rocks. { jə'ras·ik }
juvenile rift [GEOL] A stage of continental breakup before the onset of actual spreading which precedes the generation of new oceanic lithosphere. { 'jü·və·nəl 'rift }
juvenile water *See* magmatic water. { 'jü·vən·əl 'wȯd·ər }

K

K-A age [GEOL] The radioactive age of a rock determined from the ratio of potassium-40 (^{40}K) to argon-40 (^{40}A) present in the rock. { 'kā'ā' ,āj }
kaavie [METEOROL] In Scotland, a heavy fall of snow. { 'kȯ·vē }
kachchan [METEOROL] A hot, dry west or southwest wind of foehn type in the lee of the Sri Lanka (Ceylon) hills during the southwest monsoon in June and July; it is well developed at Batticaloa on the east coast, where it is strong enough to overcome the sea breeze and bring maximum temperatures of nearly 100°F (38°C). { ,käch,chän }
kal Baisakhi [METEOROL] A short-lived dusty squall at the onset of the southwest monsoon (April-June) in Bengal. { ,käl 'bī·sä·kē }
kalema [OCEANOGR] A very heavy surf breaking on the Guinea coast of Africa during the winter. { kə'lā·mə }
kame [GEOL] A low, long, steep-sided mound of glacial drift, commonly stratified sand and gravel, deposited as an alluvial fan or delta at the terminal margin of a melting glacier. { kām }
kame terrace [GEOL] A terracelike ridge deposited along the margins of glaciers by meltwater streams flowing adjacent to the valley walls. { kām ,ter·əs }
Kansan glaciation [GEOL] The second glaciation of the Pleistocene epoch in North America; began about 400,000 years ago, after the Aftonian and before the Yarmouth interglacials. { 'kan·zən ,glā·sē'ā·shən }
kaolinization [GEOL] The forming of kaolin by the weathering of aluminum silicate minerals or other clay minerals. { ,kā·ə·lə·nə'zā·shən }
karaburan [METEOROL] A violent northeast wind of Central Asia occurring during spring and summer; it carries clouds of dust (which darken the sky) instead of snow. Also known as black buran; black storm. { ,kä·rə·bu̇¦rän }
karajol [METEOROL] On the Bulgarian coast, a west wind which usually follows rain and persists 1–3 days. { ,kä·rə'yȯl }
karema [METEOROL] A violent east wind on Lake Tanganyika in Africa. { kə'rē·mə }
karif [METEOROL] A strong southwest wind on the southern shore of the Gulf of Aden, especially at Berbera, Somaliland, during the southwest monsoon. { kä'rēf }
Karnian *See* Carnian. { 'kär·nē·ən }
karoo *See* karroo. { kə'rü }
karren [GEOL] Furrows or channels formed on the surface of soluble bedrock by dissolution of a portion of the rock. Also known as lapies. { kar·ən }
karroo [GEOGR] A dry, broad, level, elevated area found especially in southern Africa, often rising to considerable elevations in terrace formations; does not support vegetation in the dry season but supports grass during the wet season. Also spelled karoo. { kə'rü }
Karroo System [GEOL] Glaciated strata formed in Permian times in southern Africa. { kə'rü ,sis·təm }
karst [GEOL] A topography formed over limestone, dolomite, or gypsum and characterized by sinkholes, caves, and underground drainage. { kärst }
karst base level [GEOL] The level below which karstification ceases in an area of karst topography. { ¦kärst ¦bās ,lev·əl }
karstbora [METEOROL] The bora of the Yugoslavian coast. { 'kärst,bȯr·ə }

karst fenster *See* karst window. { 'kärst ˌfen·stər }
karstification [GEOL] Formation of the features of karst topography by the chemical, and sometimes mechanical, action of water in a region of limestone, dolomite, or gypsum bedrock. { ˌkär·stə·fə'kā·shən }
karst plain [GEOL] A plain on which karst features are developed. { 'kärst ˌplān }
karst window [GEOL] An area over a subterranean stream that is open to the surface and appears as a depression at whose bottom the stream is visible. Also known as karst fenster. { 'kärst ˌwin·dō }
katabaric *See* katallobaric. { ¦kad·ə¦bar·ik }
katabatic wind *See* gravity wind. { ¦kad·ə¦bad·ik 'wind }
katafront [METEOROL] A front (usually a cold front) at which warm air descends the frontal surface (except, presumably, in the lowest layers). { 'kad·əˌfrənt }
katallobaric [METEOROL] Of or pertaining to a decrease in atmospheric pressure. Also known as katabaric. { kəˌtal·ə'bar·ik }
katallobaric center *See* pressure-fall center. { kəˌtal·ə'bar·ik 'sen·tər }
katazone [GEOL] The lowest depth zone of metamorphism; features include high temperatures (500–700°C), strong hydrostatic pressure, and little or no shearing stress. { 'kad·əˌzōn }
kaus [METEOROL] A moderate to gale-force southeasterly wind in the Persian Gulf, accompanied by gloomy weather, rain, and squalls; it is most frequent between December and April. Also known as cowshee; sharki. { kaůs }
kavaburd *See* cavaburd. { 'kä·vəˌbərd }
kaver *See* caver. { 'kā·vər }
kay *See* key. { kā }
Kazanian [GEOL] A European stage of geologic time: Upper Permian (above Kungurian, below Tatarian). { kə'zä·nē·ən }
K bentonite *See* potassium bentonite. { 'kā 'ben·təˌnīt }
Keewatin [GEOL] A division of the Archeozoic rocks of the Canadian Shield. { kē¦wät·ən }
Kegel karst *See* cone karst. { 'kā·gəl ˌkärst }
kelsher [METEOROL] In England, a heavy fall of rain. { 'kel·shər }
Kelvin wave [OCEANOGR] **1.** An eastward-propagating internal gravity wave that crosses the Pacific Ocean along the equator and has no north-south velocity component. **2.** A type of wave progression in relatively confined water bodies where, because of Coriolis force, the wave is higher to the right of direction of advance (in the Northern Hemisphere). { 'kel·vən ˌwāv }
Kennelly-Heaviside layer *See* E layer. { 'ken·əl·ē 'hev·ēˌsīd ˌlā·ər }
Kenoran orogeny *See* Algoman orogeny. { kə'nȯr·ən ȯ'räj·ə·nē }
kerabitumen *See* kerogen. { ¦ker·ə·bə'tü·mən }
kernel ice [HYD] In aircraft icing, an extreme form of rime ice, that is, very irregular, opaque, and of low density; it forms at temperatures of −15°C and lower. { 'kərn·əl ˌīs }
kerogen [GEOL] The complex, fossilized organic material present in sedimentary rocks, especially in shales; converted to petroleum products by distillation. Also known as kerabitumen; petrologen. { 'ker·ə·jən }
kerogen shale *See* oil shale. { 'ker·ə·jən ˌshāl }
kerosine shale *See* torbanite. { 'ker·əˌsēn ˌshāl }
kettle [GEOL] **1.** A bowl-shaped depression with steep sides in glacial drift deposits that is formed by the melting of glacier ice left behind by the retreating glacier and buried in the drift. Also known as kettle basin; kettle hole. **2.** *See* pothole. { 'ked·əl }
kettle basin *See* kettle. { 'ked·əl ˌbās·ən }
kettle hole *See* kettle. { 'ked·əl ˌhōl }
Keuper [GEOL] A European stage of geologic time, especially in Germany; Upper Triassic. { 'kȯip·ər }
Keweenawan [GEOL] The younger of two Precambrian time systems that constitute the Proterozoic period in Michigan and Wisconsin. { ¦kē·wē¦nȯ·ən }

key [GEOL] A cay, especially one of the islets off the south of Florida. Also spelled kay. { kē }

key bed [GEOL] Also known as index bed; key horizon; marker bed. **1.** A stratum or body of strata that has distinctive characteristics so that it can be easily identified. **2.** A bed whose top or bottom is employed as a datum in the drawing of structure contour maps. { 'kē ˌbed }

key day *See* control day. { 'kē ˌdā }

key horizon *See* key bed. { 'kē həˌrīz·ən }

khamsin [METEOROL] A dry, dusty, and generally hot desert wind in Egypt and over the Red Sea; it is generally southerly or southeasterly, occurring in front of depressions moving eastward across North Africa or the southeastern Mediterranean. { kam'sēn }

kibli *See* ghibli. { 'kib·lē }

kieselguhr *See* diatomaceous earth. { 'kē·zəlˌgůr }

Kimberley reefs [GEOL] Gold-bearing reefs in southern Africa that lie above the Main reef and Bird reef groups. Also known as battery reefs. { 'kim·bər·lē ˌrēfs }

Kimmeridgian [GEOL] A European stage of geologic time; middle Upper Jurassic, above Oxfordian, below Portlandian. { ˌkim·ə'rij·ē·ən }

Kinderhookian [GEOL] Lower Mississippian geologic time, above the Chautauquan of Devonian, below Osagian. { ¦kin·dər¦hůk·ē·ən }

kink band [GEOL] A deformation band in a single crystal or in foliated rocks in which the orientation is changed due to slipping on several parallel slip planes. Also known as joint drag; knick band; knick zone. { 'kiŋk ˌband }

kite observation [METEOROL] An atmospheric sounding by means of instruments carried aloft by a kite. { 'kīt ˌäb·sər'vā·shən }

klapperstein *See* rattle stone. { 'kläp·ərˌshtīn }

klint [GEOL] An exhumed coral reef or bioherm that is more resistant to the processes of erosion than the rocks that enclose it so that the core remains in relief as hills and ridges. { klint }

klintite [GEOL] The dense, hard dolomite composing a klint; gives to the core a strength and resistance to erosion. { 'klinˌtīt }

klippe [GEOL] A block of rock that is separated from underlying rocks by a fault that usually has a gentle dip. { klip }

kloof wind [METEOROL] A cold southwest wind of Simons Bay, South Africa. { 'klüf ˌwind }

knick *See* knickpoint. { nik }

knick band *See* kink band. { 'nik ˌband }

knickpoint [GEOL] A point of sharp change of slope, especially in the longitudinal profile of a stream or of its valley. Also known as break; knick; nick; nickpoint; rejuvenation head; rock step. { 'nik ˌpȯint }

knick zone *See* kink band. { 'nik ˌzōn }

knik wind [METEOROL] Local name for a strong southeast wind in the vicinity of Palmer in the Matanuska Valley of Alaska; it blows most frequently in the winter, although it may occur at any time of year. { kə'nik ˌwind }

knob [GEOL] **1.** A rounded eminence, such as a knoll, hillock, or small hill or mountain, and especially a prominent or isolated hill with steep sides. **2.** A peak or other projection at the top of a hill or mountain. { näb }

knoll [GEOL] A mound rising less than 3300 feet (1000 meters) from the sea floor. Also known as sea knoll. { ˌnōl }

Knudsen's tables [OCEANOGR] Hydrographical tables published by Martin Knudsen in 1901 to facilitate the computation of results of seawater chlorinity titrations and hydrometer temperature readings, and their conversion to salinity and density. { kə'nüd·sən ˌtā·bəlz }

koembang [METEOROL] A dry foehnlike wind from the southeast or south in Cheribon and Tegal in Java, caused by the east monsoon which develops a jet effect in passing through the gaps in the mountain ranges and descends on the leeward side. { 'kümˌbaŋ }

kona [METEOROL] A stormy, rain-bringing wind from the southwest or south-southwest in Hawaii; it blows about five times a year on the southwest slopes, which are in the lee of the prevailing northeast trade winds. { 'kō·nə }

kona cyclone [METEOROL] A slow-moving extensive cyclone which forms in subtropical latitudes during the winter season. Also known as kona storm. { 'kō·nə 'sī,klōn }

kona storm *See* kona cyclone. { 'kō·nə 'stōrm }

Köppen-Supan line [METEOROL] The isotherm connecting places which have a mean temperature of 10°C (50°F) for the warmest month of the year. { 'kep·ən sü'pän ,līn }

kossava [METEOROL] A cold, very squally wind descending from the east or southeast in the region of the Danube "Iron Gate" through the Carpathians, continuing westward over Belgrade, then spreading northward to the Rumanian and Hungarian borderlands and southward as far as Nish. { 'kȯ·sə,vä }

Krakatao winds [METEOROL] A layer of easterly winds over the tropics at an altitude of about 11 to 14.5 miles (18 to 24 kilometers), which tops the mid-tropospheric westerlies (the antitrades), is at least 3.5 miles (6 kilometers) deep, and is based at about 1.2 miles (2 kilometers) above the tropopause. { 'krak·ə,taü ,winz }

Krassowski ellipsoid of 1938 [GEOD] The reference ellipsoid of which the semimajor axis is 6,378,245 meters and the flattening or ellipticity equals 1/298.3. { kra'sȯv·skē ə'lip,sȯid əv 'nīn,tēn ,thər·dē'āt }

kremastic water *See* vadose water. { krə'mas·tik ¦wȯd·ər }

kryptoclimatology *See* cryptoclimatology. { ¦krip·tō,kli·mə'täl·ə·jē }

kukersite [GEOL] An organic sediment rich in remains of the alga *Gloexapsamorpha prisca*; found in the Ordovician of Estonia. { 'kü·kər,sīt }

Kungurian [GEOL] A European stage of geologic time; Middle Permian, above Artinskian, below Kazanian. { kùŋ'gùr·ē·ən }

Kuroshio [OCEANOGR] A fast ocean current originating off the southeast coast of Luzon, Philippines, and flowing northeastward off the coasts of China and Japan into the upper waters of the north Pacific Ocean. It carries large quantities of warm water from the tropics into the midlatitude regions, and is an important agent in redistributing global heat. { ,kü·rə'shē·ō }

Kuroshio Countercurrent [OCEANOGR] A component of the Kuroshio system flowing south and southwest between latitudes 155° and 160°E about 44 miles (70 kilometers) from the coast of Japan on the right-hand side of the Kuroshio Current. { ,kü·rə'shē·ō 'kaùnt·ər,kə·rənt }

Kuroshio extension [OCEANOGR] A general term for the warm, eastward-transitional flow that connects the Kuroshio and the North Pacific currents. { ,kü·rə'shē·ō ik 'sten·shən }

Kuroshio system [OCEANOGR] A system of ocean currents which includes part of the North Equatorial Current, the Tsushima Current, the Kuroshio Current, and the Kuroshio extension. { ,kü·rə'shē·ō ,sis·təm }

kyrohydratic point [OCEANOGR] The temperature at which a particular salt crystallizes in brine which is trapped by frozen seawater, the eutectic temperature of that salt. { ¦kī·rō·hī'drad·ik ,pȯint }

L

labbé [METEOROL] An infrequent, moderate to strong southwest wind that occurs only in March in Provence (southeastern France), bringing mild, humid, and very cloudy or rainy weather, while on the coast it raises a rough sea. { la'bā }

Labrador Current [OCEANOGR] A current that flows southward from Baffin Bay, through the Davis Strait, and southwestward along the Labrador and Newfoundland coasts. { 'lab·rə,dȯr ,kə·rənt }

laccolith [GEOL] A body of igneous rock intruding into sedimentary rocks so that the overlying strata have been notably lifted by the force of intrusion. { 'lak·ə,lith }

lacunaris *See* lacunosus. { ,lak·yə'nar·əs }

lacunosus [METEOROL] A cloud variety characterized more by the appearance of the spaces between the cloud elements than by the elements themselves, the gaps being generally rounded, often with fringed edges, and the overall appearance being that of a honeycomb or net; it is the negative of clouds composed of separate rounded elements.Formerly known as lacunaris. { ,lak·yə'nō·səs }

lacustrine [GEOL] Belonging to or produced by lakes. { lə'kəs·trən }

lacustrine sediments [GEOL] Sediments that are deposited in lakes. { lə'kəs·trən 'sed·ə·məns }

lacustrine soil [GEOL] Soil that is uniform in texture but variable in chemical composition and that has been formed by deposits in lakes which have become extinct. { lə'kəs·trən 'sȯil }

Ladinian [GEOL] A European stage of geologic time: upper Middle Triassic (above Anisian, below Carnian). { lə'din·ē·ən }

Lafond's Tables [METEOROL] A set of tables and associated information for correcting reversing thermometers and computing dynamic height anomalies, compiled by E. C. Lafond and published by the U.S. Navy Hydrographic Office. { lə'fänz ,tā·bəlz }

lag deposit [GEOL] Residual accumulation of coarse, unconsolidated rock and mineral debris left behind by the winnowing of finer material. { 'lag di,päz·ət }

lag fault [GEOL] A minor low-angle thrust fault occurring within an overthrust; it develops when one part of the mass is thrust farther than an adjacent higher or lower part. { 'lag ,fȯlt }

lag gravel [GEOL] Residual accumulations of particles that are coarser than the material that has blown away. { 'lag ,grav·əl }

lagoon [GEOGR] **1.** A shallow sound, pond, or lake generally near but separated from or communicating with the open sea. **2.** A shallow fresh-water pond or lake generally near or communicating with a larger body of fresh water. { lə'gün }

Lagrangian current measurement [OCEANOGR] Observation of the speed direction of an ocean current by means of a device, such as a parachute drogue, which follows the water movement. { lə'grän·jē·ən 'kə·rənt ,mezh·ər·mənt }

lahar [GEOL] **1.** A mudflow or landslide of pyroclastic material occurring on the flank of a volcano. **2.** The deposit of mud or land so formed. { 'lä,här }

lake [HYD] An inland body of water, small to moderately large, with its surface water exposed to the atmosphere. { lāk }

lake breeze [METEOROL] A wind, similar in origin to the sea breeze but generally weaker, blowing from the surface of a large lake onto the shores during the afternoon; it is caused by the difference in surface temperature of land and water, as in the land and sea breeze system. { 'lāk ,brēz }

lake effect [METEOROL] Generally, the effect of any lake in modifying the weather about its shore and for some distance downwind; in the United States, this term is applied specifically to the region about the Great Lakes. { 'lāk i,fekt }

lake effect storm [METEOROL] A severe snowstorm over a lake caused by the interaction between the warmer water and unstable air above it. { 'lāk i,fekt ,stórm }

lake peat [GEOL] A sedimentary peat formed near lakes. { 'lāk ,pēt }

lake plain [GEOL] One of the surfaces of the earth that represent former lake bottoms; these featureless surfaces are formed by deposition of sediments carried into the lake by streams. { 'lāk ,plān }

lamb-blasts *See* lambing storm. { 'lam,blasts }

lambing storm [METEOROL] A slight fall of snow in the spring in England. Also known as lamb-blasts; lamb-showers; lamb-storm. { 'lam·iŋ ,stórm }

lamb-showers *See* lambing storm. { 'lam ,shaú·ərz }

lamb-storm *See* lambing storm. { 'lam,stórm }

lamina [GEOL] A thin, clearly differentiated layer of sedimentary rock or sediment, usually less than 1 centimeter thick. { 'lam·ə·nə }

laminite [GEOL] Any sedimentary rock composed of millimeter- or finer-scale layers. { 'lam·ə,nīt }

Lanarkian [GEOL] A European stage of geologic time forming part of the lower Upper Carboniferous, above Lancastrian and below Yorkian, equivalent to lowermost Westphalian. { lə'när·kē·ən }

Lancastrian [GEOL] A European stage of geologic time forming part of the lower Upper Carboniferous, above Viséan and below Lanarkian. { laŋ'kas·trē·ən }

land [GEOGR] The portion of the earth's surface that stands above sea level. { land }

land and sea breeze [METEOROL] The complete cycle of diurnal local winds occurring on seacoasts due to differences in surface temperature of land and sea; the land breeze component of the system blows from land to sea, and the sea breeze blows from sea to land. { ¦land ən 'sē ,brēz }

landblink [METEOROL] A yellowish glow observed over snow-covered land in the polar regions. { 'land,bliŋk }

land breeze [METEOROL] A coastal breeze blowing from land to sea, caused by the temperature difference when the sea surface is warmer than the adjacent land; therefore, the land breeze usually blows by night and alternates with a sea breeze which blows in the opposite direction by day. { 'land ,brēz }

land bridge [GEOGR] A strip of land linking two landmasses, often subject to temporary submergence, but permitting intermittent migration of organisms. { 'land ,brij }

Landenian [GEOL] A European stage of geologic time: upper Paleocene (above Montian, below Ypresian of Eocene). { lan'den·ē·ən }

landfast ice *See* fast ice. { 'lan,fast 'īs }

landform [GEOGR] All the physical, recognizable, naturally formed features of land, having a characteristic shape; includes major forms such as a plain, mountain, or plateau, and minor forms such as a hill, valley, or alluvial fan. { 'lan,fórm }

landform map *See* physiographic diagram. { 'lan,fórm ,map }

land hemisphere [GEOGR] The half of the globe, with its pole located at 47.25°N 2.5°W, in which most of the earth's land area is concentrated. { 'land ¦hem·ə,sfir }

land ice [HYD] Any part of the earth's seasonal or perennial ice cover which has formed over land as the result, principally, of the freezing of precipitation. { 'land ,īs }

landlocked [GEOGR] Pertaining to a harbor which is surrounded or almost completely surrounded by land. { 'land,läkt }

land pebble *See* land pebble phosphate. { 'land ,peb·əl }

land pebble phosphate [GEOL] A pebble phosphate in a clay or sand bed below the ground surface; a small amount of uranium is often present and is recovered as a by-product; used as a source of phosphate fertilizer. Also known as land pebble; land rock; matrix rock. { 'land ¦peb·əl 'fäs,fāt }

land rock *See* land pebble phosphate. { 'land ,räk }

landscape [GEOGR] The distinct association of landforms that can be seen in a single view. { 'lan,skāp }

land sky [METEOROL] The relatively dark appearance of the underside of a cloud layer when it is over land that is not snow-covered, used largely in polar regions with reference to the sky map; it is brighter than water sky, but much darker than iceblink or snowblink. { 'land ,skī }

landslide [GEOL] The perceptible downward sliding or falling of a relatively dry mass of earth, rock, or combination of the two under the influence of gravity. Also known as landslip. { 'lan,slīd }

landslide track [GEOL] An exposed path in rock or earth created as the result of a landslide. { 'lan,slīd ,trak }

landslip *See* landslide. { 'lan,slip }

Langevin ion *See* large ion. { länzh·van ,ī,än }

Langmuir circulation [OCEANOGR] A form of motion found in the near-surface layer of lakes and oceans under windy conditions, and observed as streaks of bubbles, seaweed, or flotsam forming into lines running roughly parallel to the wind, called windrows. { 'laŋ·myür ,sər·kŋə,lā·shən }

lansan [METEOROL] A strong southeast trade wind of the New Hebrides and East Indies. { ¦län¦sän }

lapies *See* karren. { lə'pēz }

lapilli [GEOL] Pyroclasts that range from 0.04 to 2.6 inches (1 to 64 millimeters) in diameter. { lə'pi,lī }

lapilli-tuff [GEOL] A pyroclastic deposit that is indurated and consists of lapilli in a fine tuff matrix. { lə'pi,lī ¦təf }

lapse line [METEOROL] A curve showing the variation of temperature with height in the free air. { 'laps ,līn }

lapse rate [METEOROL] **1.** The rate of decrease of temperature in the atmosphere with height. **2.** Sometimes, the rate of change of any meteorological element with height. { 'laps ,rāt }

Laramic orogeny *See* Laramidian orogeny. { 'lar·ə·mik ȯ'räj·ə·nē }

Laramide orogeny *See* Laramidian orogeny. { 'lar·ə·məd ȯ'räj·ə·nē }

Laramide revolution *See* Laramidian orogeny. { 'lar·ə·məd ,rev·ə'lü·shən }

Laramidian orogeny [GEOL] An orogenic era typically developed in the eastern Rocky Mountains; phases extended from Late Cretaceous until the end of the Paleocene. Also known as Laramic orogeny; Laramide orogeny; Laramide revolution. { ,lar·ə'mid·ē·ən ȯ'räj·ə·nē }

lard ice *See* grease ice. { 'lärd ,īs }

lardite *See* agalmatolite. { 'lär,dīt }

large ion [METEOROL] An ion created by a small ion attaching to an Aitken nucleus; it is characterized by relatively large mass and low mobility. Also known as heavy ion; Langevin ion; slow ion. { 'lärj 'ī,än }

large nuclei [OCEANOGR] Particles of concentrated seawater or crystalline salt in the marine atmosphere having radii larger than 10^{-5} centimeter. { 'lärj 'nü·klē,ī }

large scale [METEOROL] A scale such that the curvature of the earth may not be considered negligible; this scale is applicable to the high tropospheric long-wave patterns, with four or five waves around the hemisphere in the middle latitudes. { 'lärj 'skāl }

large-scale convection [METEOROL] Organized vertical motion on a larger scale than atmospheric free convection associated with cumulus clouds; the patterns of vertical motion in hurricanes or in migratory cyclones are examples of such convection. { 'lärj ¦skāl kən'vek·shən }

latent instability [METEOROL] The state of that portion of a conditionally unstable air column lying above the level of free convection; latent instability is released only if an initial impulse on a parcel gives it sufficient kinetic energy to carry it through the layer below the level of free convection, within which the environment is warmer than the parcel. { 'lāt·ənt ,in·stə'bil·əd·ē }

lateral accretion [GEOL] The digging away of material at the outer bank of a meandering stream and the simultaneous building up to the water level by deposition of material brought there by pushing and rolling along the stream bottom. { 'lad·ə·rəl ə'krē·shən }

lateral cone *See* adventive cone. { 'lad·ə·rəl kōn }

lateral erosion [GEOL] The action of a stream in undermining a bank on one side of its channel so that material falls into the stream and disintegrates; simultaneously, the stream shifts toward the bank that is being undercut. { 'lad·ə·rəl i'rō·zhən }

lateral fault [GEOL] A fault along which there has been strike separation. Also known as strike-separation fault. { 'lad·ə·rəl 'fȯlt }

lateral moraine [GEOL] Drift material, usually thin, that was deposited by a glacier in a valley after the glacier melted. { 'lad·ə·rəl mə'rān }

lateral planation [GEOL] Reduction in land in interstream areas in a plane parallel to the stream profile; the reduction is caused by lateral movement of the stream against its banks. { 'lad·ə·rəl plā'nā·shən }

lateral secretion [GEOL] A supposed phenomenon whereby a lode's or vein's mineral content is derived from the adjacent wall rock. { 'lad·ə·rəl si'krē·shən }

laterite [GEOL] Weathered material composed principally of the oxides of iron, aluminum, titanium, and manganese; laterite ranges from soft, earthy, porous soil to hard, dense rock. { 'lad·ə,rīt }

lateritic soil [GEOL] **1.** Soil containing laterite. **2.** Any reddish soil developed from weathering. Also known as latosol. { ¦lad·ə¦rid·ik 'sȯil }

laterization [GEOL] Those conditions of weathering that lead to removal of silica and alkalies, resulting in a soil or rock with high concentrations of iron and aluminum oxides (laterite). { ,lad·ə·rə'zā·shən }

latitude [GEOD] Angular distance from a primary great circle or plane, as on the celestial sphere or the earth. { 'lad·ə,tüd }

latitude effect [GEOPHYS] The variation of a quantity with latitude; applied particularly to the increase in cosmic-ray intensity with increasing magnetic latitude. { 'lad·ə,tüd i,fect }

latitude variation [GEOPHYS] A periodic change in the latitude of any position on the earth's surface, caused by the polar variation. { 'lad·ə,tüd ,ver·ə'ā·shən }

latosol *See* lateritic soil. { 'lad·ə,sȯl }

lattice drainage pattern *See* rectangular drainage pattern. { 'lad·əs 'drān·ij ,pad·ərn }

Lattorfian *See* Tongrian. { lə'tȯr·fē·ən }

lauoho o pele *See* Pele's hair. { ,lä·ü'ō,hō ō 'pe,lē }

Laurasia [GEOL] A continent theorized to have existed in the Northern Hemisphere; supposedly it broke up to form the present northern continents about the end of the Pennsylvanian period. { lȯ'rā·zhə }

Laurentian Plateau *See* Laurentian Shield. { lȯ'ren·chən pla'tō }

Laurentian Shield [GEOL] A Precambrian plateau extending over half of Canada from Labrador southwest along Hudson Bay and northwest to the Arctic Ocean. Also known as Canadian Shield; Laurentian Plateau. { lȯ'ren·chən 'shēld }

Laurentide ice sheet [HYD] A major recurring glacier that at its maximum completely covered North America east of the Rockies from the Arctic Ocean to a line passing through the vicinity of New York, Cincinnati, St. Louis, Kansas City, and the Dakotas. { lȯr·ən,tīd 'īs ,shēt }

lava [GEOL] **1.** Molten extrusive material that reaches the earth's surface through volcanic vents and fissures. **2.** The rock mass formed by consolidation of molten rock issuing from volcanic vents and fissures, consisting chiefly of magnesium silicate; used for insulators. { 'lä·və }

lava blisters [GEOL] Small, steep-sided swellings that are hollow and raised on the surfaces of some basaltic lava flows; formed by gas bubbles pushing up the lava's viscous surface. { 'lä·və ,blis·tərz }

lava cone [GEOL] A volcanic cone that was formed of lava flows. { 'lä·və ,kōn }

lava dome *See* shield volcano. { 'lä·və ,dōm }

lava field [GEOL] A wide area of lava flow; it is commonly several square kilometers in area and forms along the base of a large compound volcano or on the flanks of shield volcanoes. { 'lä·və ˌfēld }

lava flow [GEOL] **1.** A lateral, surficial stream of molten lava issuing from a volcanic cone or from a fissure. **2.** The solidified mass of rock formed when a lava stream congeals. { 'lä·və ˌflō }

lava fountain [GEOL] A jetlike eruption of lava that issues vertically from a volcanic vent or fissure. Also known as fire fountain. { 'lä·və ˌfau̇nt·ən }

lava lake [GEOL] A lake of lava that is molten and fluid; usually contained within a summit volcanic crater or in a pit crater on the flanks of a shield volcano. { 'lä·və ˌlāk }

lava plateau [GEOL] An elevated tableland or flat-topped highland that is several hundreds to several thousands of square kilometers in area; underlain by a thick succession of lava flows. { 'lä·və plaˌtō }

lava tube [GEOL] A long, tubular opening under the crust of solidified lava. { 'lä·və ˌtüb }

law of storms [METEOROL] Historically, the general statement of the manner in which the winds of a cyclone rotate about the cyclone's center, and the way that the entire disturbance moves over the earth's surface. { 'lȯ əv 'stȯrmz }

law of superposition [GEOL] The law that strata underlying other strata must be the older if there has been neither overthrust nor inversion. { 'lȯ əv ˌsü·pər·pə'zish·ən }

layer [GEOL] A tabular body of rock, ice, sediment, or soil lying parallel to the supporting surface and distinctly limited above and below. [GEOPHYS] One of several strata of ionized air, some of which exist only during the daytime, occurring at altitudes between 30 and 250 miles (50 and 400 kilometers); the layers reflect radio waves at certain frequencies and partially absorb others. { 'lā·ər }

layer depth [OCEANOGR] **1.** The thickness of the mixed layer in an ocean. **2.** The depth to the top of the thermocline. { 'lā·ər ˌdepth }

layer depth effect [GEOPHYS] The weakening of a sound beam or seismic pulse because of abnormal spreading in passing from a positive gradient layer to an underlying negative layer. { 'lā·ər ¦depth iˌfekt }

layered complex [GEOL] An igneous rock body of large dimensions, 5–300 miles (8–480 kilometers) across and as much as 23,000 feet (7000 meters) thick, within which distinct subhorizontal stratification, or layering, is apparent and may be continuous over great distances, in some cases more than 60 miles (100 kilometers). { ¦lā·ərd 'kämˌpleks }

layer of no motion [OCEANOGR] A layer, assumed to be at rest, at some depth in the ocean. { 'lā·ər əv ¦nō 'mō·shən }

LCL *See* lifting condensation level.

leachate [GEOCHEM] A liquid that has percolated through soil and dissolved some soil materials in the process. { 'lēˌchāt }

leaching [GEOCHEM] The separation or dissolving out of soluble constituents from a rock or ore body by percolation of water. { 'lēch·iŋ }

lead [GEOL] A small, narrow passage in a cave. { led }

leader [GEOPHYS] The streamer which initiates the first phase of each stroke of a lightning discharge; it is a channel of very high ion density which propagates through the air by the continual establishment of an electron avalanche ahead of its tip. Also known as leader streamer. { 'lēd·ər }

leader stroke [GEOPHYS] The entire set of events associated with the propagation of any leader between cloud and ground in a lightning discharge. { 'lēd·ər ˌstrōk }

leaf mold [GEOL] A soil layer or compost consisting principally of decayed vegetable matter. { 'lēf ˌmōld }

leakage halo [GEOCHEM] The dispersion of elements along channels and paths followed by mineralizing solutions leading into and away from the central focus of mineralization. { 'lēk·ij ˌhā·lō }

leaking mode [GEOPHYS] A surface seismic wave which is imperfectly trapped, so that its energy leaks or escapes across a layer boundary, causing some attenuation. Also known as leaky wave. { 'lēk·iŋ ˌmōd }

leaky wave *See* leaking mode. { 'lēk·ē 'wāv }

ledge [GEOL] **1.** A narrow, shelflike ridge or rock protrusion, much longer than high, and usually horizontal, formed in a rock wall or on a cliff face. **2.** A ridge of rocks found underwater, especially one near a shore or connected with and bordering a shore. { lej }

Ledian [GEOL] Lower upper Eocene geologic time. Also known as Auversian. { 'lēd·ē·ən }

lee dune [GEOL] A dune formed to the leeward of a source of sand or of an obstacle. { 'lē ˌdün }

lee tide *See* leeward tidal current. { 'lē ˌtīd }

lee trough *See* dynamic trough. { 'lē ˌtrȯf }

leeward tidal current [OCEANOGR] A tidal current setting in the same direction as that in which the wind is blowing. Also known as lee tide; leeward tide. { 'lē·wərd ¦tīd·əl 'kə·rənt }

leeward tide *See* leeward tidal current. { 'lē·wərd ˌtīd }

left bank [GEOGR] The bank of a stream or river on the left of an observer when he is facing in the direction of flow, or downstream. { 'left ¦baŋk }

left lateral fault [GEOL] A fault in which movement is such that an observer walking toward the fault along an index plane (a bed, vein, or dike) would turn to the left to find the other part of the displaced index plane. Also known as sinistral fault. { 'left ¦lad·ə·rəl 'fȯlt }

leg [GEOPHYS] A single cycle of more or less periodic motion in a wave train on a seismogram. { leg }

length of record [CLIMATOL] The period during which observations have been maintained at a meteorological station, and which serves as the frame of reference for climatic data at that station. { 'leŋkth əv 'rek·ərd }

lens [GEOL] **1.** A geologic deposit that is thick in the middle and converges toward the edges, resembling a convex lens. **2.** An irregularly shaped formation consisting of a porous, permeable sedimentary deposit surrounded by impermeable rock. { lenz }

lenticle [GEOL] A bed or rock stratum or body that is lens-shaped. { 'len·tə·kəl }

lenticular cloud *See* lenticularis. { len'tik·yə·lər 'klau̇d }

lenticularis [METEOROL] A cloud species, the elements of which have the form of more or less isolated, generally smooth lenses; the outlines are sharp. Also known as lenticular cloud. { lenˌtik·yə'lar·əs }

Leonardian [GEOL] A North American provincial series: Lower Permian (above Wolfcampian, below Guadalupian). { ¦lā·ə¦när·dē·ən }

lentil [GEOL] **1.** A rock body that is lens-shaped and enclosed in a stratum of different material. **2.** A rock stratigraphic unit that is a subdivision of a formation and has limited geographic extent; it thins out in all directions. { 'lent·əl }

leptogeosyncline [GEOL] A deep oceanic trough that has not been filled with sedimentation and is associated with volcanism. { ˌlep·təˌjē·ō'sinˌklīn }

lesser ebb [OCEANOGR] The weaker of two ebb currents occurring during a tidal day. { 'les·ər ˌeb }

lesser flood [OCEANOGR] The weaker of two flood currents occurring during a tidal day. { 'les·ər ˌfləd }

leste [METEOROL] Spanish nautical term for east wind, specifically the hot, dry, dusty easterly or southeasterly wind which blows from the Atlantic coast of Morocco out to Madeira and the Canary Islands; it is a form of sirocco, occurring in front of depressions advancing eastward. { 'les·tā }

levante [METEOROL] The Spanish and most widely used term for an east or northeast wind occurring along the coast and inland from southern France to the Straits of Gibraltar; it is moderate or fresh, mild, very humid, and rainy, and occurs with a depression over the western Mediterranean Sea. { lə'vän·tə }

levantera [METEOROL] A persistent east wind of the Adriatic, usually accompanied by cloudy weather. { ˌle·vən'ter·ə }

Levantine Basin [OCEANOGR] A basin in the Mediterranean Ocean between Asia Minor and Egypt. { 'le·vən,tēn 'bās·ən }

leveche [METEOROL] A warm wind in Spain, either a foehn or a hot southerly wind in advance of a low-pressure area moving from the Sahara Desert. { lə'vā·chā }

levee [GEOL] **1.** An embankment bordering one or both sides of a sea channel or the low-gradient seaward part of a canyon or valley. **2.** A low ridge sometimes deposited by a stream on its sides. { 'lev·ē }

level fold *See* nonplunging fold. { 'lev·əl 'fōld }

level of free convection [METEOROL] The level at which a parcel of air lifted dry and adiabatically until saturated, and lifted saturated and adiabatically thereafter, would first become warmer than its surroundings in a conditionally unstable atmosphere. Abbreviated LFC. { 'lev·əl əv ,frē kən'vek·shən }

level of nondivergence [METEOROL] A level in the atmosphere throughout which the horizontal velocity divergence is zero; although in some meteorological situation there may be several such surfaces (not necessarily level), the level of nondivergence usually considered is that mid-tropospheric surface which separates the major regions of horizontal convergence and divergence associated with the typical vertical structure of the migratory cyclonic-scale weather systems. { 'lev·əl əv ,nän·də'vər·jəns }

level of saturation *See* water table. { 'lev·əl əv ,sach·ə'rā·shən }

level surface *See* geopotential surface. { 'lev·əl 'sər·fəs }

LFC *See* level of free convection.

Lias *See* Liassic. { 'lī·as }

Liassic [GEOL] The Lower Jurassic period of geologic time. Also known as Lias. { lī'as·ik }

Libby effect [GEOCHEM] The increase, since about 1950, in the carbon-14 content of the atmosphere, produced by the detonation of thermonuclear devices. { 'lib·ē i,fekt }

libeccio [METEOROL] Italian name for a southwest wind, used in northern Corsica for the west or southwest wind which blows throughout the year, especially in winter when it is often stormy. { li'bech·ō }

lichenometry [GEOL] Measurement of the diameter of lichens growing on exposed rock surfaces; used for dating geomorphic features, particularly of glacial origin. { ,lī·kə'näm·ə·trē }

Liesegang banding [GEOL] Colored or compositional rings or bands in a fluid-saturated rock due to rhythmic precipitation. Also known as Liesegang rings. { 'lēz·ə,gäŋ ,band·iŋ }

Liesegang rings *See* Liesegang banding. { 'lēz·ə,gäŋ ,riŋz }

lifting condensation level [METEOROL] The level at which a parcel of moist air lifted dry adiabatically would become saturated. Abbreviated LCL. Also known as isentropic condensation level (ICL). { 'lift·iŋ ,kän,den'sā·shən ,lev·əl }

light climate *See* illumination climate. { 'līt ,klī·mət }

light freeze [METEOROL] The condition when the surface temperature of the air drops to below the freezing point of water for a short time period, so that only the tenderest plants and vines are adversely affected. { 'līt ¦frēz }

light frost [HYD] A thin and more or less patchy deposit of hoarfrost on surface objects and vegetation. { 'līt ¦frȯst }

light ion *See* small ion. { 'līt 'ī,än }

lightning [GEOPHYS] An abrupt high-current electric discharge that occurs in the atmospheres of the earth and other planets and that has a path length ranging from hundreds of feet to tens of miles. Lightning occurs in thunderstorms because vertical air motions and interactions between cloud particles cause a separation of positive and negative charges. { 'līt·niŋ }

lightning channel [GEOPHYS] The irregular path through the air along which a lightning discharge occurs. { 'līt·niŋ ,chan·əl }

lightning discharge [GEOPHYS] The series of electrical processes by which charge is transferred within the atmosphere along a channel of high ion density between electric charge centers of opposite sign. { 'līt·niŋ 'dis,chärj }

lightning flash [GEOPHYS] In atmospheric electricity, the total observed luminous phenomenon accompanying a lightning discharge. { 'līt·niŋ ,flash }

lightning stroke [GEOPHYS] Any one of a series of repeated discharges comprising a single lightning discharge (or lightning flash); specifically, in the case of the cloud-to-ground discharge, a leader plus its subsequent return streamer. { 'līt·niŋ ,strōk }

light-of-the-night-sky *See* airglow. { ¦līt əv thə ¦nīt 'skī }

light pillar *See* sun pillar. { 'līt ,pil·ər }

lignite [GEOL] Coal of relatively recent origin, intermediate between peat and bituminous coal; often contains patterns from the wood from which it formed. Also known as brown coal; earth coal. { 'lig,nīt }

lignite A *See* black lignite. { 'lig,nīt 'ā }

lignite B *See* brown lignite. { 'lig,nīt 'bē }

lily-pad ice *See* pancake ice. { 'lil·ē ,pad ,īs }

limb [GEOL] One of the two sections of an anticline or syncline on either side of the axis. Also known as flank. { limb }

lime-pan playa [GEOL] A playa with a smooth, hard surface composed of calcium carbonate. { 'līm ¦pan 'plī·ə }

limestone pebble conglomerate [GEOL] A well-sorted conglomerate composed of limestone pebbles resulting from special conditions involving rapid mechanical erosion and short transport distances. { 'līm,stōn ¦peb·əl kən'gläm·ə·rət }

limit of the atmosphere [GEOPHYS] The level at which the atmospheric density becomes the same as the density of interplanetary space, which is usually taken to be about one particle per cubic centimeter. { 'lim·ət əv thə 'at·mə,sfir }

lineament [GEOL] A straight or gently curved, lengthy topographic feature expressed as depressions or lines of depressions. Also known as linear. { 'lin·ē·ə·mənt }

linear *See* lineament. { 'lin·ē·ər }

linear cleavage [GEOL] The property of metamorphic rocks of breaking into long planar fragments. { 'lin·ē·ər 'klē·vij }

lineation [GEOL] Any linear structure on or within a rock; examples are ripple marks and flow lines. { ,lin·ē'ā·shən }

line blow [METEOROL] A strong wind on the equator side of an anticyclone, probably so called because there is little shifting of wind direction during the blow, as contrasted with the marked shifting which occurs with a cyclonic windstorm. { 'līn ,blō }

line gale *See* equinoctial storm. { 'līn ,gāl }

line of strike *See* strike. { 'līn əv 'strīk }

line squall [METEOROL] A squall that occurs along a squall line. { 'līn ,skwȯl }

line storm *See* equinoctial storm. { 'līn ,stȯrm }

linguloid ripple mark *See* linguoid ripple mark. { 'liŋ·gyə,lōid 'rip·əl ,märk }

linguoid current ripple *See* linguoid ripple mark. { 'liŋ·gwȯid ¦kə·rənt ,rip·əl }

linguoid ripple mark [GEOL] An aqueous current ripple mark with tonguelike projections which are formed by action of a current of water and which point into the current. Also known as cuspate ripple mark; linguloid ripple mark; linguoid current ripple. { 'liŋ·gwȯid 'rip·əl ,märk }

Linke scale [METEOROL] A type of cyanometer; used to measure the blueness of the sky; it is simply a set of eight cards of different standardized shades of blue, numbered (evenly) 2 to 16; the odd numbers are used by the observer if the sky color lies between any of the given shades. Also known as blue-sky scale. { 'liŋk ,skāl }

Lipalian [GEOL] A hypothetical geologic period that supposedly antedated the Cambrian. { lə'pal·yən }

lipper [OCEANOGR] **1.** Slight ruffling or roughness appearing on a water surface. **2.** Light spray originating from small waves. { 'lip·ər }

liptinite *See* exinite. { 'lip·tə,nīt }

liquid-dominated hydrothermal reservoir [GEOL] Any geothermal system mainly producing superheated water (often termed brines); hot springs, fumaroles, and geysers are the surface expressions of hydrothermal reservoirs; an example is the hot-brine region in the Imperial Valley-Salton Sea area of southern California. { 'lik·wəd ¦däm·ə,nād·əd ,hī·drə¦thər·məl 'rez·əv,wär }

liquid-filled porosity [GEOL] The condition in porous rock or sand formations in which pore spaces contain fresh or salt water, liquid petroleum, pressure-liquefied butane or propane, or tar. { 'lik·wəd ¦fild pə'räs·əd·ē }

liquid limit [GEOL] The moisture content boundary that exists between the plastic and semiliquid states of a sediment. { 'lik·wəd 'lim·ət }

liquid-water content *See* water content. { 'lik·wəd ¦wȯd·ər ,kän,tent }

lithic tuff [GEOL] **1.** A tuff that is mostly crystalline rock fragments. **2.** An indurated volcanic ash deposit whose fragments are composed of previously formed rocks that first solidified in the volcanic vent and were then blown out. { 'lith·ik 'təf }

lithifaction *See* lithification. { ,lith·ə'fak·shən }

lithification [GEOL] **1.** Conversion of a newly deposited sediment into an indurated rock. Also known as lithifaction. **2.** Compositional change of coal to bituminous shale or other rock. { ,lith·ə·fə'kā·shən }

lithoclase [GEOL] A naturally produced rock fracture. { 'lith·ə,klās }

lithofacies [GEOL] A subdivision of a specified stratigraphic unit distinguished on the basis of lithologic features. { ,lith·ə'fā·shēz }

lithofacies map [GEOL] The facies map of an area based on lithologic characters; shows areal variation in all aspects of the lithology of a stratigraphic unit. { ,lith·ə'fā·shēz ,map }

lithogeochemical survey [GEOCHEM] A geochemical survey that involves the sampling of rocks. { ,lith·ō,jē·ə'kem·ə·kəl 'sər,vā }

lithographic limestone [GEOL] A dense, compact, fine-grained crystalline limestone having a pale creamy-yellow or grayish color. Also known as lithographic stone; litho stone. { ,lith·ə'graf·ik 'līm,stōn }

lithographic stone *See* lithographic limestone. { ,lith·ə'graf·ik 'stōn }

lithographic texture [GEOL] The texture of certain calcareous sedimentary rocks characterized by grain size of less than 1/256 millimeter and having a smooth appearance. { ,lith·ə'graf·ik 'teks·chər }

lithologic map [GEOL] A kind of geologic map showing the rock types of a particular area. { ¦lith·ə¦läj·ik 'map }

lithologic unit *See* rock-stratigraphic unit. { ¦lith·ə¦läj·ik 'yü·nət }

lithology [GEOL] The description of the physical character of a rock as determined by eye or with a low-power magnifier, and based on color, structures, mineralogic components, and grain size. { lə'thäl·ə·jē }

lithometeor [METEOROL] The general term for dry atmospheric suspensoids, including dust, haze, smoke, and sand. { ,lith·ə'mēd·ē·ər }

lithomorphic [GEOL] Referring to a soil whose characteristics are derived from events or conditions of a former period. { ¦lith·ə¦mȯr·fik }

lithophile [GEOCHEM] **1.** Pertaining to elements that have become concentrated in the silicate phase of meteorites or the slag crust of the earth. **2.** Pertaining to elements that have a greater free energy of oxidation per gram of oxygen than iron. Also known as oxyphile. { 'lith·ə,fīl }

lithophysa [GEOL] A large spherulitic hollow or bubble in glassy basalts and certain rhyolites. Also known as stone bubble. { ,lith·ə'fīs·ə }

lithosiderite *See* stony-iron meteorite. { ,lith·ə'sīd·ə,rīt }

lithosol [GEOL] A group of shallow soils lacking well-defined horizons and composed of imperfectly weathered fragments of rock. { 'lith·ə,sȯl }

lithosphere [GEOL] **1.** The rigid outer crust of rock on the earth about 50 miles (80 kilometers) thick, above the asthenosphere. Also known as oxysphere. **2.** Since the development of plate tectonics theory, a term referring to the rigid, upper 60 miles (100 kilometers) of the crust and upper mantle, above the asthenosphere. { 'lith·ə,sfir }

lithostatic pressure *See* ground pressure. { ¦lith·ə¦stad·ik 'presh·ər }

litho stone *See* lithographic limestone. { 'lith·ō ,stōn }

lithostratic unit *See* rock-stratigraphic unit. { ¦lith·ə¦strad·ik 'yü·nət }

lithostratigraphic unit *See* rock-stratigraphic unit. { ˌlith·əˌstrad·ə'graf·ik 'yü·nət }
lithostratigraphy [GEOL] A branch of stratigraphy concerned with the description and interpretation of sedimentary successions in terms of their lithic character. { ¦lith·ō·strə'tig·rə·fē }
lithotope [GEOL] **1.** The environment under which a sediment is deposited. **2.** An area of uniform sedimentation. { 'lith·əˌtōp }
lithotype [GEOL] A macroscopic band in humic coals, analyzed on the basis of physical characteristics rather than botanical origin. { 'lith·əˌtīp }
lit-par-lit [GEOL] Pertaining to the penetration of bedded, schistose, or other foliate rocks by innumerable narrow sheets and tongues of granitic rock. { 'lēˌpär'lē }
little brother [METEOROL] A subsidiary tropical cyclone that sometimes follows a more severe disturbance. { 'lid·əl 'brəth·ər }
Little Ice Age [GEOL] A period of expansion of mountain glaciers, marked by climatic deterioration, that began about 5500 years ago and extended to as late as A.D. 1550–1850 in some regions, as the Alps, Norway, Iceland, and Alaska. { 'lid·əl 'īs ˌāj }
littoral current [OCEANOGR] A current, caused by wave action, that sets parallel to the shore; usually in the nearshore region within the breaker zone. Also known as alongshore current; longshore current. { 'lit·ə·rəl 'kə·rənt }
littoral drift [GEOL] Materials moved by waves and currents of the littoral zone. Also known as longshore drift. { 'lit·ə·rəl 'drift }
littoral sediments [GEOL] Deposits of littoral drift. { 'lit·ə·rəl 'sed·ə·məns }
littoral transport [GEOL] The movement of littoral drift. { 'lit·ə·rəl 'tranzˌpȯrt }
L joint *See* primary flat joint. { 'el ˌjȯint }
Llandellian [GEOL] Upper Middle Ordovician geologic time. { lan'del·yən }
Llandoverian [GEOL] Lower Silurian geologic time. { ¦lan·də¦vir·ē·ən }
Llanvirnian [GEOL] Lower Middle Ordovician geologic time. { lan'vir·nē·ən }
llebetjado [METEOROL] In northeastern Spain, a hot, squally wind descending from the Pyrenees and lasting for a few hours. { ˌyā·bet'hä·dō }
load cast [GEOL] An irregularity at the base of an overlying stratum, usually sandstone, that projects into an underlying stratum, usually shale or clay. { lōd ˌkast }
load metamorphism *See* static metamorphism. { 'lōd ˌmed·ə'mȯrˌfiz·əm }
loam [GEOL] Soil mixture of sand, silt, clay, and humus. { lōm }
loaming [GEOCHEM] In geochemical prospecting, a method in which samples of material from the surface are tested for traces of a sought-after metal; its presence on the surface presumably indicates a near-surface ore body. { ˌlōm·iŋ }
lobate rill mark [GEOL] A flute cast formed by current action. { 'lōˌbāt 'ril ˌmärk }
lobe [HYD] A curved projection on the margin of a continental ice sheet. { lōb }
local angular momentum [METEOROL] Angular momentum about an arbitrarily located vertical axis which is fixed with respect to the earth. { 'lō·kəl 'aŋ·gyə·lər mə'men·təm }
local attraction *See* local magnetic disturbance. { 'lō·kəl ə'trak·shən }
local base level *See* temporary base level. { 'lō·kəl ¦bās ¦lev·əl }
local change [OCEANOGR] The time rate of change of a scalar quantity (such as temperature, salinity, pressure, or oxygen content) in a fixed locality. { 'lō·kəl 'chānj }
local extra observation [METEOROL] An aviation weather observation taken at specified intervals, usually every 15 minutes, when there are impending aircraft operations and when weather conditions are below certain operational weather limits; the observation includes ceiling, sky condition, visibility, atmospheric phenomena, and pertinent remarks. { 'lō·kəl 'ek·strə ˌäb·zər'vā·shən }
local forecast [METEOROL] Generally, any weather forecast of conditions over a relatively limited area, such as a city or airport. { 'lō·kəl 'fȯrˌkast }
local inflow [HYD] The water that enters a stream between two stream-gaging stations. { 'lō·kəl 'inˌflō }
local magnetic disturbance [GEOPHYS] An anomaly of the magnetic field of the earth, extending over a relatively small area, due to local magnetic influences. Also known as local attraction. { 'lō·kəl mag'ned·ik di'stər·bəns }

local peat [GEOL] Peat formed by groundwater. Also known as basin peat. { 'lō·kəl ˌpēt }

local relief [GEOL] The vertical difference in elevation between the highest and lowest points of a land surface within a specified horizontal distance or in a limited area. Also known as relative relief. { 'lō·kəl ri'lēf }

local storm [METEOROL] A storm of mesometeorological scale; thus, thunderstorms, squalls, and tornadoes are often put in this category. { 'lō·kəl 'stȯrm }

local winds [METEOROL] Winds which, over a small area, differ from those which would be appropriate to the general pressure distribution, or which possess some other peculiarity. { 'lō·kəl 'winz }

lode [GEOL] A fissure in consolidated rock filled with mineral; usually applied to metalliferous deposits. { lōd }

lodos [METEOROL] A southerly wind on the Black Sea coast of Bulgaria. { 'lȯ·dȯs }

lodranite [GEOL] A stony iron meteorite composed of bronzite and olivine within a fine network of nickel-iron. { 'lō·drəˌnīt }

loess [GEOL] An essentially unconsolidated, unstratified calcareous silt; commonly it is homogeneous, permeable, and buff to gray in color, and contains calcareous concretions and fossils. { les }

loess kindchen [GEOL] An irregular or spheroidal nodule of calcium carbonate that is found in loess. { 'les ˌkint·chən }

logarithmic velocity profile [METEOROL] The theoretical variation of the mean wind speed with height in the surface boundary layer under certain assumptions. { 'läg·əˌrith·mik və'läs·əd·ē ˌprōˌfīl }

lolly ice [OCEANOGR] Saltwater frazil, a heavy concentration of which is called sludge. { 'läl·ē ˌīs }

lombarde [METEOROL] An easterly wind (from Lombardy) that predominates along the French-Italian frontier, and comes from the High Alps; in winter it is violent and forms snow drifts in the mountain valleys; in the plains it is gentle and very dry. { lȯm'bär·də }

Lomonosov ridge [GEOGR] An undersea ridge which subdivides the Arctic Basin, extending from Ellesmere Land to the New Siberian Islands. { lō·mō'nȯˌsȯf ˌrij }

longitude [GEOD] Angular distance, along the Equator, between the meridian passing through a position and, usually, the meridian of Greenwich. { 'län·jəˌtüd }

longitudinal dune [GEOL] A type of linear dune ridge that extends parallel to the direction of the dominant dune-building winds. { ˌlän·jə'tüd·ən·əl 'dün }

longitudinal fault [GEOL] A fault parallel to the trend of the surrounding structure. { ˌlän·jə'tüd·ən·əl 'fȯlt }

longitudinal stream *See* subsequent stream. { ˌlän·jə'tüd·ən·əl 'strēm }

long-period tide [OCEANOGR] A tide or tidal current constituent with a period which is independent of the rotation of the earth but which depends upon the orbital movement of the moon or of the earth. { 'lȯŋ ˌpir·ē·əd 'tīd }

long-range forecast [METEOROL] A weather forecast covering periods from 48 hours to a week in advance (medium-range forecast), and ranging to even longer forecasts over periods of a month, a season, and so on. { 'lȯŋ ˌrānj 'fȯrˌkast }

longshore bar [GEOL] A ridge of sand, gravel, or mud built on the seashore by waves and currents, generally parallel to the shore and submerged by high tides. Also known as offshore bar. { 'lȯŋˌshȯr ˌbär }

longshore current *See* littoral current. { 'lȯŋˌshȯr ˌkə·rənt }

longshore drift *See* littoral drift. { 'lȯŋˌshȯr ˌdrift }

longshore trough [GEOL] A long, wide, shallow depression of the sea floor parallel to the shore. { 'lȯŋˌshȯr ˌtrȯf }

long wave [METEOROL] With regard to atmospheric circulation, a wave in the major belt of westerlies which is characterized by large length (thousands of kilometers) and significant amplitude; the wavelength is typically longer than that of the rapidly moving individual cyclonic and anticyclonic disturbances of the lower troposphere. Also known as major wave; planetary wave. { 'lȯŋ ¦wāv }

loo [METEOROL] A hot wind from the west in India. { lü }

loom [METEOROL] The glow of light below the horizon produced by greater-than-normal refraction in the lower atmosphere; it occurs when the air density decreases more rapidly with height than in the normal atmosphere. { lüm }

loop lake *See* oxbow lake. { 'lüp ˌlāk }

loop rating [HYD] A rating curve that has higher values of discharge for a certain stage when the river is rising than it does when the river is falling; thus, the curve (stage versus discharge) describes a loop with each rise and fall of the river. { 'lüp ˌrād·iŋ }

lopolith [GEOL] A large, floored intrusive body that is sunken centrally into the shape of a basin due to sagging of the underlying country rock. { 'läp·əˌlith }

losing stream *See* influent stream. { 'lüs·iŋ ˌstrēm }

lost stream [HYD] **1.** A stream that disappears from the surface into an underground channel without reappearing in the same or even a neighboring drainage basin. **2.** An evaporated stream in a desertlike region. { 'lȯst 'strēm }

Love wave [GEOPHYS] A horizontal dispersive surface wave, multireflected between internal boundaries of an elastic body, applied chiefly in the study of seismic waves in the earth's crust. { 'ləv ˌwāv }

low *See* depression. { lō }

low aloft *See* upper-level cyclone. { 'lō ə'lȯft }

low-angle fault [GEOL] A fault that dips at an angle less than 45°. { 'lō ˌaŋ·gəl ¦fȯlt }

low-angle thrust *See* overthrust. { 'lō ˌaŋ·gəl ¦thrəst }

low clouds [METEOROL] Types of clouds, the mean level of which is between the surface and 6500 feet (1980 meters); the principal clouds in this group are stratocumulus, stratus, and nimbostratus. { 'lō 'klau̇dz }

low-energy environment [GEOL] An aqueous sedimentary environment in which there is standing water with a general lack of wave or current action, permitting accumulation of very fine-grained sediments. { 'lō ˌen·ər·jē in'vī·ərn·mənt }

lower atmosphere [METEOROL] That part of the atmosphere in which most weather phenomena occur (that is, the troposphere and lower stratosphere); in other contexts, the term implies the lower troposphere. { 'lō·ər 'at·məˌsfir }

Lower Cambrian [GEOL] The earliest epoch of the Cambrian period of geologic time, ending about 540,000,000 years ago. { 'lō·ər 'kam·brē·ən }

Lower Cretaceous [GEOL] The earliest epoch of the Cretaceous period of geologic time, extending from about 140- to 120,000,000 years ago. { 'lō·ər krə'tā·shəs }

Lower Devonian [GEOL] The earliest epoch of the Devonian period of geologic time, extending from about 400- to 385,000,000 years ago. { 'lō·ər də'vō·nē·ən }

lower high water [OCEANOGR] The lower of two high tides occurring during a tidal day. { 'lō·ər 'hī ˌwȯd·ər }

Lower Jurassic [GEOL] The earliest epoch of the Jurassic period of geologic time, extending from about 185- to 170,000,000 years ago. { 'lō·ər jü'ras·ik }

lower low water [OCEANOGR] The lower of two low tides occurring during a tidal day. { 'lō·ər 'lō ˌwȯd·ər }

lower mantle [GEOL] The portion of the mantle below a depth of about 600 miles (1000 kilometers). Also known as inner mantle; mesosphere; pallasite shell. { 'lō·ər mant·əl }

Lower Mississippian [GEOL] The earliest epoch of the Mississippian period of geologic time, beginning about 350,000,000 years ago. { 'lō·ər ˌmis·ə'sip·ē·ən }

Lower Ordovician [GEOL] The earliest epoch of the Ordovician period of geologic time, extending from about 490- to 460,000,000 years ago. { 'lō·ər ˌȯr·də'vish·ən }

Lower Pennsylvanian [GEOL] The earliest epoch of the Pennsylvanian period of geologic time, beginning about 310,000,000 years ago. { 'lō·ər ˌpen·səl'vā·nyən }

Lower Permian [GEOL] The earliest epoch of the Permian period of geologic time, extending from about 275- to 260,000,000 years ago. { 'lō·ər 'pər·mē·ən }

lower plate *See* footwall. { 'lō·ər ˌplāt }

Lower Silurian [GEOL] The earliest epoch of the Silurian period of geologic time, beginning about 420,000,000 years ago. { 'lō·ər sə'lu̇r·ē·ən }

Lower Triassic [GEOL] The earliest epoch of the Triassic period of geologic time, extending from about 230- to 215,000,000 years ago. { 'lō·ər trī'as·ik }

low index [METEOROL] A relatively low value of the zonal index which, in middle latitudes, indicates a relatively weak westerly component of wind flow (usually implying stronger north-south motion), and the characteristic weather attending such motion; a circulation pattern of this type is commonly called a low-index situation. { 'lō ¦in,deks }

low-moor bog [GEOL] A bog that is at or slightly below the ground water table. { 'lō ,mür 'bäg }

low-moor peat [GEOL] Peat found in low-moor bogs or swamps and containing little or no sphagnum. Also known as fen peat. { 'lō ¦mür 'pēt }

low-rank metamorphism [GEOL] A metamorphic process that occurs under conditions of low to moderate pressure and temperature. { 'lō ,raŋk ,med·ə'mȯr·fiz·əm }

low tide *See* low water. { 'lō 'tīd }

low-tide terrace [GEOL] A flat area of a beach adjacent to the low-water line. { 'lō ¦tīd 'ter·əs }

low-velocity layer [GEOPHYS] A layer in the solid earth in which seismic wave velocity is lower than the layers immediately below or above. { 'lō və¦läs·əd·ē ,lā·ər }

low-volatile coal [GEOL] A coal that is nonagglomerating, has 78% to less than 86% fixed carbon, and 14% to less than 22% volatile matter. { 'lō ¦väl·ət·əl 'kōl }

low water [OCEANOGR] The lowest limit of the surface water level reached by the lowering tide. Also known as low tide. { 'lō 'wȯd·ər }

low-water inequality [OCEANOGR] The difference between the heights of two successive low tides. { 'lō ,wȯd·ər ,in·i'kwäl·əd·ē }

low-water interval *See* low-water lunitidal interval. { 'lō ,wȯd·ər 'in·tər·vəl }

low-water lunitidal interval [GEOPHYS] For a specific location, the interval of time between the transit (upper or lower) of the moon and the next low water. Also known as low-water interval. { 'lō ,wȯd·ər ¦lün·ə¦tīd·əl 'in·tər·vəl }

low-water neaps *See* mean low-water neaps. { 'lō ,wȯd·ər 'nēps }

low-water springs *See* mean low-water springs. { 'lō ,wȯd·ər 'spriŋz }

Ludian [GEOL] A European stage of geologic time in the uppermost Eocene, above the Bartonian and below the Tongrian of the Oligocene. { 'lü·dē·ən }

Ludlovian [GEOL] A European stage of geologic time; Upper Silurian, below Gedinnian of Devonian, above Wenlockian. { ləd'lō·vē·ən }

luganot [METEOROL] A strong south or south-southeast wind of Lake Garda, Italy. { lü'gä,nȯt }

Luisian [GEOL] A North American stage of geologic time: Miocene (above Relizian, below Mohnian). { lü'ē·shən }

lum *See* trolley. { ləm }

luminous cloud *See* sheet lightning. { 'lü·mə·nəs 'klaůd }

luminous meteor [METEOROL] According to United States weather observing practice, any one of a number of atmospheric phenomena which appear as luminous patterns in the sky, including halos, coronas, rainbows, aurorae, and their many variations, but excluding lightning (an igneous meteor or electrometeor). { 'lü·mə·nəs 'mēd·ē·ər }

lunar atmospheric tide [METEOROL] An atmospheric tide due to the gravitational attraction of the moon; the only detectable components are the 12-lunar-hour or semidiurnal component, as in the oceanic tides, and two others of very nearly the same period; the amplitude of this atmospheric tide is so small that it is detected only by careful statistical analysis of a long record. { 'lü·nər ,at·mə,sfir·ik 'tīd }

lunar inequality [GEOPHYS] A minute fluctuation of a magnetic needle from its mean position, caused by the moon. { 'lü·nər ,in·i'kwäl·əd·ē }

lunar tide [OCEANOGR] The portion of a tide produced by forces of the moon. { 'lü·nər 'tīd }

lunate bar [GEOL] A crescent-shaped bar of sand that is frequently found off the entrance to a harbor. { 'lü,nāt 'bär }

lunette [GEOL] A broad, low crescentic mound of windblown fine silt and clay. { lü'net }

lunisolar tides [OCEANOGR] Harmonic tidal constituents attributable partly to the development of both the lunar tide and the solar tide and partly to the lunisolar synodic fortnightly constituent. { ¦lü·nə'sō·lər 'tīdz }

lunitidal interval [OCEANOGR] The period between the moon's upper or lower transit over a specified meridian and a specified phase of the tidal current following the transit. { ¦lü·nə'tīd·əl 'in·tər·vəl }

Lusitanian [GEOL] Lower Jurassic geologic time. { ˌlü·sə'tan·ē·ən }

luster mottlings [GEOL] The spotted, shimmering appearance of certain rocks caused by reflection of light from cleavage faces of crystals that contain small inclusions of other minerals. { 'ləs·tər ˌmät·liŋz }

lutaceous [GEOL] Claylike. { lü'tā·shəs }

lutecite [GEOL] A fibrous, chalcedony-like quartz with optical anomalies that have led to its being considered a distinct species. { 'lüd·əˌsīt }

lutite [GEOL] A consolidated rock or sediment formed principally of clay or clay-sized particles. { 'lüˌtīt }

L wave [GEOPHYS] A phase designation for an earthquake wave that is a surface wave, without respect to type. { 'el ˌwāv }

lysocline [OCEANOGR] The level or ocean depth at which the rate of solution of calcium carbonate increases significantly. { 'lī·səˌklīn }

M

maar [GEOL] A volcanic crater that was created by violent explosion but not accompanied by igneous extrusion; frequently, it is filled by a small circular lake. { mär }
macaluba *See* mud volcano. { ˌmä·kə'lü·bə }
maceral [GEOL] The microscopic organic constituents found in coal. { ¦mas·ə¦ral }
mackerel sky [METEOROL] A sky with considerable cirrocumulus or small-element altocumulus clouds, resembling the scales on a mackerel. { 'mak·rəl 'skī }
Macky effect [METEOROL] The reduction of the effective dielectric strength of air when waterdrops are present. { 'mak·ē iˌfekt }
macroclimate [CLIMATOL] The climate of a large geographic region. { ¦mak·rō'klī·mət }
macrofacies [GEOL] A collection of sedimentary facies that are related genetically. { ¦mak·rō¦fā·shēz }
macrometeorology [METEOROL] The study of the largest-scale aspects of the atmosphere, such as the general circulation, and weather types. { ¦mak·rōˌmēd·ē·ə'räl·ə·jē }
macropore [GEOL] A pore in soil of a large enough size so that water is not held in it by capillary attraction. { 'mak·rəˌpȯr }
maculose [GEOL] Of a group of contact-metamorphosed rocks or their structures, having spotted or knotted character. { 'mak·yəˌlōs }
maelstrom [OCEANOGR] A powerful and often destructive water current caused by the combined effects of high, wind-generated waves and a strong, opposing tidal current. { 'māl·strəm }
Maestrichtian [GEOL] A European stage of geologic time: Upper Cretaceous (above Menevian, below Fastiniogian). { ma'strik·tē·ən }
maestro [METEOROL] A northwesterly wind with fine weather which blows, especially in summer, in the Adriatic, most frequently on the western shore; it is also found on the coasts of Corsica and Sardinia. { 'mī·strō }
magma [GEOL] The molten rock material from which igneous rocks are formed. { 'mag·mə }
magma chamber [GEOL] A larger reservoir in the crust of the earth that is occupied by a body of magma. { 'mag·mə ˌchām·bər }
magma geothermal system [GEOL] A geothermal system in which the dominant source of heat is a large reservoir of igneous magma within an intrusive chamber or lava pool; an example is the Yellowstone Park area of Wyoming. { 'mag·mə ¦jē·ō'thər·məl ˌsis·təm }
magma province *See* petrographic province. { 'mag·mə ˌpräv·əns }
magmatic stoping [GEOL] A process of igneous intrusion in which magma gradually works its way upward by breaking off and engulfing blocks of the country rock. Also known as stoping. { mag'mad·ik 'stōp·iŋ }
magmatic water [HYD] Water derived from or existing in molten igneous rock or magma. Also known as juvenile water. { mag'mad·ik 'wȯd·ər }
magmosphere *See* pyrosphere. { 'mag·məˌsfir }
magnafacies [GEOL] A major, continuous belt of deposits that is homogeneous in lithologic and paleontologic characteristics and that extends obliquely across time planes or through several time-stratigraphic units. { ¦mag·nə'fā·shēz }

magnetic annual change [GEOPHYS] The amount of secular change in the earth's magnetic field which occurs in 1 year. Also known as annual magnetic change. { mag'ned·ik 'an·yə·wəl 'chānj }

magnetic annual variation [GEOPHYS] The small, systematic temporal variation in the earth's magnetic field which occurs after the trend for secular change has been removed from the average monthly values. Also known as annual magnetic variation. { mag'ned·ik 'an·yə·wəl ˌver·ē'ā·shən }

magnetic bay [GEOPHYS] A small magnetic disturbance whose magnetograph resembles an indentation of a coastline; on earth, magnetic bays occur mainly in the polar regions and have a duration of a few hours. { mag'ned·ik 'bā }

magnetic character figure *See* C index. { mag'ned·ik ¦kar·ik·tər ˌfig·yər }

magnetic daily variation *See* magnetic diurnal variation. { mag'ned·ik ¦dā·lē ver·ē'ā·shən }

magnetic declination *See* declination. { mag'ned·ik ˌdek·lə'nā·shən }

magnetic dip *See* inclination. { mag'ned·ik 'dip }

magnetic direction [GEOD] Horizontal direction expressed as angular distance from magnetic north. { mag'ned·ik də'rek·shən }

magnetic diurnal variation [GEOPHYS] Oscillations of the earth's magnetic field which have a periodicity of about a day and which depend to a close approximation only on local time and geographic latitude. Also known as magnetic daily variation. { mag'ned·ik dī'ərn·əl ˌver·ē'ā·shən }

magnetic element [GEOPHYS] Magnetic declination, dip, or intensity at any location on the surface of the earth. { mag'ned·ik 'el·ə·mənt }

magnetic equator [GEOPHYS] That line on the surface of the earth connecting all points at which the magnetic dip is zero. Also known as aclinic line. { mag'ned·ik i'kwād·ər }

magnetic latitude [GEOPHYS] Angular distance north or south of the magnetic equator. { mag'ned·ik 'lad·əˌtüd }

magnetic local anomaly [GEOPHYS] A localized departure of the geomagnetic field from its average over the surrounding area. { mag'ned·ik ˌlō·kəl ə'näm·ə·lē }

magnetic meridian [GEOPHYS] A line which is at any point in the direction of horizontal magnetic force of the earth; a compass needle without deviation lies in the magnetic meridian. { mag'ned·ik mə'rid·ē·ən }

magnetic north [GEOPHYS] At any point on the earth's surface, the horizontal direction of the earth's magnetic lines of force (direction of a magnetic meridian) toward the north magnetic pole; a particular direction indicated by the needle of a magnetic compass. { mag'ned·ik 'nȯrth }

magnetic observatory [GEOPHYS] A geophysical measuring station employing some form of magnetometer to measure the intensity of the earth's magnetic field. { mag'ned·ik əb'zər·vəˌtȯr·ē }

magnetic pole [GEOPHYS] In geomagnetism, either of the two points on the earth's surface where the magnetic meridians converge, that is, where the magnetic field is vertical. Also known as dip pole. { mag'ned·ik 'pōl }

magnetic prime vertical [GEOPHYS] The vertical circle through the magnetic east and west points of the horizon. { mag'ned·ik 'prīm 'vərd·ə·kəl }

magnetic profile [GEOPHYS] A profile of a geologic structure showing magnetic anomalies. { mag'ned·ik 'prōˌfīl }

magnetic reversal [GEOPHYS] A reversal of the polarity of the earth's magnetic field that has occurred at about one-million-year intervals. { mag'ned·ik ri'vər·səl }

magnetic secular change [GEOPHYS] The gradual variation in the value of a magnetic element which occurs over a period of years. { mag'ned·ik ¦sek·yə·lər 'chānj }

magnetic station [GEOPHYS] A facility equipped with instruments for measuring local variations in the earth's magnetic field. { mag'ned·ik 'stā·shən }

magnetic storm [GEOPHYS] A worldwide disturbance of the earth's magnetic field; frequently characterized by a sudden onset, in which the magnetic field undergoes marked changes in the course of an hour or less, followed by a very gradual return to normalcy, which may take several days. Also known as geomagnetic storm. { mag'ned·ik 'stȯrm }

magnetic stratigraphy *See* paleomagnetic stratigraphy. { mag'ned·ik strə'tig·rə·fē }

magnetic survey [GEOPHYS] **1.** Magnetometer map of variations in the earth's total magnetic field; used in petroleum exploration to determine basement-rock depths and geologic anomalies. **2.** Measurement of a component of the geomagnetic field at different locations. { mag'ned·ik 'sər,vā }

magnetic temporal variation [GEOPHYS] Any change in the earth's magnetic field which is a function of time. { mag'ned·ik ¦tem·pə·rəl ,ver·ē'ā·shən }

magnetic variation [GEOPHYS] Small changes in the earth's magnetic field in time and space. { mag'ned·ik ,ver·ē'ā·shən }

magnetic wind direction [METEOROL] The direction, with respect to magnetic north, from which the wind is blowing; distinguished from true wind direction. { mag'ned·ik 'wind də,rek·shən }

magnetoionic duct [GEOPHYS] Duct along the geomagnetic lines of force which exhibits waveguide characteristics for radio-wave propagation between conjugate points on the earth's surface. { mag¦nēd·ō·ī'än·ik 'dəkt }

magnetoionic theory [GEOPHYS] The theory of the combined effect of the earth's magnetic field and atmospheric ionization on the propagation of electromagnetic waves. { mag¦nēd·ō·ī'än·ik 'thē·ə·rē }

magnetoionic wave component [GEOPHYS] Either of the two elliptically polarized wave components into which a linearly polarized wave incident on the ionosphere is separated because of the earth's magnetic field. { mag¦nēd·ō·ī'än·ik 'wāv kəm,pō·nənt }

magnetopause [GEOPHYS] A boundary that marks the transition from the earth's magnetosphere to the interplanetary medium. { mag'nēd·ə,pȯz }

magnetosheath [GEOPHYS] The relatively thin region between the earth's magnetopause and the shock front in the solar wind. { mag'nēd·ə,shēth }

magnetosphere [GEOPHYS] The region of the earth in which the geomagnetic field plays a dominant part in controlling the physical processes that take place; it is usually considered to begin at an altitude of about 60 miles (100 kilometers) and to extend outward to a distant boundary that marks the beginning of interplanetary space. { mag,'nēd·ə,sfir }

magnetospheric plasma [GEOPHYS] A low-energy plasma with particle energies less than a few electronvolts that permeates the entire region of the earth's magnetosphere. { mag'nēd·ə,sfir·ik 'plaz·mə }

magnetospheric ring current [GEOPHYS] A belt of charged particles around the earth whose perturbations give rise to ionospheric storms. { mag'nēd·ə,sfir·ik ¦riŋ ,kə·rənt }

magnetospheric substorm [GEOPHYS] A disturbance of particles and magnetic fields in the magnetosphere; occurs intermittently, lasts 1 to 3 hours, and is accompanied by various phenomena sensible from the earth's surface, such as intense auroral displays and magnetic disturbances, particularly in the nightside polar regions. { mag'nēd·ə,sfir·ik 'səb,stȯrm }

magnetostratigraphy [GEOL] A branch of stratigraphy in which sedimentary successions are described and interpreted in terms of remanent magnetization. { mag¦ned·ō·strə'tig·rə·fē }

magnetotail [GEOPHYS] The portion of the magnetosphere extending from earth in the direction away from the sun for a variable distance of the order of 1000 earth radii. { mag'ned·ō,tāl }

magnetotellurics [GEOPHYS] A geophysical exploration technique that measures natural electromagnetic fields to image subsurface electrical resistivity, providing information about the earth's interior composition and structure since naturally occurring rocks and minerals exhibit a broad range of electrical resistivities. { mag,ned·ō·tə'lür·iks }

magnitude [GEOPHYS] A measure of the amount of energy released by an earthquake. { 'mag·nə,tüd }

main joint *See* master joint. { 'mān 'jȯint }

mainland [GEOGR] A continuous body of land that constitutes the main part of a country or continent. { 'mān·lənd }

main stream [HYD] The principal or largest stream of a given area or drainage system. Also known as master stream; trunk stream. { 'mān 'strēm }

main stroke *See* return streamer. { 'mān 'strōk }

main thermocline [OCEANOGR] A thermocline that is deep enough in the ocean to be unaffected by seasonal temperature changes in the atmosphere. Also known as permanent thermocline. { 'mān 'thər·mə,klīn }

major fold [GEOL] A large-scale fold with which minor folds are usually associated. { 'mā·jər 'fōld }

major joint *See* master joint. { 'mā·jər 'jȯint }

major trough [METEOROL] A long-wave trough in the large-scale pressure pattern of the upper troposphere. { 'mā·jər 'trȯf }

malenclave [HYD] A body of contaminated groundwater surrounded by uncontaminated water. { ,mal'än,klāv }

malloseismic [GEOPHYS] Referring to an area that is likely to experience destructive earthquakes several times in a century. { ,mal·ə'sīz·mik }

malm *See* marl. { mäm }

Malm [GEOL] The Upper Jurassic geologic series, above Dogger and below Cretaceous. { mäm }

mamelon [GEOL] A small, rounded volcano which forms over a vent as a result of the slow extrusion of viscous, silicic lava. { 'mam·ə·lən }

mammillary structure *See* pillow structure. { 'ma·mə,ler·ē 'strək·chər }

mandatory layer [METEOROL] A layer of the atmosphere between two consecutive (or any two) specified mandatory levels. { 'man·də,tȯr·ē 'lā·ər }

mandatory level [METEOROL] One of several constant-pressure levels in the atmosphere for which a complete evaluation of data from upper-air observations is required. Also known as mandatory surface. { 'man·də,tȯr·ē 'lev·əl }

mandatory surface *See* mandatory level. { 'man·də,tȯr·ē 'sər·fəs }

manganese nodule [GEOL] Small, irregular black to brown concretions consisting chiefly of manganese salts and manganese oxide minerals; formed in oceans as a result of pelagic sedimentation or precipitation. { 'maŋ·gə,nēs 'naj·ül }

mantle [GEOL] The intermediate shell zone of the earth below the crust and above the core (to a depth of 2160 miles or 3480 kilometers). { 'mant·əl }

mantled gneiss dome [GEOL] A dome in metamorphic terrains that has a remobilized core of gneiss surrounded by a concordant sheath of the basal part of the overlying metamorphic sequence. { 'mant·əld ¦nīs ,dōm }

mantle rock *See* regolith. { 'mant·əl ,räk }

manto [GEOL] A sedimentary or igneous ore body occurring in flat-lying depositional layers. { 'man,tō }

map plotting [METEOROL] The process of transcribing weather information onto maps, diagrams, and so on; it usually refers specifically to decoding synoptic reports and entering those data in conventional station-model form on synoptic charts. Also known as map spotting. { 'map ,pläd·iŋ }

map spotting *See* map plotting. { 'map ,späd·iŋ }

map vertical *See* geographic vertical. { 'map 'vərd·ə·kəl }

march [METEOROL] The variation of any meteorological element throughout a specific unit of time, such as a day, month, or year; as the daily march of temperature, the complete cycle of temperature during 24 hours. { märch }

marekanite [GEOL] Rounded to subangular obsidian bodies that occur in masses of perlite. { ¦mär·ə¦ka,nīt }

mare's tail *See* precipitation trajectory. { 'merz ,tāl }

margarite [GEOL] A string of beadlike globulites; commonly found in glassy igneous rocks. { 'mär·gə,rīt }

margin [GEOGR] The boundary around a body of water. { 'mär·jən }

marginal escarpment [GEOL] A seaward slope of a marginal plateau with a gradient of 1:10 or more. { 'mär·jən·əl e'skärp·mənt }
marginal fissure [GEOL] A magma-filled fracture bordering an igneous intrusion. { 'mär·jən·əl 'fish·ər }
marginal moraine *See* terminal moraine. { 'mär·jən·əl mə'rān }
marginal plain *See* outwash plain. { 'mär·jən·əl 'plān }
marginal plateau [GEOL] A relatively flat shelf adjacent to a continent and similar topographically to, but deeper than, a continental shelf. { 'mär·jən·əl pla'tō }
marginal salt pan [GEOL] A natural, coastal salt pan. { 'mär·jən·əl 'sölt ˌpan }
marginal sea [GEOGR] A semiclosed sea adjacent to a continent and connected with the ocean at the water surface. { 'mär·jən·əl 'sē }
marginal thrust [GEOL] One of a series of faults bordering an igneous intrusion and crossing both the intrusion and the wall rock. Also known as marginal upthrust. { 'mär·jən·əl 'thrəst }
marginal upthrust *See* marginal thrust. { 'mär·jən·əl 'əpˌthrəst }
Margules equation *See* Witte-Margules equation. { mär'gü·ləs iˌkwā·zhən }
marigram [OCEANOGR] A graphic record of the rising and falling movements of the tide expressed as a curve. { 'mar·əˌgram }
marine [OCEANOGR] Pertaining to the sea. { mə'rēn }
marine abrasion [GEOL] Erosion of the ocean floor by sediment moved by ocean waves. Also known as wave erosion. { mə'rēn ə'brā·zhən }
marine arch *See* sea arch. { mə'rēn 'ärch }
marine bridge *See* sea arch. { mə'rēn 'brij }
marine cave *See* sea cave. { mə'rēn 'kāv }
marine climate [CLIMATOL] A regional climate which is under the predominant influence of the sea, that is, a climate characterized by oceanity; the antithesis of a continental climate. Also known as maritime climate; oceanic climate. { mə'rēn 'klī·mət }
marine-cut terrace [GEOL] A terrace or platform cut by wave erosion of marine origin. Also known as wave-cut terrace. { mə'rēn ¦kət 'ter·əs }
marine forecast [METEOROL] A forecast, for a specified oceanic or coastal area, of weather elements of particular interest to maritime transportation, including wind, visibility, the general state of the weather, and storm warnings. { mə'rēn 'förˌkast }
marine geology *See* geological oceanography. { mə'rēn jē'äl·ə·jē }
marine meteorology [METEOROL] That part of meteorology which deals mainly with the study of oceanic areas, including island and coastal regions; in particular, it serves the practical needs of surface and air navigation over the oceans. { mə'rēn ˌmēd·ē·ə'räl·ə·jē }
marine salina [GEOGR] A body of salt water found along an arid coast and separated from the sea by a sand or gravel barrier. { mə'rēn səl'lēn·ə }
Marinesian *See* Bartonian. { mar·ə'nē·zhē·ən }
marine snow [OCEANOGR] A concentration of living and dead organic material and inorganic debris of the sea suspended at density boundaries such as the thermocline. { mə'rēn 'snō }
marine stack *See* stack. { mə'rēn 'stak }
marine swamp [GEOGR] An area of low, salty, or brackish water found along the shore and characterized by abundant grasses, mangrove trees, and similar vegetation. Also known as paralic swamp. { mə'rēn 'swämp }
marine terrace [GEOL] A seacoast terrace formed by the merging of a wave-built terrace and a wave-cut platform. Also known as sea terrace; shore terrace. { mə'rēn 'ter·əs }
marine transgression *See* transgression. { mə'rēn tranz'gresh·ən }
marine weather observation [METEOROL] The weather as observed from a ship at sea, usually taken in accordance with procedures specified by the World Meteorological Organization. { mə'rēn 'weth·ər ˌäb·zərˌvā·shən }
maritime air [METEOROL] A type of air whose characteristics are developed over an extensive water surface and which, therefore, has the basic maritime quality of high moisture content in at least its lower levels. { 'mar·əˌtīm 'er }

maritime climate *See* marine climate. { 'mar·ə,tīm 'klī·mət }
maritime polar air [METEOROL] Polar air initially possessing similar properties to those of continental polar air, but in passing over warmer water it becomes unstable with a higher moisture content. { 'mar·ə,tīm 'pō·lər ¦er }
maritime tropical air [METEOROL] The principal type of tropical air, produced over the tropical and subtropical seas; it is very warm and humid, and is frequently carried poleward on the western flanks of the subtropical highs. { 'mar·ə,tīm 'träp·ə·kəl ¦er }
marker bed [GEOL] **1.** A stratified unit with distinctive characteristics making it an easily recognized geologic horizon. **2.** A rock layer which accounts for a characteristic portion of a seismic refraction time-distance curve. **3.** *See* key bed. { 'märk·ər ,bed }
marl [GEOL] A deposit of crumbling earthy material composed principally of clay with magnesium and calcium carbonate; used as a ertilizer for lime-deficient soils. Also known as malm. { märl }
marly [GEOL] Pertaining to, containing, or resembling marl. { 'mär·lē }
Marmor [GEOL] A North American stage of Middle Ordovician geologic time, forming the lower subdivision of Chazyan, above Whiterock and below Ashby. { 'mär,mȯr }
Marsden chart [METEOROL] A system for showing the distribution of meteorological data on a chart, especially over the oceans; using a Mercator map projection, the world between 80°N and 70°S latitudes is divided into Marsden "squares," each of 10° latitude by 10° longitude and systematically numbered to indicate position; each square may be divided into quarter squares, or into 100 one-degree subsquares numbered from 00 to 99 to give the position to the nearest degree. { 'märz·dən ,chärt }
marsh gas [GEOCHEM] Combustible gas, consisting chiefly of methane, produced as a result of decay of vegetation in stagnant water. { 'märsh ,gas }
mascaret *See* bore. { ¦mas·kə¦ret }
mascon [GEOL] A large, high-density mass concentration below a ringed mare on the surface of the moon. { 'mas,kän }
mass attraction vertical [GEOPHYS] The vertical which is a function only of the distribution of mass and is unaffected by forces resulting from the motions of the earth. { 'mas ə¦trak·shən ,verd·ə·kəl }
mass erosion [GEOL] A process in which the direct application of gravitational body stresses causes earth and rocks to fall and be carried downslope. Also known as gravity erosion. { 'mas i'rō·zhən }
mass heaving [GEOL] A comprehensive expansion of the ground due to freezing. { 'mas 'hēv·iŋ }
massif [GEOL] A massive block of rock within an erogenic belt, generally more rigid than the surrounding rocks, and commonly composed of crystalline basement or younger plutons. { ma'sēf }
massive [GEOL] Of a mineral deposit, having a large concentration of ore in one place. { 'mas·iv }
mass movement [GEOL] Movement of a portion of the land surface as a unit. { 'mas 'müv·mənt }
mass wasting [GEOL] Dislodgement and downslope transport of loose rock and soil material under the direct influence of gravitational body stresses. { 'mas ,wāst·iŋ }
master joint [GEOL] A persistent joint plane of greater than average extent, generally constituting the dominant jointing of an area. Also known as main joint; major joint. { 'mas·tər 'jȯint }
master stream *See* main stream. { 'mas·tər ,strēm }
Matanuska wind [METEOROL] A strong, gusty, northeast wind which occasionally occurs during the winter in the vicinity of Palmer, Alaska. { ,mad·ə'nüs·kə 'wind }
matched terrace *See* paired terrace. { 'macht 'ter·əs }
material unit [GEOL] A stratigraphic unit based on rocks and their fossil content without time implication. { mə'tir·ē·əl ,yü·nət }
mathematical climate [CLIMATOL] An elementary generalization of the earth's climatic pattern, based entirely on the annual cycle of the sun's inclination; this early climatic classification recognized three basic latitudinal zones (the summerless, intermediate,

and winterless), which are now known as the Frigid, Temperate, and Torrid Zones, and which are bounded by the Arctic and Antarctic Circles and the Tropics of Cancer and Capricorn. { ¦math·ə¦mad·ə·kəl 'klī·mət }

mathematical forecasting *See* numerical forecasting. { ¦math·ə¦mad·ə·kəl 'fȯr,kast·iŋ }

mathematical geography [GEOGR] The branch of geography that deals with the features and processes of the earth, and their representations on maps and charts. { ¦math·ə¦mad·ə·kəl jē'äg·rə·fē }

mathematical geology [GEOL] The branch of geology concerned with the study of probability distributions of values of random variables involved in geologic processes. { ¦math·ə¦mad·ə·kəl jē'äl·ə·jē }

matinal [METEOROL] The morning winds, that is, an east wind. { 'mat·ən·əl }

matric forces [GEOL] Forces acting on soil water that are independent of gravity but exist due to the attraction of solid surfaces for water, the attraction of water molecules for each other, and a force in the air-water interface due to the polar nature of water. { 'mā·trik ,fȯrs·əz }

matrix porosity [GEOL] Core-sample porosity determined from a small sample of the core, in contrast to total porosity, where the whole core is used. { 'mā·triks pə'räs·əd·ē }

matrix rock *See* land pebble phosphate. { 'mā·triks ,räk }

matrix velocity [GEOPHYS] The velocity of sound through a formation's rock matrix during an acoustic-velocity log. { 'mā·triks və'läs·əd·ē }

mature [GEOL] **1.** Pertaining to a topography or region, and to its landforms, having undergone maximum development and accentuation of form. **2.** Pertaining to the third stage of textural maturity of a clastic sediment. { mə'chu̇r }

matureland [GEOL] The land surface which is characteristic of the mature stage in the erosion cycle. { mə'chu̇r,land }

mature soil *See* zonal soil. { mə'chu̇r 'sȯil }

maturity [GEOL] **1.** The second stage of the erosion cycle in the topographic development of a landscape or region characterized by numerous and closely spaced mature streams, reduction of level surfaces to slopes, large well-defined drainage systems, and the absence of swamps or lakes on the uplands. Also known as topographic maturity. **2.** A stage in the development of a shore or coast that begins with the attainment of a profile of equilibrium. **3.** The extent to which the texture and composition of a clastic sediment approach the ultimate end product. **4.** The stage of stream development at which maximum vigor and efficiency has been reached. { mə'chu̇r·əd·ē }

maturity index [GEOL] A measure of the progress of a clastic sediment in the direction of chemical or mineralogic stability; for example, a high ratio of quartz + cherts to feldspar + rock fragments indicates a highly mature sediment. { mə'chu̇r·əd·ē ,in,deks }

maximum ebb [OCEANOGR] The greatest speed of an ebb current. { 'mak·sə·məm 'eb }

maximum flood [OCEANOGR] The greatest speed of a flood current. { 'mak·sə·məm 'fləd }

maximum subsidence [GEOL] The maximum amount of subsidence in a basin. { 'mak·sə·məm səb'sīd·əns }

maximum sustainable yield [OCEANOGR] **1.** In fishery management, the highest average fishing level over time that does not reduce a stock's abundance in balance with the stock's reproductive and growth capacities under a given set of environmental conditions. **2.** A level of fishing that, if approached, should signal caution rather than increased fishing. { ¦mak·sə·məm sə¦stān·ə·bul 'yēld }

maximum-wind and shear chart [METEOROL] A synoptic chart on which are plotted the altitudes of the maximum wind speed, the maximum wind velocity (wind direction optional), plus the velocity of the wind at mandatory levels both above and below the level of maximum wind. Also known as max-wind and shear chart. { 'mak·sə·məm 'wind ən 'shir ,chärt }

maximum-wind level [METEOROL] The height at which the maximum wind speed occurs, determined in a winds-aloft observation. Also known as max-wind level. { 'mak·sə·məm 'wind ˌlev·əl }

maximum-wind topography [METEOROL] The topography of the surface of maximum wind speed. Also known as max-wind topography. { 'mak·sə·məm ¦wind tə'päg·rə·fē }

maximum zonal westerlies [METEOROL] The average west-to-east component of wind over the continuous 20° belt of latitude in which this average is a maximum; it is usually found, in the winter season, in the vicinity of 40–60° north latitude. { 'mak·sə·məm ¦zōn·əl 'wes·tərˌlēz }

max-wind and shear chart *See* maximum-wind and shear chart. { 'maks 'wind ən 'shir ˌchärt }

max-wind level *See* maximum-wind level. { 'maks 'wind ˌlev·əl }

max-wind topography *See* maximum-wind topography. { 'maks 'wind tə'päg·rə·fē }

mean chart [METEOROL] Any chart on which isopleths of the mean value of a given meteorological element are drawn. Also known as mean map. { 'mēn ˌchärt }

mean depth [HYD] Average water depth in a stream channel or conduit computed by dividing the cross-sectional area by the surface width. { 'mēn 'depth }

meander [HYD] A sharp, sinuous loop or curve in a stream, usually part of a series. [OCEANOGR] A deviation of the flow pattern of a current. { mē'an·dər }

meander bar *See* point bar. { mē'an·dər ˌbär }

meander belt [GEOL] The zone along the floor of a valley across which a meandering stream periodically shifts its channel. { mē'an·dər ˌbelt }

meander core [GEOL] A hill encircled by a stream meander. Also known as rock island. { mē'an·dər ˌkȯr }

meandering stream [HYD] A stream having a pattern of successive meanders. Also known as snaking stream. { mē'an·də·riŋ 'strēm }

meander niche [GEOL] A conical or crescentic opening in the wall of a cave formed by downward and lateral stream erosion. { mē'an·dər ˌnich }

meander plain [GEOL] A plain built by the meandering process, or a plain of lateral accretion. { mē'an·dər ˌplān }

meander scar [GEOL] A crescentic, concave mark on the face of a bluff or valley wall formed by a meandering stream. { mē'an·dər ˌskär }

meander spur [GEOL] An undercut projection of high land that extends into the concave part of, and is enclosed by, a meander. { mē'an·dər ˌspər }

mean diurnal high-water inequality [OCEANOGR] Half the average difference between the heights of the two high waters of each tidal day over a 19-year period; it is obtained by subtracting the mean of all high waters from the mean of the higher high waters. { 'mēn dī'ərn·əl 'hī ˌwȯd·ər ˌin·i'kwäl·əd·ē }

mean diurnal low-water inequality [OCEANOGR] Half the average difference between the heights of the two low waters of each tidal day over a 19-year period; it is obtained by subtracting the mean of all lower low waters from the mean of the low waters. { 'mēn dī'ərn·əl 'lō ˌwȯd·ər ˌin·i·'kwal·əd·ē }

mean higher high water [OCEANOGR] The average height of higher high waters at a place over a 19-year period. { 'mēn ¦hī·ər 'hī 'wȯd·ər }

mean high water [OCEANOGR] The average height of all high waters recorded at a given place over a 19-year period. { 'mēn 'hī 'wȯd·ər }

mean high-water lunitidal interval [OCEANOGR] The average interval of time between the transit (upper or lower) of the moon and the next high water at a place. Also known as corrected establishment. { 'mēn ¦hī ¦wȯd·ər ¦lü·nə¦tīd·əl 'in·tər·vəl }

mean high-water neaps [OCEANOGR] The average height of the high waters of neap tides. Also known as neap high water. { 'mēn ¦hī ¦wȯd·ər 'nēps }

mean high-water springs [OCEANOGR] The average height of the high waters of spring tides. Also known as high-water springs; spring high water. { 'mēn ¦hī ¦wȯd·ər 'spriŋz }

mean latitude [GEOD] Half the arithmetical sum of the latitudes of two places on the same side of the equator; mean latitude is labeled N or S to indicate whether it is north or south of the equator. { 'mēn 'lad·əˌtüd }

mean lower low water [OCEANOGR] The average height of the lower low waters at a place over a 19-year period. { 'mēn ¦lō·ər 'lō 'wȯd·ər }

mean lower low-water springs [OCEANOGR] The average height of lower low-water springs at a place. { 'mēn ¦lō·ər ¦lō¦wȯd·ər 'spriŋz }

mean low water [OCEANOGR] The average height of all low waters recorded at a given place over a 19-year period. { 'mēn 'lō 'wȯd·ər }

mean low-water lunitidal interval [OCEANOGR] The average interval of time between the transit (upper or lower) of the moon and the next low water at a place. { 'mēn ¦lō ¦wȯd·ər ¦lü·nə¦tīd·əl 'in·tər·vəl }

mean low-water neaps [OCEANOGR] The average height of the low water at neap tides. Also known as low-water neaps; neap low water. { 'mēn ¦lō ¦wȯd·ər 'nēps }

mean low-water springs [OCEANOGR] The average height of the low waters of spring tides; this level is used as a tidal datum in some areas. Also known as low-water springs; spring low water. { 'mēn ¦lō ¦wȯd·ər 'spriŋz }

mean map *See* mean chart. { 'mēn ˌmap }

mean neap range *See* neap range. { 'mēn 'nēp ˌrānj }

mean neap rise [OCEANOGR] The height of mean high-water neaps above the chart datum. { 'mēn 'nēp ˌrīz }

mean range [OCEANOGR] The difference in the height between mean high water and mean low water. { 'mēn 'rānj }

mean rise interval [OCEANOGR] The average interval of time between the transit (upper or lower) of the moon and the middle of the period of rise of the tide at a place; it may be either local or Greenwich, depending on the transit to which it is referred, but the local interval is assumed unless otherwise specified. { 'mēn 'rīz ˌin·tər·vəl }

mean rise of tide [OCEANOGR] The height of mean high water above the chart datum. { 'mēn 'rīz əv 'tīd }

mean river level [HYD] The average height of the surface of a river at any point for all stages of the tide over a 19-year period. { 'mēn 'riv·ər ˌlev·əl }

mean sea level [OCEANOGR] The average sea surface level for all stages of the tide over a 19-year period, usually determined from hourly height readings from a fixed reference level. { 'mēn 'sē ˌlev·əl }

mean spring range *See* spring range. { 'mēn 'spriŋ ¦rānj }

mean spring rise [OCEANOGR] The height of mean high-water springs above the chart datum. { 'mēn 'spriŋ ¦rīz }

mean temperature [METEOROL] The average temperature of the air as indicated by a properly exposed thermometer during a given time period, usually a day, month, or year. { 'mēn 'tem·prə·chər }

mean tide *See* half tide. { 'mēn 'tīd }

mean tide level [OCEANOGR] The tide level halfway between mean high water and mean low water. { 'mēn 'tīd ˌlev·əl }

mean water level [OCEANOGR] The average surface level of a body of water. { 'mēn 'wȯd·ər ˌlev·əl }

mechanical erosion *See* corrasion. { mi'kan·ə·kəl i'rō·zhən }

mechanical instability *See* absolute instability. { mi'kan·ə·kəl ˌin·stə'bil·əd·ē }

mechanical sediment *See* clastic sediment. { mi'kan·ə·kəl 'sed·ə·mənt }

mechanical turbulence [METEOROL] Irregular air movement in the lower atmosphere resulting from obstructions, for example, tall buildings. { mi'kan·ə·kəl 'tər·byə·ləns }

mechanical weathering [GEOL] The process of weathering by which physical forces break down or reduce a rock to smaller and smaller fragments, involving no chemical change. Also known as physical weathering. { mi'kan·ə·kəl 'weth·ə·riŋ }

medial moraine [GEOL] **1.** An elongate moraine carried in or upon the middle of a glacier and parallel to its sides. **2.** A moraine formed by glacial abrasion of a rocky protuberance near the middle of a glacier. { 'mē·dē·əl mə'rān }

median mass [GEOL] A less disturbed structural block in the middle of an orogenic belt, bordered on both sides by orogenic structure, thrust away from it. Also known as betwixt mountains; Zwischengebirge. { 'mē·dē·ən 'mas }

median particle diameter [GEOL] The middlemost particle diameter of a rock or sediment, larger than 50% of the diameter in the distribution and smaller than the other 50%. { 'mē·dē·ən 'pärd·ə·kəl dī,am·əd·ər }
mediterranean *See* mesogeosyncline. { ,med·ə·tə'rā·nē·ən }
Mediterranean climate [CLIMATOL] A type of climate characterized by hot, dry, sunny summers and a winter rainy season; basically, this is the opposite of a monsoon climate. Also known as etesian climate. { ,med·ə·tə'rā·nē·ən 'klī·mət }
mediterranean sea [GEOGR] A deep epicontinental sea that is connected with the ocean by a narrow channel. { ,med·ə·tə'rā·nē·ən 'sē }
Mediterranean Sea [GEOGR] A sea that lies between Europe, Asia Minor, and Africa and is completely landlocked except for the Strait of Gibraltar, the Bosporus, and the Suez Canal; total water area is 965,000 square miles (2,501,000 square kilometers). { ,med·ə·tə'rā·nē·ən 'sē }
medium-range forecast [METEOROL] A forecast of weather conditions for a period of 48 hours to a week in advance. Also known as extended-range forecast. { 'mē·dē·əm ¦rānj 'fȯr,kast }
medium-volatile bituminous coal [GEOL] Bituminous coal consisting of 23–31% volatile matter. { 'mē·dē·əm ¦väl·ə·təl bə'tü·mə·nəs 'kōl }
megacyclothem [GEOL] A cycle of or combination of related cyclothems. { ,meg·ə'sī·klə,them }
megaripple [GEOL] A large sand wave. { 'meg·ə,rīp·əl }
megatectonics [GEOL] The tectonics of the very large structural features of the earth. { ,meg·ə,tek'tän·iks }
Melanesia [GEOGR] A group of islands in the Pacific Ocean northeast of Australia. { mel·ə'nē·zhə }
mélange [GEOL] A heterogeneous medley or mixture of rock materials; specifically, a mappable body of deformed rocks consisting of a pervasively sheared, fine-grained, commonly pelitic matrix, thoroughly mixed with angular and poorly sorted inclusions of native and exotic tectonic fragments, blocks, or slabs, of diverse origins and geologic ages, that may be as much as several kilometers in length. Also known as block clay. { mā'länzh }
melanic *See* melanocratic. { me'lan·ik }
melanocratic [GEOL] Dark-colored, referring to igneous rock containing at least 50–60% mafic minerals. Also known as chromocratic; melanic. { ¦mel·ə·nō¦krad·ik }
melting level [METEOROL] The altitude at which ice crystals and snowflakes melt as they descend through the atmosphere. { 'melt·iŋ ,lev·əl }
meltwater [HYD] Water derived from melting ice or snow, especially glacier ice. { 'melt ,wȯd·ər }
member [GEOL] A rock stratigraphic unit of subordinate rank comprising a specially developed part of a varied formation. { 'mem·bər }
mendip [GEOL] **1.** A buried hill that is exposed as an inlier. **2.** A coastal-plain hill that was originally an offshore island. { 'men,dip }
Meramecian [GEOL] A North American provincial series of geologic time: Upper Mississippian (above Osagian, below Chesterian). { ,mer·ə'mē·shən }
Mercalli scale [GEOPHYS] A 12-point scale for classifying the magnitude of an earthquake. { mer'käl·ē ,skāl }
mere [HYD] A large pond or a shallow lake. { mīr }
Merian's formula [OCEANOGR] A formula for the period of a seiche, $T = (1/n) \cdot (2L/\sqrt{gd})$, where n is the number of nodes, L is the horizontal dimension of the basin measured in the direction of wave motion, g is the acceleration of gravity, and d is the depth of the water. { 'mer·ē·ənz ,fȯr·myə·lə }
meridian [GEOD] A north-south reference line, particularly a great circle through the geographical poles of the earth. { mə'rid·ē·ən }
meridional [GEOL] Pertaining to longitudinal movements or directions, that is, northerly or southerly. { mə'rid·ē·ən·əl }
meridional cell [GEOPHYS] A very large-scale convection circulation in the atmosphere or ocean which takes place in a meridional plane, with northward and southward

currents in opposite branches of the cell, and upward and downward motion in the equatorward and poleward ends of the cell. { mə'rid·ē·ən·əl 'sel }

meridional circulation [METEOROL] An atmospheric circulation in a vertical plane oriented along a meridian; it consists, therefore, of the vertical and the meridional (north or south) components of motion only. [OCEANOGR] The exchange of water masses between northern and southern oceanic regions. { mə'rid·ē·ən·əl ˌsər·kyə'lā·shən }

meridional flow [METEOROL] A type of atmospheric circulation pattern in which the meridional (north and south) component of motion is unusually pronounced; the accompanying zonal component is usually weaker than normal. [OCEANOGR] Current moving along a meridian. { mə'rid·ē·ən·əl 'flō }

meridional front [METEOROL] A front in the South Pacific separating successive migratory subtropical anticyclones; such fronts are essentially in the form of great arcs with meridians of longitudes as chords; they have the character of cold fronts. { mə'rid·ē·ən·əl 'frənt }

meridional index [METEOROL] A measure of the component of air motion along meridians, averaged, without regard to sign, around a given latitude circle. { mə'rid·ē·ən·əl 'inˌdeks }

meridional wind [METEOROL] The wind or wind component along the local meridian, as distinguished from the zonal wind. { mə'rid·ē·ən·əl 'wind }

meromictic [HYD] Of or pertaining to a lake whose water is permanently stratified and therefore does not circulate completely throughout the basin at any time during the year. { ¦mer·ə¦mik·tik }

Mersey yellow coal *See* tasmanite. { 'mər·zē 'yel·ō 'kōl }

merzlota *See* frozen ground. { ˌmerz'lō·tə }

mesa [GEOGR] A broad, isolated, flat-topped hill bounded by a steep cliff or slope on at least one side; represents an erosion remnant. { 'mā·sə }

mesa-butte [GEOGR] A butte formed as the result of erosion and reduction of a mesa. { 'mā·sə ˌbyüt }

mesa plain [GEOGR] A flat-topped summit of a hilly mountain. { 'mā·sə ˌplān }

mesobenthos [OCEANOGR] The sea bottom at depths of 100–500 fathoms (180–900 meters). { ¦me·zō¦benˌthäs }

mesoclimate [CLIMATOL] **1.** The climate of small areas of the earth's surface which may not be representative of the general climate of the district. **2.** A climate characterized by moderate temperatures, that is, in the range 20–30°C. Also known as mesothermal climate. { ¦me·zō¦klī·mət }

mesoclimatology [CLIMATOL] The study of mesoclimates. { ˌme·zōˌklī·mə'täl·ə·jē }

mesocyclone [METEOROL] A cyclonic circulation interior to a convective storm. { ¦me·zō¦sīˌklōn }

mesogeosyncline [GEOL] A geosyncline between two continents. Also known as mediterranean. { ¦me·zōˌjē·ō'siŋˌklīn }

mesohaline [OCEANOGR] **1.** Referring to estuarine water with salinity ranging 5 – 18 parts per thousand. **2.** Referring to moderately brackish water. { ˌme·sō'ha·lēn }

mesometeorology [METEOROL] That portion of the science of meteorology concerned with the study of atmospheric phenomena on a scale larger than that of micrometeorology, but smaller than the cyclonic scale. { ˌme·zōˌmē·dē·ə'räl·ə·jē }

mesopause [METEOROL] The top of the mesosphere; corresponds to the level of minimum temperature at 50 to 60 miles (80 to 95 kilometers). { 'mez·əˌpȯz; }

mesopeak [METEOROL] The temperature maximum at about 30 miles (50 kilometers) in the mesosphere. { 'me·zōˌpēk }

mesoscale eddies *See* mode eddies. { 'me·zōˌskāl 'ed·ēz }

mesoscale motion [GEOPHYS] Motion of winds and ocean currents over regional areas with sizes of 6–60 miles (10–100 kilometers). { ˌmes·ōˌskāl 'mō·shən }

mesosiderite [GEOL] A stony-iron meteorite containing about equal amounts of silicates and nickel-iron, with considerable troilite. Also known as grahamite. { ¦me·zō'sīd·əˌrīt }

mesosphere [GEOL] *See* lower mantle. [METEOROL] The atmospheric shell between about 28–35 and 50–60 miles (45–55 and 80–95 kilometers), extending from the top of the stratosphere to the mesopause; characterized by a temperature that generally decreases with altitude. { 'mez·ə,sfir }

mesostasis [GEOL] The last-formed interstitial material, either glassy or aphanitic, of an igneous rock. { ¦me·zō'stā·səs }

mesothermal climate *See* mesoclimate. { ¦mez·ə¦thər·məl 'klī·mət }

mesotil [GEOL] A semiplastic or semifriable derivative of chemically weathered till; forms beneath a partially drained area. { 'mez·ə,til }

Mesozoic [GEOL] A geologic era from the end of the Paleozoic to the beginning of the Cenozoic; commonly referred to as the Age of Reptiles. { ¦mez·ə¦zō·ik }

metaanthracite [GEOL] Anthracite coal containing at least 98% fixed carbon. { ¦med·ə'an·thrə,sīt }

metabentonite [GEOL] Altered bentonite, formed by compaction or metamorphism; it swells very little and lacks the usual high colloidal properties of bentonite. { ¦med·ə'bent·ən,īt }

metahalloysite [GEOL] A term used in Europe for the less hydrous form of halloysite. Also known as halloysite in the United States. { ¦med·ə·hə'lȯi,sīt }

metaharmosis *See* metharmosis. { ¦med·ə·här'mō·səs }

metalimnion *See* thermocline. { ,med·ə'lim·nē,än }

metallogenic province [GEOL] A region characterized by a particular mineral assemblage, or by one or more specific types of mineralization. Also known as metallographic province. { mə¦tal·ə¦jen·ik 'präv·əns }

metallographic province *See* metallogenic province. { mə'tal·ə,graf·ik 'präv·əns }

metamorphic aureole *See* aureole. { ¦med·ə¦mȯr·fik 'ȯr·ē,ōl }

metamorphic overprint *See* overprint. { ¦med·ə¦mȯr·fik 'ō·vər,print }

metamorphic rock reservoir [GEOL] Uncommon type of formation for oil reservoir; developed when secondary porosity results from fracturing or weathering. { ¦med·ə¦mȯr·fik ¦räk 'rez·əv,wär }

metamorphic zone *See* aureole. { ¦med·ə¦mȯr·fik 'zōn }

metaripple [GEOL] An asymmetrical sand ripple. { 'med·ə,rip·əl }

metasediment [GEOL] A sediment or sedimentary rock which shows evidence of metamorphism. { ¦med·ə'sed·ə·mənt }

meteoric stone *See* stony meteorite. { ,mēd·ē'ȯr·ik 'stōn }

meteoric water [HYD] Groundwater which originates in the atmosphere and reaches the zone of saturation by infiltration and percolation. { ,mēd·ē'ȯr·ik 'wȯd·ər }

meteorite [GEOL] Any meteoroid that has fallen to the earth's surface. { 'mēd·ē·ə,rīt }

meteorite crater [GEOL] An impact crater on the surface of the earth or of a celestial body caused by a meteorite; a characteristic feature on the earth is the upturned rim, which formed as the rocks rebounded following the impact. { 'mēd·ē·ə,rīt ,krād·ər }

meteorogram [METEOROL] A chart in which meteorological variables are plotted against time. { ,med·ē'ȯr·ə,gram }

meteorolite *See* stony meteorite. { ,med·ē'ȯr·ə,līt }

meteorological [METEOROL] Of or pertaining to meteorology or weather. { ,med·ē·ə·rə'läj·ə·kəl }

meteorological chart [METEOROL] A weather map showing the spatial distribution, at an instant of time, of atmospheric highs and lows, rain clouds, and other phenomena. { ,med·ē·ə·rə'läj·ə·kəl 'chärt }

meteorological data [METEOROL] Facts pertaining to the atmosphere, especially wind, temperature, and air density. { ,med·ē·ə·rə'läj·ə·kəl 'dad·ə }

meteorological equator [METEOROL] **1.** The parallel of latitude 5° north, so called because it is the annual mean latitude of the equatorial trough. **2.** *See* equatorial trough; intertropical convergence zone. { ,med·ē·ə·rə'läj·ə·kəl i'kwād·ər }

meteorological minima [METEOROL] Minimum values of meteorological elements prescribed for specific types of flight operation. { ,med·ē·ə·rə'läj·ə·kəl 'min·ə·mə }

meteorological radar [METEOROL] A remote sensing device that transmits and receives microwave radiation for the purpose of detecting and measuring weather phenomena;

includes Doppler radar, which is used to determine air motions (to detect tornadoes), and multiparameter radar, which provides information on the phase (ice or liquid), shapes, and sizes of hydrometeors. { ˌmēd·ē·ə·rə·ˈläj·ə·kəl ˈrāˌdär }

meteorological range [METEOROL] An empirically consistant measure of the visual range of a target; a concept developed to eliminate from consideration the threshold contrast and adaptation luminance, both of which vary from observer to observer. Also known as standard visibility; standard visual range. { ˌmed·ē·ə·rəˈläj·ə·kəl ˈrānj }

meteorological solenoid [METEOROL] A hypothetical tube formed in space by the intersection of a set of surfaces of constant pressure and a set of surfaces of constant specific volume of air. Also known as solenoid. { ˌmed·ē·ə·rəˈläj·ə·kəl ˈsō·ləˌnȯid }

meteorological tide [OCEANOGR] A change in water level caused by local meteorological conditions, in contrast to an astronomical tide, caused by the attractions of the sun and moon. { ˌmed·ē·ə·rəˈläj·ə·kəl ˈtīd }

meteor trail *See* ion column. { ˈmēd·ē·ər ˌtrāl }

metharmosis [GEOL] Changes that occur in a buried sediment after uplift or consolidation but before the onset of weathering. Also spelled metaharmosis. { məˈthär·məˈsəs }

micaceous [GEOL] Pertaining to or resembling mica. { mīˈkā·shəs }

microbarm [GEOPHYS] That portion of the record of a microbarograph between any two or a specified small number of the successive crossings of the average pressure level in the same direction; analogous to microseism. { ˈmī·krəˌbärm }

microbarograph [METEOROL] A type of aneroid barograph designed to record atmospheric pressure variations of very small magnitude. { ¦mī·krōˈbar·əˌgraf }

microbreccia [GEOL] A poorly sorted sandstone containing large, angular sand particles in a fine silty or clayey matrix. { ¦mī·krōˈbrech·ə }

microburst [METEOROL] A downdraft with horizontal extent of about 2.5 miles (4 kilometers) or less, associated with atmospheric convection, often a thundershower. { ˈmī·krōˌbərst }

microclimate [CLIMATOL] The local, rather uniform climate of a specific place or habitat, compared with the climate of the entire area of which it is a part. { ¦mī·krōˈklī·mət }

microclimatology [CLIMATOL] The study of a microclimate, including the study of profiles of temperature, moisture and wind in the lowest stratum of air, the effect of the vegetation and of shelterbelts, and the modifying effect of towns and buildings. { ¦mī·krōˌklī·məˈtäl·ə·jē }

microearthquake [GEOPHYS] An earthquake with a low intensity, usually less than 3 on the Richter scale. Also known as microquake. { ˌmī·krōˈərthˌkwāk }

microgeography [GEOGR] The detailed empirical geographical study on a small scale of a specific locale. { ˌmī·krə·jēˈäg·rə·fē }

microlayer [OCEANOGR] The thin zone beneath the surface of the ocean or any free water surface within which physical processes are modified by proximity to the air-water boundary. { ˈmī·krōˌlā·ər }

micrometeorology [METEOROL] That portion of the science of meteorology that deals with the observation and explanation of the smallest-scale physical and dynamic occurrences within the atmosphere; studies are confined to the surface boundary layer of the atmosphere, that is, from the earth's surface to an altitude where the effects of the immediate underlying surface upon air motion and composition become negligible. { ¦mī·krōˌmē·dē·əˈräl·ə·jē }

micropore [GEOL] A pore small enough to hold water against the pull of gravity and to retard water flow. { ˈmī·krəˌpȯr }

micropulsation [GEOPHYS] A short-period geomagnetic variation in the range of about 0.2–600 seconds, typically exhibiting an oscillatory waveform. { ¦mī·krō·pəlˈsā·shən }

microquake *See* microearthquake.

microrelief [GEOGR] Irregularities of the land surface causing variations in elevation amounting to no more than a few feet. { ˈmī·krō·riˌlēf }

microscale motion [GEOPHYS] Local motion of winds and ocean currents over areas with sizes of 300 feet (100 meters) or less. { 'mī·krə,skāl ,mō·shən }
microseism [GEOPHYS] A weak, continuous, oscillatory motion in the earth having a period of 1–9 seconds and caused by a variety of agents, especially atmospheric agents; not related to an earthquake. { 'mī·krə,sīz·əm }
microtektite [GEOL] An extremely small tektite, 1 millimeter or less in diameter. { ,mī·krə'tek,tīt }
microthermal climate [CLIMATOL] A temperature province in both of C.W. Thornthwaite's climatic classifications, generally described as a "cool" or "cold winter" climate. { ¦mī·krə'thər·məl 'klī·mət }
microvitrain [GEOL] A coal lithotype; fine vitrain-like lenses or laminae in clarain. { ¦mī·krō'vi,trān }
mid-Atlantic ridge [GEOL] The mid-oceanic ridge in the Atlantic. { ,mid·ət'lan·tik 'rij }
Middle Cambrian [GEOL] The geologic epoch occurring between Upper and Lower Cambrian, beginning approximately 540,000,000 years ago. { 'mid·əl 'kam·brē·ən }
middle clouds [METEOROL] Types of clouds the mean level of which is between 6500 and 20,000 feet (1980 and 6100 meters); the principal clouds in this group are altocumulus and altostratus. { 'mid·əl 'klau̇dz }
Middle Cretaceous [GEOL] The geologic epoch between the Upper and Lower Cretaceous, beginning approximately 120,000,000 years ago. { 'mid·əl krə'tā·shəs }
Middle Devonian [GEOL] The geologic epoch occurring between the Upper and Lower Devonian, beginning approximately 385,000,000 years ago. { 'mid·əl di'vō·nē·ən }
Middle Jurassic [GEOL] The geologic epoch occurring between the Upper and Lower Jurassic, beginning approximately 170,000,000 years ago. { 'mid·əl jə'ras·ik }
middle-latitude westerlies *See* westerlies. { 'mid·əl ¦lad·ə,tüd 'wes·tər,lēz }
Middle Mississippian [GEOL] The geologic epoch between the Upper and Lower Mississippian. { 'mid·əl ,mis·ə'sip·ē·ən }
Middle Ordovician [GEOL] The geologic epoch occurring between the Upper and Lower Ordovician, beginning approximately 460,000,000 years ago. { 'mid·əl ,ȯr·də'vish·ən }
Middle Pennsylvanian [GEOL] The geologic epoch between the Upper and Lower Pennsylvanian. { 'mid·əl ,pen·səl'vā·nyə }
Middle Permian [GEOL] The geologic epoch occurring between the Upper and Lower Permian, beginning approximately 260,000,000 years ago. { 'mid·əl 'pər·mē·ən }
Middle Silurian [GEOL] The geologic epoch between the Upper and Lower Silurian. { 'mid·əl si'lür·ē·ən }
Middle Triassic [GEOL] The geologic epoch occurring between the Upper and Lower Triassic, beginning approximately 215,000,000 years ago. { 'mid·əl trī'as·ik }
mid-extreme tide [OCEANOGR] A level midway between the extreme high water and extreme low water occurring at a place. { 'mid ik¦strēm 'tīd }
midfan [GEOL] The portion of an alluvial fan between the fanhead and the outer, lower margins. { 'mid,fan }
mid-latitude *See* middle latitude. { 'mid¦lad·ə,tüd }
mid-latitude westerlies *See* westerlies. { 'mid¦lad·ə,tüd 'wes·tər,lēz }
mid-ocean canyon *See* deep-sea channel. { 'mid¦ō·shən 'kan·yən }
mid-oceanic ridge [GEOL] A continuous, median, seismic mountain range on the floor of the ocean, extending through the North and South Atlantic oceans, the Indian Ocean, and the South Pacific Ocean; the topography is rugged, elevation is 0.6–1.8 miles (1–3 kilometers), width is about 900 miles (1500 kilometers), and length is over 52,000 miles (84,000 kilometers). Also known as mid-ocean ridge; mid-ocean rise; oceanic ridge. { 'mid,ō·shē¦an·ik 'rij }
mid-ocean ridge *See* mid-oceanic ridge. { 'mid¦ō·shən 'rij }
mid-ocean rift *See* rift valley. { 'mid¦ō·shən 'rift }
mid-ocean rise *See* mid-oceanic ridge. { 'mid¦ō·shən 'rīs }
migma [GEOL] A mixture of solid rock materials and rock melt with mobility or potential mobility. { 'mig·mə }
migration [GEOL] **1.** Movement of a topographic feature from one place to another,

especially movement of a dune by wind action. **2.** Movement of liquid or gaseous hydrocarbons from their source into reservoir rocks. [HYD] Slow, downstream movement of a system of meanders. { mī'grā·shən }

migratory [METEOROL] Commonly applied to pressure systems embedded in the westerlies and, therefore, moving in a general west-to-east direction. { 'mī·grə,tȯr·ē }

migratory dune *See* wandering dune. { 'mī·grə,tȯr·ē 'dün }

Milankovitch cycles [GEOPHYS] Periodic variations in the earth's position relative to the sun as the earth orbits, affecting the distribution of the solar radiation reaching the earth and causing climatic changes that have profound impacts on the abundance and distribution of organisms, best seen in the fossil record of the Quaternary Period (the last 1.6 million years). { mē·lən'kō·vich ,sīk·əlz }

milky weather *See* whiteout. { 'mil·kē 'weth·ər }

Mima mound [GEOGR] A circular or oval domelike structure composed of loose silt and soil that is believed to be generated by a combination of geomorphic processes and burrowing by animals; found in northwest North America, Africa, and southern South America. { 'mē·mə ,mau̇nd }

Mindel glaciation [GEOL] The second glacial stage of the Pleistocene in the Alps. { 'min·dəl ,glā·sē'ā·shən }

Mindel-Riss interglacial [GEOL] The second interglacial stage of the Pleistocene in the Alps; follows the Mindel glaciation. { 'min·dəl 'ris ,in·tər'glā·shəl }

mineral [GEOL] A naturally occurring substance with a characteristic chemical composition expressed by a chemical formula; may occur as individual crystals or may be disseminated in some other mineral or rock; most mineralogists include the requirements of inorganic origin and internal crystalline structure. { 'min·rəl }

mineral caoutchouc *See* elaterite. { 'min·rəl 'kau̇,chu̇k }

mineral charcoal *See* fusain. { 'min·rəl 'chär,kōl }

mineral deposit [GEOL] A mass of naturally occurring mineral material, usually of economic value. { 'min·rəl di,päz·ət }

mineralization [GEOL] **1.** The process of fossilization whereby inorganic materials replace the organic constituents of an organism. **2.** The introduction of minerals into a rock, resulting in a mineral deposit. { ,min·rə·lə'zā·shən }

mineralize [GEOL] To convert to, or impregnate with, mineral material; applied to processes of ore vein deposition and of fossilization. { 'min·rə,līz }

mineralizer [GEOL] A gas or fluid dissolved in a magma that aids in the concentration and crystallization of ore minerals. { 'min·rə,līz·ər }

mineralogenetic epoch [GEOL] A geologic time period during which mineral deposits formed. { ¦min·rə·lō·jə'ned·ik 'ep·ək }

mineralogenetic province [GEOL] Geographic region where conditions were favorable for the concentration of useful minerals. { ¦min·rə·lō·jə'ned·ik 'prä·vəns }

mineral resources [GEOL] Valuable mineral deposits of an area that are presently recoverable and may be so in the future; includes known ore bodies and potential ore. { 'min·rəl ri'sȯrs·əz }

mineral soil [GEOL] Soil composed of mineral or rock derivatives with little organic matter. { 'min·rəl ,sȯil }

mineral spring [HYD] A spring whose water has a definite taste due to a high mineral content. { 'min·rəl ¦spriŋ }

mineral water [HYD] Water containing naturally or artificially supplied minerals or gases. { 'min·rəl ,wȯd·ər }

mineral wax *See* ozocerite. { 'min·rəl ¦waks }

minimum ebb [OCEANOGR] The least speed of a current that runs continuously ebb. { 'min·ə·məm 'eb }

minimum flood [OCEANOGR] The least speed of a current that runs continuously flood. { 'min·ə·məm 'fləd }

minor trough [METEOROL] A pressure trough of smaller scale than a long-wave trough; it ordinarily moves rapidly and is associated with a migratory cyclonic disturbance in the lower troposphere. { 'mīn·ər ¦trȯf }

minuano [METEOROL] A cold southwesterly wind of southern Brazil, occurring during the Southern Hemisphere winter (June to September). { ˌmin·yə'wä·nō }
minus-cement porosity [GEOL] The porosity that would characterize a sedimentary material if it contained no chemical cement. { 'mī·nəs si¦ment pə'räs·əd·ē }
Miocene [GEOL] A geologic epoch of the Tertiary period, extending from the end of the Oligocene to the beginning of the Pliocene. { 'mī·əˌsēn }
miogeocline [GEOL] A nonvolcanic (nonmagmatic) continental margin, characterized by carbonate, shale, and sandstone sediments. { ˌmī·ō'jē·əˌklīn }
miogeosyncline [GEOL] The nonvolcanic portion of an orthogeosyncline, located adjacent to the craton. { ¦mī·ō¦jē·ō'sinˌklīn }
mire [GEOL] Wet spongy earth, as of a marsh, swamp, or bog. { mīr }
misfit stream [HYD] A stream whose meanders are either too large or too small to have eroded the valley in which it flows. { 'misˌfit ¦strēm }
Mississippian [GEOL] The fifth period of the Paleozoic Era beginning about 350 million years ago and ending about 320 million years ago. The Mississippian System (referring to rocks) or Period (referring to the time during which these rocks were deposited) is employed in North America as the lower (or older) subdivision of the Carboniferous, as used in Europe and on other continents. { ¦mis·ə¦sip·ē·ən }
Missourian [GEOL] A North American provincial series of geologic time: lower Upper Pennsylvanian (above Desmoinesian, below Virgilian). { mə'zu̇r·ē·ən }
mist [METEOROL] A hydrometeor consisting of an aggregate of microscopic and more or less hygroscopic water droplets suspended in the atmosphere; it produces, generally, a thin, grayish veil over the landscape; it reduces visibility to a lesser extent than fog; the relative humidity with mist is often less than 95. { mist }
mist droplet [METEOROL] A particle of mist, intermediate between a haze droplet and a fog drop. { 'mist ˌdräp·lət }
mistral [METEOROL] A north wind which blows down the Rhone Valley south of Valence, France, and into the Gulf of Lions. Strong, squally, cold, and dry, it is the combined result of the basic circulation, a fall wind, and jet-effect wind. { mə'sträl }
mixed cloud [METEOROL] A cloud containing both water drops and ice crystals, hence a cloud whose composition is intermediate between that of a water cloud and that of an ice-crystal cloud. { 'mikst 'klau̇d }
mixed current [OCEANOGR] A type of tidal current characterized by a conspicuous difference in speed between the two flood currents or two ebb currents usually occurring each tidal day. { 'mikst 'kə·rənt }
mixed layer [OCEANOGR] The layer of water which is mixed through wave action or thermohaline convection. Also known as surface water. { 'mikst 'lā·ər }
mixed nucleus [METEOROL] A condensation nucleus of intermediate efficacy which, as a result of particle coagulation, contains both soluble hygroscopic matter and insoluble but wettable matter. { 'mikst 'nü·klē·əs }
mixed ore [GEOL] Any ore with both oxidized and unoxidized minerals. { 'mikst 'ȯr }
mixed tide [OCEANOGR] A tide in which the presence of a diurnal wave is conspicuous by a large inequality in the heights of either the two high tides or the two low tides usually occurring each tidal day. { 'mikst 'tīd }
mixing ratio [METEOROL] In a system of moist air, the dimensionless ratio of the mass of water vapor to the mass of dry air; for many purposes, the mixing ratio may be approximated by the specific humidity. { 'mik·siŋ ˌrā·shō }
mixolimnion [HYD] The upper layer of a meromictic lake, characterized by low density and free circulation; this layer is mixed by the wind. { ¦mik·sō'lim·nēˌän }
moat [GEOL] **1.** A ringlike depression around the base of a seamount. **2.** A valleylike depression around the inner side of a volcanic cone, between the rim and the lava dome. [HYD] **1.** A glacial channel in the form of a deep, wide trench. **2.** *See* oxbow lake. { mōt }
mobile belt [GEOL] A long, relatively narrow crustal region of tectonic acitivity. { 'mō·bəl ¦belt }
mobilization [GEOL] Any process by which solid rock becomes sufficiently soft and

plastic to permit it to flow or to permit geochemical migration of the mobile components. { ˌmō·bə·lə'zā·shən }

mock fog [METEOROL] A simulation of true fog by atmospheric refraction. { 'mäk ˌfäg }

mode eddies [OCEANOGR] Densely packed, irregularly oval high- and low-pressure centers roughly 240 miles (400 kilometers) in diameter in which current intensities are typically tenfold greater than the local means. Also known as mesoscale eddies. { 'mōd ˌed·ēz }

model atmosphere [METEOROL] Any theoretical representation of the atmosphere, particularly of vertical temperature distribution. { 'mäd·əl 'at·məˌsfir }

moder [GEOL] Humus consisting of plant material that is undergoing alteration from the living to the decayed state and is intermediate in acidity between mor and mull. { 'mōd·ər }

moderate breeze [METEOROL] In the Beaufort wind scale, a wind whose speed is from 11 to 16 knots (13 to 18 miles per hour or 20 to 30 kilometers per hour). { 'mäd·ə·rət 'brēz }

moderate gale [METEOROL] In the Beaufort wind scale, a wind whose speed is from 28 to 33 knots (32 to 38 miles per hour or 52 to 61 kilometers per hour). { 'mäd·ə·rət 'gāl }

modified index of refraction [METEOROL] An atmospheric index of refraction mathematically modified so that when its gradient is applied to energy propagation over a hypothetical flat earth, it is substantially equivalent to propagation over the true curved earth with the actual index of refraction. Also known as modified refractive index; refractive modulus. { 'mäd·əˌfīd 'inˌdeks əv ri'frak·shən }

modified refractive index *See* modified index of refraction. { 'mäd·əˌfīd ri¦frak·tiv 'inˌdeks }

mofette [GEOL] A small opening emitting carbon dioxide in an area of late-stage volcanic activity. { mō'fet }

Mohawkian [GEOL] A North American stage of middle Ordovician geologic time, above Chazyan and below Edenian. { mō'hȯk·ē·ən }

Mohnian [GEOL] A North American stage of geologic time: Miocene (above Luisian, below Delmontian). { 'mō·nē·ən }

Moho *See* Mohorovičić discontinuity. { 'mō·hō }

Mohole drilling [GEOL] Drilling aimed at penetration of the earth's crust, through the Mohorovičić discontinuity, to sample the mantle. { 'mōˌhōl ˌdril·iŋ }

Mohorovičić discontinuity [GEOPHYS] A seismic discontinuity that separates the earth's crust from the subjacent mantle, inferred from travel time curves indicating that seismic waves undergo a sudden increase in velocity. Also known as Moho. { ˌmō·hō'rō·vəˌchich disˌkänt·ən'ü·əd·ē }

moist adiabat *See* saturation adiabat. { 'mȯist 'ad·ē·əˌbat }

moist-adiabatic lapse rate *See* saturation-adiabatic lapse rate. { 'mȯist ¦ad·ē·ə¦bad·ik 'laps ˌrāt }

moist air [METEOROL] **1.** In atmospheric thermodynamics, air that is a mixture of dry air and any amount of water vapor. **2.** Generally, air with a high relative humidity. { 'mȯist 'er }

moist climate [CLIMATOL] In C.W. Thornthwaite's climatic classification, any type of climate in which the seasonal water surplus counteracts seasonal water deficiency; thus it has a moisture index greater than zero. { 'mȯist 'klīm·ət }

moisture [CLIMATOL] The quantity of precipitation or the precipitation effectiveness. [METEOROL] The water vapor content of the atmosphere, or the total water substance (gaseous, liquid, and solid) present in a given volume of air. { 'mȯis·chər }

moisture adjustment [METEOROL] The adjustment of observed precipitation in a storm by the ratio of the estimated probable maximum precipitable water over the basin under study to the actual precipitable water calculated for the particular storm. { 'mȯis·chər əˌjəs·mənt }

moisture factor [METEOROL] One of the simplest measures of precipitation effectiveness: moisture factor = P/T, where P is precipitation in centimeters and T is temperature degrees centigrade for the period in question. { 'mȯis·chər ˌfak·tər }

moisture film cohesion *See* apparent cohesion. { 'mȯis·chər ¦film kō,hē·zhən }

moisture inversion [METEOROL] An increase with height of the moisture content of the air; specifically, the layer through which this increase occurs, or the altitude at which the increase begins. { 'mȯis·chər in,vər·zhən }

molasse [GEOL] A paralic sedimentary facies consisting mainly of shale, subgraywacke sandstone, and conglomerate; it is more clastic and less rhythmic than the preceding flysch and is generally postorogenic. { mə'läs }

mold [GEOL] Soft, crumbling friable earth. { mōld }

moldauite *See* moldavite. { mōl'dau̇,īt }

moldavite [GEOL] A translucent, olive-to brownish-green or pale-green tektite from the western Czech Republic, characterized by surface sculpturing due to solution etching. Also known as moldauite; pseudochrysolite; vitavite. { mōl'dä,vīt }

molecular fossils *See* biomarkers. { mə¦lek·yə·lər 'fäs·əlz }

Mollisol [GEOL] An order of soils having dark or very dark, friable, thick A horizons high in humus and bases such as calcium and magnesium; most have lighter-colored or browner B horizons that are less friable and about as thick as the A horizons; all but a few have paler C horizons, many of which are calcareous. { 'mal·ə,säl }

monadnock [GEOL] A remnant hill of resistant rock rising abruptly from the level of a peneplain; commonly represents an outcrop of rock that has withstood erosion. Also known as torso mountain. { mə'nad,näk }

monimolimnion [HYD] The dense bottom stratum of a meromictic lake; it is stagnant and does not mix with the water above. { ¦man·ə·mō'lim·nē,än }

monocline [GEOL] A stratigraphic unit that dips from the horizontal in one direction only, not as part of an anticline or syncline. { 'män·ə,klīn }

monogeosyncline [GEOL] A primary geosyncline that is long, narrow, and deeply subsided; composed of the sediments of shallow water and situated along the inner margin of the borderlands. { ¦män·ō,jē·ō'sin,klīn }

monsoon [METEOROL] A large-scale wind system which predominates or strongly influences the climate of large regions, and in which the direction of the wind flow reverses from winter to summer; an example is the wind system over the Asian continent. { män'sün }

monsoon climate [CLIMATOL] The type of climate which is found in regions subject to monsoons. { män'sün ,klī·mət }

monsoon current [OCEANOGR] A seasonal wind-driven current occurring in the northern part of the Indian Ocean. { män'sün ,kə·rənt }

monsoon fog [METEOROL] An advection type of fog occurring along a coast where monsoon winds are blowing, when the air has a high specific humidity and there is a large difference in the temperature of adjacent land and sea. { män'sün ,fäg }

monsoon low [METEOROL] A seasonal low found over a continent in the summer and over the adjacent sea in the winter. { män'sün ,lō }

Montian [GEOL] A European stage of geologic time: Paleocene (above Danian, below Thanetian). { 'män·chən }

moon pillar [METEOROL] A halo consisting of a vertical shaft of light through the moon. { 'mün ,pil·ər }

moor coal [GEOL] A friable lignite or brown coal. { 'mu̇r ,kōl }

mor *See* ectohumus. { mȯr }

morainal apron *See* outwash plain. { mə'rān·əl 'ā·prən }

morainal delta *See* ice-contact delta. { mə'rān·əl 'del·tə }

morainal lake [HYD] A glacial lake filling a depression resulting from irregular deposition of drift in a terminal or ground moraine of a continental glacier. { mə'rān·əl 'lāk }

morainal plain *See* outwash plain. { mə'rān·əl 'plān }

moraine [GEOL] An accumulation of glacial drift deposited chiefly by direct glacial action and possessing initial constructional form independent of the floor beneath it. { mə'rān }

moraine bar [GEOL] A terminal moraine serving as a bar, rising out of deep water at some distance from the shore. { mə'rān ,bär }

moraine kame [GEOL] One of a group of kames characterized by the same topography, constitution, and position as a terminal moraine. { mə'rān ˌkām }
moraine plateau [GEOL] A relatively flat area within a hummocky moraine, generally at the same elevation as, or a little higher than, the summits of surrounding knobs. { mə'rān plə'tō }
morphogenetic region [GEOL] A region in which, under certain climatic conditions, the predominant geomorphic processes will contribute regional characteristics to the landscape that contrast with those of other regions formed under different climatic conditions. { ˌmȯr·fə·jə¦ned·ik 'rē·jən }
morphographic map *See* physiographic diagram. { ¦mȯr·fə¦graf·ik 'map }
mortlake *See* oxbow lake. { 'mȯrtˌlāk }
morvan [GEOL] The area where two peneplains intersect. Also known as skiou. { 'mȯr·vən }
mother lode [GEOL] A main unit of mineralized matter that may not have economic value but to which workable veins are related. { 'məth·ər ˌlōd }
mother-of-coal *See* fusain. { ¦məth·ər əv 'kōl }
mother-of-pearl clouds *See* nacreous clouds. { 'məth·ər əv 'pərl 'klau̇dz }
mother rock *See* source rock. { 'məth·ər ˌräk }
mottled [GEOL] Of a soil, irregularly marked with spots of different colors. { 'mäd·əld }
motu [GEOGR] One of a series of closely spaced coral islets separated by narrow channels; the group of islets forms a ring-shaped atoll. { 'mō·tü }
moulin [HYD] A shaft or hole in the ice of a glacier which is roughly cylindrical and nearly vertical, formed by swirling meltwater pouring down from the surface. Also known as glacial mill; glacier mill; glacier pothole; glacier well; pothole. { mü'lan }
moulin pothole *See* giant's kettle. { mü'lan 'pätˌhōl }
mound [GEOL] **1.** A low, isolated, rounded natural hill, usually of earth. Also known as tuft. **2.** A structure built by fossil colonial organisms. { mau̇nd }
mountain [GEOGR] A feature of the earth's surface that rises high above the base and has generally steep slopes and a relatively small summit area. { 'mau̇nt·ən }
mountain and valley winds [METEOROL] A system of diurnal winds along the axis of a valley, blowing uphill and upvalley by day, and downhill and downvalley by night; they prevail mostly in calm, clear weather. { 'mau̇nt·ən ən 'val·ē 'winz }
mountain breeze [METEOROL] A breeze that blows down a mountain slope due to the gravitational flow of cooled air. Also known as mountain wind. { 'mau̇nt·ən 'brēz }
mountain brown ore [GEOL] Name used in Virginia for limonite or brown iron ore. { 'mau̇nt·ən 'brau̇n 'ȯr }
mountain chain *See* mountain system. { 'mau̇nt·ən ˌchān }
mountain climate [CLIMATOL] Very generally, the climate of relatively high elevations; mountain climates are distinguished by the departure of their characteristics from those of surrounding lowlands, and the one common basis for this distinction is that of atmospheric rarefaction; aside from this, great variety is introduced by differences in latitude, elevation, and exposure to the sun; thus, there exists no single, clearly defined, mountain climate. Also known as highland climate. { 'mau̇nt·ən ¦klī·mət }
mountain-gap wind [METEOROL] A local wind blowing through a gap between mountains. { 'mau̇nt·ən ¦gap ˌwind }
mountain glacier *See* alpine glacier. { 'mau̇nt·ən 'glā·shər }
mountain mahogany *See* obsidian. { 'mau̇nt·ən mə'häg·ə·nē }
mountain meteorology [METEOROL] The branch of meteorology that studies the effects of mountains on the atmosphere, ranging over all scales of motion. { 'mau̇nt·ən ˌmēd·ē·ə'räl·ə·jē }
mountain pediment [GEOL] A plain of combined erosion and transportation at the base of and surrounding a desert mountain range; at a distance it has the appearance of a broad triangular mass. { 'mau̇nt·ən 'ped·ə·mənt }
mountain range [GEOGR] A succession of mountains or narrowly spaced mountain ridges closely related in position, direction, and geologic features. { 'mau̇nt·ən ˌrānj }

mountain slope [GEOGR] The inclined surface that forms a mountainside. { 'maủnt·ən ,slōp }

mountain system [GEOGR] A group of mountain ranges tied together by common geological features. Also known as mountain chain. { 'maủnt·ən ,sis·təm }

mountain wave [METEOROL] An undulating flow of wind on the downwind, or lee, side of a mountain ridge caused by wind blowing strongly over the ridge. { 'maủnt·ən ,wāv }

mountain wind *See* mountain breeze. { 'maủnt·ən ,wind }

mountain wood [GEOL] **1.** A compact, fibrous, gray to brown type of asbestos which has an appearance similar to dry wood. Also known as rock wood. **2.** A fibrous clay mineral; for example, sepiolite or palygorskite. { 'maủnt·ən ,wủd }

mouth [GEOGR] **1.** The place where one body of water discharges into another. Also known as influx. **2.** The entrance or exit of a geomorphic feature, such as of a cave or valley. { maủth }

Mozambique Current [OCEANOGR] The portion of the South Equatorial Current that turns and flows along the coast of Africa in the Mozambique Channel, forming one of the western boundary currents in the Indian Ocean. { ,mō·zəm'bēk 'kə·rənt }

muck [GEOL] Dark, finely divided, well-decomposed, organic matter intermixed with a high percentage of mineral matter, usually silt, forming a surface deposit in some poorly drained areas. { mək }

mud [GEOL] An unindurated mixture of clay and silt with water; it is slimy with a consistency varying from that of a semifluid to that of a soft and plastic sediment. { məd }

mud ball [GEOL] A rounded mass of mud or mudstone up to 8 inches (20 centimeters) in diameter in a sedimentary rock. Also known as chalazoidite; tuff ball. { 'məd ,bȯl }

mud cone [GEOL] A cone of sulfurous mud built around the opening of a mud volcano or mud geyser, with slopes as steep as 40° and diameters ranging upward to several hundred yards. Also known as puff cone. { 'məd ,kōn }

mud crack [GEOL] An irregular fracture formed by shrinkage of clay, silt, or mud under the drying effects of atmospheric conditions at the surface. Also known as desiccation crack; sun crack. { 'məd ,krak }

mud crack polygon *See* mud polygon. { 'məd ,krak 'päl·ə,gän }

mud flat [GEOL] A relatively level, sandy or muddy coastal strip along a shore or around an island; may be alternately covered and uncovered by the tide or may be covered by shallow water. Also known as flat. { 'məd ,flat }

mudflow [GEOL] A flowing mass of fine-grained earth material having a high degree of fluidity during movement. { 'məd,flō }

mudlump [GEOL] A diapiric sedimentary structure consisting of clay or silt and forming an island in deltaic areas; produced by the loading action of rapidly deposited delta front sands upon lighter-weight prodelta clays. { 'məd,ləmp }

mud polygon [GEOL] A nonsorted polygon whose center lacks vegetation but whose peripheral fissures contain peat and plants. Also known as mud crack polygon. { 'məd 'päl·ə,gän }

mud pot [GEOL] A type of hot spring which contains boiling mud, typically sulfurous and often multicolored; tends to be associated with geysers and other hot springs in volcanic zones. Also known as painted pot; sulfur-mud pool. { 'məd ,pät }

mudslide [GEOL] A slow-moving mudflow in which movement is mainly by sliding upon a discrete boundary shear surface. { 'məd,slīd }

mudstone [GEOL] An indurated equivalent of mud in the form of a blocky or massive, fine-grained sedimentary rock containing approximately equal proportions of silt and clay; lacks the fine lamination or fissility of shale. { 'məd,stōn }

mud volcano [GEOL] A conical accumulation of variable admixtures of sand and rock fragments, the whole resulting from eruption of wet mud and impelled upward by fluid or gas pressure. Also known as hervidero; macaluba. { 'məd väl'kā·nō }

muggy [METEOROL] Referring to warm and especially humid weather. { 'məg·ē }

mull [GEOGR] *See* headland. [GEOL] Granular forest humus that is incorporated with mineral matter. { məl }

mullion [GEOL] In folded sedimentary and metamorphic rocks, a columnar structure in which the rock columns seem to intersect. { 'məl·yən }

multicycle [GEOL] Pertaining to a landscape or landform produced by more than one cycle of erosion. { 'məl·tə,sī·kəl }

multiple-current hypothesis [OCEANOGR] The hypothesis that the Gulf Stream, instead of being composed of a single tortuous current, actually consists of many quasipermanent currents, countercurrents, and eddies. { 'məl·tə·pəl ¦kə·rənt hī,päth·ə·səs }

multiple discharge *See* composite flash. { 'məl·tə·pəl 'dis,chärj }

multiple fault *See* step fault. { 'məl·tə·pəl 'fölt }

multiple reflection [GEOPHYS] A seismic wave which has more than one reflection. Also known as repeated reflection; secondary reflection. { 'məl·tə·pəl ri'flek·shən }

multiple tropopause [GEOPHYS] A frequent condition in which the tropopause appears not as a continuous single "surface" of discontinuity between the troposphere and the stratosphere, but as a series of quasi-horizontal "leaves," which are partly overlapping in steplike arrangement. { 'məl·tə·pəl 'trō·pə,pöz }

Muschelkalk [GEOL] A European stage of geologic time equivalent to the Middle Triassic, above Bunter and below Keuper. { 'mu̇sh·əl,kälk }

mustard-seed coal [GEOL] Anthracite that will pass through circular holes in a screen which measure 3/64 inch (1.2 millimeter) in diameter. { 'məs·tərd,sēd ,kōl }

mylonitization [GEOL] Rock deformation produced by intense microbrecciation without appreciable chemical alteration of granulated materials. { mī,län·ə·tə'zā·shən }

N

nacreous clouds [METEOROL] Clouds of unknown composition, whose form resembles that of cirrus or altocumulus lenticularis, and which show very strong irisation similar to that of mother-of-pearl, especially when the sun is several degrees below the horizon; they occur at heights of about 12 or 18 miles (20 or 30 kilometers). Also known as mother-of-pearl clouds. { 'nā·krē·əs 'klaůdz }

nadir [OCEANOGR] The point on the sea floor that lies directly below the sonar during a survey. { 'nā·dər }

nailhead striation [GEOL] A glacial striation with a definite or blunt head or point of origin, generally narrowing or tapering in the direction of ice movement and coming to an indefinite end. { 'nāl,hed strī'ā·shən }

naked karst [GEOL] Karst that is developed in a region without soil cover, so that its topographic features are well exposed. { 'nā·kəd 'kärst }

nakhlite [GEOL] An achondritic stony meteorite composed of an aggregate of diopside and olivine. { 'näk,līt }

Namurian [GEOL] A European stage of geologic time; divided into a lower stage (Lower Carboniferous or Upper Mississippian) and an upper stage (Upper Carboniferous or Lower Pennsylvanian). { nə'myůr·ē·ən }

Nansen cast [OCEANOGR] A series of Nansen-bottle water samples and associated temperature observations resulting from one release of a messenger. { 'nan·sən ,kast }

Napoleonville [GEOL] A North American (Gulf Coast) stage of geologic time; a subdivision of the Miocene, above Anahauc and below Duck Lake. { nə'pōl·ē·ən,vil }

nappe [GEOL] A sheetlike, allochthonous rock unit that is formed by thrust faulting or recumbent folding or both. { nap }

narbonnais [METEOROL] A wind coming from Narbonne; a north wind in the Roussillon region of southern France resembling the tramontana; if associated with an influx of arctic air, it may be very stormy with heavy falls of rain or snow. { när·bə'nā }

nari *See* caliche. { 'när·ē }

Narizian [GEOL] A North American stage of geologic time; a subdivision of the upper Eocene, above Ulatisian and below Fresnian. { nə'rizh·ən }

narrow [GEOGR] A constricted section of a mountain pass, valley, or cave, or a gap or narrow passage between mountains. { 'nar·ō }

narrows [GEOGR] A navigable narrow part of a bay, strait, or river. { 'nar·ōz }

n'aschi [METEOROL] A northeast wind which occurs in winter on the Iranian coast of the Persian Gulf, especially near the entrance to the Gulf, and also on the Makran (West Pakistan) coast; it is probably associated with an outflow from the central Asiatic anticyclone which extends over the high land of Iran. { 'näs·chē }

Nathansohn's theory [OCEANOGR] The theory that nutrient salts in the lighted surface layers of the ocean are consumed by plants, accumulate in the deep ocean through sinking of dead plant and animal bodies, and eventually return to the euphotic layer through diffusion and vertical circulation of the water. { 'nā·thən·sənz ,thē·ə·rē }

national meridian [GEOD] A meridian chosen in a particular nation as the reference datum for determining longitude for that nation. { 'nash·ən·əl mə'rid·ē·ən }

native [GEOCHEM] Pertaining to an element found in nature in a nongaseous state. { 'nād·iv }

native asphalt [GEOL] Exudations or seepages of asphalt occurring in nature in a liquid or semiliquid state. Also known as natural asphalt. { 'nād·iv 'as,fȯlt }
native coal *See* natural coke. { 'nād·iv 'kōl }
native element [GEOL] Any of 20 elements, such as copper, gold, and silver, which occur naturally uncombined in a nongaseous state; there are three groups—metals, semimetals, and nonmetals. { 'nād·iv 'el·ə·mənt }
native metal [GEOCHEM] A metallic native element; includes silver, gold, copper, iron, mercury, iridium, lead, palladium, and platinum. { 'nād·iv 'med·əl }
native paraffin *See* ozocerite. { 'nād·iv 'par·ə·fən }
native uranium [GEOCHEM] Uranium as found in nature; a mixture of the fertile uranium-238 isotope (99.3%), the fissionable uranium-235 isotope (0.7%), and a minute percentage of other uranium isotopes. Also known as natural uranium; normal uranium. { 'nād·iv yə'rā·nē·əm }
natric horizon [GEOL] A soil horizon that has the properties of an argillic horizon, but also displays a blocky, columnar, or prismatic structure and has a subhorizon with an exchangeable-sodium saturation of over 15%. { 'nā·trik hə'rīz·ən }
natron lake *See* soda lake. { 'nā·trən ,lāk }
natural arch [GEOL] **1.** A landform similar to a natural bridge but not formed by erosive agencies. **2.** *See* natural bridge. { 'nach·rəl 'ärch }
natural asphalt *See* native asphalt. { 'nach·rəl 'as,fȯlt }
natural bitumen [GEOL] Native mineral pitch, tar, or asphalt. { 'nach·rəl bə'tü·mən }
natural bridge [GEOL] An archlike rock formation spanning a ravine or valley and formed by erosion. Also known as natural arch. { 'nach·rəl 'brij }
natural coke [GEOL] Coal that has been naturally carbonized by contact with an igneous intrusion, or by natural combustion. Also known as black coal; blind coal; carbonite; cinder coal; coke coal; cokeite; finger coal; native coal. { 'nach·rəl 'kōk }
natural glass [GEOL] An amorphous, vitreous inorganic material that has solidified from magma too quickly to crystallize. { 'nach·rəl 'glas }
natural harbor [GEOGR] A harbor where the configuration of the coast provides the necessary protection. { 'nach·rəl 'här·bər }
natural levee [GEOL] An elongate embankment compounded of sand and silt and deposited along both banks of a river channel during times of flood. { 'nach·rəl 'lev·ē }
natural load [HYD] The quantity of sediment carried by a stable stream. { 'nach·rəl 'lōd }
natural radio-frequency interference [GEOPHYS] Natural terrestrial phenomena of an electromagnetic nature, or natural electromagnetic disturbances originating outside the atmosphere, which interfere with radio communications. { 'nach·rəl ¦rād·ē·ō ¦frē·kwən·sē ,in·tər'fir·əns }
natural remanent magnetization [GEOPHYS] The magnetization of rock which exists in the absence of a magnetic field and has been acquired from the influence of the earth's magnetic field at the time of their formation or, in certain cases, at later times. Abbreviated NRM. { 'nach·rəl 'rem·ə·nənt ,mag·nə·tə'zā·shən }
natural tunnel [GEOL] A cave that is nearly horizontal and is open at both ends. Also known as tunnel cave. { 'nach·rəl 'tən·əl }
natural uranium *See* native uranium. { 'nach·rəl yu̇'rā·nē·əm }
natural well [GEOL] A sinkhole or other natural opening which resembles a well extending below the water table and from which groundwater can be withdrawn. { 'nach·rəl 'wel }
Navajo sandstone [GEOL] A fossil dune formation of Jurassic age found in the Colorado Plateau of the United States. { 'nä·və,hō 'san,stōn }
naval meteorology [METEOROL] The branch of meteorology which studies the interaction between the ocean and the overlying air mass, and which is concerned with atmospheric phenomena over the oceans, the effect of the ocean surface on these phenomena, and the influence of such phenomena on shallow and deep seawater. { 'nā·vəl ,mē·dē·ə'räl·ə·jē }
navigable semicircle [METEOROL] That half of a cyclonic storm area in which the rotary

and progressive motions of the storm tend to counteract each other, and the winds are in such a direction as to blow a vessel away from the storm track. { 'nav·i·gə·bəl 'sem·i,sər·kəl }

Navy Oceanographic and Meteorological Automatic Device [OCEANOGR] A 6-meter-long, boat-shaped, moored instrumented buoy. Abbreviated NOMAD. { ¦hāv·ē ,ō·shə·nə¦graf·ik and ,mēd·ē·ə·rə¦läj·ə·kəl ¦ȯd·ə,mad·ik di'vīs }

neap high water *See* mean high-water neaps. { 'nēp 'hī ,wȯd·ər }

neap low water *See* mean low-water neaps. { 'nēp 'lō ,wȯd·ər }

neap range [OCEANOGR] The mean semidiurnal range of tide when neap tides are occurring; the mean difference in height between neap high water and neap low water. Also known as mean neap range. { 'nēp ,rānj }

neap rise [OCEANOGR] The height of neap high water above the chart datum. { 'nēp ,rīz }

neaps *See* neap tide. { nēps }

neap tidal currents [OCEANOGR] Tidal currents of decreased speed occurring at the time of neap tides. { 'nēp 'tīd·əl ,kə·rəns }

neap tide [OCEANOGR] Tide of decreased range occurring about every 2 weeks when the moon is in quadrature, that is, during its first and last quarter. Also known as neaps. { 'nēp ,tīd }

nearshore [OCEANOGR] An indefinite zone which extends from the shoreline seaward to a point beyond the breaker zone. { 'nir,shȯr }

nearshore circulation [OCEANOGR] Ocean circulation consisting of both the nearshore currents and the coastal currents. { 'nir,shȯr ,sər·kyə'lā·shən }

nearshore current system [OCEANOGR] A current system, caused mainly by wave action in and near the breaker zone, which contains four elements: the shoreward mass transport of water; longshore currents; seaward return flow, including rip currents; and the longshore movement of the expanded heads of rip currents. { 'nir,shȯr 'kə·rənt ,sis·təm }

Nebraskan drift [GEOL] Rock material transported during the Nebraskan glaciation; it is buried below the Kansan drift in Iowa. { nə'bras·kən 'drift }

Nebraskan glaciation [GEOL] The first glacial stage of the Pleistocene epoch in North America, beginning about 1,000,000 years ago, and preceding the Aftonian interglacial stage. { nə'bras·kən glā·sē·ā·shən }

nebulosus [METEOROL] A cloud species with the appearance of a nebulous veil, showing no distinct details; found principally in the genera cirrostratus and stratus. { 'neb·yə'lō·səs }

neck [GEOGR] A narrow strip of land, especially one connecting two larger areas. [GEOL] *See* pipe. [OCEANOGR] The narrow band of water forming the part of a rip current where feeder currents converge and flow swiftly through the incoming breakers and out to the head. { nek }

neck cutoff [GEOGR] A high-angle meander cutoff formed where a stream breaks through or across a narrow meander neck, as where downstream migration of one meander has been slowed and the next meander upstream has overtaken it. { 'nek 'kət,ȯf }

needle [GEOL] A pointed, elevated, and detached mass of rock formed by erosion, such as an aiguille. [HYD] A long, slender snow crystal that is at least five times as long as it is broad. { 'nēd·əl }

needle ice *See* frazil ice; pipkrake. { 'nēd·əl ,īs }

negative area [GEOGR] An area that is almost uncultivable or uninhabitable. [GEOL] *See* negative element. { 'neg·əd·iv 'er·ē·ə }

negative element [GEOL] A large structural feature or part of the earth's crust, characterized through a long geologic time period by frequent and conspicuous downward movement (subsidence) or by extensive erosion, or by an uplift that is considerably less rapid or less frequent than that of adjacent positive elements. Also known as negative area. { 'neg·əd·iv 'el·ə·mənt }

negative landform [GEOL] **1.** A relatively depressed or low-lying topographic form, such as a valley, basin, or plain. **2.** A volcanic feature formed by a lack of material (such as a caldera). { 'neg·əd·iv 'land,fȯrm }

negative movement [GEOL] **1.** A downward movement of the earth's crust relative to an adjacent part of the crust, such as produced by subsidence. **2.** A relative lowering of the sea level with respect to the land, such as produced by a positive movement of the earth's crust or by a retreat of the sea. { 'neg·əd·iv 'müv·mənt }

negative rain [METEOROL] Rain which exhibits a net negative electric charge. { 'neg·əd·iv 'rān }

negative shoreline *See* shoreline of emergence. { 'neg·əd·iv 'shȯr,līn }

nematath [GEOL] A submarine ridge across an Atlantic-type ocean basin which is not an orogenic structure, but which is composed of otherwise undeformed continental crust that has been stretched across a sphenochasm or rhombochasm. { 'nem·ə,tath }

nemere [METEOROL] In Hungary, a stormy, cold fall wind. { 'ne,mir·ə }

neoautochthon [GEOL] A stable basement or autochthon formed where a nappe has ceased movement and has become defunct. { ¦nē·ō·ȯ'täk·thən }

Neocomian [GEOL] A European stage of Lower Cretaceous geologic time; includes Berriasian, Valanginian, Hauterivian, and Barremian. { ¦nē·ə¦kō·mē·ən }

neocryst [GEOL] An individual crystal of a secondary mineral in an evaporite. { 'nē·ə,krist }

neoformation *See* neogenesis. { ¦nē·ō·fȯr'mā·shən }

Neogene [GEOL] An interval of geologic time incorporating the Miocene and Pliocene of the Tertiary period; the Upper Tertiary. { 'nē·ə,jēn }

neogenesis [GEOL] The formation of new minerals, as by diagenesis or metamorphism. Also known as neoformation. { ¦nē·ō'jen·ə·səs }

neoglaciation [GEOL] The removal of glacier ice growth in certain mountain areas during the Little Ice Age, following its shrinkage or disappearance during the Altithermal interval. { ¦nē·ō·glā·sē'ā·shən }

neomagma [GEOL] Magma formed by partial or complete refusion of preexisting rocks under the conditions of plutonic metamorphism. { ¦nē·ō'mag·mə }

neomineralization [GEOCHEM] Chemical interchange within a rock whereby its mineral constituents are converted into entirely new mineral species. { ¦nē·ō,min·rə·lə'zā·shən }

neosome [GEOL] A geometric element of a composite rock or mineral deposit, appearing to be younger than the main rock mass. { 'nē·ə,sōm }

neostratotype [GEOL] A stratotype established after the holostratotype has been destroyed or is otherwise not usable. { ¦nē·ō'strad·ə,tīp }

neotectonic map [GEOL] A map depicting neotectonic structures. { ¦nē·ō·tek'tän·ik 'map }

neotectonics [GEOL] The study of the most recent structures and structural history of the earth's crust, after the Miocene. { ¦nē·ō·tek'tän·iks }

nephanalysis [METEOROL] The analysis of a synoptic chart in terms of the types and amount of clouds and precipitation; cloud systems are identified both as entities and in relation to the pressure pattern, fronts, and other aspects. { ,nef·ə'nal·ə·səs }

nephcurve [METEOROL] In nephanalysis, a line bounding a significant portion of a cloud system, for example, a clear-sky line, precipitation line, cloud-type line, or ceiling-height line. { 'nef,kərv }

nepheloid zone [OCEANOGR] A layer of water near the bottom of the continental rise and slope of the North Atlantic Ocean that contains suspended sediment of the clay fraction and organic matter. { 'nef·ə,lȯid ,zōn }

nephology [METEOROL] The study of clouds. { ne'fäl·ə·jē }

nephsystem *See* cloud system. { 'nef,sis·təm }

neptunian dike [GEOL] A sedimentary dike formed by infilling of sediment, generally sand, in an undersea fissure or hollow. { nep'tü·nē·ən 'dīk }

neptunianism *See* neptunism. { nep'tü·nē·ə,niz·əm }

neptunian theory *See* neptunism. { nep'tü·nē·ən 'thē·ə·rē }

neptunic rock [GEOL] A rock that is formed in the sea. { nep'tün·ik 'räk }

neptunism [GEOL] The obsolete theory that all rocks of the earth's crust were deposited from or crystallized out of water. Also known as neptunianism; neptunian theory. { 'nep·tə,niz·əm }

neritic [OCEANOGR] Of or pertaining to the region of shallow water adjoining the seacoast and extending from low-tide mark to a depth of about 660 feet (200 meters). { nə'rid·ik }

nest [GEOL] A concentration of some relatively conspicuous element of a geologic feature, such as pebbles or inclusions, within a sand layer or igneous rock. { nest }

nested [GEOL] **1.** Pertaining to volcanic cones, craters, or calderas that occur one within another. **2.** Pertaining to two or more calderas that intersect, having been formed at different times or by different explosions. { 'nes·təd }

net [GEOL] **1.** In structural petrology, coordinate network of meridians and parallels, projected from a sphere at intervals of 2°; used to plot points whose spherical coordinates are known and to study the distribution and orientation of planes and points. Also known as projection net; stereographic net. **2.** A form of horizontal patterned ground whose mesh is intermediate between a circle and a polygon. { net }

net balance [HYD] The change in mass of a glacier from the time of minimum mass in one year to the time of minimum mass in the succeeding year. Also known as net budget. { 'net 'bal·əns }

net budget *See* net balance. { 'net 'bəj·ət }

net slip [GEOL] On a fault, the distance between two formerly adjacent points on either side of the fault; defines direction and relative amount of displacement. Also known as total slip. { 'net 'slip }

neutral estuary [GEOGR] An estuary in which neither fresh-water inflow nor evaporation dominates. { 'nü·trəl 'es·chə,wer·ē }

neutral line *See* Busch lemniscate. { 'nü·trəl 'līn }

neutral point *See* col. { 'nü·trəl ,pȯint }

neutral pressure *See* neutral stress. { 'nü·trəl 'presh·ər }

neutral shoreline [GEOL] A shoreline whose essential features are independent of either the submergence of a former land surface or the emergence of a former underwater surface. { 'nü·trəl 'shȯr,līn }

neutral stability [METEOROL] The state of an unsaturated or saturated column of air in the atmosphere when its environmental lapse rate of temperature is equal to the dry-adiabatic lapse rate or the saturation-adiabatic lapse rate respectively; under such conditions a parcel of air displaced vertically will experience no buoyant acceleration. Also known as indifferent equilibrium; indifferent stability. { 'nü·trəl stə'bil·əd·ē }

neutral stress [HYD] The stress transmitted through the interstitial fluid of a soil or rock mass. Also known as neutral pressure; pore pressure; pore-water pressure. { 'nü·trəl 'stres }

neutrosphere [METEOROL] The atmospheric shell from the earth's surface upward, in which the atmospheric constituents are for the most part un-ionized, that is, electrically neutral; the region of transition between the neutrosphere and the ionosphere is somewhere between 42 and 54 miles (70 and 90 kilometers), depending on latitude and season. { 'nü·trə,sfir }

nevada [METEOROL] A cold wind descending from a mountain glacier or snowfield, for example, in the higher valleys of Ecuador. { nə'väd·ə }

Nevadan orogeny [GEOL] Orogenic episode during Jurassic and Early Cretaceous geologic time in the western part of the North American Cordillera. Also known as Nevadian orogeny; Nevadic orogeny. { nə'vad·ən ȯ'raj·ə·nē }

Nevadian orogeny *See* Nevadan orogeny. { nə'vad·ē·ən ȯ'raj·ə·nē }

Nevadic orogeny *See* Nevadan orogeny. { nə'vad·ik ȯ'raj·ə·nē }

névé [GEOGR] A geographic area of perennial snow. [HYD] An accumulation of compacted, granular snow in transition from soft snow to ice; it contains much air; the upper portions of most glaciers and ice shelves are usually composed of névé. { nā'vā }

new global tectonics [GEOL] Comprehensive theory relating the formation of mountain belts, island arcs, and ocean trenches to the relative movement of regionally extensive lithospheric plates which are delineated by the major seismic belts of the earth. { 'nü 'glō·bəl tek'tän·iks }

newly formed ice [HYD] Ice in the first stage of formation and development. Also known as fresh ice. { 'nü·lē ˌfȯrmd 'īs }

New Red Sandstone [GEOL] The red sandstone facies of the Permian and Triassic systems exposed in the British Isles. { 'nü 'red 'sanˌstōn }

new snow [METEOROL] **1.** Fallen snow whose original crystalline structure has been retained and is therefore recognizable. **2.** Snow which has fallen in a single day. { 'nü 'snō }

NEXRAD *See* next-generation radar. { 'neksˌrad }

next-generation radar [METEOROL] A Doppler radar, called WSR-88D, that enables forecasters to detect and give early warning for potentially severe weather. Abbreviated NEXRAD. { ˌnekst ˌjen·əˌrā·shən 'rāˌdär }

Niagaran [GEOL] A North American provincial geologic series, in the Middle Silurian. { nī'ag·rən }

niche [GEOL] A shallow cave or reentrant produced by weathering and erosion near the base of a rock face or cliff or beneath a waterfall. { nich }

niche glacier [HYD] A common type of small mountain glacier occupying a funnel-shaped hollow or irregular recess in a mountain slope. { 'nich ˌglā·shər }

nick *See* knickpoint. { nik }

nickpoint *See* knickpoint. { 'nikˌpȯint }

nieve penitente [GEOL] A jagged pinnacle or spike of snow or firn, up to several meters in height. Also known as penitent. { nē'ā·vā ˌpen·ə'tenˌtā }

nightglow [GEOPHYS] A subdivision of airglow in which energy comes from reactions of atomic oxygen between 42 and 60 miles (70 and 100 kilometers), and from ionic recombination around 180 miles (300 kilometers). { 'nītˌglō }

night-sky light *See* airglow. { 'nīt ˌskī 'līt }

night-sky luminescence *See* airglow. { 'nīt ˌskī ˌlü·mə'nes·əns }

night wind [METEOROL] Dry squalls which occur at night in southwest Africa and the Congo; the term is loosely applied to other diurnal local winds such as mountain wind, land breeze, and midnight wind. { 'nīt ˌwind }

nilas [HYD] A thin elastic crust of gray-colored ice formed on a calm sea; characterized by a matte surface, and easily bent by waves and thrust into a pattern of interlocking fingers. { 'nī·ləs }

nimbostratus [METEOROL] A principal cloud type, or cloud genus, gray-colored and often dark, rendered diffuse by more or less continuously falling rain, snow, or sleet of the ordinary varieties, and not accompanied by lightning, thunder, or hail; in most cases the precipitation reaches the ground. { ¦nim·bō¦strad·əs }

nimbus [METEOROL] A characteristic rain cloud; the term is not used in the international cloud classification except as a combining term, as cumulonimbus. { 'nim·bəs }

nip [GEOL] **1.** A small, low cliff or break in slope which is produced by wavelets at the high-water mark. **2.** The point on the bank of a meander lake where erosion takes place due to crowding of the stream current toward the lake. **3.** Thinning of a coal seam, particularly if caused by tectonic movements. Also known as want. { nip }

nitrogen balance [GEOCHEM] The net loss or gain of nitrogen in a soil. { 'nī·trə·jən ˌbal·əns }

nival gradient [GEOL] The angle between a nival surface and the horizon. { 'nī·vəl ˌgrād·ē·ənt }

nival surface [GEOL] The hypothetical planar surface containing all of the different snowlines of the same geologic time period. { 'nī·vəl ˌsər·fəs }

nivation [GEOL] Rock or soil erosion beneath a snowbank or snow patch, due mainly to frost action but also involving chemical weathering, solifluction, and meltwater transport of weathering products. Also known as snow patch erosion. { nī'vā·shən }

nivation cirque *See* nivation hollow. { nī'vā·shən ˌsərk }

nivation glacier [HYD] A small, newly formed glacier; represents the initial stage of glaciation. Also known as snowbank glacier. { nī'vā·shən ˌglā·shər }

nivation hollow [GEOL] A small, shallow depression formed, and occupied during part of the year, by a snow patch or snowbank that, through nivation, is thought to initiate glaciation. Also known as nivation cirque; snow niche. { nī'vā·shən,häl·ō }

nivation ridge *See* winter-talus ridge. { nī'vā·shən ˌrij }

niveal [GEOL] Property of features and effects resulting from the action of snow and ice. { 'niv·ē·əl }

niveoglacial [GEOL] Pertaining to the combined action of snow and ice. { ¦niv·ē·ō'glā·shəl }

niveolian [GEOL] Pertaining to simultaneous accumulation and intermixing of snow and airborne sand at the side of a gentle slope. { ¦niv·ē¦ō·lē·ən }

noctilucent cloud [METEOROL] A cloud of unknown composition which occurs at great heights and high altitudes; photometric measurements have located such clouds between 45 and 54 miles (75 and 90 kilometers); they resemble thin cirrus, but usually with a bluish or silverish color, although sometimes orange to red, standing out against a dark night sky. { ¦näk·tə¦lü·sənt 'klaüd }

nocturnal radiation *See* effective terrestrial radiation. { näk'tərn·əl ˌrād·ē'ā·shən }

nodal points [OCEANOGR] The no-tide points in amphidromic regions. { 'nōd·əl ˌpòins }

nodal zone [OCEANOGR] A zone in which there is a change in the prevailing direction of the littoral transport. { 'nōd·əl ˌzōn }

node [GEOL] That point along a fault at which the direction of apparent displacement changes. { nōd }

nodular chert [GEOL] Chert occurring as nodular or concretionary segregations (chert nodules). { 'näj·ə·lər 'chərt }

NOMAD *See* Navy Oceanographic and Meteorological Automatic Device. { 'nō,mad }

nodule [GEOL] A small, hard mass or lump of a mineral or mineral aggregate characterized by a contrasting composition from and a greater hardness than the surrounding sediment or rock matrix in which it is embedded. { 'näj·ül }

nominal diameter [GEOL] The diameter computed for a hypothetical sphere which would have the same volume as the calculated volume for a specific sedimentary particle. Also known as equivalent diameter. { 'näm·ə·nəl dī'am·əd·ər }

nonbanded coal [GEOL] Coal without lustrous bands, composed mainly of clarain or durain without nitrain. { 'nän,ban·dəd 'kōl }

noncaking coal [GEOL] Hard or dull coal that does not cake when heated. Also known as free-burning coal. { 'nän,kāk·iŋ 'kōl }

Noncalcic Brown soil [GEOL] A great soil group having a slightly acidic, light-pink or reddish-brown A horizon and a light-brown or dull-red B horizon, and developed under a mixture of grass and forest vegetation in a subhumid climate. Also known as Shantung soil. { ¦nän'kal·sik 'braün ˌsòil }

noncapillary porosity [GEOL] The property of a volume of large interstices in a rock or soil that do not hold water by capillarity. { ¦nän'kap·ə,ler·ē pə'räs·əd·ē }

nonclastic [GEOL] Of the texture of a sediment or sedimentary rock, formed chemically or organically and showing no evidence of a derivation from preexisting rock or mechanical deposition. Also known as nonmechanical. { ¦nän'kla,stik }

noncohesive *See* cohesionless. { ˌnän·kō'hē·siv }

nonconformity [GEOL] A type of unconformity in which rocks below the surface of unconformity are either igneous or metamorphic. { ¦nän·kən'fòr·məd·ē }

nonconservative element [OCEANOGR] An element in sea water which is so uncommon that a large proportion of its total composition enters and leaves the particulate phase. { ¦nän·kən'sər·vəd·iv 'el·ə·mənt }

noncontributing area [HYD] An area with closed drainage. { ¦nän·kən'trib·yəd·iŋ 'er·ē·ə }

noncyclic terrace [GEOL] One of a series of terraces representing previous valley floors formed during periods when continued valley deepening accompanied lateral erosion. { ¦nän'sī·klik 'ter·əs }

nondepositional unconformity *See* paraconformity. { ¦nän,dep·ə'zish·ən·əl ,ən·kən'fór·məd·ē }

nondivergent flow [OCEANOGR] Fluid flow in which the divergence of the ocean current field is zero. { 'nän·di,vər·jənt 'flō }

nonfrontal squall line *See* prefrontal squall line. { 'nän,frənt·əl 'skwól ,līn }

nongraded [GEOL] Pertaining to a soil or an unconsolidated sediment consisting of particles of essentially the same size. { ¦nän'grād·əd }

nonmechanical *See* nonclastic. { ¦nän·mi'kan·ə·kəl }

nonpenetrative [GEOL] Of a type of deformation, affecting only part of a rock, such as kink bands. { nän'pen·ə,trād·iv }

nonplunging fold [GEOL] A fold with a horizontal axial surface. Also known as horizontal fold; level fold. { ¦nän¦plən·jiŋ 'fōld }

nonrotational strain *See* irrotational strain. { ,nän·rō'tā·shən·əl 'strān }

nonsorted polygon [GEOL] A form of patterned ground which has a dominantly polygonal mesh and an unsorted appearance due to the absence of border stones, and whose borders are generally marked by wedge-shaped fissures narrowing downward. { 'nän,sórd·əd 'päl·i,gän }

nonsystematic joint [GEOL] A joint that is not part of a set. { ¦nän,sis·tə'mad·ik 'jóint }

nontidal current [OCEANOGR] Any current due to causes other than tidal, as a permanent ocean current. { 'nän,tīd·əl 'kə·rənt }

nonwetting sand [GEOL] Sand that resists infiltration of water; consists of angular particles of various sizes and occurs as a tightly packed lens. { 'nän,wed·iŋ 'sand }

Nordenskjöld line [CLIMATOL] The line connecting all places at which the mean temperature of the warmest month is equal (in degrees Celsius) to $9 - 0.1k$, where k is the mean temperature of the coldest month (in degrees Fahrenheit, it becomes $51.4 - 0.1k$). { 'nórd·ən,shēld ,līn }

nor'easter *See* northeaster. { nór'ē·stər }

Norian [GEOL] A European stage of Upper Triassic geologic time that lies above the Carnian and below the Rhaetian. { 'nór·ē·ən }

normal [METEOROL] The average value of a meteorological element over any fixed period of years that is recognized as standard for the country and element concerned. { 'nór·məl }

normal aeration [GEOL] The complete renewal of soil air to a depth of 8 inches (20 centimeters) about once each hour. { 'nór·məl e'rā·shən }

normal anticlinorium [GEOL] An anticlinorium in which axial surfaces of the subsidiary folds converge downward. { 'nór·məl,ant·i·klə'nór·ē·əm }

normal chart [METEOROL] Any chart that shows the distribution of the official normal values of a meteorological element. Also known as normal map. { 'nór·məl 'chärt }

normal consolidation [GEOL] Consolidation of a sedimentary material in equilibrium with overburden pressure. { 'nór·məl kən,säl·ə'dā·shən }

normal cycle [GEOL] A cycle of erosion whereby a region is reduced to base level by running water, especially by the action of rivers. Also known as fluvial cycle of erosion. { 'nór·məl 'sī·kəl }

normal dip *See* regional dip. { 'nór·məl 'dip }

normal dispersion [GEOPHYS] The dispersion of seismic waves in which the recorded wave period increases with time. { 'nór·məl di'spər·zhən }

normal displacement *See* dip slip. { 'nór·məl di'splās·mənt }

normal erosion [GEOL] Erosion effected by prevailing agencies of the natural environment, including running water, rain, wind, waves, and organic weathering. Also known as geologic erosion. { 'nór·məl i'rō·zhən }

normal fault [GEOL] A fault, usually of 45–90°, in which the hanging wall appears to have shifted downward in relation to the footwall. Also known as gravity fault; normal slip fault; slump fault. { 'nór·məl 'fólt }

normal fold *See* symmetrical fold. { 'nór·məl 'fōld }

normal horizontal separation *See* offset. { 'nór·məl ,hör·ə'zänt·əl ,sep·ə'rā·shən }

normal hydrostatic pressure [HYD] In porous strata or in a well, the pressure at a given point that is approximately equal to the weight of a column of water extending from the surface to that point. { 'nȯr·məl ¦hī·drə¦stad·ik 'presh·ər }

normal map *See* normal chart. { 'nȯr·məl 'map }

normal moveout [GEOPHYS] In seismic prospecting, the increase in stepout time that results from an increase in distance from source to detector when there is no dip. { 'nȯr·məl 'müv,au̇t }

normal polarity [GEOPHYS] Natural remanent magnetism nearly identical to the present ambient field. { 'nȯr·məl pə'lar·əd·ē }

normal pressure *See* standard pressure. { 'nȯr·məl 'presh·ər }

normal ripple mark [GEOL] An aqueous current ripple mark consisting of a simple asymmetrical ridge that may have various configurations. { 'nȯr·məl 'rip·əl ,märk }

normal slip fault *See* normal fault. { 'nȯr·məl 'slip ,fȯlt }

normal soil [GEOL] A soil having a profile that is more or less in equilibrium with the environment. { 'nȯr·məl 'sȯil }

normal synclinorium [GEOL] A synclinorium in which the axial surfaces of the subsidiary folds converge upward. { 'nȯr·məl ,sin·klə'nȯr·ē·əm }

normal uranium *See* native uranium. { 'nȯr·məl yu̇'rā·nē·əm }

normal water [OCEANOGR] Water whose chlorinity lies between 19.30 and 19.50 parts per thousand and has been determined to within ±0.001 per thousand. Also known as Copenhagen water; standard seawater. { 'nȯr·məl 'wȯd·ər }

norte [METEOROL] **1.** The winter north wind in Spain. **2.** A strong, cold northeasterly wind which blows in Mexico and on the shores of the Gulf of Mexico, and results from an outbreak of cold air from the north; actually, the Mexican extension of a norther. { 'nor·tā }

north [GEOD] The direction of the north terrestrial pole; the primary reference direction on the earth; the direction indicated by 000° in any system other than relative. { nȯrth }

North America [GEOGR] The northern of the two continents of the New World or Western Hemisphere, extending from narrow parts in the tropics to progressively broadened portions in middle latitudes and Arctic polar margins. { 'nȯrth ə'mer·i·kə }

North American anticyclone *See* North American high. { 'nȯrth ə'mer·i·kən ,ant·i'sī,klōn }

North American high [METEOROL] The relatively weak general area of high pressure which, as shown on mean charts of sea-level pressure, covers most of North America during winter. Also known as North American anticyclone. { 'nȯrth ə'mer·i·kən 'hī }

North Atlantic Current [OCEANOGR] A wide, slow-moving continuation of the Gulf Stream originating in the region east of the Grand Banks of Newfoundland. { 'nȯrth at'lan·tik 'kə·rənt }

North Cape Current [OCEANOGR] A warm current flowing northeastward and eastward around northern Norway, and curving into the Barents Sea. { 'nȯrth 'kāp ,kə·rənt }

Northeast Drift Current [OCEANOGR] A North Atlantic Ocean current flowing northeastward toward the Norwegian Sea, gradually widening and, south of Iceland, branching and continuing as the Irminger Current and the Norwegian Current; it is the northern branch of the North Atlantic Current. { 'nȯrth,ēst ¦drift 'kə·rənt }

northeaster [METEOROL] A northeast wind, particularly a strong wind or gale. Also spelled nor'easter. { nor'thē·stər *or* nȯ'rē·stər }

northeast storm [METEOROL] A cyclonic storm of the east coast of North America, so called because the winds over the coastal area are from the northeast; they may occur at any time of year but are most frequent and most violent between September and April. { 'nȯr'thēst 'stȯrm }

northeast trades [METEOROL] The trade winds of the Northern Hemisphere. { 'nȯr'thēst 'trādz }

North Equatorial Current [OCEANOGR] Westward ocean currents driven by the northeast trade winds blowing over tropical oceans of the Northern Hemisphere. Also known as Equatorial Current. { 'nȯrth ,ē·kwə'tȯr·ē·əl 'kə·rənt }

norther [METEOROL] A northerly wind. { 'nȯr·t͟hər }

Northern Hemisphere [GEOGR] The half of the earth north of the Equator. { 'nȯr·t͟hərn 'hem·i,sfir }

Northern Hemisphere annular mode *See* Arctic Oscillation. { ¦nȯr·*th*ərn ¦hem·ə,sfir ¦an·yə·lər 'mōd }

northern lights *See* aurora borealis. { 'nȯr·t͟hərn 'līts }

north foehn [METEOROL] A foehn condition sustained by wind flow across the Alps from north to south. { 'nȯrth 'fān }

north frigid zone [GEOGR] That part of the earth north of the Arctic Circle. { 'nȯrth 'frij·əd ,zōn }

north geographic pole *See* North Pole. { 'nȯrth ,jē·ə'graf·ik 'pōl }

north geomagnetic pole *See* North Pole. { 'nȯrth ¦jē·ō,mag¦ned·ik 'pōl }

north magnetic pole *See* North Pole. { 'nȯrth mag'ned·ik 'pōl }

North Pacific Current [OCEANOGR] The warm branch of the Kuroshio Extension flowing eastward across the Pacific Ocean. { 'nȯrth pə'sif·ik 'kə·rənt }

north pole [GEOPHYS] The geomagnetic pole in the Northern Hemisphere, at approximately latitude 78.5°N, longitude 69°W. Also known as north geomagnetic pole; north magnetic pole. { 'nȯrth 'pōl }

North Pole [GEOGR] The geographic pole located at latitude 90°N in the Northern Hemisphere of the earth; it is the northernmost point of the earth, and the northern extremity of the earth's axis of rotation. Also known as north geographic pole. { 'nȯrth 'pōl }

north-south effect [GEOPHYS] An effect whereby the intensity of cosmic radiation incident from the north is somewhat greater than that from the south in the Northern Hemisphere, and vice versa. { ¦nȯrth ¦sau̇th i,fekt }

north temperate zone [CLIMATOL] That part of the earth between the Tropic of Cancer and the Arctic Circle. { 'nȯrth 'tem·prət ,zōn }

northwester [METEOROL] A northwest wind. Also spelled nor'wester. { nȯrth'wes·tər *or* nȯr'wes·tər }

Norway Current [OCEANOGR] A continuation of the North Atlantic Current, which flows northward along the coast of Norway. Also known as Norwegian Current. { 'nȯr,wā ,kə·rənt }

Norwegian Current *See* Norway Current. { nȯr'wē·jən ,kə·rənt }

nor'wester *See* northwester. { nȯr'wes·tər }

nose [GEOL] **1.** A plunging anticline that is short and without closure. **2.** A projecting and generally overhanging buttress of rock. **3.** The projecting end of a hill, spur, ridge, or mountain. **4.** The central forward part of a parabolic dune. { nōz }

notch [GEOL] A deep, narrow cut near the high-water mark at the base of a sea cliff. [GEOGR] A narrow passage between mountains or through a ridge, hill, or mountain. { näch }

nourishment [GEOL] The replenishment of a beach, either naturally (such as by littoral transport) or artificially (such as by deposition of dredged materials). [HYD] *See* accumulation. { 'nər·ish·mənt }

novaculite [GEOL] A siliceous sedimentary rock that is dense, hard, even-textured, light-colored, and characterized by dominance of microcrystalline quartz over chalcedony. Also known as razor stone. { nə'vak·yə,līt }

novaculitic chert [GEOL] A gray chert that fragments into slightly rough, splintery pieces. { nə¦vak·yə¦lid·ik 'chərt }

nowcasting [METEOROL] **1.** The detailed description of the current weather along with forecasts obtained by extrapolation up to about 2 hours ahead. **2.** Any area-specific forecast for the period up to 12 hours ahead that is based on very detailed observational data. { 'nau̇,kast·iŋ }

NRM *See* natural remanent magnetization.

NRM wind scale [METEOROL] A wind scale adapted by the United States Forest Service for use in the forested areas of the Northern Rocky Mountains (NRM); it is an adaptation of the Beaufort wind scale; the difference between these two scales lies in the specification of the visual effects of the wind; the force numbers and the corresponding wind speeds are the same in both. { ¦en¦är'em 'wind ˌskāl }

nubbin [GEOL] **1.** One of the isolated bedrock knobs or small hills forming the last remnants of a mountain crest or mountain range that has succumbed to desert erosion. **2.** A residual boulder, commonly granitic, occurring on a desert dome or broad pediment. { 'nəb·ən }

nuclear twin-probe gage *See* profiling snow gage. { 'nü·klē·ər ¦twin ˌprōb ˌgāj }

nuclear winter [METEOROL] Predicted global-scale changes resulting from a nuclear war, in which dust raised by nuclear bursts and smoke generated in fires would cause reductions in solar energy reaching the earth's surface and reductions in surface temperatures for periods of months. { 'nü·klē·ər 'win·tər }

nucleus [HYD] A particle of any nature upon which, or a locus at which, molecules of water or ice accumulate as a result of a phase change to a more condensed state. { 'nü·klē·əs }

nuée ardente [GEOL] A turbulent, rapidly flowing, and sometimes incandescent gaseous cloud erupted from a volcano and containing ash and other pyroclastics in its lower part. Also known as glowing cloud; Pelean cloud. { ¦nü¦ā är'dänt }

nugget [GEOL] A small mass of metal found free in nature. { 'nəg·ət }

numerical forecasting [METEOROL] The forecasting of the behavior of atmospheric disturbances by the numerical solution of the governing fundamental equations of hydrodynamics, subject to observed initial conditions. Also known as dynamic forecasting; mathematical forecasting; numerical weather prediction; physical forecasting. { nü'mer·i·kəl 'fȯrˌkast·iŋ }

numerical weather prediction *See* numerical forecasting. { nü'mer·i·kəl 'weth·ər priˌdik·shən }

nunatak [GEOL] An isolated hill, knob, ridge, or peak of bedrock projecting prominently above the surface of a glacier and completely surrounded by glacial ice. { 'nən·əˌtak }

O

oasis [GEOGR] An isolated fertile area, usually limited in extent and surrounded by desert, and marked by vegetation and a water supply. { ō'ā·səs }

oberwind [METEOROL] A night wind from mountains or the upper ends of lakes; a wind of Salzkammergut in Austria. { 'ō·bər,vint }

oblique fault *See* diagonal fault. { ə'blēk 'fȯlt }

oblique joint *See* diagonal joint. { ə'blēk 'jȯint }

oblique slip fault [GEOL] A fault which has slippage along both the strike and dip of the fault plane. { ə'blēk 'slip ,fȯlt }

obscuration [METEOROL] In United States weather observing practice, the designation for the sky cover when the sky is completely hidden by surface-based obscuring phenomena, such as fog. Also known as obscured sky cover. { ,äb·skyü'rā·shən }

obscured sky cover *See* obscuration. { əb'skyürd 'skī ,kəv·ər }

obscuring phenomenon [METEOROL] In United States weather observing practice, any atmospheric phenomenon (not including clouds) which restricts the vertical visibility or slant visibility, that is, which obscures a portion of the sky from the point of observation. { əb'skyür·iŋ fə,näm·ə,nän }

obsequent [GEOL] Of a stream, valley, or drainage system, being in a direction opposite to that of the original consequent drainage. { 'äb·sə·kwənt }

obsequent fault-line scarp [GEOL] A fault-line scarp which faces in the direction opposite to that of the original fault scarp or in which the structurally upthrown block is topographically lower than the downthrown block. { 'äb·sə·kwənt 'fȯlt ,līn ,skärp }

observational day [GEOPHYS] Any 24-hour period selected as the basis for climatological or hydrological observations. { ,äb·zər'vā·shən·əl ¦dā }

obsidian [GEOL] A jet-black volcanic glass, usually of rhyolitic composition, formed by rapid cooling of viscous lava; generally forms the upper parts of lava flows. Also known as hyalopsite; Iceland agate; mountain mahogany. { äb'sid·ē·ən }

obsidianite *See* tektite. { äb'sid·ē·ə,nīt }

obstructed stream [GEOL] A stream whose valley has been blocked by a landslide, glacial moraine, sand dune, or lava flow; it frequently consists of a series of ponds or small lakes. { əb'strək·təd 'strēm }

obstruction moraine [GEOL] A moraine formed where the movement of ice is obstructed, for example, by a ridge of bedrock. { əb'strək·shən mə'rān }

obstruction to vision [METEOROL] In United States weather observing practice, one of a class of atmospheric phenomena, other than the weather class of phenomena, which may reduce horizontal visibility at the earth's surface; examples are fog, smoke, and blowing snow. { əb'strək·shən tə 'vizh·ən }

occluded cyclone [METEOROL] Any cyclone (or low) within which there has developed an occluded front. { ə'klüd·əd 'sī,klōn }

occluded front [METEOROL] A composite of two fronts, formed as a cold front overtakes a warm front or quasi-stationary front. Also known as frontal occlusion; occlusion. { ə'klüd·əd 'frənt }

occlusion *See* occluded front. { ə'klü·zhən }

ocean [OCEANOGR] The interconnected body of salt water that occupies almost three-quarters of the earth's surface. { 'ō·shən }

ocean basin [GEOL] The great depression occupied by the ocean on the surface of the lithosphere. { 'ō·shən 'bā·sən }

ocean circulation [OCEANOGR] **1.** Water current flow in a closed circular pattern within an ocean. **2.** Large-scale horizontal water motion within an ocean. { 'ō·shən ˌsər·kyə'lā·shən }

ocean current [OCEANOGR] A net transport of ocean water along a definable path. { 'ō·shən 'kə·rənt }

ocean floor [GEOL] The near-horizontal surface of the ocean basin. { 'ō·shən 'flȯr }

ocean-floor spreading *See* sea-floor spreading. { 'ō·shən ¦flȯr ˌspred·iŋ }

oceanic anticyclone *See* subtropical high. { ˌō·shē'an·ik ˌant·i'sīˌklōn }

oceanic climate *See* marine climate. { ˌō·shē'an·ik 'klī·mət }

oceanic crust [GEOL] A thick mass of igneous rock which lies under the ocean floor. { ˌō·shē'an·ik 'krəst }

oceanic heat flow [GEOPHYS] The amount of thermal energy escaping from the earth through the ocean floor per unit area and unit time. { ˌō·shē'an·ik 'hēt ˌflō }

oceanic high *See* subtropical high. { ˌō·shē'an·ik 'hī }

oceanic island [GEOL] Any island which rises from the deep-sea floor rather than from shallow continental shelves. { ˌō·shē'an·ik 'ī·lənd }

oceanicity [CLIMATOL] The degree to which a point on the earth's surface is in all respects subject to the influence of the sea; it is the opposite of continentality; oceanicity usually refers to climate and its effects; one measure for this characteristic is the ratio of the frequencies of maritime to continental types of air mass. Also known as oceanity. { ˌō·shē·ə'nis·əd·ē }

oceanic province [OCEANOGR] The water of the ocean that lies seaward of the break in the continental shelf. { ˌō·shē'an·ik 'präv·əns }

oceanic ridge *See* mid-oceanic ridge. { ˌō·shē'an·ik 'rij }

oceanic rise [GEOL] A long, broad elevation of the bottom of the ocean. { ˌō·shē'an·ik 'rīz }

oceanic stratosphere *See* cold-water sphere. { ˌō·shē'an·ik 'strad·əˌsfir }

oceanic zone [OCEANOGR] The biogeographic area of the open sea. { ˌō·shē'an·ik 'zōn }

oceanity *See* oceanicity. { ˌo·shē'an·əd·ē }

oceanization [GEOL] Process by which continental crust (sial) is converted into oceanic crust (sima). { ˌō·shə·nə'zā·shən }

oceanographic equator [OCEANOGR] **1.** The region of maximum temperature of the ocean surface. **2.** The region in which the temperature of the ocean surface is greater than 28°C. { ¦ō·shə·nə¦graf·ik i'kwād·ər }

oceanographic model [OCEANOGR] A theoretical representation of the marine environment which relates physical, chemical, geological, biological, and other oceanographic properties. { ¦ō·shə·nə¦graf·ik 'mäd·əl }

oceanographic station [OCEANOGR] A geographic location at which oceanographic observations are taken from a stationary ship. { ¦ō·shə·nə¦graf·ik 'stā·shən }

oceanographic survey [OCEANOGR] A study of oceanographic conditions with reference to physical, chemical, biological, geological, and other properties of the ocean. { ¦ō·shə·nə¦graf·ik 'sərˌvā }

oceanography [OCEANOGR] The science of the sea, including physical oceanography (the study of the physical properties of seawater and its motion in waves, tides, and currents), marine chemistry, marine geology, and marine biology. Also known as oceanology. { ˌō·shə'näg·rə·fē }

oceanology *See* oceanography. { ˌō·shə'näl·ə·jē }

ocean weather station [METEOROL] As defined by the World Meteorological Organization, a specific maritime location occupied by a ship equipped and staffed to observe weather and sea conditions and report the observations by international exchange. { 'ō·shən 'weth·ər ˌstā·shən }

Ochoan [GEOL] A North American provincial series that is uppermost in the Permian, lying above the Guadalupian and below the lower Triassic. { ō'chō·ən }

Ochrept [GEOL] A suborder of the soil order Inceptisol, with horizon below the surface, lacking clay, sesquioxides, or humus; widely distributed, occurring from the margins of the tundra region through the temperate zone, but not into the tropics. { 'ō·krept }

octahedrite [GEOL] The most common iron meteorite, containing 6–18% nickel in the metal phase and having intimate intergrowths lying parallel to the octahedral planes. { ˌäk·tə'hēˌdrīt }

offlap [GEOL] The successive lateral contraction extent of strata (in an upward sequence) due to their deposition in a shrinking sea or on the margin of a rising landmass. Also known as regressive overlap. { 'ȯfˌlap }

off-reef facies [GEOL] Facies of the inclined strata made up of reef detritus deposited along the seaward margin of a reef. { 'ȯf¦rēf 'fā·shēz }

offset [GEOL] **1.** The movement of an upcurrent part of a shore to a more seaward position than a downcurrent part. **2.** A spur from a mountain range. **3.** A level terrace on the side of a hill. **4.** The horizontal displacement component in a fault, measured parallel to the strike of the fault. Also known as normal horizontal separation. { 'ȯfˌset }

offset deposit [GEOL] A mineral deposit, especially of sulfides, formed partly by magmatic segregation and partly by hydrothermal solution and located near the source rock. { 'ȯfˌset di'päz·ət }

offset ridge [GEOL] A ridge consisting of resistant sedimentary rock that has been made discontinuous as a result of faulting. { 'ȯfˌset 'rij }

offset stream [GEOL] A stream displaced laterally or vertically by faulting. { 'ȯfˌset 'strēm }

offshore [GEOL] The comparatively flat zone of variable width extending from the outer margin of the shoreface to the edge of the continental shelf. { 'ȯf¦shȯr }

offshore bar *See* longshore bar. { 'ȯf¦shȯr 'bär }

offshore beach *See* barrier beach. { 'ȯf¦shȯr 'bēch }

offshore current [OCEANOGR] **1.** A prevailing nontidal current usually setting parallel to the shore outside the surf zone. **2.** Any current flowing away from shore. { 'ȯf¦shȯr 'kə·rənt }

offshore slope [GEOL] The frontal slope below the outer edge of an offshore terrace. { 'ȯf¦shȯr 'slōp }

offshore terrace [GEOL] A wave-built terrace in the offshore zone composed of gravel and coarse sand. { 'ȯf¦shȯr 'ter·əs }

offshore water [OCEANOGR] Water adjacent to land in which the physical properties are slightly influenced by continental conditions. { 'ȯf¦shȯr 'wȯd·ər }

offshore wind [METEOROL] Wind blowing from the land toward the sea. { 'ȯf¦shȯr 'wind }

ogive [GEOL] One of a periodically repeated series of dark, curved structures occurring down a glacier that resemble a pointed arch. { 'ōˌjīv }

oil *See* petroleum. { ȯil }

oil accumulation *See* oil pool. { 'ȯil əˌkyü·myəˌlā·shən }

oil column [GEOL] The difference in elevation between the highest and lowest portions of various producing zones of an oil-producing formation. { 'ȯil ˌkäl·əm }

oil-field brine [HYD] Connate waters, usually containing a high concentration of calcium and sodium salts and found during deep rock penetration by the drill. { 'ȯil ¦fēld ˌbrīn }

oil floor [GEOL] In a sedimentary basin, the depth below which there is no economic oil accumulation. { 'ȯil ˌflȯr }

oil pool [GEOL] An accumulation of petroleum locally confined by subsurface geologic features. Also known as oil accumulation; oil reservoir. { 'ȯil ˌpül }

oil reservoir *See* oil pool. { 'ȯil 'rez·əvˌwär }

oil-reservoir water *See* formation water. { 'ȯil ¦rez·əvˌwär ˌwȯd·ər }

oil rock [GEOL] A rock stratum containing oil. { 'ȯil ˌräk }

oil sand [GEOL] An unconsolidated, porous sand formation or sandstone containing or impregnated with petroleum or hydrocarbons. { 'ȯil ˌsand }

oil seep [GEOL] The emergence of liquid petroleum at the land surface as a result of slow migration from its buried source through minute pores or fissure networks. Also known as petroleum seep. { 'ȯil ˌsēp }

oil shale [GEOL] A finely layered brown or black shale that contains kerogen and from which liquid or gaseous hydrocarbons can be distilled. Also known as kerogen shale. { 'ȯil ˌshāl }

oil trap [GEOL] An accumulation of petroleum which, by a combination of physical conditions, is prevented from escaping laterally or vertically. Also known as trap. { 'ȯil ˌtrap }

Oiluvium *See* Pleistocene. { ȯi'lü·vē·əm }

oil-water contact *See* oil-water surface. { 'ȯil ¦wȯd·ər 'känˌtakt }

oil-water interface *See* oil-water surface. { 'ȯil ¦wȯd·ər 'in·tərˌfās }

oil-water surface [GEOL] The datum of a two-dimensional oil-water interface. Also known as oil-water contact; oil-water interface. { 'ȯil ¦wȯd·ər 'sər·fəs }

oil zone [GEOL] The formation or horizon from which oil is produced, usually immediately under the gas zone and above the water zone if all three fluids are present and segregated. { 'ȯil ˌzōn }

old age [GEOL] The last stage of the erosion cycle in the development of the topography of a region in which erosion has reduced the surface almost to base level and the land forms are marked by simplicity of form and subdued relief. Also known as topographic old age. { 'ōld 'āj }

old ice [OCEANOGR] Floating sea ice that is more than 2 years old. { 'ōld 'īs }

old lake [GEOL] **1.** A lake in an advanced stage of filling by sediments. **2.** A eutrophic or dystrophic lake. **3.** A lake whose shoreline exhibits an advanced stage of development. { 'ōld ¦lāk }

oldland [GEOL] **1.** An extensive area (as the Canadian Shield) of ancient crystalline rocks reduced to low relief by long, continuous erosion from which the materials of later sedimentary rocks were derived. **2.** A region of older land, projected above sea level behind a coastal plain, that supplied the material of which the coastal-plain strata were formed. { 'ōldˌland }

old mountain [GEOL] A mountain that was formed before the beginning of the Tertiary Period. { 'ōld ¦maůnt·ən }

Old Red Sandstone [GEOL] A Devonian formation in the United Kingdom and northwestern Europe, of nonmarine, predominantly red sedimentary rocks, consisting principally of sandstone, conglomerates, and shales. { 'ōld 'red 'sanˌstōn }

old snow [HYD] Deposited snow in which the original crystalline forms are no longer recognizable, such as firn or spring snow. Also known as firn snow. { 'ōld ¦snō }

old wives' summer [METEOROL] A period of calm, clear weather, with cold nights and misty mornings but fine warm days, which sets in over central Europe toward the end of September; comparable to Indian summer. { 'ōld ˌwīvz 'səm·ər }

Oligocene [GEOL] The third oldest of the seven geological epochs of the Cenozoic Era, beginning 34 million years ago and ending 24 million years ago. It corresponds to an interval of geological time (and rocks deposited during that time) from the close of the Eocene Epoch to the beginning of the Miocene Epoch. { ə'lig·əˌsēn }

oligomictic [HYD] Pertaining to a lake that circulates only at rare, irregular intervals during abnormal cold spells. { əˌlig·ə'mik·tik }

oligopelic [GEOL] Property of a lake bottom deposit which contains very little clay. { əˌlig·ə'pel·ik }

oligotrophic [HYD] Of a lake, lacking plant nutrients and usually containing plentiful amounts of dissolved oxygen without marked stratification. { ¦äl·ə·gō¦träf·ik }

olistolith [GEOL] An exotic block or other rock mass that has been transported by submarine gravity sliding or slumping and is included in the binder of an olistostrome. { ə'lis·təˌlith }

olistostrome [GEOL] A sedimentary deposit composed of a chaotic mass of heterogeneous material that is intimately mixed; accumulated in the form of a semifluid body by submarine gravity sliding or slumping of unconsolidated sediments. { ə'lis·təˌstrōm }

olivine-bronzite chondrite [GEOL] A type of chondritic meteorite that contains about equal amounts of olivine and bronzite. { 'äl·ə,vēn 'brän,zīt 'kän,drīt }

olivine-hypersthene chondrite [GEOL] A type of chondritic meteorite generally containing more olivine than hypersthene; the hypersthene contains 12–20% iron, giving the meteorite a relatively dark color, and the metal grains usually contain 7–12% nickel. { 'äl·ə,vēn 'hī·pər,sthēn 'kän,drīt }

olivine-pigeonite chondrite [GEOL] A type of chondritic meteorite in which olivine is the predominant mineral and pigeonite is secondary, and metal inclusions are usually rich in nickel. { 'äl·ə,vēn 'pij·ə,nīt 'kän,drīt }

omission [GEOL] The elimination or nonexposure of certain stratigraphic beds at the surface of any specified section because of disruption and displacement of the beds by faulting. { ō'mish·ən }

oncolite [GEOL] A small, variously shaped (often spheroidal), concentrically laminated, calcareous sedimentary structure resembling an oolith; formed by accretion of successive, layered masses of gelatinous sheaths of blue-green algae. { 'äŋ·kō,līt }

Onesquethawan [GEOL] A North American stage in the Lower and Middle Devonian, lying above the Deerparkian and below the Cazenovian. { ,än·ə'skweth·ə,wän }

one-year ice [OCEANOGR] Sea ice formed the previous season, not yet 1 year old. { 'wən ,yir 'īs }

onionskin weathering [GEOL] A type of spheroidal weathering in which successive shells of decayed rock resembling the layers of an onion are produced. { 'ən,yən,skin 'weth·ə·riŋ }

onlap [GEOL] A type of overlap characterized by regular and progressive pinching out of the strata toward the margins of a depositional basin; each unit transgresses and extends beyond the point of reference of the underlying unit. Also known as transgressive overlap. { 'ȯn,lap }

onshore [GEOGR] Pertaining to, in the direction toward, or located on the shore. Also known as shoreside. { 'ȯn,shȯr }

onshore wind [METEOROL] Wind blowing from the sea toward the land. { 'ȯn,shȯr ,wind }

ooze [GEOL] **1.** A soft, muddy piece of ground, such as a bog, usually resulting from the flow of a spring or brook. **2.** A marine pelagic sediment composed of at least 30% skeletal remains of pelagic organisms, the rest being clay minerals. **3.** Soft mud or slime, typically covering the bottom of a lake or river. { üz }

opacus [METEOROL] A variety of cloud (sheet, layer, or patch), the greater part of which is sufficiently dense to obscure the sun; found in the genera altocumulus, altostratus, stratocumulus, and stratus; cumulus and cumulonimbus clouds are inherently opaque. { ō'pā·kəs }

opalized wood *See* silicified wood. { 'ō·pə,līzd 'wu̇d }

opaque attritus [GEOL] Attritus that does not contain large quantities of transparent humic degradation matter. { ō'pāk ə'trīd·əs }

opaque sky cover [METEOROL] In United States weather observing practice, the amount (in tenths) of sky cover that completely hides all that might be above it; opposed to transparent sky cover. { ō'pāk 'skī ,kəv·ər }

open bay [GEOGR] An indentation between two capes or headlands which is so broad and open that waves coming directly into it are nearly as high near its center as they are in adjacent parts of the open sea. { 'ō·pən 'bā }

open coast [GEOGR] A coast that is not sheltered from the sea. { 'ō·pən 'kōst }

open fault [GEOL] A fault, or section of a fault, whose two walls have become separated along the fault surface. { 'ō·pən 'fȯlt }

open fold [GEOL] A fold having only moderately compressed limbs. { 'ō·pən ,fōld }

open harbor [GEOGR] An unsheltered harbor exposed to the sea. { 'ō·pən 'här·bər }

open ice [OCEANOGR] On navigable waters, ice that has broken apart sufficiently to permit passage of vessels. { 'ō·pən 'īs }

opening [GEOGR] A break in a coastline or a passage between shoals, and so forth. [OCEANOGR] Any break in sea ice which reveals the water. { 'ōp·ə·niŋ }

open lake [HYD] **1.** A lake that has a stream flowing out of it. **2.** A lake whose water is free of ice or emergent vegetation. { 'ō·pən 'lāk }

open pack ice [OCEANOGR] Floes of sea ice that are seldom in contact with each other, generally covering between four-tenths and six-tenths of the sea surface. { 'ō·pən 'pak ˌīs }

open rock [GEOL] Any stratum sufficiently open or porous to contain a significant amount of water or to convey water along its bed. { 'ō·pən ¦räk }

open sand [GEOL] A formation of sandstone that has porosity and permeability sufficient to provide good storage for oil. { 'ō·pən ¦sand }

open sea [GEOGR] **1.** That part of the ocean not enclosed by headlands, not within narrow straits, and so on. **2.** That part of the ocean outside the territorial jurisdiction of any country. { 'ō·pən 'sē }

open-space structure [GEOL] A structure in a carbonate sedimentary rock formed by a partial or complete occupation by internal sediments or cement. { 'ō·pən ˌspās 'strək·chər }

operational unit [GEOL] An arbitrary stratigraphic unit that is distinguished by objective criteria for some practical purpose. Also known as parastratigraphic unit. { ˌäp·ə'rā·shən·əl 'yü·nət }

open system [HYD] A condition of freezing of the ground in which additional groundwater is available either through free percolation or through capillary movement. { 'ō·pən 'sis·təm }

open water [HYD] Lake water that does not freeze during the winter. [OCEANOGR] Water less than one-tenth covered with floating ice. { 'ō·pən 'wȯd·ər }

operational weather limits [METEOROL] The limiting values of ceiling, visibility, and wind, or runway visual range, established as safety minima for aircraft landings and takeoffs. { ˌäp·ə'rā·shən·əl 'weth·ər ˌlim·əts }

opposing wind [OCEANOGR] In wave forecasting, a wind blowing in opposition to the direction that the waves are traveling. { ə'pōz·iŋ 'wind }

opposite tide [OCEANOGR] A high tide at a corresponding place on the opposite side of the earth which accompanies a direct tide. { 'äp·ə·zət 'tīd }

optical air mass [GEOPHYS] A measure of the length of the path through the atmosphere to sea level traversed by light rays from a celestial body, expressed as a multiple of the path length for a light source at the zenith. { 'äp·tə·kəl 'er ˌmas }

optical depth *See* optical thickness. { 'äp·tə·kəl 'depth }

optical horizon [GEOD] Locus of points at which a straight line from the given point becomes tangential to the earth's surface. { 'äp·tə·kəl hə'rīz·ən }

optically effective atmosphere [GEOPHYS] That portion of the atmosphere lying below the altitude (30–36 miles or 50–60 kilometers) from which scattered light at twilight still reaches the observer with sufficient intensity to be discerned. Also known as effective atmosphere. { 'äp·tə·klē i'fek·tiv 'at·məˌsfir }

optical oceanography [OCEANOGR] That aspect of physical oceanography which deals with the optical properties of sea water and natural light in sea water. { 'äp·tə·kəl ˌō·shə'näg·rə·fē }

optical thickness [METEOROL] **1.** In calculations of the transfer of radiant energy, the mass of a given absorbing or emitting material lying in a vertical column of unit cross-sectional area and extending between two specified levels. Also known as optical depth. **2.** Subjectively, the degree to which a cloud prevents light from passing through it; depends upon the physical constitution (crystals, drops, droplets), the form, the concentration of particles, and the vertical extent of the cloud. { 'äp·tə·kəl 'thik·nəs }

optimum moisture content [GEOL] The water content at which a specified compactive force can compact a soil mass to its maximum dry unit weight. { 'äp·tə·məm 'mȯis·chər ˌkänˌtent }

ora [METEOROL] A regular valley wind at Lake Garda in Italy. { 'ȯr·ə }

orbicule [GEOL] A nearly spherical body, up to 2 centimeters (0.8 inch) or more in diameter, in which the components are arranged in concentric layers. { 'ȯr·bəˌkyül }

orbit [OCEANOGR] The path of a water particle affected by wave motion; it is almost circular in deep-water waves and almost elliptical in shallow-water waves. { 'ȯr·bət }
orbital current [OCEANOGR] The flow of water which follows the orbital motion of water particles in a wave. { 'ȯr·bəd·əl 'kə·rənt }
ordinary tides [OCEANOGR] Tides which have cycles of 12 to 24 hours. { 'ȯrd·ən,er·ē 'tīdz }
ordinary-wave component [GEOPHYS] One of the two components into which an electromagnetic wave entering the ionosphere is divided under the influence of the earth's magnetic field; it has characteristics more nearly like those expected in the absence of a magnetic field. Also known as O-wave component. { 'ȯrd·ən,er·ē ¦wāv kəm,pō·nənt }
Ordovician [GEOL] The second period of the Paleozoic era, above the Cambrian and below the Silurian, from approximately 500 million to 440 million years ago. { ,ȯrd·ə'vish·ən }
ore [GEOL] **1.** The naturally occurring material from which economically valuable minerals can be extracted. **2.** Specifically, a natural mineral compound of the elements, of which one element at least is a metal. **3.** More loosely, all metalliferous rock, though it contains the metal in a free state. **4.** Occasionally, a compound of nonmetallic substances, as sulfur ore. { ȯr }
ore bed [GEOL] An economic aggregation of minerals occurring between or in rocks of sedimentary origin. { 'ȯr ,bed }
orebody [GEOL] Generally, a solid and fairly continuous mass of ore, which may include low-grade ore and waste as well as pay ore, but is individualized by form or character from adjoining country rock. { 'ȯr,bäd·ē }
ore cluster [GEOL] A group of interconnected ore bodies. { 'ȯr ,kləs·tər }
ore control [GEOL] A geologic feature that has influenced the ore deposition. { 'ȯr ,kən'trōl }
ore deposit [GEOL] Rocks containing minerals of economic value in such amount that they can be profitably exploited. { 'ȯr di,päz·ət }
ore district [GEOL] A combination of several ore deposits into one common whole or system. { 'ȯr ,dis,trikt }
ore-lead age [GEOL] An estimate of the age of the earth made by comparing the relative progress of the two radioactive decay schemes ^{235}U-^{207}Pb and ^{238}U-^{206}Pb. { 'ȯr 'led ,āj }
ore of sedimentation *See* placer. { 'ȯr əv ,sed·ə·mən'tā·shən }
ore shoot [GEOL] **1.** A large, generally vertical, pipelike ore body that is economically valuable. Also known as shoot. **2.** A large and usually rich aggregation of mineral in a vein. { 'ȯr ,shüt }
organic geochemistry [GEOCHEM] A branch of geochemistry which deals with naturally occurring carbonaceous and biologically derived substances which are of geological interest. { ȯr'gan·ik ,jē·ō'kem·ə·strē }
organic lattice *See* growth lattice. { ȯr'gan·ik 'lad·əs }
organic mound *See* bioherm. { ȯr'gan·ik 'mau̇nd }
organic reef [GEOL] A sedimentary rock structure of significant dimensions erected by, and composed almost exclusively of the remains of, corals, algae, bryozoans, sponges, and other sedentary or colonial organisms. { ȯr'gan·ik 'rēf }
organic soil [GEOL] Any soil or soil horizon consisting chiefly of, or containing at least 30% of, organic matter; examples are peat soils and muck soils. { ȯr'gan·ik 'sȯil }
organic texture [GEOL] A sedimentary texture resulting from the activity of organisms such as the secretion of skeletal material. { ȯr'gan·ik 'teks·chər }
organic weathering [GEOL] Biological processes and changes that contribute to the breakdown of rocks. Also known as biological weathering. { ȯr'gan·ik 'weth·ə·riŋ }
organogenic [GEOL] Property of a rock or sediment derived from organic substances. { ȯr¦gan·ə¦jen·ik }
organolite [GEOL] Any rock consisting mainly of organic material. { ȯr'gan·ə,līt }
orientation diagram [GEOL] Any point or contour diagram used in structural petrology. { ,ȯr·ē·ən'tā·shən ,dī·ə,gram }

oriented [GEOL] Pertaining to a specimen that is so marked as to show its exact, original position in space. { 'ȯr·ē,ent·əd }

original dip *See* primary dip. { ə'rij·ən·əl 'dip }

original valley [GEOL] A valley formed by hypogene action or by epigene action other than that of running water. { ə'rij·ən·əl 'val·ē }

orocline [GEOL] An orogenic belt with a change in horizontal direction, either a horizontal curvature or a sharp bend. Also known as geoflex. { 'ȯr·ə,klīn }

orocratic [GEOL] Pertaining to a period of time in which there is much diastrophism. { ¦ȯr·ə¦krad·ik }

orogen *See* orogenic belt. { 'ȯr·ə·jən }

orogene *See* orogenic belt. { 'ȯr·ə,jēn }

orogenesis *See* orogeny. { ,ȯr·ə'jen·ə·səs }

orogenic belt [GEOL] A linear region that has undergone folding or other deformation during the orogenic cycle. Also known as fold belt; orogen; orogene. { ¦ȯr·ə¦jen·ik 'belt }

orogenic cycle [GEOL] A time interval during which a mobile belt evolved into an orogenic belt, passing through preorogenic, orogenic, and postorogenic stages. Also known as geotectonic cycle. { ¦ȯr·ə¦jen·ik 'sī·kəl }

orogenic sediment [GEOL] Any sediment that is produced as the result of an orogeny or that is directly attributable to the orogenic region in which it is later found. { ¦ȯr·ə¦jen·ik 'sed·ə·mənt }

orogenic unconformity [GEOL] An angular unconformity produced locally in a region affected by mountain-building movements. { ¦ȯr·ə¦jen·ik ,ən·kən'fȯr·məd·ē }

orogeny [GEOL] The process or processes of mountain formation, especially the intense deformation of rocks by folding and faulting which, in many mountainous regions, has been accompanied by metamorphism, invasion of molten rock, and volcanic eruption; in modern usage, orogeny produces the internal structure of mountains, and epeirogeny produces the mountainous topography. Also known as orogenesis; tectogenesis. { ȯ'räj·ə·nē }

orogeosyncline [GEOL] A geosyncline that later became an area of orogeny. { ¦ȯr·ō¦jē·ō'sin,klīn }

orographic [GEOL] Pertaining to mountains, especially in regard to their location and distribution. { ¦ȯr·ə¦graf·ik }

orographic cloud [METEOROL] A cloud whose form and extent is determined by the disturbing effects of orography upon the passing flow of air; because these clouds are linked with the form of the terrestrial relief, they generally move very slowly, if at all, although the winds at the same level may be very strong. { ¦ȯr·ə¦graf·ik 'klau̇d }

orographic lifting [METEOROL] The lifting of an air current caused by its passage up and over surface elevations. { ¦ȯr·ə¦graf·ik 'lift·iŋ }

orographic occlusion [METEOROL] An occluded front in which the occlusion process has been hastened by the retardation of the warm front along the windward slopes of a mountain range. { ¦ȯr·ə¦graf·ik ə'klü·zhən }

orographic precipitation [METEOROL] Precipitation which results from the lifting of moist air over an orographic barrier such as a mountain range; strictly, the amount so designated should not include that part of the precipitation which would be expected from the dynamics of the associated weather disturbance, if the disturbance were over flat terrain. { ¦ȯr·ə¦graf·ik prə,sip·ə'tā·shən }

orography [GEOGR] The branch of geography dealing with mountains. { ȯ'räg·rə·fē }

orohydrography [HYD] A branch of hydrography dealing with the relations of mountains to drainage. { ¦ȯr·ō·hī'dräg·rə·fē }

orotath [GEOL] An orogenic belt that has been stretched substantially in a lengthwise direction. { 'ȯr·ə,tath }

Orthent [GEOL] A suborder of the soil order Entisol, well drained and of medium or fine texture, usually shallow to bedrock and lacking evidence of horizonation; occurs mostly on strong slopes. { 'ȯr·thənt }

Orthid [GEOL] A suborder of the soil order Aridisol, mostly well drained, gray or brownish-gray with little change from top to bottom of the soil profile; occupies younger, but not the youngest, land surfaces in deserts. { 'ȯr·thəd }

orthobituminous coal [GEOL] Bituminous coal that contains 87–89% carbon, analyzed on a dry, ash-free basis. { ¦ȯr·thō·bə'tü·mən·əs 'kōl }

orthochem [GEOCHEM] A precipitate formed within a depositional basin or within the sediment itself by direct chemical action. { 'ȯr·thə,kem }

orthochronology [GEOL] Geochronology based on a standard succession of biostratigraphically significant faunas or floras, or based on irreversible evolutionary processes. { ,ȯr·thə·krə'näl·ə·jē }

orthoconglomerate [GEOL] A conglomerate with an intact gravel framework held together by mineral cement and deposited by ordinary water currents. { ,ȯr·thə·kən'gläm·ə·rət }

Orthod [GEOL] A suborder of the soil order Spodosol having accumulations of humus, aluminum, and iron; widespread in Canada and the former Soviet Union. { 'ȯr,thäd }

orthogeosyncline [GEOL] A linear geosynclinal belt lying between continental and oceanic cratons, and having internal volcanic belts (eugeosynclinal) and external nonvolcanic belts (miogeosynclinal). Also known as geosynclinal couple; primary geosyncline. { ¦ȯr·thō,jē·ə'sin,klīn }

orthogneiss [GEOL] Gneiss originating from igneous rock. { 'ōr·thə,nīs }

orthohydrous coal [GEOL] Coal that contains 5–6% hydrogen, analyzed on a dry, ash-free basis. { ¦ȯr·thə¦hī·drəs 'kōl }

ortholignitous coal [GEOL] Coal that contains 75–80% carbon, analyzed on a dry, ash-free basis. { ¦ȯr·thō·lig'nīd·əs 'kōl }

orthomagmatic stage [GEOL] The principal stage in the crystallization of silicates from a typical magma; up to 90% of the magma may crystallize during this stage. Also known as orthotectic stage. { ¦ȯr·thō,mag'mad·ik 'stāj }

orthophotograph [GEOL] A photographic copy, prepared from a photograph formed by a perspective projection, in which the displacements due to tilt and relief have been removed. { ,ȯr·thə'fōd·ə,graf }

orthoquartzitic conglomerate [GEOL] A lithologically homogeneous, light-colored orthoconglomerate composed of quartzose residues that is commonly interbedded with pure quartz sandstone. Also known as quartz-pebble conglomerate. { ¦ȯr·thə·kwȯrt¦sid·ik kən'gläm·ə·rət }

orthostratigraphy [GEOL] Standard stratigraphy based on fossils which identify recognized biostratigraphic zones. { ¦ȯr·thō·strə'tig·rə·fē }

orthotectic stage *See* orthomagmatic stage. { ¦ȯr·thə'tek·tik ,stāj }

orthotill [GEOL] A till formed by immediate release of material from transported ice, such as by ablation and melting. { 'ȯr·thə,til }

Orthox [GEOL] A suborder of the soil order Oxisol that is moderate to low in organic matter, well drained, and moist all or nearly all year; believed to be extensive at low altitudes in the heart of the humid tropics. { 'ȯr,thäks }

Osagean [GEOL] A provincial series of geologic time in North America; Lower Mississippian (above Kinderhookian, below Meramecian). { ō'sā·jē·ən }

osar *See* esker. { 'ō,sär }

oscillation ripple *See* oscillation ripple mark. { ,äs·ə'lā·shən ,rip·əl }

oscillation ripple mark [GEOL] A symmetric ripple mark having a sharp, narrow, and relatively straight crest between broadly rounded troughs, formed by the motion of water agitated by oscillatory waves on a sandy base at a depth shallower than wave base. Also known as oscillation ripple; oscillatory ripple mark; wave ripple mark. { ,äs·ə'lā·shən 'rip·əl ,märk }

oscillatory ripple mark *See* oscillation ripple mark. { 'äs·ə·lə,tȯr·ē 'rip·əl ,märk }

Osos wind [METEOROL] In California, a strong northwest wind blowing from the Loa Osos valley to the San Luis valley. { 'ō,sōs ¦wind }

ostria [METEOROL] A warm southerly wind on the Bulgarian coast; it is considered a precursor of bad weather. Also known as auster. { 'äs·trē·ə }

ouari [METEOROL] A south wind of Somaliland, Africa; it is similar to the khamsin. { 'wä·rē }

outburst [METEOROL] Outflow from a convective event originating in cool descending air, often associated with a thunderstorm. { 'au̇t,bərst }

outcrop [GEOL] Exposed stratum or body of ore at the surface of the earth. Also known as cropout. { 'au̇t,kräp }

outcrop curvature *See* settling. { 'au̇t,kräp 'kər·və·chər }

outcrop map [GEOL] A type of geologic map that shows the distribution and shape of actual outcrops, leaving those areas without outcrops blank. { 'au̇t,kräp ,map }

outcrop water [HYD] Rain and surface water which seeps downward through outcrops of porous and fissured rock, fault planes, old shafts, or surface drifts. { 'au̇t,kräp ,wȯd·ər }

outer atmosphere [METEOROL] Very generally, the atmosphere at a great distance from the earth's surface; possibly best usage of the term is as an approximate synonym for exosphere. { 'au̇d·ər 'at·mə,sfir }

outer bar [GEOL] A bar formed at the mouth of an ebb channel of an estuary. { 'au̇d·ər 'bär }

outer beach [GEOL] The part of a beach that is ordinarily dry and reached only by the waves generated by a violent storm. { 'au̇d·ər 'bēch }

outer core [GEOL] The outer or upper zone of the earth's core, extending to a depth of 3160 miles (5100 kilometers), and including the transition zone. { 'au̇d·ər 'kȯr }

outer harbor [GEOGR] The part of a harbor toward the sea, through which a vessel enters the inner harbor. { 'au̇d·ər 'här·bər }

outer mantle *See* upper mantle. { 'au̇d·ər 'mant·əl }

outface *See* dip slope. { 'au̇t,fās }

outfall [HYD] The narrow part of a stream, lake, or other body of water where it drops away into a larger body. { 'au̇t,fȯl }

outflow cave [GEOL] A cave from which a stream issues or is known to have issued. { 'au̇t,flō ,kāv }

outlet glacier [HYD] A stream of ice from an ice cap to the sea. { 'au̇t,let ,glā·shər }

outlet head [HYD] The place where water leaves a lake and enters an effluent. { 'au̇t ,let ,hed }

outlier [GEOL] A group of rocks separated from the main mass and surrounded by outcrops of older rocks. { 'au̇t,lī·ər }

outside air temperature *See* indicated air temperature. { 'au̇t,sīd 'er ,tem·prə·chər }

outwash [GEOL] **1.** Sand and gravel transported away from a glacier by streams of meltwater and either deposited as a floodplain along a preexisting valley bottom or broadcast over a preexisting plain in a form similar to an alluvial fan. Also known as glacial outwash; outwash drift; overwash. **2.** Soil material washed down a hillside by rainwater and deposited on more gently sloping land. { 'au̇t,wäsh }

outwash apron *See* outwash plain. { 'au̇t,wäsh ,ā·prən }

outwash cone [GEOL] A cone-shaped deposit consisting chiefly of sand and gravel found at the edge of shrinking glaciers and ice sheets. { 'au̇t,wäsh ,kōn }

outwash drift *See* outwash. { 'au̇t,wäsh ,drift }

outwash fan [GEOL] A fan-shaped accumulation of outwash deposited by meltwater streams in front of the terminal moraine of a glacier. { 'au̇t,wäsh ,fan }

outwash plain [GEOL] A broad, outspread flat or gently sloping alluvial deposit of outwash in front of or beyond the terminal moraine of a glacier. Also known as apron; frontal apron; frontal plain; marginal plain; morainal apron; morainal plain; outwash apron; overwash plain; sandur; wash plain. { 'au̇t,wäsh ,plān }

outwash terrace [GEOL] A dissected and incised valley train or benchlike deposit extending along a valley downstream from an outwash plain or terminal moraine. { 'au̇t,wäsh ,ter·əs }

outwash train *See* valley train. { 'au̇t,wäsh ,trān }

oven [GEOL] **1.** A rounded, saclike, chemically weathered pit or hollow in a rock (especially a granitic rock) which has an arched roof and resembles an oven. **2.** *See* spouting horn. { 'əv·ən }

overbank deposit [GEOL] Fine-grained sediment (silt and clay) deposited from suspension on a floodplain by floodwaters from a stream channel. { ¦ō·vər¦baŋk di,päz·ət }

overbank stage [HYD] The height of the surface of a river as the river floods over its banks. { 'ō·vər,baŋk ,stāj }

overburden [GEOL] **1.** Rock material overlying a mineral deposit or coal seam. Also known as baring; top. **2.** Material of any nature, consolidated or unconsolidated, that overlies a deposit of useful materials, ores, or coal, especially those deposits that are mined from the surface by open cuts. **3.** Loose soil, sand, or gravel that lies above the bedrock. { 'ō·vər,bərd·ən }

overburdened stream *See* overloaded stream. { 'ō·vər,bərd·ənd 'strēm }

overcast [METEOROL] **1.** Pertaining to a sky cover of 1.0 (95% or more) when at least a portion of this amount is attributable to clouds or obscuring phenomena aloft, that is, when the total sky cover is not due entirely to surface-based obscuring phenomena. **2.** Cloud layer that covers most or all of the sky; generally, a widespread layer of clouds such as that which is considered typical of a warm front. { 'ō·vər,kast }

overconsolidation [GEOL] Consolidation of sedimentary material exceeding that which is normal for the existing overburden. { ¦ō·vər·kən,säl·ə'dā·shən }

overdeepening [GEOL] The erosive process by which a glacier deepens and widens an inherited preglacial valley to below the level of the subglacial surface. { ¦ō·vər'dēp·ə·niŋ }

overfalls [OCEANOGR] Short, breaking waves occurring when a strong current passes over a shoal or other submarine obstruction or meets a contrary current or wind. { 'ō·vər,fȯlz }

overflow channel [GEOL] A channel or notch cut by the overflow waters of a lake, especially the channel draining meltwater from a glacially dammed lake. { 'ō·vər,flō ,chan·əl }

overflow ice [HYD] Ice formed during high spring tides by water rising through cracks in the surface ice and then freezing. { 'ō·vər,flō ,īs }

overflow spring [HYD] A type of contact spring that develops where a permeable deposit dips beneath an impermeable mantle. { 'ō·vər,flō ,spriŋ }

overflow stream [HYD] **1.** A stream containing water that has overflowed the banks of a river or another stream. Also known as spill stream. **2.** An effluent from a lake, carrying water to a stream, a sea, or another lake. { 'ō·vər,flō ,strēm }

overfold [GEOL] A fold that is overturned. { 'ō·vər,fōld }

overhang [GEOL] The part of a salt plug that projects from the top. { 'ō·vər,haŋ }

overland flow [HYD] Water flowing over the ground surface toward a channel; upon reaching the channel, it is called surface runoff. Also known as surface flow. { 'ō·vər·lənd 'flō }

overlap [GEOL] **1.** Movement of an upcurrent part of a shore to a position extending seaward beyond a downcurrent part. **2.** Extension of strata over or beyond older underlying rocks. **3.** The horizontal component of separation measured parallel to the strike of a fault. { 'ō·vər,lap }

overlap fault [GEOL] A fault structure in which the displaced strata are doubled back upon themselves. { 'ō·vər,lap ,fȯlt }

overload [GEOL] The amount of sediment that exceeds the ability of a stream to transport it and is therefore deposited. { 'ō·vər,lōd }

overloaded stream [HYD] A stream so heavily loaded with sediment that its velocity is lessened and it is forced to deposit part of its load. Also known as overburdened stream. { ¦ō·vər¦lōd·əd 'strēm }

overprint [GEOCHEM] A complete or partial disturbance of an isolated radioactive system by thermal, igneous, or tectonic activities which results in loss or gain of radioactive or radiogenic isotopes and, hence, a change in the radiometric age that will be given the disturbed system. [GEOL] The development or superposition of metamorphic structures on original structures. Also known as imprint; metamorphic overprint; superprint. { 'ō·vər,print }

overrunning [METEOROL] A condition existing when an air mass is in motion aloft above another air mass of greater density at the surface; this term usually is applied in the case of warm air ascending the surface of a warm front or quasi-stationary front. { 'ō·və,rən·iŋ }

overseeding [METEOROL] Cloud seeding in which an excess of nucleating material is released; as the term is normally used, the excess is relative to that amount of nucleating material which would, theoretically, maximize the precipitation received at the ground. { ¦ō·vər¦sēd·iŋ }

oversteepening [GEOL] The process by which an eroding alpine glacier steepens the sides of an inherited preglacial valley. { ¦ō·vər'stēp·ə·niŋ }

overstep [GEOL] **1.** An overlap characterized by the regular truncation of older units of a complete sedimentary sequence by one or more later units of the sequence. **2.** A stratum deposited on the upturned edges of underlying strata. { 'ō·vər,step }

overthrust [GEOL] **1.** A thrust fault that has a low dip or a net slip that is large. Also known as low-angle thrust; overthrust fault. **2.** A thrust fault with the active element being the hanging wall. { 'ō·vər,thrəst }

overthrust black *See* overthrust nappe. { 'ō·vər,thrəst ,blak }

overthrust fault *See* overthrust. { 'ō·vər,thrəst ,fȯlt }

overthrust nappe [GEOL] The body of rock making up the hanging wall of a large-scale overthrust. Also known as overthrust block; overthrust sheet; overthrust slice. { 'ō·vər,thrəst ,nap }

overthrust sheet *See* overthrust nappe. { 'ō·vər,thrəst ,shēt }

overthrust slice *See* overthrust nappe. { 'ō·vər,thrəst ,slīs }

overtide [OCEANOGR] A harmonic tidal component which has a speed that is an exact multiple of the speed of one development of the tide-producing force. { 'ō·vər,tīd }

overturn [HYD] Renewal of bottom water in lakes and ponds in regions where winter temperatures are cold; in the fall, cooled surface waters become denser and sink, until the whole body of water is at 4°C; in the spring, the surface is warmed back to 4°C, and the lake is homothermous. Also known as convective overturn. { 'ō·vər,tərn }

overturned [GEOL] Of a fold or the side of a fold, tilted beyond the perpendicular. Also known as inverted; reversed. { 'ō·vər,tərnd }

overwash [GEOL] **1.** A mass of water representing the part of the wave advancing up a beach that runs over the highest part of the berm (or other structure) and that does not flow directly back to the sea or lake. **2.** *See* outwash. { 'ō·vər,wäsh }

overwash mark [GEOL] A narrow, tonguelike ridge of sand formed by overwash on the landward side of a berm. { 'ō·vər,wäsh ,märk }

overwash plain *See* outwash plain. { 'ō·vər,wäsh ,plān }

overwash pool [OCEANOGR] A tidal pool between a berm and a beach scarp which water enters only at high tide. { 'ō·vər,wäsh pül }

O-wave component *See* ordinary-wave component. { 'ō ,wāv kəm,pō·nənt }

oxbow [GEOL] The abandoned, horseshoe-shaped channel of a former stream meander after the stream formed a neck cutoff. Also known as abandoned channel. [HYD] **1.** A closely looping, U-shaped stream meander whose curvature is so extreme that only a neck of land remains between the two parts of the stream. Also known as horseshoe bend. **2.** *See* oxbow lake. { 'äks,bō }

oxbow lake [HYD] The crescent-shaped body of water located alongside a stream in an abandoned oxbow after a neck cutoff is formed and the ends of the original bends are silted up. Also known as crescentic lake; cutoff lake; horseshoe lake; loop lake; moat; mortlake; oxbow. { 'äks,bō ¦lāk }

Oxfordian [GEOL] A European stage of geologic time, in the Upper Jurassic (above Callovian, below Kimmeridgean). Also known as Divesian. { äks'fȯr·dē·ən }

oxidate [GEOL] A sediment made up of iron and manganese oxides and hydroxides crystallized from aqueous solution. { 'äk·sə,dāt }

oxidite *See* shale ball. { 'äk·sə,dīt }

oxidized zone [GEOL] A region of mineral deposits which has been altered by oxidizing surface waters. { 'äk·sə,dīzd ,zōn }

Oxisol [GEOL] A soil order characterized by residual accumulations of inactive clays, free oxides, kaolin, and quartz; mostly tropical. { 'äk·sə,sȯl }

oxoferrite [GEOL] A variety of naturally occurring iron with some ferrous oxide in solid solution. { ¦äk·sō'fe,rīt }

oxygen deficit [GEOCHEM] The difference between the actual amount of dissolved oxygen in lake or sea water and the saturation concentration at the temperature of the water mass sampled. { 'äk·sə·jən ,def·ə·sət }

oxygen distribution [OCEANOGR] The concentration of dissolved oxygen in ocean water as a function of depth, ranging from as much as 5 milliliters of oxygen per liter at the surface to a fraction of that value at great depths. { 'äk·sə·jən ,dis·trə'byü·shən }

oxygen isotope fractionation [GEOCHEM] The use of temperature-dependent variations of the oxygen-18/oxygen-16 ratio in the carbonate shells of marine organisms, to measure water temperature at the time of deposition. { 'äk·sə·jən 'īs·ə,tōp ,frak·shə'nā·shən }

oxygen minimum layer [HYD] A subsurface layer in which the content of dissolved oxygen is very low (or absent), lower than in the layers above and below. { 'äk·sə·jən 'min·ə·məm 'lā·ər }

oxygen ratio *See* acidity coefficient. { 'äk·sə·jən ,rā·shō }

oxyphile *See* lithophile. { 'äk·sə,fīl }

oxysphere *See* lithosphere. { 'äk·sə,sfir }

Oyashio [OCEANOGR] A cold current flowing from the Bering Sea southwest along the coast of Kamchatka, past the Kuril Islands, continuing close to the northeast coast of Japan, and reaching nearly 35°N. { ō'yā·shē·ō }

ozocerite [GEOL] A natural, brown to jet black paraffin wax occurring in irregular veins; consists principally of hydrocarbons, is soluble in water, and has a variable melting point. Also known as ader wax; earth wax; fossil wax; mineral wax; native paraffin; ozokerite. { ō'zäs·ə,rīt }

ozokerite *See* ozocerite. { ō'zäk·ə,rīt }

ozone hole *See* Antarctic ozone hole. { 'ō,zōn ,hōl }

ozone layer *See* stratospheric ozone. { 'ō,zōn ,lā·ər }

ozonesonde [METEOROL] A balloon-borne instrument for measuring the ozone concentration at various altitudes and transmitting the data by radio. { 'ō,zōn,sänd }

ozonosphere [METEOROL] The general stratum of the upper atmosphere in which there is an appreciable ozone concentration and in which ozone plays an important part in the radiative balance of the atmosphere; lies roughly between 6 and 30 miles (10 and 50 kilometers), with maximum ozone concentration at about 12 to 15 miles (20 to 25 kilometers). Also known as ozone layer. { ō'zō·nə,sfir }

P

paar [GEOL] A depression produced by the moving apart of crustal blocks rather than by subsidence within a crustal block. { pär }

pachoidal structure *See* flaser structure. { pə'kȯid·əl ˌstrək·chər }

Pacific anticyclone *See* Pacific high. { pə'sif·ik ˌant·i'sīˌklōn }

Pacific Equatorial Countercurrent [OCEANOGR] The Equatorial Countercurrent flowing east across the Pacific Ocean between 3° and 10°N. { pə'sif·ik ˌek·wə'tȯr·ē·əl 'kaůnt·ərˌkə·rənt }

Pacific high [METEOROL] The nearly permanent subtropical high of the North Pacific Ocean, centered, in the mean, at 30–40°N and 140–150°W. Also known as Pacific anticyclone. { pə'sif·ik 'hī }

Pacific North Equatorial Current [OCEANOGR] The North Equatorial Current which flows westward between 10° and 20°N in the Pacific Ocean. { pə'sif·ik 'nȯrth ˌek·wə'tȯr·ē·əl 'kə·rənt }

Pacific Ocean [GEOGR] The largest division of the hydrosphere, having an area of 63,690 square miles (165,000,000 square kilometers) and covering 46% of the surface of the total extent of the oceans and seas; it is bounded by Asia and Australia on the west and North and South America on the east. { pə'sif·ik 'ō·shən }

Pacific South Equatorial Current [OCEANOGR] The South Equatorial Current flowing westward between 3°N and 10°S in the Pacific Ocean. { pə'sif·ik 'saůth ˌek·wə'tȯr·ē·əl 'kə·rənt }

Pacific-type continental margin [GEOL] A continental margin typified by that of the western Pacific where oceanic lithosphere descends beneath an adjacent continent and produces an intervening island arc system. { pə'sif·ik ˌtīp ˌkänt·ən'ent·əl 'mär·jən }

pack *See* pack ice. { pak }

pack ice [OCEANOGR] Any area of sea ice, except fast ice, composed of a heterogeneous mixture of ice of varying ages and sizes, and formed by the packing together of pieces of floating ice. Also known as ice canopy; ice pack; pack. { 'pak ˌīs }

packing [GEOL] The arrangement of solid particles in a sediment or in sedimentary rock. { 'pak·iŋ }

packing density [GEOL] A measure of the extent to which the grains of a sedimentary rock occupy the gross volume of the rock in contrast to the spaces between the grains; equal to the cumulative grain-intercept length along a traverse in a thin section. { 'pak·iŋ ˌden·səd·ē }

packing proximity [GEOL] In a sedimentary rock, an estimate of the number of grains that are in contact with adjacent grains; equal to the total percentage of grain-to-grain contacts along a traverse measured on a thin section. { 'pak·iŋ präkˌsim·əd·ē }

paesa [METEOROL] A violent north-northeast wind of Lake Garda in Italy. { pī'ā·zə }

paesano [METEOROL] A northerly night breeze, blowing down from the mountains, of Lake Garda in Italy. { pī'zä·nō }

pagoda stone [GEOL] **1.** A Chinese limestone showing in section fossil orthoceratites arranged in pagodalike designs. **2.** An agate whose markings resemble pagodas. { pə'gōd·əˌstōn }

pagodite *See* agalmatolite. { 'pag·əˌdīt }

paha [GEOL] A low, elongated, rounded glacial ridge or hill which consists mainly of drift, rock, or windblown sand, silt, or clay but is capped with a thick cover of loess. { pä'hä }

pahoehoe [GEOL] A type of lava flow whose surface is glassy, smooth, and undulating; the lava is basaltic, glassy, and porous. Also known as ropy lava. { pə'hō·ē,hō·ē }

painted pot *See* mud pot. { ¦pānt·əd 'pät }

painter [METEOROL] A fog frequently experienced on the coast of Peru; the brownish deposit which it often leaves upon exposed surfaces is sometimes called Peruvian paint. Also known as Callao painter. { 'pānt·ər }

paint pot [GEOL] A mud pot containing multicolored mud. { 'pānt ,pät }

paired terrace [GEOL] One of two stream terraces that face each other at the same elevation from opposite sides of the stream valley and represent the remnants of the same floodplain or valley floor. Also known as matched terrace. { ¦perd 'ter·əs }

palagonite [GEOL] A brown to yellow altered basaltic glass found as interstitial material or amygdules in pillow lavas. { pə'lag·ə,nīt }

palasite [GEOL] The most abundant of the intermediate types of meteorites, consisting of olivine enclosed in a nickel-iron matrix. { 'pal·ə,sīt }

paleic surface [GEOL] A smooth, preglacial erosion surface. { pə'lē·ik 'sər·fəs }

paleobotanic province [GEOL] A large region defined by similar fossil floras. { ¦pāl·ē·ō·bə'tan·ik 'präv·əns }

paleoceanography [OCEANOGR] The study of the history of the circulation, chemistry, biogeography, fertility, and sedimentation of the oceans. { ¦pāl·ē·ō·shə'näg·rə·fē }

Paleocene [GEOL] The oldest of the seven geological epochs of the Cenozoic Era, spanning 65 million to 55 million years ago. Comprising the Tertiary and Quaternary periods in modern usage, it is also the oldest of the five epochs constituting the Tertiary Period. It represents an interval of geological time (and rocks deposited during that time) extending from the termination of the Cretaceous Period of the Mesozoic Era to the dawn of the Eocene Epoch. { 'pāl·ē·ə,sēn }

paleochannel [GEOL] A remnant of a stream channel cut in older rock and filled by the sediments of younger overlying rock. { ¦pāl·ē·ō¦chan·əl }

paleoclimate [GEOL] The climate of a given period of geologic time. Also known as geologic climate. { ¦pāl·ē·ō¦klī·mət }

paleoclimatic sequence [GEOL] The sequence of climatic changes in geologic time; it shows a succession of oscillations between warm periods and ice ages, but superimposed on this are numerous shorter oscillations. { ¦pāl·ē·ō·klə'mad·ik 'sē·kwəns }

paleoclimatology [GEOL] The study of climates in the geologic past, involving the interpretation of glacial deposits, fossils, and paleogeographic, isotopic, and sedimentologic data. { ¦pāl·ē·ō,klī·mə'täl·ə·jē }

paleocrystic ice [HYD] Sea ice generally considered to be at least 10 years old, especially well-weathered polar ice. { ¦pāl·ē·ō¦kris·tik 'īs }

paleocurrent [GEOL] Ancient fluid current flow whose orientation can be inferred by primary sedimentary structures and textures. { ¦pāl·ē·ō'kə·rənt }

paleoequator [GEOL] The position of the earth's equator in the geologic past as defined for a specific geologic period and based on geologic evidence. { ¦pāl·ē·ō·i'kwād·ər }

paleofluminology [GEOL] The study of ancient stream systems. { ¦pāl·ē·ō,flü·mə'näl·ə·jē }

Paleogene [GEOL] A geologic time interval comprising the Oligocene, Eocene, and Paleocene of the lower Tertiary period. Also known as Eogene. { 'pāl·ē·ō,jēn }

paleogeographic event *See* palevent. { ¦pāl·ē·ō,jē·ə'graf·ik i'vent }

paleogeographic stage *See* palstage. { ¦pāl·ē·ō,jē·ə'graf·ik 'stāj }

paleogeography [GEOL] The geography of the geologic past; concerns all physical aspects of an area that can be determined from the study of the rocks. Paleogeography is used to describe the changing positions of the continents and the ancient extent of land, mountains, shallow sea, and deep ocean basins. { ¦pāl·ē·ō·jē'äg·rə·fē }

paleogeologic map [GEOL] An areal map of the geology of an ancient surface immediately below a buried unconformity, showing the geology as it appeared at some time in the geologic past at the time the surface of unconformity was completed and before the overlapping strata were deposited. { ¦pāl·ē·ō,jē·ə'läj·ik 'map }

paleogeology [GEOL] The geology of the past, applied particularly to the interpretation of the rocks at a surface of unconformity. { ¦pāl·ē·ō·jē'äl·ə·jē }

paleogeomorphology [GEOL] A branch of geomorphology concerned with the recognition of ancient erosion surfaces and the study of ancient topographies and topographic features that are now concealed beneath the surface and have been removed by erosion. Also known as paleophysiography. { ¦pāl·ē·ō,jē·ō·mȯr'fäl·ə·jē }

paleohydrology [GEOL] The study of ancient hydrologic features preserved in rock. { ¦pāl·ē·ō·hī'dräl·ə·jē }

paleoisotherm [GEOL] The locus of points of equal temperature for some former period of geologic time. { ¦pāl·ē·ō'ī·sə,thərm }

paleokarst [GEOL] A rock or area that has undergone the karst process and subsequently been buried under sediments. { ¦pāl·ē·ō,kärst }

paleolatitude [GEOL] The latitude of a specific area on the earth's surface in the geologic past. { ¦pāl·ē·ō'lad·ə,tüd }

paleolimnology [GEOL] **1.** The study of the past conditions and processes of ancient lakes. **2.** The study of the sediments and history of existing lakes. { ¦pāl·ē·ō·lim'näl·ə·jē }

paleolithologic map [GEOL] A paleogeologic map indicating lithologic variations at a buried horizon or within a restricted zone at a specific time in the geologic past. { ¦pāl·ē·ō,lith·ə'läj·ik 'map }

paleomagnetics [GEOPHYS] The study of the direction and intensity of the earth's magnetic field throughout geologic time. { ¦pāl·ē·ō·mag'ned·iks }

paleomagnetic stratigraphy [GEOPHYS] The use of natural remanent magnetization in the identification of stratigraphic units. Also known as magnetic stratigraphy. { ¦pāl·ē·ō·mag¦ned·ik strə'tig·rə·fē }

paleometeoritics [GEOL] The study of variation of extraterrestrial debris as a function of time over extended parts of the geologic record, especially in deep-sea sediments and possibly in sedimentary rocks, and, for more recent periods, in ice. { ¦pāl·ē·ō,mēd·ē'ȯr·iks }

paleopedology [GEOL] The study of soils of past geologic ages, including determination of their ages. { ¦pāl·ē·ō·pə'däl·ə·jē }

paleophysiography *See* paleogeomorphology. { ¦pāl·ē·ō,fiz·ē'äg·rə·fē }

paleoplain [GEOL] An ancient degradational plain that is buried beneath later deposits. { 'pāl·ē·ə,plān }

paleopole [GEOL] A pole of the earth, either magnetic or geographic, in past geologic time. { 'pāl·ē·ə,pōl }

paleosalinity [GEOL] The salinity of a body of water in the geologic past, as evaluated on the basis of chemical analyses of sediment or formation water. { ¦pāl·ē·ō·sə'lin·əd·ē }

paleoslope [GEOL] The direction of initial dip of a former land surface, such as an ancient continental slope. { 'pāl·ē·ə,slōp }

paleosol [GEOL] A soil horizon that formed on the surface during the geologic past, that is, an ancient soil. Also known as buried soil; fossil soil. { 'pāl·ē·ə,sȯl }

paleosome [GEOL] A geometric element of a composite rock or mineral deposit which appears to be older than an associated younger rock element. { 'pāl·ē·ə,sōm }

paleostructure [GEOL] The geologic structure of a region or sequence of rocks in the geologic past. { ¦pāl·ē·ō'strək·chər }

paleotectonic map [GEOL] Regional map that shows the structural patterns that existed during a particular period of geologic time, for example, the Lower Cretaceous in western Canada. { ¦pāl·ē·ō·tek¦tän·ik 'map }

paleotemperature [GEOL] **1.** The temperature at which a geologic process took place in ancient past. **2.** The mean climatic temperature at a given time or place in the geologic past. { ¦pāl·ē·ō'tem·prə·chər }

paleothermal [GEOL] Pertaining to warm climates of the geologic past. { ¦pāl·ē·ō'thər·məl }

paleothermometry [GEOL] Measurement or estimation of past temperatures. { ¦pāl·ē·ō·thər'mäm·ə·trē }

paleotopography [GEOL] The topography of a given area in the geologic past. { ¦pāl·ē·ō·tə'päg·rə·fē }

Paleozoic [GEOL] The era of geologic time from the end of the Precambrian (600 million years before present) until the beginning of the Mesozoic era (225 million years before present). { ¦pāl·ē·ə¦zō·ik }

palette [GEOL] A broad sheet of calcite representing a solutional remnant in a cave. Also known as shield. { 'pal·ət }

palevent [GEOL] A relatively sudden and short-lived paleogeographic happening, such as the short, static existence of a particular depositional environment, or a rapid geographic change separating two palstages. Also known as paleogeographic event. { 'pal·ə·vənt }

palimpsest [GEOL] **1.** Referring to a kind of drainage in which a modern, anomalous drainage pattern is superimposed upon an older one, clearly indicating different topographic and possibly structural conditions at the time of development. **2.** In sedimentology, autochthonous sediment deposits which exhibit some of the attributes of the source sediment. { pə'lim·səst }

palingenetic *See* resurrected. { ¦pal·ən·jə¦ned·ik }

palinspastic map [GEOL] A paleogeographic or paleotectonic map showing restoration of the features to their original geographic positions, before thrusting or folding of the crustal rocks. { ¦pal·ən¦spas·tik 'map }

Palisade disturbance [GEOL] Appalachian orogenic episode occurring during Triassic time which produced a series of faultlike basins. { ˌpal·ə'sād di'stər·bəns }

palisades [GEOL] A series of sharp cliffs. { ˌpal·ə'sādz }

pallasite [GEOL] **1.** A stony-iron meteorite composed essentially of large single glassy crystals of olivine embedded in a network of nickel-iron. **2.** An ultramafic rock, of either meteoric or terrestrial origin, which contains more than 60% iron in the former, or more iron oxides than silica in the latter. { 'pal·əˌsīt }

pallasite shell *See* lower mantle. { 'pal·əˌsīt ˌshel }

palouser [METEOROL] A dust storm of northwestern Labrador. { pə'lüz·ər }

palstage [GEOL] A period of time when paleogeographic conditions were relatively static or were changing gradually and progressively with relation to such factors as sea level, surface relief, or the distance of the shoreline from the region in question. Also known as paleogeographic stage. { 'pal ˌstāj }

pampero [METEOROL] A wind of gale force blowing from the southwest across the pampas of Argentina and Uruguay, often accompanied by squalls, thundershowers, and a sudden drop of temperature; it is comparable to the norther of the plains of the United States. { päm'per·ō }

pan [GEOL] **1.** A shallow, natural depression or basin containing a body of standing water. **2.** A hard, cementlike layer, crust, or horizon of soil within or just beneath the surface; may be compacted, indurated, or very high in clay content. [OCEANOGR] *See* pancake ice. { pan }

panas oetara [METEOROL] A strong, warm, dry north wind in February in Indonesia. { pə'näs ˌō·ə'tär·ə }

panautomorphic rock *See* panidiomorphic rock. { ¦panˌód·ə'mór·fik 'räk }

pancake *See* pancake ice. { 'panˌkāk }

pancake ice [OCEANOGR] One or more small, newly formed pieces of sea ice, generally circular with slightly raised edges and about 1 to 10 feet (0.3 to 3 meters) across. Also known as lily-pad ice; pan; pancake; pan ice; plate ice. { 'panˌkāk ¦īs }

pan coefficient [METEOROL] The ratio of the amount of evaporation from a large body of water to that measured in an evaporation pan. { 'pan ˌkō·iˌfish·ənt }

panfan *See* pediplain. { 'panˌfan }

Pangaea [GEOL] A postulated former supercontinent supposedly composed of all the continental crust of the earth, and later fragmented by drift into Laurasia and Gondwana. Also spelled Pangea. { pan'jē·ə }

Pangea *See* Pangaea. { pan'jē·ə }

pan ice *See* pancake ice. { 'pan ˌīs }

panidiomorphic rock [GEOL] An igneous rock that is completely or predominantly idiomorphic. Also known as panautomorphic rock. { ˌpanˈid·ē·ōˈmȯr·fik ˈräk }

Pannonian [GEOL] A European stage of geologic time comprising the lower Pliocene. { pəˈnō·nē·ən }

pannus [METEOROL] Numerous cloud shreds below the main cloud; may constitute a layer separated from the main part of the cloud or attached to it. { ˈpan·əs }

panplain [GEOL] A broad, level plain formed by coalescence of several adjacent flood plains. Also spelled panplane. { ˈpanˌplān }

panplanation [GEOL] The action or process of formation or development of a panplain. { ˌpan·pləˈnā·shən }

panplane *See* panplain. { ˈpanˌplān }

Panthalassa [GEOL] The hypothetical proto-ocean surrounding Pangea, supposed by some geologists to have combined all the oceans or areas of oceanic crust of the earth at an early time in the geologic past. { ˌpan·thəˈlas·ə }

papagayo [METEOROL] A violent, northeasterly fall wind on the Pacific coast of Nicaragua and Guatemala; it consists of the cold air mass of a norte which has overridden the mountains of Central America and, being a descending wind, it brings fine, clear weather. { ˌpä·pəˈgī·yō }

paper shale [GEOL] A shale that easily separates on weathering into very thin, tough, uniform, and somewhat flexible layers or laminae suggesting sheets of paper. { ˈpā·pər ˈshāl }

paper spar [GEOL] A crystallized variety of calcite occurring in thin lamellae or paperlike plates. { ˈpā·pər ˈspär }

parabituminous coal [GEOL] Bituminous coal that contains 84–87% carbon, analyzed on a dry, ash-free basis. { ˈpar·ə·bəˈtüm·ə·nəs ˈkōl }

parabolic dune [GEOL] A long, scoop-shaped sand dune having a ground plan approximating the form of a parabola, with the horns pointing windward (upwind). Also known as blowout dune. { ˈpar·əˈbäl·ik ˈdün }

parachronology [GEOL] **1.** Practical dating and correlation of stratigraphic units. **2.** Geochronology based on fossils that supplement, or replace, biostratigraphically significant fossils. { ˌpar·ə·krəˈnäl·ə·jē }

paraclinal [GEOL] Referring to a stream or valley that is oriented in a direction parallel to the fold axes of a region. { ˈpar·əˈklīn·əl }

paraconformity [GEOL] A type of unconformity in which strata are parallel; there is little apparent erosion and the unconformity surface resembles a simple bedding plane. Also known as nondepositional unconformity; pseudoconformity. { ˈpar·ə·kənˈfȯr·məd·ē }

paraconglomerate [GEOL] A conglomerate that is not a product of normal aqueous flow but is deposited by such modes of mass transport as subaqueous turbidity currents and glacier ice; characterized by a disrupted gravel framework, often unstratified, and notable for a matrix of greater than gravel-sized fragments. { ˈpar·ə·kənˈgläm·ə·rət }

paraffin coal [GEOL] A type of light-colored bituminous coal from which oil and paraffin are produced. { ˈpar·ə·fən ˌkōl }

paraffin dirt [GEOL] A clay soil appearing rubbery or curdy and occurring in the upper several inches of a soil profile near gas seeps; probably formed by biodegradation of natural gas. { ˈpar·ə·fən ˌdərt }

parageosyncline [GEOL] An epeirogenic geosynclinal basin located within a craton or stable area. { ˈpar·əˌjē·ōˈsinˌklīn }

paraglomerate [GEOL] A conglomerate which contains more matrix than gravel-sized fragments and was deposited by subaqueous turbidity flows and glacier ice rather than normal aqueous flow. Also known as conglomeratic mudstone. { ˌpar·əˈgläm·ə·rət }

paragneiss [GEOL] A gneiss showing a sedimentary parentage. { ˈpar·əˌnīs }

paraliageosyncline [GEOL] A geosyncline developing along a present-day continental margin, such as the Gulf Coast geosyncline. { pəˌral·yəˌjē·ōˈsinˌklīn }

paralic [GEOL] Pertaining to deposits laid down on the landward side of a coast. { pə'ral·ik }
paralic coal basin [GEOL] Coal deposits formed along the margin of the sea. { pə'ral·ik 'kōl ˌbas·ən }
paralic swamp *See* marine swamp. { pə'ral·ik 'swämp }
paralimnion [HYD] The littoral part of a lake, extending from the margin to the deepest limit of rooted vegetation. { ¦par·ə'lim·nēˌän }
parallax age *See* age of parallax inequality. { 'par·əˌlaks ˌāj }
parallax inequality [OCEANOGR] The variation in the range of tide or in the speed of tidal currents due to the continual change in the distance of the moon from the earth. { 'par·əˌlaks ˌin·i'kwäl·əd·ē }
parallel [GEOD] A circle on the surface of the earth, parallel to the plane of the equator and connecting all points of equal latitude. Also known as circle of longitude; parallel of latitude. { 'par·əˌlel }
parallel drainage pattern [HYD] A drainage pattern characterized by regularly spaced streams flowing parallel to one another over a large area. { 'par·əˌlel 'drān·ij ˌpad·ərn }
parallel fold *See* concentric fold. { 'par·əˌlel 'fōld }
parallel ripple mark [GEOL] A ripple mark characterized by a relatively straight crest and an asymmetric profile. { 'par·əˌlel 'rip·əl ˌmärk }
parallel roads [GEOL] A series of horizontal beaches or wave-cut terraces occurring parallel to each other at different levels on each side of a glacial valley. { 'par·əˌlel 'rōdz }
parallochthon [GEOL] Rocks that were brought from intermediate distances and deposited near an allochthonous mass during transit. { ˌpar·ə'läkˌthän }
parametric hydrology [HYD] That branch of hydrology dealing with the development and analysis of relationships among the physical parameters involved in hydrologic events and the use of these relationships to generate, or synthesize, hydrologic events. { ¦par·ə¦me·trik hī'dräl·ə·jē }
parametric latitude *See* reduced latitude. { ¦par·ə¦me·trik 'lad·əˌtüd }
pararipple [GEOL] A large, symmetric ripple whose surface slopes gently and which shows no assortment of grains. { 'par·əˌrip·əl }
parasitic cone *See* adventive cone. { ¦par·ə¦sik·ik 'kōn }
parastratigraphic unit *See* operational unit. { ¦par·əˌstrad·ə'graf·ik 'yü·nət }
parastratigraphy [GEOL] **1.** Supplemental stratigraphy based on fossils other than those governing the prevalent orthostratigraphy. **2.** Stratigraphy based on operational units. { ¦par·ə·strə'tig·rə·fē }
parastratotype [GEOL] Another section in the original locality where a stratotype was defined. { ¦par·ə'strad·əˌtīp }
paratill [GEOL] A till formed by ice-rafting in a marine or lacustrine environment; includes deposits from ice floes and icebergs. { 'par·əˌtil }
parautochthonous [GEOL] Pertaining to a mobilized part of an autochthonous granite moved higher in the crust or into a tectonic area of lower pressure and characterized by variable and diffuse contacts with country rocks. { ¦par·ə·ȯ'täk·thə·nəs }
parcel method [METEOROL] A method of testing for instability in which a displacement is made from a steady state under the assumption that only the parcel or parcels displaced are affected, the environment remaining unchanged. { 'pär·səl ˌmeth·əd }
parental magma [GEOL] The naturally occurring mobile rock material from which a particular igneous rock solidified or from which another magma was derived. { pə'rent·əl 'mag·mə }
parent material [GEOL] The unconsolidated mineral or organic material from which the true soil develops. { 'per·ənt məˌtir·ē·əl }
parent rock [GEOL] **1.** The rock mass from which parent material is derived. **2.** *See* source rock. { 'per·ənt ˌräk }
parogenetic [GEOL] Formed previous to the enclosing rock; especially said of a concretion formed in a different (older) rock from its present (younger) host. { ˌpar·ə'jen·ik }

paroxysmal eruption *See* Vulcanian eruption. { ¦par·ək¦siz·məl i'rəp·shən }

partial-duration series [GEOPHYS] A series composed of all events during the period of record which exceed some set criterion; for example, all floods above a selected base, or all daily rainfalls greater than a specified amount. { 'pär·shəl dů'ra·shən ˌsir·ēz }

partial obscuration [METEOROL] In United States weather observing practice, the designation for sky cover when part (0.1 to 0.9) of the sky is completely hidden by surface-based obscuring phenomena. { 'pär·shəl äb·skyů'rā·shən }

partial pediment [GEOL] **1.** A broadly planate, gravel-capped, interstream bench or terrace. **2.** A broad, planate erosion surface which is formed by the coalescence of contemporaneous, valley-restricted benches developed at the same elevation in proximate valleys, and which would produce a pediment if uninterrupted planation were to continue at this level. { 'pär·shəl 'ped·ə·mənt }

partial pluton [GEOL] That part of a composite intrusion representing a single intrusive episode. { 'pär·shəl 'plü,tän }

partial potential temperature [METEOROL] The temperature that the dry-air component of an air parcel would attain if its actual partial pressure were changed to 1000 millibars (10^5 pascals). { 'pär·shəl pə'ten·chəl 'tem·prə·chər }

partial thermoremanent magnetization [GEOPHYS] The thermoremanent magnetization acquired by cooling in an ambient field over only a restricted temperature interval, as opposed to the entire temperature range from Curie point to room temperature. Abbreviated PTRM. { 'pär·shəl ¦thər·mō'rem·ə·nənt ˌmag·nə·tə'zā·shən }

partial tide [OCEANOGR] One of the harmonic components composing the tide at any point. Also known as tidal component; tidal constituent. { 'pär·shəl 'tīd }

particle diameter [GEOL] The diameter of a sedimentary particle considered as a sphere. { 'pärd·ə·kəl dī,am·əd·ər }

particle size [GEOL] The general dimensions of the particles or mineral grains in a rock or sediment based on the premise that the particles are spheres; commonly measured by sieving, by calculating setting velocities, or by determining areas of microscopic images. { 'pärd·ə·kəl ˌsīz }

particle-size analysis [GEOL] A determination of the distribution of particles in a series of size classes of a soil, sediment, or rock. Also known as size analysis; size-frequency analysis. { 'pärd·ə·kəl ¦sīz əˌnal·ə·səs }

particle velocity [OCEANOGR] In ocean wave studies, the instantaneous velocity of a water particle undergoing orbital motion. { 'pärd·ə·kəl vəˌläs·əd·ē }

parting [GEOL] **1.** A bed or bank of waste material dividing mineral veins or beds. **2.** A soft, thin sedimentary layer following a surface of separation between thicker strata of different lithology. **3.** A surface along which a hard rock can be readily separated or is naturally divided into layers. { 'pärd·iŋ }

parting cast [GEOL] A sand-filled tension crack produced by creep along the sea floor. { 'pärd·iŋ ˌkast }

parting lineation [GEOL] A small-scale primary sedimentary structure made up of a series of parallel ridges and grooves formed parallel to the current. Also known as current lineation. { 'pärd·iŋ ˌlin·ē'ā·shən }

parting plane lineation [GEOL] A parting lineation on a laminated surface, consisting of subparallel, linear, shallow grooves and ridges of low relief, generally less than 1 millimeter. { 'pärd·iŋ ¦plān ˌlin·ēˌā·shən }

parting-step lineation [GEOL] A parting lineation characterized by subparallel, steplike ridges where the parting surface cuts across several adjacent laminae. { 'pärd·iŋ ¦step 'lin·ēˌā·shən }

partiversal [GEOL] Pertaining to formations that dip in different directions roughly as far as a semicircle. { ¦pard·ə¦vər·səl }

partly cloudy [METEOROL] **1.** The character of a day's weather when the average cloudiness, as determined from frequent observations, has been from 0.1 to 0.5 for the 24-hour period. **2.** In popular usage, the state of the weather when clouds are conspicuously present, but do not completely dull the day or the sky at any moment. { 'pärt·lē 'klau̇d·ē }

parvafacies [GEOL] A body of rock constituting the part of any magnafacies that occurs between designated time-stratigraphic planes or key beds traced across the magnafacies. { ¦pär·və'fā·shēz }

pass [GEOGR] **1.** A natural break, depression, or other low place providing a passage through high terrain, such as a mountain range. **2.** A navigable channel leading to a harbor or river. **3.** A narrow opening through a barrier reef, atoll, or sand bar. { pas }

passage [GEOGR] A navigable channel, especially one through reefs or islands. { 'pas·ij }

passage bed [GEOL] A stratum marking a transition from rocks of one geological system to those of another. { 'pas·ij ˌbed }

passive fold [GEOL] A fold in which the mechanism of folding, either flow or slip, crosses the boundaries of the strata at random. { 'pas·iv 'fōld }

passive front *See* inactive front. { 'pas·iv 'frənt }

passive glacier [HYD] A glacier with sluggish movement, generally occurring in a continental environment at a high latitude, where both accumulation and ablation are minimal. { 'pas·iv 'glā·shər }

passive margin [GEOL] A continental margin formed by rifting during continental breakup. { 'pas·iv 'mär·jən }

passive permafrost [GEOL] Permafrost that will not refreeze under present climatic conditions after being disturbed or destroyed. Also known as fossil permafrost. { 'pas·iv 'pər·məˌfrȯst }

patch reef [GEOL] **1.** A small, irregular organic reef with a flat top forming a part of a reef complex. **2.** A small, thick, isolated lens of limestone or dolomite surrounded by rocks of different facies. **3.** *See* reef patch. { 'pach ˌrēf }

paternoster lake [HYD] One of a linear chain or series of small circular lakes, usually at different levels, which occupy rock basins in a glacial valley and are separated by morainal dams or riegels, but connected by streams, rapids, or waterfalls to resemble a rosary or string of beads. Also known as beaded lake; rock-basin lake; step lake. { 'päd·ərˌnäs·tər ˌlāk }

patina [GEOL] A thin, colored film produced on a rock surface by weathering. { 'pat·ən·ə *or* pə'tē·nə }

patterned ground [GEOL] Any of several well-defined, generally symmetrical forms, such as circles, polygons, and steps, that are characteristic of surficial material subject to intensive frost action. { 'pad·ərnd 'grau̇nd }

paulopost *See* deuteric. { 'pȯl·əˌpōst }

pavement [GEOL] A bare rock surface that suggests a paved road surface or other pavement in smoothness, hardness, horizontality, surface extent, or close packing of units. { 'pāv·mənt }

PCA *See* polar-cap absorption.

PDB *See* PeeDee belemnite.

pea coal [GEOL] A size of anthracite that will pass through a 13/16-inch (20.6-millimeter) round mesh but not through a 9/16-inch (14.3-millimeter) round mesh. { 'pē ˌkōl }

pea gravel [GEOL] A type of gravel whose individual particles are about the size of peas. { 'pē ˌgrav·əl }

peak [GEOL] **1.** The conical or pointed top of a hill or mountain. **2.** An individual mountain or hill taken as a whole, used especially when it is isolated or has a pointed, conspicuous summit. [METEOROL] The point of intersection of the cold and warm fronts of a mature extra-tropical cyclone. { pēk }

peak gust [METEOROL] After United States weather observing practice, the highest instantaneous wind speed recorded at a station during a specified period, usually the 24-hour observational day; therefore, a peak gust need not be a true gust of wind. { 'pēk 'gəst }

peak plain [GEOL] A high-level plain formed by a series of summits of approximately the same elevation, often described as an uplifted and fully dissected peneplain. Also known as summit plain. { 'pēk ˌplān }

pearlite *See* perlite. { 'pərˌlīt }

pearlstone *See* perlite. { 'pərlˌstōn }

pea-soup fog [METEOROL] Any particularly dense fog. { 'pē ˌsüp 'fäg }

peat [GEOL] A dark-brown or black residuum produced by the partial decomposition and disintegration of mosses, sedges, trees, and other plants that grow in marshes and other wet places. { pēt }

peat bed *See* peat bog. { 'pēt ˌbed }

peat bog [GEOL] A bog in which peat has formed under conditions of acidity. Also known as peat bed; peat moor. { 'pēt ˌbäg }

peat breccia [GEOL] Peat that has been broken up and then redeposited in water. Also known as peat slime. { 'pētˌbrech·ə }

peat coal [GEOL] A coal transitional between peat and lignite. { 'pēt ˌkōl }

peat formation [GEOCHEM] Decomposition of vegetation in stagnant water with small amounts of oxygen, under conditions intermediate between those of putrefaction and those of moldering. { 'pēt fȯr'mā·shən }

peat moor *See* peat bog. { 'pēt ˌmu̇r }

peat-sapropel [GEOL] A product of the degradation of organic matter that is transitional between peat and sapropel. Also known as sapropel-peat. { 'pēt 'sap·rəˌpel }

peat slime *See* peat breccia. { 'pēt ˌslīm }

peat soil [GEOL] Soil containing a large amount of peat; it is rich in humus and gives an acid reaction. { 'pēt ˌsȯil }

pebble [GEOL] A clast, larger than a granule and smaller than a cobble, having a diameter in the range of 0.16–2.6 inches (4–64 millimeters). Also known as pebblestone. { 'peb·əl }

pebble armor [GEOL] A desert armor made up of rounded pebbles. { 'peb·əl ˌär·mər }

pebble bed [GEOL] Any pebble conglomerate, especially one in which the pebbles weather conspicuously and become loose. Also known as popple rock. { 'peb·əl ˌbed }

pebble coal [GEOL] Coal that is transitional between peat and brown coal. { 'peb·əl ˌkōl }

pebble dike [GEOL] **1.** A clastic dike composed largely of pebbles. **2.** A tabular body containing sedimentary fragments in an igneous matrix. { 'peb·əl ˌdīk }

pebble peat [GEOL] Peat that is formed in a semiarid climate by the accumulation of moss and algae, no more than 0.25 inch (6 millimeters) in thickness, under the surface pebbles of well-drained soils. { 'peb·əl ˌpēt }

pebble phosphate [GEOL] A secondary phosphorite of either residual or transported origin, consisting of pebbles or concretions of phosphatic material. { 'peb·əl 'fäsˌfāt }

pebblestone *See* pebble. { 'peb·əlˌstōn }

pebbly mudstone [GEOL] A delicately laminated till-like conglomeratic mudstone. { 'peb·lē 'mədˌstōn }

pebbly sand [GEOL] An unconsolidated sedimentary deposit containing at least 75% sand and up to a maximum of 25% pebbles. { 'peb·lē 'sand }

pebbly sandstone [GEOL] A sandstone that contains 10–20% pebbles. { 'peb·lē 'sanˌstōn }

ped [GEOL] A naturally formed unit of soil structure. { ped }

pedalfer [GEOL] A soil in which there is an accumulation of sesquioxides; it is characteristic of a humid region. { pə'dal·fər }

pedality [GEOL] The physical nature of a soil as expressed by the features of its constituent peds. { pe'dal·əd·ē }

pedestal [GEOL] A relatively slender column of rock supporting a wider rock mass and formed by undercutting as a result of wind abrasion or differential weathering. Also known as rock pedestal. { 'ped·əst·əl }

pedestal boulder [GEOL] A rock mass supported on a rock pedestal. Also known as pedestal rock. { 'ped·əst·əl ˌbōl·dər }
pedestal rock *See* pedestal boulder. { 'ped·əst·əl ˌräk }
pediment [GEOL] A piedmont slope formed from a combination of processes which are mainly erosional; the surface is chiefly bare rock but may have a covering veneer of alluvium or gravel. Also known as conoplain. { 'ped·ə·mənt }
pedimentation [GEOL] The actions or processes by which pediments are formed. { ˌped·ə·mən'tā·shən }
pediment gap [GEOL] A broad opening formed by the enlargement of a pediment pass. { 'ped·ə·mənt ˌgap }
pediment pass [GEOL] A flat, narrow tongue that extends from a pediment on one side of a mountain to join a pediment on the other side. { 'ped·ə·mənt ˌpas }
pediocratic [GEOL] Pertaining to a period of time in which there is little diastrophism. { ˌped·ē·ə'krad·ik }
pediplain [GEOL] A rock-cut erosion surface formed in a desert by the coalescence of two or more pediments. Also known as desert peneplain; desert plain; panfan. { 'ped·əˌplān }
pediplanation [GEOL] The actions or processes by which pediplanes are formed. { ˌped·ə·plə'nā·shən }
pediplane [GEOL] Any planate erosion surface formed in the piedmont area of a desert, either bare or covered with a veneer of alluvium. { 'ped·əˌplān }
pedocal [GEOL] A soil containing a concentration of carbonates, usually calcium carbonate; it is characteristic of arid or semiarid regions. { 'ped·əˌkal }
pedogenesis *See* soil genesis. { ¦ped·ō'jen·ə·səs }
pedogenics [GEOL] The study of the origin and development of soil. { ¦ped·ō¦jen·iks }
pedogeochemical survey [GEOCHEM] A geochemical prospecting survey in which the materials sampled are soil and till. { ¦ped·ōˌjē·ō'kem·ə·kəl 'sərˌvā }
pedogeography [GEOL] The study of the geographic distribution of soils. { ¦ped·ō·jē'äg·rə·fē }
pedography [GEOL] The systematic description of soils; an aspect of soil science. { pə'däg·rə·fē }
pedolith [GEOL] A surface formation that has undergone one or more pedogenic processes. { 'ped·əˌlith }
pedologic age [GEOL] The relative maturity of a soil profile. { ¦ped·ō¦läj·ik 'āj }
pedologic unit [GEOL] A soil considered without regard to its stratigraphic relations. { ¦ped·ō¦läj·ik 'yü·nət }
pedology *See* soil science. { pe'däl·ə·jē }
pedon [GEOL] The smallest unit or volume of soil that represents or exemplifies all the horizons of a soil profile; it is usually a horizontal, hexagonal area of about 1 square meter, or possibly larger. { 'peˌdän }
pedorelic [GEOL] Referring to a soil feature that is derived from a preexisting soil horizon. { ¦ped·ō¦rel·ik }
pedosphere [GEOL] That shell or layer of the earth in which soil-forming processes occur. { 'ped·əˌsfir }
pedotubule [GEOL] A soil feature consisting of skeleton grains, or skeleton grains plus plasma, and having a tubular external form (either single tubes or branching systems of tubes) characterized by relatively sharp boundaries and relatively uniform cross-sectional size and shape (circular or elliptical). { ¦ped·ō'tüb·yül }
PeeDee belemnite [GEOCHEM] Limestone from the PeeDee Formation in South Carolina (derived from the Cretaceous marine fossil *Belemnitella americana*), the carbon and oxygen isotope ratios of which are used as an international reference standard. Abbreviated PDB. { ˌpēˌdē bə'lemˌnīt }
peel thrust [GEOL] A sedimentary sheet peeled off a sedimentary sequence, usually along a bedding plane. { 'pēl ˌthrəst }
peesweep storm [METEOROL] An early-spring storm in Scotland and England. { 'pēz ˌwēp ˌstȯrm }

pegmatitic stage [GEOL] A stage in the normal sequence of crystallization of magma containing volatiles when the residual fluid is sufficiently enriched in volatile materials to permit the formation of coarse-grained rocks, that is pegmatites. { ¦peg·mə¦tid·ik ˌstāj }

pegmatitization [GEOL] Formation of or replacement by a pegmatite. { ˌpeg·məˌtīd·ə'za·shən }

peg model [GEOL] Three-dimensional model used to illustrate and study stratigraphic and structural conditions of subsurface geology; consists of a flat platform onto which vertical pegs of varying heights are mounted to represent the contours of various strata. { 'peg ˌmäd·əl }

pelagic [GEOL] Pertaining to regions of a lake at depths of 33–66 feet (10–20 meters) or more, characterized by deposits of mud or ooze and by the absence of vegetation. Also known as eupelagic. [OCEANOGR] Pertaining to water of the open portion of an ocean, above the abyssal zone and beyond the outer limits of the littoral zone. { pə'laj·ik }

pelagic limestone [GEOL] A fine-textured limestone formed in relatively deep water by the concentration of calcareous tests of pelagic Foraminifera. { pə'laj·ik 'līmˌstōn }

pelagochthonous [GEOL] Referring to coal derived from a submerged forest or from driftwood. { ˌpel·ə'gäk·thə·nəs }

pelagosite [GEOL] A superficial calcareous crust a few millimeters thick, generally white, gray, or brownish with a pearly luster, formed in the intertidal zone by ocean spray and evaporation, and composed of calcium carbonate with higher contents of magnesium carbonate, strontium carbonate, calcium sulfate, and silica than are found in normal limy sediments. { pə'lag·əˌsīt }

Pelean cloud *See* nuée ardente. { pə'lē·ən 'klau̇d }

pelelith [GEOL] Vesicular or pumiceous lava in the throat of a volcano. { pə'lāˌlith }

Pele's hair [GEOL] A spun volcanic glass formed naturally by blowing out during quiet fountaining of fluid lava. Also known as capillary ejecta; filiform lapilli; lauoho o pele. { ˌpāˌlāz 'her }

Pele's tears [GEOL] Volcanic glass in the form of small, solidified drops which precede pendants of Pele's hair. { 'pāˌlāz 'tirz }

pelite [GEOL] A sediment or sedimentary rock, such as mudstone, composed of fine, clay- or mud-size particles. Also spelled pelyte. { 'pēˌlīt }

pelitic [GEOL] Pertaining to, characteristic of, or derived from pelite. { pə'lid·ik }

pellet [GEOL] A fine-grained, sand-size, spherical to elliptical aggregate of clay-sized calcareous material, devoid of internal structure, and contained in the body of a well-sorted carbonate rock. { 'pel·ət }

pellicular water [HYD] Films of groundwater adhering to particles or cavities above the water table. { pə'lik·yə·lər 'wȯd·ər }

pell-mell structure [GEOL] A sedimentary structure characterized by absence of bedding in a coarse deposit of waterworn material; it may occur where deposition is too rapid for sorting or where slumping has destroyed the layered arrangement. { 'pel¦mel 'strək·chər }

pellodite *See* pelodite. { 'pel·əˌdīt }

pelmicrite [GEOL] A limestone containing less than 25% each of intraclasts and ooliths, having a volume ratio of pellets to fossils greater than 3 to 1, and with the micrite matrix more abundant than the sparry-calcite cement. { 'pel·məˌkrīt }

pelodite [GEOL] A lithified glacial rock flour which is composed of glacial pebbles in a silt or clay matrix and which was formed by redeposition of the fine fraction of a till. Also spelled pellodite. { 'pel·əˌdīt }

pelogloea [GEOL] Marine detrital slime from settled plankton. { ¦pel·ə¦glē·ə }

pelphyte [GEOL] A lake-bottom deposit consisting mainly of fine, nonfibrous plant remains. { 'pelˌfīt }

pelyte *See* pelite. { 'peˌlīt }

pencil cleavage [GEOL] Cleavage in which fracture produces long, slender pieces of rock. { 'pen·səl ˌklē·vij }

pencil gneiss [GEOL] A gneiss that splits into thin, rodlike quartz-feldspar crystal aggregates. { 'pen·səl ˌnīs }
pencil ore [GEOL] Hard, fibrous masses of hematite that can be broken up into splinters. { 'pen·səl ˌȯr }
pendant *See* roof pendant. { 'pen·dənt }
pendant cloud *See* tuba. { 'pen·dənt ˌklau̇d }
pendent terrace [GEOL] A connecting ribbon of sand that joins an isolated point of rock with a neighboring coast. { 'pen·dənt ˌter·əs }
pendular water [HYD] Capillary water ringing the contact points of adjacent rock or soil particles in the zone of areation. { 'pen·jə·lər ˌwȯd·ər }
penecontemporaneous [GEOL] Of a geologic process or the structure or mineral that is formed by the process, occurring immediately following deposition but before consolidation of the enclosing rock. { ¦pēn·ē·kənˌtem·pə'rā·nē·əs }
peneplain *See* base-leveled plain. { 'pēn·əˌplān }
peneplanation [GEOL] The actions or processes by which peneplains are formed. { ˌpēn·ə·plə'nā·shən }
penetration funnel [GEOL] An impact crater, generally funnel-shaped, formed by a small meteorite striking the earth at a relatively low velocity and containing nearly all the impacting mass within it. { ˌpen·ə'trā·shən ˌfən·əl }
penetrative [GEOL] Referring to a texture of deformation that is uniformly distributed in a rock, without notable discontinuities; for example, slaty cleavage. { 'pen·əˌtrā·div }
peninsula [GEOGR] A body of land extending into water from the mainland, sometimes almost entirely separated from the mainland except for an isthmus. { pə'nin·sə·lə }
penitent *See* nieve penitente. { 'pen·ə·tənt }
penitent ice [HYD] A jagged spike or pillar of compacted firn caused by differential melting and evaporation; necessary for this formation are air temperature near freezing, dew point much below freezing, and strong insolation. { 'pen·ə·tənt 'īs }
penitent snow [HYD] A jagged spike or pillar of compacted snow caused by differential melting and evaporation. { 'pen·ə·tənt 'snō }
pennant [METEOROL] A means of representing wind speed in the plotting of a synoptic chart; it is a triangular flag, drawn pointing toward lower pressure from a wind-direction shaft. { 'pen·ənt }
Pennsylvanian [GEOL] A division of late Paleozoic geologic time, extending from 320 to 280 million years ago, varyingly considered to rank as an independent period or as an epoch of the Carboniferous period; named for outcrops of coal-bearing rock formations in Pennsylvania. { ¦pen·sal¦vā·nyən }
Penokean *See* Animikean. { pə'nō·kē·ən }
pentad [CLIMATOL] A period of 5 consecutive days, often preferred to the week for climatological purposes since it is an exact factor of the 365-day year. { 'penˌtad }
Penutian [GEOL] A North American stage of geologic time: lower Eocene (above Bulitian, below Ulatasian). { pə'nü·shən }
peperite [GEOL] A breccialike material in marine sedimentary rock, considered to be either a mixture of lava with sediment, or shallow intrusions of magma into wet sediment. { 'pep·əˌrīt }
P-E ratio *See* precipitation-evaporation ratio. { ¦pē'ē ˌrā·shō }
perbituminous [GEOL] Referring to bituminous coal containing more than 5.8% hydrogen, analyzed on a dry, ash-free basis. { ¦pər·bə'tü·mə·nəs }
perched aquifer [HYD] An aquifer that is separated from another water-bearing stratum by an impermeable layer. { 'pərcht 'ak·wə·fər }
perched block [GEOL] A large, detached rock fragment presumed to have been transported and deposited by a glacier, and perched in a conspicuous and precarious position on the side of a hill. Also known as balanced rock; perched boulder; perched rock. { 'pərcht 'bläk }
perched boulder *See* perched block. { 'pərcht 'bōl·dər }
perched groundwater *See* perched water. { 'pərcht 'grau̇ndˌwȯd·ər }
perched lake [HYD] A perennial lake whose surface level lies at a considerably higher

elevation than those of other bodies of water, including aquifers, directly or closely associated with the lake. { 'pərcht 'lāk }

perched rock *See* perched block. { 'pərcht 'räk }

perched spring [HYD] A spring that arises from a body of perched water. { 'pərcht 'spriŋ }

perched stream [HYD] A stream whose surface level is above that of the water table and that is separated from underlying groundwater by an impermeable bed in the zone of aeration. { 'pərcht 'strēm }

perched water [HYD] Groundwater that is unconfined and separated from an underlying main body of groundwater by an unsaturated zone. Also known as perched groundwater. { 'pərcht 'wȯd·ər }

perched water table [HYD] The water table or upper surface of a body of perched water. Also known as apparent water table. { 'pərcht 'wȯd·ər ˌtā·bəl }

perching bed [GEOL] A body of rock, generally stratiform, that supports a body of perched water. { 'pərch·iŋ ˌbed }

percolation [HYD] Gravity flow of groundwater through the pore spaces in rock or soil. { pər·kə'lā·shən }

percolation zone [HYD] The area on a glacier or ice sheet where a limited amount of surface melting occurs, but the meltwater refreezes in the same snow layer and the snow layer is not completely soaked or brought up to the melting temperature. { pər·kə'lā·shən ˌzōn }

percussion mark [GEOL] A small, crescent-shaped scar produced on a hard, dense pebble by a blow. { pər'kəsh·ən ˌmärk }

pereletok [GEOL] A frozen layer of ground, at the base of the active layer, which may persist for one or several years. Also known as intergelisol. { ˌper·ə·lə'täk }

perennial lake [HYD] A lake that retains water in its basin throughout the year and is not usually subject to extreme water-level fluctuations. { pə'ren·ē·əl 'lāk }

perennial spring [HYD] A spring that flows continuously, as opposed to an intermittent spring or a periodic spring. { pə'ren·ē·əl 'spriŋ }

perennial stream [HYD] A stream which contains water at all times except during extreme drought. { pə'ren·ē·əl 'strēm }

perezone [GEOL] A zone in which sediments accumulate along coastal lowlands; includes lagoons and brackish-water bays. { 'per·əˌzōn }

perfect prognostic [METEOROL] The observed pressure pattern at the verifying time of a forecast of some element other than pressure; used in objective forecast studies in which a forecast of the element is based on a simultaneous relation between this element and the pressure pattern plus a forecast of the pressure pattern at some future time. { 'pər·fikt präg'näs·tik }

perfemic rock [GEOL] An igneous rock in which the ratio of salicalic to femic minerals is less than 1:7. { pər'fem·ik 'räk }

perforated crust [HYD] A type of snow crust containing pits and hollows produced by ablation. { 'pər·fəˌrād·əd 'krəst }

perforation deposit [GEOL] An isolated kame consisting of material that accumulated in a vertical shaft which pierced a glacier and afforded no outlet for water at the bottom. { ˌpər·fə'rā·shən diˌpäz·ət }

pergelation [HYD] The act or process of forming permafrost. { ˌpər·jə'lā·shən }

pergelic [GEOL] Referring to a soil temperature regime in which the mean annual temperature is less than 0°C and there is permafrost. { pər'jel·ik }

pergelisol table *See* permafrost table. { ¦pər'jel·əˌsȯl }

perhumid climate [CLIMATOL] As defined by C. W. Thornthwaite in his climatic classification, a type of climate which has humidity index values of +100 and above; this is his wettest type of climate (designated A), and compares closely to the "wet climate" which heads his 1931 grouping of humidity provinces. { ¦pər'hyü·məd 'klī·mət }

perhydrous coal [GEOL] Coal that contains more than 6% hydrogen, analyzed on a dry, ash-free basis. { ¦pər'hī·drəs 'kōl }

periblinite [GEOL] A variety of provitrinite consisting of cortical tissue. { pə'rib·ləˌnīt }

periclinal [GEOL] Referring to strata and structures that dip radially outward from, or inward toward, a center, forming a dome or a basin. { ¦per·ə¦klīn·əl }

pericline [GEOL] A fold characterized by central orientation of the dip of the beds. { 'per·ə‚klīn }

pericline ripple mark [GEOL] A ripple mark arranged in an orthogonal pattern either parallel to or transverse to the current direction and having a wavelength up to 80 centimeters and amplitude up to 30 centimeters. { 'per·ə‚klīn 'ripəl ‚märk }

peridotite shell *See* upper mantle. { pə'rid·ə‚tīt ‚shel }

perigean range [OCEANOGR] The average range of tide at the time of perigean tides, when the moon is near perigee; the perigean range is greater than the mean range. { ¦per·ə¦jē·ən 'rānj }

perigean tidal currents [OCEANOGR] Tidal currents of increased speed occurring at the time of perigean tides. { ¦per·ə¦jē·ən 'tīd·əl ‚kə·rəns }

perigean tide [OCEANOGR] Tide of increased range occurring when the moon is near perigee. { ¦per·ə¦jē·ən 'tīd }

perigenic [GEOL] Referring to a rock constituent or mineral formed at the same time as the rock it is part of, but not formed at the specific location it now occupies in the rock. { ¦per·ə¦jen·ik }

periglacial [GEOL] Of or pertaining to the outer perimeter of a glacier, particularly to the fringe areas immediately surrounding the great continental glaciers of the geologic ice ages, with respect to environment, topography, areas, processes, and conditions influenced by the low temperature of the ice. { ¦per·ə'glā·shəl }

periglacial climate [CLIMATOL] The climate which is characteristic of the regions immediately bordering the outer perimeter of an ice cap or continental glacier; the principal climatic feature is the high frequency of very cold and dry winds off the ice area; it is also thought that these regions offer ideal conditions for the maintenance of a belt of intense cyclonic activity. { ¦per·ə'glā·shəl 'klī·mət }

perimagmatic [GEOL] Referring to a hydrothermal mineral deposit located near its magmatic source. { ¦per·ə·mag'ned·ik }

period [GEOL] A unit of geologic time constituting a subdivision of an era; the fundamental unit of the standard geologic time scale. { 'pir·ē·əd }

periodic current [OCEANOGR] Current produced by the tidal influence of moon and sun or by any other oscillatory forcing function. { ¦pir·ē¦äd·ik 'kə·rənt }

periodic spring [HYD] A spring that ebbs and flows periodically, apparently due to natural siphon action. { ¦pir·ē¦äd·ik 'spriŋ }

peripediment [GEOL] The segment of a pediplane extending across the younger rocks or alluvium of a basin which is always beyond but adjacent to the segment developed on the older upland rocks. { ¦per·ə'ped·ə·mənt }

peripheral depression *See* ring depression. { pə'rif·ə·rəl di'presh·ən }

peripheral faults [GEOL] Arcuate faults bounding an elevated or depressed area such as a diapir. { pə'rif·ə·rəl 'fȯls }

peripheral sink *See* rim syncline. { pə'rif·ə·rəl 'siŋk }

peripheral stream [HYD] A stream that flows parallel to the edge of a glacier, usually just beyond the moraine. { pə'rif·ə·rəl 'strēm }

perlite [GEOL] A rhyolitic glass with abundant spherical or convolute cracks that cause it to break into small pearllike masses or pebbles, usually less than a centimeter across; it is commonly gray or green with a pearly luster and has the composition of rhyolite. Also known as pearlite; pearlstone. { 'pər‚līt }

perlucidus [METEOROL] A cloud variety, usually of the species stratiformis, in which distinct spaces between its elements permit the sun, moon, blue sky, or higher clouds to be seen. { pər'lü·səd·əs }

permafrost [GEOL] Perennially frozen ground, occurring wherever the temperature remains below 0°C for several years, whether the ground is actually consolidated by ice or not and regardless of the nature of the rock and soil particles of which the earth is composed. { 'pər·mə‚frȯst }

permafrost island [GEOL] A small, shallow, isolated patch of permafrost surrounded by unfrozen ground. { 'pər·mə‚frȯst 'ī·lənd }

permafrost line [GEOL] A line on a map representing the border of the arctic permafrost. { 'pər·mə,fròst ,līn }

permafrost table [GEOL] The upper limit of permafrost. Also known as pergelisol table. { 'pər·mə,fròst ,tā·bəl }

permanent aurora *See* airglow. { 'pər·mə·nənt ò'ròr·ə }

permanent current [OCEANOGR] A current which continues with relatively little periodic or seasonal change. { 'pər·mə·nənt 'kə·rənt }

permanent extinction [GEOL] The extinction of a lake by destruction of the lake basin, because of such processes as deposition of sediments, erosion of the basin rim, filling with vegetation, or catastrophic events. { 'pər·mə·nənt ik'stiŋk·shən }

permanent ice foot [HYD] An ice foot that does not melt completely in summer. { 'pər·mə·nənt 'īs ,fu̇t }

permanent thermocline *See* main thermocline. { 'pər·mə·nənt 'thər·mə,klīn }

permanent water [HYD] A source of water that remains constant throughout the year. { 'pər·mə·nənt 'wȯd·ər }

permeability [GEOL] The capacity of a porous rock, soil, or sediment for transmitting a fluid without damage to the structure of the medium. Also known as conductivity; perviousness. { ,pər·mē·ə'bil·əd·ē }

permeability trap [GEOL] An oil trap formed by lateral variation within a reservoir bed which seals the contained hydrocarbons through a change of permeability. { ,pər·mē·ə'bil·əd·ē ,trap }

permeable bed [GEOL] A porous reservoir formation through which hydrocarbon fluids (oil or gas) or water (waterflood or interstitial) can flow. { 'pər·mē·ə·bəl 'bed }

Permian [GEOL] The last period of geologic time in the Paleozoic era, from 280 to 225 million years ago. { 'pər·mē·ən }

permineralization [GEOL] A fossilization process whereby additional minerals are deposited in the pore spaces of originally hard animal parts. { pər,min·rə·lə'zā·shən }

Permo-Carboniferous [GEOL] **1.** The Permian and Carboniferous periods considered as one unit. **2.** The Permian and Pennsylvanian periods considered as a single unit. **3.** The rock unit, or the period of geologic time, transitional between the Upper Pennsylvanian and the Lower Permian periods. { ¦pər·mō,kär·bə'nif·ə·rəs }

perpendicular slip [GEOL] The component of a fault slip measured at right angles to the trace of the fault on any intersecting surface. { ¦pər·pən¦dik·yə·lər 'slip }

perpendicular slope [GEOL] A very steep slope or precipitous face, as on a mountain. { ¦pər·pən¦dik·yə·lər 'slōp }

perpendicular throw [GEOL] The distance between two points which were formerly adjacent in a faulted bed, vein, or other surface, measured at right angles to the surface. { ¦pər·pən¦dik·yə·lər 'thrō }

perpetual frost climate [CLIMATOL] The climate of the ice cap regions of the world; thus, it requires temperatures sufficiently cold so that the annual accumulation of snow and ice is never exceeded by ablation. Also known as ice-cap climate. { pər'pech·ə·wəl 'fròst ,klī·mət }

Perret phase [GEOL] That stage of a volcanic eruption that is characterized by the emission of much high-energy gas that may significantly enlarge the volcanic conduit. { 'per·ət ,fāz }

perry [METEOROL] In England, a sudden, heavy fall of rain; a squall, sometimes referred to as "half a gale." { 'per·ē }

persalic rock [GEOL] An igneous rock in which the ratio of salic to femic minerals is greater than 7:1. { pər'sal·ik 'räk }

persistence [METEOROL] With respect to the long-term nature of the wind at a given location, the ratio of the magnitude of the mean wind vector to the average speed of the wind without regard to direction. Also known as constancy; steadiness. { pər'sis·təns }

persistence forecast [METEOROL] A forecast that the future weather condition will be the same as the present condition; often used as a standard of comparison in

measuring the degree of skill of forecasts prepared by other methods. { pər'sis·təns ,fȯr,kast }
perthite [GEOL] A parallel to subparallel intergrowth of potassium and sodium feldspar; the potassium-rich phase is usually the host from which the sodium-rich phase evolves. { 'pər,thīt }
perthitic [GEOL] Of a texture produced by perthite, exhibiting sodium feldspar as small strings, blebs, films, or irregular veinlets in a host of potassium feldspar. { pər'thid·ik }
Peru Current [OCEANOGR] The cold ocean current flowing north along the coasts of Chile and Peru. Also known as Humboldt Current. { pə'rü 'kə·rənt }
perviousness *See* permeability. { 'pər·vē·əs·nəs }
petrifaction [GEOL] A fossilization process whereby inorganic matter dissolved in water replaces the original organic materials, converting them to a stony substance. { ,pe·trə'fak·shən }
petrified wood *See* silicified wood. { 'pe·trə,fīd 'wu̇d }
petrochemistry [GEOCHEM] An aspect of geochemistry that deals with the study of the chemical composition of rocks. { ¦pe·trō'kem·ə·strē }
petrofacies *See* petrographic facies. { ¦pe·trō'fā·shēz }
petrographer [GEOL] An individual who does petrography. { pə'träg·rə·fər }
petrographic facies [GEOL] Facies distinguished principally by composition and appearance. Also known as petrofacies. { ¦pe·trə¦graf·ik 'fā·shēz }
petrographic period [GEOL] The extension in time of a rock association. { ¦pe·trə¦graf·ik 'pir·ē·əd }
petrographic province [GEOL] A broad area in which similar igneous rocks are formed during the same period of igneous activity. Also known as comagmatic region; igneous province; magma province. { ¦pe·trə¦graf·ik 'präv·əns }
petrography [GEOL] The branch of geology that deals with the description and systematic classification of rocks, especially by means of microscopic examination. { pə'träg·rə·fē }
petroleum [GEOL] A naturally occurring complex liquid hydrocarbon which after distillation yields combustible fuels, petrochemicals, and lubricants; can be gaseous (natural gas), liquid (crude oil, crude petroleum), solid (asphalt, tar, bitumen), or a combination of states. { pə'trō·lē·əm }
petroleum geology [GEOL] The branch of economic geology dealing with the origin, occurrence, movement, accumulation, and exploration of hydrocarbon fuels. { pə'trō·lē·əm jē'äl·ə·jē }
petroleum seep *See* oil seep. { pə'trō·lē·əm ,sēp }
petroleum trap [GEOL] Stable underground formation (geological or physical) of such nature as to trap and hold liquid or gaseous hydrocarbons; usually consists of sand or porous rock surrounded by impervious rock or clay formations. { pə'trō·lē·əm ,trap }
petroliferous [GEOL] Containing petroleum. { ,pe·trə'lif·ə·rəs }
petrologen *See* kerogen. { pə'träl·ə·jən }
petrologist [GEOL] An individual who studies petrology. { pə'träl·ə·jəst }
petrology [GEOL] The branch of geology concerned with the origin, occurrence, structure, and history of rocks, principally igneous and metamorphic rock. { pə'träl·ə·jē }
petromict [GEOL] Of a sediment, composed of metastable rock fragments. { 'pe·trə,mikt }
petromorph [GEOL] A speleothem or cave formation that is exposed to the surface by erosion of the limestone in which the cave was formed. { 'pe·trə,mȯrf }
petrophysics [GEOL] Study of the physical properties of reservoir rocks. { ¦pe·trō¦fiz·iks }
petrotectonics [GEOL] Extension of the field of structural petrology to include analysis of the movements that produced the rock's fabric. Also known as tectonic analysis. { ¦pe·trō·tek'tän·iks }
pezograph *See* regmaglypt. { 'pez·ə,graf }

phacolith [GEOL] A minor, concordant, lens-shaped, and usually granitic intrusion into folded sedimentary strata. { 'fak·ə,lith }

Phanerozoic [GEOL] The part of geologic time for which there is abundant evidence of life, especially higher forms, in the corresponding rock, essentially post-Precambrian. { ¦fan·ə·rō¦zō·ik }

phantom [GEOL] A bed or member that is absent from a specific stratigraphic section but is usually present in a characteristic position in a sequence of similar geologic age. { 'fan·təm }

phantom bottom [OCEANOGR] A false bottom indicated by an echo sounder, some distance above the actual bottom; such an indication, quite common in the deeper parts of the ocean, is due to large quantities of small organisms. { 'fan·təm 'bäd·əm }

phantom horizon [GEOL] In seismic reflection prospecting, a line constructed so that it is parallel to the nearest actual dip segment at all points along a profile. { 'fan·təm hə'rīz·ən }

phase age *See* age of phase inequality. { 'fāz ,āj }

phase inequality [OCEANOGR] Variations in the tide or tidal currents associated with changes in the phase of the moon. { 'fāz ,in·i'kwäl·əd·ē }

phase lag [OCEANOGR] Angular retardation of the maximum of a constituent of the observed tide behind the corresponding maximum of the same constituent of the hypothetical equilibrium tide. Also known as tidal epoch. { 'fāz ,lag }

phenology [CLIMATOL] The science which treats of periodic biological phenomena with relation to climate, especially seasonal changes; from a climatologic viewpoint, these phenomena serve as bases for the interpretation of local seasons and the climatic zones, and are considered to integrate the effects of a number of bioclimatic factors. { fə'näl·ə·jē }

phi grade scale [GEOL] A logarithmic transformation of the Wentworth grade scale in which the diameter value of the particle is replaced by the negative logarithm to the base 2 of the particle diameter (in millimeters). { 'fī 'grād ,skāl }

phorogenesis [GEOL] The shifting or slipping of the earth's crust relative to the mantle. { ¦fȯr·ə'jen·ə·səs }

phosphatic nodule [GEOL] A dark, usually black, earthy mass or pebble of variable size and shape, having a hard shiny surface and occurring in marine strata. { fä'sfad·ik 'naj·yül }

phosphatization [GEOCHEM] Conversion to a phosphate or phosphates; for example, the diagenetic replacement of limestone, mudstone, or shale by phosphate-bearing solutions, producing phosphates of calcium, aluminum, or iron. { ,fäs·fəd·ə'zā·shən }

phosphorization [GEOCHEM] Impregnation or combination with phosphorus or a compound of phosphorus; for example, the diagenetic process of phosphatization. { ,fäs·fə·rə'zā·shən }

phosphorus-nitrogen ratio [OCEANOGR] The proportion, by weight, of phosphorus to nitrogen in seawater or in plankton; the ratio is approximately 7:1. { 'fäs·fə·rəs 'nī·trə·jən 'rā·shō }

photochemical smog [METEOROL] Chemical pollutants in the atmosphere resulting from chemical reactions involving hydrocarbons and nitrogen oxides in the presence of sunlight. { ¦fōd·ō'kem·ə·kəl 'smäg }

photoclinometry [GEOL] A technique for ascertaining slope information from an image brightness distribution, used especially for studying the amount of slope to a lunar crater wall or ridge by measuring the density of its shadow. { ¦fōd·ō·klə'näm·ə·trē }

photogeologic anomaly [GEOL] Any systematic deviation of a photogeologic factor from the expected norm in a given area. { ¦fōd·ō,jē·ə'läj·ik ə'näm·ə·lē }

photogeologic map [GEOL] A compilation of interpretations of a series of aerial photographs, including annotations of geologic features. { ¦fōd·ō,jē·ə'läj·ik 'map }

photogeology [GEOL] The geologic interpretation of landforms by means of aerial photographs. { ¦fōd·ō,jē'äl·ə·jē }

photogeomorphology [GEOL] The study of landforms by means of aerial photographs. { ¦fōd·ō,jē·ō·mȯr'fäl·ə·jē }
phreatic [GEOL] Of a volcanic explosion of material such as steam or mud, not being incandescent. { frē'ad·ik }
phreatic cycle [HYD] The period of time during which the water table rises and then falls. { frē'ad·ik 'sī·kəl }
phreatic gas [GEOL] A gas formed by the contact of atmospheric or surface water with ascending magma. { frē'ad·ik 'gas }
phreatic surface *See* water table. { frē'ad·ik 'sər·fəs }
phreatic water [HYD] Groundwater in the zone of saturation. { frē'ad·ik 'wȯd·ər }
phreatic-water discharge *See* groundwater discharge. { frē'ad·ik ¦wȯd·ər 'dis,chärj }
phreatic zone *See* zone of saturation. { frē'ad·ik ,zōn }
phreatomagmatic [GEOL] Pertaining to a volcanic explosion that extrudes both magmatic gases and steam; it is caused by the contact of the magma with groundwater or ocean water. { frē¦ad·ō·mag'mad·ik }
phyllofacies [GEOL] A facies differentiated on the basis of stratification characteristics, especially the stratification index. { ,fil·ō'fā·shēz }
phyllomorphic stage [GEOL] The most advanced geochemical stage of diagenesis, characterized by authigenic development of micas, feldspars, and chlorites at the expense of clays. { ¦fil·ə¦mȯr·fik ,stāj }
physical climate [CLIMATOL] The actual climate of a place, as distinguished from a hypothetical climate, such as the solar climate or mathematical climate. { 'fiz·ə·kəl 'klī·mət }
physical climatology [CLIMATOL] The major branch of climatology, which deals with the explanation of climate, rather than with presentation of it (climatography). { 'fiz·ə·kəl ,klī·mə'täl·ə·jē }
physical exfoliation [GEOL] A type of exfoliation caused by physical forces; for example, by the freezing of water that has penetrated fine cracks in rock or by the removal of overburden concealing deeply buried rocks. { 'fiz·ə·kəl eks,fō·lē'ā·shən }
physical forecasting *See* numerical forecasting. { 'fiz·ə·kəl 'fȯr,kast·iŋ }
physical geography [GEOGR] The study of the earth's surface features and associated processes. { 'fiz·ə·kəl jē'äg·rə·fē }
physical geology [GEOL] That branch of geology concerned with understanding the composition of the earth and the physical changes occurring in it, based on the study of rocks, minerals, and sediments, their structures and formations, and their processes of origin and alteration. { 'fiz·ə·kəl jē'äl·ə·jē }
physical meteorology [METEOROL] That branch of meteorology which deals with optical, electrical, acoustical, and thermodynamic phenomena of the atmosphere, its chemical composition, the laws of radiation, and the explanation of clouds and precipitation. { 'fiz·ə·kəl ,mēd·ē·ə'räl·ə·jē }
physical oceanography [OCEANOGR] The study of the physical aspects of the ocean, the movements of the sea, and the variability of these factors in relationship to the atmosphere and the ocean bottom. { 'fiz·ə·kəl ,ō·shə'näg·rə·fē }
physical residue [GEOL] A residue which results from physical, as opposed to chemical, weathering processes. { 'fiz·ə·kəl 'rez·ə,dü }
physical stratigraphy [GEOL] Stratigraphy based on the physical aspects of rocks, especially the sedimentologic aspects. { 'fiz·ə·kəl strə'tig·rə·fē }
physical time [GEOL] Geologic time as measured by some physical process, such as the radioactive decay of elements. { 'fiz·ə·kəl 'tīm }
physical weathering *See* mechanical weathering. { 'fiz·ə·kəl 'weth·ə·riŋ }
physiographic diagram [GEOL] A small-scale map showing landforms by the systematic application of a standardized set of simplified pictorial symbols that represent the appearance such forms would have if viewed obliquely from the air at an angle of about 45°. Also known as landform map; morphographic map. { ¦fiz·ē·ə¦graf·ik 'dī·ə,gram }
physiographic feature [GEOL] A prominent or conspicuous physiographic form or noticeable part thereof. { ¦fiz·ē·ə¦graf·ik 'fē·chər }

physiographic form [GEOL] A landform considered with regard to its origin, cause, or history. { ¦fiz·ē·ə¦graf·ik 'fórm }
physiographic province [GEOL] A region having a pattern of relief features or landforms that differs significantly from that of adjacent regions. { ¦fiz·ē·ə¦graf·ik 'präv·əns }
phyteral [GEOL] Morphologically recognizable forms of vegetal matter in coal. { 'fīd·ə·rəl }
phytoclimatology [CLIMATOL] The study of the microclimate in the air space occupied by plant communities, on the surfaces of the plants themselves and, in some cases, in the air spaces within the plants. { ¦fīd·ō,klī·mə'täl·ə·jē }
phytocollite [GEOL] A black, gelatinous, nitrogenous humic body occurring beneath or within peat deposits. { fī'täk·ə,līt }
Piacention *See* Plaisancian. { ,pē·ə'sen·chən }
pibal *See* pilot-balloon observation. { 'pī,bal }
piecemeal stoping [GEOL] Magmatic stoping in which only isolated blocks of roof rock are assimilated. { 'pēs,mēl ,stōp·iŋ }
piedmont [GEOL] Lying or formed at the base of a mountain or mountain range, as a piedmont terrace or a piedmont pediment. { 'pēd,mänt }
piedmont alluvial plain *See* bajada. { 'pēd,mänt ə'lüv·ē·əl 'plān }
piedmont angle [GEOL] The sharp break of slope between a hill and a plain, such as the angle at the junction of a mountain front and the pediment at its base. { 'pēd ,mänt ¦aŋ·gəl }
piedmont bench *See* piedmont step. { 'pēd,mänt ¦bench }
piedmont benchland [GEOL] One of several successions or systems of piedmont steps. Also known as piedmont stairway; piedmont treppe. { 'pēd,mänt 'bench,land }
piedmont bulb [HYD] The lobe or fan of ice formed when a glacier spreads out on a plain at the lower end of a valley. { 'pēd,mänt ¦bəlb }
piedmont flat *See* piedmont step. { 'pēd,mänt ,flat }
piedmont glacier [HYD] A thick, continuous ice sheet formed at the base of a mountain range by the spreading out and coalescing of valley glaciers from higher mountain elevations. { 'pēd,mänt ¦glā·shər }
piedmont gravel [GEOL] Coarse gravel derived from high ground by mountain torrents and spread out on relatively flat ground where the velocity of the water is decreased. { 'pēd,mänt ¦grav·əl }
piedmont ice [HYD] An ice sheet formed by the joining of two or more glaciers on a comparatively level plain at the base of the mountains down which the glaciers descended; it may be partly afloat. { 'pēd,mänt ¦īs }
piedmont lake [HYD] An oblong lake occupying a partly overdeepened basin excavated from rock by a piedmont glacier, or dammed by a glacial moraine. { 'pēd,mänt ¦lāk }
piedmont plain *See* bajada. { 'pēd,mänt ¦plān }
piedmont plateau [GEOL] A plateau lying between the mountains and the plains or the ocean. { 'pēd,mänt pla'tō }
piedmont scarp [GEOL] A small, low cliff formed in alluvium on a piedmont slope at the foot of a steep mountain range; due to dislocation of the surface, especially by faulting. Also known as scarplet. { 'pēd,mänt ¦skärp }
piedmont slope *See* bajada. { 'pēd,mänt ¦slōp }
piedmont stairway *See* piedmont benchland. { 'pēd,mänt 'ster,wā }
piedmont step [GEOL] A terracelike or benchlike piedmont feature that slopes outward or downvalley. Also known as piedmont bench; piedmont flat. { 'pēd,mänt ¦step }
piedmont treppe *See* piedmont benchland. { 'pēd,mänt 'trēp·ə }
piercement *See* diapir. { 'pirs·mənt }
piercement dome *See* diapir. { 'pirs·mənt ,dōm }
piercing fold *See* diapir. { 'pirs·iŋ ,fōld }
piezocrystallization [GEOL] Crystallization of a magma under pressure, such as the pressure associated with orogeny. { pē¦ā·zō,krist·əl·ə'zā·shən }
piezogene [GEOL] Pertaining to the formation of minerals primarily under the influence of pressure. { pē'ā·zō,jēn }

piezoglypt *See* regmaglypt. { pē'ā·zō,glipt }
piezometric surface *See* potentiometric surface. { pē¦ā·zō¦me·trik 'sər·fəs }
pike [GEOL] A mountain or hill which has a peaked summit. { pīk }
pileus [METEOROL] An accessory cloud of small horizontal extent, often cirriform, in the form of a cap, hood, or scarf, which occurs above or attached to the top of a cumuliform cloud that often pierces it; several pileus clouds fairly often are observed above each other. Also known as scarf cloud. { 'pil·ē·əs }
pillar [GEOL] **1.** A natural formation shaped like a pillar. **2.** A joint block produced by columnar jointing. **3.** *See* stalacto-stalagmite. { 'pil·ər }
pillow lava [GEOL] Any lava characterized by pillow structure and presumed to have formed in a subaqueous environment. Also known as ellipsoidal lava. { 'pil·ō ,läv·ə }
pillow structure [GEOL] A primary sedimentary structure that resembles a pillow in size and shape. Also known as mammillary structure. { 'pil·ō ,strək·chər }
pilmer [METEOROL] In England, a heavy shower of rain. { 'pil·mər }
pilotaxitic [GEOL] Pertaining to the texture of the groundmass of a holocrystalline igneous rock in which lath-shaped microlites (usually of plagioclase) are arranged in a glass-free felty mesh, often aligned along the flow lines. { ¦pī·lō·tak'sid·ik }
pilot-balloon observation [METEOROL] A method of winds-aloft observation, that is, the determination of wind speeds and directions in the atmosphere above a station; involves reading the elevation and azimuth angles of a theodolite while visually tracking a pilot balloon. Also known as pibal. { 'pī·lət bə,lün ,äb·zər'vā·shən }
pilot briefing [METEOROL] Oral comment on the observed and forecast weather conditions along a route, given by a forecaster to the pilot, navigator, or other air crew member prior to takeoff. Also known as briefing; flight briefing; flight-weather briefing. { 'pī·lət ,brēf·iŋ }
pilot report [METEOROL] A report of in-flight weather by an aircraft pilot or crew member; a complete pilot report includes the following information in this order: location or extent of reported weather phenomena, time of observation, description of phenomena, altitude of phenomena, type of aircraft (only with reports of turbulence or icing). Also known as aircraft report; pirep. { 'pī·lət ri,pȯrt }
pilot streamer [GEOPHYS] A relatively slow-moving, nonluminous lightning streamer, the existence of which has been postulated to help account for the observed mode of advance of a stepped leader as it initiates a lightning discharge. { 'pī·lət ,strēm·ər }
pimple mound [GEOL] A low, flattened, roughly circular or elliptical dome consisting of sandy loam that is entirely distinct from the surrounding soil; peculiar to the Gulf coast of eastern Texas and southwestern Louisiana. { 'pim·pəl ,maủnd }
pimple plain [GEOL] A plain distinguished by the presence of numerous, conspicuous pimple mounds. { 'pim·pəl ,plān }
pinch [GEOL] Thinning of a rock layer, as where a vein narrows. { pinch }
pinch-and-swell structure [GEOL] A structural condition common in pegmatites and veins of quartz in metamorphosed rocks; the vein is pinched at frequent intervals, leaving expanded parts between. { ¦pinch ən 'swel ,strək·chər }
piner [METEOROL] In England, a rather strong breeze from the north or northeast. { 'pīn·ər }
pingo [HYD] A frost mound resembling a volcano, being a relatively large and conical mound of soil-covered ice, elevated by hydrostatic pressure of water within or below the permafrost of arctic regions. { 'piŋ·gō }
pingo ice [HYD] Clear or relatively clear ice that occurs in permafrost; originates from groundwater under pressure. { 'piŋ·gō ,īs }
pingo remnant [GEOL] A rimmed depression formed by the rupturing of a pingo summit which results in the exposure of the ice core to melting followed by partial or total collapse. Also known as pseudokettle. { 'piŋ·gō ¦rem·nənt }
pinnacle [GEOL] **1.** A sharp-pointed rock rising from the bottom, which may extend above the surface of the water, and may be a hazard to surface navigation; due to the sheer rise from the sea floor, no warning is given by sounding. **2.** Any high tower or spire-shaped pillar of rock, alone or cresting a summit. { 'pin·ə·kəl }

pinnacled iceberg [OCEANOGR] An iceberg weathered in such manner as to produce spires or pinnacles. Also known as irregular iceberg; pyramidal iceberg. { 'pin·ə·kəld 'īs,bərg }

pinnate drainage [HYD] A dendritic drainage pattern in which the main stream receives many closely spaced, subparallel tributaries that join it at acute angles; resembles a feather in plan view. { 'pi,nāt 'drā·nij }

pipe [GEOL] **1.** A vertical, cylindrical ore body. Also known as chimney; neck; ore chimney; ore pipe; stock. **2.** A tubular cavity of varying depth in calcareous rocks, often filled with sand and gravel. **3.** A vertical conduit through the crust of the earth below a volcano, through which magmatic materials have passed. Also known as breccia pipe. { pīp }

pipe amygdule [GEOL] An elongate amygdule occurring toward the base of a lava flow, probably formed by the generation of gases or vapor from the underlying material. { 'pīp ə'mig,dyül }

pipe clay [GEOL] A mass of fine clay, usually lens-shaped, which forms the surface of bedrock and upon which often rests the gravel of old river beds. { 'pīp ,klā }

pipernoid texture [GEOL] The eutaxitic texture of certain extrusive igneous rocks in which dark patches and stringers occur in a light-colored groundmass. { 'pī·pər,nȯid ,teks·chər }

pipe vesicle [GEOL] A slender vertical cavity, a few centimeters or tens of centimeters in length, extending upward from the base of a lava flow. { 'pīp ,ves·ə·kəl }

piping [HYD] Erosive action of water passing through or under a dam, which may result in leakage or failure. { 'pīp·iŋ }

pipkrake [HYD] A small, thin needlelike crystal of ice formed just below ground level and growing perpendicular to the soil surface. Also known as needle ice. { 'pip,krāk }

pirep *See* pilot report. { 'pī,rep }

pisolith [GEOL] Small, more or less spherical particles found in limestones and dolomites, having a diameter of 2–10 millimeters and often formed of calcium carbonate. { 'pī·zə,lith }

pisolitic tuff [GEOL] Of a tuff, composed of accretionary lapilli or pisolites. { ¦pī·zə¦lid·ik 'təf }

pitch *See* plunge. { pich }

pitch coal *See* bituminous lignite. { 'pich ,kōl }

pitching fold *See* plunging fold. { 'pich·iŋ ,fōld }

pitchstone [GEOL] A type of volcanic glass distinguished by a waxy, dull, resinous, pitchy luster. Also known as fluolite. { 'pich,stōn }

pit-run gravel [GEOL] A natural deposit of a mixture of gravel, sand, and foreign materials. { 'pit ,rən ,grav·əl }

pitted outwash plain [GEOL] An outwash plain characterized by numerous depressions such as kettles, shallow pits, and potholes. { 'pid·əd 'au̇t,wäsh ,plān }

pitted pebble [GEOL] A pebble having marked concavities not related to the texture of the rock in which it appears or to differential weathering. { 'pid·əd 'peb·əl }

pivotal fault *See* rotary fault. { 'piv·əd·əl 'fȯlt }

placanticline [GEOL] A gentle, anticlinal-like uplift of the continental platform, usually asymmetric and without a typical outline. { plak'ant·i,klīn }

placer [GEOL] A mineral deposit at or near the surface of the earth, formed by mechanical concentration of mineral particles from weathered debris. Also known as ore of sedimentation. { 'plās·ər }

placic horizon [GEOL] A black to dark red soil horizon that is usually cemented with iron and is not very permeable. { 'plā·sik hə'rīz·ən }

Plaggept [GEOL] A suborder of the soil order Inceptisol, with very thick surface horizons of mixed mineral and organic materials resulting from manure or human wastes added over long periods of time. { 'plä·gept }

plain [GEOGR] An extensive, broad tract of level or rolling, almost treeless land with a shrubby vegetation, usually at a low elevation. [GEOL] A flat, gently sloping region of the sea floor. Also known as submarine plain. { plān }

plain of denudation [GEOL] A surface that has been reduced to sea level or to just above sea level by the agents of erosion (usually considered to be of subaerial origin). { 'plān əv ˌdē·nü'dā·shən }

plain of lateral planation [GEOL] An extensive, smooth, apronlike surface developed at the base of a mountain or escarpment by the widening of valleys and the coalescence of floodplains as a result of lateral planation. { 'plān əv ¦lad·ə·rəl plā'nā·shən }

plain of marine denudation [GEOL] A plane or nearly plane surface worn down by the gradual encroachment of ocean waves upon the land; or a plane or nearly plane imaginary surface representing such a plain after uplift and partial subaerial erosion. Also known as plain of submarine denudation. { 'plān əv mə¦rēn ˌdē·nü'dā·shən }

plain of marine erosion [GEOL] A theoretical platform representing a plane surface of unlimited width produced below sea level by the complete cutting away of the land by marine processes acting over a very long period of stillstand. { 'plān əv mə¦rēn i'rō·zhən }

plain of submarine denudation *See* plain of marine denudation. { 'plān əv ¦səb·məˌrēn ˌdē·nü'dā·shən }

plains-type fold [GEOL] An anticlinal or domelike structure of the continental platform which has no typical outline and for which there is no corresponding synclinal structure. { 'plānz ¦tīp ˌfōld }

plain tract [GEOL] The lower part of a stream, characterized by a low gradient and a wide floodplain. { 'plān ¦trakt }

Plaisancian [GEOL] A European stage of geologic time: lower Pliocene (above Pontian of Miocene, below Astian). Also known as Piacention; Plaisanzian. { plā'zän·chən }

Plaisanzian *See* Plaisancian. { plā'zän·zhən }

plaiting [GEOL] A texture in some schists that results from the intersection of relict bedding planes with well-developed cleavage planes. Also known as gaufrage. { 'plād·iŋ }

planar cross-bedding [GEOL] Cross-bedding characterized by planar surfaces of erosion in the lower bounding surface. { 'plā·nər 'krȯs ˌbed·iŋ }

planate [GEOL] Referring to a surface that has been flattened or leveled by planation. { 'plāˌnāt }

planation [GEOL] Erosion resulting in flat surfaces, caused by meandering streams, waves, ocean currents, wind, or glaciers. { plā'nā·shən }

planation stream piracy [HYD] Capture effected by the lateral planation of a stream invading and diverting the upper part of a smaller stream. { plā'nā·shən ¦strēm ˌpī·rə·sē }

plane atmospheric wave [METEOROL] An atmospheric wave represented in two-dimensional rectangular cartesian coordinates, in contrast to a wave considered on the spherical earth. { 'plān ¦at·mə¦sfir·ik 'wāv }

plane bed [GEOL] A sedimentary bed without elevations or depressions larger than the maximum size of the bed material. { 'plān ˌbed }

plane jet [HYD] A stream flow pattern characteristic of hyperpycnal inflow, in which the inflowing water spreads as a parabola whose width is about three times the square root of the distance downstream from the mouth. { 'plān ˌjet }

plane of saturation *See* water table. { 'plān əv ˌsach·ə'rā·shən }

planetary boundary layer [METEOROL] That layer of the atmosphere from the earth's surface to the geostrophic wind level, including, therefore, the surface boundary layer and the Ekman layer; above this layer lies the free atmosphere. { 'plan·əˌter·ē 'bau̇n·drē ˌlā·ər }

planetary circulation *See* general circulation. { 'plan·əˌter·ē ˌsər·kyə'lā·shən }

planetary geology [GEOL] A science that applies geologic principles and techniques to the study of planets and their natural satellites. Also know as planetary geoscience. { 'plan·əˌter·ē jē'äl·ə·jē }

planetary geoscience *See* planetary geology. { 'plan·əˌter·ē ¦jē·ō'sī·əns }

planetary vorticity effect [GEOPHYS] The effect of the variation of the earth's vorticity with latitude in altering the relative vorticity of a flow with a meridional component; a fluid with a free surface in a rotating cylinder exhibits a corresponding effect, owing to the shrinking or stretching of radially displaced columns. { 'plan·ə,ter·ē vȯr'tis·əd·ē i,fekt }

planetary wave *See* Rossby wave. { 'plan·ə,ter·ē 'wāv }

planetary wind [METEOROL] Any wind system of the earth's atmosphere which owes its existence and direction to solar radiation and to the rotation of the earth. { 'plan·ə,ter·ē 'wind }

planform [GEOGR] A body of water's outline or morphology as defined by the still water line. { 'plan,fȯrm }

planoconformity [GEOL] The relation between conformable strata that are approximately uniform in thickness and sensibly parallel throughout. { ¦plā·nō·kən'fȯr·məd·ē }

Planosol [GEOL] An intrazonal, hydromorphic soil having a clay pan or hardpan covered with a leached surface layer; developed in a humid to subhumid climate. { 'plan·ə,sȯl }

plash [HYD] A shallow, standing, usually short-lived pool or small pond resulting from a flood, heavy rain, or melting snow. { plash }

plasma [GEOL] The part of a soil material that can be, or has been, moved, reorganized, or concentrated by soil-forming processes. { 'plaz·mə }

plasma mantle [GEOPHYS] A thick layer of plasma just inside the magnetopause characterized by a tailward bulk flow with a speed of 60 to 120 miles (100 to 200 kilometers) per second and by a gradual decrease of density, temperature, and speed as the depth inside the magnetosphere increases. { 'plaz·mə 'mant·əl }

plasmapause [GEOPHYS] The sharp outer boundary of the plasmasphere, at which the plasma density decreases by a factor of 100 or more. { 'plaz·mə,pȯz }

plasma sheet [GEOPHYS] A region of relatively hot plasma outside the plasmasphere, which reaches, during quiet times, from an altitude of about 30,000 miles (50,000 kilometers) to at least past the moon's orbit in a long tail extending away from the sun; composed of particles with typical thermal energies of 2 to 4 kiloelectronvolts. { 'plaz·mə ,shēt }

plasmasphere [GEOPHYS] A region of relatively dense, cold plasma surrounding the earth and extending out to altitudes of approximately 2 to 6 earth radii, composed predominantly of electrons and protons, with thermal energies not exceeding several electronvolts. { 'plaz·mə,sfir }

plaster conglomerate [GEOL] A conglomerate composed entirely of boulders derived from a partially exhumed monadnock forming a wedgelike mass of its flank. { 'plas·tər kən'gläm·ə·rət }

plastic equilibrium [GEOL] State of stress within a soil mass or a portion thereof that has been deformed to such an extent that its ultimate shearing resistance is mobilized. { 'plas·tik ,ē·kwə'lib·rē·əm }

plasticity index [GEOL] The percent difference between moisture content of soil at the liquid and plastic limits. { plas'tis·əd·ē ,in,deks }

plasticlast [GEOL] An intraclast consisting of calcareous mud that has been torn up while still soft. { 'plas·tə,klast }

plastic limit [GEOL] The water content of a sediment, such as a soil, at the point of transition between the plastic and semisolid states. { 'plas·tik 'lim·ət }

plastic zone [GEOL] A region located adjacent to the rupture zone of an explosion crater and at an increased distance from the shot site, differing from the rupture zone by having less fracturing and only small permanent deformations. { 'plas·tik ,zōn }

plate [GEOL] **1.** A smooth, thin, flat fragment of rock, such as a flagstone. **2.** A large rigid, but mobile, block involved in plate tectonics; thickness ranges from 30 to 150 miles (50 to 250 kilometers) and includes both crust and a portion of the upper mantle. { plāt }

plateau [GEOGR] An extensive, flat-surfaced upland region, usually more than 45–90 meters (150–300 feet) in elevation and considerably elevated above the adjacent country and limited by an abrupt descent on at least one side. [GEOL] A broad, comparatively flat and poorly defined elevation of the sea floor, commonly over 60 meters (200 feet) in elevation. { pla'tō }

plateau basalt [GEOL] One or a succession of high-temperature basaltic lava flows from fissure eruptions which accumulate to form a plateau. Also known as flood basalt. { pla'tō bə'sölt }

plateau glacier [HYD] A highland glacier that overlies a generally flat mountain tract; usually overflows its edges in hanging glaciers. { pla'tō ¦glā·shər }

plateau gravel [GEOL] A sheet, spread, or patch of surficial gravel, often compacted, occupying a flat area on a hilltop, plateau, or other high region at a height above that normally occupied by a stream terrace gravel. { pla'tō ¦grav·əl }

plateau mountain [GEOL] A pseudomountain produced by the dissection of a plateau. { pla'tō ¦maunt·ən }

plateau plain [GEOL] An extensive plain surmounted by a sublevel summit area and bordered by escarpments. { pla'tō ,plān }

plate crystal [HYD] An ice crystal exhibiting typical hexagonal (rarely triangular) symmetry and having comparatively little thickness parallel to its principal axis (*c* axis); as such crystals fall through the clouds in which they form, they may encounter conditions causing them to develop dendritic extensions, that is, to become plane-dendritic crystals. { 'plāt ¦krist·əl }

plate ice *See* pancake ice. { 'plāt ,īs }

platelet [HYD] A small ice crystal which, when united with other such crystals, forms a layer of floating ice, especially sea ice, and serves as seed crystals for further thickening of the ice cover. { 'plāt·lət }

plate tectonics [GEOL] Global tectonics based on a model of the earth characterized by a small number (10–25) of semirigid plates which float on some viscous underlayer in the mantle; each plate moves more or less independently and grinds against the others, concentrating most deformation, volcanism, and seismic activity along the periphery. Also known as raft tectonics. { 'plāt tek'tän·iks }

platform [GEOL] **1.** Any level or almost level surface; a small plateau. **2.** A continental area covered by relatively flat or gently tilted, mainly sedimentary strata which overlay a basement of rocks consolidated during earlier deformations; platforms and shields together constitute cratons. { 'plat,fȯrm }

platform beach [GEOL] A looped bar or ridge of sand and gravel formed on a wave-cut platform. { 'plat,fȯrm ,bēch }

platform facies *See* shelf facies. { 'plat,fȯrm ,fā·shēz }

platform reef [GEOL] An organic reef, generally small but more extensive than a patch reef, with a flat upper surface. { 'plat,fȯrm ,rēf }

platte [GEOL] A resistant knob of rock in a glacial valley or rising in the midst of an existing glacier, often causing a glacier to split near its snout. { 'plad·ə }

platy [GEOL] **1.** Referring to a sedimentary particle whose length is more than three times its thickness. **2.** Referring to a sandstone or limestone that splits into laminae having thicknesses in the range of 2 to 10 millimeters. { 'plad·ē }

playa [GEOL] **1.** A low, essentially flat part of a basin or other undrained area in an arid region. **2.** A small, generally sandy land area at the mouth of a stream or along the shore of a bay. **3.** A flat, alluvial coastland, as distinguished from a beach. { 'plī·ə }

playa lake [HYD] A shallow temporary sheet of water covering a playa in the wet season. { 'plī·ə ,lāk }

Playfair's law [GEOL] The law that each stream cuts its own valley, the valley being proportional in size to its stream, and the stream junctions in the valley are accordant in level. { 'plā,ferz ,lȯ }

Pleistocene [GEOL] The older of the two epochs of the Quaternary Period, spanning about 1.8 million to 10,000 years ago. It represents the interval of geological time (and rocks accumulated during that time) extending from the end of the Pliocene Epoch (and the end of Tertiary Period) to the start of the Holocene Epoch. It is commonly characterized as an epoch when the earth entered its most recent phase of widespread glaciation. Also known as Ice Age; Oiluvium. { 'plī·stə,sēn }

plerotic water [HYD] That part of subsurface water that forms the zone of saturation, including underground streams. { plə'räd·ik 'wȯd·ər }

plexus [GEOL] An area on a subglacial deposit that encloses a giant's kettle. { 'plek·səs }

plication [GEOL] Intense, small-scale folding. { plī'kā·shən }

Pliensbachian [GEOL] A European stage of geologic time: Lower Jurassic (above Sinemurian, below Toarcian). { plēnz'bäk·ē·ən }

Plinian eruption *See* Vulcanian eruption. { 'plin·ē·ən i'rəp·shən }

plinth [GEOL] The lower and outer part of a seif dune, beyond the slip-face boundaries, that has never been subjected to sand avalanches. { plinth }

plinthite [GEOL] In a soil, a material consisting of a mixture of clay and quartz with other diluents, that is rich in sesquioxides, poor in humus, and highly weathered. { 'plin,thīt }

Pliocene [GEOL] The youngest of the five geological epochs of the Tertiary Period. The Pliocene represents the interval of geological time (and rocks deposited during that time) extending from the end of the Miocene Epoch to the beginning of the Pleistocene Epoch of the Quaternary Period. Modern time scales assign the duration of 5.0 million to 1.8 million years ago to the Pliocene. { 'plī·ə,sēn }

pliothermic [GEOL] Pertaining to a period in geologic history characterized by more than average climatic warmth. { ¦plī·ō¦thər·mik }

plough wind *See* plow wind. { 'plau̇ ,wind }

plowshare [HYD] A wedge-shaped feature developed on a snow surface by further ablation of foam crust. { 'plau̇,sher }

plow sole [GEOL] A pressure pan representing a layer of soil compacted by repeated plowing to the same depth. { 'plau̇ ,sōl }

plow wind [METEOROL] A term used in the midwestern United States to describe strong, straight-line winds associated with squall lines and thunderstorms; resulting damage is usually confined to narrow zones like that caused by tornadoes; however, the winds are all in one direction. Also spelled plough wind. Also known as derecho. { 'plau̇ ,wind }

plucking [GEOL] A process of glacial erosion which involves the penetration of ice or rock wedges into subglacial niches, crevices, and joints in the bedrock; as the glacier moves, it plucks off pieces of jointed rock and incorporates them. Also known as glacial plucking; quarrying. { 'plək·iŋ }

plug [GEOL] **1.** A vertical pipelike magmatic body representing the conduit to a former volcanic vent. **2.** A crater filling of lava, the surrounding material of which has been removed by erosion. **3.** A mass of clay, sand, or other sediment filling the part of a stream channel abandoned by the formation of a cutoff. { pləg }

plug dome [GEOL] A volcanic dome characterized by an upheaved, consolidated conduit filling. { 'pləg ,dōm }

plug reef [GEOL] A small, triangular reef that grows with its apex pointing seaward through openings between linear shelf-edge reefs. { 'pləg ,rēf }

plum [GEOL] A clast embedded in a matrix of a different kind, especially a pebble in a conglomerate. { pləm }

plumb line [GEOPHYS] A continuous curve to which the direction of gravity is everywhere tangential. { 'pləm ,līn }

plume structure [GEOL] On the surface of a master joint, a ridgelike tracing in a plumelike pattern, usually oriented parallel to the upper and lower surfaces of the constituent rock unit. Also known as plumose structure. { 'plüm ,strək·chər }

plumose structure *See* plume structure. { 'plü,mōs ,strək·chər }

plunge [GEOL] The inclination of a geologic structure, especially a fold axis, measured by its departure from the horizontal. Also known as pitch; rake. { plənj }

plunge basin [GEOL] A deep, large hollow or cavity scoured in the bed of a stream at the foot of a waterfall or cataract by the force and eddying effect of the falling water. { 'plənj ,bās·ən }

plunge point [OCEANOGR] The point at which a plunging wave curls over and falls as it moves toward the shore. { 'plənj ,pȯint }

plunge pool [HYD] **1.** The water in a plunge basin. **2.** A deep, circular lake occupying a plunge basin after the waterfall has ceased to exist or the stream has been diverted. Also known as waterfall lake. **3.** A small, deep plunge basin. { 'plənj ˌpül }

plunging breaker [OCEANOGR] A breaking wave whose crest curls over and collapses suddenly. Also known as spilling breaker; surging breaker. { 'plənj·iŋ 'brāk·ər }

plunging cliff [GEOL] A sea cliff bordering directly on deep water, having a base that lies well below water level. { 'plənj·iŋ 'klif }

plunging fold [GEOL] A fold having a relatively steep plunge. Also known as pitching fold. { 'plənj·iŋ 'fōld }

plutology [GEOL] The study of the interior of the earth. { plü'täl·ə·jē }

pluton [GEOL] **1.** An igneous intrusion. **2.** A body of rock formed by metasomatic replacement. { 'plü,tän }

plutonian *See* plutonic. { plü'tō·nē·ən }

plutonic [GEOL] Pertaining to rocks formed at a great depth. Also known as abyssal; deep-seated; plutonian. { plü'tän·ik }

plutonic breccia [GEOL] Breccia consisting of older annular rock fragments enclosed in younger plutonic rock. { plü'tän·ik 'brech·ə }

plutonic metamorphism [GEOL] Deep-seated regional metamorphism at high temperatures and pressures, often accompanied by strong deformation. { plü'tän·ik ˌmed·ə'mȯr,fiz·əm }

plutonic rock [GEOL] A rock formed at considerable depth by crystallization of magma or by chemical alteration. { plü'tän·ik 'räk }

plutonic water [HYD] Juvenile water in magma, or derived from magma, at a considerable depth, probably several kilometers. { plü'tän·ik 'wȯd·ər }

plutonism [GEOL] **1.** Pertaining to the processes associated with pluton formation. **2.** The theory that the earth formed by solidification of a molten mass. { 'plüt·ən,iz·əm }

pluvial [GEOL] Of a geologic process or feature, effected by rain action. [METEOROL] Pertaining to rain, or more broadly, to precipitation, particularly to an abundant amount thereof. { 'plü·vē·əl }

pluvial lake [GEOL] A lake formed during a period of exceptionally heavy rainfall; specifically, a Pleistocene lake formed during a period of glacial advance and now either extinct or only a remnant. { 'plü·vē·əl 'lāk }

pluviofluvial [GEOL] Pertaining to the combined action of rainwater and streams. { ¦plü·vē·ō¦flü·vē·əl }

pluviometric coefficient [METEOROL] For any month at a given station, the ratio of the monthly normal precipitation to one-twelfth of the annual normal precipitation. Also known as hyetal coefficient. { ¦plü·vē·ə¦me·trik ˌkō·i'fish·ənt }

pneumatogenic [GEOL] Referring to a rock or mineral deposit formed by a gaseous agent. { ¦nü·məd·ō¦jen·ik }

pneumatolysis [GEOL] Rock alteration or mineral crystallization effected by gaseous emanations from solidifying magma. { ˌnü·mə'täl·ə·səs }

pneumatolytic [GEOL] Formed by gaseous agents. { ¦nü·məd·ō¦lid·ik }

pneumatolytic stage [GEOL] The stage in the cooling of a magma in which the solid and gaseous phases are in equilibrium. { ¦nü·məd·ō¦lid·ik 'stāj }

pneumotectic [GEOL] Referring to processes and products of magmatic consolidation affected to some degree by the gaseous constituents of the magma. { ¦nü·mō¦tek·tik }

pocket [GEOL] **1.** A cavity that contains a deposit such as a gas or an ore. **2.** An enclosed or sheltered place along a coast, such as a reentrant between rocky, cliffed headlands or a bight on a lee shore. { 'päk·ət }

pocket beach [GEOL] A small, narrow beach formed in a pocket, commonly crescentic in plan, with the concave edge toward the sea, and displaying well-sorted sands. { 'päk·ət ˌbēch }

pocket valley [GEOL] A valley whose head is enclosed by steep walls at the base of which underground water emerges as a spring. { 'päk·ət ˌval·ē }

pod [GEOL] An orebody of elongate, lenticular shape. Also known as podiform orebody. { päd }

Podzol [GEOL] A soil group characterized by mats of organic matter in the surface layer and thin horizons of organic minerals overlying gray, leached horizons and dark-brown illuvial horizons; found in coal forests to temperate coniferous or mixed forests. { 'päd,zȯl }

podzolic soil *See* red-yellow podzolic soil. { päd¦zäl·ik 'sȯil }

podzolization [GEOL] The process by which a soil becomes more acid because of the depletion of bases, and develops surface layers that have been leached of clay. { ,päd·zə·lə'zā·shən }

pogonip *See* ice fog. { 'päg·ə,nip }

poikiloblast [GEOL] A large crystal (xenoblast) formed by recrystallization during metamorphism and containing numerous inclusions of small idioblasts. { pȯi'kil·ə,blast }

poikilocrystallic *See* poikilotopic. { pȯi¦kil·ə·kri¦stal·ik }

poikilophitic [GEOL] Referring to ophitic texture characterized by lath-shaped feldspar crystals completely included in large, anhedral pyroxene crystals. { pȯi¦kil·ə¦fid·ik }

poikilotope [GEOL] A large crystal enclosing smaller crystals of another mineral in a sedimentary rock showing poikilotopic fabric. { pȯi'kil·ə,tōp }

poikilotopic [GEOL] Referring to the fabric of a crystalline sedimentary rock in which the constituent crystals are multisized and larger crystals enclose smaller crystals of another mineral. Also known as poikilocrystallic. { pȯi¦kil·ə¦täp·ik }

point [GEOGR] A tapering piece of land projecting into a body of water; it is generally less prominent than a cape. { pȯint }

point bar [GEOL] One of a series of low, arcuate sand and gravel ridges formed on the inside of a growing meander by the gradual addition of accretions. Also known as meander bar. { 'pȯint ,bär }

point rainfall [METEOROL] The rainfall during a given time interval (or often one storm) measured in a rain gage, or an estimate of the amount which might have been measured at a given point. { 'pȯint 'rān,fȯl }

points of the compass *See* compass points. { 'pȯins əv thə 'käm·pəs }

poised stream [HYD] A stream that is neither eroding nor depositing sediment. { 'pȯizd 'strēm }

Poisson relation [GEOPHYS] A model of elastic behavior used in experimental structural geology that takes the Poisson ratio as equal to 0.25. { pwä'sōn ri,lā·shən }

polacke [METEOROL] A cold, dry, northeasterly katabatic wind in Bohemia descending from the Sudeten Mountains (from the direction of Poland). { pō'läk·ə }

polar air [METEOROL] A type of air whose characteristics are developed over high latitudes; there are two types: continental polar air and maritime polar air. { 'pō·lər 'er }

polar anticyclone *See* arctic high; subpolar high. { 'pō·lər ,ant·i'sī,klōn }

polar automatic weather station [METEOROL] An automatic weather station which measures meteorological elements and transmits them by radio; the station is designed to function primarily in frigid or polar climates in order to fill the need for weather reports from inaccessible regions where manned stations are not practicable; since the equipment is designed to operate on ice or slush, the main structure is in the form of a sled with external pontoons for added stability. { 'pō·lər ¦ȯd·ə¦mad·ik 'weth·ər ,stā·shən }

polar cap [HYD] An ice sheet centered at one of the poles of the earth. { 'pō·lər ,kap }

polar-cap absorption [GEOPHYS] Very strong attenuation of radio waves over the polar regions during strong solar flares, due to extremely heavy ionization of the upper atmosphere. Abbreviated PCA. { 'pō·lər ,kap əb,sȯrp·shən }

polar-cap ice *See* polar ice. { 'pō·lər ,kap 'īs }

polar circle [GEOD] A parallel of latitude whose distance from the pole is equal to the obliquity of the ecliptic (approximately 23°27′). { 'pō·lər 'sər·kəl }

polar climate [CLIMATOL] The climate of a geographical polar region, most commonly taken to be a climate which is too cold to support the growth of trees. Also known as arctic climate; snow climate. { 'pō·lər 'klī·mət }

polar continental air [METEOROL] Air of an air mass that originates over land or frozen ocean areas in the polar regions; characterized by low temperature, stability, low specific humidity, and shallow vertical extent. { 'pō·lər ˌkant·ən'ent·əl 'er }

polar convergence [OCEANOGR] The line of convergence of polar and subpolar water masses in the ocean. { 'pō·lər kən'vər·jəns }

polar cyclone *See* polar vortex. { 'pō·lər 'sīˌklōn }

polar desert [GEOGR] A high-latitude desert where the existing moisture is frozen in ice sheets and is thus unavailable for plant growth. Also known as arctic desert. { 'pō·lər 'dez·ərt }

polar easterlies [METEOROL] The rather shallow and diffuse body of easterly winds located poleward of the subpolar low-pressure belt; in the mean in the Northern Hemisphere, these easterlies exist to an appreciable extent only north of the Aleutian low and Icelandic low. { 'pō·lər 'ēs·tərˌlēz }

polar-easterlies index [METEOROL] A measure of the strength of the easterly wind between the latitudes of 55° and 70°N; the index is computed from the average sea-level pressure difference between these latitudes and is expressed as the east to west component of geostrophic wind in meters and tenths of meters per second. { 'pō·lər 'ēs·tərˌlēz 'inˌdeks }

polar electrojet [GEOPHYS] An intense current that flows in a relatively narrow band of the auroral zone ionosphere during disturbances of the magnetosphere. { 'pō·lər i'lek·trəˌjet }

polar firn [HYD] Firn formed at low temperatures with no melting or liquid water present. Also known as dry firn. { 'pō·lər 'fərn }

polar front [METEOROL] The semipermanent, semicontinuous front separating air masses of tropical and polar origin; this is the major front in terms of air mass contrast and susceptibility to cyclonic disturbance. { 'pō·lər 'frənt }

polar-front theory [METEOROL] A theory whereby a polar front, separating air masses of polar and tropical origin, gives rise to cyclonic disturbances which intensify and travel along the front, passing through various phases of a characteristic life history. { 'pō·lər ¦frənt ˌthē·ə·rē }

polar glacier [HYD] A glacier whose temperature is below freezing throughout its mass, and on which there is no melting during any season. { 'pō·lər 'glā·shər }

polar high *See* arctic high; subpolar high. { 'pō·lər 'hī }

polar ice [OCEANOGR] Sea ice that is more than 1 year old; the thickest form of sea ice. Also known as polar-cap ice. { 'pō·lər 'īs }

polarity epoch [GEOPHYS] A period of time during which the earth's magnetic field was predominantly of a single polarity. { pə'lar·əd·ē ˌep·ək }

polarity event [GEOPHYS] A period of no more than about 100,000 years when the earth's magnetic polarity was opposite to the predominant polarity of that polarity epoch. { pə'lar·əd·ē iˌvent }

polarity zone [GEOL] In stratigraphy, a material unit that is defined in terms of magnetic polarity, that is, reversals of the earth's magnetic field. { pə'lar·əd·ē ˌzōn }

polarization isocline [METEOROL] A locus of all points at which the inclination to the vertical of the plane of polarization of the diffuse sky radiation has the same value. { ˌpō·lə·rə'zā·shən 'īs·əˌklīn }

polar lake [HYD] A lake whose surface temperature never exceeds 4°C. { 'pō·lər 'lāk }

polar low *See* polar vortex. { 'pō·lər 'lō }

polar maritime air [METEOROL] Air of an air mass that originates in the polar regions and is then modified by passing over a relatively warm ocean surface; characterized by moderately low temperature, moderately high surface specific humidity, and a considerable degree of vertical instability. { 'pō·lər 'mar·əˌtīm 'er }

polar meteorology [METEOROL] The application of meteorological principles to a study of atmospheric conditions in the earth's high latitudes or polar-cap regions, northern and southern. { 'pō·lər ˌmē·dē·ə'räl·ə·jē }

polar migration *See* polar wandering. { 'pō·lər mī'grā·shən }

polar outbreak [METEOROL] The movement of a cold air mass from its source region; almost invariably applied to a vigorous equatorward thrust of cold polar air, a rapid equatorward movement of the polar front. Also known as cold-air outbreak. { 'pō·lər 'au̇t,brāk }

polar regions [GEOGR] The regions near the geographic poles; no definite limit for these regions is recognized. { 'pō·lər ,rē·jənz }

polar trough [METEOROL] In tropical meteorology, a wave trough in the circumpolar westerlies having sufficient amplitude to reach the tropics in the upper air; at the surface it is reflected as a trough in the tropical easterlies, but at moderate elevations it is characterized by westerly winds. { 'pō·lər 'tro̊f }

polar variation [GEOPHYS] A small movement of the earth's axis of rotation relative to the geoid, the resultant of the Chandler wobble and other smaller movements. { 'pō·lər ,ver·ē'ā·shən }

polar vortex [METEOROL] The large-scale cyclonic circulation in the middle and upper troposphere centered generally in the polar regions; specifically, the vortex has two centers in the mean, one near Baffin Island and another over northeastern Siberia; the associated cyclonic wind system comprises the westerlies of middle latitudes. Also known as Antarctic vortex; circumpolar whirl; polar cyclone; polar low. { 'pō·lər 'vo̊r,teks }

polar wandering [GEOL] Migration during geologic time of the earth's poles of rotation and magnetic poles. Also known as Chandler motion; polar migration. { 'pō·lər 'wan·də·riŋ }

polar westerlies *See* westerlies. { 'pō·lər 'wes·tər,lēz }

pole of inaccessibility *See* ice pole. { 'pōl əv ,in·ak,ses·ə'bil·əd·ē }

pole tide [OCEANOGR] An ocean tide, theoretically 6 millimeters in amplitude, caused by the Chandler wobble of the earth; has a period of 428 days. { 'pōl ,tīd }

polyclinal fold [GEOL] One of a group of adjacent folds, the axial surfaces of which are oriented randomly, but which have similar surface axes. { ¦päl·i¦klīn·əl 'fōld }

polygene [GEOL] An igneous rock composed of two or more minerals. Also known as polymere. { 'päl·i,jēn }

polygenetic [GEOL] **1.** Resulting from more than one process of formation or derived from more than one source, or originating or developing at various places and times. **2.** Consisting of more than one type of material, or having a heterogeneous composition. Also known as polygenic. { ¦päl·i·jə'ned·ik }

polygenic *See* polygenetic. { ¦päl·i¦jen·ik }

polygeosyncline [GEOL] A geosynclinal-geoanticlinal belt that lies along the continental margin and receives sediments from a borderland on its oceanic side. { ¦päl·i,jē·ō'sin,klīn }

polygonal ground [GEOL] A ground surface consisting of polygonal arrangements of rock, soil, and vegetation formed on a level or gently sloping surface by frost action. Also known as cellular soil. { pə'lig·ən·əl 'grau̇nd }

polygonal karst [GEOL] A karst pattern that is characteristic of tropical types such as cone karsts, with the surface completely divided into a polygonal network. { pə'lig·ən·əl 'kärst }

polymetamorphic diaphthoresis [GEOL] Retrograde changes during a second phase of metamorphism that is clearly separated from a previous, higher-grade metamorphic period. { ¦päl·i,med·ə'mo̊r·fik dī,af·thə'rē·səs }

polymetamorphism [GEOL] Polyphase or multiple metamorphism whereby two or more successive metamorphic events have left their imprint upon the same rocks. { ¦päl·i,med·ə'mo̊r,fiz·əm }

polymictic [HYD] Pertaining to or characteristic of a lake having no stabile thermal stratification. { ¦päl·i¦mik·tik }

polyn'ya [OCEANOGR] A Russian term for a water area, other than a lead, lane, or crack, which is surrounded by sea ice; the term "window" is sometimes used for a similar open area in river ice. Also known as ice clearing. { ,päl·ən,yä }

polytropic atmosphere [METEOROL] A model atmosphere in hydrostatic equilibrium with a constant nonzero lapse rate. { ¦päl·i¦träp·ik 'at·mə,sfir }

pond [GEOGR] A small natural body of standing fresh water filling a surface depression, usually smaller than a lake. { pänd }
pondage [HYD] Water held in a reservoir for short periods to regulate natural flow, usually for hydroelectric power. { 'pän·dij }
pondage land [GEOL] Land on which water is stored as dead water during flooding, and which does not contribute to the downstream passage of flow. Also known as flood fringe. { 'pän·dij ˌland }
ponded stream [HYD] A stream in which a pond forms due to an interruption of the normal streamflow. { 'pän·dəd ˌstrēm }
ponding [HYD] The natural formation of a pond in a stream by an interruption of the normal streamflow. { 'pänd·iŋ }
ponente [METEOROL] A west wind on the French Mediterranean coast, the northern Roussillon region, and Corsica. { pȯ'nȯnt }
poniente [METEOROL] The west wind in the Straits of Gibraltar. { ˌpō·nē'en·tē }
Pontian [GEOL] A European stage of geologic time in the uppermost Miocene, above the Sarmatian and below the Plaisancian of the Pliocene; it has also been regarded as the lowermost Pliocene. { 'pän·chən }
pontic [GEOL] Pertaining to sediments or facies deposited in comparatively deep and motionless water, such as an association of black shales and dark limestones deposited in a stagnant basin. { 'pän·tik }
pool [GEOL] Underground accumulation of petroleum. [HYD] A small deep body of water, often fed by a spring. { pül }
pool stage [HYD] As used along the Ohio and upper Mississippi Rivers of the United States, a low-water condition with the navigation dams up so that the river is a series of shallow pools; when this condition exists, the river is said to be "in pool"; river depth is regulated by the dams so as to be adequate for navigation. { 'pül ˌstāj }
popple rock *See* pebble bed. { 'päp·əl ˌräk }
porcelain jasper [GEOL] A hard, naturally baked, impure clay (or porcellanite) which because of its red color had long been considered a variety of jasper. { 'pȯr·slən 'jas·pər }
porcelaneous [GEOL] Resembling unglazed porcelain. { ¦pȯr·sə¦lā·nē·əs }
pore [GEOL] An opening or channelway in rock or soil. { pȯr }
pore compressibility [GEOL] The fractional change in reservoir-rock pore volume with a unit change in pressure upon that rock. { 'pȯr kəmˌpres·ə'bil·əd·ē }
pore ice [HYD] Ice which fills or partially fills pore spaces in permafrost; forms by freezing soil water in place, with no addition of water. { 'pȯr ˌīs }
pore pressure *See* neutral stress. { 'pȯr ˌpresh·ər }
pore-size distribution [GEOL] Variations in pore sizes in reservoir formations; each type of rock has its own typical pore size and related permeability. { 'pȯr¦sīz ˌdis·trə'byü·shən }
pore space [GEOL] The pores in a rock or soil considered collectively. Also known as pore volume. { 'pȯr ˌspās }
pore volume *See* pore space. { 'pȯr ˌväl·yəm }
pore-water pressure *See* neutral stress. { 'pȯr ¦wȯd·ər ˌpresh·ər }
poriaz [METEOROL] Violent northeast winds on the Black Sea near the Bosporus. { 'pȯr·ēˌäz }
Porlezzina [METEOROL] An east wind on Lake Lugano (Italy and Switzerland), blowing from the Gulf of Porlezza. { ˌpȯr·let'sē·nə }
porosity trap *See* stratigraphic trap. { pə'räs·əd·ē ˌtrap }
porphyrocrystallic *See* porphyrotopic. { pȯr¦fir·ō·kri'stal·ik }
porphyroskelic [GEOL] Pertaining to an arrangement in a soil fabric whereby the plasma occurs as a dense matrix in which skeleton grains are set like phenocrysts in a porphyritic rock. { pȯr¦fir·ə¦skel·ik }
porphyrotope [GEOL] A large crystal enclosed in a finer-grained matrix in a sedimentary rock showing porphyrotopic fabric. { pȯr'fir·əˌtōp }
porphyrotopic [GEOL] Referring to the fabric of a crystalline sedimentary rock in which the constituent crystals are of more than one size and in which larger crystals are

enclosed in a finer-grained matrix. Also known as porphyrocrystallic. { pȯr¦fir·ə¦täp·ik }

port *See* harbor. { pȯrt }

Porterfield [GEOL] A North American geologic stage of the Middle Ordovician, forming the lower division of the Mohawkian, and lying above Ashby and below Wilderness. { 'pȯrd·ər,fēld }

Portlandian [GEOL] A European geologic stage of the Upper Jurassic, above Kimmeridgian, below Berriasian of Cretaceous. { pȯrt'land·ē·ən }

positive area *See* positive element. { 'päz·əd·iv 'er·ē·ə }

positive axis [METEOROL] In tropical synoptic analysis, a locus of maximum streamline curvature in an easterly wave; used primarily in the analysis of waves that span the equatorial trough (equatorial waves); a positive axis corresponds to a trough line in the Northern Hemisphere and a ridge line in the Southern Hemisphere. { 'päz·əd·iv 'ak·səs }

positive element [GEOGR] A large structural feature of the earth's crust characterized by long-term upward movement (uplift, emergence) or subsidence less rapid than that of adjacent negative elements. Also known as archibole; positive area. { 'päz·əd·iv 'el·ə·mənt }

positive estuary [HYD] An estuary in which there is a measurable dilution of seawater by land drainage. { 'päz·əd·iv 'es·chə,wer·ē }

positive landform [GEOL] An upstanding topographic form, such as a mountain, hill, plateau, or cinder cone. { 'päz·əd·iv 'land,fȯrm }

positive movement [GEOL] **1.** Uplift or emergence of the earth's crust relative to an adjacent area of the crust. **2.** A relative rise in sea level with respect to land level. { 'päz·əd·iv 'müv·mənt }

positive shoreline *See* shoreline of submergence. { 'päz·əd·iv 'shȯr,līn }

Postglacial *See* Holocene. { pōst'glā·shəl }

posthumous structure [GEOL] Folds, faults, and other structural features in covering strata which revive or mimic the structure of older underlying rocks that are generally more deformed. { 'päs·chə·məs 'strək·chər }

postmagmatic [GEOL] Pertaining to geologic reactions or events occurring after the bulk of the magma has crystallized. { ,pōst·mag'mad·ik }

postobsequent stream [HYD] A strike stream developed after the obsequent stream into which it flows. { ¦pōst·äb'sē·kwənt 'strēm }

postorogenic [GEOL] Of a geologic process or event, occurring after a period of orogeny. { ,pōst,ȯr·ə'jen·ik }

potamology [HYD] The scientific study of rivers. { ,päd·ə'mäl·ə·jē }

potash bentonite *See* potassium bentonite. { 'päd,ash 'bent·ən,īt }

potash kettle *See* giant's kettle. { 'päd,ash 'ked·əl }

potash lake [HYD] An alkali lake whose waters contain a high content of dissolved potassium salts. { 'päd,ash 'lāk }

potassium-argon dating [GEOL] Dating of archeological, geological, or organic specimens by measuring the amount of argon accumulated in the matrix rock through decay of radioactive potassium. { pə'tas·ē·əm 'är,gän 'dād·iŋ }

potassium bentonite [GEOL] A clay of the illite group that contains potassium and is formed by alteration of volcanic ash. Also known as K bentonite; potash bentonite. { pə'tas·ē·əm 'bent·ən,īt }

potato stone [GEOL] A potato-shaped geode, especially one consisting of hard, silicified limestone with an internal lining of quartz crystals. { pə'tā·dō ,stōn }

potential evaporation *See* evaporative power. { pə'ten·chəl i,vap·ə'rā·shən }

potential evapotranspiration [HYD] Generally, the amount of moisture which, if available, would be removed from a given land area by evapotranspiration; expressed in units of water depth. { pə'ten·chəl i,vap·ō,tranz·pə'rā·shən }

potential index of refraction [METEOROL] An atmospheric index of refraction so formulated that it would have no height variation in an adiabatic atmosphere. Also known as potential refractive index. { pə'ten·chəl ¦in,deks əv ri'frak·shən }

potential instability *See* convective instability. { pə'ten·chəl ,in·stə'bil·əd·ē }

potential refractive index *See* potential index of refraction. { pə'ten·chəl ri'frak·tiv 'in,deks }

potentiometric map [HYD] A map showing the elevation of a potentiometric surface of an aquifer by means of contour lines or other symbols. Also known as pressure-surface map. { pə¦ten·chē·ə¦me·trik 'map }

potentiometric surface [HYD] An imaginary surface that represents the static head of groundwater and is defined by the level to which water will rise. Also known as isopotential level; piezometric surface; pressure surface. { pə¦ten·chē·ə¦me·trik 'sər·fəs }

pothole [GEOL] **1.** A shaftlike cave opening upward to the surface. **2.** Any bowl-shaped, cylindrical, or circular hole formed by the grinding action of a stone in the rocky bed of a river or stream. Also known as churn hole; colk; eddy mill; evorsion hollow; kettle; pot. **3.** A vertical, or nearly vertical shaft in limestone. Also known as aven; cenote. **4.** A small depression with steep sides in a coastal marsh; contains water at or below low-tide level. Also known as rotten spot. [HYD] *See* moulin. { 'pät,hōl }

potrero [GEOL] An elongate, islandlike beach ridge, surrounded by mud flats and separated from the coast by a lagoon and barrier island, made up of a series of accretionary dune ridges. { pə'trer·ō }

Poulter seismic method [GEOPHYS] A type of air shooting in which the explosive is set on poles above the ground. { 'pōl·tər 'sīz·mik ,meth·əd }

powder avalanche [GEOL] Loose powder snow rapidly descending a mountainside. { ¦pau̇d·ər ¦av·ə,lanch }

powder snow [HYD] A cover of dry snow that has not been compacted in any way. { 'pau̇d·ər ,snō }

power-law profile [METEOROL] A formula for the variation of wind with height in the surface boundary layer. { 'pau̇·ər ,lȯ ,prō,fīl }

pozzolan [GEOL] A finely ground burnt clay or shale resembling volcanic dust, found near Pozzuoli, Italy; used in cement because it hardens underwater. { 'pät·sə·lən }

praecipitatio [METEOROL] Precipitation falling from a cloud and apparently reaching the earth's surface; this supplementary cloud feature is mostly encountered in altostratus, nimbostratus, stratocumulus, stratus, cumulus, and cumulonimbus. { prē¦sip·ə¦tā·shō }

prairie [GEOGR] An extensive level-to-rolling treeless tract of land in the temperate latitudes of central North America, characterized by deep, fertile soil and a cover of coarse grass and herbaceous plants. { 'prer·ē }

prairie climate *See* subhumid climate. { 'prer·ē ,klī·mət }

prairie soil [GEOL] A group of zonal soils having a surface horizon that is dark or grayish brown, which grades through brown soil into lighter-colored parent material; it is 2–5 feet (0.6–1.5 meters) thick and develops under tall grass in a temperate and humid climate. { 'prer·ē ,sȯil }

prealpine facies [GEOL] A geosynclinal facies characteristic of neritic areas, displaying thick limestone deposits and coarse terrigenous material and resembling epicontinental platform sediments. { prē'al,pīn 'fā·shēz }

Precambrian [GEOL] All geologic time prior to the beginning of the Paleozoic era (before 600,000,000 years ago); equivalent to about 90% of all geologic time. { prē 'kam·brē·ən }

precipice [GEOL] A very steeply inclined, vertical, or overhanging wall or surface of rock. { 'pres·ə·pəs }

precipitable water [METEOROL] The total atmospheric water vapor contained in a vertical column of unit cross-sectional area extending between any two specified levels, commonly expressed in terms of the height to which that water substance would stand if completely condensed and collected in a vessel of the same unit cross section. Also known as precipitable water vapor. { pri'sip·əd·ə·bəl 'wȯd·ər }

precipitable water vapor *See* precipitable water. { pri'sip·əd·ə·bəl 'wȯd·ər ,vā·pər }

precipitation [METEOROL] **1.** Any or all of the forms of water particles, whether liquid or solid, that fall from the atmosphere and reach the ground. **2.** The amount, usually expressed in inches of liquid water depth, of the water substance that has fallen at a given point over a specified period of time. { prə,sip·ə'tā·shən }

precipitation area [METEOROL] **1.** On a synoptic surface chart, an area over which precipitation is falling. **2.** In radar meteorology, the region from which a precipitation echo is received. { prə,sip·ə'tā·shən ,er·ē·ə }

precipitation ceiling [METEOROL] After United States weather observing practice, a ceiling classification applied when the ceiling value is the vertical visibility upward into precipitation; this is necessary when precipitation obscures the cloud base and prevents a determination of its height. { prə,sip·ə'tā·shən ,sēl·iŋ }

precipitation cell [METEOROL] In radar meteorology, an element of a precipitation area over which the precipitation is more or less continuous. { prə,sip·ə'tā·shən ,sel }

precipitation current [METEOROL] The downward transport of charge, from cloud region to earth, that occurs in a fall of electrically charged rain or other hydrometeors. { prə,sip·ə'tā·shən ,kə·rənt }

precipitation echo [METEOROL] A type of radar echo returned by precipitation. { prə,sip·ə'tā·shən ,ek·ō }

precipitation effectiveness *See* precipitation-evaporation ratio. { prə,sip·ə'tā·shən i¦fek·tiv·nəs }

precipitation electricity [GEOPHYS] **1.** That branch of the study of atmospheric electricity concerned with the electric charges carried by precipitation particles and with the manner in which these charges are acquired. **2.** The electric charge borne by precipitation particles. { prə,sip·ə'tā·shən ,i,lek'tris·əd·ē }

precipitation-evaporation ratio [CLIMATOL] For a given locality and month, an empirical expression devised for the purpose of classifying climates numerically on the basis of precipitation and evaporation. Abbreviated P-E ratio. Also known as precipitation effectiveness. { prə,sip·ə'tā·shən i,vap·ə'rā·shən ,rā·shō }

precipitation excess [HYD] The volume of water from precipitation that is available for direct runoff. { prə,sip·ə'tā·shən 'ek,ses }

precipitation facies [GEOL] Facies characteristics that provide evidence of depositional conditions; revealed mainly by sedimentary structures (such as cross-bedding and ripple marks) and by primary constituents (especially fossils). { prə,sip·ə'tā·shən ,fā·shēz }

precipitation-generating element [METEOROL] In radar meteorology, a relatively small volume of supercooled cloud droplets in which ice crystals form and grow much more rapidly than in a lower, larger cloud mass. { prə,sip·ə'tā·shən 'jen·ə,rād·iŋ ,el·ə·mənt }

precipitation intensity [METEOROL] The rate of precipitation, expressed in inches or millimeters per hour. Also known as rainfall intensity. { prə,sip·ə'tā·shən in,ten·səd·ē }

precipitation inversion [METEOROL] As found in some mountain areas, a decrease of precipitation with increasing elevation of ground above sea level. Also known as rainfall inversion. { prə,sip·ə'tā·shən in,vər·zhən }

precipitation physics [METEOROL] The study of the formation and precipitation of liquid and solid hydrometeors from clouds; a branch of cloud physics and of physical meteorology. { prə,sip·ə'tā·shən ,fiz·iks }

precipitation station [METEOROL] A station at which only precipitation observations are made. { prə,sip·ə'tā·shən ,stā·shən }

precipitation trails *See* virga. { prə,sip·ə'tā·shən ,trālz }

precipitation trajectory [METEOROL] In radar meteorology, a characteristic echo observed on range-height indicator scopes and time-height sections which represents the height-range pattern of snow falling from isolated precipitation-generating elements of a few miles in diameter. Also known as mare's tail. { prə,sip·ə'tā·shən trə,jek·trē }

pre-cold-frontal squall line *See* prefrontal squall line. { ¦prē 'kōld ,frənt·əl 'skwȯl ,līn }

preconsolidation pressure [GEOL] The greatest effective stress exerted on a soil; result of this pressure from overlying materials is compaction. Also known as prestress. { ¦prē·kən,säl·ə'dā·shən ,presh·ər }

predict *See* forecast. { pri'dikt }

prediction [METEOROL] **1.** The act of making a weather forecast. **2.** The forecast itself. { prə'dik·shən }

prefrontal squall line [METEOROL] A squall line or instability line located in the warm sector of a wave cyclone, about 50 to 300 miles (80 to 480 kilometers) in advance of the cold front, usually oriented roughly parallel to the cold front, and moving in about the same manner as the cold front. Also called nonfrontal squall line; precold-frontal squall line. { prē'frənt·əl 'skwȯl ˌlīn }

preglacial [GEOL] **1.** Pertaining to the geologic time immediately preceding the Pleistocene epoch. **2.** Of material, underlying glacial deposits. { prē'glā·shəl }

preliminary waves [GEOPHYS] The body of waves of an earthquake, including both P waves and S waves. { pri'lim·əˌner·ē ¦wāvz }

preorogenic [GEOL] The initial phase of an orogenic cycle during which geosynclines form. { ¦prēˌȯr·ə'jen·ik }

presque isle [GEOGR] A promontory or peninsula extending into a lake, nearly or almost forming an island; its head or end section is connected with the shore by a sag or low gap only slightly above water level or by a strip of lake bottom exposed as a land surface by a drop in lake level. { ¦pres'kīl }

pressolved [GEOL] Referring to a sedimentary bed or rock in which the grains have undergone pressure solution. { pri'zälvd }

pressure altitude [METEOROL] The height above sea level at which the existing atmospheric pressure would be duplicated in the standard atmosphere; atmospheric pressure expressed as height according to a standard scale. { 'presh·ər ˌal·təˌtüd }

pressure-altitude variation [METEOROL] The pressure difference, in feet or meters, between mean sea level and the standard datum plane. { 'presh·ər ¦al·təˌtüd ˌver·ē'ā·shən }

pressure center [METEOROL] **1.** On a synoptic chart (or on a mean chart of atmospheric pressure), a point of local minimum or maximum pressure; the center of a low or high. **2.** A center of cyclonic or anticyclonic circulation. { 'presh·ər ˌsen·tər }

pressure-change chart [METEOROL] A chart indicating the change in atmospheric pressure of a constant-height surface over some specified interval of time. Also known as pressure-tendency chart. { 'presh·ər ¦chānj ˌchärt }

pressure contour [METEOROL] A line connecting points of equal height of a given barometric pressure; the intersection of a constant pressure surface by a plane parallel to mean sea level. { 'presh·ər ˌkänˌtür }

pressure depth [OCEANOGR] The depth at which an ocean sample was taken, as inferred from the difference in readings on protected and unprotected thermometers on the sampler; the higher reading is on the unprotected thermometer due to the effect of pressure on the mercury column at the sampling depth. { 'presh·ər ˌdepth }

pressure-fall center [METEOROL] A point of maximum decrease in atmospheric pressure over a specified interval of time; on synoptic charts, a point of greatest negative pressure tendency. Also known as center of falls; isallobaric low; isallobaric minimum; katallobaric center. { 'presh·ər ¦fȯl ˌsen·tər }

pressure field [OCEANOGR] A representation of a pressure gradient as isobar contours, parallel to which ocean currents flow. { 'presh·ər ˌfēld }

pressure force *See* pressure-gradient force. { 'presh·ər ˌfȯrs }

pressure gradient [METEOROL] The change in atmospheric pressure per unit horizontal distance, usually measured along a line perpendicular to the isobars. { 'presh·ər ˌgrād·ē·ənt }

pressure-gradient force [METEOROL] The force due to differences of pressure within the atmosphere; it usually refers only to the horizontal component of the force. Also known as pressure force. { 'presh·ər ¦grād·ē·ənt ˌfȯrs }

pressure ice [OCEANOGR] Ice, especially sea ice, which has been deformed or altered by the lateral stresses of any combination of wind, water currents, tides, waves, and surf; may include ice pressed against the shore, or one piece of ice upon another. { 'presh·ər ˌīs }

pressure ice foot [HYD] An ice foot formed along a shore by the freezing together of stranded pressure ice. { 'presh·ər ¦īs ˌfu̇t }

pressure jump [METEOROL] A steady-state propagation of a sudden finite change of inversion height, in analogy to the shock wave in a compressible fluid or to a hydraulic jump; the prefrontal squall line has been interpreted as a pressure jump, with the cold front providing the initial pistonlike impetus. { 'presh·ər ˌjəmp }

pressure-jump line [METEOROL] A fast-moving line of sudden rise in atmospheric pressure, followed by a higher pressure level than that which preceded the jump; under suitable moisture conditions, sudden instability of the atmosphere conducive to the formation of thunderstorms can result. { 'presh·ər ¦jəmp ˌlīn }

pressure pan [GEOL] An induced soil pan which has a higher bulk density and a lower total porosity than the soil directly above or below it and is produced as a result of pressure applied by normal tillage operations or by other artificial means. { 'presh·ər ˌpan }

pressure pattern [METEOROL] The general geometric characteristics of atmospheric pressure distribution as revealed by isobars on a constant-height chart; usually applied to cyclonic-scale features of a surface chart. { 'presh·ər ˌpad·ərn }

pressure penitente [GEOL] A nieve penitente composed of brilliantly white ice which is shaped into a slender ridge by lateral pressure of converging morainal streams and by melting of the adjacent debris-covered ice. { 'presh·ər ˌpen·ə'ten·tā }

pressure plateau [GEOL] An uplifted area of a thick lava flow, measuring up to 10 or 13 feet (3 or 4 meters), the uplift of which is due to the intrusion of new lava from below that does not reach the surface. { 'presh·ər plaˌtō }

pressure release [GEOPHYS] The outward-expanding force of pressure which is released within rock masses by unloading, as by erosion of superincumbent rocks or by removal of glacial ice. { 'presh·ər riˌlēs }

pressure-release jointing [GEOL] Exfoliation that occurs in once deeply buried rock that erosion has brought nearer the surface, thus releasing its confining pressure. { 'presh·ər ri¦lēs ˌjȯint·iŋ }

pressure ridge [GEOL] **1.** A seismic feature resulting from transverse pressure and shortening of the land surface. **2.** An elongate upward movement of the congealing crust of a lava flow. **3.** A ridge of glacier ice. [OCEANOGR] A ridge or wall of hummocks where one ice floe has been pressed against another. { 'presh·ər ˌrij }

pressure-rise center [METEOROL] A point of maximum increase in atmospheric pressure over a specified interval of time; on synoptic charts, a point of maximum positive pressure tendency. Also known as anallobaric center; center of rises; isallobaric high; isallobaric maximum. { 'presh·ər ¦rīz ˌsen·tər }

pressure surface *See* potentiometric surface. { 'presh·ər ˌsər·fəs }

pressure-surface map *See* potentiometric map. { 'presh·ər ¦sər·fəs ˌmap }

pressure system [METEOROL] An individual cyclonic-scale feature of atmospheric circulation, commonly used to denote either a high or a low, less frequently a ridge or a trough. { 'presh·ər ˌsis·təm }

pressure tendency [METEOROL] The character and amount of atmospheric pressure change for a 3-hour or other specified period ending at the time of observation. Also known as barometric tendency. { 'presh·ər ˌten·dən·sē }

pressure-tendency chart *See* pressure-change charty. { 'presh·ər ¦ten·dən·sē ˌchärt }

pressure topography *See* height pattern. { 'presh·ər təˌpäg·rə·fē }

pressure tube [HYD] A deep, slender, cylindrical hole formed in a glacier by the sinking of an isolated stone that has absorbed more solar radiation than the surrounding ice. { 'presh·ər ˌtüb }

pressure wave [METEOROL] A wave or periodicity which exists in the variation of atmospheric pressure on any time scale, usually excluding normal diurnal or seasonal trends. { 'presh·ər ˌwāv }

prester [METEOROL] A whirlwind or waterspout accompanied by lightning in the Mediterranean Sea and Greece. { 'pres·tər }

prestress *See* preconsolidation pressure. { ¦prē'stres }

presuppression [GEOPHYS] In seismic prospecting, the suppression of the early events on a seismic record for control of noise and reflections on that portion of the record. { ¦prē·sə'presh·ən }

prevailing current [OCEANOGR] The ocean current most frequently observed during a given period, such as a month, a season, or a year. { pri'vāl·iŋ ,kə·rənt }

prevailing visibility [METEOROL] In United States weather observing practice, the greatest horizontal visibility equaled or surpassed throughout half of the horizon circle; in the case of rapidly varying conditions, it is the average of the prevailing visibility while the observation is being taken. { pri'vāl·iŋ ,viz·ə'bil·əd·ē }

prevailing westerlies [METEOROL] The prevailing westerly winds on the poleward sides of the subtropical high-pressure belts. { pri'vāl·iŋ 'wes·tər·lēz }

prevailing wind *See* prevailing wind direction. { pri'vāl·iŋ 'wind }

prevailing wind direction [METEOROL] The wind direction most frequently observed during a given period; the periods most often used are the observational day, month, season, and year. Also known as prevailing wind. { pri'vāl·iŋ 'wind di,rek·shən }

previtrain [GEOL] The woody lenses in lignite that are equivalent to vitrain in coal of higher rank. { prē'vi,trān }

Priabonian [GEOL] A European stage of geologic time in the upper Eocene, believed to consist of Auversian and Bartonian. { ,prē·ə'bō·nē·ən }

primary [GEOL] **1.** A young shoreline whose features are produced chiefly by nonmarine agencies. **2.** Of a mineral deposit, unaffected by supergene enrichment. { 'prī ,mer·ē }

primary arc [GEOL] **1.** A curved segment of elongated mountain zones that are the areas of the earth's major and most recent tectonic activity. **2.** *See* internides. { 'prī,mer·ē 'ärk }

primary circle *See* primary great circle. { 'prī,mer·ē 'sər·kəl }

primary circulation [METEOROL] The prevailing fundamental atmospheric circulation on a planetary scale which must exist in response to radiation differences with latitude, to the rotation of the earth, and to the particular distribution of land and oceans, and which is required from the viewpoint of conservation of energy. { 'prī,mer·ē ,sər·kyə'lā·shən }

primary clay *See* residual clay. { 'prī,mer·ē 'klā }

primary crater [GEOL] **1.** An impact crater produced directly by the high-velocity impact of a meteorite or other projectile. **2.** *See* true crater. { 'prī,mer·ē 'krād·ər }

primary cyclone [METEOROL] Any cyclone (or low), especially a frontal cyclone, within whose circulation one or more secondary cyclones have developed. Also known as primary low. { 'prī,mer·ē 'sī,klōn }

primary dip [GEOL] The slight dip assumed by a bedded deposit at its moment of deposition. Also known as depositional dip; initial dip; original dip. { 'prī,mer·ē 'dip }

primary flat joint [GEOL] An approximately horizontal joint plane in igneous rocks. Also known as L joint. { 'prī,mer·ē 'flat ,jȯint }

primary front [METEOROL] The principal, and usually original, front in any frontal system in which secondary fronts are found. { 'prī,mer·ē 'frənt }

primary geosyncline *See* orthogeosyncline. { 'prī,mer·ē ¦jē·ō'sin,klīn }

primary great circle [GEOD] A great circle used as the origin of measurement of a coordinate; particularly, such a circle 90° from the poles of a system of spherical coordinates, as the equator. Also known as fundamental circle; primary circle. { 'prī,mer·ē 'grāt 'sər·kəl }

primary low *See* primary cyclone. { 'prī,mer·ē 'lō }

primary magma [GEOL] A magma that originates below the earth's crust. { 'prī,mer·ē 'mag·mə }

primary orogeny [GEOL] Orogeny that is characteristic of the internides and that involves deformation, regional metamorphism, and granitization. { 'prī,mer·ē ȯ'räj·ə·nē }

primary pollutant [METEOROL] A pollutant that enters the air directly from a source. { 'prī,mer·ē pə'lüt·ənt }

primary porosity [GEOL] Natural porosity in petroleum reservoir sands or rocks. { 'prī,mer·ē pə'räs·əd·ē }

primary sedimentary structure [GEOL] A sedimentary structure produced during deposition, such as ripple marks and graded bedding. { 'prī,mer·ē ,sed·ə'men·trē ,strək·chər }

primary stratification [GEOL] Stratification which develops when sediments are first deposited. Also known as direct stratification. { 'prī,mer·ē ,strad·ə·fə'kā·shən }

primary stratigraphic trap [GEOL] A stratigraphic trap formed by the deposition of clastic materials (such as shoestring sands, lenses, sand patches, bars, or cocinas) or through chemical deposition (such as organic reefs or biostromes). { 'prī,mer·ē ¦strad·ə¦graf·ik 'trap }

primary stress field *See* ambient stress field. { 'prī,mer·ē 'stres ,fēld }

primary structure [GEOL] A structure, in an igneous rock, that formed at the same time as the rock, but before its final consolidation. { 'prī,mer·ē 'strək·chər }

primary wave [GEOPHYS] The first seismic wave that reaches a station from an earthquake. { 'prī,mer·ē 'wāv }

prime meridian [GEOD] The meridian of longitude 0°, used as the origin for measurement of longitude; the meridian of Greenwich, England, is almost universally used for this purpose. { 'prīm mə,rid·ē·ən }

prime vertical circle [GEOD] The vertical circle through the east and west points of the horizon. { 'prīm 'vərd·ə·kəl 'sər·kəl }

priming of the tides [OCEANOGR] The acceleration in the times of occurrence of high and low tides when the sun's tidal effect comes before that of the moon. { 'prīm·iŋ əv t͟hə 'tīdz }

primitive water [HYD] Water that has been imprisoned in the earth's interior, in either molecular or dissociated form, since the formation of the earth. { 'prim·əd·iv 'wȯd·ər }

principal vertical circle [GEOD] The vertical circle through the north and south points of the horizon, coinciding with the celestial meridian. { 'prin·sə·pəl 'vərd·ə·kəl ,sər·kəl }

principle of uniformity *See* uniformitarianism. { 'prin·sə·pəl əv ,yü·nə'fȯr·məd·ē }

prism [GEOL] A long, narrow, wedge-shaped sedimentary body with a width-thickness ratio greater than 5 to 1 but less than 50 to 1. { 'priz·əm }

prismatic jointing *See* columnar jointing. { priz'mad·ik 'jȯint·iŋ }

prismatic structure *See* columnar jointing. { priz'mad·ik 'strək·chər }

prism crack [GEOL] A mud crack that develops in regular or irregular polygonal patterns on the surface of drying mud puddles and that breaks the sediment into prisms standing normal to the bedding. { 'priz·əm ,krak }

private stream [HYD] Any stream which diverts part or all of the drainage of another stream. { 'prī·vət 'strēm }

probability forecast [METEOROL] A forecast of the probability of occurrence of one or more of a mutually exclusive set of weather contingencies, as distinguished from a series of categorical statements. { ,präb·ə'bil·əd·ē ,fȯr,kast }

probable maximum precipitation [METEOROL] The theoretically greatest depth of precipitation for a given duration that is physically possible over a particular drainage area at a certain time of year; in practice, this is derived over flat terrain by storm transposition and moisture adjustment to observed storm patterns. { 'präb·ə·bəl 'mak·sə·məm pri,sip·ə'tā·shən }

Procellarian [GEOL] Pertaining to lunar lithologic map units and topographic forms constituting, or closely associated with, the maria. { ,prō·sə'lar·ē·ən }

process lapse rate [METEOROL] The rate of decrease of the temperature of an air parcel as it is lifted, expressed as $-dT/dz$, where z is the altitude, or occasionally dT/dp, where p is pressure; the concept may be applied to other atmospheric variables, such as the process lapse rate of density. { 'prä,səs 'laps ,rāt }

prod cast [GEOL] The cast of a prod mark. Also known as impact cast. { 'präd ,kast }

prodelta [GEOL] The part of a delta lying beyond the delta front, and sloping gently down to the basin floor of the delta; it is entirely below the water level. { 'prō,del·tə }

prodelta clay [GEOL] Fine sand, silt, and clay transported by the river and deposited on the floor of a sea or lake beyond the main body of a delta. { 'prō,del·tə ,klā }

prod mark [GEOL] A short tool mark oriented parallel to the current and gradually deepening downcurrent. Also known as impact mark. { 'präd ,märk }

profile [GEOL] **1.** The outline formed by the intersection of the plane of a vertical section and the ground surface. Also known as topographic profile. **2.** Data recorded by a single line of receivers from one shot point in seismic prospecting. [GEOPHYS] A graphic representation of the variation of one property, such as gravity, usually as ordinate, with respect to another property, usually linear, such as distance. [HYD] A vertical section of a potentiometric surface, such as a water table. { 'prō,fīl }

profile line [GEOL] The top line of a profile section, representing the intersection of a vertical plane with the surface of the ground. { ¦prō,fīl ,līn }

profile of equilibrium [GEOL] **1.** The slope of the floor of a sea, ocean, or lake, taken in a vertical plane, when deposition of sediment is balanced by erosion. **2.** The longitudinal profile of a graded stream. Also known as equilibrium profile; graded profile. { ¦prō,fīl əv ,ē·kwə'lib·rē·əm }

profile section [GEOL] A diagram or drawing that shows along a given line the configuration or slope of the surface of the ground as it would appear if it were intersected by a vertical plane. { ¦prō,fīl ,sek·shən }

profiling snow gage [HYD] A type of radioactive gage for measuring the water equivalent and density/depth distribution of a snowpack, consisting of a radioactive source and a radioactivity detector which move up and down in two adjacent vertical pipes surrounded by snow. Also known as nuclear twin-probe gage. { 'prō,fīl·iŋ 'snō ,gāj }

proglacial [GEOL] Of streams, deposits, and other features, being immediately in front of or just beyond the outer limits of a glacier or ice sheet, and formed by or derived from glacier ice. { prō'glā·shəl }

prognostic chart [METEOROL] A chart showing, principally, the expected pressure pattern (or height pattern) of a given synoptic chart at a specified future time; usually, positions of fronts are also included, and the forecast values of other meteorological elements may be superimposed. { präg'näs·tik ¦chärt }

prognostic equation [METEOROL] Any equation governing a system which contains a time derivative of a quantity and therefore can be used to determine the value of that quantity at a later time when the other terms in the equation are known (for example, the vorticity equation). { präg'näs·tik i'kwā·zhən }

progradation [GEOL] Seaward buildup of a beach, delta, or fan by nearshore deposition of sediments transported by a river, by accumulation of material thrown up by waves, or by material moved by longshore drifting. { ¦prō·grə'dā·shən }

prograde metamorphism [GEOL] Metamorphic changes in response to a higher pressure or temperature than that to which the rock was last adjusted. { ¦prō'grād ,med·ə'mȯr,fiz·əm }

prograding shoreline [GEOL] A shoreline that is being built seaward by accumulation or deposition. { ¦prō'grād·iŋ 'shȯr,līn }

progressive metamorphism [GEOL] Systematic change in metamorphic grade from lower to higher in any metamorphic terrain. { prə'gres·iv ,med·ə'mȯr,fiz·əm }

progressive sand wave [GEOL] A sand wave characterized by downcurrent migration. { prə'gres·iv 'sand ,wāv }

progressive sorting [GEOL] Sorting of sedimentary particles in the downcurrent direction, resulting in a systematic downcurrent decrease in the mean grain size of the sediment. { prə'gres·iv 'sȯrd·iŋ }

progressive wave [METEOROL] A wave or wavelike disturbance which moves relative to the earth's surface. { prə'gres·iv 'wāv }

prolapsed bedding [GEOL] Bedding characterized by a series of flat folds with near-horizontal axial planes contained entirely within a bed which has undisturbed boundaries. { 'prō,lapst 'bed·iŋ }

proluvium [GEOL] A complex, friable, deltaic sediment accumulated at the foot of a slope as a result of an occasional torrential washing of fragmental material. { prō'lü·vē·əm }

promontory [GEOL] **1.** A high, prominent projection or point of land, or a rock cliff, jutting out boldly into a body of water. **2.** A cape, either low-lying or of considerable height, with a bold termination. **3.** A bluff or prominent hill overlooking or projecting into a lowland. { 'präm·ən,tȯr·ē }

prong reef [GEOL] A wall reef that has developed irregular buttresses normal to its axis in both leeward and (to a smaller degree) seaward directions. { präŋ ,rēf }

propaedeutic stratigraphy *See* prostratigraphy. { ,prō·pi'düd·ik strə'tig·rə·fē }

propagation forecasting [METEOROL] Forecasting in which the known or predicted vertical distribution of the index of refraction over an area is used to forecast the propagation performance of radars or any microwave radio equipment operating in that area. { ,präp·ə'gā·shən 'fȯr,kast·iŋ }

prostratigraphy [GEOL] Preliminary stratigraphy, including lithologic and paleontologic studies, without consideration of the time factor. Also known as propaedeutic stratigraphy; protostratigraphy. { ,prä·strə'tig·rə·fē }

protactinium-ionium age method [GEOL] A method of calculating the ages of deep-sea sediments formed during the last 150,000 years from measurements of the ratio of protactinium-231 to ionium (thorium-230), based on the gradual change of this ratio over time because of the difference in half-lives. { ¦prōd,ak'tin·ē·əm ī'ō·nē·əm 'āj ,meth·əd }

protalus rampart [GEOL] An arcuate ridge consisting of boulders and other coarse debris marking the downslope edge of an existing or melted snowbank. { prō'tal·əs 'ram,pärt }

Proterozoic [GEOL] Geologic time between the Archean and Paleozoic eras, that is, from 2500 million to 550 million years ago. Also known as Algonkian. { ¦präd·ə·rə¦zō·ik }

protointraclast [GEOL] A limestone component that resulted from a premature attempt at resedimentation while it was still in an unconsolidated and viscous or plastic state, and that never existed as a free clastic entity. { ,prōd·ō'in·trə,klast }

protostratigraphy *See* prostratigraphy. { ¦prōd·ō·strə'tig·rə·fē }

provenance [GEOL] The location, topography, and composition of the source area for any sedimentary rock. Also known as source area; sourceland. { 'präv·ə·nəns }

province [OCEANOGR] An area composed of a grouping of like bathymetric elements whose features are in obvious contrast with surrounding regions. { 'präv·əns }

provincial series [GEOL] A time-stratigraphic series recognized only in a particular region and involving a major division of time within a period. { prə'vin·chəl 'sir·ēz }

provitrain [GEOL] Vitrain in which some plant structure can be discerned by microscope. Also known as telain. { prō'vi,trān }

provitrinite [GEOL] A variety of vitrinite characteristic of provitrain and including the varieties periblinite, suberinite, and xylinite. { prō'vi·trə,nīt }

proximal [GEOL] Of a sedimentary deposit, composed of coarse clastics and formed near the source. { 'präk·sə·məl }

Psamment [GEOL] A suborder of the soil order Entisol, characterized by a texture of loamy fine sand or coarser sand, and by a coarse fragment content of less than 35. { 'sa,ment }

psammitic *See* arenaceous. { sə'mid·ik }

psephicity [GEOL] A coefficient of roundability of a pebble- or sand-size mineral fragment, expressed as the ratio of specific gravity to hardness (as measured in the air) or the quotient of specific gravity minus one divided by hardness (as measured in water). { sə'fis·əd·ē }

psephite [GEOL] A sediment or sedimentary rock composed of fragments that are coarser than sand and which are set in a qualitatively and quantitatively varying matrix; equivalent to a rudite or, generally, a conglomerate. { sē,fīt }

psephyte [GEOL] A lake-bottom deposit consisting mainly of coarse, fibrous plant remains. { sē,fīt }

pseudoadiabat [METEOROL] On a thermodynamic diagram, a line representing a pseudoadiabatic expansion of an air parcel; in practice, approximate computations are employed, and the resulting lines represent, ambiguously, pseudoadiabats and saturation adiabats. { ¦sü·dō'ad·ē·ə,bat }

pseudoadiabatic chart *See* Stuve chart. { ¦sü·dō,ad·ē·ə'bad·ik 'chärt }

pseudoadiabatic expansion [GEOPHYS] A saturation-adiabatic process in which the condensed water substance is removed from the system, and which therefore is best treated by the thermodynamics of open systems; meteorologically, this process corresponds to rising air from which the moisture is precipitating. { ¦sü·dō,ad·ē·ə'bad·ik ik'span·shən }

pseudoallochem [GEOL] An object resembling an allochem but produced in place within a calcareous sediment by a secondary process such as recrystallization. { ,sü·dō'al·ə,kem }

pseudocannel coal [GEOL] Cannel coal that contains much humic matter. Also known as humic-cannel coal. { ¦sü·dō¦kan·əl 'kōl }

pseudochrysolite *See* moldavite. { ¦sü·dō'kris·ə,līt }

pseudocol [GEOL] A landform represented by a constriction of a stream valley diverted by a glacial ponding, formed by the cutting through of a cover of drift and subsequent exposure of a former col. { 'süd·ə,kòl }

pseudo cold front *See* pseudo front. { 'sü·dō 'kōld ,frənt }

pseudoconcretion [GEOL] A subspherical, secondary sedimentary structure resembling a true concretion but not formed by orderly precipitation of mineral matter in the pores of a sediment. { ¦sü·dō·kän'krē·shən }

pseudoconformity *See* paraconformity. { ¦sü·dō·kən'fȯr·məd·ē }

pseudoconglomerate [GEOL] A rock that resembles, or may easily be mistaken for, a true or normal (sedimentary) conglomerate. { ¦sü·dō·kən'gläm·ə·rət }

pseudo cross-bedding [GEOL] **1.** An inclined bedding produced by deposition in response to ripple-mark migration and characterized by foreset beds that appear to dip into the current. **2.** A structure resembling cross-bedding, caused by distortion-free slumping and sliding of a semiconsolidated mass of sediments (such as sandy shales). { 'sü·dō 'krȯs ,bed·iŋ }

pseudodiffusion [GEOL] Mixing of thin superpositioned layers of slowly accumulated marine sediments by the action of water motion or subsurface organisms. { ¦sü·dō·di'fyü·zhən }

pseudoequivalent temperature *See* equivalent temperature. { ¦sü·dō·i'kwiv·ə·lənt 'tem·prə·chər }

pseudofault [GEOL] A faultlike feature resulting from weathering along joint, shrinkage, or bedding planes. { 'süd·ə,fȯlt }

pseudofibrous peat [GEOL] Peat that is fibrous in texture but is plastic and incoherent. { ¦sü·dō'fī·brəs 'pēt }

pseudo front [METEOROL] A small-scale front, formed in association with organized severe convective activity, between a mass of rain-cooled air from the thunderstorm clouds and the warm surrounding air. Also known as pseudo cold front. { 'sü·dō 'frənt }

pseudogley [GEOL] A densely packed, silty soil that is alternately waterlogged and rapidly dried out. { 'sü·dō,glā }

pseudogradational bedding [GEOL] A structure in metamorphosed sedimentary rock in which the original textural graduation (coarse at the base, finer at the top) appears to be reversed because of the formation of porphyroblasts in the finer-grained part of the rock. { ¦sü·dō·grā'dā·shən·əl 'bed·iŋ }

pseudokarst [GEOL] A topography that resembles karst but that is not formed by the dissolution of limestone; usually a rough-surfaced lava field in which ceilings of lava tubes have collapsed. { 'süd·ə,kärst }

pseudokettle *See* pingo remnant. { ¦sü·dō'ked·əl }

pseudomicroseism [GEOPHYS] A microseism due to instrumental effects. { ,sü·dō'mī·krə,sīz·əm }

pseudomountain [GEOL] A mountain formed by differential erosion, in contrast to one produced by uplift. { ¦sü·dō¦maůnt·ən }

pseudonodule [GEOL] A primary sedimentary structure consisting of a ball-like mass of sandstone enclosed in shale or mudstone; characterized by a rounded base with upturned or inrolled edges and resulting from the settling of sand into underlying clay or mud which has welled up between isolated sand masses. Also known as sand roll. { ¦sü·dō'näj,ül }

pseudo-oolith [GEOL] A spherical or roundish pellet or particle (generally less than 1 millimeter in diameter) in a sedimentary rock, externally resembling an oolith in size or shape but of secondary origin and amorphous or crypto- or microcrystalline, and lacking the radial or concentric internal structure of an oolith. Also known as false oolith. { ¦sü·dō'ō,ō,lith }

pseudo ripple mark [GEOL] A bedding-plane feature that resembles a ripple mark but is formed by lateral pressure caused by slumping or by local, small-scale tectonic deformation. { ¦sü·dō 'rip·əl ,märk }

pseudostratification *See* sheeting structure. { ¦sü·dō,strad·ə·fə'kā·shən }

pseudotillite [GEOL] A nonglacial tillite-like rock, such as a pebbly mudstone, formed on land by the flow of nonglacial mud or deposited by a subaqueous turbidity flow. { ¦sü·dō'täd·əl,īt }

pseudounconformity [GEOL] A stratigraphic relationship that appears unconformable but is characterized by a superabundance or an excess accumulation of sediment, due to factors like submarine slumping which occurs penecontemporaneously with sedimentation off the sides of a rising anticline or dome. { ¦sü·dō·kən'fór·məd·ē }

pseudovitrinite [GEOL] A maceral of coal that is superficially similar to vitrinite but that is higher in reflectance from polished surfaces in oil immersion and has slitted structure, remnant cellular structures, uncommon fracture patterns, higher relief, and paucity or absence of pyrite inclusions. { ¦sü·dō'vi·trə,nīt }

pseudovitrinoid [GEOL] Pseudovitrinite occurring in bituminous coal. { ¦sü·dō'vi·trə,nòid }

pseudovolcano [GEOL] A large crater or circular hollow believed not to be associated with recent volcanic activity, such as a crater which is the result of cauldron subsidence or of a phreatic explosion in the distant past. { ¦sü·dō·väl'kā·nō }

psychrosphere [OCEANOGR] The cold deep layer of the ocean, 100–700 meters (330–2300 feet) below the surface, where the water temperature is typically less than 10°C (50°F). { 'sī·krə,sfir }

pteropod ooze [GEOL] A pelagic sediment containing at least 45% calcium carbonate in the form of tests of marine animals, particularly pteropods. { 'ter·ə,päd 'üz }

PTRM *See* partial thermoremanent magnetization.

ptygma [GEOL] Pegmatitic material with migmatite or gneiss, resembling disharmonic folds. Also known as ptygmatic fold. { 'tig·mə }

ptygmatic fold *See* ptygma. { tig'mad·ik 'fōld }

pudding ball *See* armored mud ball. { 'pùd·iŋ ,bòl }

puddingstone [GEOL] In the United Kingdom, a conglomerate consisting of rounded pebbles whose colors are in marked contrast with the matrix, giving a section of the rock the appearance of a raisin pudding. { 'pùd·iŋ,stōn }

puelche [METEOROL] An east wind which has crossed the Andes; the Andean foehn of the South American west coast. { 'pwel·chē }

puff cone *See* mud cone. { 'pəf ,kōn }

puff of wind [METEOROL] A slight local breeze which causes a patch of ripples on the surface of the sea. { 'pəf əv 'wind }

pull-apart [GEOL] A precompaction sedimentary structure having the appearance of boudinage and consisting of beds that have been stretched and pulled apart into relatively short slabs. { 'pùl ə,pärt }

pumice [GEOL] A rock froth, formed by the extreme puffing up of liquid lava by expanding gases liberated from solution in the lava prior to and during solidification. Also known as foam; pumice stone; pumicite; volcanic foam. { 'pəm·əs }

pumice fall [GEOL] Pumice falling from a volcano eruption cloud. { 'pəm·əs ,fòl }

pumiceous [GEOL] Pertaining to the texture of a pyroclastic rock, such as pumice, characterized by numerous small cavities presenting a spongy, frothy appearance. { pyü'mish·əs }

pumice stone *See* pumice. { 'pəm·əs ˌstōn }

pumicite *See* pumice. { 'pəm·əˌsīt }

pumilith [GEOL] A lithified deposit of volcanic ash. { 'pəm·əˌlith }

Purbeckian [GEOL] A stage of geologic time in Great Britain: uppermost Jurassic (above Bononian, below Cretaceous). { pər'bek·ē·ən }

pure coal *See* vitrain. { 'pyůr ¦kōl }

purga [METEOROL] A severe storm similar to the blizzard and buran, which rages in the tundra regions of northern Siberia in winter. { 'pu̇r·gə }

purl [HYD] A swirling or eddying stream or rill, moving swiftly around obstructions. { pərl }

purple light [GEOPHYS] The faint purple glow observed on clear days over a large region of the western sky after sunset and over the eastern sky before sunrise. { 'pər·pəl 'līt }

push moraine [GEOL] A broad, smooth, arc-shaped ridge consisting of material mechanically pushed or shoved along by an advancing glacier. Also known as push-ridge moraine; shoved moraine; thrust moraine; upsetted moraine. { 'pu̇sh məˌrān }

push-ridge moraine *See* push moraine. { 'pu̇sh ¦rij məˌrān }

puy [GEOL] A small, remnant volcanic cone. { pwē }

P wave [GEOPHYS] A body wave that can pass through all layers of the earth. It is fastest of all seismic waves, traveling at a velocity of 3–4 miles (5–7 kilometers) per second in the crust and 5–6 miles (8–9 kilometers) per second in the upper mantle. Also known as compressional wave; longitudinal wave; primary wave. { 'pē ˌwāv }

pycnocline [GEOPHYS] A change in density of ocean or lake water or rock with displacement in some direction, especially a rapid change in density with vertical displacement. [OCEANOGR] A region in the ocean where water density increases relatively rapidly with depth. { 'pik·nəˌklīn }

pyramidal iceberg *See* pinnacled iceberg. { ¦pir·ə¦mid·əl 'īsˌbərg }

Pyrenean orogeny [GEOL] A short-lived orogeny that occurred during the late Eocene, between the Bartonian and Ludian stages. { ˌpir·ə'nē·ən ȯ'räj·ə·nē }

pyritization [GEOL] A common process of hydrothermal alteration involving introduction of or replacement by pyrite. { ˌpīˌrīd·ə'zā·shən }

pyritobitumen [GEOL] Any of various dark-colored, relatively hard, nonvolatile hydrocarbon substances often associated with mineral matter, which decompose upon heating to yield bitumens. Also known as pyrobitumen. { pə¦rīd·ō·bə'tü·mən }

pyrobitumen *See* pyritobitumen. { ¦pī·rō·bə'tü·mən }

pyroclast [GEOL] An individual pyroclastic fragment or clast. { 'pī·rəˌklast }

pyroclastic flow [GEOL] Ash flow not involving high-temperature conditions. { ¦pī·rə¦klas·tik 'flō }

pyroclastic ground surge [GEOL] The relatively thin mantle of rock found around a volcanic vent; the thickness is not uniform, the internal stratification is not parallel to the top and bottom of the layer, and the extent is a few kilometers from the source. { ¦pī·rə¦klas·tik 'grau̇nd ˌsərj }

pyrogenesis [GEOL] The intrusion and extrusion of magma and its derivatives. { ˌpī·rō'jen·ə·səs }

pyroheliometer [METEOROL] An instrument that measures the sun's radiation output. { ˌpī·rōˌhē·lē'äm·əd·ər }

pyromagma [GEOL] A highly mobile lava, oversaturated with gases, that exists at shallower depths than hypomagma. { ˌpī·rō'mag·mə }

pyrosphere [GEOL] The zone of the earth below the lithosphere, consisting of magma. Also known as magmosphere. { 'pī·rəˌsfir }

pyroxene alkali syenite [GEOL] A quartz-poor (less than 20%) member of the charnockite series, characterized by the presence of microperthite. { pə'räkˌsēn 'al·kəˌlī 'sī·əˌnīt }

pyroxene monzonite [GEOL] A quartz-poor (less than 20%) member of the charnockite series, containing approximately equal amounts of microperthite and plagioclase. { pə'räk,sēn 'män·zə,nīt }

pyroxene syenite [GEOL] A quartz-poor (less than 20%) member of the charnockite series, containing more microperthite than plagioclase. { pə'räk,sēn 'sī·ə,nīt }

quadrantal point *See* intercardinal point. { kwä'drant·əl ˌpȯint }

quake sheet [GEOL] A well-defined bed resembling a slump sheet but produced by an earthquake and resulting in the formation of a load cast without horizontal slip. { kwāk ˌshēt }

quaking bog [GEOL] A peat bog floating or growing over water-saturated land which shakes or trembles when walked on. { 'kwāk·iŋ 'bäg }

quality of snow [METEOROL] The amount of ice in a snow sample expressed as a percent of the weight of the sample. Also known as thermal quality of snow. { 'kwäl·əd·ē əv 'snō }

quantitative geomorphology [GEOL] The assignment of dimensions of mass, length, and time to all descriptive parameters of landform geometry and geomorphic processes, followed by the derivation of empirical mathematical relationships and formulation of rational mathematical models relating these parameters. { 'kwän·ə·tād·iv ˌjē·ō·mȯr'fäl·ə·jē }

quaquaversal [GEOL] Of strata and geologic structures, dipping outward in all directions away from a central point. { ¦kwä·kwə¦vər·səl }

quarrying *See* plucking. { 'kwär·ē·iŋ }

quartzose [GEOL] Referring to a substance which contains quartz as a principal constituent. { 'kwȯrt‚sōs }

quartz-pebble conglomerate *See* orthoquartzitic conglomerate. { 'kwȯrts ˌpeb·əl kən'gläm·ə·rət }

quasi-cratonic [GEOL] Pertaining to a part of oceanic crust marginal to the continent which is considered to be former continental material that stretched and foundered during expansion. Also known as semicratonic. { ¦kwä·zē krə'tän·ik }

quasi-equilibrium [GEOL] The state of balance or grade in a stream cross section, whereby conditions of approximate equilibrium tend to be established in a reach of the stream as soon as a rather smooth longitudinal profile has been established in that reach, even though downcutting may go on. { ¦kwä·zē ˌē·kwə'lib·rē·əm }

quasi-hydrostatic approximation [METEOROL] The use of the hydrostatic equation as the vertical equation of motion, thus implying that the vertical accelerations are small without constraining them to be zero. Also known as quasi-hydrostatic assumption. { ¦kwä·zē ¦hī·drə¦stad·ik ə‚präk·sə'mā·shən }

quasi-hydrostatic assumption *See* quasi-hydrostatic approximation. { ¦kwä·zē ¦hī·drə¦stad·ik ə'səm·shən }

quasi-stationary front [METEOROL] A front which is stationary or nearly so; conventionally, a front which is moving at a speed less than about 5 knots (0.26 meter per second) is generally considered to be quasi-stationary. Commonly known as stationary front. { ¦kwä·zē 'stā·shə‚ner·ē 'frənt }

Quaternary [GEOL] The second period of the Cenozoic geologic era, following the Tertiary, and including the last 2–3 million years. { 'kwät·ən‚er·ē }

queenslandite *See* Darwin glass. { 'kwēnz·lən‚dīt }

Queenston shale [GEOL] A red bed series from the Ordovician found in Niagara Gorge; it is composed of deltaic red shale. { 'kwēnz·tən 'shāl }

queenstownite *See* Darwin glass. { 'kwēn·stə‚nīt }

quenched water [OCEANOGR] Ocean water which produces an abnormally large propagation loss in the sound passing through it; it is usually found in shallow water or near shores where there are strong currents accompanied by considerable turbulence. { 'kwencht 'wȯd·ər }

quick [GEOL] **1.** Referring to a sediment that, when mixed with or absorbing water, becomes extremely soft, incoherent, or loose, and is capable of flowing easily under load or by force of gravity. **2.** Referring to a soil in which a decrease in effective stress allows water to flow upward with sufficient velocity to reduce significantly the soil's bearing capacity. **3.** Referring to a highly porous soil that readily absorbs heat. { kwik }

quick clay [GEOL] Clay that loses its shear strength after being disturbed. { 'kwik ˌklā }

quicksand [GEOL] A highly mobile mass of fine sand consisting of smooth, rounded grains with little tendency to mutual adherence, usually thoroughly saturated with upward-flowing water; tends to yield under pressure and to readily swallow heavy objects on the surface. Also known as running sand. { 'kwikˌsand }

quickwater [HYD] The part of a stream characterized by a strong current. { 'kwik ˌwȯd·ər }

quilted surface [GEOL] A land surface characterized by broad, rounded, uniformly convex hills separating valleys that are comparatively narrow. { 'kwil·təd 'sər·fəs }

R

rabal [METEOROL] A method of winds-aloft observation, that is, the determination of wind speeds and directions in the atmosphere above a station; it is accomplished by recording the elevation and azimuth angles of the balloon at specified time intervals while visually tracking a radiosonde balloon with the theodolite. { 'rā,bäl }

race [OCEANOGR] A rapid current, or a constricted channel in which such a current flows; the term is usually used only in connection with a tidal current, which may be called a tide race. { rās }

radar climatology [CLIMATOL] The statistics in time and space of radar weather echoes. { 'rā,där ,klī·mə'täl·ə·jē }

radar interferometry [GEOPHYS] A microwave remote sensing method for combining imagery collected over time by radar systems on board airplane or satellite platforms to map the elevations, movements, and changes of the earth's surface. Such detectable changes include earthquakes, volcanoes, glaciers, landslides, and underground explosions, as well as fires, floods, forestry operations, moisture changes, and vegetation growth. { ¦rā,där ,in·tər·fə'räm·ə·trē }

radar meteorological observation [METEOROL] Evaluation of the echoes appearing on the indicator of a weather radar, in terms of orientation, coverage, intensity, tendency of intensity, height, movement, and unique characteristics of echoes, that may be indicative of certain types of severe storms (such as hurricanes, tornadoes, or thunderstorms) and of anomalous propagation. Also known as radar weather observation. { 'rā,där ,mēd·ē·ə·rə'läj·ə·kəl ,äb·sər'vā·shən }

radar meteorology [METEOROL] The study of the scattering of radar waves by all types of atmospheric phenomena and the use of radar for making weather observations and forecasts. { 'rā,där ,mēd·ē·ə'räl·ə·jē }

radar report [METEOROL] The encoded and transmitted report of a radar meteorological observation; these reports usually give the azimuth, distance, altitude, intensity, shape and movement, and other characteristics of precipitation echoes observed by the radar. Also known as rain area report. Abbreviated RAREP. { 'rā,där ri,pȯrt }

radar storm detection [METEOROL] The detection of certain storms or stormy conditions by means of radar; liquid or frozen water drops within the storm reflect radar echoes. { 'rā,där 'stȯrm di,tek·shən }

radar storm-detection equation [METEOROL] The equation which relates the variables involved in the radar detection of precipitation. { 'rā,där 'stȯrm di,tek·shən i,kwā·zhən }

radar upper band *See* upper bright band. { 'rā,där 'əp·ər ,band }

radar weather observation *See* radar meteorological observation. { 'rā,där 'weth·ər ,äb·zər,vā·shən }

radar wind [METEOROL] Wind of which the movement, speed, and direction is observed or determined by a radar tracking of a balloon carrying a radiosonde, a radio transmitter, or a radar reflector. { 'rā,där ,wind }

radial drainage pattern [GEOL] A drainage pattern characterized by radiating streams diverging from a high central area. Also known as centrifugal drainage pattern. { 'rād·ē·əl 'drān·ij ,pad·ərn }

radial faults [GEOL] Faults arranged like the spokes of a wheel, radiating from a central point. { 'rād·ē·əl 'fȯls }

radiational cooling [METEOROL] The cooling of the earth's surface and adjacent air, accomplished (mainly at night) whenever the earth's surface suffers a net loss of heat due to terrestrial radiation. { ˌrād·ē'ā·shən·əl 'kül·iŋ }

radiational inversion [METEOROL] An inversion at the land surface resulting from rapid radiational cooling of lower air; usually occurs on cold winter nights. { ˌrad·ē'ā·shən·əl in'vər·zhən }

radiation budget [GEOPHYS] A quantitative statement of the amounts of radiation entering and leaving a given region of the earth. { ˌrād·ē'ā·shən ˌbəj·ət }

radiation chart [GEOPHYS] Any chart or diagram which permits graphical solution of the (generally unintegrable) flux integrals arising in problems of atmospheric infrared radiation transfer. { ˌrād·ē'ā·shən ˌchärt }

radiation fog [METEOROL] A major type of fog, produced over a land area when radiational cooling reduces the air temperature to or below its dew point; thus, strictly, a nighttime occurrence, although the fog may begin to form by evening twilight and often does not dissipate until after sunrise. { ˌrād·ē'ā·shən ˌfäg }

radiative diffusivity [METEOROL] A characteristic property of a given layer of the atmosphere which governs the rate at which that layer will warm or cool as a result of the transfer, within it, of infrared radiation; the radiative diffusivity is dependent upon the temperature and water-vapor content of the layer of air and upon the pressure within the layer. { 'rād·ēˌād·iv ˌdiˌfyü'siv·əd·ē }

radiative forcing [METEOROL] The relative effectiveness of greenhouse gases to restrict long-wave radiation from escaping back into space. For a particular greenhouse gas, radiative forcing is measured as the change in average net radiation (in watts per square meter) at the top of the troposphere, and depends on the wavelength at which the gas absorbs the radiation, the strength of absorption per molecule, and the concentration of the gas. { ˌrād·ēˌād·iv 'fȯrs·iŋ }

radiatus [METEOROL] A cloud variety whose elements are arranged in straight parallel bands; owing to the effect of perspective, these bands seem to converge toward a point on the horizon, or, when the bands cross the entire sky, toward two opposite points. { ˌrād·ē'äd·əs }

radiochronology [GEOL] An absolute-age dating method based on the existing ratio between radioactive parent elements (such as uranium-238) and their radiogenic daughter isotopes (such as lead-206). { ¦rad·ē·ō·krə'näl·ə·jē }

radio climatology [CLIMATOL] The study of regional and seasonal variations in the manner of propagation of radio energy through the atmosphere. { 'rād·ē·ō 'klī·mə'täl·ə·jē }

radio duct [GEOPHYS] An atmospheric layer, typically shallow and almost horizontal, in which radio waves propagate in an anomalous fashion; ducts occur when, due to sharp inversions of temperature or humidity, the vertical gradient of the radio index of refraction exceeds a critical value. { 'rād·ē·ō ˌdəkt }

radioelectric meteorology *See* radio meteorology. { ¦rād·ē·ō·i'lek·trik ˌmēd·ē·ə'räl·ə·jē }

radiogeology [GEOCHEM] The study of the distribution patterns of radioactive elements in the earth's crust and the role of radioactive processes in geologic phenomena. { ¦rād·ē·ō·jē'äl·ə·jē }

radioglaciology [GEOPHYS] The study of glacier ice by means of radar, especially the sounding of ice depth. { ¦rād·ē·ōˌglā·sē'äl·ə·jē }

radio hole [GEOPHYS] Strong fading of the radio signal at some position in space along an air-to-air or air-to-ground path; the effect is caused by the abnormal refraction of radio waves. { 'rād·ē·ō ˌhōl }

radiolarian chert [GEOL] A homogeneous cryptocrystalline radiolarite with a well-developed matrix. { ¦rād·ē·ō¦lar·ē·ən 'chərt }

radiolarian earth [GEOL] A porous, unconsolidated siliceous sediment formed from the opaline silica skeletal remains of Radiolaria; formed from radiolarian ooze. { ¦rād·ē·ō¦lar·ē·ən 'ərth }

radiolarian ooze [GEOL] A siliceous ooze containing the skeletal remains of the Radiolaria. { ¦rād·ē·ō¦lar·ē·ən 'üz }

radiolarite [GEOL] **1.** A whitish, hard, consolidated equivalent of radiolarian earth. **2.** Radiolarian ooze that has been indurated. { ˌrād·ē·ō'laˌrīt }

radio meteorology [METEOROL] That branch of the science of meteorology which embraces the propagation of radio energy through the atmosphere, and the use of radio and radar equipment in meteorology; this is the most general term and includes radar meteorology. Also known as radioelectric meteorology. { 'rād·ē·ō ˌmēd·ē·ə'räl·ə·jē }

radiometric age [GEOL] Geologic age expressed in years determined by quantitatively measuring radioactive elements and their decay products. { ¦rād·ē·ō¦me·trik 'āj }

radiosonde observation [METEOROL] An evaluation in terms of temperature, relative humidity, and pressure aloft, of radio signals received from a balloon-borne radiosonde; the height of each mandatory and significant pressure level of the observation is computed from these data. Also known as raob. { 'rād·ē·ōˌsänd ˌäb·zərˌvā·shən }

radio window [GEOPHYS] A band of frequencies extending from about 6 to 30,000 megahertz, in which radiation from the outer universe can enter and travel through the atmosphere of the earth. { 'rād·ē·ō ˌwin·dō }

Radstockian [GEOL] A European stage of geologic time forming the upper Upper Carboniferous, above Staffordian and below Stephanian, equivalent to uppermost Westphalian. { rad'stäk·ē·ən }

raffiche [METEOROL] In the Mediterranean region, gusts from the mountains; violent gusts of the bora. { rä'fēsh }

raft [GEOL] **1.** A rock fragment caught up in a magma and drifting freely, more or less vertically. **2.** *See* float coal. [HYD] An accumulation or jam of floating logs, driftwood, dislodged trees, or other debris, formed naturally in a stream by caving of the banks. { raft }

rafted ice [OCEANOGR] A form of pressure ice composed of overlying pieces of ice floe. { 'raf·təd 'īs }

rafting [GEOL] Transporting of rock by floating ice or floating organic materials (such as logs) to places not reached by water currents. [OCEANOGR] The process of forming rafted ice. { 'raft·iŋ }

raft lake [HYD] A relatively short-lived body of water impounded along a stream by a raft. { 'raft ˌlāk }

raft tectonics *See* plate tectonics. { 'raft tekˌtän·iks }

ragged ceiling *See* indefinite ceiling. { 'rag·əd'sēl·iŋ }

raggiatura [METEOROL] Land squalls descending with great force from ravines and valleys in high land in Italy; they extend only a short distance off the west coast. { ˌrä·jə'tu̇r·ə }

rain [METEOROL] Precipitation in the form of liquid water drops with diameters greater than 0.5 millimeter, or if widely scattered the drops may be smaller; the only other form of liquid precipitation is drizzle. { rān }

rain and snow mixed [METEOROL] Precipitation consisting of a mixture of rain and wet snow; usually occurs when the temperature of the air layer near the ground is slightly above freezing. { 'rān ən 'snō 'mikst }

rain area report *See* radar report. { 'rān ¦er·ē·ə riˌpȯrt }

rain cloud [METEOROL] Any cloud from which rain falls; a popular term having no technical denotation. { 'rān ˌklau̇d }

rain crust [HYD] A type of snow crust, formed by refreezing after surface snow crystals have been melted and wetted by liquid precipitation; composed of individual ice particles such as firn. { 'rān ˌkrəst }

raindrop [METEOROL] A drop of water of diameter greater than 0.5 millimeter falling through the atmosphere. { 'rānˌdräp }

raindrop impressions *See* rain prints. { 'rānˌdräp im'presh·ənz }

raindrop imprints *See* rain prints. { 'rānˌdräp 'imˌprins }

rain factor [HYD] A coefficient designed to measure the combined effect of temperature and moisture on the formation of soil humus; it is obtained by dividing the annual rainfall (in millimeters) by the mean annual temperature (in degrees Celsius). { 'rān ˌfak·tər }

rainfall [METEOROL] The amount of precipitation of any type; usually taken as that amount which is measured by means of a rain gage (thus a small, varying amount of direct condensation is included). { 'rānˌfȯl }

rainfall frequency [CLIMATOL] The number of times, during a specified period of years, that precipitation of a certain magnitude or greater occurs or will occur at a station; numerically, the reciprocal of the frequency is usually given. { 'rānˌfȯl ˌfrē·kwən·sē }

rainfall intensity *See* precipitation intensity. { 'rānˌfȯl in'ten·səd·ē }

rainfall inversion *See* precipitation inversion. { 'rānˌfȯl inˌvər·zhən }

rainfall penetration [HYD] The depth below the soil surface to which water from a given rainfall has been able to infiltrate. { 'rānˌfȯl ˌpen·əˌtrā·shən }

rainfall regime [CLIMATOL] The character of the seasonal distribution of rainfall at any place; the chief rainfall regimes, as defined by W. G. Kendrew, are equatorial, tropical, monsoonal, oceanic and continental westerlies, and Mediterranean. { 'rānˌfȯl rəˌzhēm }

rainforest climate *See* wet climate. { 'rānˌfär·əst ˌklī·mət }

rain gush *See* cloudburst. { 'rān ˌgəsh }

rain gust *See* cloudburst. { 'rān ˌgəst }

raininess [METEOROL] Generally, the quantitative character of rainfall for a given place. { 'rān·ē·nəs }

rainmaking [METEOROL] Popular term applied to all activities designed to increase, through any artificial means, the amount of precipitation released from a cloud. { 'rānˌmāk·iŋ }

rain pillar [GEOL] A minor landform consisting of a column of soil or soft rock capped and protected by pebbles or concretions, produced by the differential erosion from the impact of falling rain. { 'rān ˌpil·ər }

rain prints [GEOL] Small, shallow depressions formed in soft sediment or mud by the impact of falling raindrops. Also known as raindrop impressions; raindrop imprints. { 'rān ˌprins }

rain shadow [METEOROL] An area of diminished precipitation on the lee side of mountains or other topographic obstacles. { 'rān ˌshad·ō }

rainsquall [METEOROL] A squall associated with heavy convective clouds, frequently the cumulonimbus type; usually sets in shortly before the thunderstorm rain, blowing outward from the storm and generally lasting only a short time. Also known as thundersquall. { 'rānˌskwȯl }

rain stage [METEOROL] The thermodynamic process of condensation of water from moist air in an idealized saturation-adiabatic or pseudoadiabatic lifting, at temperatures above the freezing point; begins at the condensation level. { 'rān ˌstāj }

rainwash [GEOL] **1.** The washing away of loose surface material by rainwater after it has reached the ground but before it has been concentrated into definite streams. **2.** Material transported and accumulated, or washed away, by rainwater. { 'rānˌwäsh }

rainwater [HYD] Water that has fallen as rain and is quite soft, as it has not yet collected soluble matter from the soil. { 'rānˌwȯd·ər }

rainy climate [CLIMATOL] In W. Koppen's climatic classification, any climate type other than the dry climates; however, it is generally understood that this refers principally to the tree climates and not the polar climates. { 'rān·ē 'klī·mət }

rainy season [CLIMATOL] In certain types of climate, an annually recurring period of one or more months during which precipitation is a maximum for that region. Also known as wet season. { 'rān·ē ˌsēz·ən }

raised beach [GEOL] An ancient beach raised to a level above the present shoreline by uplift or by lowering of the sea level; often bounded by inland cliffs. { 'rāzd 'bēch }

rake *See* plunge. { rāk }

ram [HYD] An underwater ledge or projection from an ice wall, ice front, iceberg, or floe, usually caused by the more intensive melting and erosion of the unsubmerged part. Also known as apron; spur. { ram }

rambla [GEOL] A dry ravine, or the dry bed of an ephemeral stream. { 'ram·blə }

rampart [GEOL] **1.** A narrow, wall-like ridge, 3–7 feet (1–2 meters) high, built up by waves along the seaward edge of a reef flat, and consisting of boulders, shingle, gravel, or reef rubble, commonly capped by dune sand. **2.** A wall-like ridge of unconsolidated material formed along a beach by the action of strong waves and current. **3.** A crescentic or ringlike deposit of pyroclastics around the top of a volcano. { 'ram,pärt }

ramp [HYD] An accumulation of snow forming an inclined plane between land or land ice and sea ice or shelf ice. Also known as drift ice foot. { ramp }

rampart wall [GEOL] A rimming wall formed along the outer or seaward margin of a terrace, as on various high limestone Pacific islands. { 'ram,pärt ¦wȯl }

ramp valley [GEOL] A trough between faults, forced downward by lateral pressure. { 'ramp ,val·ē }

Rancholabrean [GEOL] A stage of geologic time in southern California, in the upper Pleistocene, above the Irvingtonian. { ,ran·chō·lə'brā·ən }

randkluft [HYD] A crevasse at the head of a mountain glacier, separating the moving ice and snow from the surrounding rock wall of the valley, where no ice apron is present. { 'ränt,klu̇ft }

random forecast [METEOROL] A forecast in which one of a set of meteorological contingencies is selected on the basis of chance; it is often used as a standard of comparison in determining the degree of skill of another forecast method. { 'ran·dəm 'fȯr,kast }

range of tide [OCEANOGR] The difference in height between consecutive high and low tides at a place. { 'rānj əv 'tīd }

range zone [GEOL] Formal biostratigraphic zone made up of a body of strata comprising the total horizontal (geographic) and vertical (stratigraphic) range of occurrence of a specified taxon of a group of taxa. { 'rānj ,zōn }

rank [GEOL] **1.** A coal classification based on degree of metamorphism. **2.** *See* stack. { raŋk }

raob *See* radiosonde observation. { 'rā,äb }

rapid [HYD] A portion of a stream in swift, disturbed motion, but without cascade or waterfall; usually used in the plural. { 'rap·əd }

rapid flow [HYD] Water flow whose velocity exceeds the velocity of propagation of a long surface wave in still water. Also known as supercritical flow. { 'rap·əd 'flō }

RAREP *See* radar report. { 'rer,ep }

rate-of-change map [GEOL] A derived stratigraphic map that shows the rate of change of structure, thickness, or composition of a given stratigraphic unit. { 'rāt əv 'chānj ,map }

rate of sedimentation [GEOL] The amount of sediment accumulated in an aquatic environment over a given period of time, usually expressed as thickness of accumulation per unit time. Also known as sedimentation rate. { 'rāt əv ,sed·ə·mən'tā·shən }

rating curve [HYD] For a given point on a stream, a graph of discharge versus stage. { 'rād·iŋ ,kərv }

ratio map [GEOL] A facies map that depicts the ratio of thicknesses between rock types in a given stratigraphic unit. { 'rā·shō ,map }

rational formula [HYD] The expression of peak discharge as equal to the product of rainfall, drainage area, and a runoff coefficient depending on drainage-basin characteristics. { 'rash·ən·əl 'fȯr·myə·lə }

ratio of rise [OCEANOGR] The ratio of the height of tide at two places. { 'rā·shō əv 'rīz }

rattlesnake ore [GEOL] A gray, black, and yellow mottled ore of carnotite and vanoxite; its spotted appearance resembles that of a rattlesnake. { 'rad·əl,snāk ,ȯr }

rattle stone [GEOL] A concretion composed of concentric laminae of different compositions, in which the more soluble layers have been removed by solution, leaving the central part detached from the outer part, such as a concretion of iron oxide filled with loose sand that rattles on shaking. Also known as klapperstein. { 'rad·əl ,stōn }

Rauracian [GEOL] A substage of Upper Jurassic geologic time in Great Britain forming the middle Lusitanian, above the Argovian and below the Sequanian. { rau̇'rā·shən }

ravelly ground [GEOL] Rock that breaks into small pieces when drilled and tends to cave or slough into the hole when the drill string is pulled, or binds the drill string by becoming wedged or locked between the drill rod and the borehole wall. { 'rav·lē 'grau̇nd }
ravine [GEOGR] A small and narrow valley with steeply sloping sides. { rə'vēn }
ravinement [GEOL] **1.** The formation of a ravine or ravines. **2.** An irregular junction which marks a break in sedimentation, such as an erosion line occurring where shallow-water marine deposits have cut down into slightly eroded underlying beds. { rə'vēn·mənt }
raw [METEOROL] Colloquially descriptive of uncomfortably cold weather, usually meaning cold and damp, but sometimes cold and windy. { ró }
raw humus *See* ectohumus. { 'ró 'hyü·məs }
rawin [METEOROL] A method of winds-aloft observation, that is, the determination of wind speeds and directions in the atmosphere above a station; accomplished by tracking a balloon-borne radar target, responder, or radiosonde transmitter with either radar or a radio direction finder. { 'rā,win }
rawinsonde [METEOROL] A method of upper-air observation consisting of an evaluation of the wind speed and direction, temperature, pressure, and relative humidity aloft by means of a balloon-borne radiosonde tracked by a radar or radio direction finder. { 'rā·wən,sänd }
raw map [GEOPHYS] A seismic map in which the z coordinate is time. { 'ró 'map }
Rayleigh atmosphere [METEOROL] An idealized atmosphere consisting of only those particles, such as molecules, that are smaller than about one-tenth of the wavelength of all radiation incident upon that atmosphere; in such an atmosphere, simple Rayleigh scattering would prevail. { 'rā·lē ,at·mə,sfir }
Rayleigh wave [GEOPHYS] In seismology, a surface wave with a retrograde, elliptical motion at the free surface. Also known as R wave. { 'rā·lē ,wāv }
ray parameter [GEOPHYS] A function p that is constant along a given seismic ray, given by $p = rv^{-1} \sin i$, where r is the distance from the center O of the earth, v is the velocity, and i is the angle that the ray at a point P makes with the radius OP. { 'rā pə,ram·əd·ər }
razorback [GEOL] A sharp, narrow ridge. { 'rā·zər,bak }
razor stone *See* novaculite. { 'rā·zər ,stōn }
reach [GEOGR] **1.** A continuous, unbroken surface of land or water. Also known as stretch. **2.** A bay, estuary, or other arm of the sea extending up into the land. [HYD] A straight, continuous, or extended part of a river, stream, or restricted waterway. { rēch }
rebat [METEOROL] The lake breeze of Lake Geneva, Switzerland; it blows from about 10 a.m. to 4 p.m. { re'bä }
rebound [GEOL] The isostatic readjustment upward of a landmass depressed by glacial loading. { 'rē,bau̇nd }
reboyo [METEOROL] A persistent (day-long) storm from the southwest during the rainy season on the Brazilian coast. { rə'bói·ō }
Recent *See* Holocene. { 'rē·sənt }
recess [GEOL] **1.** An indentation occurring in a surface, bounded by a straight line. **2.** An area having the axial traces of folds concave toward the outer edge of the folded belt. { 'rē,ses }
recession [GEOL] **1.** The backward movement, or retreat, of an eroded escarpment. **2.** A continuing landward movement of a shoreline or beach undergoing erosion. Also known as retrogression. **3.** The withdrawal of a body of water (as a sea or lake), thereby exposing formerly submerged areas. [HYD] The gradual upstream retreat of a waterfall. { ri'sesh·ən }
recessional moraine [GEOL] **1.** An end moraine formed during a temporary halt in the final retreat of a glacier. **2.** A moraine formed during a minor readvance of the ice front during a period of glacial recession. Also known as stadial moraine. { ri'sesh·ən·əl mə'rān }

recession curve [HYD] A hydrograph showing the decrease of the runoff rate after rainfall or the melting of snow. { ri'sesh·ən ˌkərv }

recharge [HYD] **1.** The processes involved in the replenishment of water to the zone of saturation. **2.** The amount of water added or absorbed. Also known as groundwater increment; groundwater recharge; groundwater replenishment; increment; intake. { rē'chärj }

recharge area [HYD] An area in which water is absorbed that eventually reaches the zone of saturation in one or more aquifers. Also known as intake area. { 'rēˌchärj ˌer·ē·ə }

recharge well [HYD] A well used as a source of water in the process of artificial recharge. Also known as injection well. { 'rēˌchärj ˌwel }

reclined fold *See* recumbent fold. { ˌri'klīnd 'fōld }

reconstitution [GEOL] The formation of new chemicals, minerals, or structures under the influence of metamorphism. { rēˌkän·stə'tü·shən }

record observation [METEOROL] A type of aviation weather observation; the most complete of all such observations and usually taken at regularly specified and equal intervals (hourly, usually on the hour). Also known as hourly observation. { 'rek·ərd ˌäb·zərˌvā·shən }

recovery [HYD] The rise in static water level in a well, occurring upon the cessation of discharge from that well or a nearby well. { ri'kəv·ə·rē }

recrystallization flow [GEOL] Flow in which there is molecular rearrangement by solution and redeposition, solid diffusion, or local melting. { rēˌkrist·əl·ə'zā·shən ˌflō }

rectangular cross ripple mark [GEOL] An oscillation cross ripple mark consisting of two sets of ripples which intersect at right angles, enclosing a rectangular pit. { rek'taŋ·gyə·lər ¦krȯs 'rip·əl ˌmärk }

rectangular drainage pattern [GEOL] A drainage pattern characterized by many right-angle bends in both the main streams and their tributaries. Also known as lattice drainage pattern. { rek'taŋ·gyə·lər 'drān·ij ˌpad·ərn }

rectification [GEOL] The simplification and straightening of the outline of an initially irregular and crenulate shoreline through the cutting back of headlands and offshore islands by marine erosion, and through deposition of waste from erosion or of sediment brought down by neighboring rivers. { ˌrek·tə·fə'kā·shən }

rectilinear shoreline [GEOL] A long, relatively straight shoreline. { ¦rek·tə'lin·ē·ər 'shȯrˌlīn }

rectorite [GEOL] A white clay-mineral mixture with a regular interstratification of two mica layers (pyrophyllite and vermiculite) and one or more water layers. Also known as allevardite. { 'rek·təˌrīt }

recumbent fold [GEOL] An overturned fold with a nearly horizontal axial surface. Also known as reclined fold. { ri'kəm·bənt 'fōld }

recurrence interval [HYD] The average time interval between occurrences of a hydrologic event, such as a flood, of a given or greater magnitude. { ri'kər·əns 'int·ər·vəl }

recurrent folding [GEOL] A type of folding due to periodic deformation or subsidence and characterized by thinning or possible disappearance of formations at the crest. Also known as revived folding. { ri'kər·ənt 'fōld·iŋ }

recurvature [METEOROL] With respect to the motion of severe tropical cyclones (hurricanes and typhoons), the change in direction from westward and poleward to eastward and poleward; such recurvature of the path frequently occurs as the storm moves into middle latitudes. { rē'kər·və·chər }

recurved spit *See* hook. { rē'kərvd 'spit }

redbed [GEOL] Continentally deposited sediment composed principally of sandstone, siltsone, and shale; red in color due to the presence of ferric oxide (hematite). Also known as red rock. { 'redˌbed }

red clay [GEOL] A fine-grained, reddish-brown pelagic deposit consisting of relatively large proportions of windblown particles, meteoric and volcanic dust, pumice, shark teeth, manganese nodules, and debris transported by ice. Also known as brown clay. { 'red 'klā }

Reddish-Brown Lateritic soil [GEOL] One of a zonal, lateritic group of soils developed from a mottled red parent material and characterized by a reddish-brown surface horizon and underlying red clay. { 'red·ish ¦braün ,lad·ə'rid·ik 'sȯil }

Reddish-Brown soil [GEOL] A group of zonal soils having a reddish, light brown surface horizon overlying a heavier, more reddish horizon and a light-colored lime horizon. { 'red·ish ¦braün 'sȯil }

red earth [GEOL] Leached, red, deep, clayey soil that is characteristic of a tropical climate. Also known as red loam. { 'red ¦ərth }

redeposition [GEOL] Formation into a new accumulation, such as the deposition of sedimentary material that has been picked up and moved (reworked) from the place of its original deposition, or the solution and reprecipitation of mineral matter. { rē,dep·ə'zish·ən }

red loam *See* red earth. { 'red 'lōm }

red magnetism [GEOPHYS] The magnetism of the north-seeking end of a freely suspended magnet; this is the magnetism of the earth's south magnetic pole. { 'red 'mag·nə,tiz·əm }

red mud [GEOL] A reddish terrigenous mud composed of up to 25% calcium carbonate and deriving its color from the presence of ferric oxide; found on the sea floor near deserts and near the mouths of large rivers. { 'red 'məd }

redoxomorphic stage [GEOCHEM] The earliest geochemical stage of diagenesis characterized by mineral changes primarily due to oxidation and reduction reactions. { ri¦däk·sə¦mȯr·fik ,stāj }

red rock *See* redbed. { 'red ,räk }

Red Sea [GEOGR] A body of water that lies between Arabia and northeastern Africa, about 1200 miles (2000 kilometers) long, 180 miles (300 kilometers) wide, and a maximum depth of about 7600 feet (2300 meters). { 'red 'sē }

red snow [HYD] A snow surface of reddish color caused by the presence within it of certain microscopic algae or particles of red dust. { 'red 'snō }

reduced latitude [GEOD] The angle at the center of a sphere tangent to a reference ellipsoid along the equator, between the plane of the equator and a radius to the point intersected on the sphere by a straight line perpendicular to the plane of the equator. Also known as geometric latitude; parametric latitude. { ri'düst 'lad·ə,tüd }

reduced pressure [METEOROL] The calculated value of atmospheric pressure at mean sea level or some other specified level, as derived (reduced) from station pressure or actual pressure; thus, sea level pressure is nearly always a reduced pressure. { ri'düst 'presh·ər }

reduction [GEOL] The lowering of a land surface by erosion. { ri'dək·shən }

reduction index [GEOL] The rate of wear of a sedimentary particle subject to abrasion, expressed as the difference between the mean weight of the particle before and after transport divided by the product of mean weight before transport and the distance traveled. { ri'dək·shən ,in,deks }

reduction of tidal current [OCEANOGR] The processing of observed tidal current data to obtain mean values of tidal current constants. { ri'dək·shən əv 'tīd·əl ,kə·rənt }

reduction of tides [OCEANOGR] The processing of observed tidal data to obtain mean values of tidal constants. { ri'dək·shən əv 'tīdz }

reduction sphere [GEOL] A white, leached, spheroidal mass produced in a reddish or brownish sandstone by a localized reducing environment, commonly surrounding an organic nucleus or a pebble and ranging in size from a poorly defined speck to a large, perfect sphere more than 10 inches (25 centimeters) in diameter. { ri'dək·shən ,sfir }

reduzate [GEOL] A sediment accumulated under reducing conditions and consequently rich in organic carbon and in iron sulfide minerals; examples are coal and black shale. { 'rej·yə,zāt }

Red-Yellow Podzolic soil [GEOL] Any of a group of acidic, zonal soils having a leached, light-colored surface layer and a subsoil containing clay and oxides of aluminum and iron, varying in color from red to yellowish red to a bright yellowish brown. { 'red 'yel·ō päd'zäl·ik 'sȯil }

reef [GEOL] **1.** A ridge- or moundlike layered sedimentary rock structure built almost exclusively by organisms. **2.** An offshore chain or range of rock or sand at or near the surface of the water. { rēf }
reef cap [GEOL] A deposit of fossil-reef material overlying or covering an island or mountain. { 'rēf ˌkap }
reef cluster [GEOL] A group of reefs of wholly or partly contemporaneous growth, found within a circumscribed area or geologic province. { 'rēf ˌkləs·tər }
reef complex [GEOL] The solid reef core and the heterogeneous and contiguous fragmentary material derived from it by abrasion. { 'rēf ˌkämˌpleks }
reef conglomerate *See* reef talus. { 'rēf kənˌgläm·ə·rət }
reef core [GEOL] The rock mass constructed in place, and within the rigid growth lattice formed by reef-building organisms. { 'rēf ˌkȯr }
reef debris *See* reef detritus. { 'rēf dəˌbrē }
reef detritus [GEOL] Fragmental material derived from the erosion of an organic reef. Also known as reef debris. { 'rēf diˌtrīd·əs }
reef edge [GEOL] The seaward margin of the reef flat, commonly marked by surge channels. { 'rēf ˌej }
reef flank [GEOL] The part of the reef that surrounds, interfingers with, and locally overlies the reef core, often indicated by massive or medium beds of reef talus dipping steeply away from the reef core. { 'rēf ˌflaŋk }
reef flat [GEOL] A flat expanse of dead reef rock which is partly or entirely dry at low tide; shallow pools, potholes, gullies, and patches of coral debris and sand are features of the reef flat. { 'rēf ˌflat }
reef front [GEOL] The upper part of the outer or seaward slope of a reef, extending to the reef edge from above the dwindle point of abundant living coral and coralline algae. { 'rēf ˌfrənt }
reef-front terrace [GEOL] A shelflike or benchlike eroded surface, sometimes veneered with organic growth, sloping seaward to a depth of 8–15 fathoms (15–27 meters). { 'rēf ¦frənt ˌter·əs }
reef knoll [GEOL] **1.** A bioherm or fossil coral reef represented by a small, prominent, rounded hill, up to 330 feet (100 meters) high, consisting of resistant reef material, being either a local exhumation of an original reef feature or a feature produced by later erosion. **2.** A present-day reef in the form of a knoll; a small reef patch developed locally and built upward rather than outward. { 'rēf ˌnōl }
reef milk [GEOL] A very-fine-grained matrix material of the back-reef facies, consisting of white, opaque microcrystalline calcite derived from abrasion of the reef core and reef flank. { 'rēf ˌmilk }
reef patch [GEOL] A single large colony of coral formed independently on a shelf at depths less than 220 feet (70 meters) in the lagoon of a barrier reef or of an atoll. Also known as patch reef. { 'rēf ˌpach }
reef pinnacle [GEOL] A small, isolated spire of rock or coral, especially a small reef patch. { 'rēf ˌpin·ə·kəl }
reef segment [GEOL] A part of an organic reef lying between passes, gaps, or channels. { 'rēf ˌseg·mənt }
reef slope [GEOL] The face of a reef rising from the sea floor. { 'rēf ˌslōp }
reef talus [GEOL] Massive inclined strata composed of reef detritus deposited along the seaward margin of an organic reef. Also known as reef conglomerate. { 'rēf ˌtā·ləs }
reef tufa [GEOL] Drusy, prismatic, fibrous calcite deposited directly from supersaturated water upon the void-filling internal sediment of the calcite mudstone of a reef knoll. { 'rēf ˌtüf·ə }
reef wall [GEOL] A wall-like upgrowth of living coral and the skeletal remains of dead coral and other reef-building organisms, which reaches an intertidal level and acts as a partial barrier between adjacent environments. { 'rēf ˌwȯl }
reentrant [GEOL] A prominent, generally angular indentation into a coastline. { rē'en·trənt }
reference level [OCEANOGR] **1.** Level of no motion. **2.** A level for which current is

known; allows determination of absolute current from relative current. { 'ref·rəns ˌlev·əl }

reference locality [GEOL] A locality containing a reference section, established to supplement the type locality. { 'ref·rəns lō'kal·əd·ē }

reference section [GEOL] A rock section, or group of sections, designated to supplement the type section, or sometimes to supplant it (as where the type section is no longer exposed), and to afford a standard for correlation for a certain part of the geologic column. { 'ref·rəns ˌsek·shən }

reference spheroid [GEOD] An ellipsoid of revolution, chosen to approximate the geoid, on which geodetic triangulation measurements are computed. { 'ref·rəns 'sfirˌȯid }

reference station [OCEANOGR] **1.** A place for which independent daily predictions are given in the tide or current tables, from which corresponding predictions are obtained for other stations by means of differences or factors. **2.** A place for which tidal or tidal current constants have been determined and which is used as a standard for the comparison of simultaneous observations at a second station. Also known as standard station. { 'ref·rəns ˌstā·shən }

reflected buried structure [GEOL] The distortion of surface beds that reflect a similar structural distortion of underlying formations. { ri'flek·təd 'ber·ēd 'strək·chər }

reflector [GEOPHYS] A layer or horizon that reflects seismic waves. { ri'flek·tər }

refolding [GEOL] A process by which folds of one generation are subjected to and stressed by a force of different orientation. { rē'fōld·iŋ }

refoliation [GEOL] A foliation that is subsequent to and oriented differently from an earlier foliation. { riˌfō·lē'ā·shən }

refraction coefficient [OCEANOGR] The square root of the ratio of the spacing between orthogonals in deep water and in shallow water; it is a measure of the effect of refraction in diminishing wave height by increasing the length of the wave crest. { ri'frak·shən ˌkō·iˌfish·ənt }

refraction diagram [OCEANOGR] A chart showing the position of the wave crests at a particular time, or the successive positions of a particular wave crest as it moves shoreward. { ri'frak·shən ˌdī·əˌgram }

refractive modulus *See* modified index of refraction. { ri'frak·tiv 'mäj·ə·ləs }

Refugian [GEOL] A North American stage of geologic time in the Eocene and Oligocene, above the Fresnian and below the Zemorrian. { rə'fyü·jē·ən }

reg [GEOL] An extensive, nearly level, low desert plain from which fine sand has been removed by wind, leaving a sheet of coarse, smoothly angular, wind-polished gravel and small stones lying on an alluvial soil, strongly cemented by mineralized solutions to form a broad desert pavement. Also known as gravel desert. { reg }

regelation [HYD] Phenomenon in which ice melts at the bottom of droplets of highly concentrated saline solution that are trapped in ice which has frozen over polar waters, and freezes at the top of these droplets, so that the droplets move downward through the ice, leaving it hard and clear. { ¦rē·jə'lā·shən }

regenerated flow control [HYD] Control of glacial drainage by modified morainal features, resulting from the readvance of a previously stagnant glacier. { rē'jen·əˌrād·əd 'flō kənˌtrōl }

regenerated glacier [HYD] A glacier that becomes active after a period of stagnation. { rē'jen·əˌrād·əd 'glā·shər }

regime [GEOL] The existence in a stream channel of a balance between erosion and deposition over a period of years. { rə'zhēm }

regimen [HYD] **1.** The behavior characteristic of the total amount of water involved in a drainage basin. **2.** Analysis of the total volume of water involved with a lake, including water losses and gains, over a period of a year. **3.** The flow characteristics of a stream with respect to velocity, volume, form of and alterations in the channel, capacity to transport sediment, and the amount of material supplied for transportation. { 'rej·ə·mən }

regional dip [GEOL] The nearly uniform and generally low-angle inclination of strata over a wide area. Also known as normal dip. { 'rēj·ən·əl 'dip }

regional forecast *See* area forecast. { 'rēj·ən·əl 'fȯr,kast }

regional geology [GEOL] The geology of a large region, treated from the viewpoint of the spatial distribution and position of stratigraphic units, structural features, and surface forms. { 'rēj·ən·əl jē'äl·ə·jē }

regional metamorphism [GEOL] Geological metamorphism affecting an extensive area. { 'rēj·ən·əl ,med·ə'mȯr,fiz·əm }

regional metasomatism [GEOL] Metasomatic processes affecting extensive areas whereby the introduced material may be derived from partial fusion of the rocks involved from deep-seated magmatic sources. { 'rēj·ən·əl ¦med·ə'sō·mə,tiz·əm }

regional slope [GEOL] The generally uniform dip of rock strata or land surface over a wide area. { 'rēj·ən·əl 'slōp }

regional slope deposit [GEOL] A sedimentary deposit widely distributed as a thin sheet over a regional slope. { 'rēj·ən·əl ¦slōp di,päz·ət }

regional snowline [HYD] The level above which, averaged over a large area, snow accumulation exceeds ablation year after year. { 'rēj·ən·əl 'snō,līn }

regional unconformity [GEOL] A continuous unconformity extending throughout a wide region that may be nearly continentwide, and usually represents a long period of time. { 'rēj·ən·əl ,ən·kən'fȯr·məd·ē }

region of escape *See* exosphere. { 'rē·jən əv i'skāp }

regmagenesis [GEOL] Diastrophic production of regional strike-slip displacements. { ¦reg·mə'jen·ə·səs }

regmaglypt [GEOL] Any of various small, well-defined, characteristic indentations or pits on the surface of meteorites, frequently resembling the imprints of fingertips in soft clay. Also known as pezograph; piezoglypt. { 'reg·mə,glipt }

regolith [GEOL] The layer rock or blanket of unconsolidated rocky debris of any thickness that overlies bedrock and forms the surface of the land. Also known as mantle rock. { 'reg·ə,lith }

Regosol [GEOL] In early United States soil classification systems, one of an azonal group of soils that form from deep, unconsolidated deposits and have no definite genetic horizons. { 'reg·ə,säl }

regradation [GEOL] The formation by a stream of a new profile of equilibrium, as when the former profile, after gradation, became deformed by crustal movements. { ,rē·grā'dā·shən }

regression [GEOL] The theory that some rivers have sources on the rainier sides of mountain ranges and gradually erode backward until the ranges are cut through. [OCEANOGR] Retreat of the sea from land areas, and the consequent evidence of such withdrawal. { ri'gresh·ən }

regression conglomerate [GEOL] A coarse sedimentary deposit formed during a retreat (recession) of the sea. { ri'gresh·ən kən,gläm·ə·rət }

regressive overlap *See* offlap. { ri'gres·iv 'o·vər,lap }

regressive reef [GEOL] One of a series of nearshore reefs or bioherms superimposed on basinal deposits during the rising of a landmass or the lowering of the sea level, and developed more or less parallel to the shore. { ri'gres·iv 'rēf }

regressive ripple [GEOL] An asymmetric ripple mark formed by a current but oriented in a direction opposite to the general movement of current flow (steep side facing upcurrent). { ri'gres·iv 'rip·əl }

regressive sediment [GEOL] A sediment deposited during the retreat or withdrawal of water from a land area or during the emergence of the land, and characterized by an offlap arrangement. { ri'gres·iv 'sed·ə·mənt }

regur [GEOL] One of a group of calcareous, intrazonal soils characterized by dark color and a high clay content. Also known as black cotton soil. { 'reg·ər }

Reichenbach's lamellae [GEOL] Thin, platy inclusions of foreign minerals (usually troilite, schreibersite, or chromite) occurring in iron meteorites. { 'rī·kən,bäks lə'mel·ē }

rejected recharge [HYD] Water that infiltrates to the water table but then discharges because the aquifer is full and cannot accept it. { ri'jek·təd 'rē,chärj }

rejuvenate [GEOL] The act of stimulating a stream to renewed erosive activity either by tectonic uplift or a drop in sea level. { ri'jü·və,nāt }

rejuvenated fault scarp [GEOL] A fault scarp revived by renewed movement along an old fault line after partial dissection or erosion of the initial scarp. Also known as revived fault scarp. { ri'jü·və,nād·əd 'fólt ,skärp }

rejuvenated stream [HYD] A mature stream that has reverted to the behavior and forms of a more youthful stage due to rejuvenation, usually as a result of uplift. Also known as revived stream. { ri'jü·və,nād·əd 'strēm }

rejuvenated water [HYD] Water returned to the terrestrial water supply as a result of compaction and metamorphism. { ri'jü·və,nād·əd 'wód·ər }

rejuvenation [GEOL] The restoration of youthful features to fluvial landscapes; the renewal of youthful vigor to low-gradient streams is usually caused by regional upwarping of broad areas formerly at or near base level. [HYD] **1.** The stimulation of a stream to renew erosive activity. **2.** The renewal of youthful vigor in a mature stream. { ri,jü·və'nā·shən }

rejuvenation head *See* knickpoint. { ri,jü·və'nā·shən ,hed }

relative age [GEOL] The geologic age of a fossil organism, rock, or geologic feature or event defined relative to other organisms, rocks, or features or events rather than in terms of years. { 'rel·əd·iv 'āj }

relative chronology [GEOL] Geochronology in which the time order is based on superposition or fossil content rather than on an age expressed in years. { 'rel·əd·iv krə'näl·ə·jē }

relative contour *See* thickness line. { 'rel·əd·iv 'kän,tür }

relative current [OCEANOGR] The current which is a function of the dynamic slope of an isobaric surface and which is determined from an assumed layer of no motion. { 'rel·əd·iv 'kə·rənt }

relative dating [GEOL] The proper chronological placement of a feature, object, or happening in the geologic time scale without reference to its absolute age. { 'rel·əd·iv 'dād·iŋ }

relative deflection *See* astrogeodetic deflection.

relative divergence *See* development index. { 'rel·əd·iv di'vər·jəns }

relative geologic time [GEOL] Nonabsolute geological time in which events may be placed relatively to one another. { 'rel·əd·iv ,jē·ə,läj·ik 'tīm }

relative humidity [METEOROL] The (dimensionless) ratio of the actual vapor pressure of the air to the saturation vapor pressure. Abbreviated RH. { 'rel·əd·iv hyü'mid·əd·ē }

relative hypsography *See* thickness pattern. { 'rel·əd·iv hip'säg·rə·fē }

relative isohypse *See* thickness line. { 'rel·əd·iv 'ī·sə,hips }

relative permeability [GEOL] Specific permeability of a porous rock formation to a particular phase (oil, water, gas) at a particular saturation and a particular saturation distribution; for example, ratio of effective permeability to a specified phase to the rock's absolute permeability. { 'rel·əd·iv ,pər·mē·ə'bil·əd·ē }

relative relief *See* local relief. { 'rel·əd·iv ri'lēf }

relative time [GEOL] Geologic time determined by the placing of events in a chronologic order of occurrence, especially time as determined by organic evolution or superposition. { 'rel·əd·iv 'tīm }

relative topography *See* thickness pattern. { 'rel·əd·iv tə'päg·rə·fē }

relaxation [GEOL] In experimental structural geology, the diminution of applied stress with time, as the result of any of various creep processes. { ,rē,lak'sā·shən }

release fracture [GEOL] A fracture formed as a result of a decrease in the maximum principal stress. { ri'lēs ,frak·chər }

release joint *See* sheeting structure. { ri'lēs ,jóint }

relic [GEOL] **1.** A landform that remains intact after decay or disintegration or that remains after the disappearance of the major portion of its substance. **2.** A vestige of a particle in a sedimentary rock, such as a trace of a fossil fragment. { 'rel·ik }

relict [GEOL] **1.** Referring to a topographic feature that remains after other parts of the feature have been removed or have disappeared. **2.** Pertaining to a mineral, structure, or feature of a rock which represents features of an earlier rock and which persists in spite of processes tending to destroy it, such as metamorphism. { 'rel·ikt }

relict dike [GEOL] In a granitized mass, a tabular, crystalloblastic body that represents a dike which was emplaced prior to, and which was relatively resistant to, the granitization process. { 'rel·ikt 'dīk }

relict glacier [HYD] A remnant of an older and larger glacier. { 'rel·ikt 'glā·shər }

reliction [HYD] The slow and gradual withdrawal or recession of the water in a sea, a lake, or a stream, leaving the former bottom as permanently exposed and uncovered dry land. { rə'lik·shən }

relict lake [HYD] A lake that survives in an area formerly covered by the sea or a larger lake, or a lake that represents a remnant resulting from a partial extinction of the original body of water. { 'rel·ikt 'lāk }

relict permafrost [GEOL] Permafrost formed in the past which persists in areas where it would not form today. { 'rel·ikt 'pər·mə,fròst }

relict sediment [GEOL] A sediment which was in equilibrium with its environment when first deposited but which is unrelated to its present environment even though it is not buried by later sediments, such as a shallow-marine sediment on the deep ocean floor. { 'rel·ikt 'sed·ə·mənt }

relict soil [GEOL] A soil formed on a preexisting landscape but not subsequently buried under younger sediments. { 'rel·ikt 'sòil }

relict texture [GEOL] In mineral deposits, an original texture that persists after partial replacement. { 'rel·ikt 'teks·chər }

relief [GEOD] The configuration of a part of the earth's surface, with reference to altitude and slope variations and to irregularities of the land surface. { ri'lēf }

Relizean stage [GEOL] A subdivision of the Miocene in the California-Oregon-Washington area. { rə'lē·zē·ən ,stāj }

remanent magnetization [GEOPHYS] That component of a rock's magnetization whose direction is fixed relative to the rock and which is independent of moderate, applied magnetic fields. { 'rem·ə·nənt ,mag·nə·tə'zā·shən }

remolded soil [GEOL] Soil that has had its natural internal structure modified or disturbed by manipulation so that it lacks shear strength and gains compressibility. { rē'mōl·dəd 'sòil }

remolding index [GEOL] The ratio of the modulus of deformation of a soil in the undisturbed state to that of a soil in the remolded state. { rē'mōld·iŋ ,in,deks }

remotely operated vehicle [OCEANOGR] A crewless submersible vehicle that is tethered to a vessel on the surface by a cable; it has a video camera, lights, thrusters that generally provide three-dimensional maneuverability, depth sensors, and a wide array of manipulative and acoustic devices, as well as special instrumentation to perform a variety of work tasks. Abbreviated ROV. { rə,mōt·lē ,äp·ə,rād·əd 'vē·ə·kəl }

Rendoll [GEOL] A suborder of the soil order Mollisol, formed in highly calcareous parent materials, mostly restricted to humid, temperate regions; the soil profile consists of a dark upper horizon grading to a pale lower horizon. { 'ren,däl }

Rendzina [GEOL] One of an intrazonal, calcimorphic group of soils characterized by a brown to black, friable surface horizon and a light-gray or yellow, soft underlying horizon; found under grasses or forests in humid to semiarid climates. { rent'sin·ə }

repeated reflection *See* multiple reflection. { ri'pēd·əd ri'flek·shən }

repetition [GEOL] The duplication of certain stratigraphic beds at the surface or in any specified section owing to disruption and displacement of the beds by faulting or intense folding. { ,rep·ə'tish·ən }

Repettian [GEOL] A North American stage of lower Pliocene geologic time, above the Delmontian and below the Venturian. { rə'pesh·ən }

repi [HYD] A lake, pond, or other standing water body associated with a sink or subsidence of land surface. { 'rep·ē }

replacement [GEOL] Growth of a new or chemically different mineral in the body of an old mineral by simultaneous capillary solution and deposition. { ri'plās·mənt }

replacement dike [GEOL] A dike which is made by gradual transformation of wall rock by solutions along fractures or permeable zones. { ri'plās·mənt ˌdīk }
replacement texture [GEOL] The texture exhibited where one mineral has replaced another. { ri'plās·mənt ˌteks·chər }
replacement vein [GEOL] A mineral vein formed by the gradual transformation of an original vein by secondary fluids. { ri'plās·mənt vān }
replenishment [GEOL] The stage in development of a cavern in which the presence of air in the passages allows the deposition of speleothems. { ri'plen·ish·mənt }
réseau [METEOROL] The term adopted by the World Meteorological Organization for the worldwide network of meteorological stations which have been chosen to represent the meteorology of the globe (*réseau mondial*). { rā'zō }
resedimentation [GEOL] **1.** Sedimentation of material derived from a preexisting sedimentary rock, that is, redeposition of sedimentary material. **2.** Mechanical deposition of material in cavities of postdepositional age, such as the deposition of carbonate muds and silts by internal mechanical erosion or solution of a limestone. **3.** The general process of subaqueous, downslope movement of sediment under the influence of gravity, such as the formation of a turbidity-current deposit. { rēˌsed·ə·mən'tā·shən }
resequent [GEOL] Referring to a geologic or topographic feature that resembles or agrees with a consequent feature but that developed from the feature at a later date. { rē'sē·kwənt }
resequent fault-line scarp [GEOL] A fault-line scarp which faces in the same direction as the original fault scarp or in which the downthrown block is topographically lower than the upthrown block. { rē'sē·kwənt 'fólt ˌlīn ˌskärp }
resequent stream [HYD] A stream whose direction follows an original consequent stream but is generally lower; resequent streams are generally tributary to a subsequent stream. { rē'sē·kwənt 'strēm }
reservoir [GEOL] **1.** A subsurface accumulation of crude oil or natural gas under adequate trap conditions. **2.** An area covered by névé where snow collects to form a glacier. **3.** A space within the earth that is occupied by magma. { 'rez·əvˌwär }
reservoir fluid [GEOL] The subterranean fluid trapped by a reservoir formation; can include natural gas, liquid and vapor petroleum hydrocarbons, and interstitial water. { 'rez·əvˌwär ˌflü·əd }
reservoir pressure [GEOL] **1.** The pressure on fluids (water, oil, gas) in a subsurface formation. Also known as formation pressure. **2.** The pressure under which fluids are confined in rocks. { 'rez·əvˌwär ˌpresh·ər }
reservoir rock [GEOL] Friable, porous sandstone containing deposits of oil or gas. { 'rez·əvˌwär ˌräk }
reshabar [METEOROL] A strong, very turbulent, dry northeast wind of bora type which blows down mountain ranges in southern Kurdistan in Persia; it is dry and hot in summer and cold in winter. { ¦rā·shə¦bär }
residual [GEOL] **1.** Of a mineral deposit, formed by either mechanical or chemical concentration. **2.** Pertaining to a residue left in place after weathering of rock. **3.** Of a topographic feature, representing the remains of a formerly great mass or area and rising above the surrounding surface. { rə'zij·ə·wəl }
residual anticline [GEOL] In salt tectonics, a relative structural high resulting from the depression of two adjacent rim synclines. Also known as residual dome. { rə'zij·ə·wəl 'ant·iˌklīn }
residual clay [GEOL] Very finely divided clay material formed in place by weathering of rock. Also known as primary clay. { rə'zij·ə·wəl 'klā }
residual compaction [GEOL] The difference between the amount of compaction that will ultimately occur for a given increase in applied stress, and that which has occurred at a specified time. { rə'zij·ə·wəl kəm'pak·shən }
residual dome *See* residual anticline. { rə'zij·ə·wəl ¦dōm }
residual kame [GEOL] A ridge or mound of sand or gravel formed by the denudation of glaciofluvial material that had been deposited in glacial lakes or on the flanks of hills of till. { rə'zij·ə·wəl 'kām }

residual liquid [GEOL] The volatile components of a magma that remain in the magma chamber after much crystallization has taken place. { rə'zij·ə·wəl 'lik·wəd }
residual liquor *See* rest magma. { rə'zij·ə·wəl 'lik·ər }
residual map [GEOL] A stratigraphic map that displays the small-scale variations (such as local features in the sedimentary environment) of a given stratigraphic unit. { rə'zij·ə·wəl ¦map }
residual material [GEOL] Unconsolidated or partly weathered parent material of a soil, presumed to have developed in place (by weathering) from the consolidated rock on which it lies. { rə'zij·ə·wəl mə'tir·ē·əl }
residual mineral [GEOL] A mineral that has been concentrated in place by weathering and leaching of rock. { rə'zij·ə·wəl ¦min·rəl }
residual ochre [GEOL] An earthy, red, yellow, or brownish iron oxide powder of iron oxide (usually the mineral limonite) produced during chemical weathering. { rə'zij·ə·wəl 'ō·kər }
residual sediment *See* resistate. { rə'zij·ə·wəl 'sed·ə·mənt }
residual stress field *See* ambient stress field. { rə'zij·ə·wəl 'stres ˌfēld }
residual swelling [GEOL] The difference between the original prefreezing level of the ground and the level reached by the settling after the ground is completely thawed. { rə'zij·ə·wəl 'swel·iŋ }
residual valley [GEOL] An intervening trough between uplifted mountains. { rə'zij·ə·wəl 'val·ē }
residue [GEOL] The in-place accumulation of rock debris which remains after weathering has removed all but the least soluble constituent. { 'rez·əˌdü }
resinite [GEOL] A variety of exinite composed of resinous compounds, often in elliptical or spindle-shaped bodies. { 'rez·ənˌīt }
resinous coal [GEOL] Coal in which large proportions of resinous material are contained in the attritus. { 'rez·ən·əs 'kōl }
resinous luster [GEOL] The luster on the fractured surfaces of certain minerals (such as opal, sulfur, amber, and sphalerite) and rocks (such as pitchstone) that resemble the appearance of resin. { 'rez·ən·əs 'ləs·tər }
resistate [GEOL] A sediment consisting of minerals that are chemically resistant and are enriched in the residues of weathering processes. Also known as residual sediment. { ri'zisˌtāt }
resistivity factor *See* formation factor. { ˌrēˌzis'tiv·əd·ē ˌfak·tər }
resonance trough [METEOROL] A large-scale pressure trough which forms at an appropriate wavelength away from a dominant trough; for example, the mean trough over the Mediterranean in winter is often considered a resonance trough between the two more dynamically active troughs along the east coasts of North America and Asia. { 'rez·ən·əns ˌtrȯf }
resorbed reef [GEOL] A reef characterized by embayed margins and by the numerous isolated patches of reef that are closely distributed about the main mass. { rē'sȯrbd 'rēf }
rest hardening [GEOL] The increase of strength, with time, of a clay subsequent to its deposition, remolding, or modification by the application of shear stress. { 'rest ˌhärd·ən·iŋ }
rest magma [GEOL] The part of magma that remains after many minerals have crystallized from it during a long series of differentiations. Also known as residual liquor. { 'rest ˌmag·mə }
restricted [GEOL] Referring to tectonic transport or movement in which elongation of particles is transverse to the direction of movement. { ri'strik·təd }
restricted basin [GEOL] A depression in the floor of the ocean in which the water circulation is topographically restricted and therefore generally is oxygen-depleted. Also known as barred basin; silled basin. { ri'strik·təd 'bās·ən }
resultant wind [CLIMATOL] The vectorial average of all wind directions and speeds for a given level at a given place for a certain period, such as a month. { ri'zəlt·ənt 'wind }
resurgence [HYD] The point where an underground stream reappears at the surface

to become a surface stream. Also known as emergence; exsurgence; rise. { ri'sər·jəns }

resurgent [GEOL] Referring to magmatic water or gases that were derived from sources on the earth's surface, from its atmosphere, or from country rock of the magma. { ri'sər·jənt }

resurgent cauldron [GEOL] A cauldron in which the cauldron block has been uplifted following subsidence, usually in the form of a structural dome. { ri'sər·jənt 'kȯl·drən }

resurrected [GEOL] Pertaining to a surface, landscape, or feature (such as a mountain, peneplain, or fault scarp) that has been restored by exhumation to its previous status in the existing relief. Also known as exhumed. [HYD] Pertaining to a stream that follows an earlier drainage system after a period of brief submergence has slightly masked the old course by a thin film of sediments. Also known a palingenetic. { ¦rez·ə¦rek·təd }

retained water [HYD] The water remaining in rock or soil after gravity groundwater has been drained out. { ri'tānd 'wȯd·ər }

retardation [OCEANOGR] The amount of time by which corresponding tidal phases grow later day by day, averaging approximately 50 minutes. { ˌrēˌtär'dā·shən }

reticulate [GEOL] **1.** Referring to a vein or lode with netlike texture. **2.** Referring to rock texture in which crystals are partly altered to a secondary material, forming a network that encloses the remnants of the original mineral. Also known as mesh texture; reticular; reticulated. { rə'tik·yə·lət }

reticulated bar [GEOL] One of a group of slightly submerged sandbars in two sets, both of which are diagonal to the shoreline, forming a crisscross pattern. { rə'tik·yəˌlād·əd 'bär }

retrograde reservoir [GEOL] Hydrocarbon reservoir in which hydrocarbons are initially in the vapor phase; as pressure is reduced, the bubble-point line is passed and liquids are formed; upon further pressure reduction, a vapor phase is again formed. { 're·trəˌgrād 'rez·əvˌwär }

retrograde wave [METEOROL] An atmospheric wave which moves in a direction opposite to that of the flow in which the wave is embedded; retrogression of a particular wave on daily charts is rarely seen, but is frequently observed on 4-day or monthly mean charts. { 're·trəˌgrād 'wāv }

retrograding shoreline [GEOL] A shoreline that is being moved landward by wave erosion. { 're·trəˌgrād·iŋ 'shȯrˌlīn }

retrogression [GEOL] *See* recession. [METEOROL] The motion of an atmospheric wave or pressure system in a direction opposite to that of the basic flow in which it is embedded. { ˌre·trə'gresh·ən }

return [GEOPHYS] Any of those surface waves on the record of a large earthquake which have traveled around the earth's surface by the long (greater than 180°) arc between epicenter and station, or which have passed the station and returned after traveling the entire circumference of the earth. { ri'tərn }

return flow [HYD] Irrigation water not consumed by evapotranspiration but returned to its source or to another body of ground or surface water. Also known as return water. { ri'tərn ˌflō }

return streamer [GEOPHYS] The intensely luminous streamer which propagates upward from earth to cloud base in the last phase of each lightning stroke of a cloud-to-ground discharge. Also known as main stroke; return stroke. { ri'tərn ˌstrēm·ər }

return stroke *See* return streamer. { ri'tərn ˌstrōk }

return water *See* return flow. { ri'tərn ˌwȯd·ər }

reversal of dip [GEOL] Change in the dip direction of bedding near a fault such that the beds curve toward the fault surface in a direction exactly opposite that of the drag folds. Also known as dip reversal. { ri'vər·səl əv 'dip }

reverse cell [METEOROL] A circulating fluid system in which the circulation in a vertical plane is thermally indirect; that is, cooler air rises relative to warmer air. { ri'vərs 'sel }

reversed *See* overturned. { ri'vərst }

reversed arc [GEOL] A curved belt of islands which is concave toward the open ocean, the opposite of most island arcs. { ri'vərst 'ärk }

reversed consequent stream [HYD] A consequent stream whose direction of flow is contrary to that normally consistent with the geologic structure. { ri'vərst 'kän·sə·kwənt 'strēm }

reversed polarity [GEOPHYS] Natural remanent magnetism opposite that of the present geomagnetic field. { ri'vərst pə'lar·əd·ē }

reversed stream [HYD] A stream whose direction of flow has been reversed, as by glacial action, landsliding, gradual tilting of a region, or capture. { ri'vərst 'strēm }

reversed tide [OCEANOGR] An oceanic tide that is out of phase with the apparent motions of the tide-producing body, so that low tide is directly under the tide-producing body and is accompanied by a low tide on the opposite side of the earth. Also known as inverted tide. { ri'vərst 'tīd }

reverse fault *See* thrust fault. { ri'vərs 'fȯlt }

reverse-flowage fold [GEOL] A fold in which flow from deformation has thickened the anticlinal crests and thinned the synclinal troughs, contrary to the normal flow pattern of a flow fold. { ri'vərs ¦flō·ij 'fōld }

reverse saddle [GEOL] A mineral deposit associated with the trough of a synclinal fold and following the bedding plane. Also known as trough reef. { ri'vərs ¦sad·əl }

reverse similar fold [GEOL] A fold whose strata are thickened on the limbs and thinned on the axes, contrary to the pattern of a similar fold. { ri'vərs 'sim·ə·lər 'fōld }

reverse slip fault *See* thrust fault. { ri'vərs 'slip ˌfȯlt }

reverse slope [GEOL] A hill descending away from a ridge. { ri'vərs ¦slōp }

reversing current [OCEANOGR] Any current that changes direction, with a period of slack water at each reversal of direction. { ri'vərs·iŋ ˌkə·rənt }

reversing dune [GEOL] A dune that tends to develop unusual height but migrates only a limited distance because seasonal shifts in dominant wind direction cause it to move alternately in nearly opposite directions. { ri'vərs·iŋ ˌdün }

revet-crag [GEOL] One of a series of narrow, pointed outliers or ridges of eroded strata inclined like a revetment against a mountain spur. { rə'vet ˌkrag }

revived fault scarp *See* rejuvenated fault scarp. { ri'vīvd 'fȯlt ˌskärp }

revived folding *See* recurrent folding. { ri'vīvd 'fōld·iŋ }

revived stream *See* rejuvenated stream. { ri'vīvd 'strēm }

revolution [GEOL] A little-used term to describe a time of profound crustal movements, on a continentwide or worldwide scale, which led to abrupt geographic, climatic, and environmental changes that were related to changes in forms of life. { ˌrev·ə'lü·shən }

revolving storm [METEOROL] A cyclonic storm, or one in which the wind revolves about a central low-pressure area. { ri'välv·iŋ 'stȯrm }

rework [GEOL] Any geologic material that has been removed or displaced by natural agents from its origin and incorporated in a younger formation. { 'rēˌwərk }

Reynolds effect [METEOROL] A process of drop growth in clouds which involves net evaporation from cloud drops warmer than others and net condensation on the cooler drops. { 'ren·əlz iˌfekt }

Reynolds model [OCEANOGR] A laboratory model of ocean currents in which inertial forces and frictional forces predominate, and in which the Reynolds number is used extensively in calculations. { 'ren·əlz ˌmäd·əl }

RH *See* relative humidity.

Rhaetian [GEOL] A European stage of geologic time; the uppermost Triassic (above Norian, below Hettangian of Jurassic). Also known as Rhaetic. { 'rē·shən }

Rhaetic *See* Rhaetian. { 'rēd·ik }

rhegmagenesis [GEOL] Orogeny characterized by the development of large-scale strike-slip faults. { ¦reg·mə'jen·ə·səs }

rheid [GEOL] A substance (below its melting point) which deforms by viscous flow during applied stress at an order of magnitude at least three times that of elastic deformation under similar circumstances. { 'rē·əd }

rheid fold [GEOL] A fold whose strata deform by viscous flow as if they were fluid. { 'rē·əd ˌfōld }
rheidity [GEOL] Relaxation time of a substance, divided by 1000. { rē'id·əd·ē }
rheoignimbrite [GEOL] An ignimbrite, on the slope of a volcanic crater, that has developed secondary flowage due to high temperatures. { ¦rē·ō'ig·nim,brīt }
rhexistasy [GEOL] The mechanical breaking up and transport of old soils or other surface residual materials. { rek'sis·tə·sē }
rhizic water *See* soil water. { 'rīz·ik ˌwȯd·ər }
rhizoconcretion *See* root cast. { ¦rī·zō·kän'krē·shən }
rhizosphere [GEOL] The soil region subject to the influence of plant roots and characterized by a zone of increased microbiological activity. { 'rī·zəˌsfir }
Rhodanian orogeny [GEOL] A short-lived orogeny that occurred at the end of the Miocene Period. { rō'dān·ē·ən ȯ'räj·ə·nē }
rhombochasm [GEOL] A parallel-sided gap in the sialic crust occupied by simatic crust, probably caused by spreading and separation. { 'räm·bəˌkaz·əm }
rhomboid ripple mark [GEOL] An aqueous current ripple mark characterized by a reticular arrangement of diamond-shaped tongues of sand, with each tongue having two acute angles, one pointing upcurrent and the other pointing downcurrent. { 'räm ˌbȯid 'rip·əl ˌmärk }
rhourd [GEOL] A pyramid-shaped sand dune, formed by the intersection of other dunes. { rȯrd }
rhyolitic glass [GEOL] Volcanic glass that is chemically equivalent to rhyolite. { ¦rī·ə¦lid·ik 'glas }
rhyolitic lava [GEOL] A highly viscous, silica-rich lava. { ¦rī·ə¦lid·ik 'lä·və }
rhyolitic tuff [GEOL] A tuff composed of fragments of rhyolitic lava. { ¦rī·ə¦lid·ik 'təf }
rhythmic accumulations [GEOL] Regular patterns of ripples and cusps in sediment on the beach or the sea floor, formed by currents and waves. { 'rith·mik əˌkyü·mə'lā·shənz }
rhythmic layering [GEOL] A type of layering in an igneous intrusion which is easily observable and in which there is repetition of zones of varying composition. { 'rith·mik 'lā·ər·iŋ }
rhythmic sedimentation [GEOL] A repetitious, regular sequence of rock units formed by sedimentary succession and indicating a frequent, predictable recurrence of the same sequence of conditions. { 'rith·mik ˌsed·ə·men'tā·shən }
rhythmic stratification [GEOL] The occurrence of sediment layers in repetitive patterns, such as a regular alternation of layers of lime and clay. { 'rith·mik ˌstrad·ə·fə'kā·shən }
rhythmic succession [GEOL] A succession of rock units showing continual and repeated changes of lithology. { 'rith·mik sək'sesh·ən }
rhythmite [GEOL] An independent unit of a rhythmic succession or of beds that were developed by rhythmic sedimentation. { 'rith·mīt }
ria [GEOGR] **1.** Any broad, estuarine river mouth. **2.** A long, narrow coastal inlet, except a fjord, whose depth and width gradually and uniformly diminish inland. { 'rē·ə }
ria coast [GEOGR] A coast with several parallel rias extending far inland and alternating with ridgelike promontories. { 'rē·ə 'kōst }
ria shoreline [GEOGR] A type of coastline developed along a drowning landmass in which numerous long and narrow arms of the sea extend inland parallel with one another and perpendicular to the coastline. { 'rē·ə 'shȯrˌlīn }
rib [GEOL] A layer or dike of rock forming a small ridge on a steep mountainside. { rib }
rib-and-furrow [GEOL] The bedding-plane expression for micro-cross-bedding, consisting of sets of small, transverse arcuate markings confined to long, narrow, parallel grooves oriented parallel to the current flow and separated by narrow ridges. { 'rib ən 'fər·ō }
riband jasper *See* ribbon jasper. { 'rib·ənd 'jas·pər }
ribbed moraine [GEOL] One of a group of irregularly subparallel, locally branching,

generally smoothly rounded and arcuate ridges that are convex in the downstream direction of a glacier but that curve upstream adjacent to eskers. { 'ribd mə'rān }

ribble *See* ripple till. { 'rib·əl }

ribbon bomb [GEOL] An elongate and flattened volcanic bomb derived from ropes of lava. { 'rib·ən ˌbäm }

ribbon diagram [GEOL] A continuous geologic cross section that is drawn in perspective along a curved or sinuous line. { 'rib·ən ¦dī·əˌgram }

ribbon jasper [GEOL] Banded jasper with parallel, ribbonlike stripes of alternating colors or shades of color. Also known as riband jasper. { 'rib·ən ˌjas·pər }

ribbon lightning [GEOPHYS] Ordinary streak lightning that appears to be spread horizontally into a ribbon of parallel luminous streaks when a very strong wind is blowing at right angles to the observer's line of sight; successive strokes of the lightning flash are then displaced by small angular amounts and may appear to the eye or camera as distinct paths. Also known as band lightning; fillet lightning. { 'rib·ən ˌlīt·niŋ }

ribbon reef [GEOL] A linear reef within the Great Barrier Reef off the northeast coast of Australia, having inwardly curved extremities, and forming a festoon along the precipitous edge of the continental shelf. { 'rib·ən ˌrēf }

ribbon structure [GEOL] A succession of thin layers of different mineralogy and texture often contorted and deformed. { 'rib·ən ˌstrək·chər }

ribbon vein *See* banded vein. { 'rib·ən ˌvān }

ribut [METEOROL] Sharp, short squalls during comparatively calm winds from May to November in Malaya. { ri'bət }

rice coal [GEOL] Anthracite that will pass through circular holes in a screen, the holes measuring 5/16 inch (7.9 millimeters), but not 3/16 inch (4.8 millimeters), in diameter. { 'rīs ˌkōl }

Richmondian [GEOL] A North American stage of geologic time: Upper Ordovician (above Maysvillian, below Lower Silurian). { rich'mən·dē·ən }

Richter scale [GEOPHYS] A scale of numerical values of earthquake magnitude ranging from 1 to 9. { 'rik·tər ˌskāl }

rideau [GEOL] A small ridge or mound of earth, or a slightly elevated piece of ground. { ri'dō }

ridge [GEOL] An elongate, narrow, steep-sided elevation of the earth's surface or the ocean floor. [METEOROL] An elongated area of relatively high atmospheric pressure, almost always associated with, and most clearly identified as, an area of maximum anticyclonic curvature of wind flow. Also known as wedge. { rij }

ridge aloft *See* upper-level ridge. { 'rij ə'lȯft }

ridged ice [OCEANOGR] Sea ice having readily observed surface features in the form of one or more pressure ridges. { 'rijd 'īs }

ridge fault [GEOL] A fault structure that is a set of two faults bounding a horst. { 'rij ˌfȯlt }

ridge-top trench [GEOL] A trench, occasionally found at or near the crest of high, steep-sided mountain ridges, formed by the creep displacement of a large slab of rock along shear surfaces more or less parallel with the side slope of the ridge. { 'rij ˌtäp ˌtrench }

ridging [OCEANOGR] A form of deformation of floating ice, caused by lateral pressure, whereby ice is forced or piled haphazardly to form ridged ice. { 'rij·iŋ }

riebungsbreccia [GEOL] A breccia developed during folding. { 'rē·bəŋzˌbrech·ə }

riegel [GEOL] A low, traverse ridge of bedrock on the floor of a glacial valley. Also known as rock bar; threshold; verrou. { 'rē·gəl }

riffle [HYD] **1.** A shallows across a stream bed over which water flows swiftly and is broken into waves by submerged obstructions. **2.** Shallow water flowing over a riffle. { 'rif·əl }

rift [GEOL] **1.** A narrow opening in a rock caused by cracking or splitting. **2.** A high, narrow passage in a cave. { rift }

rift-block mountain [GEOL] A mountain range which is a horst block bounded by normal faults. { 'rift ¦bläk 'mau̇nt·ən }

rift-block valley [GEOL] A valley which occupies a graben. { 'rift ¦bläk 'val·ē }
rift lake *See* sag pond. { 'rift ,lāk }
rift valley [GEOL] A deep, central cleft with a mountainous floor in the crest of a midoceanic ridge. Also known as central valley; midocean rift. { 'rift ,val·ē }
rift-valley lake *See* sag pond. { 'rift ¦val·ē 'lāk }
right-lateral fault *See* dextral fault. { 'rīt ¦lad·ə·rəl 'fölt }
right-lateral slip fault *See* dextral fault. { 'rīt ¦lad·ə·rəl 'slip ,fölt }
right side up *See* right way up. { 'rīt 'sīd 'əp }
right-slip fault *See* dextral fault. { 'rīt ,slip ,fölt }
right way up [GEOL] The state of strata where the present upward succession of layers is the original (normal) order of deposition. Also known as right side up. { 'rīt 'wā 'əp }
rill [GEOL] A small, transient runnel. [HYD] A small brook or stream. { ril }
rillenstein [GEOL] A pattern of tiny solution grooves of about 1 millimeter or less in width, formed on the limestone surface of a karstic region. { 'ril·ən,stīn }
rill erosion [GEOL] The formation of numerous, closely spaced rills due to the uneven removal of surface soil by streamlets of running water. Also known as rilling; rill wash; rillwork. { 'ril i'rō·zhən }
rill flow [HYD] Surface runoff flowing in small irregular channels too small to be considered rivulets. { 'ril ,flō }
rilling *See* rill erosion. { 'ril·iŋ }
rill mark [GEOL] A small, dendritic channel formed on beach mud or sand by a rill, especially if on the lee side of a partially buried obstruction. { 'ril ,märk }
rillstone *See* ventifact. { 'ril,stōn }
rill wash *See* rill erosion. { 'ril ,wäsh }
rillwork *See* rill erosion. { 'ril,wərk }
rima [GEOL] A long, narrow aperture, cleft, or fissure. { 'rī·mə }
rim cement [GEOL] A thin layer of calcium carbonate, hematite, or silica developed on the surface of detrital grains during diagenesis. { 'rim si,mənt }
rime [HYD] A white or milky and opaque granular deposit of ice formed by the rapid freezing of supercooled water drops as they impinge upon an exposed object; composed essentially of discrete ice granules, and has densities as low as 0.2–0.3 gram per cubic centimeter. { rīm }
rime fog *See* ice fog. { 'rīm ,fäg }
rim gypsum [GEOCHEM] Gypsum in thin films between anhydrite crystals, believed to have been introduced in solution rather than produced by replacement. { 'rim ,jip·səm }
rimmed kettle [GEOL] A morainal depression with raised edges. { 'rimd 'ked·əl }
rimmed solution pool [GEOL] A pool in rock with a hardened rim resulting from deposition of lime during evaporation at low tide. { 'rimd sə'lü·shən ,pül }
rimming wall [GEOL] A steep, ridgelike erosional remnant of continuous layers of porous, permeable, poorly cemented, detrital limestones, believed to form under tropical or subtropical conditions by surface-controlled secondary cementation of an original steep slope and followed by differential erosion that brings the cemented zone into relief. { 'rim·iŋ ,wöl }
rim ridge [GEOL] A minor ridge of till defining the edge of a moraine plateau. { 'rim ,rij }
rimrock [GEOL] A top layer of resistant rock on a plateau outcropping with vertical or near vertical walls. { 'rim,räk }
rimstone [GEOL] A calcium-containing deposit ringing an overflowing basin such as a hot spring. { 'rim,stōn }
rim syncline [GEOL] In salt tectonics, a local depression that develops as a border around a salt dome, as the salt in the underlying strata is displaced toward the dome. Also known as peripheral sink. { 'rim 'sin,klīn }
rincon [GEOL] **1.** A small, secluded valley. **2.** A bend in a stream. { riŋ'kōn }
ring complex [GEOL] An association of two ring-shaped igneous intrusive forms, ring dikes and cone sheets. { 'riŋ ,käm,pleks }

ring current [GEOPHYS] A westward electric current which is believed to circle the earth at an altitude of several earth radii during the main phase of geomagnetic storms, resulting in a large worldwide decrease in the geomagnetic field horizontal component at low latitudes. { 'riŋ ˌkə·rənt }

ring depression [GEOL] The annular, structurally depressed area surrounding the central uplift of a cryptoexplosion structure; faulting and folding may be involved in its formation. Also known as peripheral depression; ring syncline. { 'riŋ diˌpresh·ən }

ring dike [GEOL] A roughly circular dike that is vertical or inclined away from the center of the arc. Also known as ring-fracture intrusion. { 'riŋ ˌdīk }

ring fault [GEOL] **1.** A fault that bounds a rift valley. **2.** A steep-sided fault pattern that is cylindrical in outline and associated with cauldron subsidence. Also known as ring fracture. { 'riŋ ˌfȯlt }

ring fissure [GEOL] A roughly circular desiccation crack formed on a playa around a point source (generally a phreatophyte). { 'riŋ ˌfish·ər }

ring fracture *See* ring fault. { 'riŋ ˌfrak·chər }

ring-fracture intrusion *See* ring dike. { 'riŋ ¦frak·chər inˌtrü·zhən }

ring-fracture stoping [GEOL] Large-scale magmatic stoping that is associated with cauldron subsidence. { 'riŋ ¦frak·chər ˌstōp·iŋ }

ringite [GEOL] An igneous rock formed by the mixing of silicate and carbonatite magmas. { 'riŋˌīt }

ring structure [GEOL] A formation on the surface of the earth, moon, or a planet, having a ring-shaped trace in plan. { 'riŋ ˌstrək·chər }

ring syncline *See* ring depression. { 'riŋ 'sinˌklīn }

rip [OCEANOGR] A turbulent agitation of water generally caused by the interaction of currents and wind. { rip }

riparian water loss [HYD] Discharge of water through evapo-transpiration along a watercourse, especially water transpired by vegetation growing along the watercourse. { rə'per·ē·ən 'wȯd·ər ˌlȯs }

rip channel [GEOL] A channel, often more than 2 meters (6.6 feet) deep, carved on the shore by a rip current. { 'rip ˌchan·əl }

rip current [OCEANOGR] The return flow of water piled up on shore by incoming waves and wind. { 'rip ˌkə·rənt }

ripe [GEOL] Referring to peat, in an advanced state of decay. [HYD] Descriptive of snow that is in a condition to discharge meltwater; ripe snow usually has a coarse crystalline structure, a snow density near 0.5, and a temperature near 32°F (0°C). { rīp }

ripple [GEOL] A very small ridge of sand resembling or suggesting a ripple of water and formed on the bedding surface of a sediment. [OCEANOGR] A small curling or undulating wave controlled to a significant degree by both surface tension and gravity. { 'rip·əl }

ripple bedding [GEOL] A bedding surface characterized by ripple marks. { 'rip·əl ˌbed·iŋ }

ripple biscuit [GEOL] A bedding structure produced by lenticular lamination of sand in a bay or lagoon. { 'rip·əl ˌbis·kət }

ripple drift [GEOL] A pattern of cross-lamination formed by sedimentary deposits on both sides of a migrating ripple. { 'rip·əl ˌdrift }

ripple index [GEOL] On a rippled surface, the ratio of the crest-to-crest distance to the crest-to-trough distance. { 'rip·əl ˌinˌdeks }

ripple lamina [GEOL] An internal sedimentary structure formed in sand or silt by currents or waves, as opposed to a ripple mark formed externally on a surface. { 'rip·əl ˌlam· ə·nə }

ripple load cast [GEOL] A load cast of a ripple mark showing evidence of penecontemporaneous deformation in the accumulation of its trough and crest and in the oversteepening of the component laminae. { 'rip·əl 'lōd ˌkast }

ripple mark [GEOL] **1.** A surface pattern on incoherent sedimentary material, especially loose sand, consisting of alternating ridges and hollows formed by wind or water action. **2.** One of the ridges on a ripple-marked surface. { 'rip·əl ˌmärk }

ripple scour [GEOL] A shallow, linear trough with transverse ripple marks. { 'rip·əl ˌskau̇r }

ripple symmetry index [GEOL] A measure of the degree of symmetry of a ripple mark, equal to the ratio of the length of the gentle (upcurrent) side to the steep (downcurrent) side. { 'rip·əl 'sim·ə·trē ˌinˌdeks }

ripple till [GEOL] A till sheet containing low, winding smooth-topped ridges lying at right angles to the direction of ice movement, and grouped into narrow belts up to 48 miles (80 kilometers) long that are generally parallel to the direction of ice movement. Also known as ribble. { 'rip·əl ¦til }

rips [OCEANOGR] A turbulent agitation of water, generally caused by the interaction of currents and wind; in nearshore regions they may be currents flowing swiftly over an irregular bottom; sometimes referred to erroneously as tide rips. { rips }

rise [GEOL] A long, broad elevation which rises gently from its surroundings, such as the sea floor. [HYD] *See* resurgence. { rīz }

rise of tide [OCEANOGR] Vertical distance from the chart datum to a higher water datum. { 'rīz əv 'tīd }

rise pit [GEOL] A pit through which an underground stream rises to the surface with a calm and steady flow. { 'rīz ˌpit }

riser [GEOL] A steplike topographic feature, such as a steep slope between terraces. { 'rīz·ər }

rising limb [HYD] The rising portion of the hydrograph resulting from runoff of rainfall or snowmelt. { 'rīz·iŋ 'lim }

rising tide [OCEANOGR] The portion of the tide cycle between low water and the following high water. { 'rīz·iŋ 'tīd }

Riss [GEOL] **1.** A European stage of geologic time: Pleistocene (above Mindel, below Würm). **2.** The third stage of glaciation of the Pleistocene in the Alps. { ris }

Riss-Würm [GEOL] The third interglacial stage of the Pleistocene in the Alps, following the Riss glaciation and preceding the Würm glaciation. { 'ris'virm }

river [HYD] A large, natural freshwater surface stream having a permanent or seasonal flow and moving toward a sea, lake, or another river in a definite channel. { 'riv·ər }

river bar [GEOL] A ridgelike accumulation of alluvium in the channel, along the banks, or at the mouth of a river. { 'riv·ər ˌbär }

river basin [GEOL] The area drained by a river and all of its tributaries. { 'riv·ər ˌbās·ən }

riverbed [GEOL] The channel which contains, or formerly contained, a river. { 'riv·ərˌbed }

river bottom [GEOL] The low-lying alluvial land along a river. Also known as river flat. { 'riv·ər ˌbäd·əm }

river breathing [HYD] Fluctuation of the water level of a river. { 'riv·ər ˌbrēth·iŋ }

river-deposition coast [GEOL] A deltaic coast characterized by lobate seaward bulges crossed by river distributaries and bordered by lowlands. { 'riv·ər ˌdep·ə'zish·ən ˌkōst }

river drift [GEOL] Rock material deposited by a river in one place after having been moved from another. { 'riv·ər ˌdrift }

river end [HYD] The lowest point of a river with no outlet to the sea, situated where its water disappears by percolation or evaporation. { 'riv·ər ˌend }

river flat *See* river bottom. { 'riv·ər ˌflat }

river forecast [HYD] A forecast of the expected stage or discharge at a specified time, or of the total volume of flow within a specified time interval, at one or more points along a stream. { 'riv·ər ˌfórˌkast }

river ice [HYD] Any ice formed in or carried by a river. { 'riv·ər ˌīs }

river morphology [GEOL] The study of the channel pattern and the channel geometry at several points along a river channel, including the network of tributaries within the drainage basin. Also known as channel morphology; fluviomorphology; stream morphology. { 'riv·ər mór'fäl·ə·jē }

river plain *See* alluvial plain. { 'riv·ər ˌplān }

river run gravel [GEOL] Natural gravel as found in deposits that have been subjected to the action of running water. { 'riv·ər ¦rən ˌgrav·əl }

river system [HYD] The aggregate of stream channels draining a river basin. { 'riv·ər ˌsis·təm }

river terrace *See* stream terrace. { 'riv·ər ˌter·əs }

river tide [HYD] A tide that occurs in rivers emptying directly into the sea, showing three characteristic modifications of ocean tides: the speed at which the tide travels upstream depends on the depth of the channel, the further upstream the longer the duration of the falling tide and shorter the duration of the rising tide, and the range of the tide decreases with distance upstream. { 'riv·ər ˌtīd }

riverwash [GEOL] **1.** Soil material that has been transported and deposited by rivers. **2.** An alluvial deposit in a river bed or flood channel, subject to erosion and deposition during recurring flood periods. { 'riv·ərˌwäsh }

river water [HYD] Water having carbonate, sulfate, and calcium as its main dissolved constituents; distinguished from seawater by its chloride and sodium content. { 'riv·ər ˌwȯd·ər }

riving [GEOL] The splitting off, cracking, or fracturing of rock, especially by frost action. { 'rīv·iŋ }

rivulet [HYD] A small stream; a brook. { 'riv·yə·lət }

road [GEOL] One of a series of erosional terraces in a glacial valley, formed as the water level dropped in an ice-dammed lake. { rōd }

roadstead [GEOGR] An area near the shore, where vessels can anchor in safety; usually a shallow indentation in the coast. { 'rōdˌsted }

roaring forties [METEOROL] A popular nautical term for the stormy ocean regions between 40° and 50° latitude; it usually refers to the Southern Hemisphere, where there is an almost completely uninterrupted belt of ocean with strong prevailing westerly winds. { 'rȯr·iŋ 'fȯr·dēz }

roaring sand [GEOL] A sounding sand, found on a desert dune, that sets up a low roaring sound that sometimes can be heard for a distance of 1200 feet (400 meters). { 'rȯr·iŋ 'sand }

Robin Hood's wind [METEOROL] In the United Kingdom, saturated air with temperatures near freezing; it is raw and penetrating. { 'räb·ən 'hüdz 'wind }

rocdrumlin *See* rock drumlin. { 'räk¦drəm·lən }

roche moutonnée [GEOL] A small, elongate hillock of bedrock sculptured by a large glacier so that its long axis is oriented in the direction of ice movement; the upstream side is gently inclined, smoothly rounded, but striated, and the downstream side is steep, rough, and hackly. { 'rōch ¦müt·ən¦ā }

rock asphalt *See* asphalt rock. { 'räk 'asˌfȯlt }

rock awash [OCEANOGR] In U.S. Coast and Geodetic Survey terminology, a rock exposed at any stage of the tide between the datum of mean high water and the sounding datum, or one just bare at these data. { 'räk ə'wäsh }

rock bar *See* riegel. { 'räk ˌbär }

rock-basin lake *See* paternoster lake. { 'räk ¦bas·ən ˌlāk }

rock bench *See* structural bench. { 'räk ˌbench }

rock-bulk compressibility [GEOL] One of three types of rock compressibility (matrix, bulk, and pore); the fractional change in volume of the bulk volume of the rock with a unit change in pressure. { 'räk ¦bəlk kəmˌpres·ə'bil·əd·ē }

rock cave *See* shelter cave. { 'räk ˌkāv }

rock control [GEOL] The influences of differences in earth materials on development of landforms. { 'räk kənˌtrōl }

rock creep [GEOL] A form of slow flowage in rock materials evident in the downhill bending of layers of bedded or foliated rock and in the slow downslope migration of large blocks of rock away from their parent outcrop. { 'räk ˌkrēp }

rock cycle [GEOL] The interrelated sequence of events by which rocks are initially formed, altered, destroyed, and reformed as a result of magmatism, erosion, sedimentation, and metamorphism. { 'räk ˌsī·kəl }

rock-defended terrace [GEOL] **1.** A river terrace having a ledge or outcrop of resistant rock at its base which serves as protection against undermining. **2.** A marine terrace having a mass of resistant rock at the base of the cliff which protects against wave erosion. { 'räk di¦fen·dəd 'ter·əs }

rock desert [GEOL] An upland desert in which bedrock is either exposed or is covered with a thin veneer of coarse rock fragments. { 'räk ˌdez·ərt }

rock drum *See* rock drumlin. { 'räk ¦drəm }

rock drumlin [GEOL] A smooth, streamlined hill modeled by glacial erosion, which has a core of bedrock usually veneered with a layer of glacial till and which resembles a true drumlin in outline and form but is generally less symmetrical and less regularly shaped. Also known as drumlinoid; false drumlin; rocdrumlin; rock drum. { 'räk ¦drəm·lən }

rocket lightning [GEOPHYS] A rare form of lightning whose luminous channel seems to advance through the air with only the speed of a skyrocket. { 'räk·ət ˌlīt·niŋ }

rock failure [GEOL] Fracture of a rock that has been stressed beyond its ultimate strength. { 'räk ˌfāl·yər }

rockfall [GEOL] **1.** The fastest-moving landslide; free fall of newly detached bedrock segments from a cliff or other steep slope; usually occurs during spring thaw. **2.** The rock material moving in or moved by a rockfall. { 'räkˌfȯl }

rock fan [GEOL] A fan-shaped bedrock surface whose apex is where a mountain stream debouches upon a piedmont slope, and which occupies an area where a pediment meets the mountain slope. { 'räk ˌfan }

rock-floor robbing [GEOL] A form of sheetflood erosion in which sheetfloods remove crumbling debris from rock surfaces in desert mountains. { 'räk ¦flȯr ˌräb·iŋ }

rock flour [GEOL] A fine, chemically unweathered powder of rock-forming minerals produced by pulverization of rock fragments during natural transport or crushing. Also known as glacial flour. { 'räk ˌflaü·ər }

rockforming [GEOL] Referring to any minerals which commonly occur in important proportions in common rocks. { 'räkˌfȯrm·iŋ }

rock glacier [GEOL] Boulders and fine material cemented by ice about a meter below the surface. Also known as talus glacier. { 'räk ˌglā·shər }

rock-glacier creep [GEOL] A rapid talus creep of tongues of debris in a cold region, caused by the expansive force of the alternate freeze and thaw of ice in the interstices of the debris. { 'räk ¦glā·shər ˌkrēp }

rocking stone [GEOL] A stone or boulder, often of great size, so finely poised upon its foundation (as on the side of a hill or cliff) that it can be moved slightly backward and forward with little force (as with the hand) and still retain its original position. Also known as roggan. { 'räk·iŋ ˌstōn }

rock island *See* meander core. { 'räk ˌī·lənd }

rock magnetism [GEOPHYS] The natural remanent magnetization of igneous, metamorphic, and sedimentary rocks resulting from the presence of iron oxide minerals. { 'räk 'mag·nəˌtiz·əm }

rock matrix compressibility [GEOL] One of three types of rock compressibility (matrix, bulk, and pore); the fractional change in volume of the solid rock material (grains) with a unit change in pressure. { 'räk ¦mā·triks kəmˌpres·ə'bil·əd·ē }

rock mechanics [GEOPHYS] Application of the principles of mechanics and geology to quantify the response of rock when it is acted upon by environmental forces, particularly when human-induced factors alter the original ambient forces. { 'räk miˌkan·iks }

rock pediment [GEOL] A pediment formed on the surface of bedrock. { 'räk ˌped·ə·mənt }

rock permeability [GEOL] The ability of a rock to receive, hold, or pass fluid materials (oil, water, and gas) by nature of the interconnections of its internal porosity. { 'räk ˌpər·mē·ə'bil·əd·ē }

rock pillar [GEOL] **1.** A column of rock produced by differential weathering or erosion, as along a joint plane. **2.** In a cave, a pillar-type structure that is residual bedrock rather than a stalactostalagmite. { 'räk ˌpil·ər }

rock pool [GEOL] A tidal pool formed along a rocky shoreline. { 'räk ˌpül }

rock pressure [GEOPHYS] **1.** Stress in underground geologic material due to weight of overlying material, residual stresses, and pressures resulting from swelling clays. **2.** *See* ground pressure. { 'räk ˌpresh·ər }

rock river [GEOL] A very long and narrow rock stream. { 'räk ˌriv·ər }

rock shelter [GEOL] A cave that is formed by a ledge of overhanging rock. { 'räk ˌshel·tər }

rockslide [GEOL] The sudden, rapid downward movement of newly detached bedrock segments over a surface of weakness, such as of bedding, jointing, or faulting. Also known as rock slip. { 'räkˌslīd }

rock slip *See* rockslide. { 'räk ˌslip }

rock stack [GEOL] A rocky crag that has been uplifted from an old sea floor. { 'räk ˌstak }

rock step *See* knickpoint. { 'räk ˌstep }

rock-stratigraphic unit [GEOL] A lithologically homogeneous body of strata characterized by certain observable physical features, or by the dominance of a certain rock type or combination of rock types; rock-stratigraphic units include groups, formations, members, and beds. Also known as geolith; lithologic unit; lithostratic unit; lithostratigraphic unit; rock unit. { 'räk ¦strad·ə¦graf·ik 'yü·nət }

rock stream [GEOL] Rocks moving (or already moved) in a mass down a slope under the influence of their own weight. { 'räk ˌstrēm }

rock system [GEOPHYS] In rock mechanics, all natural environmental factors that can influence the behavior of that portion of the earth's crust that will become part of an engineering structure. { 'räk ˌsis·təm }

rock terrace [GEOL] A stream terrace on the side of a valley composed of resistant bedrock which remains during erosion of weaker overlying and underlying beds. { 'räk ˌter·əs }

rock unit *See* rock-stratigraphic unit. { 'räk ˌyü·nət }

rock varnish [GEOL] A dark coating on rock surfaces exposed to the atmosphere. It is composed of about 30% manganese and iron oxides, up to 70% clay minerals, and over a dozen trace and rare-earth minerals. Although found in all terrestrial environments, it is mostly developed and best preserved in arid regions. Also know as desert varnish. { 'räk ˌvär·nəsh }

rock wood *See* mountain wood. { 'räk ˌwu̇d }

rod [GEOL] A rodlike sedimentary particle characterized by a width-length ratio less than 2/3 and a thickness-width ratio more than 2/3. Also known as roller. { räd }

rofla [GEOL] An extremely narrow, tortuous gorge, frequently formed by meltwater streams flowing from a glacier. { 'rō·flə }

ROFOR [METEOROL] An international code word used to indicate a route forecast (along an air route). { 'rōˌfȯr }

ROFOT [METEOROL] An international code word used to indicate a route forecast, with units in the English system. { 'rōˌfät }

rogenstein [GEOL] An oolite in which the ooliths are united by argillaceous cement. { 'rō·gənˌstīn }

roggan *See* rocking stone. { 'räg·ən }

roil [HYD] A small section of a stream, characterized by swiftly flowing, turbulent water. { 'ròil }

roily water [HYD] **1.** Muddy or sediment-filled water. **2.** Turbulent, agitated, or swirling water. { 'ròil·ē 'wȯd·ər }

roll [GEOL] A primary sedimentary structure produced by deformation involving subaqueous slump or vertical foundering. { rōl }

roll cloud *See* rotor cloud. { 'rōl ˌklau̇d }

roller [GEOL] *See* rod. [OCEANOGR] A long, massive wave which usually retains its form without breaking until it reaches the beach or a shoal. { 'rō·lər }

rollers [OCEANOGR] Swells coming from a great distance and forming large breakers on exposed coasts. { 'rō·lərz }

roll flattening *See* flattening. { 'rōl ˌflat·ən·iŋ }

rolling beach [GEOL] At the base of a sea cliff, the upper part of an accumulation of boulder sand pebbles which is being ground to sand and finer particles. { 'rōl·iŋ 'bēch }

Romanche trench [GEOL] A 24,320-foot-deep (7370-meter) trench in the Mid-Atlantic Ridge near the equator. { rō'mänsh 'trench }

ROMET [METEOROL] An international code word denoting route forecast, with units in the metric system. { 'rō,met }

rondada [METEOROL] In Spain, a wind that shifts diurnally from northwest through north, east, south, and west. { ròn'däd·ə }

rongstockite [GEOL] A medium- to fine-grained plutonic rock composed of zoned plagioclase, orthoclase, some cancrinite, augite, mica, hornblende, magnetite, sphene, and apatite. { raŋ'stä,kīt }

roof [GEOL] **1.** The rock above an orebody. **2.** The country rock bordering the upper surface of an igneous intrusion. { rüf }

roofed dike [GEOL] A dike that has an upward termination. { ''rüft 'dīk }

roof foundering [GEOL] Collapse of overlying rock into a magma chamber following excavation of a large quantity of magma. { 'rüf ,faůn·driŋ }

roof pendant [GEOL] Downward projection or sag into an igneous intrusion of the country rock of the roof. Also known as pendant. { 'rüf ,pen·dənt }

room [GEOL] An open area in a cave. { rüm }

rooster tail [HYD] A plumelike form of water and sometimes spray that occurs at the intersection of two crossing waves. { 'rüs·tər ,tāl }

root [GEOL] **1.** The lower limit of an orebody. Also known as bottom. **2.** The part of a fold nappe that was originally linked to its root zone. { rüt }

root cast [GEOL] A slender, tubular, near-vertical, and commonly downward-branching sedimentary structure formed by the filling of a tubular opening left by a root. Also known as rhizoconcretion. { 'rüt ,kast }

root clay *See* underclay. { 'rüt ,klā }

rootless vent [GEOL] A source of lava that is not directly connected to a volcanic vent or magma source. { 'rüt·ləs 'vent }

root sheath [GEOL] A hollow root cast. { 'rüt ,shēth }

root zone [GEOL] **1.** The area where a low-angle thrust fault steepens and descends into the crust. **2.** The source of the root of a fold nappe. { 'rüt ,zōn }

ropak [OCEANOGR] An ice cake standing on edge as a result of excessive pressure. Also known as turret ice. { 'rō,pak }

ropy lava *See* pahoehoe. { 'rō·pē 'lä·və }

rose diagram [GEOL] A circular graph indicating values in several classes of vector properties of rocks such as cross-bedding direction. { 'rōz 'dī·ə,gram }

Ross Barrier [OCEANOGR] A wall of shelf ice bordering on the Ross Sea. { 'ròs 'bar·ē·ər }

Rossby wave [METEOROL] A large, slow-moving, planetary-scale wave generated in the troposphere by ocean-land temperature contrasts and topographic forcing (winds flowing over mountains), and affected by the Coriolis effect due to the earth's rotation. Rossby waves have also been observed in the ocean. Also known as planetary wave. { 'ròs·bē ,wāv }

Rossel Current [OCEANOGR] A seasonal Pacific Ocean current flowing westward and north-westward along both the southern and northeastern coasts of New Guinea, the southern part flowing through Torres Strait and losing its identity in the Arafura Sea, and the northern part curving northeastward to join the equatorial countercurrent of the Pacific Ocean. { 'ròs·əl ,kə·rənt }

Ross Sea [GEOGR] Arm of the South Pacific Ocean off Antarctica. { 'ròs 'sē }

rotary current [OCEANOGR] A current with the direction of flow rotating through all points of the compass. { 'rōd·ə·rē 'kə·rənt }

rotary fault [GEOL] A fault in which displacement is downward at one point and upward at another point. Also known as pivotal fault; rotational fault. { 'rōd·ə·rē 'fòlt }

rotating models [OCEANOGR] Laboratory models for studying ocean currents, the models being rotated to simulate in part the earth's rotation. { 'rō,tād·iŋ ¦mäd·əlz }

rotational bomb [GEOL] A bomb whose shape is formed by spiral motion or rotation during flight. { rō'tā·shən·əl 'bäm }
rotational fault *See* rotary fault. { rō'tā·shən·əl 'fȯlt }
rotational landslide [GEOL] A landslide in which shearing takes place on a well-defined, curved shear surface, concave upward in cross section, producing a backward rotation in the displaced mass. { rō'tā·shən·əl 'lan,slīd }
rotational movement [GEOL] Apparent fault-block displacement in which the blocks have rotated relative to one another, so that alignment of formerly parallel features is disturbed. { rō'tā·shən·əl 'müv·mənt }
rotational wave *See* shear wave; S wave. { rō'tā·shən·əl 'wāv }
rotenturm wind [METEOROL] A warm south wind blowing through Rotenturm Pass in the Transylvanian Alps. { 'rōt·ən,tu̇rm ,wind }
Rotliegende [GEOL] A European series of geologic time: Lower and Middle Permian. { 'rōt,lē·gən·də }
rotor cloud [METEOROL] Turbulent, altocumulus-type cloud formation found in the lee of some large mountain barriers, particularly in the Sierra Nevadas near Bishop, California; the air in the cloud rotates around an axis parallel to the range. Also known as roll cloud. { 'rōd·ər ,klau̇d }
rotten ice [HYD] Any piece, body, or area of ice which is in the process of melting or disintegrating; it is characterized by honeycomb structure, weak bonding between crystals, or the presence of meltwater or sea water between grains. Also known as spring sludge. { 'rät·ən ¦īs }
rotten spot *See* pothole. { 'rät·ən ,spät }
rougemontite [GEOL] A coarse-grained igneous rock composed of anorthite, titanaugite, and small amounts of olivine and iron ore. { 'rüzh,män,tīt }
rough ice [HYD] An expanse of ice having an uneven surface caused by formation of pressure ice or by growlers frozen in place. { 'rəf 'īs }
roughness elements [OCEANOGR] Structures attached to laboratory models to simulate the roughness of the ocean floor. { 'rəf·nəs ,el·ə·məns }
roughness length *See* dynamic roughness. { 'rəf·nəs ,leŋkth }
roundness [GEOL] The degree of abrasion of sedimentary particles; expressed as the radius of the average radius of curvature of the edges or corners to the radius of curvature of the maximum inscribed sphere. { 'rau̇nd·nəs }
roundstone [GEOL] Any naturally rounded rock fragment of any size larger than a sand grain (diameter greater than 2 millimeters), such as a boulder, cobble, pebble, or granule. { 'rau̇nd,stōn }
round wind [METEOROL] A wind that gradually changes direction through approximately 180° during the daylight hours. { 'rau̇nd ,wind }
route component [METEOROL] The average forecast wind component parallel to the flight path at flight level for an entire route; it is positive if helping (tailwind), and negative if retarding (headwind). { 'rüt kəm,pō·nənt }
route forecast [METEOROL] An aviation weather forecast for one or more specified air routes. { 'rüt ,fȯr,kast }
routivarite [GEOL] A fine-grained igneous rock containing orthoclase, plagioclase, quartz, and garnet. { ¦rüd·ə¦va,rīt }
rouvillite [GEOL] A light-colored theralite composed predominantly of labradorite and nepheline, with small amounts of titanaugite, hornblende, pyrite, and apatite. { 'rüv·ə,līt }
ROV *See* remotely operated vehicle.
rubber ice [OCEANOGR] Newly formed sea ice which is weak and elastic. { 'rəb·ər ,īs }
rubble [GEOL] **1.** A loose mass of rough, angular rock fragments, coarser than sand. **2.** *See* talus. [HYD] Fragments of floating or grounded sea ice in hard, roughly spherical blocks measuring 0.5–1.5 meters (1.5–4.5 feet) in diameter, and resulting from the breakup of larger ice formations. Also known as rubble ice. { 'rəb·əl }
rubble drift [GEOL] **1.** A rubbly deposit (or congeliturbate) formed by solifluction under periglacial conditions. **2.** A coarse mass of angular debris and large blocks set in an earthy matrix of glacial origin. { 'rəb·əl ,drift }

rubble ice *See* rubble. { 'rəb·əl ,īs }

rubble tract [GEOL] The part of the reef flat immediately behind and on the lagoon side of the reef front, paved with cobbles, pebbles, blocks, and other coarse reef fragments. { 'rəb·əl ,trakt }

rubidium-strontium dating [GEOL] A method for determining the age of a mineral or rock based on the decay rate of rubidium-87 to strontium-87. { rü'bid·ē·əm 'strän·chəm 'dād·iŋ }

rudite [GEOL] A sedimentary rock composed of fragments coarser than sand grains. { 'rü,dīt }

Rudzki anomaly [GEOPHYS] A gravity anomaly calculated by replacing the surface topography by its mirror image within the geoid. { 'rüd·skē a,näm·ə·lē }

ruffle [GEOL] A ripple mark produced by an eddy. { 'rəf·əl }

ruffled groove cast [GEOL] A groove cast with a feather pattern, consisting of a groove with lateral wrinkles that join the main cast in the downcurrent direction at an acute angle. { 'rəf·əld ¦grüv ,kast }

ruggedness number [GEOL] A dimensionless number that expresses the geometric characteristics of a drainage system; derived from the product of maximum basin relief and drainage density within the drainage basin. { 'rəg·əd·nəs ,nəm·bər }

rule of V's [GEOL] The outcrop of a formation that crosses a valley forms an acute angle (a V) that points in the direction in which the formation lies underneath the stream. { 'rül əv 'vēz }

run [GEOL] **1.** A ribbonlike, flat-lying, irregular orebody following the stratification of the host rock. **2.** A branching or fingerlike extension of the feeder of an igneous intrusion. { rən }

runnel [GEOL] A troughlike hollow on a tidal sand beach which carries water drainage off the beach as the tide retreats. { 'rən·əl }

running sand *See* quicksand. { 'rən·iŋ ,sand }

run-of-bank gravel *See* bank-run gravel. { ¦rən əv 'baŋk ,grav·əl }

runoff [HYD] **1.** Surface streams that appear after precipitation. **2.** The flow of water in a stream, usually expressed in cubic feet per second; the net effect of storms, accumulation, transpiration, meltage, seepage, evaporation, and percolation. { 'rən,ȯf }

runoff coefficient [HYD] The percentage of precipitation that appears as runoff. { 'rən,ȯf ,kō·i,fish·ənt }

runoff cycle [HYD] The part of the hydrologic cycle involving water between the time it reaches the land as precipitation and its subsequent evapotranspiration or runoff. { 'rən,ȯf ,sī·kəl }

runoff intensity [HYD] The excess of rainfall intensity over infiltration capacity, usually expressed in inches of rainfall per hour. Also known as runoff rate. { 'rən,ȯf in,ten·səd·ē }

runoff rate *See* runoff intensity. { 'rən,ȯf ,rāt }

run of the coast [GEOGR] The trend of the coast. { 'rən əv thə 'kōst }

runout [HYD] The location where an avalanche slows down or stops, depositing the avalanche debris. { 'rən,au̇t }

run-up *See* swash. { 'rən,əp }

runway [GEOL] The channel of a stream. { 'rən,wā }

runway observation [METEOROL] An evaluation of certain meteorological elements observed at a specified point on or near an airport runway; temperature, wind speed and direction, ceiling, and visibility are among the elements frequently observed at such locations, because of the importance of these data to aircraft landing and takeoff operations. { 'rən,wā ,äb·zər'vā·shən }

runway temperature [METEOROL] The temperature of the air just above the runway at an airport (usually at about 4 feet or 1.2 meters but ideally at engine or wing height), used in the determination of density altitude; therefore, runway temperature observations are made and reported at airports when critical values of density altitude prevail. { 'rən,wā ,tem·prə·chər }

runway visibility [METEOROL] The visibility along an identified runway, determined from a specified point on the runway with the observer facing in the same direction as a pilot using the runway. { 'rən,wā ,viz·ə'bil·əd·ē }

runway visual range [METEOROL] The maximum distance along the runway at which the runway lights are visible to a pilot after touchdown. { 'rən,wā 'vizh·ə·wəl 'rānj }

Rupelian [GEOL] A European stage of middle Oligocene geologic time, above the Tongrian and below the Chattian. Also known as Stampian. { rü'pel·yən }

rupture *See* fracture. { 'rəp·chər }

rupture zone [GEOL] The region immediately adjacent to the boundary of an explosion crater, characterized by excessive in-place crushing and fracturing where the stresses produced by the explosion have exceeded the ultimate strength of the medium. { 'rəp·chər ,zōn }

rusting [GEOL] The formation of red, yellow, or brown iron oxide minerals by oxidation of mineral deposits. { 'rəst·iŋ }

R wave *See* Rayleigh wave. { 'är ,wāv }

S

Saalic orogeny [GEOL] A short-lived orogeny that occurred early in the Permian period, between the Autunian and Saxonian stages. { 'sä·lik ȯ'räj·ə·nē }

sabach *See* caliche. { ˌsäˌbäk }

Sabinas [GEOL] A North American (Gulf Coast) provincial series in Upper Jurassic geologic time, below the Coahuilan. { sə'bēn·əs }

sabkha *See* sebkha. { 'sab·kə }

sabulous *See* arenaceous. { 'sab·yə·ləs }

sackungen [GEOL] Deep-seated rock creep which has produced a ridge-top trench by gradual settlement of a slablike mass into an adjacent valley. { 'saˌku̇ŋ·ən }

saddle [GEOL] **1.** A gap that is broad and gently sloping on both sides. **2.** A relatively flat ridge that connects the peaks of two higher elevations. **3.** That part along the surface axis or axial trend of an anticline that is a low point or depression. { 'sad·əl }

saddleback [GEOL] A hill or ridge with a concave outline along its crest. [METEOROL] The cloudless air between the "towers" of two cumulus congestus or cumulonimbus clouds and above a lower cloud mass. { 'sad·əlˌbak }

saddle fold [GEOL] A flexural fold perpendicular to the parent fold and having an additional flexure at its crest. { 'sad·əl ˌfōld }

saddle point *See* col. { 'sad·əl ˌpȯint }

saddle reef [GEOL] A mineral deposit associated with the crest of an anticlinal fold and following the bedding plane, usually found in vertical succession. Also known as saddle vein. { 'sad·əl ˌrēf }

saddle vein *See* saddle reef. { 'sad·əl ˌvān }

sag [GEOL] **1.** A pass or gap in a ridge or mountain range shaped like a saddle. **2.** A shallow depression in a relatively flat land surface. **3.** A regional basin with gently sloping sides. { sag }

sag-and-swell topography [GEOGR] An undulating surface characteristic of till sheets, for example, the landscape of the midwestern United States. { ¦sag ən ¦swel tə'päg·rə·fē }

sagenitic [GEOL] Containing acicular minerals. { ˌsaj·ə'nid·ik }

sag pond [GEOL] A small body of water occupying an enclosed depression or sag formed where active or recent fault movement has impounded drainage. Also known as fault-trough lake; rift lake; rift-valley lake. { 'sag ˌpänd }

sahel [METEOROL] A strong dust-bearing desert wind in Morocco. { sə'hel }

Saint Peter sandstone [GEOL] An artesian aquifer of early Lower Paleozoic age which underlies part of Minnesota, Wisconsin, Iowa, Illinois, and Indiana. { 'sānt 'pēd·ər 'sanˌstōn }

Sakmarian [GEOL] A European stage of geologic time; the lowermost Permian, above Stephanian of Carboniferous and below Artinskian. { säk'mär·ē·ən }

Salado formation [GEOL] A red-bed formation from the Permian found in southeast New Mexico; contains rock salt and potash salts. { sə'lä·dō fȯrˌmā·shən }

salcrete [GEOL] A thin, hard crust of salt-cemented sand grains, occurring on a marine beach that is occasionally or periodically saturated by saline water. { 'salˌkrēt }

salfemic rock [GEOL] An igneous rock in which the ratio of salic to femic minerals is greater than 3:5 and less than 5:3. { sal'fē·mik 'räk }

salic [GEOL] A soil horizon enriched with secondary salts, at least 2 percent, and measuring at least 6 inches (15 centimeters) in thickness. { 'sal·ik }

salient [GEOL] **1.** A landform that projects or extends outward or upward from its surroundings. **2.** An area in which the axial traces of folds are convex toward the outer edge of the folded belt. { 'sāl·yənt }

saliferous stratum [GEOL] A stratum that contains, produces, or is impregnated with salt. Also known as saliniferous stratum. { sə'lif·ə·rəs 'strad·əm }

salina [GEOL] An area, such as a salt flat, in which deposits of crystalline salts are formed or found. [HYD] A body of water containing high concentrations of salt. { sə'lē·nə }

salinastone [GEOL] A sedimentary rock composed mostly of saline minerals which are usually precipitated but may be fragmental. { sə'lē·nə,stōn }

saline-alkali soil [GEOL] A salt-affected soil with a content of exchangeable sodium greater than 15, with much soluble salts, and with a pH value usually less than 9.5. { 'sā,lēn 'al·kə,lī ,sȯil }

salinelle [GEOL] A mud volcano erupting saline mud. { ,sa·lə'nel }

saline soil [GEOL] A nonalkali, salt-affected soil with a high content of soluble salts, with exchangeable sodium of less than 15, and with a pH value less than 8.5. { 'sā,lēn ,sȯil }

saliniferous stratum *See* saliferous stratum. { ,sal·ə'nif·ə·rəs 'strad·əm }

salinity [OCEANOGR] The total quantity of dissolved salts in sea water, measured by weight in parts per thousand. { sə'lin·əd·ē }

salinity current [OCEANOGR] A density current in the ocean whose flow is caused, controlled, or maintained by its relatively greater density due to excessive salinity. { sə'lin·əd·ē ,kə·rənt }

salinization [GEOL] In a soil of an arid, poorly drained region, the accumulation of soluble salts by the evaporation of the waters that bore them to the soil zone. { ,sal·ən·ə'zā·shən }

salt-affected soil [GEOL] A general term for a soil that is not suitable for the growth of crops because of an excess of salts, exchangeable sodium, or both. { 'sȯlt i¦fek·təd 'sȯil }

salt-and-pepper sand [GEOL] A sand composed of a mixture of light- and dark-colored grains. { ¦sȯlt ən ¦pep·ər 'sand }

salt anticline [GEOL] A structure like a salt dome but with a linear salt core. Also known as salt wall. { 'sȯlt 'ant·i,klīn }

saltation [GEOL] Transport of a sediment in which the particles are moved forward in a series of short intermittent bounces from a bottom surface. { sȯl'tā·shən }

saltation load [GEOL] The part of the bed load that is bouncing along the stream bed or is moved, directly or indirectly, by the impact of bouncing particles. { sȯl'tā·shən ,lōd }

salt bottom [GEOL] A flat piece of relatively low-lying ground encrusted with salt. { 'sȯlt ,bäd·əm }

salt burst [GEOL] Rock destruction caused by crystallization of soluble salts that enter the pores. { 'sȯlt ,bərst }

salt crust [HYD] A salt deposit formed on an ice surface by crystal growth forcing salt out of young sea ice and pushing it upward. { 'sȯlt ,krəst }

salt dome [GEOL] A diapiric or piercement structure in which there is a central, equidimensional salt plug. { 'sȯlt ,dōm }

salt-dome breccia [GEOL] A breccia found in deep shale sequences and occurring as a dome-shaped mass in a broad zone surrounding a salt plug. { 'sȯlt ¦dōm 'brech·ə }

salt field [GEOL] An area overlying a usually workable salt deposit of economic value. { 'sȯlt ,fēld }

salt flat [GEOL] The level, salt-encrusted bottom of a lake or pond that is temporarily or permanently dried up. { 'sȯlt ,flat }

salt flowers *See* ice flowers. { 'sȯlt ,flau̇·ərz }

salt glacier [GEOL] A gravitational flow of salt down the slopes of a salt plug, following the preexisting structure. { 'sȯlt ,glā·shər }

salt haze [METEOROL] A haze created by the presence of finely divided particles of sea salt in the air, usually derived from the evaporation of sea spray. { 'sȯlt ˌhāz }

salt hill [GEOL] An abrupt hill of salt, with sinkholes and pinnacles at its summit. { 'sȯlt ˌhil }

saltierra [GEOL] A deposit of salt left by evaporation of a shallow salt lake. { ˌsal·tē'er·ə }

salt lake [HYD] A confined inland body of water having a high concentration of salts, principally sodium chloride. { 'sȯlt ˌlāk }

salt pan [GEOL] **1.** An undrained, usually small and shallow, natural depression or hollow in which water accumulates and evaporates, leaving a salt deposit. **2.** A shallow lake of brackish water occupying such a depression. { 'sȯlt ˌpan }

saltpeter cave [GEOL] A cave in which there are deposits of saltpeter earth. { sȯlt'pēd·ər ˌkāv }

saltpeter earth [GEOL] A deposit containing calcium nitrate and found in caves. { sȯlt 'pēd·ər ˌərth }

salt pillow [GEOL] An embryonic salt dome rising from its source bed, still at depth. { 'sȯlt ˌpil·ō }

salt pit [GEOL] A pit in which sea water is received and evaporated and from which salt is obtained. { 'sȯlt ˌpit }

salt plug [GEOL] The salt core of a salt dome. { 'sȯlt ˌpləg }

salt polygon [GEOL] A surface of salt on a playa, having three to eight sides marked by ridges of material formed as a result of the expansive forces of crystallizing salt, and ranging in width from an inch or so to 100 feet (30 meters). { 'sȯlt 'päl·iˌgän }

salt stock [GEOL] An immature salt dome comprising a pluglike salt diapir that has pierced the overlying strata. { 'sȯlt ˌstäk }

salt tectonics [GEOL] The study of the structure and mechanism of emplacement of salt domes. Also known as halokinesis. { 'sȯlt tek'tän·iks }

salt wall *See* salt anticline. { 'sȯlt ˌwȯl }

salt water *See* seawater. { 'sȯlt ¦wȯd·ər }

salt-water front [OCEANOGR] The interface between fresh and salt water in a coastal aquifer or in an estuary. { 'sȯlt ¦wȯd·ər ˌfrənt }

salt-water intrusion [HYD] Displacement of fresh surface water or groundwater by salt water due to its greater density. { 'sȯlt ¦wȯd·ər inˌtrü·zhən }

salt-water underrun [OCEANOGR] A type of density current occurring in a tidal estuary, due to the greater salinity of the bottom water. { 'sȯlt ¦wȯd·ər 'ən·dəˌrən }

salt-water wedge [OCEANOGR] A wedge-shaped intrusion of salty ocean water into a fresh-water estuary or tidal river; it slopes downward in the upstream direction, and salinity increases with depth. { 'sȯlt ¦wȯd·ər ˌwej }

salt weathering [GEOL] The granular disintegration or fragmentation of rock material produced by saline solutions or by salt-crystal growth. { 'sȯlt ˌweth·ə·riŋ }

sand [GEOL] Unconsolidated granular material consisting of mineral, rock, or biological fragments between 63 micrometers and 2 millimeters in diameter, usually produced primarily by the chemical or mechanical breakdown of older source rocks, but may also be formed by the direct chemical precipitation of mineral grains or by biological processes. { sand }

sand apron [GEOL] A deposit of sand along the shore of a lagoon of a reef. { 'sand ˌā·prən }

sand auger *See* dust whirl. { 'sand ˌȯg·ər }

sand avalanche [GEOL] Movement of large masses of sand down a dune face when the angle of repose is exceeded or when the dune is disturbed. { 'sand ˌav·əˌlanch }

sandbag [GEOL] In the roof of a coal seam, a deposit of glacial debris formed by scour and fill subsequent to coal formation. { 'sanˌbag }

sandbank [GEOL] A deposit of sand forming a mound, hillside, bar, or shoal. { 'sanˌbaŋk }

sandbar [GEOL] A bar or low ridge of sand bordering the shore and built up, or near, to the surface of the water by currents or wave action. Also known as sand reef. { 'sanˌbär }

sandblasting [GEOL] Abrasion affected by the action of hard, windblown mineral grains. { 'san,blast·iŋ }
sand cay *See* sandkey. { 'san ,kē }
sand cone [GEOL] **1.** A cone-shaped deposit of sand, produced especially in an alluvial cone. **2.** A low debris cone whose protective veneer consists of sand. { 'san ,kōn }
sand crystal [GEOL] A large crystal loaded up to 60% with detrital sand inclusions formed in a sandstone during or as a result of cementation. { 'san ,krist·əl }
sand devil *See* dust whirl. { 'san ,dev·əl }
sand dike [GEOL] A sedimentary dike consisting of sand that has been squeezed or injected upward into a fissure. { 'san ,dīk }
sand drift [GEOL] **1.** Movement of windblown sand along the surface of a desert or shore. **2.** An accumulation of sand against the leeward side of a fixed obstruction. { 'san ,drift }
sand drip [GEOL] A rounded or crescentic surface form on a beach sand, resulting from the sudden absorption of overwash. { 'san ,drip }
sand dune [GEOL] A mound of loose windblown sand commonly found along low-lying seashores above high-tide level. { 'san ,dün }
sandfall *See* slip face. { 'san,fȯl }
sand flat [GEOL] A sandy tidal flat barren of vegetation. { 'san ,flat }
sand flood [GEOL] A vast body of sand moving or borne along a desert, as in the Arabian deserts. { 'san ,fləd }
sand gall *See* sand pipe. { 'san ,gȯl }
sand glacier [GEOL] **1.** An accumulation of sand that is blown up the side of a hill or mountain and through a pass or saddle, and then spread out on the opposite side to form a wide, fan-shaped plain. **2.** A horizontal plateau of sand terminated by a steep talus slope. { 'san ,glā·shər }
sand hill [GEOL] A ridge of sand, especially a sand dune in a desert region. { 'san ,hil }
sand hole [GEOL] A small pit (7–8 millimeters in depth and a little less wide than deep) with a raised margin, formed on a beach by waves expelling air from a formerly saturated mass of sand. { 'san ,hōl }
sand horn [GEOL] A pointed sand deposit extending from the shore into shallow water. { 'san ,hȯrn }
sandkey [GEOL] A small sandy island parallel with the shore. Also known as sand cay. { 'san,kē }
sand levee *See* whaleback dune. { 'san ,lev·ē }
sand lobe [GEOL] A rounded sand deposit extending from the shore into shallow water. { 'san ,lōb }
sand pavement [GEOL] A sandy surface derived from coarse-grained sand ripples, developed on the lower, windward slope of a dune or rolling sand area during a period of intermittent light, variable winds. { 'san ,pāv·mənt }
sand pipe [GEOL] A pipe formed in sedimentary rocks, filled with considerable sand and some gravel. Also known as sand gall. { 'san ,pīp }
sand plain [GEOL] A small outwash plain formed by deposition of sand transported by meltwater streams flowing from a glacier. { 'san ,plān }
sand reef *See* sandbar. { 'san ,rēf }
sand ridge [GEOL] **1.** Any low ridge of sand formed at some distance from the shore, and either submerged or emergent, such as a longshore bar or a barrier beach. **2.** One of a series of long, wide, extremely low, parallel ridges believed to represent the eroded stumps of former longitudinal sand dunes. **3.** A crescent-shaped land-form found on a sandy beach, such as a beach cusp. **4.** *See* sand wave. { 'san ,rij }
sand river [GEOL] A river that deposits much of its sand load along its middle course, to be subsequently removed by the wind. { 'san ,riv·ər }
sandrock [GEOL] A field term for a sandstone that is not firmly cemented. { 'san,räk }
sand roll *See* pseudonodule. { 'san ,rōl }
sand run [GEOL] **1.** A fluidlike motion of dry sand. **2.** A mass of dry sand in motion. { 'san ,rən }
sand sea [GEOL] **1.** An extensive assemblage of sand dunes of several types in an area

where a great supply of sand is present; characterized by an absence of travel lines, or directional indicators, and by a wavelike appearance of dunes separated by troughs. **2.** The flat, rain-smoothed plain of volcanic ash and other pyroclastics on the floor of a caldera. { 'san 'sē }

sand shadow [GEOL] A lee-side accumulation of sand, as a small turret-shaped dune, formed in the shelter of, and immediately behind, a fixed obstruction, such as clumps of vegetation. { 'san ˌshad·ō }

sandshale [GEOL] A sedimentary deposit consisting of thin alternating beds of sandstone and shale. { 'sanˌshāl }

sand-shale ratio [GEOL] The ratio between the thickness or percentage of sandstone and that of shale in a geologic section. { 'san 'shāl 'rā·shō }

sand sheet [GEOL] A thin accumulation of coarse sand or fine gravel having a flat surface. { 'san ˌshēt }

sand snow [HYD] Snow that has fallen at very cold temperatures (of the order of −25°C); as a surface cover, it has the consistency of dust or light dry sand. { 'san ˌsnō }

sandspit [GEOL] A spit consisting principally of sand. { 'sanˌspit }

sand splay [GEOL] A floodplain splay consisting of coarse sand particles. { 'san ˌsplā }

sandstone dike [GEOL] A dike made of sandstone or lithified sand. { 'sanˌstōn 'dīk }

sandstone sill [GEOL] A tabular mass of sandstone that has been emplaced by sedimentary injection parallel to the structure or by bedding of preexisting rock in the manner of an igneous sill. { 'sanˌstōn 'sil }

sandstorm [METEOROL] A strong wind carrying sand through the air, the diameter of most particles ranging from 0.08 to 1 millimeter; in contrast to a duststorm, the sand particles are mostly confined to the lowest 7 feet (2 meters) above ground, rarely rising more than 36 feet (11 meters). { 'sanˌstȯrm }

sand streak [GEOL] A low, linear ridge formed at the interface of sand and air or water, oriented parallel to the direction of flow, and having a symmetric cross section. { 'san ˌstrēk }

sand stream [GEOL] A small sand delta spread out at the mouth of a gully, or a deposit of sand along the bed of a small creek, formed by a torrential rain. { 'san ˌstrēm }

sand strip [GEOL] A long, narrow ridge of sand extending for a long distance downwind from each horn of a dune. { 'san ˌstrip }

sandur *See* outwash plain. { 'san·dər }

sandwash [GEOL] A sandy or gravel stream bed, devoid of vegetation, containing water only during a sudden and heavy rainstorm. { 'sanˌwäsh }

sand wave [GEOL] A large, ridgelike primary structure resembling a water wave on the upper surface of a sedimentary bed that is formed by high-velocity air or water currents. Also known as sand ridge. { 'san ˌwāv }

sand wedge [GEOL] A wedge-shaped accumulation of sand with the apex downward formed by the filling in of winter contraction cracks. { 'san ˌwej }

Sangamon [GEOL] The third interglacial stage of the Pleistocene epoch in North America, following the Illinoian glacial and preceding the Wisconsin. { 'saŋ·gəˌmən }

sanidal [GEOL] Pertaining to the continental shelf. { 'san·əd·əl }

sansar [METEOROL] A northwest wind of Persia. { 'sän·sər }

sansicl [GEOL] An unconsolidated sediment, consisting of a mixture of sand, silt, and clay, in which no component forms 50% or more of the whole aggregate. { 'san ˌsik·əl }

Santa Ana [METEOROL] A hot, dry, foehnlike desert wind, generally from the northeast or east, especially in the pass and river valley of Santa Ana, California, where it is further modified as a mountain-gap wind. { 'san·tə ¦an·ə }

Santa Rosa storm [METEOROL] In Argentina, an annual storm near the end of August. { 'san·tə 'rō·zə 'stȯrm }

Santonian [GEOL] A European stage of geologic time in the Upper Cretaceous, above the Coniacian and below the Campanian. { san'tō·nē·ən }

sapping [GEOL] Erosion along the base of a cliff by the wearing away of softer layers, thus removing the support for the upper mass which breaks off into large blocks and falls from the cliff face. Also known as undermining. { 'sap·iŋ }

Saprist [GEOL] A suborder of the soil order Histosol consisting of residues in which plant structures have been largely obliterated by decay; saturated with water most of the time. { 'sa,prist }

saprogenous ooze [GEOL] Ooze formed of putrefying organic matter. { sə'präj·ə·nəs 'üz }

saprolite [GEOL] A soft, earthy red or brown, decomposed igneous or metamorphic rock that is rich in clay and formed in place by chemical weathering. Also known as saprolith; sathrolith. { 'sap·rə,līt }

saprolith *See* saprolite. { 'sap·rə,lith }

sapropel [GEOL] A mud, slime, or ooze deposited in more or less open water. { 'sap·rə,pel }

sapropel-clay [GEOL] A sedimentary deposit in which the amount of clay is greater than that of sapropel. { 'sap·rə,pel ,klā }

sapropelic coal [GEOL] Coal formed by putrefaction of organic matter under anaerobic conditions in stagnant or standing bodies of water. Also known as sapropelite. { ¦sap·rə¦pel·ik 'kōl }

sapropelite *See* sapropelic coal. { 'sap·rə,pe,līt }

sapropel-peat *See* peat-sapropel. { 'sap·rə,pel ,pēt }

sàrca [METEOROL] A violent north wind of Lake Garda in Italy. { 'sär·kə }

Sardic orogeny [GEOL] A short-lived orogeny that occurred near the end of the Cambrian period. { 'sär·dik ȯ'räj·ə·nē }

Sargasso Sea [GEOGR] A region of the North Atlantic Ocean; boundaries are defined in the west and north by the Gulf Stream, in the east by longitude 40°W, and in the south by latitude 20°N. { sär'ga·sō 'sē }

Sarmatian [GEOL] A European stage of geologic time: the upper Miocene, above Tortonian, below Pontian. { sär'mā·shən }

sarnaite [GEOL] A feldspathoid-bearing syenite composed of cancrinite and acmite. { 'sär·nə,īt }

sarospatakite [GEOL] A micaceous clay mineral composed of mixed layers of illite and montmorillonite. { ,sar·ə'spād·ə,kīt }

sastruga [HYD] A ridge of snow up to 2 inches (5 centimeters) high formed by wind erosion and aligned parallel to the wind. Also known as skavl; zastruga. { 'zas·trə·gə }

satellite meteorology [METEOROL] That branch of meteorological science that employs sensing elements on meteorological satellites to define the state of the atmosphere. { 'sad·əl,īt ,mēd·ē·ə'räl·ə·jē }

satellitic crater *See* secondary crater. { ¦sad·ə¦lid·ik 'krād·ər }

sathrolith *See* saprolite. { 'sath·rə,lith }

satin ice *See* acicular ice. { 'sat·ən 'īs }

saturated air [METEOROL] Moist air in a state of equilibrium with a plane surface of pure water or ice at the same temperature and pressure; that is, air whose vapor pressure is the saturation vapor pressure and whose relative humidity is 100. { 'sach·ə,rād·əd 'er }

saturated permafrost [GEOL] Permafrost that contains no more ice than the ground could hold if the water were in the liquid state. { 'sach·ə,rād·əd 'pər·mə,frȯst }

saturated surface *See* water table. { 'sach·ə,rād·əd 'sər·fəs }

saturated zone *See* zone of saturation. { 'sach·ə,rād·əd 'zōn }

saturation [METEOROL] The maximum water vapor per unit volume that a parcel of air can contain at a given temperature. { ,sach·ə'rā·shən }

saturation adiabat [METEOROL] On a thermodynamic diagram, a line of constant wet-bulb potential temperatures; in practice, approximate computations are usually employed, and the resulting lines represent, ambiguously, saturation adiabats and pseudoadiabats. Also known as moist adiabat; wet adiabat. { ,sach·ə'rā·shən 'ad·ē·ə,bat }

saturation-adiabatic lapse rate [METEOROL] A special case of process lapse rate, defined as the rate of decrease of temperature with height of an air parcel lifted in a saturation-adiabatic process through an atmosphere in hydrostatic equilibrium. Also known as moist-adiabatic lapse rate. { ˌsach·əˈrā·shən ¦ad·ē·ə¦bad·ik ˈlaps ˌrāt }

saturation-adiabatic process [METEOROL] An adiabatic process in which the air is maintained at saturation by the evaporation or condensation of water substance, the latent heat being supplied by or to the air respectively; the ascent of cloudy air, for example, is often assumed to be such a process. { ˌsach·əˈrā·shən ¦ad·ē·ə¦bad·ik ˈprä·səs }

saturation curve [GEOL] A curve showing the weight of solids per unit volume of a saturated soil mass as a function of water content. { ˌsach·əˈrā·shən ¦kərv }

saturation deficit [METEOROL] **1.** The difference between the actual vapor pressure and the saturation vapor pressure at the existing temperature. **2.** The additional amount of water vapor needed to produce saturation at the current temperature and pressure, expressed in grams per cubic meter. Also known as vapor-pressure deficit. { ˌsach·əˈrā·shən ˈdef·ə·sət }

saturation mixing ratio [METEOROL] A thermodynamic function of state; the value of the mixing ratio of saturated air at the given temperature and pressure; this value may be read directly from a thermodynamic diagram. { ˌsach·əˈrā·shən ˈmik·siŋ ˌrā·shō }

saturation ratio [METEOROL] The ratio of the actual specific humidity to the specific humidity of saturated air at the same temperature. { ˌsach·əˈrā·shən ¦rā·shō }

Saucesian [GEOL] A North American stage of geologic time in the Oligocene and Miocene, above the Zemorrian and below the Relizian. { sȯˈsē·zhən }

sault [HYD] A waterfall or rapids in a stream. { sü }

saussuritization [GEOL] A metamorphic process involving replacement of plagioclase in basalts and gabbros by a fine-grained aggregate of zoisite, epidote, albite, calcite, sericite, and zeolites. { sȯˈsu̇r·əd·əˈzā·shən }

savanna climate *See* tropical savanna climate. { səˈvan·ə ˌklī·mət }

savic orogeny [GEOL] A short-lived orogeny that occurred in late Oligocene geologic time, between the Chattian and Aquitanian stages. { ˈsav·ik ȯˈräj·ə·nē }

saw-cut [GEOL] A large canyon that cuts abruptly across a terrace, so that it is visible only from locations near its edge. { ˈsȯ ˌkət }

Saxonian [GEOL] A European stage of geologic time in the Middle Permian, above the Autonian and below the Thuringian. { sakˈsō·nē·ən }

scabland [GEOL] Elevated land that is essentially flat-lying and covered with basalt and has only a thin soil cover, sparse vegetation, and usually deep, dry channels. { ˈskabˌland }

scabrock [GEOL] **1.** An outcropping of scabland. **2.** Weathered material of a scabland surface. { ˈskabˌräk }

scaglia [GEOL] A dark, very-fine-grained, somewhat calcareous shale usually developed in the Upper Cretaceous and Lower Tertiary periods of the northern Apennines. { ˈskal·yə }

scale height [GEOPHYS] A measure of the decrease of atmospheric pressure with height; when the atmospheric temperature is constant with height, the pressure varies exponentially with height, and the scale height is the height interval over which the pressure changes by a factor of $1/e$. Also known as e-folding height. { ˈskāl ˌhīt }

scales of motion [OCEANOGR] A series of increasing characteristic magnitudes of motion, ranging from tiny eddies of turbulence to oceanwide currents, each member of the series interacting with the adjacent members. { ˈskālz əv ˈmō·shən }

scallop *See* scalloping. { ˈskäl·əp }

scalloped upland [GEOL] The region near or at the divide of an upland into which glacial cirques have cut from opposite sides. { ˈskäl·əpt ˈəp·lənd }

scalloping [GEOL] A sedimentary structure superficially resembling an oscillation ripple mark, and having a concave side that is always oriented toward the top of the bed. Also known as scallop. { ˈskäl·ə·piŋ }

scalped anticline *See* breached anticline. { ˈskalpt ˈant·iˌklīn }

scapolitization [GEOL] Introduction of or replacement by scapolite. { skap·əˌlid·əˈzā·shən }

scar [GEOL] **1.** A steep, rocky eminence, such as a cliff or precipice, where bare rock is well exposed. Also known as scaur; scaw. **2.** *See* shore platform. { skär }

scarp *See* escarpment. { skärp }

scarped plain [GEOL] A terrain characterized by a succession of faintly inclined or gently folded strata. { 'skärpt 'plān }

scarp face *See* scarp slope. { 'skärp ˌfās }

scarp-foot spring [HYD] A spring that flows onto the land surface at or near the foot of an escarpment. { 'skärp ¦fu̇t ˌspriŋ }

scarpland [GEOGR] A region marked by a succession of nearly parallel cuestas separated by lowlands. { 'skärp·lənd }

scarplet *See* piedmont scarp. { 'skärp·lət }

scarpline [GEOL] A relatively straight line of cliffs of considerable extent, produced by faulting or erosion along a fault. { 'skärpˌlīn }

scarp slope [GEOL] The steep face of a cuesta, or asymmetric ridge, facing in an opposite direction to the dip of the strata. Also known as front slope; inface; scarp face. { 'skärp ˌslōp }

scarp stream [HYD] An obsequent stream flowing down a scarp, such as down the scarp slope of a cuesta. { 'skärp ˌstrēm }

scattered [METEOROL] Descriptive of a sky cover of 0.1 to 0.5 (5 to 54%), applied only when clouds or obscuring phenomena aloft are present, not applied for surface-based obscuring phenomena. { 'skad·ərd }

scattering layer [OCEANOGR] A layer of organisms in the sea which causes sound to scatter and to return echoes. { 'skad·ə·riŋ ˌlā·ər }

scaur *See* scar. { skär }

scavenger well [HYD] A well located between a good well (or group of wells) and a source of potential contamination, which is pumped (or allowed to flow) as waste to prevent the contaminated water from reaching the good well. { 'skav·ən·jər ˌwel }

scaw *See* scar. { skȯ }

scharnitzer [METEOROL] A cold, northerly wind of long duration in Tyrol, Austria. { 'shär·nit·sər }

schist [GEOL] A large group of coarse-grained metamorphic rocks which readily split into thin plates or slabs as a result of the alignment of lamellar or prismatic minerals. { shist }

schistose [GEOL] Pertaining to rocks exhibiting schistosity. { 'shisˌtōs }

schistosity [GEOL] A type of cleavage characteristic of metamorphic rocks, notably schists and phyllites, in which the rocks tend to split along parallel planes defined by the distribution and parallel arrangement of platy mineral crystals. { shis'täs·əd·ē }

Schlernwind [METEOROL] East wind blowing down from the Schlern near Bozen in Tyrol, Austria. { 'shlərnˌvint }

schlieren arch [GEOL] An intrusive igneous body with flow layers which occur along its borders but which are poorly developed or absent in its interior. { 'shlir·ən ˌärch }

schlieren dome [GEOL] An intrusive body more or less completely outlined by flow layers which culminate in one central area. { 'shlir·ən ˌdōm }

Schmidt net [GEOL] A coordinate or reference system used to plot a Schmidt projection. { 'shmit ˌnet }

Schmidt projection [GEOL] A Lambert azimuthal equal-area projection of the lower hemisphere of a sphere onto the plane of a meridian; used in structural geology. { 'shmit prəˌjek·shən }

schott [GEOGR] A shallow saline lake in southern Tunisia or on the plateaus of northern Algeria, which is usually dry during the summer. { shät }

schrund line [GEOL] The base of the bergschrund, or deep crevasse, at a late stage in the excavation of a cirque; the schrund line separates the steep slope of the cirque wall from the gentler slope below. { 'shru̇nt ˌlīn }

Schumann resonance [GEOPHYS] A resonance created by lightning-induced electromagnetic radiation trapped in the spherical waveguide formed between the ionosphere and the earth. { 'shü·mən ˌrez·ən·əns }

schungite [GEOL] Amorphous carbon-rich material occurring in Precambrian schists. { 'shu̇ŋ,gīt }

schuppen structure *See* imbricate structure. { 'shu̇p·ən ,strək·chər }

scissors fault [GEOL] A fault on which the offset or separation along the strike increases in one direction from an initial point and decreases in the other direction. Also known as differential fault. { 'siz·ərz ,fȯlt }

sclerotinite [GEOL] A variety of inertinite composed of fungal sclerotia. { 'skler·ə·tə,nīt }

scolite [GEOL] Any of the small tubes in rock believed to be the fossilized burrows of worms. { 'skō,līt }

scopulite [GEOL] A crystallite in the form of a rod with terminal brush or plume. { 'skäp·yə,līt }

score *See* scoring. { skȯr }

scoria [GEOL] Vesicular, cindery, dark lava formed by the escape and expansion of gases in basaltic or andesitic magma; generally denser and darker than pumice. { 'skȯr·ē·ə }

scoria cone [GEOL] A volcanic cone composed of a vesicular, cindery crust on the surface of lava that is basaltic or andesitic in nature. { 'skȯr·ē·ə ,kōn }

scoria mound [GEOL] A volcanic knoll composed of vesicular, cindery crust on the surface of lava that is basaltic or andesitic in nature. { 'skȯr·ē·ə ,mau̇nd }

scoria tuff [GEOL] A deposit of fragmented scoria in a fine-grained tuff matrix. { 'skȯr·ē·ə ,təf }

scoring [GEOL] **1.** The formation of parallel scratches, lines, or grooves in a bedrock surface by the abrasive action of rock fragments transported by a moving glacier. **2.** A scratch, line, or groove produced by this process. Also known as score. { 'skȯr·iŋ }

Scotch mist [METEOROL] A combination of thick mist (or fog) and heavy drizzle occurring frequently in Scotland and in parts of England. { 'skäch 'mist }

Scotch-type volcano [GEOL] A volcanic form characterized by concentric cuestas and produced by cauldron subsidence. { 'skäch ¦tīp väl'kā·nō }

scour *See* tidal scour. { 'skau̇·ər }

scour and fill [GEOL] The process of first digging out and then refilling a channel instigated by the action of a stream or tide; refers particularly to the process that occurs during a period of flood. { 'skau̇·ər ən 'fil }

scour channel [GEOL] A large, groovelike erosional feature produced in sediments by scour. { 'skau̇·ər ,chan·əl }

scour depression [GEOL] A crescent-shaped hollow in the stream bed near the outside of the stream's bend, caused by water that scours below the grade of the stream. { 'skau̇·ər di,presh·ən }

scouring [GEOL] **1.** An erosion process resulting from the action of the flow of air, ice, or water. **2.** *See* glacial scour. { 'skau̇r·iŋ }

scouring velocity [GEOL] The velocity of water which is necessary to dislodge stranded solids from the stream bed. { 'skau̇r·iŋ və,läs·əd·ē }

scour lineation [GEOL] A smooth, low, narrow (2–5 centimeters or 1–2 inches wide) ridge formed on a sedimentary surface and believed to result from the scouring action of a current of water. { 'skau̇·ər ,lin·ē,ā·shən }

scour mark [GEOL] A mark produced by the cutting or scouring action of a current flowing over the bottom of a river or body of water. { 'skau̇·ər ,mark }

scourway [GEOL] A channel created by a powerful water current, particularly the temporary channels formed by streams on the edge of a Pleistocene ice sheet. { 'skau̇·ər,wā }

scree [GEOL] **1.** A mound of loose, angular material, less than 4 inches (10 centimeters). **2.** *See* talus. { skrē }

screened pan [METEOROL] An evaporation pan the top of which is covered by wire-mesh screening (1/4-inch or 6-millimeter mesh); the screening reduces air circulation and insolation, and results in a pan coefficient nearer to unity than that for unscreened pans. { 'skrēnd 'pan }

screw ice [HYD] **1.** Small ice fragments in heaps or ridges, produced by the collision of ice cakes. **2.** A small formation of pressure ice. { 'skrü ˌīs }

scroll [GEOL] One of a series of crescent-shaped sediments on the inner bank of a moving channel, deposited there by the stream. { skrōl }

scroll meander [GEOL] A type of forced-cut meander, in which the scrolls built on the inner bank cause erosion of the outer bank. { 'skrōl mē'an·dər }

scud [METEOROL] Ragged low clouds, usually stratus fractus; most often applied when such clouds are moving rapidly beneath a layer of nimbostratus. { skəd }

Scythian stage [GEOL] A stage in the lesser Triassic series of the alpine facies. Also known as Werfenian stage. { 'sith·ē·ən ˌstāj }

sea [GEOGR] A usually salty lake lacking an outlet to the ocean. [OCEANOGR] **1.** A major subdivision of the ocean. **2.** A heavy swell or ocean wave still under the influence of the wind that produced it. **3.** *See* ocean. { sē }

sea arch [GEOL] An opening through a headland, formed by wave erosion or solution (as by the enlargement of a sea cave, or by the meeting of two sea caves from opposite sides), which leaves a bridge of rock over the water. Also known as marine arch; marine bridge; sea bridge. { 'sē ˌärch }

sea ball [OCEANOGR] A spherical mass of somewhat fibrous material of living or fossil vegetation (especially algae), produced mechanically in shallow waters along a seashore by the compacting effect of wave movement. { 'sē ˌbȯl }

seabeach [GEOL] A beach along the margin of the sea. { 'sēˌbēch }

seabed *See* sea floor. { 'sēˌbed }

sea bottom *See* sea floor. { 'sē ˌbäd·əm }

sea breeze [METEOROL] A coastal, local wind that blows from sea to land, caused by the temperature difference when the sea surface is colder than the adjacent land; it usually blows on relatively calm, sunny summer days, and alternates with the oppositely directed, usually weaker, nighttime land breeze. { 'sē ˌbrēz }

sea breeze of the second kind *See* cold-front-like sea breeze. { 'sē ˌbrēz əv t͟hə 'sek·ənd ˌkīnd }

sea bridge *See* sea arch. { 'sē ˌbrij }

sea-captured stream [HYD] A stream, flowing parallel to the seashore, that is cut in two as a result of marine erosion and that may enter the sea by way of a waterfall. { 'sē ¦kap·chərd ˌstrēm }

sea cave [GEOL] A split or hollow opening, usually at sea level, in the base of a sea cliff, formed by waves acting on weak parts of the weathered rock. Also known as marine cave; sea chasm. { 'sē ˌkāv }

sea channel [GEOL] A long, narrow, U-shaped or V-shaped shallow depression of the sea floor, usually occurring on a gently sloping plain or fan. { 'sē ˌchan·əl }

sea chasm *See* sea cave. { 'sē ˌkaz·əm }

sea cliff [GEOL] An erosional landform, produced by wave action, which is either at the seaward edge of the coast or at the landward side of a wave-cut platform and which denotes the inner limit of the beach erosion. { 'sē ˌklif }

seacoast [GEOGR] The land adjacent to the sea. { 'sēˌkōst }

sea fan *See* submarine fan. { 'sē ˌfan }

sea floor [GEOL] The bottom of the ocean. Also known as seabed; sea bottom. { 'sē ˌflȯr }

sea-floor spreading [GEOL] The hypothesis that the ocean floor is spreading away from the midoceanic ridges and is being conveyed landward by convective cells in the earth's mantle, carrying the continental blocks as passive passengers; the ocean floor moves away from the midoceanic ridge at the rate of 0.4 to 4 inches (1 to 10 centimeters) per year and provides the source of power in the hypothesis of plate tectonics. Also known as ocean-floor spreading; spreading concept; spreading floor hypothesis. { 'sē ¦flȯr ˌspred·iŋ }

sea fog [METEOROL] A type of advection fog formed over the ocean as a result of any of a variety of processes, as when air that has been lying over a warm water surface is transported over a colder water surface, resulting in a cooling of the lower layer of air below its dew point. { 'sē ˌfäg }

sea front [GEOGR] An area partly bounded by the sea. { 'sē ˌfrənt }
sea gate [GEOGR] A way giving access to the sea such as a gate, channel, or beach. { 'sē ˌgāt }
sea glow [OCEANOGR] The luminous, cobalt-blue appearance of very clear water in the open ocean, caused by upward-scattered light from which much of the red has been absorbed. { 'sē ˌglō }
sea gully *See* slope gully. { 'sē ˌgəl·ē }
sea ice [OCEANOGR] **1.** Ice formed from seawater. **2.** Any ice floating in the sea. { 'sē ˌīs }
sea-ice shelf [OCEANOGR] Sea ice floating in the vicinity of its formation and separated from fast ice, of which it may have been a part, by a tide crack or a family of such cracks. { 'sē ¦īs ˌshelf }
sea knoll *See* knoll. { 'sē ˌnōl }
sea level [GEOL] The level of the surface of the ocean; especially, the mean level halfway between high and low tide, used as a standard in reckoning land elevation or sea depths. { 'sē ˌlev·əl }
sea-level chart *See* surface chart. { 'sē ¦lev·əl ˌchärt }
sea-level pressure [METEOROL] The atmospheric pressure at mean sea level, either directly measured or, most commonly, empirically determined from the observed station pressure. { 'sē ¦lev·əl ˌpresh·ər }
sea-level pressure chart *See* surface chart. { 'sē ¦lev·əl ¦presh·ər ˌchärt }
sealing-wax structure [GEOL] A primary sedimentary flow structure produced by slumping, characterized by the lack of a sharply defined slip plane at the base or a contemporaneous erosion plane at the top, and occupying a zone of highly fluid contortion in an otherwise normal sedimentary succession. { 'sēl·iŋ ¦waks ˌstrək·chər }
seam [GEOL] **1.** A stratum or bed of coal or other mineral. **2.** A thin layer or stratum of rock. **3.** A very narrow coal vein. { sēm }
sea meadow [OCEANOGR] Any of the upper layers of the open ocean that have such an abundance of phytoplankton that they provide food for marine organisms. { 'sē ˌmed·ō }
sea mist *See* steam fog. { 'sē ˌmist }
seamount [GEOL] A mountain rising from the ocean floor as a result of submarine volcanism. { 'sēˌmau̇nt }
seamount chain [GEOL] Several seamounts in a line with bases separated by a relatively flat sea floor. { 'sēˌmau̇nt ˌchān }
seamount group [GEOL] Several closely spaced seamounts not in a line. { 'sē ˌmau̇nt ˌgrüp }
seamount range [GEOL] Three or more seamounts having connected bases and aligned along a ridge or rise. { 'sēˌmau̇nt ˌrānj }
sea mud [GEOL] A rich, slimy deposit in a salt marsh or along a seashore, sometimes used as a manure. Also known as sea ooze. { 'sē ˌməd }
sea ooze *See* sea mud. { 'sē ˌüz }
sea peak [GEOL] A peaked elevation of the sea floor, rising 3300 feet (1000 meters) or more from the floor. { 'sē ˌpēk }
seaquake [GEOPHYS] An earth tremor whose epicenter is beneath the ocean and can be felt only by ships in the vicinity of the epicenter. Also known as submarine earthquake. { 'sēˌkwāk }
sea salt [OCEANOGR] The salt remaining after the evaporation of seawater, containing sodium and magnesium chlorides and magnesium and calcium sulfates. { 'sē ˌsȯlt }
sea-salt nucleus [OCEANOGR] A condensation nucleus of a highly hygroscopic nature produced by partial or complete desiccation of particles of sea spray or of seawater droplets derived from breaking bubbles. { 'sē ¦sȯlt ˌnü·klē·əs }
seascape [OCEANOGR] The surrounding sea as it appears to an observer. { 'sēˌskāp }
seascarp [GEOL] A submarine cliff that is relatively long, high, and straight. { 'sēˌskärp }
seashore [GEOL] **1.** The strip of land that borders a sea or ocean. Also known as

seaside; shore. **2.** The ground between the usual tide levels. Also known as seastrand. { 'sē,shȯr }

seashore lake [GEOGR] A lake, containing either fresh or salt water, which lies along a seashore; it is separated from the sea by a river, a delta, or a wall of sediment. { 'sē,shȯr ,lāk }

seaside *See* seashore. { 'sē,sīd }

sea slope [GEOL] The slope of land toward the sea. { 'sē ,slōp }

sea smoke *See* steam fog. { 'sē ,smōk }

season [CLIMATOL] A division of the year according to some regularly recurrent phenomena, usually astronomical or climatic. { 'sēz·ən }

seasonal current [OCEANOGR] An ocean current which has large changes in speed or direction due to seasonal winds. { 'sēz·ən·əl 'kə·rənt }

seasonally frozen ground [GEOL] Ground that is frozen during low temperatures and remains so only during the winter season. Also known as frost zone. { 'sēz·ən·lē ¦frō·zən 'grau̇nd }

seasonal recovery [HYD] Recharge of groundwater during and after a wet season, with a rise in the level of the water table. { 'sēz·ən·əl ri'kəv·ə·rē }

seasonal stream [HYD] A stream whose flow is not constant because it has water in its course only during certain seasons. { 'sēz·ən·əl 'strēm }

seasonal thermocline [OCEANOGR] A thermocline which develops in the oceans in summer at relatively shallow depths due to surface heating and downward transport of heat caused by mixing of water generated by summer winds. { 'sēz·ən·əl 'thər·mə,klīn }

seasonal variation [GEOPHYS] The variation of any parameter of the upper atmosphere with season; for example, the variation of ion densities of different parts of the ionosphere, and the resulting variation in transmission of radio signals over large distances. { 'sēz·ən·əl ,ver·ē'ā·shən }

sea state [OCEANOGR] The numerical or written description of ocean-surface roughness. { 'sē ,stāt }

seastrand *See* seashore. { 'sē,strand }

sea-surface slope [OCEANOGR] A gradual change in the level of the sea surface with distance, caused by Coriolis and wind forces. { 'sē ¦sər·fəs ,slōp }

seat clay *See* underclay. { 'sēt ,klā }

seat earth *See* underclay. { 'sēt ,ərth }

sea terrace *See* marine terrace. { 'sē ,ter·əs }

sea turn [METEOROL] A wind coming from the sea, often bringing mist; the term is limited mainly to the New England section of the United States. { 'sē ,tərn }

sea valley [GEOL] A relatively shallow, wide depression with gentle slopes in the sea floor, the bottom of which grades continuously downward. { 'sē ,val·ē }

seawall [GEOL] A steep-faced, long embankment situated by powerful storm waves along a seacoast at high-water mark. { 'sē,wȯl }

seawater [OCEANOGR] Water of the seas, distinguished by high salinity. Also known as salt water. { 'sē,wȯd·ər }

sebcha *See* sebkha. { 'seb·kə }

sebka *See* sebkha. { 'seb·kə }

sebkha [GEOL] A geologic feature, in North Africa, which is a smooth, flat, plain usually high in salt; after a rain the plain may become a marsh or a shallow lake until the water evaporates. Also known as sabkha; sebcha; sebka; sibjet. { 'seb·kə }

seca [METEOROL] A drought, or dry wind, in Brazil. { 'sā·kə }

sechard [METEOROL] A dry, warm foehn wind over Lake Geneva in Switzerland. { se'chär }

seclusion [METEOROL] A special case of the process of occlusion, where the point at which the cold front first overtakes the warm front (or quasi-stationary front) is at some distance from the apex of the wave cyclone. { si'klü·zhən }

secondary [GEOL] A term with meanings that changed from early to late in the 19th century, when the term was confined to the entire Mesozoic era; it was finally replaced by Mesozoic era. { 'sek·ən,der·ē }

secondary circle *See* secondary great circle. { 'sek·ən,der·ē 'sər·kəl }
secondary clay [GEOL] A clay that has been transported from its place of formation and redeposited elsewhere. { 'sek·ən,der·ē 'klā }
secondary coast [GEOL] A relatively stable seacoast or shoreline whose features are the result of present-day marine processes. { 'sek·ən,der·ē 'kōst }
secondary cold front [METEOROL] A front which forms behind a frontal cyclone and within a cold air mass, characterized by an appreciable horizontal temperature gradient. { 'sek·ən,der·ē 'kōld ,frənt }
secondary consequent stream [HYD] A tributary of a subsequent stream, flowing parallel to or down the same slope as the original consequent stream; it is usually developed after the formation of a subsequent stream, but in a direction consistent with that of the original consequent stream. Also known as subconsequent stream. { 'sek·ən,der·ē 'kän·sə·kwənt 'strēm }
secondary consolidation [GEOL] Consolidation of sedimentary material, at essentially constant pressure, resulting from internal processes such as recrystallization. { 'sek·ən,der·ē kən,säl·ə'dā·shən }
secondary cosmic rays [GEOPHYS] Radiation produced when primary cosmic rays enter the atmosphere and collide with atomic nuclei and electrons. { 'sek·ən,der·ē 'käz·mik 'rāz }
secondary crater [GEOL] An impact crater produced by the relatively low-velocity impact of fragments ejected from a large primary crater. Also known as satellitic crater. { 'sek·ən,der·ē 'krād·ər }
secondary cyclone [METEOROL] A cyclone which forms near or in association with a primary cyclone. Also known as secondary low. { 'sek·ən,der·ē 'sī,klōn }
secondary enrichment [GEOL] The addition to a vein or ore body of material that originated later in time from the oxidation of decomposed ore masses that overlie the vein. { 'sek·ən,der·ē in'rich·mənt }
secondary front [METEOROL] A front which may form within a baroclinic cold air mass which itself is separated from a warm air mass by a primary frontal system; the most common type is the secondary cold front. { 'sek·ən,der·ē 'frənt }
secondary geosyncline [GEOL] A geosyncline appearing at the culmination of or after geosynclinal orogeny. { 'sek·ən,der·ē ,jē·ō'sin,klīn }
secondary glacier [HYD] A small valley glacier that joins a larger trunk glacier as a tributary glacier. { 'sek·ən,der·ē 'glā·shər }
secondary great circle [GEOD] A great circle perpendicular to a primary great circle, as a meridian. Also known as secondary circle. { 'sek·ən,der·ē 'grāt 'sər·kəl }
secondary interstices [GEOL] Openings in a rock that formed after the enclosing rock was formed. { 'sek·ən,der·ē in'tər·stə,sēz }
secondary low *See* secondary cyclone. { 'sek·ən,der·ē 'lō }
secondary pollutant [METEOROL] An air pollutant produced by the reaction of a primary pollutant with some other component in the air. { ¦sek·ən,der·ē pə'lüt·ənt }
secondary porosity [GEOL] The interstices that appear in a rock formation after it has formed, because of dissolution or stress distortion taking place naturally or artificially as a result of the effect of acid treatment or the injection of coarse sand. { 'sek·ən,der·ē pə'räs·əd·ē }
secondary reflection *See* multiple reflection; shoot. { 'sek·ən,der·ē ri'flek·shən }
secondary stratification [GEOL] The layering that occurs when sediments that were at one time deposited are resuspended and redeposited. Also known as indirect stratification. { 'sek·ən,der·ē ,strad·ə·fə'kā·shən }
secondary stratigraphic trap *See* stratigraphic trap. { 'sek·ən,der·ē ¦strad·ə¦graf·ik 'trap }
secondary structure [GEOL] A structure such as a fault, fold, or joint resulting from tectonic movement that started after the rock in which it is found was emplaced. { 'sek·ən,der·ē 'strək·chər }
secondary tectonite [GEOL] A tectonite having a deformation fabric. { 'sek·ən,der·ē 'tek·tə,nīt }
secondary wave *See* S wave. { 'sek·ən,der·ē 'wāv }

second bottom [GEOL] The first terrace rising over a floodplain. { 'sek·ənd 'bäd·əm }

second-derivative map [GEOPHYS] A map of the second vertical derivative of a potential field such as the earth's gravity or magnetic field. { 'sek·ənd də¦riv·əd·iv 'map }

second-foot [HYD] A contraction of cubic foot per second (cfs), the unit of stream discharge commonly used in the United States. { 'sek·ənd ¦fu̇t }

second-foot day [HYD] The volume of water represented by a flow of 1 cubic foot per second for 24 hours; equal to 86,400 cubic feet (approximately 2446.58 cubic meters); used extensively as a unit of runoff volume or reservoir capacity, particularly in the eastern United States. { 'sek·ənd ¦fu̇t 'dā }

second-order climatological station [CLIMATOL] A station at which observations of atmospheric pressure, temperature, humidity, winds, clouds, and weather are made at least twice daily at fixed hours, and at which the daily maximum and minimum of temperature, the daily amount of precipitation, and the duration of bright sunshine are observed. { 'sek·ənd ¦ȯr·dər ˌklī·mət·əl'äj·ə·kəl 'stā·shən }

second-order relief [GEOGR] Extensive relief features consisting of major mountain systems and other surface formations of subcontinental extent. { 'sek·ənd ¦ȯr·dər ri'lēf }

second-order station [METEOROL] After U.S. Weather Bureau practice, a station operated by personnel certified to make aviation weather observations or synoptic weather observations. { 'sek·ənd ¦ȯr·dər 'stā·shən }

second-year ice [OCEANOGR] Sea ice that has survived only one summer's melt. Also known as two-year ice. { 'sek·ənd ¦yir 'īs }

secretion [GEOL] A secondary structure formed of material deposited (from solution) within an empty cavity in any rock, especially a deposit formed on or parallel to the walls of the cavity, the first layer being the outer one. { si'krē·shən }

section [GEOL] **1.** An inclined or vertical surface that is uncovered either naturally (as a sea cliff or stream bank) or artificially (as a strip mine or road cut) through a part of the earth's crust. **2.** A description or scale drawing of the successive rock units or geologic structures shown by the exposed surface, or their appearance if cut through by any intersecting plane. **3.** *See* columnar section; geologic section; thin section; type section. { 'sek·shən }

sector [METEOROL] Something resembling the sector of a circle, as a warm sector between the warm and cold fronts of a cyclone. { 'sek·tər }

sector wind [METEOROL] The average observed or computed wind (direction and speed) at flight level for a given sector of an air route; sectors for over-ocean flights usually consist of 10° of longitude. { 'sek·tər ˌwind }

secular variation [GEOPHYS] The changes, measured in hundreds of years, in the magnetic field of the earth. Also known as geomagnetic secular variation. { 'sek·yə·lər ˌver·ē'ā·shən }

sedimentation [GEOL] **1.** The act or process of accumulating sediment in layers. **2.** The process of deposition of sediment. { ˌsed·ə·mən'tā·shən }

secundine dike [GEOL] A dike which has been intruded into hot country rock. { 'sek·ənˌdīn 'dīk }

sedentary soil [GEOL] Soil that still lies on the rock from which it was formed. { 'sed·ənˌter·ē 'sȯil }

sedifluction [GEOL] The subaquatic or subaerial movement of material in unconsolidated sediments, occurring in the primary stages of diagenesis. { ˌsed·ə'flək·shən }

sediment [GEOL] **1.** A mass of organic or inorganic solid fragmented material, or the solid fragment itself, that comes from weathering of rock and is carried by, suspended in, or dropped by air, water, or ice; or a mass that is accumulated by any other natural agent and that forms in layers on the earth's surface such as sand, gravel, silt, mud, fill, or loess. **2.** A solid material that is not in solution and either is distributed through the liquid or has settled out of the liquid. { 'sed·ə·mənt }

sedimentary cycle *See* cycle of sedimentation. { ¦sed·ə¦men·trē 'sī·kəl }

sedimentary differentiation [GEOL] The progressive separation (by erosion and transportation) of a well-defined rock mass into physically and chemically unlike products that are resorted and deposited as sediments in more or less separate areas. { ¦sed·ə¦men·trē ˌdif·əˌren·chē'ā·shən }

sedimentary dike [GEOL] A tabular mass of sedimentary material that cuts across the structure or bedding of preexisting rock in the manner of an igneous dike and that is formed by the filling of a crack or fissure by forcible injection or intrusion of sediments under abnormal pressure, or by simple infilling of sediments. { ¦sed·ə¦men·trē 'dīk }

sedimentary facies [GEOL] A stratigraphic facies differing from another part or parts of the same unit in both lithologic and paleontologic characters. { ¦sed·ə¦men·trē 'fā·shēz }

sedimentary insertion [GEOL] The emplacement of sedimentary material among deposits or rocks already formed, such as by infilling, injection, or intrusion, or through localized subsidence due to solution of underlying rock. { ¦sed·ə¦men·trē in'sər·shən }

sedimentary intrusion *See* intrusion. { ¦sed·ə¦men·trē in'trü·zhən }

sedimentary laccolith [GEOL] An intrusion of plastic sedimentary material (such as clayey salt breccia) forced up under high pressure and penetrating parallel or nearly parallel to the bedding planes of the invaded formation; characterized by a very irregular thickness. { ¦sed·ə¦men·trē 'lak·əˌlith }

sedimentary lag [GEOL] Delay between the formation of potential sediment by weathering and its removal and deposition. { ¦sed·ə¦men·trē 'lag }

sedimentary structure [GEOL] A structure in sedimentary rocks, such as cross-bedding, ripple marks, and sandstone dikes, produced either contemporaneously with deposition (primary sedimentary structures) or shortly after deposition (secondary sedimentary structures). { ¦sed·ə¦men·trē 'strək·chər }

sedimentary tectonics [GEOL] Folding and deformation in geosynclinal basins caused by subsidence and buckling of strata. { ¦sed·ə¦men·trē tek'tän·iks }

sedimentary trap [GEOL] An area in which sedimentary material accumulates instead of being transported farther, as in an area between high-energy and low-energy environments. { ¦sed·ə¦men·trē 'trap }

sedimentary tuff [GEOL] A tuff containing a small amount of nonvolcanic detrital material. { ¦sed·ə¦men·trē 'təf }

sedimentary volcanism [GEOL] The expelling, extruding, or breaking through of overlying formations by a mixture of sediment, water, and gas, driven by the gas under pressure. { ¦sed·ə¦men·trē 'väl·kəˌniz·əm }

sedimentation basin [GEOL] A depression in the ocean floor with a wide, flat bottom in which sediment accumulates. { ˌsed·ə·mən'tā·shən ˌbās·ən }

sedimentation curve [GEOL] A curve showing cumulatively, and in successive units of time, the amount of sediment accumulated or removed from an originally uniform suspension. { ˌsed·ə·mən'tā·shən ˌkərv }

sedimentation diameter [GEOL] The diameter of a sedimentary particle, determined from the measurement of a hypothetical sphere of the same gravity and settling velocity as those of a given sedimentary particle in the same fluid. { ˌsed·ə·mən'tā·shən dīˌam·əd·ər }

sedimentation radius [GEOL] One-half of the sedimentation diameter. { ˌsed·ə·mən'tā·shən ˌrād·ē·əs }

sedimentation rate *See* rate of sedimentation. { ˌsed·ə·mən'tā·shən ˌrāt }

sedimentation trend [GEOL] The direction in which sediments were laid down. { ˌsed·ə·mən'tā·shən ˌtrend }

sedimentation trough [GEOL] A depression in the ocean floor with a narrow U- or V-shaped bottom in which sediment accumulates. { ˌsed·ə·mən'tā·shən ˌtrȯf }

sedimentation unit [GEOL] A sedimentary deposit formed during one distinct act of sedimentation. { ˌsed·ə·mən'tā·shən ˌyü·nət }

sediment charge [HYD] In a stream, the ratio of the weight or volume of sediment to the weight or volume of water passing a given cross section per unit of time. { 'sed·ə·mənt ˌchärj }

sediment concentration [HYD] The ratio of the dry weight of the sediment in a water-sediment mixture (obtained from a stream or other body of water) to the total weight of the mixture. { 'sed·ə·mənt ˌkän·sən'trā·shən }

sediment-delivery ratio [GEOL] The ratio of sediment yield of a drainage basin to the total amount of sediment moved by sheet erosion and channel erosion. { 'sed·ə·mənt di¦liv·ə·rē ˌrā·shō }

sediment discharge [HYD] The amount of sediment moved by a stream in a given time, measured by dry weight or by volume. Also known as sediment-transport rate. { 'sed·ə·mənt 'disˌchärj }

sediment discharge rating [HYD] A relationship between the discharge of sediment and the total discharge of the stream. Also known as silt discharge rating. { 'sed·ə·mənt 'disˌchärj ˌrād·iŋ }

sediment load [HYD] The solid material that is transported by a natural agent, especially by a stream. { 'sed·ə·mənt ˌlōd }

sedimentology [GEOL] The science concerned with the description, classification, origin, and interpretation of sediments and sedimentary rock. { ˌsed·ə·mən'täl·ə·jē }

sediment-production rate [GEOL] Sediment yield per unit of drainage area, derived by dividing the annual sediment yield by the area of the drainage basin. { 'sed·ə·mənt prə'dək·shən ˌrāt }

sediment station [HYD] A vertical cross-sectional plane of a stream, usually normal to the mean direction of flow, where samples of suspended load are collected on a systematic basis for determining concentration, particle-size distribution, and other characteristics. { 'sed·ə·mənt ˌstā·shən }

sediment-transport rate *See* sediment discharge. { 'sed·ə·mənt 'tranzˌpȯrt ˌrāt }

sediment vein [GEOL] A sedimentary dike formed by the filling of a fissure from above with sedimentary material. { 'sed·ə·mənt ˌvān }

sediment yield [GEOL] The amount of material eroded from the land surface by runoff and delivered to a stream system. { 'sed·ə·mənt ˌyēld }

Seelandian [GEOL] A European stage of geologic time in the lowermost Paleocene. { zā'län·dē·ən }

seep [GEOL] An area, generally small, where water, or another liquid such as oil, percolates slowly to the land surface. { sēp }

seepage [HYD] The slow movement of water through small openings and spaces in the surface of unsaturated soil into or out of a body of surface or subsurface water. { 'sēp·ij }

segregation [GEOL] The formation of a secondary feature within a sediment after deposition due to chemical rearrangement of minor constituents. { ˌseg·rə'gā·shən }

seepage face [GEOL] A belt on a slope, such as the bank of a stream, along which water emerges at atmospheric pressure and flows down the slope. { 'sēp·ij ˌfās }

seepage lake [HYD] **1.** A closed lake that loses water mainly by seepage through the walls and floor of its basin. **2.** A lake that receives its water mainly from seepage. { 'sēp·ij ˌlāk }

segregated ice [HYD] Ice films, seams, lenses, rods, or layers generally 0.04 to 6 inches (1 to 150 millimeters) thick that grow in permafrost by drawing in water as the ground freezes. Also known as Taber ice. { 'seg·rəˌgād·əd 'īs }

segregated vein [GEOL] A fissure filled with mineral matter derived from country rock by the action of percolating water. Also known as exudation vein. { 'seg·rəˌgād·əd 'vān }

seiche [OCEANOGR] A standing-wave oscillation of an enclosed or semienclosed water body, continuing pendulum-fashion after cessation of the originating force, which is usually considered to be strong winds or barometric pressure changes. { sāsh }

seif dune [GEOL] A large, tapering, longitudinal dune or chain of sand dunes with a sharp crest that in profile consists of a succession of peaks and cols. { 'sāf ˌdün }

seismic activity *See* seismicity. { 'sīz·mik akˌtiv·əd·ē }

seismic anisotropy [GEOPHYS] The dependence of seismic velocity on the direction of propagation. { 'sīz·mik ˌan·ə'sä·trə·pē }

seismic area *See* earthquake zone. { 'sīz·mik ,er·ē·ə }
seismic belt [GEOPHYS] An elongate seismic zone, such as that in the Circum-Pacific. { 'sīz·mik ,belt }
seismic discontinuity [GEOPHYS] **1.** A surface at which velocities of seismic waves change abruptly. **2.** A boundary between seismic layers of the earth. Also known as interface; velocity discontinuity. { 'sīz·mik ,dis·känt·ən'ü·əd·ē }
seismic efficiency [GEOPHYS] The proportion of the total available strain energy which is radiated as seismic waves. { 'sīz·mik i'fish·ən·sē }
seismic-electric effect [GEOPHYS] The variation of resistivity with elastic deformation of rocks. { 'sīz·mik i'lek·trik i,fekt }
seismic event [GEOPHYS] An earthquake or a somewhat similar transient earth motion caused by an explosion. { 'sīz·mik i,vent }
seismic gradient [GEOPHYS] The variation of seismic velocity with distance in a specified direction. Also known as velocity gradient. { 'sīz·mik 'grād·ē·ənt }
seismic hazard [GEOPHYS] Any physical phenomenon, such as ground shaking or ground failure, that is associated with an earthquake and that may produce adverse effects on human activities. { 'sīz·mik 'haz·ərd }
seismic intensity [GEOPHYS] The average rate of flow of seismic-wave energy through a unit section perpendicular to the direction of propagation. { 'sīz·mik in'ten·səd·ē }
seismicity [GEOPHYS] The phenomena of earth movements. Also known as seismic activity. { sīz'mis·əd·ē }
seismic map [GEOPHYS] A contour map constructed from seismic data, the *z* coordinate of which could be either time or depth. { 'sīz·mik 'map }
seismic prospecting [GEOPHYS] Geophysical prospecting based on the analysis of elastic waves generated in the earth by artificial means. { 'sīz·mik 'präs,pek·tiŋ }
seismic ray [GEOL] The path along which seismic energy travels. { 'sīz·mik 'rā }
seismic reflector [GEOPHYS] A subsurface profile that is generated by seismic data and indicates a distinctive type of sediment geometry produced by sea-level changes; used to correlate stratigraphic sequences. { ¦sīz·mik ri'flek·tər }
seismic risk [GEOPHYS] **1.** An assortment of earthquake effects that range from ground shaking, surface faulting, and landsliding to economic loss and casualties. **2.** The probability that social or economic consequences of earthquakes will equal or exceed specified values at a site, at several sites, or in an area, during a specified exposure time. { 'sīz·mik 'risk }
seismic stratigraphy [GEOL] A branch of stratigraphy in which sediments and sedimentary rocks are interpreted in a geometrical context from seismic reflectors. { 'sīz·mik strə'tig·rə·fē }
seismic tomography [GEOPHYS] The estimation of seismic wave velocities throughout a region of interest from the travel times of either transmitted or reflected waves, generally through numerical models and iterative procedures. { 'sīz,mik tō'mäg·rə·fē }
seismic velocity [GEOPHYS] The rate of propagation of an elastic wave, usually measured in kilometers per second. { 'sīz·mik və'läs·əd·ē }
seismic vertical [GEOL] **1.** The point on the earth's surface directly over the point within the earth from which an earthquake impulse originates. **2.** The vertical line between the surface point and the point of origin. { 'sīz·mik 'vərd·ə·kəl }
seismology [GEOPHYS] **1.** The study of earthquakes. **2.** The science of strain-wave propagation in the earth. { sīz'mäl·ə·jē }
seistan [METEOROL] A strong wind of monsoon origin which blows from between the northwest and north-northwest and sets in about the end of May or early June in the historic Seistan district of eastern Iran and Afghanistan; it continues almost without cessation until about the end of September; because of its duration it is known as the wind of 120 days (bad-i-sad-o-bistroz). { 'sā,stän }
sejunction water [HYD] Capillary water bounded by menisci, and in static equilibrium in the soil above the capillary fringe. { sə'jəŋk·shən ,wȯd·ər }
selatan [METEOROL] Strong, dry, southerly winds of the southeast monsoon in the Netherlands East Indies and the Celebes. { sā'lä,tän }

selective fusion [GEOL] The fusion of only a portion of a mixture, such as a rock. { si'lek·tiv 'fyü·zhən }
selective replacement [GEOL] The replacement of one mineral by another, preferentially within an altered rock mass. { si'lek·tiv ri'plās·mənt }
selenite butte [GEOL] A small tabular mound, rising 3.3–10 feet (1–3 meters) above a playa, composed of lake sediments capped with a veneer of selenite formed by deflation of the playa or by the effects of rising groundwater. { 'sel·ə,nīt 'byüt }
self-reversal [GEOPHYS] Acquisition by a rock of a natural remanent magnetization opposite to the ambient magnetic field direction at the time of rock formation. { ¦self ri¦vər·səl }
self-rising ground [GEOL] The puffy, irregular, surface or near-surface zone of certain playas, formed by the effects of capillary rise of groundwater. { 'self ¦rīz·iŋ 'grau̇nd }
semianthracite [GEOL] Coal which is between bituminous coal and anthracite in metamorphic rank, and which has a fixed-carbon content of 86–92%. { ,sem·ē'an·thrə,sīt }
semiarid climate *See* steppe climate. { ¦sem·ē'ar·əd 'klī·mət }
semibituminous coal [GEOL] Coal that is harder and more brittle than bituminous coal, has a high fuel ratio, contains 10–20% volatile matter, and burns without smoke; ranks between bituminous and semianthracite coals. { ¦sem·i·bə'tü·mə·nəs 'kōl }
semibolson [GEOL] A wide desert basin or valley whose central playa is absent or poorly developed, and which is drained by an intermittent stream that flows through canyons at each end and reaches a surface outlet. { ¦sem·i'bōls·ən }
semibright coal [GEOL] A type of banded coal defined microscopically as consisting of between 80 and 61% bright ingredients such as vitrain, clarain, and fusain, with clarodurain and durain composing the remainder. { 'sem·i,brīt 'kōl }
semicratonic *See* quasi-cratonic. { ¦sem·i·krə'tän·ik }
semidiurnal [METEOROL] Pertaining to a meteorological event that occurs twice a day. { ¦sem·i·dī'ərn·əl }
semidiurnal current [OCEANOGR] A tidal current in which the tidal-day current cycle consists of two flood currents and two ebb currents, separated by slack water, or of two changes in direction of 360° of a rotary current; this is the most common type of tidal current throughout the world. { ¦sem·i·dī'ərn·əl 'kə·rənt }
semidiurnal tide [OCEANOGR] A tide having two high waters and two low waters during a tidal day. { ¦sem·i·dī'ərn·əl 'tīd }
semidull coal [GEOL] A type of banded coal consisting mainly of clarodurain and durain, with from 40 to 21% bright ingredients such as vitrain, clarain, and fusain. { 'sem·i,dəl 'kōl }
semifusinite [GEOL] A coal maceral with a well-defined woody structure and optical properties intermediate between those of nitrinite and those of fusinite. { ¦sem·i'fyüz·ən,īt }
semisplint coal [GEOL] Banded coal that is intermediate between bright-banded and splint coal, and has 20–30% opaque attritus and more than 5% anthraxylon. { ¦sem·i'splint 'kōl }
Senecan [GEOL] A North American provincial series of geologic time, forming the lower part of the Upper Devonian, above the Erian and below the Chautauquan. { 'sen·i·kən }
senescence [GEOL] The part of the erosion cycle at which the stage of old age begins. { si'nes·əns }
senescent lake [HYD] A lake that is approaching extinction—for example, from filling by remains of aquatic vegetation. { si'nes·ənt 'lāk }
senesland [GEOL] A land surface intermediate between a matureland and a peneplain. { 'sen·əs,land }
senile [GEOL] Pertaining to the stage of senility of the cycle or erosion. { 'sē,nīl }
senility [GEOL] The stage of the cycle of erosion in which erosion of a land surface has reached a minimum, most of the hills have disappeared, and base level has been approached. { si'nil·əd·ē }

Senonian [GEOL] A European stage of geologic time, forming the Upper Cretaceous, above the Turonian and below the Danian. { sə'nō·nē·ən }
sensible atmosphere [METEOROL] That part of the atmosphere that offers resistance to a body passing through it. { 'sen·sə·bəl 'at·mə,sfir }
sensible-heat flow [METEOROL] In the atmosphere, the poleward transport of sensible heat (enthalpy) across a given latitude belt by fluid flow. { 'sen·sə·bəl ¦hēt 'flō }
sensible temperature [METEOROL] The temperature at which air with some standard humidity, motion, and radiation would provide the same sensation of human comfort as existing atmospheric conditions. { 'sen·sə·bəl 'tem·prə·chər }
sensitive clay [GEOL] A clay whose shear strength is reduced to a very small fraction of its former value on remolding at constant moisture content. { 'sen·səd·iv 'klā }
sensitivity [GEOL] The effect of remolding on the consistency of a clay or cohesive soil, regardless of the physical nature of the causes of the change. { ,sen·sə'tiv·əd·ē }
separate *See* soil separate. { 'sep·rət }
separation [GEOL] The apparent relative displacement on a fault, measured in any given direction. { ,sep·ə'rā·shən }
septarian [GEOL] Pertaining to the irregular polygonal pattern of internal cracks developed in septaria. { sep'tar·ē·ən }
septarian boulder *See* septarium. { sep'tar·ē·ən 'bōl·dər }
septarian nodule *See* septarium. { sep'tar·ē·ən 'näj·ül }
septarium [GEOL] A large (32–36 inches or 80–90 centimeters in diameter), spheroidal concretion, usually composed of argillaceous carbonate, characterized by internal cracking into irregular polygonal blocks that become cemented together by crystalline minerals. Also known as beetle stone; septarian boulder; septarian nodule; turtle stone. { sep'tar·ē·əm }
Sequanian [GEOL] Upper Lower Jurassic (Upper Lusitanian) geologic time. Also known as Astartian. { sə'kwā·nē·ən }
sequence [GEOL] **1.** A sequence of geologic events, processes, or rocks, arranged in chronological order. **2.** A geographically discrete, major informal rock-stratigraphic unit of greater than group or supergroup rank. Also known as stratigraphic sequence. **3.** A body of rock deposited during a complete cycle of sea-level change. [METEOROL] *See* collective. { 'sē·kwəns }
sequence of current [OCEANOGR] The order of occurrence of the tidal current strengths of a day, with special reference to whether the greater flood immediately precedes or follows the greater ebb. { 'sē·kwəns əv 'kə·rənt }
sequence of tide [OCEANOGR] The order in which the tides of a day occur, with special reference to whether the higher high water immediately precedes or follows the lower low water. { 'sē·kwəns əv 'tīd }
sequence stratigraphy [GEOL] A branch of stratigraphy that subdivides the sedimentary record along continental margins and in interior basins into a succession of depositional sequences as regional and interregional correlative units. { 'sē·kwəns strə'tig·rə·fē }
sequential landform [GEOL] One of an orderly succession of smaller landforms that are developed by the erosion, weathering, and mass wasting of larger initial landforms. { si'kwen·chəl 'land,fȯrm }
serac [HYD] A sharp ridge or pinnacle of ice among the crevasses of a glacier. { sə'rak }
serein [METEOROL] The doubtful phenomenon of fine rain falling from an apparently clear sky, the clouds, if any, being too thin to be visible; frequently, fine rain is observed with a clear sky overhead, but clouds to windward clearly indicate the source of the drops. { sə'ran }
serial observation [OCEANOGR] The procurement of water samples and temperature readings at a number of levels between the surface and the bottom of an ocean. { 'sir·ē·əl ,äb·zər'vā·shən }
serial station [OCEANOGR] An oceanographic station consisting of one or more Nansen casts. { 'sir·ē·əl ,stā·shən }
seriate [GEOL] Having crystals that vary gradually in size. { 'sir·ē,āt }

sericitization [GEOL] A hydrothermal or metamorphic process involving the introduction of or replacement by sericite. { ˌser·əˌsīd·əˈzā·shən }

series [GEOL] **1.** A number of rocks, minerals, or fossils that can be arranged in a natural sequence due to certain characteristics, such as succession, composition, or occurrence. **2.** A time-stratigraphic unit, below system and above stage, composed of rocks formed during an epoch of geologic time. { ˈsir·ēz }

serpentine spit [GEOGR] A spit that is extended in more than one direction due to variable or periodically shifting currents. { ˈsər·pənˌtēn ˈspit }

serpentinization [GEOL] A hydrothermal process by which magnesium-rich silicate minerals are converted into or replaced by serpentine minerals. { ˌsər·pənˌtē·nəˈzā·shən }

serpent kame *See* esker. { ˈsər·pənt ˈkām }

serrate [GEOL] Pertaining to topographic features having a notched or toothed edge, or a saw-edge profile. { ˈseˌrāt }

serrate ridge *See* arête. { ˈseˌrāt ˈrij }

seston [OCEANOGR] Minute living organisms and particles of nonliving matter which float in water and contribute to turbidity. { ˈseˌstän }

set [GEOL] A group of essentially conformable strata or cross-strata, separated from other sedimentary units by surfaces of erosion, nondeposition, or abrupt change in character. [OCEANOGR] The direction toward which an oceanic current flows. { set }

settled [METEOROL] Pertaining to weather, devoid of storms for a considerable period. { ˈsed·əld }

settled snow [HYD] An old snow that has been strongly metamorphosed and compacted. { ˈsed·əld ˈsnō }

settlement [GEOL] The subsidence of surficial material (such as coastal sediments) due to compaction. { ˈsed·əl·mənt }

settling [GEOL] The sag in outcrops of layered strata, caused by rock creep. Also known as outcrop curvature. { ˈset·liŋ }

severe storm [METEOROL] In general, any destructive storm, but usually applied to a severe local storm, that is, an intense thunderstorm, hail storm, or tornado. { siˈvir ˈstȯrm }

severe-storm observation [METEOROL] An observation (and report) of the occurrence, location, time, and direction of movement of severe local storms. { siˈvir ¦stȯrm ˌäb·zərˈvā·shən }

severe weather [METEOROL] A more general term for severe storm. { siˈvir ˈweth·ər }

Sevier orogeny [GEOL] The deformation that occurred along the eastern edge of the Great Basin in Utah (eastern edge of the Cordilleran miogeosyncline) during times intermediate between the Nevadan orogeny to the west and the Laramide orogeny to the east, culminating early in the Late Cretaceous. { seˈvyā ȯˈräj·ə·nē }

sferics *See* atmospheric interference. { ˈsfir·iks }

sferics fix [METEOROL] The estimated location of a source of atmospherics, presumably a lightning discharge. { ˈsfir·iks ˌfiks }

sferics observation [METEOROL] An evaluation, from one or more sferics receivers, of the location of weather conditions with which lightning is associated; such observations are more commonly obtained from networks of two or three widely spaced stations; simultaneous observations of the azimuth of the discharge are made at all stations, and the location of the storm is determined by triangulation. { ˈsfir·iks ˌäb·zərˌvā·shən }

shadow zone [GEOPHYS] The zone, between 103 and 143° from the epicenter of an earthquake, in which direct seismic waves do not arrive because of refraction and absorption by the earth's core. { ˈshad·ō ˌzōn }

shaft [GEOL] A passage in a cave that is vertical or nearly vertical. { shaft }

shake wave *See* S wave. { ˈshāk ˌwāv }

shale ball [GEOL] A meteorite partly or wholly converted to iron oxides by weathering. Also known as oxidite. { ˈshāl ˌbȯl }

shale break [GEOL] A thin layer or parting of shale between harder strata or within a bed of sandstone or limestone. { ˈshāl ˌbrāk }

shale crescent [GEOL] A crescent formed by the filling of a ripple-mark trough by shale. { 'shāl ˌkres·ənt }

shale ice [HYD] A mass of thin and brittle plates of river or lake ice formed when sheets of skim ice break up into small pieces. { 'shāl ˌīs }

shale reservoir [GEOL] Underground hydrocarbon reservoir in which the reservoir rock is a brittle, siliceous, fractured shale. { 'shāl 'rez·əvˌwär }

shalification [GEOL] The formation of shale. { ˌshāl·ə·fə'kā·shən }

shallow-focus earthquake [GEOPHYS] An earthquake whose focus is located within 70 kilometers of the earth's surface. { 'shal·ō ¦fō·kəs 'ərthˌkwāk }

shallow fog [METEOROL] In weather-observing terminology, low-lying fog that does not obstruct horizontal visibility at a level 6 feet (1.8 meters) or more above the surface of the earth; this is, almost invariably, a form of radiation fog. { 'shal·ō 'fäg }

shallow inland seas [GEOL] Epeiric seas which periodically cover cratonic areas as a result of continental subsidence or eustatic rises in sea level. { 'shal·ō 'in·lənd 'sēz }

shallow marginal seas [GEOL] Epeiric seas along the cratonic margins. { 'shal·ō 'märj·ən·əl 'sēz }

shallows [HYD] A shallow place or area in a body of water, or an expanse of shallow water. { 'shal·ōz }

shallow water [HYD] Water of such a depth that bottom topography affects surface waves. { 'shal·ō 'wȯd·ər }

shallow-water wave [HYD] A progressive gravity wave in water whose depth is much less than the wavelength. { 'shal·ō ¦wȯd·ər ˌwāv }

shallow well [HYD] **1.** A water well, generally dug up by hand or by excavating machinery, or put down by driving or boring, that taps the shallowest aquifer in the vicinity. **2.** A well whose water level is shallow enough to permit use of a suction pump, the practical lift of which is taken as 22 feet (6.7 meters). { 'shal·ō 'wel }

shaluk [METEOROL] Any hot desert wind other than simoom. { shä'lək }

shaly [GEOL] Pertaining to, composed of, containing, or having the properties of shale, especially readily split along close-spaced bedding planes. { 'shāl·ē }

shaly bedding [GEOL] Laminated bedding varying between 2 and 10 millimeters in thickness. { 'shāl·ē 'bed·iŋ }

shamal [METEOROL] The northwest wind in the lower valley of the Tigris and Euphrates and the Persian Gulf; it may set in suddenly at any time, and generally lasts from 1 to 5 days, dying down at night and freshening again by day; but in June and early July it continues almost without cessation (the great or 40-day shamal). { shə'mäl }

shantung [GEOL] A monadnock in the process of burial by huangho deposits. { shan'təŋ }

Shantung soil *See* Noncalcic Brown soil. { shan'təŋ ˌsȯil }

shard [GEOL] A vitric fragment in pyroclastics, having a characteristically curved surface of fracture. { shärd }

sharki *See* kaus. { 'shär·kē }

sharkskin pahoehoe [GEOL] A type of pahoehoe displaying numerous tiny spines or spicules on the surface. { 'shärkˌskin pə'hō·ēˌhō·ē }

shark-tooth projection [GEOL] Sharp pointed projections several centimeters in length, formed by the pulling apart of plastic lava. { 'shärk ¦tüth prə'jek·shən }

sharp-edged gust [METEOROL] A gust that represents an instantaneous change in wind direction or speed. { 'shärp ¦ejd 'gəst }

sharp sand [GEOL] An angular-grain sand free of clay, loam, and other foreign particles. { 'shärp 'sand }

sharpstone [GEOL] Any rock fragment having angular edges and corners and being more than 2 millimeters in diameter. { 'shärpˌstōn }

shatter cone [GEOL] A striated conical rock fragment along which fracturing has occurred. { 'shad·ər ˌkōn }

shatter zone [GEOL] An area of randomly fissured or cracked rock that may be filled by mineral deposits, forming a network pattern of veins. { 'shad·ər ˌzōn }

sheaf structure [GEOL] A bundled arrangement of crystals that is characteristic of certain fibrous minerals, such as stibnite. { 'shēf ˌstrək·chər }

shear cleavage *See* slip cleavage. { 'shir ,klē·vij }

shear fold [GEOL] A similar fold whose mechanism is shearing or slipping along closely spaced planes that are parallel to the fold's axial surface. Also known as glide fold; slip fold. { 'shir ,fōld }

shear-gravity wave [GEOPHYS] A combination of gravity waves and a Helmholtz wave on a surface of discontinuity of density and velocity. { 'shir 'grav·əd·ē ,wāv }

shear joint [GEOL] A joint that is a shear fracture; it is a potential plane of shear. Also known as slip joint. { 'shir ,jȯint }

shear line [METEOROL] A line or narrow zone across which there is an abrupt change in the horizontal wind component parallel to this line; a line of maximum horizontal wind shear. { 'shir ,līn }

shear moraine [GEOL] A debris-laden surface or zone found along the margin of any ice sheet or ice cap, dipping in toward the center of the ice sheet but becoming parallel to the bed at the base. { 'shir mə'rān }

shear plane *See* shear surface. [HYD] A planar surface in a glacier, usually laden with rock debris, attributed to discontinuous shearing or overthrusting. { 'shir ,plān }

shear slide [GEOL] A landslide, especially a slump, produced by shear failure usually along a plane of weakness such as a bedding or cleavage plane. { 'shir ,slīd }

shear sorting [GEOL] Sorting of sediments in which the smaller grains tend to move toward the zone of greatest shear strain, and the larger grains toward the zone of least shear. { 'shir ,sȯrd·iŋ }

shear structure [GEOL] A local structure in which earth stresses have been relieved by many small, closely spaced fractures. { 'shir ,strək·chər }

shear surface [GEOL] A surface along which differential movement has taken place parallel to the surface. Also known as shear plane. { 'shir ,sər·fəs }

shear wave *See* S wave. { 'shir ,wāv }

shear zone [GEOL] A tabular area of rock that has been crushed and brecciated by many parallel fractures resulting from shear strain; often becomes a channel for underground solutions and the seat of ore deposition. { 'shir ,zōn }

sheer [GEOL] A steep face of a cliff. { shir }

sheet [GEOL] **1.** A thin flowstone coating of calcite in a cave. **2.** A tabular igneous intrusion, especially when concordant or only slightly discordant. [HYD] *See* sheetflood. { shēt }

sheet crack [GEOL] A planar crack attributed to shrinkage of sediment due to dewatering. { 'shēt ,krak }

sheet deposit [GEOL] A stratiform mineral deposit that is more or less horizontal and extensive relative to its thickness. { 'shēt di,päz·ət }

sheet drift [GEOL] An evenly spread deposit of glacial drift that did not significantly alter the form of the underlying rock surface. { 'shēt ,drift }

sheeted fissure [GEOL] A closely spaced fissure. { 'shēd·əd 'fish·ər }

sheeted vein [GEOL] A vein filling a shear zone. { 'shēd·əd 'vān }

sheeted zone [GEOL] An area of mineral deposits consisting of sheeted veins. { 'shēd·əd 'zōn }

sheet erosion [GEOL] Erosion of thin layers of surface materials by continuous sheets of running water. Also known as sheetflood erosion; sheetwash; surface wash; unconcentrated wash. { 'shēt i,rō·zhən }

sheetflood [HYD] A broad expanse of moving, storm-borne water that spreads as a thin, continuous, relatively uniform film over a large area for a short distance and duration. Also known as sheet; sheetwash. { 'shēt,fləd }

sheetflood erosion *See* sheet erosion. { 'shēd,fləd i,rō·zhən }

sheet flow [HYD] An overland flow or downslope movement of water taking the form of a thin, continuous film over relatively smooth soil or rock surfaces and not concentrated into channels larger than rills. { 'shēt,flō }

sheet frost [HYD] A thick coating of rime formed on windows and other surfaces. { 'shēt ,frȯst }

sheet ice [HYD] A smooth, thin layer of ice formed by rapid freezing of the surface layer of a body of water. { 'shēt ,īs }

sheeting [GEOL] The process by which thin sheets, slabs, scales, plates, or flakes of rock are successively broken loose or stripped from the outer surface of a large rock mass in response to release of load. Also known as exfoliation. { 'shēd·iŋ }

sheeting structure [GEOL] A fracture or joint formed by pressure-release jointing or exfoliation. Also known as exfoliation joint; expansion joint; pseudostratification; release joint; sheet joint; sheet structure. { 'shēd·iŋ ˌstrək·chər }

sheet joint *See* sheeting structure. { 'shēt ˌjȯint }

sheet lightning [GEOPHYS] A diffuse, but sometimes fairly bright, illumination of those parts of a thundercloud that surround the path of a lightning flash, particularly a cloud discharge or cloud-to-cloud discharge. Also known as luminous cloud. { 'shēt¦līt·niŋ }

sheet sand *See* blanket sand. { 'shēt ˌsand }

sheet sandstone [GEOL] A thin, blanket-shaped deposit of sandstone of regional extent. { 'shēt 'sanˌstōn }

sheet spar [GEOL] A sheet crack filled with spar. { 'shēt ˌspär }

sheet structure *See* sheeting structure. { 'shēt ˌstrək·chər }

sheetwash [GEOL] **1.** The detritus deposited by a sheetflood. **2.** *See* sheet erosion. [HYD] **1.** A wide, moving expanse of water on an arid plain; the combined result of many streams issuing from the mountains. **2.** *See* sheetflood. { 'shētˌ wäsh }

shelf [GEOL] **1.** Solid rock beneath alluvial deposits. **2.** A flat, projecting ledge of rock. **3.** *See* continental shelf. { shelf }

shelf break [GEOL] An obvious steepening of the gradient between the continental shelf and the continental slope. { 'shelf ˌbrāk }

shelf channel [GEOL] A valley formed in a shelf by erosion. { 'shelf ˌchan·əl }

shelf edge [GEOL] The demarcation, without dramatic change in gradient, between continental shelf and continental slope. { 'shelf ˌej }

shelf facies [GEOL] A sedimentary facies characterized by carbonate rocks and fossil shells and produced in the neritic environments of marginal shelf seas. Also known as foreland facies; platform facies. { 'shelf ˌfā·shēz }

shelf ice [HYD] The ice of an ice shelf. Also known as barrier ice. { 'shelf ˌīs }

shelf sea [OCEANOGR] A shallow marginal sea located on the continental shelf, usually less than 150 fathoms (275 meters) in depth; an example is the North Sea. { 'shelf ˌsē }

shelfstone [GEOL] A speleothem formed at the water's edge as a horizontally projecting ledge. { 'shelfˌstōn }

shell [GEOL] **1.** The crust of the earth. **2.** A thin hard layer of rock. { shel }

shell ice [HYD] Ice, on a body of water, that remains as an unbroken surface when the water level drops so that a cavity is formed between the water surface and the ice. { 'shel ˌīs }

shell marl [GEOL] A light-colored calcareous deposit formed on the bottoms of small fresh-water lakes, composed largely of uncemented mollusk shells and precipitated calcium carbonate, along with the hard parts of minute organisms. { 'shel ˌmärl }

shell sand [GEOL] A loose aggregate that is largely composed of shell fragments of sand size. { 'shel ˌsand }

shelly [GEOL] **1.** Pertaining to a sediment or sedimentary rock containing the shells of animals. **2.** Pertaining to land abounding in or covered with shells. { 'shel·ē }

shelly facies [GEOL] A nongeosynclinal sedimentary facies that is commonly characterized by abundant calcareous fossil shells, dominant carbonate rocks (limestones and dolomites), mature orthoquartzitic sandstones, and a paucity of shales. { 'shel·ē 'fā·shēz }

shelly pahoehoe [GEOL] A type of pahoehoe characterized by open tubes and blisters on the surface. { 'shel·ē pə'hō·ēˌhō·ē }

shelter cave [GEOL] A cave which extends only a short way underground, and whose roof of overlying rock usually extends beyond its sides. Also known as rock cave. { 'shel·tər ˌkāv }

shelter porosity [GEOL] A type of primary interparticle porosity created by the sheltering effect of relatively large sedimentary particles which prevent the infilling of pore space by finer clastic particles. { 'shel·tər pə'räs·əd·ē }

shergottite [GEOL] An achondritic stony meteorite that is composed chiefly of pigeonite and maskelynite. { 'shər·gə,tīt }

shield [GEOL] **1.** The very old, rigid core of relatively stable rocks within a continent around which younger sedimentary rocks have been deposited. Also known as continental shield. **2.** *See* palette. { shēld }

shield basalt [GEOL] A basaltic lava flow from a group of small, close-spaced shield-volcano vents that coalesced to form a single unit. { shēld bə'sȯlt }

shield cone [GEOL] A cone or dome-shaped volcano built up by successive outpourings of lava. { shēld ,kōn }

shielding factor [GEOPHYS] The ratio of the strength of the magnetic field at a directional compass to its strength if there were no disturbing material; usually expressed as a decimal. { 'shēld·iŋ ,fak·tər }

shielding layer [METEOROL] The layer of air nearest the earth, with reference to the manner in which this layer shields the earth from activity in the free atmosphere above, or vice versa. { 'shēld·iŋ ,lā·ər }

shield volcano [GEOL] A broad, low volcano shaped like a flattened dome and built of basaltic lava. Also known as basaltic dome; lava dome. { 'shēld väl,kā·nō }

shift [GEOL] The relative displacement of the units affected by a fault but outside the fault zone itself. { shift }

shifting [GEOL] The movement of the crest of a divide away from a more actively eroding stream (as on the steeper slope of an asymmetric ridge) toward a weaker stream on the gentler slope. { 'shift·iŋ }

shimmer [METEOROL] To appear tremulous or wavering, due to varying atmospheric refraction in the line of sight. { 'shim·ər }

shingle [GEOL] Pebbles, cobble, and other beach material, coarser than ordinary gravel but roughly the same size and occurring typically on the higher parts of a beach. { 'shiŋ·gəl }

shingle barchan [GEOL] A dunelike ridge formed of shingle perpendicular to the beach in shallow water. { 'shiŋ·gəl bär'kän }

shingle beach [GEOL] A narrow beach composed of shingle and commonly having a steep slope on both its landward and seaward sides. Also known as cobble beach. { 'shiŋ·gəl ,bēch }

shingle-block structure *See* imbricate structure. { 'shiŋ·gəl ¦bläk 'strək·chər }

shingle rampart [GEOL] A rampart of shingle built along a reef on the seaward edge. { 'shiŋ·gəl 'ram,pärt }

shingle ridge [GEOL] A steeply sloping bank of shingle heaped upon and parallel with the shore. { 'shiŋ·gəl ,rij }

shingle structure *See* imbricate structure. { 'shin·gəl ,strək·chər }

shingling *See* imbrication. { 'shiŋ·gliŋ }

ship drift [OCEANOGR] A method of measuring ocean currents; the ship itself is used as a current tracer, its motions being measured by navigating equipment on board. { 'ship ,drift }

ship report [METEOROL] The encoded and transmitted report of a marine weather observation. { 'ship ri,pȯrt }

ship synoptic code [METEOROL] A synoptic code for communicating marine weather observations; it is a modification of the international synoptic code. { 'ship si'näp·tik 'kōd }

shoal [GEOL] A submerged elevation that rises from the bed of a shallow body of water and consists of, or is covered by, unconsolidated material, and may be exposed at low water. { 'shōl }

shoaling [OCEANOGR] The bottom effect which influences the height of waves moving from deep to shallow water. { 'shōl·iŋ }

shoal patches [OCEANOGR] Individual and scattered elevations of the bottom, with depths of 10 fathoms (18 meters) or less, but composed of any material except rock or coral. { 'shōl ˌpach·əz }

shoal reef [GEOL] A reef formed in irregular masses amid submerged shoals of calcareous reef detritus. { 'shōl ˌrēf }

shoal water [OCEANOGR] Shallow water; over a shoal. { 'shōl ˌwȯd·ər }

shock lithification [GEOL] The conversion of originally loose fragmental materials into coherent aggregates by the action of shock waves, such as those generated by explosions or meteorite impacts. { 'shäk ˌlith·ə·fəˌkā·shən }

shock loading [GEOPHYS] The process of subjecting material to the action of high-pressure shock waves generated by artificial explosions or by meteorite impact. { 'shäk ˌlōd·iŋ }

shock melting [GEOPHYS] Fusion of material as a result of the high temperatures produced by the action of high-pressure shock waves. { 'shäk ˌmelt·iŋ }

shock zone [GEOL] A volume of rock in or around an impact or explosion crater in which a distinctive shock-metamorphic deformation or transformation effect is present. { 'shäk ˌzōn }

shoestring [GEOL] A long, relatively straight and narrow sedimentary body having a width/thickness ratio of less than 5:1, usually 1:1. { shüˌstriŋ }

shoestring rill [GEOL] One of several long, narrow, uniform channels, closely spaced and roughly parallel with one another, that merely score the homogeneous surface of a relatively steep slope of bare soil or weak, clay-rich bedrock, and that develop wherever overland flow is intense. { 'shüˌstriŋ ˌril }

shoestring sand [GEOL] A shoestring composed of sand and usually buried in mud or shale, usually a sandbar or channel fill. { 'shüˌstriŋ ˌsand }

shoot [GEOL] *See* ore shoot. [GEOPHYS] The energy that goes up through the strata from a seismic profiling shot and is reflected downward at the surface or at the base of the weathering; appears either as a single wave or unites with a wave train that is traveling downward. Also known as secondary reflection. [HYD] **1.** A place where a stream flows or descends swiftly. **2.** A natural or artificial channel, passage, or trough through which water is moved to a lower level. **3.** A rush of water down a steep place or a rapids. { shüt }

shore [GEOL] **1.** The narrow strip of land immediately bordering a body of water. **2.** *See* seashore. { shȯr }

shore current [HYD] A water current near a shoreline, often flowing parallel to the shore. { 'shȯr ˌkə·rənt }

shore drift *See* littoral drift. { 'shȯr ˌdrift }

shoreface [GEOL] The narrow, steeply sloping zone between the seaward limit of the shore at low water and the nearly horizontal offshore zone. { 'shȯrˌfās }

shoreface terrace [GEOL] A wave-built terrace in the shoreface region, composed of gravel and coarse sand swept from the wave-cut bench into deeper water. { 'shȯrˌfās ˌter·əs }

shore ice [OCEANOGR] Sea ice that has been beached by wind, tides, currents, or ice pressure; it is a type of fast ice, and may sometimes be rafted ice. { 'shȯr ˌīs }

shore lead [OCEANOGR] A lead between pack ice and fast ice or between floating ice and the shore; it may be closed by wind or currents so that only a tide crack remains. { 'shȯr ˌlēd }

shoreline [GEOL] The intersection of a specified plane of water, especially mean high water, with the shore; a limit which changes with the tide or water level. Also known as strandline; waterline. { 'shȯrˌlīn }

shoreline cycle [GEOL] The cycle of changes through which sequential forms of coastal features pass during shoreline development, from the establishment of a water level to the time when the water can do no more work. { 'shȯrˌlīn ˌsī·kəl }

shoreline-development ratio [GEOL] A ratio indicating the degree of irregularity of a lake shoreline, given as the length of the shoreline to the circumference of a circle whose area is equal to that of the lake. { 'shȯrˌlīn di'vel·əp·mənt ˌrā·shō }

shoreline of depression [GEOL] A shoreline of submergence that implies an absolute subsidence of the land. { 'shȯrˌlīn əv di'presh·ən }

shoreline of elevation [GEOL] A shoreline of emergence that implies an absolute rise of the land. { 'shȯr,līn əv ,el·ə'vā·shən }

shoreline of emergence [GEOL] A straight or gently curving shoreline formed by the dominant relative emergence of the floor of an ocean or a lake. Also known as emerged shoreline; negative shoreline. { 'shȯr,līn əv i'mər·jəns }

shoreline of submergence [GEOL] A shoreline, characterized by bays, promontories, and other minor features, formed by the dominant relative submergence of a landmass. Also known as positive shoreline; submerged shoreline. { 'shȯr,līn əv səb'mər·jəns }

shore platform [GEOL] The horizontal or gently sloping surface produced along a shore by wave erosion. Also known as scar. { 'shȯr ,plat,fȯrm }

shore polyn'ya [OCEANOGR] A polyn'ya between pack ice and the coast, or between pack ice and an ice front, formed by a current or by wind. { 'shȯr ¦pal·ən¦yä }

shoreside *See* onshore. { 'shȯr,sīd }

shore terrace [GEOL] **1.** A terrace produced along the shore by wave and current action. **2.** *See* marine terrace. { 'shȯr ,ter·əs }

short-crested wave [OCEANOGR] An ocean wave whose crest is of finite length; that is, the type actually found in nature. { 'shȯrt ¦kres·təd 'wāv }

short-range forecast [METEOROL] A weather forecast made for a time period generally not greater than 48 hours in advance. { 'shȯrt ¦rānj 'fȯr,kast }

short wave *See* deep-water wave. { 'shȯrt ¦wāv }

shot copper [GEOL] Small, rounded particles of native copper, molded by the shape of vesicles in basaltic host rock, and resembling shot in size and shape. { 'shät ,käp·ər }

shoulder [GEOL] **1.** A short, rounded spur protruding laterally from the slope of a mountain or hill. **2.** The sloping segment below the summit of a mountain or hill. **3.** A bench on the flanks of a glaciated valley, located at the sharp change of slope where the steep sides of the inner glaciated valley meet the more gradual slope above the level of glaciation. **4.** A joint structure on a joint face produced by the intersection of plume-structure ridges with fringe joints. { 'shōl·dər }

shoved moraine *See* push moraine. { 'shəvd mə'rān }

Showalter stability index [METEOROL] A measure of the local static stability of the atmosphere, expressed as a numerical index. { 'shō,wȯl·tər stə,bil·əd·ē ,in,deks }

shower [METEOROL] Precipitation from a convective cloud; characterized by the suddenness with which it starts and stops, by the rapid changes of intensity, and usually by rapid changes in the appearance of the sky. { 'shaü·ər }

shrinkage [GEOL] The decrease in volume of soil, sediment, fill, or excavated earth due to the reduction of voids by mechanical compaction, superimposed loads, natural consolidation, or drying. { 'shriŋ·kij }

shrinkage crack [GEOL] A small crack produced in fine-grained sediment or rock by the loss of contained water during drying or dehydration. { 'shriŋ·kij ,krak }

shrinkage index [GEOL] The numerical difference between the plastic limit of a material and its shrikage limit. { 'shriŋ·kij ,in,deks }

shrinkage limit [GEOL] That moisture content of a soil below which a decrease in moisture content will not cause a decrease in volume, but above which an increase in moisture will cause an increase in volume. { 'shriŋ·kij ,lim·ət }

shrinkage pore [GEOL] An irregular pore formed in muddy sediment by shrinkage. { 'shriŋ·kij ,pȯr }

shrinkage ratio [GEOL] The ratio of a volume change to the moisture-content change above the shrinkage limit. { 'shriŋ·kij ,rā·shō }

shrub-coppice dune [GEOL] A small dune formed on the leeward side of bush-and-clump vegetation. { 'shrəb ¦käp·əs 'dün }

shuga [OCEANOGR] A spongy, rather opaque, whitish chunk of ice which forms instead of pancake ice if the freezing takes place in sea water which is considerably agitated. { 'shü·gə }

shungite [GEOL] A hard, black, amorphous, coallike material composed of more than 98% carbon. { 'shəŋ,īt }

shutterridge [GEOL] A ridge formed by vertical, lateral, or oblique displacement of a fault traversing a ridge-and-valley topography with the displaced part of a ridge shutting in the adjacent ravine or canyon. { 'shəd·ə,rij }

Siberian anticyclone *See* Siberian high. { sī'bir·ē·ən 'ant·i,sī,klōn }

Siberian high [METEOROL] An area of high pressure which forms over Siberia in winter, and which is particularly apparent on mean charts of sea-level pressure; centered near lake Baikal. Also known as Siberian anticyclone. { sī'bir·ē·ən 'hī }

sibjet *See* sebkha. { 'sib·jət }

SID *See* sudden ionospheric disturbance.

side canyon [GEOL] A ravine or other valley smaller than a canyon, through which a tributary flows into the main stream. { 'sīd ,kan·yən }

sideraerolite *See* stony-iron meteorite. { ¦sid·ə·rə¦er·ə,līt }

sideroferrite [GEOL] A variety of native iron occurring as grains in petrified wood. { ,sid·ə·rə'fe,rīt }

siderolite *See* stony-iron meteorite. { 'sid·ə·rə,līt }

sideronitic texture [GEOL] In mineral deposits, a mesh of silicate minerals so shattered and pressed as to force out solutions and other volatiles. { ¦sid·ə·rə¦nid·ik 'teks·chər }

siderophyre [GEOL] A stony-iron meteorite containing bronzite and tridymite crystals in a nickel-iron network. Also known as siderophyry. { 'sid·ə·rə,fīr }

siderophyry *See* siderophyre. { ,sid·ə'räf·ə·rē }

siderosphere *See* inner core. { 'sid·ə·rə,sfir }

side stream *See* tributary. { 'sīd,strēm }

sideswipe [GEOPHYS] **1.** A phenomenon wherein two cross reflections come from a single seismograph, due to the almost simultaneous arrival of reflection energy from both limbs of a syncline or from two nearby, steeply dipping fault scarps. **2.** In refraction shooting, the lateral deflection of a minimum-time path to include a nearby, steeply dipping, high-velocity boundary such as a flank of a salt dome. { 'sīd,swīp }

sierozem [GEOL] A soil found in cool to temperate arid regions, characterized by a brownish-gray surface on a lighter layer based on a carbonate or hardpan layer. { 'sir·ə,zem }

sierra [GEOGR] A high range of hills or mountains with irregular peaks that give a sawtooth profile. { sē'er·ə }

sieve deposition [GEOL] The formation of coarse-grained lobate masses on an alluvial fan whose material is sufficiently coarse and permeable to permit complete infiltration of water before it reaches the toe of the fan. { 'siv ,dep·ə,zish·ən }

sieve lobe [GEOL] A coarse-grained lobate mass produced by sieve deposition on an alluvial fan. { 'siv ,lōb }

siffanto [METEOROL] A southwest wind of the Adriatic Sea; it is often violent. { si'fän·tō }

sigma-T [OCEANOGR] An abbreviated value of the density of a sea-water sample of temperature T and salinity S: $\sigma T = [\rho(S,T) - 1] \times 10^3$, where $\rho(S,T)$ is the value of the sea-water density in centimeter-gram-second units at standard atmospheric pressure. { 'sig·mə ¦tē }

sigmoidal dune [GEOL] A dune with an S-shaped ridge crest formed by the merger of crescentic dunes. { sig'mȯid·əl 'dün }

sigmoidal fold [GEOL] A recumbent fold having an axial surface which resembles the Greek letter sigma. { sig'mȯid·əl 'fōld }

significant wave [OCEANOGR] Statistically, a wave with the average height of the highest third of the waves of a given wave group. { sig'nif·i·kənt ,wāv }

sigua [METEOROL] A straight-blowing monsoon gale of the Philippines. { 'sē,wä }

sikussak [OCEANOGR] Very old sea ice trapped in fjords; it resembles glacier ice because snowfall and snow drifts contribute to its formation. { sə'kü,säk }

silcrete [GEOL] A conglomerate of sand and gravel cemented by silica. { 'sil,krēt }

silexite [GEOL] Chert occurring in calcareous beds. { sī'lek,sīt }

silica sand [GEOL] Sand having a very high percentage of silicon dioxide; a source of silicon. { 'sil·ə·kə ¦sand }

silication [GEOL] The conversion to or the replacement by silicates. { ˌsil·əˈkā·shən }
siliceous earth [GEOL] A loose, friable, soft, porous, lightweight, fine-grained, and usually white siliceous sediment, usually derived from the remains of organisms. { səˈlish·əs ˈərth }
siliceous ooze [GEOL] An ooze composed of siliceous skeletal remains of organisms, such as radiolarians. { səˈlish·əs ˈüz }
siliceous sediment [GEOL] Fine-grained sediment and sedimentary rock mainly composed of the microscopic remains of the unicellular, silica-secreting plankton diatoms and radiolarians. Minor constituents include extremely small shards of sponge spicules and other microorganisms such as silicoflagellates. Siliceous sedimentary rock sequences are often highly porous and can form excellent petroleum source and reservoir rocks. { səˈlish·əs ˈsed·ə·mənt }
silicification [GEOL] Introduction of or replacement by silica. Also known as silification. { səˌlis·ə·fəˈkā·shən }
silicified wood [GEOL] A material formed by the silicification of wood, generally in the form of opal or chalcedony, in such a manner as to preserve the original form and structure of the wood. Also known as agatized wood; opalized wood; petrified wood; woodstone. { səˈlis·əˌfīd ˈwu̇d }
silicinate [GEOL] Pertaining to the silica cement of a sedimentary rock. { səˈlis·ənˌāt }
silification *See* silicification. { ˌsil·ə·fəˈkā·shən }
silk [GEOL] Microscopic needle-shaped crystalline inclusions of rutile in a natural gem from which subsurface reflections produce a whitish sheen resembling that of a silk fabric. { silk }
sill [GEOL] **1.** Submarine ridge in relatively shallow water that separates a partly closed basin from another basin or from an adjacent sea. **2.** A tabular igneous intrusion that is oriented parallel to the planar structure of surrounding rock. { sil }
sill depth [OCEANOGR] The maximum depth at which there is horizontal communication between an ocean basin and the open ocean. Also known as threshold depth. { ˈsil ˌdepth }
silled basin *See* restricted basin. { ˈsild ˈbās·ən }
silt [GEOL] **1.** A rock fragment or a mineral or detrital particle in the soil having a diameter of 0.002–0.05 millimeter that is, smaller than fine sand and larger than coarse clay. **2.** Sediment carried or deposited by water. **3.** Soil containing at least 80% silt and less than 12% clay. { silt }
silt discharge rating *See* sediment discharge rating. { ˈsilt ˈdisˌchärj ˌrād·iŋ }
silting [GEOL] The deposition or accumulation of stream-deposited silt that is suspended in a body of standing water. { ˈsilt·iŋ }
siltite *See* siltstone. { ˈsilˌtīt }
silt loam [GEOL] A soil containing 50–88% silt, 0–27% clay, and 0–50% sand. { ˈsilt ˌlōm }
silt soil [GEOL] A soil containing 80% or more of silt, and not more than 12% of clay and 20% of sand. { ˈsilt ˌsȯil }
siltstone [GEOL] Indurated silt having a shalelike texture and composition. Also known as siltite. { ˈsiltˌstōn }
silttil [GEOL] A chemically decomposed and eluviated till consisting of a friable, brownish, open-textured silt that contains a few small siliceous pebbles. { ˈsilˌtil }
Silurian [GEOL] **1.** A period of geologic time of the Paleozoic era, covering a time span of between 430–440 and 395 million years ago. **2.** The rock system of this period. { siˈlu̇r·ē·ən }
silver frost [METEOROL] A deposit of glaze built up on trees, shrubs, and other exposed objects during a fall of freezing precipitation; the product of an ice storm. Also known as silver thaw. { ˈsil·vər ˈfrȯst }
silver storm *See* ice storm. { ˈsil·vər ˈstȯrm }
silver thaw *See* silver frost. { ˈsil·vər ˈthȯ }
similar fold [GEOL] A fold in deformed beds in which the successive folds resemble each other. { ˈsim·ə·lər ˈfōld }
simoom [METEOROL] A strong, dry, dust-laden desert wind which blows in the Sahara,

Israel, Syria, and the desert of Arabia; its temperature may exceed 130°F (54°C), and the humidity may fall below 10. { sə'müm }

simple crater [GEOL] A meteorite impact crater of relatively small diameter, characterized by a uniformly concave-upward shape and a maximum depth in the center, and lacking a central uplift. { 'sim·pəl 'krād·ər }

simple cross-bedding [GEOL] Cross-bedding in which the lower bounding surfaces are nonerosional surfaces. { 'sim·pəl 'krȯs ˌbed·iŋ }

simple ore [GEOL] An ore of a single metal. { 'sim·pəl 'ȯr }

simple shear [GEOPHYS] Strain caused by differential movements on one set of parallel planes which results in internal rotation of fabric elements. { 'sim·pəl 'shir }

simple spit [GEOGR] A spit, either straight or recurved, without the development of minor spits at its end or along its inner side. { 'sim·pəl 'spit }

simple valley [GEOL] A valley that maintains a constant relation to the general structure of the underlying strata. { 'sim·pəl 'val·ē }

Sinemurian [GEOL] A European stage of geologic time; Lower Jurassic, above Hattangian and below Pliensbachian. { sin·ə'myu̇r·ē·ən }

singing sand *See* sounding sand. { 'siŋ·iŋ ¦sand }

single-cycle mountain [GEOL] A fold mountain that has been destroyed without reelevation of any of its important parts. { 'siŋ·gəl ¦sī·kəl 'mau̇nt·ən }

single-station analysis [METEOROL] The analysis or reconstruction of the weather pattern from more or less continuous meteorological observations made at a single geographic location, or the body of techniques employed in such an analysis. { 'siŋ·gəl ¦stā·shən ə'nal·ə·səs }

single-theodolite observation [METEOROL] The usual type of pilot-balloon observation, that is, using one theodolite. { 'siŋ·gəl thē¦äd·əlˌīt ˌäb·zər'vā·shən }

singular corresponding point [METEOROL] A center of elevation or depression on a constant-pressure chart (or a center of high or low pressure on a constant-height chart) considered as a reappearing characteristic of successive charts. { 'siŋ·gyə·lər ¦kär·ə¦spänd·iŋ 'pȯint }

singularity [METEOROL] A characteristic meteorological condition which tends to occur on or near a specific calendar date more frequently than chance would indicate; an example is the January thaw. { ˌsiŋ·gyə'lar·əd·ē }

sinistral fault *See* left lateral fault. { 'sin·əs·trəl 'fȯlt }

sinistral fold [GEOL] An asymmetric fold whose long limb, when viewed along its dip, appears to have a leftward offset. { 'sin·əs·trəl 'fōld }

sink [GEOL] **1.** A circular or ellipsoidal depression formed by collapse on the flank of or near to a volcano. **2.** A slight, low-lying desert depression containing a central playa or saline lake with no outlet, as where a desert stream comes to an end or disappears by evaporation. { siŋk }

sinkhole [GEOL] Closed surface depressions in regions of karst topography produced by solution of surface limestone or the collapse of cavern roofs. { 'siŋkˌhōl }

sinkhole plain [GEOL] A regionally extensive plain or plateau characterized by well-developed karst features. { 'siŋkˌhōl ˌplān }

sinking [OCEANOGR] The downward movement of surface water generally caused by converging currents or when a water mass becomes denser than the surrounding water. Also known as downwelling. { 'siŋk·iŋ }

siphon [GEOL] A passage in a cave system that connects with a water trap. { 'sī·fən }

sirocco [METEOROL] A warm south or southeast wind in advance of a depression moving eastward across the southern Mediterranean Sea or North Africa. { sə'rä·kō }

skarn [GEOL] A lime-bearing silicate derived from nearly pure limestone and dolomite with the introduction of large amounts of silicon, aluminum, iron, and magnesium. { skärn }

skauk [HYD] An extensive field of crevasses in a glacier. { skȯk }

skavl *See* sastruga. { 'skav·əl }

skeleton grain [GEOL] A relatively stable and not readily translocated grain of soil material, concentrated or reorganized by soil-forming processes. { 'skel·ət·ən ˌgrān }

skeleton layer [OCEANOGR] The structure that is formed at the bottom of sea ice while freezing, and consists of vertically oriented platelets of ice separated by layers of brine. { 'skel·ət·ən ˌlā·ər }

skerry [GEOL] A low, small, rugged and rocky island or reef. { 'sker·ē }

skid boulder [GEOL] An isolated angular block of stone resting on the floor of a playa, derived from an outcrop near the playa margin, and associated with a trail or mark indicating that the boulder has recently slid across the mud surface. { 'skid ˌbōl·dər }

Skiddavin *See* Arenigian. { skə'dav·ən }

skill score [METEOROL] In synoptic meteorology, an index of the degree of skill of a set of forecasts, expressed with reference to some standard such as forecasts based upon chance, persistence, or climatology. { 'skil ˌskór }

skim ice [HYD] First formation of a thin layer of ice on the water surface. { 'skim ˌīs }

skimming [HYD] **1.** Diversion of water from a stream or conduit by shallow overflow in order to avoid diverting sand, silt, or other debris carried as bottom load. **2.** Withdrawal of fresh groundwater from a thin body or lens floating on salt water by means of shallow wells or infiltration galleries. { 'skim·iŋ }

skin-friction coefficient [METEOROL] A dimensionless drag coefficient expressing the proportionality between the frictional force per unit area, or the shearing stress exerted by the wind at the earth's surface, and the square of the surface wind speed. { 'skin ¦frik·shən ˌkō·i'fish·ənt }

skiou *See* morvan. { skyō }

skip cast [GEOL] The cast of a skip mark. { 'skip ˌkast }

skip mark [GEOL] A crescent-shaped mark that is one of a linear pattern of regularly spaced marks made by an object that skipped along the bottom of a stream. { 'skip ˌmärk }

skiron [METEOROL] The Greek name for the northwest wind, which is cold in winter but hot and dry in summer. { 'skē·rón }

sky cover [METEOROL] In surface weather observations, the amount of sky covered but not necessarily concealed by clouds or by obscuring phenomena aloft, the amount of sky concealed by obscuring phenomena that reach the ground, or the amount of sky covered or concealed by a combination of the two phenomena. { 'skī ˌkəv·ər }

sky map [METEOROL] A pattern of variable brightness observable on the underside of a cloud layer, and caused by the different reflectivities of material on the earth's surface immediately beneath the clouds; this term is used mainly in polar regions. { 'skī ˌmap }

slab [GEOL] A cleaved or finely parallel jointed rock, which splits into tabular plates from 1 to 4 inches (2.5 to 10 centimeters) thick. Also known as slabstone. [HYD] A layer in, or the whole-thickness of, a snowpack that is very hard and has the ability to sustain elastic deformation under stress. { slab }

slab jointing [GEOL] Jointing produced in rock by the formation of numerous cleaved or closely spaced parallel fissures dividing the rock into thin slabs. { 'slab ˌjóint·iŋ }

slab pahoehoe [GEOL] A pahoehoe whose surface consists of a jumbled arrangement of slabs of flow crust. { 'slab pə'hō·ēˌhō·ē }

slack [GEOL] A hollow or depression between lines of shore dunes or in a sandbank or mudbank on a shore. { slak }

slack ice [HYD] Ice fragments on still or slow-moving water. { 'slak 'īs }

slack water [OCEANOGR] The interval when the speed of the tidal current is very weak or zero; usually refers to the period of reversal between ebb and flood currents. { 'slak 'wód·ər }

slaking [GEOL] **1.** Crumbling and disintegration of earth materials when exposed to air or moisture. **2.** The breaking up of dried clay when saturated with water. { 'slāk·iŋ }

slate ribbon [GEOL] A relict ribbon sructure on the cleavage surface of slate, in which varicolored and straight, wavy, or crumpled stripes cross the cleavage surface. { 'slāt ˌrib·ən }

slaty cleavage *See* flow cleavage. { 'slād·ē 'klē·vij }

sleet [METEOROL] Colloquially in some parts of the United States, precipitation in the form of a mixture of snow and rain. { slēt }

slice [GEOL] An arbitrary section of some uniform standard, such as thickness of a stratigraphic unit that is otherwise indivisible for purposes of analytic study. { slīs }
slice method [METEOROL] A method of evaluating the static stability over a limited area at any reference level in the atmosphere; unlike the parcel method, the slice method takes into account continuity of mass by considering both upward and downward motion. { 'slīs ˌmeth·əd }
slick [OCEANOGR] Area in which capillary waves are absent or suppressed. { slik }
slickens [GEOL] A layer of fine silt deposited by a flooding stream. { 'slik·ənz }
slickenside [GEOL] A surface that is polished and smoothly striated and results from slippage along a fault plane. { 'slik·ənˌsīd }
slickolite [GEOL] A vertically discontinuous slip-scratch surface made by slippage and shearing and developed on sharply dipping bedding planes of limestone that shapes the wall of a solution cavity. { 'slik·əˌlīt }
slide [GEOL] **1.** A vein of clay intersecting and dislocating a vein vertically, or the vertical dislocation itself. **2.** A rotational or planar mass movement of earth, snow, or rock resulting from failure under shear stress along one or more surfaces. { slīd }
sliding *See* gravitational sliding. { 'slīd·iŋ }
sliding scale [METEOROL] A set of combinations of ceilings and visibilities which constitute the operational weather limits at an airport; as the observed value of one element increases, the limiting value of the other element decreases, and vice versa. { 'slīd·iŋ 'skāl }
sliming [OCEANOGR] The formation of films of algae on submerged structures. { 'slīm·iŋ }
slip [GEOL] The actual relative displacement along a fault plane of two points which were formerly adjacent on either side of the fault. Also known as actual relative movement; total displacement. { slip }
slip bedding [GEOL] Convolute bedding formed as the result of subaqueous sliding. { 'slip ˌbed·iŋ }
slip block [GEOL] A separate rock mass that has slid away from its original position and come to rest down the slope without undergoing much deformation. { 'slip ˌbläk }
slip cleavage [GEOL] Cleavage that is superposed on slaty cleavage or schistosity, characterized by spaced cleavage with thin tabular bodies of rock between the cleavage planes. Also known as close-joints cleavage; crenulation cleavage; shear cleavage; strain-slip cleavage. { 'slip ˌklē·vij }
slip face [GEOL] The steeply sloping leeward surface of a sand dune. Also known as sandfall. { 'slip ˌfās }
slip fold *See* shear fold. { 'slip ˌfōld }
slip joint *See* shear joint. { 'slip ˌjȯint }
slip-off slope [GEOL] The long, low, gentle slope on the inside of the downstream face of a stream meander. { 'slip ˌȯf ˌslōp }
slip plane [GEOL] A planar slip surface. { 'slip ˌplān }
slip sheet [GEOL] A stratum or rock on the limb of an anticline that has slid down and away from the anticline; a gravity collapse structure. { 'slip ˌshēt }
slip surface [GEOL] The displacement surface of a landslide. { 'slip ˌsər·fəs }
slope [GEOL] The inclined surface of any part of the earth's surface. { slōp }
slope correction [GEOL] A tape correction applied to a distance measured on a slope in order to reduce it to a horizontal distance, between the vertical lines through its end points. Also known as grade correction. { 'slōp kəˌrek·shən }
slope current *See* gradient current. { 'slōp ˌkə·rənt }
slope failure [GEOL] The downward and outward movement of a mass of soil beneath a natural slope or other inclined surface; four types of slope failure are rockfall, rock flow, plane shear, and rotational shear. { 'slōp ˌfāl·yər }
slope gully [GEOL] A small, discontinuous submarine valley, usually formed by slumping along a fault scarp or the slope of a river delta. Also known as sea gully. { 'slōp ˌgəl·ē }
slope stability [GEOL] The resistance of an inclined surface to failure by sliding or collapsing. { 'slōp stəˌbil·əd·ē }

slope wash [GEOL] **1.** The mass-wasting process, assisted by nonchanneled running water, by which rock and soil is transported down a slope, specifically, sheet erosion. **2.** The material that is or has been transported. { 'slōp ˌwäsh }

slough [HYD] A minor marshland or tidal waterway which usually connects other tidal areas; often more or less equivalent to a bayou. { slaủ }

slough ice [HYD] Slushy ice or snow. { 'slaủ ˌīs }

slow ion *See* large ion. { 'slō 'īˌän }

slud [GEOL] **1.** Muddy material which has moved downslope by solifluction. **2.** Ground that behaves as a viscous fluid, including material moved by solifluction and by mechanisms not limited to gravitational flow. { sləd }

sludge [GEOL] A soft or muddy bottom deposit as on tideland or in a stream bed. [OCEANOGR] A dense, soupy accumulation of new sea ice consisting of incoherent floating frazil crystals. Also known as cream ice; sludge ice; slush. { sləj }

sludge cake [OCEANOGR] An accumulation of sludge hardened into a cake strong enough to bear the weight of a man. { 'sləj ˌkāk }

sludge floe [OCEANOGR] Sludge that is hardened into a floe strong enough to bear the weight of a person. { 'sləj ˌflō }

sludge ice *See* sludge. { 'sləj ˌīs }

sludge lump [OCEANOGR] An irregular mass of sludge formed as a result of strong winds. { 'sləj ˌləmp }

sludging *See* solifluction. { 'sləj·iŋ }

slump [GEOL] A type of landslide characterized by the downward slipping of a mass of rock or unconsolidated debris, moving as a unit or several subsidiary units, characteristically with backward rotation on a horizontal axis parallel to the slope; common on natural cliffs and banks and on the sides of artificial cuts and fills. { sləmp }

slump ball [GEOL] A relatively flattened mass of sandstone resembling a large concretion, measuring from 0.8 inch to 10 feet (2 centimeters to 3 meters) across, commonly thinly laminated with internal contortions and a smooth or lumpy external form, and formed by subaqueous slumping. { 'sləmp ˌbȯl }

slump basin [GEOL] A shallow basin near the base of a canyon wall and on a shale hill or ridge, formed by small, irregular slumps. { 'sləmp ˌbās·ən }

slump bedding [GEOL] Also known as slurry bedding. **1.** Any disturbed bedding. **2.** Convolute bedding produced by subaqueous slumping or lateral movement of newly deposited sediment. { 'sləmp ˌbed·iŋ }

slump fault *See* normal fault. { 'sləmp ˌfȯlt }

slump fold [GEOL] An intraformational fold produced by slumping of soft sediments, as at the edge of the continental shelf. { 'sləmp ˌfōld }

slump overfold [GEOL] A fold consisting of hook-shaped masses of sandstone produced during slumping. { 'sləmp 'ō·vərˌfōld }

slump scarp [GEOL] A low cliff or rim of thin solidified lava occurring along the margins of a lava flow and against the valley walls or around steptoes after the central part of the lava crust collapsed due to outflow of still-molten underlying layers. { 'sləmp ˌskärp }

slump sheet [GEOL] A well-defined bed of limited thickness and wide horizontal extent, containing slump structures. { 'sləmp ˌshēt }

slump structure [GEOL] Any sedimentary structure produced by subaqueous slumping. { 'sləmp ˌstrək·chər }

slurry bedding *See* slump bedding. { 'slər·ē ˌbed·iŋ }

slurry slump [GEOL] A slump in which the incoherent sliding mass is mixed with water and disintegrates into a quasiliquid slurry. { 'slər·ē ˌsləmp }

slush [HYD] Snow or ice on the ground that has been reduced to a soft, watery mixture by rain, warm temperature, or chemical treatment. [OCEANOGR] *See* sludge. { sləsh }

slush avalanche [GEOL] A rapid and far-reaching downslope transport of rock debris released by snow supersaturated with meltwater and marking the catastrophic opening of ice- and snow-dammed brooks to the spring flood. { 'sləsh 'av·əˌlanch }

slush ball [HYD] An extremely compact accretion of snow, frazil, and ice particles. { 'sləsh ˌbȯl }

slush field [HYD] An area of water-saturated snow having a soupy consistency. Also known as snow swamp. { 'sləsh ˌfēld }

slushflow [HYD] **1.** A mudflow-like outburst of water-saturated snow along a stream course, commonly occurring in the Arctic Zone after intense thawing has produced more meltwater than can drain through the snow, and having a width generally several times greater than that of the stream channel. **2.** A flow of clear slush on a glacier, as in Greenland. { 'shəshˌflō }

slush icing [METEOROL] The accumulation of ice and water on exposed surfaces of aircraft when the craft is flown through wet snow or snow and liquid drops at temperatures near 0°C. { 'sləsh ˌīs·iŋ }

slush pond [HYD] A pool or lake containing slush, on the ablation surface of a glacier. { 'sləsh ˌpänd }

small circle [GEOD] A circle on the surface of the earth, the plane of which does not pass through the earth's center. { 'smȯl 'sər·kəl }

small-craft warning [METEOROL] A warning, for marine interests, of impending winds up to 28 knots (32 miles per hour or 52 kilometers per hour). { 'smȯl ¦kraft 'wȯrn·iŋ }

small diurnal range [OCEANOGR] The difference in height between mean lower high water and mean higher low water. { 'smȯl dī¦ərn·əl 'rānj }

small hail [METEOROL] Frozen precipitation consisting of small, semitransparent, roundish grains, each grain consisting of a snow pellet surrounded by a very thin ice covering, giving it a glazed appearance. { 'smȯl 'hāl }

small ice floe [OCEANOGR] An ice floe of sea ice 30 to 600 feet (9 to 180 meters) across. { 'smȯl 'īs ˌflō }

small ion [METEOROL] An atmospheric ion of the type that has the greatest mobility; and hence, collectively, it is the principal agent of atmospheric conduction; evidence indicates that each ion is a singly charged atmospheric molecule (or, rarely, an atom) about which a few other neutral molecules are held by the electrical attraction of the central ionized molecule; estimates of the number of satellite molecules range as high as 12. Also known as fast ion; light ion. { 'smȯl 'īˌän }

small-ion combination [METEOROL] Either of two processes by which small ions disappear: the union of a small ion and a neutral Aitken nucleus to form a new large ion, or the neutralization of a large ion by the small ion. { 'smȯl 'īˌän ˌkäm·bə'nā·shən }

small tropic range [OCEANOGR] The difference in height between tropic lower high water and tropic higher low water. { 'smȯl 'träp·ik 'rānj }

smog [METEOROL] Air pollution consisting of smoke and fog. { smäg }

smoke horizon [METEOROL] The top of a smoke layer which is confined by a low-level temperature inversion in such a way as to give the appearance of the horizon when viewed from above against the sky; in such instances the true horizon is usually obscured by the smoke layer. { 'smōk həˌrīz·ən }

smokes [METEOROL] Dense white haze and dust clouds common in the dry season on the Guinea coast of Africa, particularly at the approach of the harmattan. { smōks }

smooth [OCEANOGR] Comparatively calm water between heavy seas. { smüth }

smooth chert [GEOL] A hard, dense, homogeneous chert (insoluble residue) characterized by a conchoidal-to-even fracture surface that is devoid of roughness and by a lack of crystallinity, granularity, or other distinctive structure. { 'smüth 'chərt }

smooth phase [GEOL] The part of stream traction whereby a mass of sediment travels as a sheet with gradually increasing density from the surface downward. { 'smüth ˌfāz }

smooth sea [OCEANOGR] Sea with waves no higher than ripples or small wavelets. { 'smüth 'sē }

smothered bottom [GEOL] A sedimentary surface on which complete, well-preserved, and commonly very fragile and delicate fossils were saved by an influx of mud that buried them instantly. { 'sməth·ərd 'bäd·əm }

SMOW *See* standard mean ocean water. { smaůw *or* ¦es¦em¦ō'dəb·əlˌyü }

snaking stream *See* meandering stream. { 'snāk·iŋ ˌstrēm }

snap [METEOROL] A brief period of extreme (generally cold) weather setting in suddenly, as in a "cold snap." { snap }
SNC group [GEOL] A group of meteorites comprising the shergottites, nakhlites, and chassignites, which are all believed to have originated from Mars. { ¦es¦en'sē ˌgrüp }
snezhura *See* snow slush. { 'snezh·ə·rə }
sno *See* elvegust. { snȯ }
snout [GEOGR] A promontory or protruding mass of rock. [HYD] The protruding lower extremity of a glacier. { snau̇t }
snow [METEOROL] The most common form of frozen precipitation, usually flakes of starlike crystals, matted ice needles, or combinations, and often rime-coated. { snō }
snow accumulation [METEOROL] The actual depth of snow on the ground at any instant during a storm, or after any single snowstorm or series of snowstorms. { 'snō əˌkyü·myəˌlā·shən }
snow avalanche [HYD] An avalanche of relatively pure snow; some rock and earth material may also be carried downward. Also known as snowslide. { 'snō ˌav·əˌlanch }
snowbank glacier *See* nivation glacier. { 'snōˌbaŋk ˌglā·shər }
snow banner [METEOROL] Snow being blown from a mountain crest. Also known as snow plume; snow smoke. { 'snō ˌban·ər }
snow barchan [HYD] A crescentic or horseshoe-shaped snow dune of windblown snow with the ends pointing downwind. Also known as snow medano. { 'snō bär'kän }
snow blink [METEOROL] A bright, white glare on the underside of clouds, produced by the reflection of light from a snow-covered surface; this term is used in polar regions with reference to the sky map. Also known as snow sky. { 'snō ˌbliŋk }
snowbridge [HYD] Snow bridging a crevasse in a glacier. { 'snōˌbrij }
snow cap [HYD] **1.** Snow covering a mountain peak when no snow exists at lower elevations. **2.** Snow on the surface of a frozen lake. { 'snō ˌkap }
snow climate *See* polar climate. { 'snō ˌklī·mət }
snow cloud [METEOROL] A popular term for any cloud from which snow falls. { 'snō ˌklau̇d }
snow concrete [HYD] Snow that is compacted at low temperatures by heavy objects (as by a vehicle) and that sets into a tough substance of considerably greater strength than uncompressed snow. Also known as snowcrete. { 'snō ˌkänˌkrēt }
snow course [HYD] An established line, usually from several hundred feet to as much as a mile long, traversing representative terrain in a mountainous region of appreciable snow accumulation; along this course, measurements of snow cover are made to determine its water equivalent. { 'snō ˌkȯrs }
snow cover [HYD] **1.** All accumulated snow on the ground, including that derived from snowfall, snowslides, and drifting snow. Also known as snow mantle. **2.** The extent, expressed as a percentage, of snow cover in a particular area. { 'snō ˌkəv·ər }
snow-cover chart [METEOROL] A synoptic chart showing areas covered by snow and contour lines of snow depth. { 'snō ¦kəv·ər ˌchärt }
snowcreep [HYD] The slow internal deformation of a snowpack resulting from the stress of its own weight and metamorphism of snow crystals. { 'snōˌkrēp }
snowcrete *See* snow concrete. { 'snōˌkrēt }
snow crust [HYD] A crisp, firm, outer surface upon snow. { 'snō ˌkrəst }
snow crystal [METEOROL] Any of several types of ice crystal found in snow; a snow crystal is a single crystal, in contrast to a snowflake which is usually an aggregate of many single snow crystals. { 'snō ˌkrist·əl }
snow cushion [HYD] An accumulation of snow, commonly deep, soft, and unstable, deposited in the lee of a cornice on a steep mountain slope. { 'snō ku̇sh·ən }
snow density [HYD] The ratio of the volume of meltwater that can be derived from a sample of snow to the original volume of the sample; strictly speaking, this is the specific gravity of the snow sample. { 'snō ˌden·səd·ē }
snowdrift [HYD] Snow deposited on the lee of obstacles, lodged in irregularities of a surface, or collected in heaps by eddies in the wind. { 'snōˌdrift }
snowdrift glacier [HYD] A semipermanent mass of firn, formed by drifted snow in

depressions in the ground or behind obstructions. Also known as catchment glacier; drift glacier. { 'snō,drift ,glā·shər }

snowdrift ice [HYD] Permanent or semipermanent masses of ice, formed by the accumulation of drifted snow in the lee of projections, or in depressions of the ground. Also known as glacieret. { 'snō,drift ,īs }

snow dune [HYD] An accumulation of wind-transported snow resembling the forms of sand dunes. { 'snō ,dün }

snow dust [METEOROL] Fine snow crystals fragmented or driven by the wind. { 'snō ,dəst }

snow eater [METEOROL] **1.** Any warm wind blowing over a snow surface; usually applied to a foehn wind. **2.** A fog over a snow surface; so called because of the frequently observed rapidity with which a snow cover disappears after a fog sets in. { 'snō ,ēd·ər }

snowfall [METEOROL] **1.** The rate at which snow falls; in surface weather observations, this is usually expressed as inches of snow depth per 6-hour period. **2.** A snow storm. { 'snō,fȯl }

snowfield [HYD] **1.** A broad, level, relatively smooth and uniform snow cover on ground or ice at high altitudes or in mountainous regions above the snow line. **2.** The accumulation area of a glacier. **3.** A small glacier or accumulation of perennial ice and snow too small to be designated a glacier. { 'snō,fēld }

snowflake [METEOROL] An ice crystal or, much more commonly, an aggregation of many crystals which falls from a cloud; simple snowflakes (single crystal) exhibit beautiful variety of form, but the symmetrical shapes reproduced so often in photomicrographs are not actually found frequently in snowfalls; broken single crystals, fragments, or clusters of such elements are much more typical of actual snows. { 'snō,flāk }

snow flurry [METEOROL] Popular term for snow shower, particularly of a very light and brief nature. { 'snō ,flər·ē }

snowflush [GEOL] An accumulation of drifted snow, windblown soil, and wind-transported seeds on a lee slope, characteristically marked during the winter by a dark patch of soil. { 'snō,fləsh }

snow forest climate [CLIMATOL] A major category in W. Köppen's climatic classification, defined by a coldest-month mean temperature of less than 26.6°F (3°C) and a warmest-month mean temperature of greater than 50°F (10°C). { 'snō ¦fär·əst ,klī·mət }

snow gage [HYD] An instrument for measuring the amount of water equivalent in a snowpack. Also known as snow sampler. { 'snō ,gāj }

snow garland [HYD] A rare phenomenon in which snow is festooned from trees, fences, and so on, in the form of a rope of snow, several feet long and several inches in diameter; produced by surface tension acting in thin films of water bonding individual crystals; such garlands form only when the surface temperature is close to the melting point, for only then will the requisite films of slightly supercooled water exist. { 'snō ,gär·lənd }

snow geyser [METEOROL] Fine, powdery snow blown upward by a snow tremor. { 'snō ,gī·zər }

snow glide [HYD] The slow slip of a snowpack over the ground surface caused by the stress of its own weight. { 'snō ,glīd }

snow grains [METEOROL] Precipitation in the form of very small, white opaque particles of ice; the solid equivalent of drizzle; the grains resemble snow pellets in external appearance, but are more flattened and elongated, and generally have diameters of less than 1 millimeter; they neither shatter nor bounce when they hit a hard surface. Also known as granular snow. { 'snō ,grānz }

snow ice [HYD] Ice crust formed from snow, either by compaction or by the refreezing of partially thawed snow. { 'snō ,īs }

snow line [GEOGR] **1.** A transient line delineating a snow-covered area or altitude. **2.** An area with more than 50% snow cover. **3.** The altitude or geographic line separating areas in which snow melts in summer from areas having perennial ice and snow. { 'snō ,līn }

snow mantle *See* snow cover. { 'snō ˌmant·əl }
snow medano *See* snow barchan. { 'snō mə'dä·nō }
snowmelt [HYD] The water resulting from the melting of snow; it may evaporate, seep into the ground, or become a part of runoff. { 'snōˌmelt }
snow niche *See* nivation hollow. { 'snō ˌnich }
snowpack [HYD] The amount of annual accumulation of snow at higher elevations in the western United States, usually expressed in terms of average water equivalent. { 'snōˌpak }
snow patch erosion *See* nivation. { 'snō ¦pach iˌrō·zhən }
snow pellets [METEOROL] Precipitation consisting of white, opaque, approximately round (sometimes conical) ice particles which have a snowlike structure and are about 2 to 5 millimeters in diameter; snow pellets are crisp and easily crushed, differing in this respect from snow grains, and they rebound when they fall on a hard surface and often break up. Also known as graupel; soft hail; tapioca snow. { 'snō ˌpel·əts }
snow plume *See* snow banner. { 'snō ˌplüm }
snowquake *See* snow tremor. { 'snōˌkwāk }
snow ripple *See* wind ripple. { 'snō ˌrip·əl }
snow roller [HYD] A mass of snow, shaped somewhat like a lady's muff, rather common in mountainous or hilly regions; it occurs when snow, moist enough to be cohesive, is picked up by wind blowing down a slope and rolled onward and downward until either it becomes too large or the ground levels off too much for the wind to propel it further; snow rollers vary in size from very small cylinders to some as large as 4 feet (1.2 meters) long and 7 feet (2.1 meters) in circumference. { 'snō ˌrō·lər }
snow sampler *See* snow gage. { 'snō ˌsam·plər }
snowshed [HYD] A drainage basin primarily supplied by snowmelt. { 'snōˌshed }
snow sky *See* snow blink. { 'snō ˌskī }
snowslide *See* snow avalanche. { 'snōˌslīd }
snow sludge [OCEANOGR] Sludge formed mainly from snow. { 'snō ˌsləj }
snow slush [HYD] Slush formed from snow that has fallen into water that is at a temperature below that of the snow. Also known as snezhura. { 'snō ˌsləsh }
snow smoke *See* snow banner. { 'snō ˌsmōk }
snow stage [METEOROL] The thermodynamic process of sublimation of water vapor into snow in an idealized saturation-adiabatic or pseudoadiabatic expansion (lifting) of moist air; the snow stage begins at the condensation level when it is higher than the freezing level. { 'snō ˌstāj }
snowstorm [METEOROL] A storm in which snow falls. { 'snōˌstȯrm }
snow survey [HYD] The process of determining depth and water content of snow at representative points, for example, along a snow course. { 'snō ˌsərˌvā }
snow swamp *See* slush field. { 'snō ˌswämp }
snow tremor [HYD] A disturbance in a snowfield, caused by the simultaneous settling of a large area of thick snow crust or surface layer. Also known as snowquake. { 'snō ˌtrem·ər }
socked in [METEOROL] In the early days of aviation, pertaining to weather at an airport when ceiling or visibility were of such low values that the airport was effectively closed to aircraft operations. { 'säkt 'in }
soda lake [HYD] An alkali lake rich in dissolved sodium salts, especially sodium carbonate, sodium chloride, and sodium sulfate. Also known as natron lake. { 'sōd·ə ˌlāk }
soffione [GEOL] A jet of steam and other vapors issuing from the ground in a volcanic area. { ˌsä·fē'ō·nē }
soffosian knob *See* frost mound. { sə'fō·zhən 'näb }
soft coal *See* bituminous coal. { 'sȯft 'kōl }
soft hail *See* snow pellets. { 'sȯft 'hāl }
soft rime [HYD] A white, opaque coating of fine rime deposited chiefly on vertical surfaces, especially on points and edges of objects, generally in supercooled fog. { 'sȯft 'rīm }
Sohm Abyssal Plain [GEOL] A basin in the North Atlantic, about 2400 fathoms (4390

meters) deep, between Newfoundland and the Mid-Atlantic Ridge. { 'sōm ə'bis·əl 'plān }

soil [GEOL] **1.** Unconsolidated rock material over bedrock. **2.** Freely divided rock-derived material containing an admixture of organic matter and capable of supporting vegetation. { sȯil }

soil air [GEOL] The air and other gases in spaces in the soil; specifically, that which is found within the zone of aeration. Also known as soil atmosphere. { ¦sȯil ¦er }

soil atmosphere *See* soil air. { ¦sȯil ¦at·mə,sfir }

soil blister *See* frost mound. { 'sȯil ,blis·tər }

soil chemistry [GEOCHEM] The study and analysis of the inorganic and organic components and the life cycles within soils. { ¦sȯil ¦kem·ə·strē }

soil colloid [GEOL] Colloidal complex of soils composed principally of clay and humus. { ¦sȯil ¦kä,lȯid }

soil complex [GEOL] A mapping unit used in detailed soil surveys; consists of two or more recognized classifications. { ¦sȯil ¦käm,pleks }

soil creep [GEOL] The slow, steady downhill movement of soil and loose rock on a slope. Also known as surficial creep. { 'sȯil ,krēp }

soil element [GEOL] A unit that represents an arbitrarily small volume of soil within a soil mass. { 'sȯil ,el·ə·mənt }

soil erosion [GEOL] The detachment and movement of topsoil by the action of wind and flowing water. { 'sȯil i,rōzh·ən }

soil flow *See* solifluction. { 'sȯil ,flō }

soil fluction *See* solifluction. { 'sȯil ,flək·shən }

soil formation *See* soil genesis. { 'sȯil ,fȯr·mā·shən }

soil genesis [GEOL] The mode by which soil originates, with particular reference to processes of soil-forming factors responsible for the development of true soil from unconsolidated parent material. Also known as pedogenesis; soil formation. { 'sȯil ,jen·ə·səs }

soil moisture *See* soil water. { ¦sȯil ¦mȯis·chər }

soil physics [GEOPHYS] The study of the physical characteristics of soils; concerned also with the methods and instruments used to determine these characteristics. { ¦sȯil ¦fiz·iks }

soil profile [GEOL] A vertical section of a soil, showing horizons and parent material. { ¦sȯil ¦prō,fīl }

soil science [GEOL] The study of the formation, properties, and classification of soil; includes mapping. Also known as pedology. { 'sȯil ,sī·əns }

soil separate [GEOL] Any of a group of rock or mineral particles, separated from a soil sample, having diameters less than 0.8 inch (2 millimeters) and ranging within the limits of one of the standard classifications of soil particle size. Also known as separate. { 'sȯil ,sep·rət }

soil series [GEOL] A family of soils having similar profiles, and developing from similar original materials under the influence of similar climate and vegetation. { 'sȯil ,sir·ēz }

soil shear strength [GEOL] The maximum resistance of a soil to shearing stresses. { 'sȯil 'shir ,streŋkth }

soil stripes [GEOL] Alternating bands of fine and coarse material in a soil structure. { 'sȯil ,strīps }

soil structure [GEOL] Arrangement of soil into various aggregates, each differing in the characteristics of its particles. { 'sȯil ,strək·chər }

soil survey [GEOL] The systematic examination of soils, their description and classification, mapping of soil types, and the assessment of soils for various agricultural and engineering uses. { 'sȯil 'sər,vā }

soil water [HYD] Water in the belt of soil water. Also known as rhizic water; soil moisture. { 'sȯil ,wȯd·ər }

soil-water belt *See* belt of soil water. { 'sȯil ¦wȯd·ər ,belt }

soil-water zone *See* belt of soil water. { 'sȯil ¦wȯd·ər ,zōn }

solaire [METEOROL] A name generally applied to winds from an easterly direction (that is, from the rising sun) in central and southern France. { sō'ler }

sol-air temperature [METEOROL] The temperature which, under conditions of no direct solar radiation and no air motion, would cause the same heat transfer into a house as that caused by the interplay of all existing atmospheric conditions. { 'säl'er ˌtem·prə·chər }

solano [METEOROL] A southeasterly or easterly wind on the southeast coast of Spain in summer; usually an extension of the sirocco; it is hot and humid and sometimes brings rain; when dry, it is dusty. { sō'lä·nō }

solar absorption index [GEOPHYS] A relation of the sun's angle at various latitudes and local times with the ionospheric absorption. { 'sō·lər əp'sȯrp·shən ˌinˌdeks }

solar air mass [METEOROL] The optical air mass penetrated by light from the sun for any given position of the sun. { 'sō·lər 'er ˌmas }

solar atmospheric tide [GEOPHYS] An atmospheric tide due to the thermal or gravitational action of the sun. { 'sō·lər 'at·məˌsfir·ik 'tīd }

solar climate [CLIMATOL] The hypothetical climate which would prevail on a uniform solid earth with no atmosphere; thus, it is a climate of temperature alone and is determined only by the amount of solar radiation received. { 'sō·lər 'klī·mət }

solar constant [METEOROL] The rate at which energy from the sun is received just outside the earth's atmosphere on a surface normal to the incident radiation and at the earth's mean distance from the sun; it is approximately 1367 watts per square meter. { 'sō·lər 'kän·stənt }

solar evaporation [HYD] The evaporation of water due to the sun's heat. { 'sō·lər iˌvap·ə'rā·shən }

solar-radiation observation [GEOPHYS] An evaluation of the radiation from the sun that reaches an observation point; the observing instrument is usually a pyrheliometer or pyranometer. { 'sō·lər ˌrād·ē'ā·shən ˌäb·zərˌvā·shən }

solar-terrestrial phenomena [GEOPHYS] All observed physical effects that are caused by solar activity; the phenomena may be in the atmosphere or on the earth's surface; an example is the aurora borealis. { 'sō·lər tə'res·trē·əl fə'näm·əˌnä }

solar tide [OCEANOGR] The tide caused solely by the tide-producing forces of the sun. { 'sō·lər 'tīd }

solar-topographic theory [CLIMATOL] The theory that the changes of climate through geologic time (the paleoclimates) have been due to changes of land and sea distribution and orography, combined with fluctuations of solar radiation of the order of 10–20% on either side of the mean. { 'sō·lər ¦täp·ə¦graf·ik 'thē·ə·rē }

solar wind [GEOPHYS] The supersonic flow of gas, composed of ionized hydrogen and helium, which continuously flows from the sun out through the solar system with velocities of 180 to 600 miles (300 to 1000 kilometers) per second; it carries magnetic fields from the sun. { 'sō·lər 'wind }

sole [GEOGR] The lowest part of a valley. [GEOL] **1.** The bottom of a sedimentary stratum. **2.** The middle and lower portion of the shear surface of a landslide. **3.** The underlying fault plane of a thrust nappe. Also known as sole plane. [HYD] The basal ice of a glacier, often dirty in appearance due to contained rock fragments. { sōl }

sole injection [GEOL] An igneous intrusion that was put in place along a thrust plane. { 'sōl inˌjek·shən }

sole mark [GEOL] An irregularity or penetration on the undersurface of a sedimentary stratum. { 'sōl ˌmärk }

solenoid *See* meteorological solenoid. { 'säl·əˌnȯid }

solenoidal index [METEOROL] The difference between the mean virtual temperature from the surface to some specified upper level averaged around the earth at 55° latitude, and the mean virtual temperature for the corresponding layer averaged at 35° latitude. { ¦säl·ə¦nȯid·əl 'inˌdeks }

sole plane *See* sole. { 'sōl ˌplān }

solfatara [GEOL] A fumarole from which sulfurous gases are emitted. { ˌsäl·fə'tär·ə }

solifluction [GEOL] A rapid soil creep, especially referring to downslope soil movement in periglacial areas. Also known as sludging; soil flow; soil fluction. { ¦säl·ə'flək·shən }

solifluction lobe [GEOL] An isolated, tongue-shaped feature of the land surface with a steep front and a smooth upper surface formed by more rapid solifluction on certain sections of the slope. Also known as solifluction tongue. { ¦säl·ə'flək·shən ¦lōb }

solifluction mantle [GEOL] The locally derived, unsorted material moved downslope by solifluction. Also known as flow earth. { ¦säl·ə'flək·shən ¦mant·əl }

solifluction sheet [GEOL] A broad deposit of a solifluction mantle. { ¦säl·ə'flək·shən ¦shēt }

solifluction stream [GEOL] A narrow, streamlike deposit of a solifluction mantle. { ¦säl·ə'flək·shən ¦strēm }

solifluction tongue *See* solifluction lobe. { ¦säl·ə'flək·shən ¦təŋ }

solodize [GEOL] To improve a soil by removing alkalies from it. { 'sō·lə,dīz }

Solod soil *See* Soloth soil. { 'sō·ləd ,sȯil }

Solonchak soil [GEOL] One of an intrazonal, balamorphic group of light-colored soils rich in soluble salts. { ¦säl·ən¦chäk ,sȯil }

Solonetz soil [GEOL] One of an intrazonal group of black alkali soils having a columnar structure. { ¦säl·ə¦nets ,sȯil }

solore [METEOROL] A cold, night wind of the mountains following the course of the Drome River in southeastern France. { sə'lȯr }

Soloth soil [GEOL] One of an intrazonal halomorphic group of soils formed from saline material; the surface layer is soft and friable, and overlies a light-colored leached horizon which, in turn, overlies a dark horizon. Also known as Solod soil. { 'sō·lət ,sȯil }

solstitial tidal currents [OCEANOGR] Tidal currents of especially large tropic diurnal inequality occurring at the time of solstitial tides. { sälz'tish·əl 'tīd·əl ,kə·rəns }

solstitial tides [OCEANOGR] Tides occurring near the times of the solstices, when the tropic range is especially large. { sälz'tish·əl 'tīdz }

solum [GEOL] The upper part of a soil profile, composed of A and B horizons in mature soil. Also known as true soil. { 'sō·ləm }

solution groove [GEOL] One of a series of continuous, subparallel furrows developed on an inclined or vertical surface of a soluble and homogeneous rock (such as the limestone walls of a cave) by the slow corroding action of trickling water. { sə'lü·shən ,grüv }

solution pool [GEOL] A pool in a rock that is formed by the dissolution of the rock in ocean water. { sə'lü·shən ,pül }

solution potholes [GEOL] Potholes produced in carbonate rocks by dissolution. { sə'lü·shən ,pät,hōlz }

solution transfer [GEOL] A process whereby pressure solution of detrital mineral grains at contact areas is followed by recrystallization on the less strained parts of the grain surfaces. { sə'lü·shən ,tranz·fər }

Somali Current *See* East Africa Coast Current. { sə'mäl·ē 'kə·rənt }

somma [GEOL] The rim of a volcano. { 'säm·ə }

sonora [METEOROL] A summer thunderstorm in the mountains and deserts of southern California and Baja California. { sə'nȯr·ə }

sordawalite *See* tachylite. { sȯr'dä·wə,līt }

sorotiite [GEOL] A type of meteorite similar to the pallasites, with troilite substituting for olivine. { sə'räd·ē,īt }

sorted [GEOL] **1.** Pertaining to a nongenetic group of patterned-ground features displaying a border of stones, including boulders, commonly alternating with very small particles, including silt, sand, and clay. **2.** Pertaining to an unconsolidated sediment or a cemented detrital rock consisting of particles of essentially uniform size or of particles lying within the limits of a single grade. { 'sȯrd·əd }

sorted polygon [GEOL] A patterned ground having a sorted appearance due to a border of stones and characterized by a polygonal mesh. Also known as stone polygon. { 'sȯrd·əd 'päl·i,gän }

sorting [GEOL] The process by which similar in size, shape, or specific gravity sedimentary particles are selected and separated from associated but dissimilar particles by the agent of transportation. { 'sȯrd·iŋ }

sorting coefficient [GEOL] A sorting index equal to the square root of the ratio of the larger quartile (the diameter having 25% of the cumulative size-frequency distribution larger than itself) to the smaller quartile (the diameter having 75% of the cumulative size-frequency distribution larger than itself). { 'sȯrd·iŋ ˌkō·iˌfish·ənt }

sorting index [GEOL] A measure of the degree of sorting in a sediment based on the statistical spread of the frequency curve of particle sizes. { 'sȯrd·iŋ ˌinˌdeks }

sou'easter *See* southeaster. { 'saủˌēs·tər }

sounding *See* upper-air observation. { 'saủnd·iŋ }

sounding sand [GEOL] Sand that emits musical, humming, or crunching sounds when disturbed. Also known as singing sand. { 'saủnd·iŋ ˌsand }

source area *See* provenance. { 'sȯrs ˌer·ē·ə }

source bed [GEOL] The original stratigraphic horizon from which secondary sulfide minerals were derived. { 'sȯrs ˌbed }

sourceland *See* provenance. { 'sȯrsˌland }

source region [METEOROL] An extensive area of the earth's surface characterized by essentially uniform surface conditions and so situated with respect to the general atmospheric circulation that an air mass may remain over it long enough to acquire its characteristic properties. { 'sȯrs ˌrē·jən }

source rock [GEOL] **1.** Rock from which fragments have been derived which form a later, usually sedimentary rock. Also known as mother rock; parent rock. **2.** Sedimentary rock, usually shale and limestone, deposited together with organic matter which was subsequently transformed to liquid or gaseous hydrocarbons. { 'sȯrs ˌräk }

south [GEOD] The direction 180° from north. { saủth }

South America [GEOGR] The southernmost of the Western Hemisphere continents, three-fourths of which lies within the tropics. { 'saủth ə'mer·ə·kə }

South Atlantic Current [OCEANOGR] An eastward-flowing current of the South Atlantic Ocean that is continuous with the northern edge of the West Wind Drift. { 'saủth at'lan·tik 'kə·rənt }

Southeast Drift Current [OCEANOGR] A North Atlantic Ocean current flowing southeastward and southward from a point west of the Bay of Biscay toward southwestern Europe and the Canary Islands, where it continues as the Canary Current. { saủ'thēst 'drift 'kə·rənt }

southeaster [METEOROL] A southeasterly wind, particularly a strong wind or gale; for example, the winter southeast storms of the Bay of San Francisco. Also spelled sou'easter. { saủ'thēs·tər }

South Equatorial Current [OCEANOGR] Any of several ocean currents, flowing westward, driven by the southeast trade winds blowing over the tropical oceans of the Southern Hemisphere and extending slightly north of the equator. Also known as Equatorial Current. { 'saủth ˌek·wə'tȯr·ē·əl 'kə·rənt }

souther [METEOROL] A south wind, especially a strong wind or gale. { 'saủth·ər }

southerly burster [METEOROL] A cold wind from the south in Australia. { 'səth·ər·lē 'bər·stər }

southern lights *See* aurora australis. { 'səth·ərn 'līts }

Southern Polar Front *See* Antarctic Convergence. { 'səth·ərn 'pō·lər 'frənt }

south foehn [METEOROL] A foehn condition sustained by a strong south-to-north airflow across a transverse mountain barrier; the south foehn of the Alps may well be the most striking foehn in the world. { 'saủth 'fān }

south frigid zone [GEOGR] That part of the earth south of the Antarctic Circle. { 'saủth 'frij·əd ˌzōn }

south geographical pole [GEOGR] The geographical pole in the Southern Hemisphere, at latitude 90°S. Also known as South Pole. { 'saủth ¦jē·ə¦graf·ə·kəl ˌpōl }

south geomagnetic pole [GEOPHYS] The geomagnetic pole in the Southern Hemisphere at approximately 78.5°S, longitude 111°E, 180° from the north geomagnetic pole. Also known as south pole. { 'sau̇th ¦jē·ō·mag'ned·ik ˌpōl }

South Indian Current [OCEANOGR] An eastward-flowing current of the southern Indian Ocean that is continuous with the northern edge of the West Wind Drift. { 'sau̇th 'in·dē·ən 'kə·rənt }

South Pacific Current [OCEANOGR] An eastward-flowing current of the South Pacific Ocean that is continuous with the northern edge of the West Wind Drift. { 'sau̇th pə'sif·ik 'kə·rənt }

south pole *See* south geomagnetic pole. { 'sau̇th 'pōl }

South Pole *See* south geographical pole. { 'sau̇th 'pōl }

south temperate zone [GEOGR] That part of the earth between the Tropic of Capricorn and the Antarctic Circle. { 'sau̇th 'tem·prət ˌzōn }

southwester [METEOROL] A southwest wind, particularly a strong wind or gale. Also spelled sou'wester. { sau̇th'wes·tər }

sou'wester *See* southwester. { sau̇'wes·tər }

space charge [GEOPHYS] In atmospheric electricity, the preponderance of either negative or positive ions within any given portion of the atmosphere. { 'spās ˌchärj }

space weather [GEOPHYS] The conditions on the sun and in the solar wind, magnetosphere, ionosphere, and thermosphere that can influence the performance and reliability of space-borne and ground-based technological systems and endanger human life or health. { 'spās ˌweth·ər }

spall [GEOL] **1.** A fragment removed from the surface of a rock by weathering. **2.** A relatively thin, sharp-edged fragment produced by exfoliation. **3.** A rock fragment produced by chipping with a hammer. { spȯl }

spalling [GEOL] The chipping or fracturing with an upward heaving, of rock caused by a compressional wave at a free surface. { 'spȯl·iŋ }

sparagmite [GEOL] Late Precambrian fragmental rocks of Scandinavia, characterized by high proportions of microcline. { spə'rag,mīt }

Sparnacean [GEOL] A European stage of geologic time; upper upper Paleocene, above Thanetian, below Ypresian of Eocene. { spär'nāsh·ən }

sparry cement [GEOL] Clear, relatively coarse-grained calcite in the interstices of any sedimentary rock. { 'spär·ē si'ment }

spasmodic turbidity current [GEOPHYS] A single, rapidly developed turbidity current. { spaz'mäd·ik tər'bid·əd·ē ˌkə·rənt }

spatial autocorrelation [GEOGR] In mathematical geography, the degree of interdependence among data arranged on a three-dimensional grid. { 'spā·shəl ¦ȯd·ō,kä·rə'lā·shən }

spatial dendrite [METEOROL] A complex ice crystal with fernlike arms that extend in many directions (spatially) from a central nucleus; its form is roughly spherical. Also known as spatial dendritic crystal. { 'spā·shəl 'den,drīt }

spatial dendritic crystal *See* spatial dendrite. { 'spā·shəl den'drid·ik 'krist·əl }

spatter cone [GEOL] A low, steep-sided cone of small pyroclastic fragments built up on a fissure or vent. Also known as agglutinate cone; volcanello. { 'spad·ər ˌkōn }

spatter rampart [GEOL] A low, circular ridge of pyroclastics built up around the margins of small volcanoes. { 'spad·ər ˌram,pärt }

special observation [METEOROL] A category of aviation weather observation taken to report significant changes in one or more of the observed elements since the last previous record observation. { 'spesh·əl ˌäb·zər'vā·shən }

special weather report [METEOROL] The encoded and transmitted weather report of a special observation. { 'spesh·əl 'weth·ər ri,pȯrt }

species number [OCEANOGR] The first number in the argument number in a Doodson tide schedule; indicates approximately the period of a component of tidal potential. { 'spē·shēz ˌnəm·bər }

specific energy [HYD] The energy at any cross section of an open channel, measured above the channel bottom as datum; numerically the specific energy is the sum of the water depth plus the velocity head, $v^2/2g$, where v is the velocity of flow and g the acceleration of gravity. { spə'sif·ik 'en·ər·jē }

specific humidity [METEOROL] In a system of moist air, the (dimensionless) ratio of the mass of water vapor to the total mass of the system. { spə'sif·ik hyü'mid·əd·ē }

specific retention [GEOL] The ratio of the volume of water that a given body of rock or soil will retain after saturation, and the pull of gravity to the volume of the body itself. { spə¦sif·ik ri'ten·chən }

specific-volume anomaly [OCEANOGR] The excess of the actual specific volume of the sea water at any point in the ocean over the specific volume of sea water of salinity 35 parts per thousand (+) and temperature 0°C at the same pressure. Also known as steric anomaly. { spə'sif·ik ¦väl·yəm ə'näm·ə·lē }

specific yield [HYD] The quantity of water which a unit volume of aquifer, after being saturated, will yield by gravity; it is expressed either as a ratio or as a percentage of the volume of the aquifer; specific yield is a measure of the water available to wells. { spə'sif·ik 'yēld }

spelean [GEOL] Of or pertaining to a feature in a cave. { spə'lē·ən }

speleology [GEOL] The study and exploration of caves. { ˌspē·lē'äl·ə·jē }

speleothem [GEOL] A secondary mineral deposited in a cave by the action of water. Also known as cave formation. { 'spē·lē·əˌthem }

spending beach [GEOL] In a wave basin, the beach on which the entering waves spend themselves, except for the small remainder entering the inner harbor. { 'spend·iŋ ˌbēch }

spergenite [GEOL] A biocalcarenite containing ooliths and fossil debris and having a maximum quartz content of 10%. Also known as Bedford limestone; Indiana limestone. { 'spər·jəˌnīt }

sphenochasm [GEOL] A triangular gap of oceanic crust separating two continental blocks and converging to a point. { ˌsfē·nə'kaz·əm }

sphenolith [GEOL] A wedgelike igneous intrusion that is partly concordant and partly discordant. { 'sfēn·əlˌith }

spherical weathering *See* spheroidal weathering. { 'sfir·ə·kəl 'weth·ə·riŋ }

spheroidal recovery [GEOPHYS] The hypothetical return of the earth to spheroid form after it has been distorted. { sfir'ȯid·əl ri'kəv·ə·rē }

spheroidal weathering [GEOL] Chemical weathering in which concentric or spherical shells of decayed rock are successively separated from a block of rock; commonly results in the formation of a rounded boulder of decomposition. Also known as concentric weathering; spherical weathering. { sfir'ȯid·əl 'weth·ə·riŋ }

spherulite [GEOL] A spherical body or coarsely crystalline aggregate having a radial internal structure arranged about one or more centers. { 'sfir·əˌlīt }

spilling [OCEANOGR] The process by which steep waves break on approaching the shore; white water appears on the crest and the wave top gradually rolls over, without a crash. { 'spil·iŋ }

spilling breaker *See* plunging breaker. { 'spil·iŋ ˌbrāk·ər }

spillover [METEOROL] That part of orographic precipitation which is carried along by the wind so that it reaches the ground in the nominal rain shadow on the lee side of the barrier. { 'spilˌō·vər }

spill stream *See* overflow stream. { 'spil ˌstrēm }

spiral band [METEOROL] Spiral-shaped radar echoes received from precipitation areas within intense tropical cyclones (hurricanes or typhoons); they curve cyclonically in toward the center of the storm and appear to merge to form the wall around the eye of the storm. Also known as hurricane band; hurricane radar band. { 'spī·rəl 'band }

spiral layer *See* Ekman layer. { 'spī·rəl ¦lā·ər }

spit [GEOGR] A small point of land commonly consisting of sand or gravel and which terminates in open water. { spit }

Spitsbergen Current [OCEANOGR] An ocean current flowing northward and westward from a point south of Spitsbergen, and gradually merging with the East Greenland

Current in the Greenland Sea; the Spitsbergen Current is the continuation of the northwestern branch of the Norwegian Current. { 'spits,bər·gən 'kə·rənt }

splash erosion [GEOL] Erosion resulting from the impact of falling raindrops. { 'splash i,rōzh·ən }

splent coal *See* splint coal. { 'splent ¦kōl }

spliced [GEOL] Relating to veins that pinch out and are overlapped at that point by another parallel vein. { splīst }

splint *See* splint coal. { splint }

splint coal [GEOL] A hard, dull, blocky, grayish-black, banded bituminous coal characterized by an uneven fracture and a granular texture; burns with intense heat. Also known as splent coal; splint. { 'splint ,kōl }

split [GEOL] A coal seam that cannot be mined as a single unit because it is separated by a parting of other sedimentary rock. Also known as coal split; split coal. { split }

spodic horizon [GEOL] A soil horizon characterized by illuviation of amorphous substances. { 'späd·ik hə'rīz·ən }

Spodosol [GEOL] A soil order characterized by accumulations of amorphous materials in subsurface horizons. { 'späd·ə,sól }

spongework [GEOL] A pattern of small irregular interconnecting cavities on walls of limestone caves. { 'spənj,wərk }

spongolite [GEOL] A rock or sediment composed chiefly of the remains of sponges. Also known as spongolith. { 'späŋ·gə,līt }

spongolith *See* spongolite. { 'spaŋ·gə,lith }

spontaneous nucleation [METEOROL] The nucleation of a phase change of a substance without the benefit of any seeding nuclei within or otherwise in contact with that substance; examples of such systems are a pure vapor condensing to its pure liquid state, a pure liquid freezing to its pure solid state, and a pure solution crystallizing to yield pure solute crystals. { spän'tā·nē·əs ,nü·klē'ā·shən }

sporadic E layer [GEOPHYS] A layer of intense ionization that occurs sporadically within the E layer; it is variable in time of occurrence, height, geographical distribution, penetration frequency, and ionization density. { spə'rad·ik 'ē ,lā·ər }

sporinite [GEOL] A variety of exinite composed of spore exines which have been compressed parallel to the stratification. { 'spór·ə,nīt }

spot wind [METEOROL] In air navigation, wind direction and speed, either observed or forecast if so specified, at a designated altitude over a fixed location. { 'spät ,wind }

spouting horn [GEOL] A sea cave with a rearward or upward opening through which water spurts or sprays after waves enter the cave. Also known as chimney; oven. { 'spaúd·iŋ ¦hórn }

spray region *See* fringe region. { 'sprā ,rē·jən }

spreading concept *See* sea-floor spreading. { 'spred·iŋ ,kän,sept }

spreading-floor hypothesis *See* sea-floor spreading. { 'spred·iŋ ¦flór hī,päth·ə·səs }

spring [HYD] A general name for any discharge of deep-seated, hot or cold, pure or mineralized water. { spriŋ }

spring crust [HYD] A type of snow crust, formed when loose firn is recemented by a decrease in temperature; it is most common in late winter and spring. { 'spriŋ 'krəst }

spring high water *See* mean high-water springs. { 'spriŋ 'hī ,wód·ər }

spring low water *See* mean low-water springs. { 'spriŋ 'lō ,wód·ər }

spring range [OCEANOGR] The mean semidiurnal range of tide when spring tides are occurring; the mean difference in height between spring high water and spring low water. Also known as mean spring range. { 'spriŋ 'rānj }

spring rise [OCEANOGR] The height of mean high-water springs above the chart datum. { 'spriŋ 'rīz }

spring seepage [HYD] A spring of small discharge. Also known as weeping spring. { 'spriŋ ,sēp·ij }

spring sludge *See* rotten ice. { 'spriŋ 'sləj }

spring snow [HYD] A coarse, granular snow formed during spring by alternate freezing and thawing. Also known as corn snow. { 'spriŋ 'snō }

spring tidal currents [OCEANOGR] Tidal currents of increased speed occurring at the time of spring tides. { 'spriŋ 'tīd·əl ˌkə·rəns }
spring tide [OCEANOGR] Tide of increased range which occurs about every 2 weeks when the moon is new or full. { 'spriŋ 'tīd }
spring velocity [OCEANOGR] The average speed of the maximum flood and maximum ebb of a tidal current at the time of spring tides. { 'spriŋ və'läs·əd·ē }
sprinkle [METEOROL] A very light shower of rain. { 'spriŋ·kəl }
sprite [METEOROL] A transient illumination that can appear over a laterally extensive thunderstorm, with a red body about 20 kilometers (12 miles) in diameter, extending up to an altitude of 85–90 kilometers (51–54 miles), and blue tendrils extending down to an altitude of about 45 kilometers (27 miles). { sprīt }
spur [GEOL] A ridge or rise projecting from a larger elevational feature. [HYD] *See* ram. { spər }
squall [METEOROL] A strong wind with sudden onset and more gradual decline, lasting for several minutes; in the United States observational practice, a squall is reported only if a wind speed of 16 knots (8.23 meters per second) or higher is sustained for at least 2 minutes. { skwȯl }
squall cloud [METEOROL] A small eddy cloud sometimes formed below the leading edge of a thunderstorm cloud, between the upward and downward currents. { 'skwȯl ˌklau̇d }
squall line [METEOROL] A line of thunderstorms near whose advancing edge squalls occur along an extensive front; the region of thunderstorms is typically 12 to 30 miles (20 to 50 kilometers) wide and a few hundred to 1200 miles (2000 kilometers) long. { 'skwȯl ˌlīn }
stability [GEOL] **1.** The resistance of a structure, spoil heap, or clay bank to sliding, overturning, or collapsing. **2.** Chemical durability, resistance to weathering. { stə'bil·əd·ē }
stability chart [METEOROL] A synoptic chart that shows the distribution of a stability index. { stə'bil·əd·ē ˌchärt }
stability index [METEOROL] An indication of the local static stability of a layer of air. { stə'bil·əd·ē ˌinˌdeks }
stack [GEOL] An erosional, coastal landform that is a steep-sided, pillarlike rocky island or mass that has been detached by wave action from a shore made up of cliffs; applies particularly to a stack that is columnar in structure and has horizontal stratifications. Also known as marine stack; rank. { stak }
stade [GEOL] A substage of a glacial stage marked by a secondary advance of glaciers. { stād }
stadial moraine *See* recessional moraine. { 'stād·ē·əl mə'rān }
Staffordian [GEOL] A European stage of geologic time forming the middle Upper Carboniferous, above Yorkian and below Radstockian, equivalent to part of the upper Westphalian. { sta'fȯrd·ē·ən }
stage [GEOL] **1.** A developmental phase of an erosion cycle in which landscape features have distinctive characteristic forms. **2.** A phase in the historical development of a geologic feature. **3.** A major subdivision of a glacial epoch. **4.** A time-stratigraphic unit ranking below series and above chronozone, composed of rocks formed during an age of geologic time. [HYD] The elevation of the water surface in a stream as measured by a river gage with reference to some arbitrarily selected zero datum. Also knonw as stream stage. { stāj }
stagnant glacier [HYD] A glacier which has ceased to move. { 'stag·nənt 'glā·shər }
stagnant water [HYD] Motionless water, not flowing in a stream or current. Also known as standing water. { 'stag·nənt 'wȯd·ər }
stagnation [HYD] **1.** The condition of a body of water unstirred by a current or wave. **2.** The condition of a glacier that has stopped flowing. { stag'nā·shən }
stagnum [HYD] A pool of water with no outlet. { 'stag·nəm }
stalactite [GEOL] A conical or roughly cylindrical speleothem formed by dripping water and hanging from the roof of a cave; usually composed of calcium carbonate. { stə'lakˌtīt }

stalacto-stalagmite [GEOL] A columnar deposit formed by the union of a stalactite with its complementary stalagmite. Also known as column; pillar. { stə¦lak·tō stə'lag,mīt }

stalagmite [GEOL] A conical speleothem formed upward from the floor of a cave by the action of dripping water; usually composed of calcium carbonate. { stə'lag,mīt }

Stampian *See* Rupelian. { 'stam·pē·ən }

stamukha [OCEANOGR] An individual piece of stranded ice. { ,sta,mü·kə }

stand [OCEANOGR] The interval at high or low water when there is no appreciable change in the height of the tide. Also known as tidal stand. { stand }

standard artillery atmosphere [METEOROL] A set of values describing atmospheric conditions on which ballistic computations are based, namely: no wind, a surface temperature of 15°C, a surface pressure of 1000 millibars, a surface relative humidity of 78%, and a lapse rate which yields a prescribed density-altitude relation. { 'stan·dərd är'til·ə·rē 'at·mə,sfir }

standard artillery zone [METEOROL] A vertical subdivision of the standard artillery atmosphere; it may be considered a layer of air of prescribed thickness and altitude. { 'stan·dərd är'til·ə·rē ,zōn }

standard atmosphere [METEOROL] A hypothetical vertical distribution of atmospheric temperature, pressure, and density which is taken to be representative of the atmosphere for purposes of pressure altimeter calibrations, aircraft performance calculations, aircraft and missile design, and ballistic tables; the air is assumed to obey the perfect gas law and hydrostatic equation, which, taken together, relate temperature, pressure, and density variations in the vertical; it is further assumed that the air contains no water vapor, and that the acceleration of gravity does not change with height. { 'stan·dərd 'at·mə,sfir }

standard mean ocean water [GEOL] An international reference standard used to determine oxygen and hydrogen isotopic content. Abbreviated SMOW. { ¦stan·dərd ,mēn 'ō·shən ,wȯd·ər }

standard meridian [GEOD] The meridian used for reckoning standard time; throughout most of the world the standard meridians are those whose longitudes are exactly divisible by 15°. { 'stan·dərd mə'rid·ē·ən }

standard parallel [GEOD] The parallel or parallels of latitude used as control lines in the computation of a map projection. { 'stan·dərd 'par·ə,lel }

standard pressure [METEOROL] The arbitrarily selected atmospheric pressure of 1000 millibars to which adiabatic processes are referred for definitions of potential temperature, equivalent potential temperature, and so on. { 'stan·dərd 'presh·ər }

standard project flood [HYD] The volume of streamflow expected to result from the most severe combination of meteorological and hydrologic conditions which are reasonably characteristic of the geographic region involved, excluding extremely rare combinations. { 'stan·dərd 'präj,ekt ,fləd }

standard seawater *See* normal water. { 'stan·dərd 'sē,wȯd·ər }

standard station *See* reference station. { 'stan·dərd 'stā·shən }

standard visibility *See* meteorological range. { 'stan·dərd ,viz·ə'bil·əd·ē }

standard visual range *See* meteorological range. { 'stan·dərd 'vizh·ə·wəl 'rānj }

standing cloud [METEOROL] Any stationary cloud maintaining its position with respect to a mountain peak or ridge. { 'stand·iŋ 'klaủd }

standing water *See* stagnant water. { ¦stand·iŋ 'wȯd·ər }

starved basin [GEOL] A sedimentary basin in which rate of subsidence exceeds rate of sedimentation. { 'stärvd 'bās·ən }

state of the sea [OCEANOGR] A description of the properties of the wind-generated waves on the surface of the sea. { 'stāt əv thə 'sē }

state of the sky [METEOROL] The aspect of the sky in reference to the cloud cover; the state of the sky is fully described when the amounts, kinds, directions of movement, and heights of all clouds are given. { 'stāt əv thə 'skī }

static level [HYD] The height to which water will rise in an artesian well; the static level of a flowing well is above the ground surface. { 'stad·ik 'lev·əl }

static metamorphism [GEOL] Regional metamorphism caused by heat and solvents at high lithostatic pressures. Also known as load metamorphism. { 'stad·ik ˌmed·ə'mȯrˌfiz·əm }

static oceanography [OCEANOGR] Branch of oceanography that deals with the physical and chemical nature of water in the ocean and with the shape and composition of the ocean bottom. { 'stad·ik ˌō·shə'näg·rə·fē }

static stability [METEOROL] The stability of an atmosphere in hydrostatic equilibrium with respect to vertical displacements, usually considered by the parcel method. Also known as convectional stability; convection stability; hydrostatic stability; vertical stability. { 'stad·ik stə'bil·əd·ē }

stationary front *See* quasi-stationary front. { 'stā·shəˌner·ē 'frənt }

station continuity chart [METEOROL] A chart or graph on which time is one coordinate, and one or more of the observed meteorological elements at that station is the other coordinate. { 'stā·shən ˌkan·tə'nü·əd·ē ˌchärt }

station elevation [METEOROL] The vertical distance above mean sea level that is adopted as the reference datum level for all current measurements of atmospheric pressure at the station. { 'stā·shən ˌel·əˌvā·shən }

station model [METEOROL] A specified pattern for entering, on a weather map, the meteorological symbols that represent the state of the weather at a particular observation station. { 'stā·shən ˌmäd·əl }

station pressure [METEOROL] The atmospheric pressure computed for the level of the station elevation. { 'stā·shən ˌpresh·ər }

statistical forecast [METEOROL] A weather forecast based upon a systematic statistical examination of the past behavior of the atmosphere, as distinguished from a forecast based upon thermodynamic and hydrodynamic considerations. { stə'tis·tə·kəl 'fȯrˌkast }

steadiness *See* persistence. { 'sted·ē·nəs }

steam fog [METEOROL] Fog formed when water vapor is added to air which is much colder than the vapor's source; most commonly, when very cold air drifts across relatively warm water. Also known as frost smoke; sea mist; sea smoke; steam mist; water smoke. { 'stēm ˌfäg }

steam mist *See* steam fog. { 'stēm ˌmist }

steatization [GEOL] Introduction of or replacement by talc or steatite. { stēˌad·ə'zā·shən }

steering [METEOROL] Loosely used for any influence upon the direction of movement of an atmospheric disturbance exerted by another aspect of the state of the atmosphere; for example, a surface pressure system tends to be steered by isotherms, contour lines, or streamlines aloft, or by warm-sector isobars or the orientation of a warm front. { 'stir·iŋ }

steering level [METEOROL] A hypothetical level, in the atmosphere, where the velocity of the basic flow bears a direct relationship to the velocity of movement of an atmospheric disturbance embedded in the flow. { 'stir·iŋ ˌlev·əl }

Stefan's formula [OCEANOGR] A formula for the growth of thickness h of an ice cover on the ocean at various freezing temperatures, expressed as

$$h \approx \sqrt{\left(\frac{2}{\lambda_i \rho_i}\right) \psi}$$

where l is the coefficient of thermal conductivity, λ_i is the latent heat of fusion, ρ_i is the density of ice, and ψ is the cold sum (in degree days below 0°C). { 'shteˌfänz ˌfȯr·myə·lə }

steinkern [GEOL] **1.** Rock material formed from consolidated mud or sediment that filled a hollow organic structure, such as a fossil shell. **2.** The fossil formed after dissolution of the mold. Also known as endocast; internal cast. { 'shtīnˌkərn }

stellar lightning [GEOPHYS] Lightning consisting of several flashes seeming to radiate from a single point. { 'stel·ər 'līt·niŋ }

step [GEOL] A hitch or dislocation of the strata. { step }

step fault [GEOL] One of a set of closely spaced, parallel faults. Also known as distributive fault; multiple fault. { 'step ˌfȯlt }

Stephanian [GEOL] A European stage of Upper Carboniferous geologic time, forming the Upper Pennsylvanian, above the Westphalian and below the Sakmarian of the Permian. { stə'fān·ē·ən }

step lake *See* paternoster lake. { 'step ˌlāk }

step-out time [GEOPHYS] In seismic prospecting, the time differentials in arrivals of a given peak or trough of a reflected or refracted event for successive detector positions on the earth's surface. { 'step ¦au̇t ˌtīm }

steppe [GEOGR] An extensive grassland in the semiarid climates of southeastern Europe and Asia; it is similar to but more arid than the prairie of the United States. { step }

steppe climate [CLIMATOL] The type of climate in which precipitation though very slight, is sufficient for growth of short, sparse grass; typical of the steppe regions of south-central Eurasia. Also known as semiarid climate. { 'step ˌklī·mət }

stepped leader [GEOPHYS] The initial streamer of a lightning discharge; an intermittently advancing column of high ion density which established the channel for subsequent return streamers and dart leaders. { 'stept 'lēd·ər }

steptoe [GEOL] An isolated protrusion of bedrock, such as the summit of a hill or mountain, in a lava flow. { 'stepˌtō }

steric anomaly *See* specific-volume anomaly. { 'ster·ik ə'näm·ə·lē }

stewartite [GEOL] A steel-gray, iron-containing variety of bort that has magnetic properties. { 'stü·ərˌtīt }

stillstand [GEOL] A period during which a land area, a continent, or an island remains stationary with respect to the interior of the earth or to sea level. { 'stilˌstand }

still water [HYD] A portion of a stream having a very slight gradient and no visible current. { 'stil 'wȯd·ər }

still-water level [OCEANOGR] The level that the sea surface would assume in the absence of wind waves. { 'stil ¦wȯd·ər ˌlev·əl }

still well [METEOROL] A device, used in evaporation pan measurements, which provides an undisturbed water surface and support for the hook gage; the U.S. Weather Bureau model consists of a brass cylinder, 8 inches (20.32 centimeters) high and 3.5 inches (8.89 centimeters) in diameter, mounted over a hole in a triangular galvanized iron base which is provided with leveling screws. { 'stil 'wel }

stinkstone [GEOL] A stone containing decomposing organic matter that gives off an offensive odor when rubbed or struck. { 'stiŋkˌstōn }

stock *See* pipe. { stäk }

stockwork [GEOL] A mineral deposit in the form of a network of veinlets diffused in the country rock. { 'stäkˌwərk }

stone [GEOL] **1.** A small fragment of rock or mineral. **2.** *See* stony meteorite. { stōn }

stone bubble *See* lithophysa. { 'stōn ¦bəb·əl }

stone ice *See* ground ice. { 'stōn ˌīs }

stone polygon *See* sorted polygon. { 'stōn 'päl·iˌgän }

stone ring [GEOL] A ring of stones surrounding a central area of finer material; characteristic of sorted circle and sorted polygon. { 'stōn 'riŋ }

stony-iron meteorite [GEOL] Any of the rare meteorites containing at least 25% of both nickel-iron and heavy basic silicates. Also known as iron-stony meteorite; lithosiderite; sideraerolite; siderolite; syssiderite. { 'stō·nē ¦ī·ərn 'mēd·ē·əˌrīt }

stony meteorite [GEOL] Any meteorite composed principally of silicate minerals, especially olivine, pyroxene, and plagioclase. Also known as aerolite; asiderite; meteoric stone; meteorolite; stone. { 'stō·nē 'mēd·ē·əˌrīt }

stooping [METEOROL] An atmospheric refraction phenomenon; a special case of sinking in which the curvature of light rays due to atmospheric refraction decreases with elevation so that the visual image of a distant object is foreshortened in the vertical. { 'stüp·iŋ }

storage equation [HYD] The equation of continuity applied to unsteady flow; it states that the fluid inflow to a given space during an interval of time minus the outflow during the same interval is equal to the change in storage; it is applied in hydrology to the routing of floods through a reservoir or a reach of a stream; the moisture continuity equation applied to the atmosphere is a modification of this. { 'stȯr·ij i͵kwā·zhən }

storage routing *See* flood routing. { 'stȯr·ij ͵rüd·iŋ }

storm [METEOROL] An atmospheric disturbance involving perturbations of the prevailing pressure and wind fields on scales ranging from tornadoes (0.6 mile or 1 kilometer across) to extratropical cyclones (up to 1800 miles or 3000 kilometers across); also the associated weather (rain storm or blizzard) and the like. { stȯrm }

storm beach [GEOL] A ridge composed of gravel or shingle built up by storm waves at the inner margin of a beach. { 'stȯrm ͵bēch }

storm center [METEOROL] The area of lowest atmospheric pressure of a cyclone; this is a more general expression than eye of the storm, which refers only to the center of a well-developed tropical cyclone, in which there is a tendency of the skies to clear. { 'stȯrm ͵sen·tər }

storm delta *See* washover. { 'stȯrm ͵del·tə }

storm detection [METEOROL] Any of the methods and techniques used to ascertain the formation of storms, including procedures for locating, tracking, and forecasting; special tools adapted to this purpose are radar and satellites to supplement meteorological charts and visual observations. { 'stȯrm di͵tek·shən }

storm ice foot [OCEANOGR] An ice foot produced by the breaking of a heavy sea or the freezing of wind-driven spray. { 'stȯrm 'īs ͵fu̇t }

storm microseism [GEOPHYS] A microseism lasting 25 or more seconds, caused by ocean waves. { 'stȯrm 'mī·krə͵sīz·əm }

storm model [METEOROL] A physical, three-dimensional representation of the inflow, outflow, and vertical motion of air and water vapor in a storm. { 'stȯrm ͵mäd·əl }

storm surge [OCEANOGR] A rise above normal water level on the open coast due only to the action of wind stress on the water surface; includes the rise in level due to atmospheric pressure reduction as well as that due to wind stress. Also known as storm wave; surge. { 'stȯrm ͵sərj }

storm tide [OCEANOGR] Height of a storm surge or hurricane wave above the astronomically predicted sea level. { 'stȯrm ͵tīd }

storm track [METEOROL] The path followed by a center of low atmospheric pressure. { 'stȯrm ͵trak }

storm transposition [METEOROL] The transfer of precipitation patterns or DDA (depth-duration-area) values from the areas where they actually occurred to areas where they could occur; if necessary, the precipitation values are modified to account for differences in elevation or intervening barriers, and restrictions on change in shape or orientation of the storm may be imposed. { 'stȯrm ͵tranz·pə'zish·ən }

storm warning [METEOROL] A specially worded forecast of severe weather conditions, designed to alert the public to impending dangers; usually, this refers to a warning of potentially dangerous wind conditions for marine interests. { 'stȯrm ͵wȯrn·iŋ }

storm-warning signal [METEOROL] An arrangement of flags or pennants (by day) and lanterns (by night) displayed on a coastal storm-warning tower. { 'stȯrm ¦wȯrn·iŋ ͵sig·nəl }

storm-warning tower [METEOROL] A tower, generally constructed of steel, for displaying coastal storm-warning signals. { 'stȯrm ¦wȯrn·iŋ ͵tau̇·ər }

storm wave *See* storm surge. { 'stȯrm ͵wāv }

storm wind [METEOROL] In the Beaufort wind scale, a wind whose speed is from 56 to 63 knots (64 to 72 miles per hour or 104 to 117 kilometers per hour). { 'stȯrm ͵wind }

stoss [GEOL] Of the side of a hill, knob, or prominent rock, facing the upstream side of a glacier. { stäs }

stoss-and-lee topography [GEOL] A type of glaciated landscape in which small hills or other landforms exhibit gentle eroded slopes on the up-glacier or upstream side and less eroded, steeper slopes on the lee side. { ¦stäs ənd ¦lē tə'päg·rə·fē }

strain-slip [GEOL] A rock fracture resulting in a slight displacement. { 'strān ͵slip }

strain-slip cleavage *See* slip cleavage. { 'strān ¦slip ˌklē·vij }

strait [GEOGR] **1.** A neck of land. **2.** A narrow waterway connecting two larger bodies of water. { strāt }

strand [GEOL] A beach bordering a sea or an arm of an ocean. { strand }

stranded-floe ice foot *See* stranded ice foot. { 'stran·dəd ¦flō 'īs ˌfu̇t }

stranded ice [OCEANOGR] Ice held in place by virtue of being grounded. Also known as grounded ice. { 'stran·dəd 'īs }

stranded ice foot [OCEANOGR] An ice foot formed by the stranding of floes or small icebergs along a shore; it may be built up by freezing spray or breaking seas. Also known as stranded-floe ice foot. { 'stran·dəd 'īs ˌfu̇t }

strand flat *See* wave-cut platform. { 'strand ˌflat }

strandline [GEOL] **1.** A beach raised above the present sea level. **2.** The level at which a body of standing water meets the land. **3.** *See* shoreline. { 'strandˌlīn }

strath [GEOL] **1.** A broad, elongate depression with steep sides on the continental shelf. **2.** An extensive remnant of a broad, flat valley floor that has undergone degradation following uplift. { strath }

strath terrace [GEOL] An extensive remnant of a strath from a former erosion cycle. { 'strath ˌter·əs }

stratification [GEOL] An arrangement or deposition of sedimentary material in layers, or of sedimentary rock in strata. [HYD] **1.** The arrangement of a body of water, as a lake, into two or more horizontal layers of differing characteristics, especially densities. **2.** The formation of layers in a mass of snow, ice, or firn. { ˌstrad·ə·fə'kā·shən }

stratification index [GEOL] A measure of the beddedness of a stratigraphic unit, expressed as the number of beds in the unit per 100 feet (30 meters) of section. { ˌstrad·ə·fə'kā·shən ˌinˌdeks }

stratification plane [GEOL] A demarcation between two layers of sedimentary rock, often signifying that the layers were deposited under different conditions. { ˌstrad·ə·fə'kā·shən ˌplān }

stratified drift [GEOL] Fluvioglacial drift composed of material deposited by a meltwater stream or settled from suspension. { 'strad·əˌfīd 'drift }

stratified ocean [OCEANOGR] An ocean where there is a vertical gradient of density. { 'strad·əˌfīd 'ō·shən }

stratiform [GEOL] **1.** Descriptive of a layered mineral deposit of either igneous or sedimentary origin. **2.** Consisting of parallel bands, layers, or sheets. [METEOROL] Description of clouds of extensive horizontal development, as contrasted to the vertically developed cumuliform types. { 'strad·əˌfȯrm }

stratiformis [METEOROL] A cloud species consisting of a very extensive horizontal layer or layers which need not be continuous; this species is the most common form of the genera altocumulus and stratocumulus and is occasionally found in cirrocumulus. { ¦strad·ə¦fȯr·məs }

stratigrapher [GEOL] A geologist who deals with stratified rocks, for example, the classification, nomenclature, correlation, and interpretation of rocks. { strə'tig·rə·fər }

stratigraphic geology *See* stratigraphy. { ¦strad·ə¦graf·ik jē'äl·ə·jē }

stratigraphic map [GEOL] A map showing the areal distribution, configuration, or aspect of a stratigraphic unit or surface, such as an isopach map or a lithofacies map. { ¦strad·ə¦graf·ik 'map }

stratigraphic oil fields [GEOL] Hydrocarbon reserves in stratigraphic (sedimentary) traps formed by the positioning of clastic materials through chemical deposition. { ¦strad·ə¦graf·ik 'ȯil ˌfēlz }

stratigraphic separation *See* stratigraphic throw. { ¦strad·ə¦graf·ik ˌsep·ə'rā·shən }

stratigraphic throw [GEOL] The thickness of the strata which originally separated two beds brought into contact at a fault. Also known as stratigraphic separation. { ¦strad·ə¦graf·ik 'thrō }

stratigraphic trap [GEOL] Sealing of a reservoir bed due to lithologic changes rather than geologic structure. Also known as porosity trap; secondary stratigraphic trap. { ¦strad·ə¦graf·ik 'trap }

stratigraphic unit [GEOL] A stratum of rock or a body of strata classified as a unit on the basis of character, property, or attribute. { ¦strad·ə¦graf·ik 'yü·nət }

stratigraphy [GEOL] A branch of geology concerned with the form, arrangement, geographic distribution, chronologic succession, classification, correlation, and mutual relationships of rock strata, especially sedimentary. Also known as stratigraphic geology. { strə'tig·rə·fē }

stratocumulus [METEOROL] A principal cloud type predominantly stratiform, in the form of a gray or whitish layer of patch, which nearly always has dark parts. { ¦strad·ō'kyü·myə·ləs }

stratopause [METEOROL] The boundary or zone of transition separating the stratosphere and the mesosphere; it marks a reversal of temperature change with altitude. { 'strad·ə,pȯz }

stratosphere [METEOROL] The atmospheric shell above the troposphere and below the mesosphere; it extends, therefore, from the tropopause to about 33 miles (55 kilometers), where the temperature begins again to increase with altitude. { 'strad·ə,sfir }

stratosphere radiation [GEOPHYS] Any infrared radiation involved in the complex infrared exchange continually proceeding within the stratosphere. { 'strad·ə,sfir ,rād·ē'ā·shən }

stratospheric coupling [METEOROL] The interaction between disturbances in the stratosphere and those in the troposphere. { ¦strad·ə¦sfir·ik 'kəp·liŋ }

stratospheric ozone [METEOROL] Atmospheric ozone that is relatively concentrated in the lower stratosphere in a layer between 9 and 18 miles (15 and 30 kilometers) above the earth's surface, and plays a critical role for the biosphere by absorbing the damaging ultraviolet radiation with wavelengths 320 nanometers and lower. Also known as ozone layer. { ,strad·ə,sfir·ik 'ō,zōn }

stratospheric steering [METEOROL] The steering of lower-level atmospheric disturbances along the contour lines of the tropopause, which lines are presumably roughly parallel to the direction of the wind at the tropopause level. { ¦strad·ə¦sfir·ik 'stir·iŋ }

stratotype [GEOL] A specifically bounded type section of rock strata to which a time-stratigraphic unit is ascribed, ideally consisting of a complete and continuously exposed and deposited sequence of correlatable strata, and extending from a readily identifiable basal boundary to a readily identifiable top boundary. { 'strad·ə,tīp }

stratovolcano [GEOL] A volcano constructed of lava and pyroclastics, deposited in alternating layers. Also known as composite volcano. { ¦strad·ō·väl'kā·nō }

stratum [GEOL] A mass of homogeneous or gradational sedimentary material, either consolidated rock or unconsolidated soil, occurring in a distinct layer and visually separable from other layers above and below. { 'strad·əm }

stratus [METEOROL] A principal cloud type in the form of a gray layer with a rather uniform base; a stratus does not usually produce precipitation, but when it does occur it is in the form of minute particles, such as drizzle, ice crystals, or snow grains. { 'strad·əs }

stray [GEOL] A lenticular rock formation encountered unexpectedly in drilling an oil or a gas well; it differs from an adjacent persistent formation in lithology and hardness. { strā }

strays *See* atmospheric interference. { strāz }

stray sand [GEOL] A stray composed of sandstone. { 'strā 'sand }

streak lightning [GEOPHYS] Ordinary lightning, of a cloud-to-ground discharge, that appears to be entirely concentrated in a single, relatively straight lightning channel. { 'strēk ,līt·niŋ }

stream [HYD] A body of running water moving under the influence of gravity to lower levels in a narrow, clearly defined natural channel. { strēm }

stream-built terrace *See* alluvial terrace. { 'strēm ¦bilt 'ter·əs }

stream capacity [GEOL] The ability of a stream to carry detritus, measured at a given point per unit of time. { 'strēm kə,pas·əd·ē }

stream channel [GEOL] A long, narrow, sloping troughlike depression where a natural stream flows or may flow. Also known as streamway. { 'strēm ˌchan·əl }

stream-channel form ratio [GEOL] The mathematical relationship between a stream channel width, depth, and channel perimeter. { 'strēm ˌchan·əl 'fȯrm ˌrā·shō }

stream current [HYD] A steady current in a stream or river. [OCEANOGR] A deep, narrow, well-defined fast-moving ocean current. { 'strēm ˌkə·rənt }

streamer [GEOPHYS] A sinuous channel of very high ion-density which propagates itself through a gas by continual establishment of an electron avalanche just ahead of its advancing tip; in lightning discharges, the stepped leader, and return streamer all constitute special types of streamers. { 'strē·mər }

stream erosion [GEOL] The progressive removal of exposed matter from the surface of a stream channel by a stream. { 'strēm iˌrō·zhən }

streamflow [HYD] A type of channel flow, applied to surface runoff moving in a stream. { 'strēmˌflō }

streamflow routing *See* flood routing. { 'strēmˌflō 'rüd·iŋ }

stream frequency [GEOL] A measure of topographic texture expressed as the ratio of the number of streams in a drainage basin to the area of the basin. Also known as channel frequency. { 'strēm ˌfrē·kwən·sē }

stream gradient [GEOL] The angle, measured in the direction of flow, between the water surface (for large streams) or the channel flow (for small streams) and the horizontal. Also known as stream slope. { 'strēm ˌgrād·ē·ənt }

stream-gradient ratio [GEOL] Ratio of the stream gradient of a stream channel of one order to the stream gradient of the next higher order channel in the same drainage basin. Also known as channel gradient ratio. { 'strēm ˌgrād·ē·ənt ˌrā·shō }

stream-length ratio [HYD] Ratio of the mean length of a stream of a given order to the mean length of the next lower order stream in the same basin. { 'strēm ¦leŋkth ˌrā·shō }

stream load [GEOL] Solid material transported by a stream. { 'strēm ˌlōd }

stream morphology *See* river morphology. { 'strēm mȯr'fäl·ə·jē }

stream order [HYD] The designation by a dimensionless integer series (1, 2, 3, . . .) of the relative position of stream segments in the network of a drainage basin. Also known as channel order. { 'strēm ¦ȯr·dər }

stream profile [HYD] The longitudinal profile of a stream. { 'strēm ˌprōˌfīl }

stream segment [HYD] The part of a stream extending between designated tributary junctions. Also known as channel segment. { 'strēm ˌseg·mənt }

streamsink [GEOL] An opening in the surface of the earth down which a stream disappears underground. { 'strēmˌsiŋk }

stream slope *See* stream gradient. { 'strēm ˌslōp }

stream terrace [GEOL] One of a series of level surfaces on a stream valley flanking and parallel to a stream channel and above the stream level, representing the uneroded remnant of an abandoned floodplain or stream bed. Also known as river terrace. { 'strēm ˌter·əs }

stream tin [GEOL] The mineral cassiterite occurring as pebbles in alluvial deposits. { 'strēm ˌtin }

stream transport [GEOL] Movement of rock material in and by a stream. { 'strēm 'tranzˌpȯrt }

streamway *See* stream channel. { 'strēmˌwā }

strength of current [OCEANOGR] **1.** The phase of a tidal current at which the speed is a maximum. **2.** The velocity of the current at this time. { 'streŋkth əv 'kə·rənt }

strength of ebb [OCEANOGR] **1.** The ebb current at the time of maximum speed. **2.** The speed of the current at this time. { 'streŋkth əv 'eb }

strength-of-ebb interval [OCEANOGR] The time interval between the transit (upper or lower) of the moon and the next maximum ebb current at a place. { 'streŋkth əv 'eb 'in·tər·vəl }

strength of flood [OCEANOGR] **1.** The flood current at the time of maximum speed. **2.** The speed of the current at this time. { 'streŋkth əv 'fləd }

strength-of-flood interval [OCEANOGR] The time interval between the transit (upper or lower) of the moon and the next maximum flood current at a place. { 'streŋkth əv 'fləd 'in·tər·vəl }

stretch *See* reach. { strech }

stretched pebbles [GEOL] Pebbles in a sedimentary rock which have been elongated from their original shape by deformation. { 'strecht 'peb·əlz }

stretch fault *See* stretch thrust. { 'strech ˌfȯlt }

stretch thrust [GEOL] A reverse fault developed as a result of shear in the middle limb of an overturned fold. Also known as stretch fault. { 'strech ˌthrəst }

striated ground *See* striped ground. { 'strīˌād·əd 'grau̇nd }

striation [GEOL] One of a series of parallel or subparallel scratches, small furrows, or lines on the surface of a rock or rock fragment; usually inscribed by rock fragments embedded at the base of a moving glacier. { strī'ā·shən }

strike [GEOL] The direction taken by a structural surface, such as a fault plane, as it intersects the horizontal. Also known as line of strike. { strīk }

strike fault [GEOL] A fault whose strike is parallel with that of the strata involved. { 'strīk ˌfȯlt }

strike joint [GEOL] A joint that strikes parallel to the bedding or cleavage of the constituent rock. { 'strīk ˌjȯint }

strike separation [GEOL] The distance of separation on either side of a fault surface of two formerly adjacent beds. { 'strīk ˌsep·əˌrā·shən }

strike-separation fault *See* lateral fault. { 'strīk ˌsep·ə¦rā·shən ˌfȯlt }

strike-shift fault *See* strike-slip fault. { 'strīk ¦shift ˌfȯlt }

strike slip [GEOL] The component of the slip of a fault that is parallel to the strike of the fault. Also known as horizontal displacement; horizontal separation. { 'strīk ˌslip }

strike-slip fault [GEOL] A fault whose direction of movement is parallel to the strike of the fault. Also known as strike-shift fault. { 'strīk ¦slip ˌfȯlt }

strike stream *See* subsequent stream. { 'strīk ˌstrēm }

string [GEOL] A very small vein, either independent or occurring as a branch of a larger vein. Also known as stringer. { striŋ }

stringer *See* string. { 'striŋ·ər }

stringer lode [GEOL] A lode that consists of many narrow veins in a mass of country rock. { 'striŋ·ər ˌlōd }

striped ground [GEOL] A pattern of alternating stripes formed by frost action on a sloping surface. Also known as striated ground; striped soil. { 'strīpt 'grau̇nd }

striped soil *See* striped ground. { 'strīpt 'sȯil }

stripped plain [GEOL] The upper, exposed surface of a resistant stratum that forms a stripped structural surface when extended over a considerable area. { 'stript 'plān }

stripped structural surface [GEOL] An erosion surface formed in an area underlain by horizontal or gently sloping strata of unequal resistance where the overlying softer beds have been removed by erosion. Also known as stripped surface. { 'stript ¦strək·chə·rəl 'sər·fəs }

stripped surface *See* stripped structural surface. { 'stript 'sər·fəs }

stroke density [GEOPHYS] The areal density of lightning discharges over a given region during some specified period of time, as number per square mile per year. { 'strōk ˌden·səd·ē }

stromatite [GEOL] Chorismite having flat or folded parallel layers of two or more textural elements. Also known as stromatolith. { 'strō·məˌtīt }

stromatolite [GEOL] A structure in calcareous rocks consisting of concentrically laminated masses of calcium carbonate and calcium-magnesium carbonate which are believed to be of calcareous algal origin; these structures are irregular to columnar and hemispheroidal in shape, and range from 1 millimeter to many meters in thickness. Also known as callenia. { strə'mad·əlˌīt }

stromatolith [GEOL] **1.** A complex sill-like igneous intrusion interfingered with sedimentary strata. **2.** *See* stromatite. { strə'mad·əlˌith }

strombolian [GEOL] A type of volcanic eruption characterized by fire fountains of lava from a central crater. { sträm'bō·lē·ən }

strong breeze [METEOROL] In the Beaufort wind scale, a wind whose speed is from 22 to 27 knots (25 to 31 miles per hour or 41 to 50 kilometers per hour). { 'strȯŋ 'brēz }
strong gale [METEOROL] In the Beaufort wind scale, a wind whose speed is from 41 to 47 knots (47 to 54 miles per hour or 76 to 87 kilometers per hour). { 'strȯŋ 'gāl }
structural bench [GEOL] A bench typifying the resistant edge of a terrace that is being reduced by erosion. Also known as rock bench. { 'strək·chə·rəl 'bench }
structural contour map [GEOL] A map representation of a subsurface stratigraphic unit; depicts the configuration of a rock surface by means of elevation contour lines. { 'strək·chə·rəl 'kän,tu̇r ,map }
structural geology [GEOL] A branch of geology concerned with the form, arrangement, and internal structure of the rocks. { 'strək·chə·rəl jē'äl·ə·jē }
structural high [GEOL] Any of various structural features such as a crest, culmination, anticline, or dome. { 'strək·chə·rəl 'hī }
structural low [GEOL] Any of various structural features such as a basin, a syncline, a saddle, or a sag. { 'strək·chə·rəl 'lō }
structural terrace [GEOL] A terracelike landform developed where generally steeply inclined and otherwise uniformly dipping strata locally flatten. { 'strək·chə·rəl 'ter·əs }
structural trap [GEOL] Containment in a reservoir bed of oil or gas due to flexure or fracture of the bed. { 'strək·chə·rəl 'trap }
structural valley [GEOL] A valley whose form and origin is attributable to the underlying geologic structure. { 'strək·chə·rəl 'val·ē }
structure [GEOL] **1.** An assemblage of rocks upon which erosive agents have been or are acting. **2.** The sum total of the structural features of an area. { 'strək·chər }
structure contour [GEOL] A contour that portrays a structural surface, such as a fault. Also known as subsurface contour. { 'strək·chər ,kän,tu̇r }
structure-contour map [GEOL] A map that uses structure contour lines to portray subsurface configuration. Also known as structure map. { 'strək·chər ¦kän,tu̇r ,map }
structure map *See* structure-contour map. { 'strək·chər ,map }
structure section [GEOL] A vertical section showing the observed or inferred geologic structure on a vertical surface or plane. { 'strək·chər ,sek·shən }
Stuve chart [METEOROL] A thermodynamic diagram with atmospheric temperature as the x axis and atmospheric pressure to the power 0.286 as the y ordinate, increasing downward; named after G. Stuve. Also known as adiabatic chart; pseudoadiabatic chart. { 'stüv·ə ,chärt }
stylolite [GEOL] An irregular surface, generally parallel to a bedding plane, in which small toothlike projections on one side of the surface fit into cavities of complementary shape on the other surface; interpreted to result diagenetically by pressure solution. { 'stī·lə,līt }
S-type magma [GEOL] Magma formed from sedimentary source material. { 'es ¦tīp 'mag·mə }
subaerial [GEOL] Pertaining to conditions and processes occurring beneath the atmosphere or in the open air, that is, on or adjacent to the land surface. { ¦səb'er·ē·əl }
subage [GEOL] A subdivision of a geologic age. { 'səb,āj }
subalkaline [GEOCHEM] Pertaining to a soil in which the pH is 8.0 to 8.5, usually in a limestone or salt-marsh region. { ¦səb'al·kə,līn }
Subantarctic Intermediate Water [OCEANOGR] A layer of water above the deep-water layer in the South Atlantic. { ¦səb·ant'ärd·ik ,in·tər,mēd·ē·ət 'wȯd·ər }
subaqueous [HYD] Pertaining to conditions and processes occurring in, under, or beneath the surface of water, especially fresh water. { ¦səb'ā·kwē·əs }
subaqueous dune [GEOL] A dune resulting from entrainment of grains by the flow of moving water. { ¦səb'ā·kwē·əs 'dün }
subarctic [GEOGR] Pertaining to regions adjacent to the Arctic Circle or having characteristics somewhat similar to those of these regions. { ¦səb'ärd·ik }
subarctic climate *See* taiga climate. { ¦səb'ärd·ik 'klī·mət }

subarid [CLIMATOL] Pertaining to regions that are moderately or slightly arid. { ¦səb'ar·əd }
subarkose [GEOL] Sandstone that is intermediate in composition between arkose and pure quartz sandstone; it contains less feldspar than arkose. { səb'är,kōs }
subartesian well [HYD] A well that requires artificial pumping to raise water to the surface because confining pressure forces the water only part of the distance up the well shaft. { ¦səb·är'tē·zhən 'wel }
subbituminous coal [GEOL] Black coal intermediate in rank between lignite and bituminous coal; has more carbon and less moisture than lignite. { ¦səb·bə'tü·mə·nəs 'kōl }
subbottom reflection [GEOPHYS] The return of sound energy from a discontinuity in material below the surface of the sea bottom. { ¦səb'bäd·əm ri'flek·shən }
subcapillary interstice [GEOL] An interstice in which the molecular attraction of its walls extends across the entire opening; it is smaller than a capillary interstice. { ¦səb'kap·ə,ler·ē in'tər·stəs }
subconchoidal [GEOL] Pertaining to a fracture that is partly or vaguely conchoidal in shape. { ¦səb·kən'kȯid·əl }
subconsequent stream *See* secondary consequent stream. { ¦səb'kän·sə·kwənt 'strēm }
subcontinent [GEOGR] **1.** A landmass such as Greenland that is large but not as large as the generally recognized continents. **2.** A large subdivision of a continent (for example, the Indian subcontinent) distinguished geologically or geomorphically from the rest of the continent. { ¦səb'känt·ən·ənt }
subcrop [GEOL] An occurrence of strata beneath the subsurface of an inclusive stratigraphic unit that succeeds an unconformity on which there is marked overstep. { 'səb,kräp }
subduction [GEOL] The process by which one crustal block descends beneath another, such as the descent of the Pacific plate beneath the Andean plate along the Andean Trench. { səb'dək·shən }
subduction zones [GEOL] Regions where portions of the earth's tectonic plates are diving beneath other plates, into the earth's interior. They are defined by deep oceanic trenches, lines of volcanoes parallel to the trenches, and zones of large earthquakes that extend from the trenches landward. { səb'dək·shən ,zōnz }
suberinite [GEOL] A variety of provitrinite composed of corky tissue. { sü'ber·ə,nīt }
subfeldspathic [GEOL] Referring to mature lithic wacke or arenite containing an abundance of quartz grains with less than 10% feldspar grains. { ¦səb·fel'spath·ik }
subgelisol [GEOL] Unfrozen ground beneath permafrost. { ¦səb'jel·ə,sȯl }
subgeostrophic wind [METEOROL] Any wind of lower speed than the geostrophic wind required by the existing pressure gradient. { ¦səb¦jē·ō¦sträf·ik 'wind }
subglacial [GEOL] Pertaining to the area in or at the bottom of, or immediately beneath, a glacier. { ¦səb'glā·shəl }
subglacial moraine *See* ground moraine. { ¦səb'glā·shəl mə'rān }
subgradient wind [METEOROL] A wind of lower speed than the gradient wind required by the existing pressure gradient and centrifugal force. { ¦səb'grād·ē·ənt 'wind }
subhumid climate [CLIMATOL] A humidity province based on its typical vegetation. Also known as grassland climate; prairie climate. { ¦səb'hyü·məd 'klī·mat }
subjacent [GEOL] Being lower than but not directly underneath. { ,səb'jās·ənt }
subjacent igneous body [GEOL] An igneous intrusion without a known floor, and which presumably enlarges downward. { ,səb'jās·ənt 'ig·nē·əs 'bäd·ē }
sublacustrine [GEOL] Existing or formed on the bottom of a lake. { ¦səb·lə'kəs·trən }
sublacustrine channel [GEOL] A channel eroded in a lake bed either before the lake existed or by a strong current in the lake. { ¦səb·lə'kəs·trən ,chan·əl }
sublimation nucleus [METEOROL] Any particle upon which an ice crystal may grow by the process of sublimation. { ,səb·lə'mā·shən ¦nü·klē·əs }
sublimation vein [GEOL] A vein of mineral that has condensed from a vapor. { ,səb·lə'mā·shən ¦vān }
sublittoral zone [OCEANOGR] The benthic region extending from mean low water (2–3 fathoms or 40–60 meters, according to some authorities) to a depth of about

110 fathoms (200 meters), or the edge of a continental shelf, beyond which most abundant attached plants do not grow. { ¦səb'lid·ə·rəl ˌzōn }

submarine [OCEANOGR] Being or functioning in the sea. { ¦səb·mə'rēn }

submarine canyon [GEOL] Steep-sided valleys winding across the continental shelf or continental slope, probably originally produced by Pleistocene stream erosion, but presently the site of turbidity flows. { ¦səb·mə'rēn 'kan·yən }

submarine cave *See* submarine fan. { ¦səb·mə'rēn 'kāv }

submarine delta *See* submarine fan. { ¦səb·mə'rēn 'del·tə }

submarine earthquake *See* seaquake. { ¦səb·mə'rēn 'ərthˌkwāk }

submarine fan [GEOL] A shallow marine sediment that is fan- or cone-shaped and lies off the seaward opening of large rivers and submarine canyons. Also known as abyssal cave; abyssal fan; sea fan; submarine cave; submarine delta; subsea apron. { ¦səb·mə'rēn 'fan }

submarine geology *See* geological oceanography. { ¦səb·mə'rēn jē'äl·ə·jē }

submarine isthmus [GEOL] A submarine elevation joining two land areas and separating two basins or depressions by a depth less than that of the basins. { ¦səb·mə'rēn 'is·məs }

submarine peninsula [GEOL] An elevated portion of the submarine relief resembling a peninsula. { ¦səb·mə'rēn pə'nin·sə·lə }

submarine pit [GEOL] A cavity on the bottom of the sea. Also known as submarine well. { ¦səb·mə'rēn 'pit }

submarine plain *See* plain. { ¦səb·mə'rēn 'plān }

submarine relief [GEOL] Relative elevations of the ocean bed, or the representation of them on a chart. { ¦səb·mə'rēn ri'lēf }

submarine spring [HYD] A spring of water issuing from the bottom of the sea. { ¦səb·mə'rēn 'spriŋ }

submarine station [OCEANOGR] **1.** One of the places for which tide or tidal current predictions are determined by applying a correction to the predictions of a reference station. **2.** A tide or tidal current station at which a short series of observations have been made; these observations are reduced by comparison with simultaneous observations at a reference station. { ¦səb·mə'rēn 'stā·shən }

submarine topography [GEOL] Configuration of a surface such as the sea bottom or of a surface of given characteristics within the water mass. { ¦səb·mə'rēn tə'päg·rə·fē }

submarine trench *See* trench. { ¦səb·mə'rēn 'trench }

submarine trough *See* trough. { ¦səb·mə'rēn 'tróf }

submarine valley *See* valley. { ¦səb·mə'rēn 'val·ē }

submarine weathering [GEOL] A slow alteration of the form, texture, and composition of the sea floor from chemical, thermal, and biological causes. { ¦səb·mə'rēn 'weth·ə·riŋ }

submarine well *See* submarine pit. { ¦səb·mə'rēn 'wel }

submerged breakwater [OCEANOGR] A breakwater with its top below the still water level; when struck by a wave, part of the wave energy is reflected seaward and the remaining energy is largely dissipated in a breaker, transmitted shoreward as a multiple crest system, or transmitted shoreward as a simple wave system. { səb'mərjd 'brākˌwód·ər }

submerged coastal plain [GEOL] The continental shelf as the seaward extension of a coastal plain on the land. Also known as coast shelf. { səb'mərjd 'kōst·əl 'plān }

submerged lands [GEOL] Lands covered by water at any stage of the tide, as distinguished from tidelands which are attached to the mainland or an island and are covered or uncovered with the tide; tidelands presuppose a high-water line as the upper boundary, submerged lands do not. { səb'mərjd 'lanz }

submerged shoreline *See* shoreline of submergence. { səb'mərjd 'shórˌlīn }

submergence [GEOL] A change in the relative levels of water and land either from a sinking of the land or a rise of the water level. { səb'mər·jəns }

subpolar anticyclone *See* subpolar high. { ¦səb'pō·lər ¦ant·i'sīˌklōn }

subpolar glacier [HYD] A polar glacier with 30 to 60 feet (10 to 20 meters) of firn in the accumulation area where some melting occurs. { ¦səb'pō·lər 'glā·shər }
subpolar high [METEOROL] A high that forms over the cold continental surfaces of subpolar latitudes, principally in Northern Hemisphere winters; these highs typically migrate eastward and southward. Also known as polar anticyclone; polar high; subpolar anticyclone. { ¦səb'pō·lər 'hī }
subpolar low-pressure belt [METEOROL] A belt of low pressure located, in the mean, between 50 and 70° latitude; in the Northern Hemisphere, this belt consists of the Aleutian low and the Icelandic low; in the Southern Hemisphere, it is supposed to exist around the periphery of the Antarctic continent. { ¦səb'pō·lər ¦lō 'presh·ər ˌbelt }
subpolar westerlies *See* westerlies. { ¦səb'pō·lər 'wes·tərˌlēz }
subsea apron *See* submarine fan. { 'səbˌsē 'ā·prən }
subsequent [GEOL] Referring to a geologic feature that followed in time the development of a consequent feature of which it is a part. { 'səb·sə·kwənt }
subsequent drainage [HYD] Drainage by a stream developed subsequent to the system of which it is a part; drainage follows belts of weak rocks. { 'səb·sə·kwənt 'drā·nij }
subsequent fold *See* cross fold. { 'səb·sə·kwənt 'fōld }
subsequent stream [HYD] A stream that flows in the general direction of the strike of the underlying strata and is subsequent to the formation of the consequent stream of which it is a tributary. Also known as longitudinal stream; strike stream. { 'səb·sə·kwənt 'strēm }
subsequent valley [GEOL] A valley eroded by a stream developed subsequent to the system of which it is a part. { 'səb·sə·kwənt 'val·ē }
subsidence [METEOROL] A descending motion of air in the atmosphere, usually with the implication that the condition extends over a rather broad area. { səb'sīd·əns }
subsidence inversion [METEOROL] A temperature inversion produced by the adiabatic warming of a layer of subsiding air; this inversion is enhanced by vertical mixing in the air layer below the inversion. { səb'sīd·əns inˌvər·zhən }
subsidiary fracture *See* tension fracture. { səb'sid·ēˌer·ē 'frak·chər }
subsoil [GEOL] **1.** Soil underlying surface soil. **2.** *See* B horizon. { 'səbˌsȯil }
subsoil ice *See* ground ice. { 'səbˌsȯil 'īs }
substratosphere [METEOROL] A region of indefinite lower limit just below the stratosphere. { ¦səb'strad·əˌsfir }
substratum [GEOL] Any layer underlying the true soil. { ¦səb'strad·əm }
subsurface contour *See* structure contour. { ¦səb'sər·fəs 'känˌtu̇r }
subsurface current [OCEANOGR] An underwater current which is not present at the surface or whose core (region of maximum velocity) is below the surface. { ¦səb'sər·fəs 'kə·rənt }
subsurface flow [HYD] Interflow plus groundwater flow. { ¦səb'sər·fəs 'flō }
subsurface geology [GEOL] The study of geologic features beneath the land or seafloor surface. Also known as underground geology. { ¦səb'sər·fəs jē'äl·ə·jē }
subterranean ice *See* ground ice. { ¦səb·tə'rā·nē·ən 'īs }
subterranean stream [HYD] A subsurface stream that flows through a cave or a group of communicating caves. { ¦səb·tə'rā·nē·ən 'strēm }
subtropic [METEOROL] An indefinite belt in each hemisphere between the tropic and temperate regions; the polar boundaries are considered to be roughly 35–40° northern and southern latitudes, but vary greatly according to continental influence, being farther poleward on the western coasts of continents and farther equatorward on the eastern coasts. { ¦səb'träp·ik }
subtropical anticyclone *See* subtropical high. { ˌsəb'träp·ə·kəl ¦ant·i'sīˌklōn }
Subtropical Convergence [OCEANOGR] The zone of converging currents, generally located in midlatitudes. { ˌsəb'träp·ə·kəl kən'vər·jəns }
subtropical cyclone [METEOROL] The low-level (surface chart) manifestation of a cutoff low. { ˌsəb'träp·ə·kəl 'sīˌklōn }
subtropical easterlies *See* tropical easterlies. { ˌsəb'träp·ə·kəl 'ēs·tərˌlēz }
subtropical easterlies index [METEOROL] A measure of the strength of the easterly

wind between the latitudes of 20° and 35°N; the index is computed from the average sea-level pressure difference between these latitudes and is expressed as the east to west component of the corresponding geostrophic wind in meters and tenths of meters per second. { ˌsəb'träp·ə·kəl 'ēs·tərˌlēz ˌinˌdeks }

subtropical high [METEOROL] One of the semipermanent highs of the subtropical high-pressure belt; these highs appear as centers of action on mean charts of surface pressure; they lie over oceans and are best developed in the summer season. Also known as oceanic anticyclone; oceanic high; subtropical anticyclone. { ˌsəb'träp·ə·kəl 'hī }

subtropical high-pressure belt [METEOROL] One of the two belts of high atmospheric pressure that are centered, in the mean, near 30°N and 30°S latitudes; these belts are formed by the subtropical highs. { ˌsəb'träp·ə·kəl ¦hī 'presh·ər ˌbelt }

subtropical westerlies *See* westerlies. { ˌsəb'träp·ə·kəl 'wes·tərˌlēz }

succession [GEOL] A group of rock units or strata that succeed one another in chronological order. { sək'sesh·ən }

sudburite [GEOL] A basic basalt composed of hypersthene, augite, and magnetite, among other minerals. { 'səd·bəˌrīt }

sudden commencement [GEOPHYS] Magnetic storms which start suddenly (within a few seconds) and simultaneously all over the earth. { 'səd·ən kə'mens·mənt }

sudden ionospheric disturbance [GEOPHYS] A complex combination of sudden changes in the condition of the ionosphere following the appearance of solar flares, and the effects of these changes. Abbreviated SID. { 'səd·ən ī¦än·ə¦sfir·ik di'stər·bəns }

suestada [METEOROL] Strong southeast winds occurring in winter along the coast of Argentina, Uruguay, and southern Brazil; they cause heavy seas and are accompanied by fog and rain; the counterpart of the northeast storm in North America. { swā 'städ·ə }

suevite [GEOL] A grayish or yellowish fragmental rock associated with meteorite impact craters; resembles tuff breccia or pumiceous tuff but is of nonvolcanic origin. { 'swāˌvīt }

sugar berg [OCEANOGR] An iceberg of porous glacier ice. { 'shu̇g·ər ˌbərg }

sugarloaf sea [OCEANOGR] A sea characterized by waves that rise into sugarloaf shapes, with little wind, possibly resulting from intersecting waves. { 'shu̇g·ərˌlōf ˌsē }

sugar snow *See* depth hoar. { 'shu̇g·ər ˌsnō }

sukhovei [METEOROL] Literally dry wind; a dry, hot, dusty wind in the south Russian steppes, which blows principally from the east and frequently brings a prolonged drought and crop damage. { sü·kō·vā }

sulfofication [GEOCHEM] Oxidation of sulfur and sulfur compounds into sulfates, occurring in soils by the agency of bacteria. { ˌsəl·fə·fə'kā·shən }

sulfophile element [GEOCHEM] An element occurring preferentially in an oxygen-free mineral. Also known as thiophile element. { ¦səl·fə'fīl ˌel·ə·mənt }

sulfur ball [GEOL] A bubble of hot volcanic gas encased in a sulfurous mud skin that solidified on contact with air. { 'səl·fər ˌbȯl }

sulfur-mud pool *See* mud pot. { 'səl·fər ¦məd ˌpül }

sulfur spring [HYD] A spring containing sulfur compounds such as hydrogen sulfide. { 'səl·fər ˌspriŋ }

sullage [GEOL] Mud, silt, or other sediments carried and deposited by flowing water. { 'səl·ij }

sultriness [METEOROL] An oppressively uncomfortable state of the weather which results from the simultaneous occurrence of high temperature and high humidity, and often enhanced by calm air and cloudiness. { 'səl·trē·nəs }

sumatra [METEOROL] A squall, with wind speeds occasionally exceeding 30 miles (48 kilometers) per hour, in the Malacca Strait between Malay and Sumatra during the southwest monsoon (April through November). { sü'mä·trə }

summation principle [METEOROL] In United States weather observing practice, the rule which governs the assignment of sky cover amount to any layer of cloud or obscuring phenomenon, and to the total sky cover; in essence, this principle states that the sky cover at any level is equal to the summation of the sky cover of the lowest layer plus the additional sky cover provided at all successively higher layers up to and including the layer in question; thus, no layer can be assigned a sky cover less than a lower layer, and no sky cover can be greater than 1.0 (10/10). { sə'mā·shən ˌprin·sə·pəl }

summit plain *See* peak plain. { 'səm·ət ˌplān }

sun crack *See* mud crack. { 'sən ˌkrak }

sun cross [METEOROL] A rare halo phenomenon in which bands of white light intersect over the sun at right angles; it appears probable that most of such observed crosses appear merely as a result of the superposition of a parhelic circle and a sun pillar. { 'sən ˌkrȯs }

sun crust [HYD] A type of snow crust, formed by refreezing of surface snow crystals after having been melted by the sun. { 'sən ˌkrəst }

sun drawing water [METEOROL] Popular designation for a phenomenon of the sun showing through scattered openings in a layer of clouds into a layer of turbid air that is hazy or dusty; bright bands are seen where the several beams of sunlight pass down through the subcloud layer; sailors called the phenomenon the backstays of the sun. { 'sən 'drȯ·iŋ 'wȯd·ər }

sunlit aurora [GEOPHYS] An aurora which occurs in the part of the upper atmosphere which is in the sunlight, above the earth's shadow. { 'sənˌlit ə'rȯr·ə }

sun pillar [METEOROL] A luminous streak of light, white or slightly reddened, extending above and below the sun, most frequently observed near sunrise or sunset; it may extend to about 20° above the sun, and generally ends in a point. Also known as light pillar. { 'sən ˌpil·ər }

superadiabatic lapse rate [METEOROL] An environmental lapse rate greater than the dry-adiabatic lapse rate, such that potential temperature decreases with height. { ¦sü·pərˌad·ē·ə'bad·ik 'laps ˌrāt }

supercapillary interstice [GEOL] An interstice that is too large to hold water above the free water surface by surface tension; it is larger than a capillary interstice. { ¦sü·pər'kap·əˌler·ē in'tər·stəs }

supercell [METEOROL] A thunderstorm with a persistent rotating updraft. While rare, it produces the most severe weather such as tornadoes, strong winds, and hail. { 'sü·pərˌsel }

supercontinent [GEOL] A large continental mass, such as Pangea, that existed early in geologic time and from which smaller continents formed and separated by fragmentation and drifting. { 'sü·pərˌkänt·ən·ənt }

supercooled cloud [METEOROL] A cloud composed of supercooled liquid waterdrops. { ¦sü·pər'küld 'klaůd }

supercritical flow *See* rapid flow. { ¦sü·pər'krid·ə·kəl 'flō }

superficial deposit *See* surficial deposit. { ¦sü·pər¦fish·əl di'päz·ət }

supergeostrophic wind [METEOROL] Any wind of greater speed than the geostrophic wind required by the pressure gradient. { ¦sü·pər¦jē·ō¦sträf·ik 'wind }

superglacial [HYD] Of or pertaining to the upper surface of a glacier or ice sheet. { ¦sü·pər'glā·shəl }

supergradient wind [METEOROL] A wind of greater speed than the gradient wind required by the existing pressure gradient and centrifugal force. { ¦sü·pər'grād·ē·ənt 'wind }

supergroup [GEOL] A lithostratigraphic material unit of the highest order. { 'sü·pərˌgrüp }

superimposed [GEOL] Pertaining to layered or stratified rocks. { ¦sü·pər·im'pōzd }

superimposed drainage [HYD] A naturally evolved drainage system that became established on a preexisting surface, now eroded, and whose course is unrelated to the present underlying geological structure. { ¦sü·pər·im'pōzd 'drā·nij }

superimposed fan [GEOL] An alluvial fan developed on, and having a steeper gradient than, an older fan. { ¦sü·pər·im'pōzd 'fan }

superimposed fold *See* cross fold. { ¦sü·pər·im'pōzd 'fōld }

superimposed glacier [GEOL] A glacier whose course is maintained despite different preexisting structures and lithologies as the glacier erodes downward. { ¦sü·pər·im'pōzd 'glā·shər }

superimposed stream [HYD] A stream, started on a new surface, that kept its course through the different preexisting lithologies and structures encountered as it eroded downward into the underlying rock. Also known as superinduced stream. { ¦sü·pər·im'pōzd 'strēm }

superimposed valley [GEOL] A valley eroded by or containing a superimposed stream. { ¦sü·pər·im'pōzd 'val·ē }

superincumbent [GEOL] Pertaining to a superjacent layer, especially one that is situated so as to exert pressure. { ¦sü·pər·in'kəm·bənt }

superinduced stream *See* superimposed stream. { ¦sü·pər·in'düst 'strēm }

superior air [METEOROL] An exceptionally dry mass of air formed by subsidence and usually found aloft but occasionally reaching the earth's surface during extreme subsidence processes. { sə'pir·ē·ər 'er }

superior tide [OCEANOGR] The tide in the hemisphere in which the moon is above the horizon. { sə'pir·ē·ər 'tīd }

superjacent [GEOL] Pertaining to a stratum situated immediately upon or over a particular lower stratum or above an unconformity. { ¦sü·pər'jā·sənt }

superjacent waters [OCEANOGR] The waters above the continental shelf. { ¦sü·pər'jā·sənt 'wȯd·ərz }

supermature [GEOL] Pertaining to a texturally mature clastic sediment whose grains have become rounded. { ¦sü·pər·mə'chu̇r }

superposed stream *See* consequent stream. { ¦sü·pər'pōzd 'strēm }

superposition [GEOL] **1.** The order in which sedimentary layers are deposited, the highest being the youngest. **2.** The process by which the layering occurs. { ˌsü·pər·pə'zish·ən }

superprint *See* overprint. { 'sü·pərˌprint }

superrefraction *See* ducting. { ¦sü·pər·ri'frak·shən }

superresolution [OCEANOGR] Separation of tides into components of different frequencies, without taking measurements for the full extent of the longest-period component. { ¦sü·pərˌrez·ə'lü·shən }

supersaturation [METEOROL] The condition existing in a given portion of the atmosphere when the relative humidity is greater than 100%, in respect to a plane surface of pure water or pure ice. { ¦sü·pərˌsach·ə'rā·shən }

supply current [GEOPHYS] The electrical current in the atmosphere which is required to balance the observed air-earth current of fair-weather regions by transporting positive charge upward or negative charge downward. { sə'plī ˌkə·rənt }

supracrustal rocks [GEOL] Rocks that overlie basement rock. { ¦sü·prə'krəst·əl 'räks }

supragelisol *See* suprapermafrost layer. { ¦sü·prə'jel·əˌsȯl }

supralateral tangent arcs [METEOROL] Two oblique luminous arcs, concave to the sun and tangent to the halo of 46° at points above the altitude of the sun. { ¦sü·prə'lad·ə·rəl 'tan·jənt 'ärks }

suprapermafrost layer [HYD] The layer of ground above permafrost; it includes the active layer and possibly occurrences of talik and perelotok. Also known as supragelisol. { ¦sü·prə'pər·məˌfrȯst ˌlā·ər }

supratidal sediment [GEOL] The sediment deposited immediately above the high-tide level. { ¦sü·prə'tīd·əl 'sed·ə·mənt }

supratidal zone [GEOL] Pertaining to the shore area immediately marginal to and above the high-tide level. { ¦sü·prə'tīd·əl zōn }

surf [OCEANOGR] Wave activity in the area between the shoreline and the outermost limit of breakers, that is, in the surf zone. { sərf }

surface boundary layer [METEOROL] That thin layer of air adjacent to the earth's surface, extending up to the so-called anemometer level (the base of the Ekman layer); within this layer the wind distribution is determined largely by the vertical temperature gradient and the nature and contours of the underlying surface, and shearing stresses are approximately constant. Also known as atmospheric boundary layer; friction layer; ground layer; surface layer. { 'sər·fəs 'bau̇n·drē ˌlā·ər }

surface chart [METEOROL] An analyzed synoptic chart of surface weather observations; essentially, a surface chart shows the distribution of sea-level pressure (therefore, the positions of highs, lows, ridges, and troughs) and the location and nature of fronts and air masses, plus the symbols of occurring weather phenomena, analysis of pressure tendency (isallobars), and indications of the movement of pressure systems and fronts. Also known as sea-level chart; sea-level-pressure chart; surface map. { 'sər·fəs ˌchärt }
surface creep [GEOL] A stage of the wind erosion process in which grains of sand move each other along the surface. { 'sər·fəs ˌkrēp }
surface current [OCEANOGR] **1.** Water movement which extends to depths of 3–10 feet (1–3 meters) below the surface in nearshore areas, and to about 33 feet (10 meters) in deep-ocean areas. **2.** Any current whose maximum velocity core is at or near the surface. { 'sər·fəs ˌkə·rənt }
surface deposit *See* surficial deposit. { 'sər·fəs diˌpäz·ət }
surface detention [HYD] Water in temporary storage as a thin sheet over the soil surface during the occurrence of overland flow. { 'sər·fəs diˌten·chən }
surface drainage [HYD] Natural or artificial removal of excess groundwater. { 'sər·fəs ˌdrā·nij }
surface duct [GEOPHYS] Atmospheric duct for which the lower boundary is the surface of the earth. { 'sər·fəs ˌdəkt }
surface flow *See* overland flow. { 'sər·fəs ˌflō }
surface friction [GEOPHYS] The drag or skin friction of the earth on the atmosphere, usually expressed in terms of the shearing stress of the wind on the earth's surface. { 'sər·fəs ˌfrik·shən }
surface geology [GEOL] The scientific study of the features at the surface of the earth. { 'sər·fəs jēˌäl·ə·jē }
surface hoar [HYD] **1.** Fernlike ice crystals formed directly on a snow surface by sublimation; a type of hoarfrost. **2.** Hoarfrost that has grown primarily in two dimensions, as on a window or other smooth surface. { 'sər·fəs ˌhȯr }
surface inversion [METEOROL] A temperature inversion based at the earth's surface; that is, an increase of temperature with height beginning at ground level. Also known as ground inversion. { 'sər·fəs inˌvər·zhən }
surface layer *See* surface boundary layer. { 'sər·fəs ˌlā·ər }
surface map *See* surface chart. { 'sər·fəs ˌmap }
surface of discontinuity [METEOROL] An interface, applied to the atmosphere; for example, an atmospheric front is represented ideally by a surface of discontinuity of velocity, density, temperature, and pressure gradient. { ¦sər·fəs əv ˌdisˌkänt·ən'ü·əd·ē }
surface phase [GEOCHEM] A thin rock layer differing in geochemical properties from those of the volume phases on either side. Also known as volume phase. { 'sər·fəs ˌfāz }
surface pressure [METEOROL] The atmospheric pressure at a given location on the earth's surface; the expression is applied loosely and about equally to the more specific terms: station pressure and sea-level pressure. { 'sər·fəs ˌpresh·ər }
surface retention *See* surface storage. { 'sər·fəs riˌten·chən }
surface runoff [HYD] Runoff that moves over the soil surface to the nearest surface stream. { 'sər·fəs 'rənˌȯf }
surface soil [GEOL] The soil extending 5 to 8 inches (13 to 20 centimeters) below the surface. { 'sər·fəs ˌsȯil }
surface storage [HYD] The part of precipitation retained temporarily at the ground surface as interception or depression storage so that it does not appear as infiltration or surface runoff either during the rainfall period or shortly thereafter. Also known as initial detention; surface retention. { 'sər·fəs ˌstȯr·ij }
surface temperature [METEOROL] Temperature of the air near the surface of the earth. [OCEANOGR] Temperature of the layer of seawater nearest the atmosphere. { 'sər·fəs ˌtem·prə·chər }

surface visibility [METEOROL] The visibility determined from a point on the ground, as opposed to control-tower visibility. { 'sər·fəs ˌviz·ə'bil·əd·ē }
surface wash *See* sheet erosion. { 'sər·fəs ˌwäsh }
surface water [HYD] All bodies of water on the surface of the earth. [OCEANOGR] *See* mixed layer. { 'sər·fəs ˌwȯd·ər }
surface wave [OCEANOGR] A progressive gravity wave in which the disturbance is of greatest amplitude at the air-water interface. { 'sər·fəs ˌwāv }
surface weather observation [METEOROL] An evaluation of the state of the atmosphere as observed from a point at the surface of the earth, as opposed to an upper-air observation, and applied mainly to observations which are taken for the primary purpose of preparing surface synoptic charts. { 'sər·fəs 'weth·ər ˌäb·zərˌvā·shən }
surface wind [METEOROL] The wind measured at a surface observing station; customarily, it is measured at some distance above the ground itself to minimize the distorting effects of local obstacles and terrain. { 'sər·fəs ˌwind }
surf beat [OCEANOGR] Oscillations of water level near shore, associated with groups of high breakers. { 'sərf ˌbēt }
surficial creep *See* soil creep. { sər'fish·əl 'krēp }
surficial deposit [GEOL] Unconsolidated alluvial, residual, or glacial deposits overlying bedrock or occurring on or near the surface of the earth. Also known as superficial deposit; surface deposit. { sər'fish·əl di'päz·ət }
surficial geology [GEOL] The scientific study of surficial deposits, including soils. { sər'fish·əl jē'äl·ə·jē }
surf ripple [GEOL] A ripple mark formed on a sandy beach by wave-generated currents. { 'sərf ˌrip·əl }
surf zone [OCEANOGR] The area between the landward limit of wave uprush and the farthest seaward breaker. { 'sərf ˌzōn }
surge [OCEANOGR] **1.** Wave motion of low height and short period, from about 1/2 to 60 minutes. **2.** *See* storm surge. { sərj }
surge line [METEOROL] A line along which a discontinuity in the wind speed occurs. { 'sərj ˌlīn }
surging breaker *See* plunging breaker. { 'sərj·iŋ 'brā·kər }
surging glacier [HYD] A glacier that alternates periodically between surges (brief periods of rapid flow) and stagnation. { 'sərj·iŋ 'glā·shər }
suroet [METEOROL] A persistent, rain-bearing southwest wind on the west coast of France. { sər·ə'wā }
suspended load [GEOL] The part of the stream load that is carried for a long time in suspension. Also known as suspension load. { sə'spen·dəd 'lōd }
suspended water *See* vadose water. { sə'spen·dəd 'wȯd·ər }
suspension current *See* turbidity current. { sə'spen·shən ˌkə·rənt }
suspension load *See* suspended load. { sə'spen·shən ¦lōd }
swale [GEOL] **1.** A slight depression, sometimes swampy, in the midst of generally level land. **2.** A shallow depression in an undulating ground moraine due to uneven glacial deposition. **3.** A long, narrow, generally shallow, troughlike depression which lies between two beach ridges and is aligned roughly parallel to the coastline. { swāl }
swallow hole [GEOL] An opening that occurs occasionally at the bottom of a sinkhole which permits direct drainage from the surface into an underground channel. { 'swäl·ō ˌhōl }
swash [GEOL] **1.** A narrow channel or ground within a sand bank, or between a sand bank and the shore. **2.** A bar over which the sea washes. [OCEANOGR] The rush of water up onto the beach following the breaking of a wave. Also known as run-up; uprush. { swäsh }
swash mark [GEOL] A fine, wavy or arcuate line or minute ridge consisting of fine sand, seaweed, and other debris on a beach; marks the farthest advance of wave uprush. Also known as debris line; wave line; wavemark. { 'swäsh ˌmärk }
S wave [GEOPHYS] A seismic body wave propagated in the crust or mantle of the earth by a shearing motion of material; speed is 1.9–2.5 miles (3–4 kilometers) per second

in the crust and 2.7–2.9 miles (4.4–4.6 kilometers) in the mantle. Also known as distortional wave; equivoluminal wave; rotational wave; secondary wave; shake wave; shear wave; tangential wave; transverse wave. { 'es ˌwāv }

swell [GEOL] **1.** The volumetric increase of soils on being removed from their compacted beds due to an increase in void ratio. **2.** A local enlargement or thickening in a vein or ore deposit. **3.** A low dome or quaquaversal anticline of considerable areal extent; long and generally symmetrical waves contribute to the mixing processes in the surface layer and thus to its sound transmission properties. **4.** Gently rising ground, or a rounded hill above the surrounding ground or ocean floor. [OCEANOGR] Ocean waves which have traveled away from their generating area; these waves are of relatively long length and period, and regular in character. { swel }

swell-and-swale topography [GEOGR] A low-relief, undulating landscape characterized by gentle slopes and rounded hills interspersed with shallow depressions. { ¦swel ən ¦swāl tə'päg·rə·fē }

swell direction [OCEANOGR] The direction from which swell is moving. { 'swel diˌrek·shən }

swelled ground [GEOL] A soil or rock that expands when wetted. { 'sweld 'graủnd }

swell forecast [OCEANOGR] Prediction of the frequency and height of swell waves in a remote area from the characteristics of the waves at their origin. { 'swel ˌfȯrˌkast }

swelling clay [GEOL] Clay that can absorb large amounts of water, such as bentonite. { 'swel·iŋ ¦klā }

symmetrical fold [GEOL] A fold whose limbs have approximately the same angle of dip relative to the axial surface. Also known as normal fold. { sə'me·trə·kəl 'fōld }

symmetric ripple mark [GEOL] A ripple mark whose cross-section profile is symmetric. { sə'me·trik 'rip·əl ˌmärk }

symmict [GEOL] Referring to a sedimentation unit that is structureless and in which coarse- and fine-grained particles are mixed more extensively in the lower part. { 'sim·ikt }

symon fault *See* horseback. { 'sī·mən ˌfȯlt }

synantexis [GEOL] Deuteric alteration. { ¦sin·ən¦tek·səs }

synchronous [GEOL] Geological rock units or features formed at the same time. { 'siŋ·krə·nəs }

synchronous pluton [GEOL] Any pluton whose time of emplacement coincides with a major orogeny. { 'siŋ·krə·nəs 'plüˌtän }

synclinal axis *See* trough surface. { sin'klīn·əl 'ak·səs }

synclinal valley [GEOL] Pertaining to a topographic valley whose sides coincide with a synclinal fold. { sin'klīn·əl 'val·ē }

syncline [GEOL] A fold having stratigraphically younger rock material in its core; it is concave upward. { 'sinˌklīn }

synclinorium [GEOL] A composite synclinal structure in a region of lesser folds. { ˌsin·klə'nȯr·ē·əm }

syngenesis [GEOL] In place formation of unconsolidated sediments. { sin'jen·ə·səs }

syngenetic [GEOL] **1.** Pertaining to a primary sedimentary structure formed contemporaneously with sediment deposition. **2.** Pertaining to a mineral deposit formed contemporaneously with the enclosing rock. Also known as ideogenous. { ¦sin·jə¦ned·ik }

synkinematic *See* syntectonic. { ¦sinˌkin·ə'mad·ik }

synoptic [METEOROL] Refers to the use of meteorological data obtained simultaneously over a wide area for the purpose of presenting a comprehensive and nearly instantaneous picture of the state of the atmosphere. { sə'näp·tik }

synoptic chart [METEOROL] Any chart or map on which data and analyses are presented that describe the state of the atmosphere over a large area at a given moment in time. { sə'näp·tik 'chärt }

synoptic climatology [CLIMATOL] The study and analysis of climate in terms of synoptic weather information, principally in the form of synoptic charts; the information thus obtained gives the climate (that is, average weather) of a given locality in a given synoptic situation rather than the usual climatic parameters which represent averages over all synoptic conditions. { sə'näp·tik ˌklī·mə'täl·ə·jē }

synoptic code [METEOROL] In general, any code by which synoptic weather observations are communicated; among the synoptic codes in use are the international synoptic code, ship synoptic code, U.S. Airways code, and RECCO code. { sə'näp·tik 'kōd }

synoptic meteorology [METEOROL] The study and analysis of synoptic weather information. { sə'näp·tik ˌmēd·ē·ə'räl·ə·jē }

synoptic model [METEOROL] Any model specifying a space distribution of some meteorological elements; the distribution of clouds, precipitation, wind, temperature, and pressure in the vicinity of a front is an example of a synoptic model. { sə'näp·tik 'mäd·əl }

synoptic oceanography [OCEANOGR] The study of the physical spatial parameters of the ocean through analysis of simultaneous observations from many stations. { sə'näp·tik ˌō·shə'näg·rə·fē }

synoptic report [METEOROL] An encoded and transmitted synoptic weather observation. { sə'näp·tik ri'pȯrt }

synoptic scale *See* cyclonic scale. { sə'näp·tik 'skāl }

synoptic wave chart [OCEANOGR] A chart of an ocean area on which is plotted synoptic wave reports from vessels, along with computed wave heights for areas where reports are lacking; atmospheric fronts, highs, and lows are also shown; isolines of wave height and the boundaries of areas having the same dominant wave direction are drawn. { sə'näp·tik 'wāv ˌchärt }

synoptic weather observation [METEOROL] A surface weather observation, made at periodic times (usually at 3- and 6-hourly intervals specified by the World Meteorological Organization), of sky cover, state of the sky, cloud height, atmospheric pressure reduced to sea level, temperature, dew point, wind speed and direction, amount of precipitation, hydrometeors and lithometeors, and special phenomena that prevail at the time of the observation or have been observed since the previous specified observation. { sə'näp·tik 'weth·ər ˌäb·zərˌvā·shən }

synorogenic [GEOL] Referring to a geologic process occurring at the same time as orogenic activity. { ¦sinˌȯr·ə'jen·ik }

syntectic *See* syntexis. { sin'tek·tik }

syntectonic [GEOL] Refers to a geologic process or event occurring during tectonic activity. Also known as synkinematic. { ¦sin·tek'tän·ik }

syntexis [GEOL] Magma made by the melting of two or more rock types and the assimilation of country rock. Also known as syntectic. { sin'tek·səs }

synthem [GEOL] A chronostratigraphic unit that defines an unconformity-bounded regional body of sediments and represents a cycle of sedimentation in response to changes in relative sea level or tectonics. { 'sinˌthem }

syssiderite *See* stony-iron meteorite. { sə'sid·əˌrīt }

system [GEOL] **1.** A major time-stratigraphic unit of worldwide significance, representing the basic unit of Phanerozic rocks. **2.** A group of related structures, such as joints. **3.** A chronostratigraphic unit, below erathem and above series. { 'sis·təm }

systematic joints [GEOL] Joints occurring in patterns or sets and oriented perpendicular to the boundaries of the constituent rock unit. { ˌsis·tə'mad·ik 'jȯins }

systems tract [GEOL] A discrete package of distinctive sediment types (facies) that are laid down during different phases of a cycle of sea-level change. { 'sis·təmz ˌtrakt }

T

Taber ice *See* segregated ice. { 'tā·bər ˌīs }
tabetisol *See* talik. { tə'bed·əˌsȯl }
table iceberg *See* tabular iceberg. { 'tā·bəl 'īsˌbərg }
table knoll [GEOGR] A knoll with a comparatively smooth, flat top. { 'tā·bəl ˌnōl }
tableland [GEOGR] A broad, elevated, nearly level, and extensive region of land that has been deeply cut at intervals by valleys or broken by escarpments. Also known as continental plateau. { 'tā·bəlˌand }
tablemount *See* guyot. { 'tā·bəlˌmau̇nt }
table mountain [GEOGR] A flat-topped mountain. { 'tā·bəl ˌmau̇nt·ən }
table reef [GEOL] A small, isolated organic reef which has a flat top and does not enclose a lagoon. { 'tā·bəl ˌrēf }
tabular [GEOL] Referring to a sedimentary particle whose length is two to three times its thickness. { 'tab·yə·lər }
tabular berg *See* tabular iceberg. { 'tab·yə·lər 'bərg }
tabular iceberg [OCEANOGR] An iceberg with clifflike sides and a flat top; usually arises by detachment from an ice shelf. Also known as table iceberg; tabular berg. { 'tab·yə·lər 'īsˌbərg }
tachylite [GEOL] A black, green, or brown volcanic glass formed from basaltic magma. Also known as basalt glass; basalt obsidian; hyalobasalt; jaspoid; sordawalite; wichtisite. { 'tak·əˌlīt }
Taconian orogeny [GEOL] A process of formation of mountains in the latter part of the Ordovician period, particularly in the northern Appalachians. Also known as Taconic orogeny. { tə'kō·nē·ən ȯ'räj·ə·nē }
Taconic orogeny *See* Taconian orogeny. { tə'kän·ik ȯ'räj·ə·nē }
taconite [GEOL] The siliceous iron formation from which high-grade iron ores of the Lake Superior district have been derived; consists chiefly of fine-grained silica mixed with magnetite and hematite. { 'tak·əˌnīt }
taele *See* frozen ground. { 'tā·lə }
Tahuian [GEOL] A local Eocene time subdivision in Australia whose identification is based on foraminiferans. { tə'wī·ən }
taiga climate [CLIMATOL] In general, a climate which produces taiga vegetation, that is, too cold for prolific tree growth but milder than the tundra climate and moist enough to promote appreciable vegetation. Also known as subarctic climate. { 'tī·gə ˌklī·mət }
tailwind [METEOROL] A wind which assists the intended progress of an exposed, moving object, for example, rendering an airborne object's ground speed greater than its airspeed; the opposite of a headwind. Also known as following wind. { 'tālˌwind }
taino [METEOROL] A tropical cyclone (hurricane) in parts of the Greater Antilles. { 'tī·nō }
taku wind [METEOROL] A strong, gusty, east-northeast wind, occurring in the vicinity of Juneau, Alaska, between October and March; it sometimes attains hurricane force at the mouth of the Taku River, after which it is named. { 'tä·kü ˌwind }
talik [GEOL] A Russian term applied to permanently unfrozen ground in regions of permafrost; usually applies to a layer which lies above the permafrost but below

the active layer, that is, when the permafrost table is deeper than the depth reached by winter freezing from the surface. Also known as tabetisol. { 'tä·lik }

talus [GEOL] Also known as rubble; scree. **1.** Coarse and angular rock fragments derived from and accumulated at the base of a cliff or steep, rocky slope. **2.** The accumulated heap of such fragments. { 'tal·əs }

talus creep [GEOL] The slow, downslope movement of talus. { 'tal·əs ˌkrēp }

talus glacier *See* rock glacier. { 'tal·əs ˌglā·shər }

talus slope [GEOL] A steep, concave slope consisting of an accumulation of talus. Also known as debris slope. { 'tal·əs ˌslōp }

tangent arc [METEOROL] Generic name for several types of halo arcs that form as loci tangent to other halos; the halo of 22° occasionally exhibits the horizontal and vertical tangent arcs, and the halo of 46° exhibits the infralateral tangent arcs and the supralateral tangent arcs. { 'tan·jənt ¦ärk }

tangential wave *See* S wave. { tan'jen·chəl 'wāv }

taphrogenesis *See* taphrogeny. { ˌtaf·rə'jen·ə·səs }

taphrogeny [GEOL] The formation of rift or trench phenomena, characterized by block faulting and associated subsidence. Also known as taphrogenesis. { tə'fräj·ə·nē }

taphrogeosyncline [GEOL] A geosyncline formed as a rift basin between faults. { ¦taf·rōˌjē·ō'sinˌklīn }

tapioca snow *See* snow pellets. { ˌtap·ē'ō·kə 'snō }

tarantata [METEOROL] A strong breeze from the northwest in the Mediterranean region. { ˌtär·ən'täd·ə }

tarn [GEOGR] A landlocked pool or small lake that may occur in a marsh or swamp, or that may occupy a basin amid mountain ranges. { tärn }

tar sand [GEOL] A type of oil sand; a sand whose interstices are filled with asphalt that remained after the escape of the lighter fractions of crude oil. { 'tär ˌsand }

tar seep [GEOL] Natural tar that, because of its close proximity to the ground surface, seeps from cracks in the earth or from between rocks, often forming pits or pools. { 'tär ˌsēp }

tasmanite [GEOL] An impure coal, transitional between cannel coal and oil shale. Also known as combustible shale; Mersey yellow coal; white coal; yellow coal. { 'taz·məˌnīt }

tau-value [METEOROL] The time rate of change of D value at a fixed point defined by the relation $\tau = (\Delta_t D)/(\Delta t)$, where Δt is the change in time and $\Delta_t D$ is the change in D value during this time interval; tau-values are expressed in terms of feet per hour; tau-value lines are drawn on 4-D charts and constitute the time dimension of these charts. { 'tau̇ ˌval·yü }

taylorite *See* bentonite. { 'tā·ləˌrīt }

Tchernozem *See* Chernozem. { 'chər·nəˌzem }

tear fault [GEOL] A very steep to vertical fault associated with and perpendicular to the strike of an overthrust fault. { 'tar ˌfȯlt }

tectite *See* tektite. { 'tekˌtīt }

tectofacies [GEOL] A lithofacies that is interpreted tectonically. { ¦tek·tə'fā·shēz }

tectogene [GEOL] A long, relatively narrow downward fold of sialic crust considered to be an early phase in mountain-building processes. Also known as geotectogene. { 'tek·təˌjēn }

tectogenesis *See* orogeny. { ¦tek·tə'jen·ə·səs }

tectonic analysis *See* petrotectonics. { tek'tän·ik ə'nal·ə·səs }

tectonic conglomerate *See* crush conglomerate. { tek'tän·ik kən'gläm·ə·rət }

tectonic cycle [GEOL] The orogenic cycle which relates larger crustal features, such as mountain belts, to a series of stages of development. Also known as geosynclinal cycle. { tek'tän·ik 'sī·kəl }

tectonic framework [GEOL] The relationship in space and time of subsiding, stable, and rising tectonic elements in a sedimentary source area. { tek'tän·ik 'frāmˌwərk }

tectonic land [GEOL] Linear fold ridges and volcanic islands which existed for a short time in the interior sections of an orogenic belt during the geosynclinal phase. { tek'tän·ik 'land }

tectonic lens [GEOL] An elongate, sausage-shaped body of rock formed by distortion of a continuous incompetent layer enclosed between competent layers, similar to a boudin, but genetically distinct. { tek'tän·ik 'lenz }

tectonic map [GEOL] A map which shows the architecture of the upper portion of the earth's crust. { tek'tän·ik 'map }

tectonic moraine [GEOL] An aggregation of boulders incorporated in the base of an overthrust mass. { tek'tän·ik mə'rān }

tectonic patterns [GEOL] The arrangement of the large structural units of the earth's crust, such as mountain systems, shields or stable areas, basins, arches, and volcanic archipelagoes. { tek'tän·ik 'pad·ərnz }

tectonic plate [GEOL] Any one of the internally rigid crustal blocks of the lithosphere which move horizontally across the earth's surface relative to one another. Also known as crustal plate. { tek'tän·ik 'plāt }

tectonic rotation [GEOL] Internal rotation of a tectonite in the direction of transport. { tek'tän·ik rō'tā·shən }

tectonics [GEOL] A branch of geology that deals with regional structural and deformational features of the earth's crust, including the mutual relations, origin, and historical evolution of the features. Also known as geotectonics. { tek'tän·iks }

tectonoeustatism [OCEANOGR] Fluctuations of sea level due to changes in the capacities of the ocean basins resulting from earth movements. { ¦tek·tə·nō'yü·stə,tiz·əm }

tectonomagnetism [GEOPHYS] Study of magnetic anomalies due to tectonic stress. { ¦tek·tə·nō'mag·nə,tiz·əm }

tectonophysicist [GEOPHYS] One who studies elastic deformation of flow and rupture of constituent materials of the earth's crust and makes deductions concerning the forces that cause these deformations. { ¦tek·tə·nō'fiz·ə,sist }

tectonophysics [GEOPHYS] A branch of geophysics dealing with the physical processes involved in forming geological structures. { ¦tek·tə·nō'fiz·iks }

tectosome [GEOL] A body of strata representing a tectotope. { 'tek·tə,sōm }

tectosphere [GEOL] The region of the earth's crust occupied by the tectonic plates. { 'tek·tə,sfir }

teeth of the gale [METEOROL] An old nautical term for the direction from which the wind is blowing (upwind, windward); to sail into the teeth of the gale is to sail to windward. { 'tēth əv thə 'gāl }

tehuantepecer [METEOROL] A violent squally wind from north or north-northeast in the Gulf of Tehuantepec in winter; it originates in the Gulf of Mexico, as a norther which crosses the isthmus and blows through the gap between the Mexican and Guatemalan mountains. { tə'wän·tə,pek·ər }

tektite [GEOL] A collective term applied to certain objects of natural glass of debatable origin that are widely strewn over the land and in sediments under the oceans; composition and size vary, and overall shapes resemble splash forms; most tektites are believed to be of extraterrestrial origin. Also known as obsidianite; tectite. { 'tek,tīt }

telain *See* provitrain. { 'te,lān }

telemagmatic [GEOL] Pertaining to a hydrothermal mineral deposit that is distant from its magmatic source. { ¦tel·ə·mag'mad·ik }

telemeteorometry [METEOROL] The study of making meteorological observations at a distance. { ¦tel·ə,mēd·ē·ə'räm·ə·trē }

telescope structure [GEOL] An alluvial fan structure characterized by younger fans with flatter gradients spreading out between older fans with steeper gradients. { 'tel·ə,skōp ,strək·chər }

teleseism [GEOPHYS] An earthquake that is far from the recording station. { 'tel·ə,sīz·əm }

teleseismology [GEOPHYS] The aspect of seismology dealing with records made at a distance from the source of the impulse. { ,tel·ə·sīz'mäl·ə·jē }

telethermal [GEOL] Pertaining to a hydrothermal mineral deposit precipitated at a shallow depth and at a mild temperature. { ¦tel·ə'thər·məl }

telinite [GEOL] A variety of provitrinite composed of plant cell-wall material. { 'tē·lə,nīt }

telluric current *See* earth current. { tə'lu̇r·ik ,kə·rənt }

tellurics [GEOPHYS] A geophysical exploration technique that measures variations in the conductivity (or resistivity) of rocks; often used for metallic mineral prospecting. { tə'lu̇r·iks }

temperate belt [CLIMATOL] A belt around the earth within which the annual mean temperature is less than 20°C (68°F) and the mean temperature of the warmest month is higher than 10°C (50°F). { 'tem·prət ,belt }

temperate climate [CLIMATOL] The climate of the middle latitudes; the climate between the extremes of tropical climate and polar climate. { 'tem·prət 'klī·mət }

temperate glacier [HYD] A glacier which, at the end of the melting season, is composed of firn and ice at the melting point. { 'tem·prət 'glā·shər }

temperate rainy climate [CLIMATOL] One of the major categories in W. Kippen's climatic classification; the coldest-month mean temperature is less than 64.4°F (18°C) and greater than 26.6°F (-3°C), and the warmest-month mean temperature is more than 50°F (10°C). { 'tem·prət 'rān·ē ,klī·mət }

temperate westerlies *See* westerlies. { 'tem·prət 'wes·tər·lēz }

temperate-westerlies index [METEOROL] A measure of the strength of the westerly wind between latitudes 35°N and 55°N; the index is computed from the average sea-level pressure difference between these latitudes and is expressed as the west to east component of geostrophic wind in meters and tenths of meters per second. { 'tem·prət ¦wes·tər·lez 'in,deks }

Temperate Zone [CLIMATOL] Either of the two latitudinal zones on the earth's surface which lie between 23°27′ and 66°32′ N and S (the North Temperate Zone and South Temperate Zone, respectively). { 'tem·prət ,zōn }

temperature belt [METEOROL] The belt which may be drawn on a thermograph trace or other temperature graph by connecting the daily maxima with one line and the daily minima with another. { 'tem·prə·chər ,belt }

temperature-humidity index [METEOROL] An index which gives a numerical value, in the general range of 70–80, reflecting outdoor atmospheric conditions of temperature and humidity as a measure of comfort (or discomfort) during the warm season of the year; equal to 15 plus 0.4 times the sum of the dry-bulb and wet-bulb temperatures in degrees Fahrenheit. Also known as comfort index; discomfort index. Abbreviated CI; DI; THI. { 'tem·prə·chər hyü'mid·ədē ,in,deks }

temperature inversion [METEOROL] A layer in the atmosphere in which temperature increases with altitude; the principal characteristic of an inversion layer is its marked static stability, so that very little turbulent exchange can occur within it; strong wind shears often occur across inversion layers, and abrupt changes in concentrations of atmospheric particulates and atmospheric water vapor may be encountered on ascending through the inversion layer. Also known as thermal inversion. [OCEANOGR] A layer of a large body of water in which temperature increases with depth. { 'tem·prə·chər in,vər·zhən }

temperature province [CLIMATOL] A major division of C.W. Thornthwaite's schemes of climatic classification, determined as a function of the temperature-efficiency index or the potential evapotranspiration. { 'tem·prə·chər ,präv·əns }

temperature-salinity diagram [OCEANOGR] The plot of temperature versus salinity data of a water column; the resulting diagram identifies the water masses within the column, the column's stability, indicates the σ_T value via lines of constant σ_T printed on paper, and allows an estimate of the accuracy of the temperature and salinity measurements. Also known as T-S curve; T-S diagram; T-S relation. { 'tem·prə·chər sə'lin·əd·ē ,dī·ə,gram }

temperature zone [CLIMATOL] A portion of the earth's surface defined by relatively uniform temperature characteristics, and usually bounded by selected values of some measure of temperature or temperature effect. { 'tem·prə·chər ,zōn }

temporale [METEOROL] A rainy wind from the southwest to west resulting from a deflection of the southeast trades of the eastern South Pacific onto the Pacific coast of Central America. { ˌtem·pȯ'rä·lē }

temporal unit [GEOL] A stratigraphic unit defined in terms of time-related characteristics. { 'tem·prəl ˌyü·nət }

temporary base level [GEOL] Any base level, other than sea level, below which a land area temporarily cannot be reduced by erosion. Also known as local base level. { 'tem·pəˌrer·ē 'bās ˌlev·əl }

tendency [METEOROL] The local rate of change of a vector or scalar quality with time at a given point in space. { 'ten·dən·sē }

tendency chart *See* change chart. { 'ten·dən·sē ˌchärt }

tendency equation [METEOROL] An equation for the local change of pressure at any point in the atmosphere, derived by combining the equation of continuity with an integrated form of the hydrostatic equation. { 'ten·dən·sē iˌkwā·zhən }

tendency interval [METEOROL] The finite increment of time over which a change of the value of a meteorological element is measured in order to estimate its tendency; the most familiar example is the three-hour time interval over which local pressure differences are measured in determining pressure tendency. { 'ten·dən·sē ˌin·tər·vəl }

tenggara [METEOROL] A strong, dry, hazy, east or southeast wind during the east monsoon in the Spermunde Archipelago. { teŋ'gär·ə }

tension crack [GEOL] An extension fracture caused by tensile stress. { 'ten·chən ˌkrak }

tension fault [GEOL] A fault in which crustal tension is a factor, such as a normal fault. Also known as extensional fault. { 'ten·chən ˌfȯlt }

tension fracture [GEOL] A minor rock fracture developed at right angles to the direction of maximum tension. Also known as subsidiary fracture. { 'ten·chən ˌfrak·chər }

tension joint [GEOL] A joint that is a tension fracture. { 'ten·chən ˌjȯint }

tented ice [OCEANOGR] A type of pressure ice formed when ice is pushed up vertically, producing a flat-sided arch with a cavity between the raised ice and the water beneath. { 'ten·təd 'īs }

tenting [OCEANOGR] The vertical displacement upward of ice under pressure to form a flat-sided arch with a cavity beneath. { 'tent·iŋ }

tepee butte [GEOGR] A tepeelike hill or knoll, especially one comprising soft material capped by more resistant rock. { 'tē·pē ˌbyüt }

tepee structure [GEOL] A disharmonic sedimentary structure consisting of a fold that resembles an inverted depressed V in cross section. { 'tē·pē ˌstrək·chər }

tepetate *See* caliche. { ˌtep·ə'täd·ē }

tephigram [METEOROL] A thermodynamic diagram designed by Napier Shaw with temperature and logarithm of potential temperature as coordinates; isobars are gently curved lines and the chart is rotated so that pressure increases downward; vapor lines and saturation adiabats are curved; on this chart, energy is proportional to the area enclosed by the curve representing the process. { 'tef·əˌgram }

tephra [GEOL] All pyroclastics of a volcano. { 'tef·rə }

tephrochronology [GEOL] The dating of different layers of volcanic ash for the establishment of a sequence of geologic and archeologic occurrences. { ¦tef·rō·krə'näl·ə·jē }

tephrostratigraphy [GEOL] The use of pyroclastic layers, in particular volcanic ash, as a correlational tool in the study of stratigraphic sequences. { ˌtef·rō·strə'tig·rə·fē }

terdiurnal [METEOROL] Pertaining to a meteorological event that occurs three times a day. { ˌtər·dī'ərn·əl }

terminal forecast [METEOROL] An aviation weather forecast for one or more specified air terminals. { 'tər·mən·əl ¦fȯrˌkast }

terminal moraine [GEOL] An end moraine that extends as an arcuate or crescentic ridge across a glacial valley; marks the farthest advance of a glacier. Also known as marginal moraine. { 'tər·mən·əl mə'rān }

terraced pool [GEOGR] A shallow, rimmed pool on the surface of a reef. { 'ter·əst 'pül }

terrace [GEOL] **1.** A horizontal or gently sloping embankment of earth along the contours of a slope to reduce erosion, control runoff, or conserve moisture. **2.** A narrow coastal strip sloping gently toward the water. **3.** A long, narrow, nearly level surface bounded by a steeper descending slope on one side and by a steeper ascending slope on the other side. **4.** A benchlike structure bordering an undersea feature. { 'ter·əs }

terracette [GEOL] A small steplike form developed on the surface of a slumped soil mass along a steep grassy incline. { ¦ter·ə¦set }

terral levante [METEOROL] A land breeze of Spain and Brazil, sometimes a northwest squall of foehn character. { tə'räl lə'vän·tā }

terrane [GEOL] A rock formation, a cluster of rock formations, or the general area of outcrops. { tə'rān }

terra rossa [GEOL] A reddish-brown soil overlying limestone bedrock. { 'ter·ə 'rós·ə }

terrestrial coordinates *See* geographical coordinates. { tə'res·trē·əl kō'ȯrd·ən·əts }

terrestrial electricity [GEOPHYS] Electric phenomena and properties of the earth; used in a broad sense to include atmospheric electricity. Also known as geoelectricity. { tə'res·trē·əl ˌiˌlek'tris·əd·ē }

terrestrial environment [GEOGR] The earth's land area, including its human-made and natural surface and subsurface features, and its interfaces and interactions with the atmosphere and the oceans. { tə'res·trē·əl in'vī·rən·mənt }

terrestrial equator *See* astronomical equator. { tə'res·trē·əl i'kwād·ər }

terrestrial frozen water [HYD] Seasonally or perennially frozen waters of the earth, exclusive of the atmosphere. { tə'res·trē·əl 'frō·zən 'wȯd·ər }

terrestrial gravitation [GEOPHYS] The effect of gravitational attraction of the earth. { tə'res·trē·əl ˌgrav·ə'tā·shən }

terrestrial magnetism *See* geomagnetism. { tə'res·trē·əl 'mag·nəˌtiz·əm }

terrestrial meridian *See* astronomical meridian. { tə'res·trē·əl mə'rid·ē·ən }

terrestrial radiation [GEOPHYS] Electromagnetic radiation originating from the earth and its atmosphere at wavelengths determined by their temperature. Also known as earth radiation; eradiation. { tə'res·trē·əl ˌrād·ē'ā·shən }

terrestrial sediment [GEOL] A sedimentary deposit on land above tidal reach. { tə'res·trē·əl 'sed·ə·mənt }

terrigenous sediment [GEOL] Shallow marine sedimentary deposits composed of eroded terrestrial material. { tə'rij·ə·nəs 'sed·ə·mənt }

Tertiary [GEOL] The older major subdivision (period) of the Cenozoic era, extending from the end of the Cretaceous to the beginning of the Quaternary, from 70,000,000 to 2,000,000 years ago. { 'tər·shēˌer·ē }

tertiary circulation [METEOROL] The generally small, localized atmospheric circulations, represented by such phenomena as local winds, thunderstorms, and tornadoes. { 'tər·shēˌer·ē ˌsər·kyə'lā·shən }

Tethys [GEOL] **1.** A sea which existed for extensive periods of geologic time between the northern and southern continents of the Eastern Hemisphere. **2.** A composite geosyncline from which many structures of the present Alpine-Himalayan orogenic belt were formed. { 'tē·thəs }

texture [GEOL] The physical nature of the soil according to composition and particle size. { 'teks·chər }

thalassic [OCEANOGR] Of or pertaining to the smaller seas. { thə'las·ik }

thalassocratic [GEOL] **1.** Pertaining to a thalassocraton. **2.** Referring to a period of high sea level in the geologic past. { thə¦las·ə¦krad·ik }

thalassocraton [GEOL] A craton that is part of the oceanic crust. { ˌthal·ə'säk·rəˌtān }

thalassophile element [GEOCHEM] An element that is relatively more abundant in sea water than in normal continental waters, such as sodium and chlorine. { thə'las·əˌfīl ˌel·ə·mənt }

thalweg [GEOGR] The middle of the principal navigable waterway which serves as a boundary between two states. [GEOL] **1.** A line connecting the lowest points along a stream bed or a valley. Also known as valley line. **2.** A line crossing all contour lines on a land surface perpendicularly. [HYD] Water seeping through the ground below the surface in the same direction as a surface stream course. { 'täl,veg }

Thanetian [GEOL] A European stage of geologic time; uppermost Paleocene, above Montian, below Ypresian of Eocene. { thə'nē·shən }

thaw [CLIMATOL] A warm spell during which ice and snow melt, as a January thaw. { thȯ }

thawing index [CLIMATOL] The number of degree days, above and below 32°F, between the lowest and highest points on the cumulative degree-days time curve for one thawing season. { 'thȯ·iŋ ,in,deks }

thawing season [CLIMATOL] The period of time between the lowest point and the succeeding highest point on the time curve of cumulative degree days above and below 32°F. { 'thȯ·iŋ ,sē·zən }

thermal [METEOROL] A relatively small-scale, rising current of air produced when the atmosphere is heated enough locally by the earth's surface to produce absolute instability in its lower layers. { 'thər·məl }

thermal aureole *See* aureole. { 'thər·məl 'ȯr·ē,ōl }

thermal climate [CLIMATOL] Climate as defined by temperature, and divided regionally into temperature zones. { 'thər·məl 'klī·mət }

thermal convection [METEOROL] Atmospheric currents, predominantly vertical, arising from the release of gravitational visibility; commonly produced by solar heating of the ground; the cause of convective (cumulus) clouds. Also known as free convection; gravitational convection. { 'thər·məl kən'vek·shən }

thermal equator *See* heat equator. { 'thər·məl i'kwād·ər }

thermal gradient [GEOPHYS] The rate of temperature change with distance; for example, its increase with depth below the surface of the earth. { 'thər·məl 'grād·ē·ənt }

thermal high [METEOROL] A high resulting from the cooling of air by a cold underlying surface, and remaining relatively stationary over the cold surface. { 'thər·məl 'hī }

thermal inversion *See* temperature inversion. { 'thər·məl in'vər·zhən }

thermal jet [METEOROL] A region in the atmosphere where isotherms or thickness lines are closely packed; therefore, a region of very strong thermal wind. { 'thər·məl 'jet }

thermal low [METEOROL] An area of low atmospheric pressure due to high temperatures caused by intensive heating at the earth's surface; common to the continental subtropics in summer, thermal lows remain stationary over the area that produces them, their cyclonic circulation is generally weak and diffuse, and they are nonfrontal. Also known as heat low. { 'thər·məl 'lō }

thermal quality of snow *See* quality of snow. { 'thər·məl 'kwäl·əd·ē əv 'snō }

thermal spring [HYD] A spring whose water temperature is higher than the local mean annual temperature of the atmosphere. { 'thər·məl 'spriŋ }

thermal steering [METEOROL] The steering of an atmospheric disturbance in the direction of the thermal wind in its vicinity; equivalent to steering along thickness lines; for this purpose the thermal wind is usually taken from the earth's surface to a level in the middle troposphere. { 'thər·məl 'stir·iŋ }

thermal stratification [HYD] Horizontal layers of differing densities produced in a lake by temperature changes at different depths. { 'thər·məl ,strad·ə·fə'kā·shən }

thermal tide [METEOROL] A variation in atmospheric pressure due to the diurnal differential heating of the atmosphere by the sun; so-called in analogy to the conventional gravitational tide. { 'thər·məl 'tīd }

thermal vorticity [METEOROL] The vorticity of a thermal wind. { 'thər·məl vȯr'tis·əd·ē }

thermal vorticity advection [METEOROL] The advection or transport of the thermal vorticity by the thermal wind, in analogy to the advection of the vorticity by the wind. { 'thər·məl vȯr¦tis·əd·ē ad'vek·shən }

thermal wind [METEOROL] The mean wind-shear vector in geostrophic balance with the gradient of mean temperature of a layer bounded by two isobaric surfaces. { 'thər·məl 'wind }

thermal wind equation [METEOROL] An equation for the vertical variation of the geostrophic wind in hydrostatic equilibrium which may be written $-(\partial \mathbf{V}/\partial p) = (R/pf)\ \mathbf{k} \times \nabla_p T$, where $\mathbf{V}$ is the vector geostrophic wind, p the pressure (used here as the vertical coordinate), R the gas constant for air, *f* the Coriolis parameter, **k** a vertically directed unit vector, and ∇_p the isobaric del operator. { 'thər·məl ¦wind i,kwā·zhən }

thermocline [GEOPHYS] **1.** A temperature gradient as in a layer of sea water, in which the temperature decrease with depth is greater than that of the overlying and underlying water. Also known as metalimnion. **2.** A layer in a thermally stratified body of water in which such a gradient occurs. { 'thər·mə,klīn }

thermocyclogenesis [METEOROL] A theory of cyclogenesis by G. Stüve, in which the disturbance is initiated in the stratosphere and is reflected in the development of a disturbance in the lower troposphere. { ¦thər·mō,sī·klə'jen·ə·səs }

thermohaline [OCEANOGR] Pertaining to the joint activity of salinity and temperature in the oceans. { ¦thər·mō'hā,līn }

thermohaline convection [OCEANOGR] Vertical water movement observed when sea water, due to conditions of decreasing temperature or increasing salinity, becomes heavier than the water beneath it. { ¦thər·mō'hā,līn kən'vek·shən }

thermoisopleth [CLIMATOL] An isopleth of temperature; specifically, a line on a climatic graph showing the variation of temperature in relation to two coordinates. { ¦thər·mō'īs·ə,pleth }

thermokarst topography [GEOL] An irregular land surface formed in a permafrost region by melting ground ice. { 'thər·mə,kärst tə'päg·rə·fē }

thermometric depth [OCEANOGR] The ocean depth, in meters, deduced from the difference between the paired protected and unprotected reversing thermometer readings; the unprotected reversing thermometer indicates higher temperature due to pressure effects on the instrument. { ¦thər·mə¦me·trik 'depth }

thermoremanent magnetization [GEOPHYS] The permanent magnetization of igneous rocks, acquired at the time of cooling from the molten state. { ¦thər·mō'rem·ə·nənt ,mag·nəd·ə'zā·shən }

thermosphere [METEOROL] The atmospheric shell extending from the top of the mesosphere to outer space; it is a region of more or less steadily increasing temperature with height, starting at 40 to 50 miles (70 to 80 kilometers); the thermosphere includes, therefore, the exosphere and most or all of the ionosphere. { 'thər·mə,sfir }

thermosteric anomaly [OCEANOGR] Component of the specific volume anomaly for a parcel of sea water at a pressure of 1 atmosphere due to its temperature being other than the standard temperature of 0°C. { ¦thər·mə¦ster·ik ə'näm·ə·lē }

thermotropic model [METEOROL] A model atmosphere used in numerical forecasting, in which the parameters are the height of one constant-pressure surface (usually 500 millibars) and one temperature (usually the mean temperature between 100 and 500 millibars). { ¦thər·mō¦träp·ik 'mad·əl }

THI *See* temperature-humidity index.

thick-bedded [GEOL] Pertaining to a sedimentary bed that ranges in thickness from 60 to 120 centimeters (2–4 feet). { 'thik ,bed·əd }

thickness [METEOROL] The vertical depth, measured in geometric or geopotential units, of a layer in the atmosphere bounded by surfaces of two different values of the same physical quantity, usually constant-pressure surfaces. { 'thik·nəs }

thickness chart [METEOROL] A type of synoptic chart showing the thickness of a certain physically defined layer in the atmosphere; it almost always refers to an isobaric thickness chart, that is, a chart of vertical distance between two constant-pressure surfaces. { 'thik·nəs ,chärt }

thickness line [METEOROL] A line drawn through all geographic points at which the thickness of a given atmospheric layer is the same. Also known as relative contour; relative isohypse. { 'thik·nəs ,līn }

thickness pattern [METEOROL] The general geometric distribution of thickness lines on a thickness chart. Also known as relative hypsography; relative topography. { 'thik·nəs ,pad·ərn }

thick-skinned structure [GEOL] Any large-scale structure, such as a fold or fault, believed to have originated as a result of basement movement beneath overlying rocks. { 'thik ¦skind 'strək·chər }

thick-thin chart *See* isentropic thickness chart. { 'thik 'thin ˌchärt }

Thiessen polygon method [METEOROL] A method of assigning areal significance to point rainfall values: perpendicular bisectors are constructed to the lines joining each measuring station with those immediately surrounding it; the bisectors form a series of polygons, each polygon containing one station; the value of precipitation measured at a station is assigned to the whole area covered by the enclosing polygon. { 'tē·sən 'päl·iˌgän ˌmeth·əd }

thill *See* underclay. { thil }

thin [METEOROL] In aviation weather observations, the description of a sky cover that is predominantly transparent. { thin }

thin-bedded [GEOL] Pertaining to a sedimentary bed that ranges in thickness from 2 inches to 2 feet (5 to 60 centimeters). { 'thin ˌbed·əd }

thin-out [GEOL] Gradual thinning of a stratum, vein, or other body of rock until the upper and lower surfaces meet and the rock disappears. { 'thin'au̇t }

thin section [GEOL] A piece of rock or mineral specifically prepared to study its optical properties; the sample is ground to 0.03-millimeter thickness, then polished and placed between two microscope slides. Also known as section. { 'thin 'sek·shən }

thin-skinned structure [GEOL] Any large-scale structure, such as a fold or fault, confined to and originating within a thin layer of rocks above a surface of décollement. { 'thin ¦skind 'strək·chər }

thiophile element *See* sulfophile element. { 'thī·əˌfīl ˌel·ə·mənt }

third-order climatological station [CLIMATOL] As defined by the World Meteorological Organization, a station, other than a precipitation station, at which the observations are of the same kind as those at a second-order climatological station, but are not so comprehensive, are made once a day only, and are made at other than the specified hours. { 'thərd ¦ȯr·dər ˌklī·mə·tə'läj·ə·kəl 'stā·shən }

third-order relief [GEOGR] Specific landform complexes that are smaller in extent and size than formations of subcontinental extent. { 'thərd ¦ȯr·dər ri'lēf }

thirty-day forecast [METEOROL] A weather forecast for a period of 30 days; as issued by the U.S. Weather Bureau, the forecast concerns expected departures of temperature and precipitation from normal. { ¦thər·dē ¦dā 'fȯrˌkast }

thirty-two nucleus [METEOROL] An unidentified type of freezing nucleus which first becomes active when a supercooled cloud is cooled to about −32°C. { ¦thər·dē¦tü 'nü·klē·əs }

thixotropic clay [GEOL] A clay that weakens when disturbed and increases in strength upon standing. { ¦thik·sə¦träp·ik 'klā }

thread [GEOL] An extremely small vein, even thinner than a stringer. { thred }

thread-lace scoria [GEOL] Scoria whose vesicle walls have collapsed and are represented only by a network of threads. { 'thred ¦lās 'skȯr·ē·ə }

three-point method [GEOL] A method used to determine the dip and strike of a structural surface from three points of varying elevation along the surface. { 'thrē ¦pȯint 'meth·əd }

threshold *See* riegel. { 'threshˌhōld }

threshold depth *See* sill depth. { 'threshˌhōld ˌdepth }

threshold velocity [GEOPHYS] The minimum velocity at which wind or water begins to move particles of soil, sand, or other material at a given place under specified conditions. { 'threshˌhōld vəˌläs·əd·ē }

through glacier [HYD] A two-ended glacier, consisting of two valley glaciers in a depression, flowing in opposite directions. { 'thrü ˌglā·shər }

through valley [GEOL] **1.** A depression eroded across a divide by glacier ice or meltwater streams. **2.** A valley excavated by a through glacier. { 'thrü ˌval·ē }

throw [GEOL] The vertical component of dip separation on a fault, or generally the amount of vertical displacement on any fault. { thrō }

thrust [GEOL] Overriding movement of one crystal unit over another. Also known as mountain thrust. { thrəst }

thrust block *See* thrust nappe. { 'thrəst ˌbläk }

thrust fault [GEOL] A low-angle (less than a 45° dip) fault along which the hanging wall has moved up relative to the footwall. Also known as reverse fault; reverse slip fault; thrust slip fault. { 'thrəst ˌfȯlt }

thrust moraine *See* push moraine. { 'thrəst mə·rān }

thrust nappe [GEOL] The body of rock that makes up the hanging wall of a thrust fault. Also known as thrust block; thrust plate; thrust sheet; thrust slice. { 'thrəst ˌnap }

thrust plate *See* thrust nappe. { 'thrəst ˌplāt }

thrust sheet *See* thrust nappe. { 'thrəst ˌshēt }

thrust slice *See* thrust nappe. { 'thrəst ˌslīs }

thrust slip fault *See* thrust fault. { 'thrəst 'slip ˌfȯlt }

thucolite [GEOL] Concentrations of carbonaceous matter in ancient sedimentary rocks. { 'thü·kəˌlīt }

thunder [GEOPHYS] The sound emitted by rapidly expanding gases along the channel of a lightning discharge. { 'thən·dər }

thunderbolt [GEOPHYS] In mythology, a lightning flash accompanied by a material bolt or dart and which causes great damage; it is still used as a popular term for a single lightning discharge accompanied by thunder. { 'thən·dərˌbōlt }

thundercloud [METEOROL] A convenient and often used term for the cloud mass of a thunderstorm, that is, a cumulonimbus. { 'thən·dərˌklau̇d }

thunderhead *See* incus. { 'thən·dərˌhed }

thundersquall *See* rainsquall. { 'thən·dərˌskwȯl }

thunderstorm [METEOROL] A convective storm accompanied by lightning and thunder and rain, rarely snow showers but often hail, and gusty squall winds at the onset of precipitation; the characteristic cloud is the cumulonimbus. { 'thən·dərˌstȯrm }

thunderstorm cell [METEOROL] The convection cell of a cumulonimbus cloud. { 'thən·dərˌstȯrm ˌsel }

thunderstorm charge separation [GEOPHYS] **1.** The process by which the large electric field found within thunderclouds is generated. **2.** The processes by which particles bearing opposite electrical charges are given the charges and are transported to different regions of the active cloud. { 'thən·dərˌstȯrm 'chärj ˌsep·əˌrā·shən }

thunderstorm day [METEOROL] An observational day during which thunder is heard at the station; precipitation need not occur. { 'thən·dərˌstȯrm 'dā }

Thuringian [GEOL] A European stage of Upper Permian geologic time, above the Saxonian and below the Triassic. { thə'rin·jē·ən }

tidal bore *See* bore. { 'tīd·əl 'bȯr }

tidal channel [OCEANOGR] A major channel followed by tidal currents, extending from the ocean into a tidal marsh or tidal flat. { 'tīd·əl 'chan·əl }

tidal component *See* partial tide. { 'tīd·əl kəm'pō·nənt }

tidal constants [OCEANOGR] Tidal relations that remain essentially constant for any particular locality. { 'tīd·əl 'kän·stəns }

tidal constituent *See* partial tide. { 'tīd·əl kən'stich·ə·wənt }

tidal correction [GEOPHYS] A correction made in gravity observations to remove the effect of the earth's tides. { 'tīd·əl kə'rek·shən }

tidal current [OCEANOGR] The alternating horizontal movement of water associated with the rise and fall of the tide caused by the astronomical tide-producing forces. { 'tīd·əl 'kə·rənt }

tidal-current chart [OCEANOGR] A chart showing by arrows and numbers the average direction and speed of tidal currents at a particular part of the current cycle. { 'tīd·əl ¦kə·rənt ˌchärt }

tidal-current tables [OCEANOGR] Tables issued annually which give daily predictions of the times of slack water and the times and velocities of the strength of flood and ebb currents for a number of reference stations, together with differences and constants for obtaining predictions at subordinate stations. { 'tīd·əl ¦kə·rənt ˌtā·bəlz }

tidal cycle *See* tide cycle. { 'tīd·əl ˌsī·kəl }

tidal datum [OCEANOGR] A level of the sea, defined by some phase of the tide, from which water depths and heights of tide are reckoned. Also known as tidal datum plane. { 'tīd·əl 'dad·əm }

tidal datum plane *See* tidal datum. { 'tīd·əl ¦dad·əm ˌplān }
tidal day [OCEANOGR] The interval between two consecutive high waters of the tide at a given place, averaging 24 hours 51 minutes. { 'tīd·əl 'dā }
tidal delta [GEOL] A sand bar or shoal formed in the entrance of an inlet by the action of reversing tidal currents. { 'tīd·əl 'del·tə }
tidal difference [OCEANOGR] The difference in time or height of a high or low water at a subordinate station and at a reference station for which predictions are given in the tide tables; the difference applied to the prediction at the reference station gives the corresponding time or height for the subordinate station. { 'tīd·əl 'dif·rəns }
tidal energy [OCEANOGR] The energy in a tide flowing from a basin into an open sea. { 'tīd·əl 'en·ər·jē }
tidal epoch *See* phase lag. { 'tīd·əl 'ep·ik }
tidal excursion [OCEANOGR] The net horizontal distance over which a water particle moves during one tidal cycle of flood and ebb; the distances traversed during ebb and flood are rarely equal in nature, since there is usually a layered circulation in an estuary, with a net surface flow in one direction compensated by an opposite flow at depth. { 'tīd·əl ik'skər·zhən }
tidal flat [GEOL] A marshy, sandy, or muddy nearly horizontal coastal flatland which is alternately covered and exposed as the tide rises and falls. { 'tīd·əl 'flat }
tidal frequency [OCEANOGR] The rate of travel, in degrees per day, of a component of a tide, the component being created by a particular juxtaposition of forces in the sun-earth-moon system. { 'tīd·əl 'frē·kwən·sē }
tidal friction [OCEANOGR] The frictional effect of the tidal wave particularly in shallow waters that lengthens the tidal epoch and tends to slow the rotational velocity of the earth, thus increasing very slowly the length of the day. { 'tīd·əl 'frik·shən }
tidal glacier *See* tidewater glacier. { 'tīd·əl 'glā·shər }
tidal harbor [OCEANOGR] A harbor affected by the tides, in distinction to a harbor in which the water level is maintained by caissons or gates. { 'tīd·əl 'här·bər }
tidal inlet [GEOL] A natural inlet maintained by tidal currents. { 'tīd·əl 'in·lət }
tidalite [GEOL] Any sediment transported and deposited by tidal currents. { 'tīd·əlˌīt }
tidal marsh [GEOGR] Any marsh whose surface is covered and uncovered by tidal flow. { 'tīd·əl 'märsh }
tidal platform ice foot [OCEANOGR] An ice foot between high and low water levels, produced by the the rise and fall of the tide. { 'tīd·əl ¦platˌfȯrm 'īs ˌfu̇t }
tidal pool [OCEANOGR] An accumulation of sea water remaining in a depression on a beach or reef after the tide recedes. { 'tīd·əl 'pül }
tidal potential [OCEANOGR] Tidal forces expressed as components of a vector field. { 'tīd·əl pəˌten·chəl }
tidal prism [OCEANOGR] The difference between the mean high-water volume and the mean low-water volume of an estuary. { 'tīd·əl 'priz·əm }
tidal range *See* tide range. { 'tīd·əl 'rānj }
tidal scour [GEOL] Sea-floor erosion caused by strong tidal currents, resulting in removal of inshore sediments and formation of deep holes and channels. Also known as scour. { 'tīd·əl 'skau̇r }
tidal water [OCEANOGR] Any water whose level changes periodically due to tidal action. { 'tīd·əl ˌwȯd·ər }
tidal wave [OCEANOGR] **1.** Any unusually high and generally destructive sea wave or water level along a shore. **2.** *See* tide wave. { 'tīd·əl ˌwāv }
tidal wind [METEOROL] A very light breeze which occurs in calm weather in inlets where the tide sets strongly; it blows onshore with rising tide and offshore with ebbing tide. { 'tīd·əl 'wind }
tide [OCEANOGR] The periodic rising and falling of the oceans resulting from lunar and solar tide-producing forces acting upon the rotating earth. { tīd }
tide amplitude [OCEANOGR] One-half of the difference in height between consecutive high water and low water; half the tide range. { 'tīd 'am·pləˌtüd }
tide bulge *See* tide wave. { 'tīd ˌbəlj }
tide crack [OCEANOGR] A crack in sea ice, parallel to the shore, caused by the vertical

movement of the water due to tides; several such cracks often appear as a family. { 'tīd ˌkrak }

tide curve [OCEANOGR] Any graphic representation of the rise and fall of the tide; time is generally represented by the abscissas, and the height of the tide by the ordinates; for normal tides the curve so produced approximates a sine curve. { 'tīd ˌkərv }

tide cycle [OCEANOGR] A period which includes a complete set of tide conditions or characteristics, such as a tidal day or a lunar month. Also known as tidal cycle. { 'tīd ˌsī·kəl }

tidehead [OCEANOGR] The inland limit of water affected by a tide. { 'tīdˌhed }

tide hole [OCEANOGR] A hole made in ice to observe the height of the tide. { 'tīd ˌhōl }

tideland [GEOGR] Land which is under water at high tide and uncovered at low tide. { 'tīd·lənd }

tidemark [OCEANOGR] **1.** A high-water mark left by tidal water. **2.** The highest point reached by a high tide. { 'tīdˌmärk }

tide notes [OCEANOGR] Notes included on nautical charts which give information on the mean range or the diurnal range of the tide, mean tide level, and extreme low water at key places on the chart. { 'tīd ˌnōts }

tide prediction [OCEANOGR] The mathematical process by which the times and heights of the tide are determined in advance from the harmonic constituents at a place. { 'tīd priˌdik·shən }

tide-producing force [GEOPHYS] The slight local difference between the gravitational attraction of two astronomical bodies and the centrifugal force that holds them apart. { 'tīd prəˈdüs·iŋ 'fȯrs }

tide race [OCEANOGR] A strong tidal current or a channel in which such a current flows. { 'tīd ˌrās }

tide range [OCEANOGR] The difference in height between consecutive high and low waters. Also known as tidal range. { 'tīd ˌrānj }

tide rips *See* rips. { 'tīd ˌrips }

tide station [OCEANOGR] A place where observations of the tides are obtained. { 'tīd ˌstā·shən }

tide table [OCEANOGR] A table giving daily predictions, usually a year in advance, of the times and heights of the tide for a number of reference stations. { 'tīd ˌtā·bəl }

tidewater [OCEANOGR] **1.** A body of water, such as a river, affected by tides. **2.** Water inundating land at flood tide. { 'tīdˌwȯd·ər }

tidewater glacier [HYD] A glacier that descends into the sea and usually has a terminal ice cliff. Also known as tidal glacier. { 'tīdˌwȯd·ər 'glā·shər }

tide wave [OCEANOGR] A long-period wave associated with the tide-producing forces of the moon and sun, and identified with the rising and falling of the tide. Also known as tidal wave; tide bulge. { 'tīd ˌwāv }

tideway [OCEANOGR] A channel through which a tidal current runs. { 'tīdˌwā }

tie bar *See* tombolo. { 'tī ˌbär }

tight fold *See* closed fold. { 'tīt 'fōld }

tight sand [GEOL] A sand whose interstices are filled with finer grains of the matrix material, thus effectively destroying porosity and permeability. Also known as close sand. { 'tīt 'sand }

till [GEOL] Unsorted and unstratified drift consisting of a heterogeneous mixture of clay, sand, gravel, and boulders which is deposited by and underneath a glacier. Also known as boulder clay; glacial till; ice-laid drift. { til }

till billow [GEOL] An undulating mass of glacial drift that is disposed in an irregular pattern with regard to the direction of movement of the ice. { 'til ˌbil·ō }

tilloid [GEOL] A nonglacial till-like deposit. { 'tiˌlȯid }

till plain [GEOL] An extensive, relatively flat area overlying a till. { 'til ˌplān }

till sheet [GEOL] A sheet, layer, or bed of till. { 'til ˌshēt }

tilt [METEOROL] The inclination to the vertical of a significant feature of the circulation (or pressure) pattern or of the field of temperature or moisture; for example, troughs in the westerlies usually display a westward tilt with altitude in the lower and middle troposphere. { tilt }

tilt block [GEOL] A tilted fault block. { 'tilt ˌbläk }

tilted iceberg [OCEANOGR] A tabular iceberg that has become unbalanced, so that the flat, level top is inclined. { 'til·təd 'īsˌbərg }

tilted interface [GEOL] Oil-water interface in which water moves in a generally linear direction under an oil accumulation which is, for instance, in an anticline. { 'til·təd 'in·tərˌfās }

tilth [GEOL] The physical condition of a soil as expressed in terms of fitness for growth of specified plants or crops. { tilth }

time correlation [GEOL] A correlation of age or mutual time relations between stratigraphic units in separated areas. { 'tīm ˌkär·ə'lā·shən }

time line [GEOL] **1.** A line that indicates equal geologic age in a correlation diagram. **2.** A rock unit represented by a time line. { 'tīm ˌlīn }

time-rock unit *See* time-stratigraphic unit. { 'tīm 'räk ˌyü·nət }

time-stratigraphic facies [GEOL] A stratigraphic facies based on the amount of geologic time during which deposition and nondeposition of sediment occurred. { 'tīm ¦strad·ə¦graf·ik ˌfā·shēz }

time-stratigraphic unit [GEOL] A stratigraphic unit based on geologic age or time of origin. Also known as chronolith; chronolithologic unit; chronostratic unit; chronostratigraphic unit; time-rock unit. { 'tīm ¦strad·ə¦graf·ik ˌyü·nət }

time-transgressive *See* diachronous. { 'tīm tranzˌgres·iv }

Tithonian [GEOL] Southern European equivalent of the Portlandian stage (uppermost Jurassic) of geologic time. { ti'thō·nē·ən }

tivano [METEOROL] A night breeze blowing down the valley at Lake Como in Italy. { ti'vä·nō }

tjaele *See* frozen ground. { 'chā·lē }

Toarcian [GEOL] A European stage of geologic time; Lower Jurassic (above Pliensbachian, below Bajocian). { tō'är·shən }

todorokite [GEOL] A hydrated manganese oxide mineral containing calcium, barium, potassium, sodium, and sometimes magnesium; a major constituent of manganese nodules, which occur in large quantities ($>10^{12}$ tons) on the ocean floors. { tə'dȯr·əˌkīt }

toe [GEOL] The leading edge of a thrust nappe. { tō }

tofan [METEOROL] A violent spring storm common in the mountains of Indonesia. { tō'fän }

toise [GEOD] A unit of length equal to about 6.4 feet (1.95 meters); used in early geodetic surveys. { 'tȯiz }

tombolo [GEOL] A sand or gravel bar or spit that connects an island with another island or an island with the mainland. Also known as connecting bar; tie bar; tying bar. { 'täm·bəˌlō }

tombolo cluster *See* complex tombolo. { 'täm·bəˌlō ˌkləs·tər }

tombolo series *See* complex tombolo. { 'täm·bəˌlō ˌsir·ēz }

tongara [METEOROL] A hazy, southeast wind in the Macassar Strait. { täŋ'gär·ə }

Tongrian [GEOL] A European stage of geologic time; lower Oligocene (above Ludian of Eocene, below Rupelian). Also known as Lattorfian. { 'täŋ·grē·ən }

tongue [GEOL] **1.** A minor rock-stratigraphic unit of limited geographic extent; it disappears laterally in one direction. **2.** A lava flow branching from a larger flow. [OCEANOGR] **1.** A protrusion of water into a region of different temperature, or salinity, or dissolved oxygen concentrating. **2.** A protrusion of one water mass into a region occupied by a different water mass. { təŋ }

tonstein [GEOL] Kaolinitic bands in certain coalfields which have characteristic fossil fauna from short-lived but widespread marine invasions. { 'tänˌshtīn }

tool mark [GEOL] Any of the wide variety of current marks, such as groove marks, prod marks, and skip marks, produced by the continuous contact or intermittent impact of solid, current-borne objects against a muddy bottom. { 'tül ˌmärk }

top *See* overburden. { täp }

topographical latitude *See* geodetic latitude. { ¦täp·ə¦graf·ə·kəl 'lad·əˌtüd }

topographic curl effect [OCEANOGR] A term in Ekman's differential equation for the effects of variable wind stress, variable depth, variable friction, and variable latitude on the deep current; tends to make the curl G (velocity of deep current) positive when the current flows over increasing depth and negative when the depth decreases in the direction of the current. { ¦täp·ə¦graf·ik 'kərl i,fəkt }

topographic infancy *See* infancy. { ¦täp·ə¦graf·ik 'in·fən·sē }

topographic maturity *See* maturity. { ¦täp·ə¦graf·ik mə'chür·əd·ē }

topographic old age *See* old age. { ¦täp·ə¦graf·ik 'ōld 'āj }

topographic passage [OCEANOGR] A pass or gap through a sea-floor feature that possesses high topography, such as a ridge or a plateau. { ¦täp·ə,graf·ik 'pas·ij }

topographic profile *See* profile. { ¦täp·ə¦graf·ik 'prō,fīl }

topographic unconformity [GEOGR] A lack of harmony or conformity between two parts of a landscape or two kinds of topography. { ¦täp·ə¦graf·ik ,ən·kən'fór·məd·ē }

topographic youth *See* youth. { ¦täp·ə¦graf·ik 'yüth }

topography [GEOGR] **1.** The general configuration of a surface, including its relief; may be a land or water-bottom surface. **2.** The natural surface features of a region, treated collectively as to form. { tə'päg·rə·fē }

topset bed [GEOL] One of the nearly horizontal sedimentary layers deposited on the top surface of an advancing delta. { 'täp,set ¦bed }

topsoil [GEOL] **1.** Soil presumed to be fertile and used to cover areas of special planting. **2.** Surface soil, usually corresponding with the A horizon, as distinguished from subsoil. { 'täp,sóil }

tor [GEOGR] An isolated, rough pinnacle or rocky peak. { tór }

torbanite [GEOL] A variety of coal that resembles a carbonaceous shale in outward appearance; it is fine-grained, black to brown, and tough. Also known as bitumenite; kerosine shale. { 'tór·bə,nīt }

tornado [METEOROL] An intense rotary storm of small diameter, the most violent of weather phenomena; tornadoes always extend downward from the base of a convective-type cloud, generally in the vicinity of a severe thunderstorm. { tór'nād·ō }

tornado belt [METEOROL] The district of the United States in which tornadoes are most frequent; it encompasses the great lowland areas of the central and upper Mississippi, the Ohio, and lower Missouri River valleys. { tór'nād·ō ,belt }

tornado cloud *See* tuba. { tór'nād·ō ,klaüd }

tornado echo [METEOROL] A type of radar precipitation echo which has been observed in connection with a number of tornadoes; it frequently appears, on plan-position-indicator scopes, in the form of the figure 6 in the southwest sector of the storm; this echo has not been noted with all radar-observed tornadoes. { tór'nād·ō ,ek·ō }

torose load cast [GEOL] One of a group of elongate load casts with alternate contractions and swellings, which may terminate down current in bulbous, teardrop, or spiral forms. { 'tó,rōs 'lōd ,kast }

Torrert [GEOL] A suborder of the soil order Vertisol; it is the driest soil of the order and forms cracks that tend to remain open; occurs in arid regions. { 'tór·ərt }

Torrid Zone [CLIMATOL] The zone of the earth's surface which lies between the Tropics of Cancer and Capricorn. { 'tär·əd ,zōn }

Torrox [GEOL] A suborder of the soil order Oxisol that is low in organic matter, well drained, and dry most of the year; believed to have been formed under rainier climates of past eras. { 'tór,äks }

torsion fault *See* wrench fault. { 'tór·shən ,fólt }

torso mountain *See* monadnock. { 'tór·sō ,maünt·ən }

Tortonian [GEOL] A European stage of geologic time: Miocene (above Helvetian, below Sarmatian). { tór'tō·nē·ən }

tosca [METEOROL] A southwest wind on Lake Garda in Italy. { 'tós·kə }

total allowable catch [OCEANOGR] A fishery management approach to assign an annual quota that, if exceeded, will terminate the fishery for that year, the total allowable catch is set at a level to prevent a catch so large that the stock will be overfished. { ¦tōd·əl ə,laü·ə·bəl 'kach }

total conductivity [GEOPHYS] In atmospheric electricity, the sum of the electrical conductivities of the positive and negative ions found in a given portion of the atmosphere. { 'tōd·əl ˌkän·dək'tiv·əd·ē }

total displacement *See* slip. { 'tōd·əl di'splās·mənt }

total evaporation *See* evapotranspiration. { 'tōd·əl iˌvap·ə'rā·shən }

total porosity [GEOL] The ratio of total void space in porous oil-reservoir rock to the bulk volume of the rock itself. { 'tōd·əl pə'räs·əd·ē }

total slip *See* net slip. { 'tōd·əl 'slip }

touriello [METEOROL] A south wind of foehn type descending from the Pyrenees in the Ariège valley, France; it is especially violent in February and March, when it melts the snow, flooding the rivers and sometimes causing avalanches. { ˌtu̇r·ē'el·ō }

Tournaisian [GEOL] European stage of lowermost Carboniferous time. { tu̇r'nā·zhən }

Toussaint's formula [METEOROL] A rule for the linear decrease of temperature with height in an atmosphere for which the temperature at mean sea level is 15°C, and given by the formula $t = 15 - 0.0065z$, where t is the temperature in degrees Celsius, and z is the geometric height in meters above mean sea level. { tü'sanz ˌfȯr·myə·lə }

towering [METEOROL] A refraction phenomenon; a special case of looming in which the downward curvature of the light rays due to atmospheric refraction increases with elevation so that the visual image of a distant object appears to be stretched in the vertical direction. { 'tau̇·ə·riŋ }

towering cumulus [METEOROL] A descriptive term, used mostly in weather observing, for the cloud type cumulus congestus. { 'tau̇·ə·riŋ 'kyü·myə·ləs }

trace [GEOL] The intersection of two geological surfaces. [METEOROL] A precipitation of less than 0.005 inch (0.127 millimeter). { trās }

trace element [GEOCHEM] An element found in small quantities (usually less than 1.0%) in a mineral. Also known as accessory element; guest element. { 'trās ˌel·ə·mənt }

trace fossil [GEOL] A trail, track, or burrow made by an animal and found in ancient sediments such as sandstone, shale, or limestone. Also known as ichnofossil. { 'trās ˌfäs·əl }

trace slip [GEOL] That component of the net slip in a fault which is parallel to the trace of an index plane on a fault plane. { 'trās ˌslip }

trace-slip fault [GEOL] A fault whose net slip is trace slip. { 'trās ˌslip ˌfȯlt }

trachytoid texture [GEOL] The texture of a phaneritic extrusive igneous rock in which the microlites of a mineral, not necessarily feldspar, in the groundmass have a subparallel or randomly divergent alignment. { 'trak·əˌtȯid ˌteks·chər }

traction [GEOL] Transport of sedimentary particles along and parallel to a bottom surface of a stream channel by rolling, sliding, dragging, pushing, or saltation. { 'trak·shən }

trade air [METEOROL] The type of air of which the trade winds consist, and whose chief thermodynamic characteristic is the presence of the trade-wind inversion. { 'trād ¦er }

trade cumulus *See* trade-wind cumulus. { 'trād ¦kyü·myə·ləs }

trade-wind [METEOROL] The wind system, occupying most of the tropics, which blows from the subtropical highs toward the equatorial trough; a major component of the general circulation of the atmosphere; the winds are northeasterly in the Northern Hemisphere and southeasterly in the Southern Hemisphere; hence they are known as the northeast trades and southeast trades, respectively. { 'trād ¦wind }

trade-wind cumulus [METEOROL] The characteristic cumulus cloud of the trade winds over the oceans in average, undisturbed weather conditions; the individual cloud usually exhibits a blocklike appearance since its vertical growth ends abruptly in the lower stratum of the trade-wind inversion; a group of fully grown clouds shows considerable uniformity in size and shape. Also known as trade cumulus. { 'trād ¦wind ˌkyü·myə·ləs }

trade-wind desert [CLIMATOL] **1.** An area of very little rainfall and high temperature which occurs where the trade winds or their equivalent (such as the harmattan) blow over land; the best examples are the Sahara and Kalahari deserts. **2.** The arid cold-water coasts on the western shores of North and South America and Africa. { 'trād ¦wind ˌdez·ərt }

trade-wind inversion [METEOROL] A characteristic temperature inversion usually present in the trade-wind streams over the eastern portions of the tropical oceans: it is formed by broad-scale subsidence of air from high altitudes in the eastern extremities of the subtropical highs; while descending, the current meets the opposition of the low-level maritime air flowing equatorward; the inversion forms at the meeting point of these two strata which flow horizontally in the same direction. { 'trād ¦wind inˌvər·zhən }

traersu [METEOROL] A violent east wind of Lake Garda in Italy. { 'trerˌzü }

trail [GEOL] A line of rock fragments that were picked up by glacial ice at a localized outcropping and left scattered along a fairly well-defined tract during the movement of a glacier. { trāl }

trajectory [GEOPHYS] The path followed by a seismic wave. { trə'jek·trē }

tramontana [METEOROL] A cold wind from the northeast or north, particularly on the west coast of Italy and northern Corsica, but also in the Balearic Islands and the Ebro valley in Catalonia. { ˌträ·mōn'tä·nə }

transcurrent fault [GEOL] A strike-slip fault characterized by a steeply inclined surface. Also known as transverse thrust. { ¦tranz¦kə·rənt 'fȯlt }

transform fault [GEOL] A strike-slip fault with offset ridges characteristic of a midoceanic ridge. { 'tranzˌfȯrm ˌfȯlt }

transgression [GEOL] Geologic evidence of landward extension of the sea. Also known as invasion; marine transgression. [OCEANOGR] Extension of the sea over land areas. { tranz'gresh·ən }

transgressive deposit [GEOL] Sediment deposited during transgression of the sea or during subsidence of the land. { tranz'gres·iv di'päz·ət }

transgressive overlap *See* onlap.

transition zone [GEOL] **1.** A region within the upper mantle bordering the lower mantle, at a depth of 246–600 miles (410–1000 kilometers), characterized by a rapid increase in density of about 20% and an increase in seismic wave velocities. **2.** A region within the outer core, transitional to the inner core. { tran'zish·ən ˌzōn }

translational fault [GEOL] A fault in which there has been uniform movement in one direction and no rotational component of movement. Also known as translatory fault. { tran'slā·shən·əl 'fȯlt }

translational movement [GEOL] Movement, as of fault blocks, that is uniform, without rotation, so that parallel features maintain their orientation. { tran'slā·shən·əl 'müv·mənt }

translatory fault *See* translational fault. { 'tran·sləˌtȯr·ē 'fȯlt }

translucent attritus [GEOL] Attritus composed principally of transparent humic degradation matter. Also known as humodurite. { tran'slüs·əns ə'trīd·əs }

translucidus [METEOROL] A cloud variety occurring in a layer, patch, or extensive sheet, the greater part of which is sufficiently translucent to reveal the position of the sun, or through which higher clouds may be discerned; this variety is found in the general altocumulus, altostratus, stratocumulus, and stratus. { tran'slüs·əd·əs }

transmission function [GEOPHYS] A mathematical formulation of relationships between infrared transmission in the atmosphere, the path length, and the concentration of absorbing gases. { tranz'mish·ən ˌfəŋk·shən }

transparent sky cover [METEOROL] In United States weather-observing practice, that portion of sky cover through which higher clouds and blue sky may be observed; opposed to opaque sky cover. { tranz'par·ənt 'skī ˌkəv·ər }

transportation [GEOL] A phase of sedimentation concerned with movement by natural agents of sediment or any loose or weathered material from one place to another. { ˌtranz·pər'tā·shən }

transverse bar [GEOL] A slightly submerged sand bar extending perpendicular to the shoreline. { trans¦vərs 'bär }

transverse basin *See* exogeosyncline. { trans¦vərs 'bās·ən }

transverse dune [GEOL] A sand dune with a nearly straight ridge crest formed by the merger of crescentic dunes; elongated at right angles to the direction of prevailing winds, with a gentle windward slope and a steep leeward slope. { trans¦vərs 'dün }
transverse fault [GEOL] A fault whose strike is more or less perpendicular to the general structural trend of the region. { trans¦vərs 'fȯlt }
transverse fold *See* cross fold. { trans¦vərs 'fōld }
transverse joint *See* cross joint. { trans¦vərs 'jȯint }
transverse ripple mark [GEOL] A ripple mark formed nearly perpendicular to the direction of the current. { trans¦vərs 'rip·əl ˌmärk }
transverse thrust *See* transcurrent fault. { trans¦vərs 'thrəst }
transverse valley [GEOL] **1.** A valley perpendicular to the general strike of the underlying strata. **2.** A valley cutting perpendicularly across a ridge, range, or chain of mountains. Also known as cross valley. { trans¦vərs 'val·ē }
transverse wave *See* S wave. { trans¦vərs 'wāv }
trap *See* oil trap. { trap }
trapdoor fault [GEOL] A circular fault that is hinged at one end. { 'trap¦dȯr ˌfȯlt }
trapped radiation [GEOPHYS] Radiation from space that has become trapped in the magnetic field of the earth, as in the Van Allen belt. { 'trapt ˌrād·ē'ā·shən }
traveling dune *See* wandering dune. { 'trav·əl·iŋ 'dün }
travel-time curve [GEOPHYS] A plot of P-, S-, and L-wave travel times used by seismologists to locate earthquakes. { 'trav·əl ˌtīm ˌkərv }
traverse [GEOL] A line of survey or sampling across a thin section of geological region. [METEOROL] A westerly wind in central France; it is moderate to strong, generally squally, humid and thundery in summer, especially on slopes facing west; it is cold in winter and spring and brings snow or hail showers. { tra'vərs }
traversia [METEOROL] A South American nautical term (especially Chile) for a west wind from the sea. { ˌtra·vər'sē·ə }
traversier [METEOROL] In the Mediterranean, a dangerous wind blowing directly into port. { trəˌver·sē'ā }
travertine [GEOL] Concretionary limestone deposited at the mouth of a hot spring. { 'trav·ərˌtēn }
tree climate [CLIMATOL] Any type of climate which supports the growth of trees, including the tropical rainy climates, temperate rainy climates, and snow-forest climates. { 'trē ˌklī·mət }
tree-ring hydrology *See* dendrohydrology. { ¦trē ¦riŋ hī'dräl·ə·jē }
trellis drainage [HYD] A drainage pattern characterized by parallel main streams and secondary tributaries intersected at right angles by tributaries. Also known as espalier drainage; grapevine drainage. { 'trel·əs ˌdrā·nij }
tremor [GEOPHYS] A minor earthquake. Also known as earthquake tremor; earth tremor. { 'trem·ər }
trench [GEOGR] **1.** A narrow, straight, elongate, U-shaped valley between two mountain ranges. **2.** A narrow stream-eroded canyon, gulley, or depression with steep sides. [GEOL] A long, narrow, deep depression of the sea floor, with relatively steep sides. Also known as submarine trench. { trench }
trend [GEOL] The direction of an outcrop of a layer, vein, fold, or other kind of geologic feature. Also known as direction. { trend }
Trentonian [GEOL] A North American stage of geologic time; Middle Ordovician (above Wilderness, below Edenian); equivalent to the upper Mohawkian. { tren'tō·nē·ən }
triangular facet [GEOL] A triangular-shaped steep-sloped hill or cliff formed usually by the erosion of a fault-truncated hill. { trī'aŋ·gyə·lər 'fas·ət }
Triassic [GEOL] The first period of the Mesozoic era, lying above Permian and below Jurassic, 180–225 million years ago. { trī'aˌsik }
tributary [HYD] A stream that feeds or flows into or joins a larger stream or a lake. Also known as contributory; feeder; side stream; tributary stream. { 'trib·yəˌter·ē }
tributary glacier [GEOL] A glacier that flows into a larger glacier. { 'trib·yəˌter·ē 'glā·shər }
tributary stream *See* tributary. { 'trib·yəˌter·ē 'strēm }

tributary waterway [HYD] Any body of water that flows into a larger body, that is, a creek in relation to a river, a river in relation to a bay, and a bay in relation to the open sea. { 'trib·yə,ter·ē 'wȯd·ər,wā }

tripoli [GEOL] A lightweight, porous, friable, siliceous sedimentary rock that may have a white, gray, pink, red, or yellow color; used for polishing metals and stones. { 'trip·ə·lē }

tripolite *See* diatomaceous earth. { 'trip·ə,līt }

trolley [GEOL] A basin-shaped depression in strata. Also known as lum. { 'träl·ē }

Tropept [GEOL] A suborder of the order Inceptisol, characterized by moderately dark A horizons with modest additions of organic matter, B horizons with brown or reddish colors, and slightly pale C horizons; restricted to tropical regions with moderate or high rainfall. { 'trä,pept }

tropical air [METEOROL] A type of air whose characteristics are developed over low latitudes. { 'träp·ə·kəl 'er }

tropical climate [CLIMATOL] A climate which is typical of equatorial and tropical regions, that is, one with continually high temperatures and with considerable precipitation, at least during part of the year. { 'träp·ə·kəl 'klī·mət }

tropical cyclone [METEOROL] The general term for a cyclone that originates over tropical oceans; at maturity, the tropical cyclone is one of the most intense storms of the world; winds exceeding 175 knots (324 kilometers per hour) have been measured, and the rain is torrential. { 'träp·ə·kəl 'sī,klōn }

tropical disturbance [METEOROL] A cyclonic wind system of the tropics, of lesser intensity than a tropical cyclone. { 'träp·ə·kəl di'stər·bəns }

tropical easterlies [METEOROL] The trade winds when shallow and exhibiting a strong vertical shear; at about 500 feet (152 meters) the easterlies give way to the upper westerlies, which are sufficiently strong and deep to govern the course of cloudiness and weather. Also known as subtropical easterlies. { 'träp·ə·kəl 'ēs·tər·lēz }

tropical front *See* intertropical front. { 'träp·ə·kəl 'frənt }

tropical meteorology [METEOROL] The study of the tropical atmosphere; the dividing lines, in each hemisphere, between the tropical easterlies and the mid-latitude westerlies in the middle troposphere roughly define the poleward boundaries of this region. { 'träp·ə·kəl ,mēd·ē·ə'räl·ə·jē }

tropical monsoon climate [CLIMATOL] One of the tropical rainy climates; it is sufficiently warm and rainy to produce tropical rainforest vegetation, but it does exhibit the monsoon climate influences in that it has a winter dry season. { 'träp·ə·kəl män'sün ,klī·mət }

Tropical Rainfall Measuring Mission [METEOROL] A meteorological satellite used for mapping tropical precipitation in order to better understand the earth's climate system and to verify climate models. Abbreviated TRMM. { ¦träp·ə·kəl 'rān,fȯl ,mezh·ər·iŋ ,mish·ən }

tropical rainforest climate [CLIMATOL] In general, the climate which produces tropical rainforest vegetation, that is, a climate of unbroken warmth, high humidity, and heavy annual precipitation. Also known as tropical wet climate. { 'träp·ə·kəl 'rān,fär·əst ,klī·mət }

tropical rainy climate [CLIMATOL] A major category in W. Köppen's climatic classification, characterized by a mean temperature of the coldest month of 64.4°F (18°C) or higher, and by a mean annual precipitation, in inches, greater than $0.44(t - a)$, where t is the mean annual temperature in degrees Fahrenheit, and a equals 32 for precipitation chiefly in winter, 19.4 for evenly distributed precipitation, and 6.8 for precipitation chiefly in summer. { 'träp·ə·kəl ¦rān·ē ,klī·mət }

tropical savanna climate [CLIMATOL] In general, the type of climate which produces the vegetation of the tropical and subtropical savanna; thus, a climate with a winter dry season, a relatively short but heavy rainy summer season, and high year-round temperatures. Also known as savanna climate; tropical wet and dry climate. { 'träp·ə·kəl sə'van·ə ,klī·mət }

tropical wet and dry climate *See* tropical savanna climate. { 'träp·ə·kəl ¦wet ən ¦drī ,klī·mət }

tropical wet climate *See* tropical rainforest climate. { 'träp·ə·kəl 'wet ˌklī·mət }

tropic higher-high-water interval [OCEANOGR] The lunitidal interval pertaining to the higher high waters at the time of tropic tides. { 'träp·ik 'hī·ər ¦hī ˌwȯd·ər ˌin·tər·vəl }

tropic high-water inequality [OCEANOGR] The average difference between the heights of the two high waters of the tidal day at the time of tropic tides. { 'träp·ik 'hī ˌwȯd·ər ˌin·i'kwäl·əd·ē }

tropic lower-low-water interval [OCEANOGR] The lunitidal interval pertaining to the lower low waters at the time of tropic tides. { 'träp·ik 'lō·ər ¦lō ˌwȯd·ər ˌin·tər·vəl }

tropic low-water inequality [OCEANOGR] The average difference between the heights of the two low waters of the tidal day at the time of tropic tides. { 'träp·ik 'lō ˌwȯd·ər ˌin·i'kwäl·əd·ē }

Tropic of Cancer [GEOD] A parallel of latitude 23.45° north of the equator, marking the northernmost latitude at which the sun reaches its zenith. { 'träp·ik əv 'kan·sər }

Tropic of Capricorn [GEOD] A parallel of latitude 23.45° south of the equator, marking the southernmost latitude at which the sun reaches its zenith. { 'träp·ik əv 'kap·riˌkȯrn }

tropics [CLIMATOL] Any portion of the earth characterized by a tropical climate. { 'träp·iks }

tropic tidal currents [OCEANOGR] Tidal currents of increased diurnal inequality occurring at the time of tropic tides. { 'träp·ik 'tīd·əl 'kə·rəns }

tropic tide [OCEANOGR] A tide occurring when the moon is near maximum declination; the diurnal inequality is then at a maximum. { 'träp·ik 'tīd }

tropic velocity [OCEANOGR] The speed of the greater flood or greater ebb at the time of tropic currents. { 'träp·ik və'läs·əd·ē }

tropopause [METEOROL] The boundary between the troposphere and stratosphere, usually characterized by an abrupt change of lapse rate; the change is in the direction of increased atmospheric stability from regions below to regions above the tropopause; its height varies from 9 to 12 miles (15 to 20 kilometers) in the tropics to about 6 miles (10 kilometers) in polar regions. { 'trōp·əˌpȯz }

tropopause chart [METEOROL] A synoptic chart showing the contour lines of the tropopause and tropopause break lines. { 'trōp·əˌpȯz ˌchärt }

tropopause fold [METEOROL] A phenomenon occurring in the stratosphere in which a tapering cone of dry, ozone-rich air intrudes into the troposphere. { 'trōp·əˌpȯz 'fōld }

tropopause inversion [METEOROL] The decrease in the lapse rate of temperature encountered at the level of the tropopause. Also known as upper inversion. { 'trōp·əˌpȯz in'vər·zhən }

troposphere [METEOROL] That portion of the atmosphere from the earth's surface to the tropopause, that is, the lowest 10 to 20 kilometers of the atmosphere. { 'trōp·əˌsfir }

tropospheric duct *See* duct. { ¦trōp·ə¦sfir·ik 'dəkt }

tropospheric ducting *See* ducting. { ¦trōp·ə¦sfir·ik 'dəkt·iŋ }

tropospheric superrefraction [GEOPHYS] Phenomenon occurring in the troposphere whereby radio waves are bent sufficiently to be returned to the earth. { ¦trōp·ə¦sfir·ik ¦sü·pər·ri'frak·shən }

trough [GEOL] **1.** A small, straight depression formed just offshore on the bottom of a sea or lake and on the landward side of a longshore bar. **2.** Any narrow, elongate depression in the surface of the earth. **3.** An elongate depression on the sea floor that is wider and shallower than a trench. Also known as submarine trench. **4.** The line connecting the lowest points of a fold. [METEOROL] An elongated area of relatively low atmospheric pressure; the opposite of a ridge. { trȯf }

trough aloft *See* upper-level trough. { 'trȯf ə'lōft }

trough crossbedding [GEOL] A variety of crossbedding in which the lower crossbedding surfaces are smoothly curved, rather than planar. { 'trȯf 'krȯsˌbed·iŋ }

trough fault [GEOL] One of a set of two faults bounding a graben. { 'trȯft ˌfȯlt }

trough plane *See* trough surface. { 'trȯf ˌplān }

trough reef *See* reverse saddle. { 'trȯf ˌrēf }

trough surface [GEOL] A surface or plane connecting the troughs of the bed of a syncline. Also known as synclinal axis; trough plane. { 'trȯf ˌsər·fəs }
trough valley *See* U-shaped valley. { 'trȯf ˌval·ē }
true [GEOD] Related to true north. { trü }
true air temperature [METEOROL] Basic air temperature corrected for heat of compression error due to high-speed motion of the thermometer through the air, as on an aircraft. { 'trü 'er ˌtem·prə·chər }
true altitude *See* corrected altitude. { 'trü 'al·təˌtüd }
true convergence [GEOD] The angle at which one meridian is inclined to another on the surface of the earth. { 'trü kən'vər·jəns }
true crater [GEOL] The primary depression formed by impact or explosion before modification by slumping or by deposition of ejected material. Also known as primary crater. { 'trü 'krād·ər }
true dip *See* dip. { 'trü 'dip }
true formation resistivity [GEOPHYS] Electrical resistivity of a clean (nonshaly) porous reservoir formation containing hydrocarbons and formation water; value is greater than the resistivity when there is added water incursion. { 'trü fȯr'mā·shən ˌrē·zis'tiv·əd·ē }
true mean temperature [METEOROL] As adopted by the International Meteorological Organization, a monthly or annual mean air temperature based upon hourly observations at a given place, or on some combination of less frequent observations designed to represent this mean as nearly as possible. { 'trü 'mēn 'tem·prə·chər }
true soil *See* solum. { 'trü 'sȯil }
true wind [METEOROL] Wind relative to a fixed point on the earth. { 'trü 'wind }
true wind direction [METEOROL] The direction, with respect to true north, from which the wind is blowing. { 'trü 'wind dəˌrek·shən }
truncated landform [GEOGR] A landform which has been cut off by erosion, creating a steep side or cliff. { 'trəŋˌkād·əd 'landˌfȯrm }
trunk stream *See* main stream. { 'trəŋk ˌstrēm }
T-S curve *See* temperature-salinity diagram. { ¦tē¦es ˌkərv }
T-S diagram *See* temperature-salinity diagram. { ¦tē¦es ˌdī·əˌgram }
T-S relation *See* temperature-salinity diagram. { ¦tē¦es riˌlā·shən }
tsunami [OCEANOGR] An ocean wave or series of waves generated by any large, abrupt disturbance of the sea-surface by an earthquake in marine and coastal regions, as well as by a suboceanic landslide, volcanic eruption, or asteroid impact. { tsü'nä·mē }
tsunamiite [GEOL] **1.** A sedimentary deposit resulting from a tsunami generated by an asteroid or comet impact. **2.** Rock deposited by a tsunami. Also known as tsunamite. { tsü'näm·ēˌīt }
tsunamite *See* tsunamiite. { 'tsü·nəˌmīt }
Tsushima Current [OCEANOGR] That part of the Kuroshio Current flowing northeastward through the Korea Strait and along the Japanese coast in the Sea of Japan. { 'tsü·shēˌmä 'kə·rənt }
tuba [METEOROL] A cloud column or inverted cloud cone, pendant from a cloud base; this supplementary feature occurs mostly with cumulus and cumulonimbus; when it reaches the earth's surface it constitutes the cloudy manifestation of an intense vortex, namely, a tornado or waterspout. Also known as pendant cloud; tornado cloud. { 'tü·bə }
tube [GEOL] A passage in a cave having smooth sides and an elliptical to nearly circular cross section. { 'tüb }
tufa [GEOL] A spongy, porous limestone formed by precipitation from evaporating spring and river waters, often onto leaves and stems of neighboring plants. Also known as calcareous sinter; calcareous tufa. { 'tü·fə }
tufaceous [GEOL] Pertaining to or similar to tufa. { tü'fā·shəs }
tuff [GEOL] Consolidated volcanic ash, composed largely of fragments (less than 4 millimeters) produced directly by volcanic eruption; much of the fragmented material represents finely comminuted crystals and rocks. { təf }
tuffaceous [GEOL] Pertaining to sediments which contain up to 50% tuff. { tə'fā·shəs }

tuff ball *See* mud ball. { 'təf ˌbȯl }
tuft *See* mound. { təft }
tumuli lava [GEOL] A type of lava flow forming ovoid mounds, a few feet high and a few tens of feet long, caused by buckling up of the crust. { 'tü·myəˌlī 'lä·və }
tundra climate [CLIMATOL] The climate which produces tundra vegetation; it is too cold for the growth of trees but does not have a permanent snow-ice cover. { 'tən·drə ˌklī·mət }
tunnel cave *See* natural tunnel. { 'tən·əl ˌkāv }
turbidite [GEOL] Any sediment or rock transported and deposited by a turbidity current, generally characterized by graded bedding, large amounts of matrix, and commonly exhibiting a Bouma sequence. { 'tər·bəˌdīt }
turbidity [METEOROL] Any condition of the atmosphere which reduces its transparency to radiation, especially to visible radiation. { tər'bid·əd·ē }
turbidity current [OCEANOGR] A highly turbid, relatively dense current carrying large quantities of clay, silt, and sand in suspension which flows down a submarine slope through less dense sea water. Also known as density current; suspension current. { tər'bid·əd·ē ˌkə·rənt }
turbidity factor [GEOPHYS] A measure of the atmospheric transmission of incident solar radiation; if I_0 is the flux density of the solar beam just outside the earth's atmosphere, I the flux density measured at the earth's surface with the sun at a zenith distance which implies an optical air mass *m*, and $I_{m,w}$ the intensity which would be observed at the earth's surface for a pure atmosphere containing 1 centimeter of precipitable water viewed through the given optical air mass, then turbidity factor θ is given by $\theta = (\ln I_0 - \ln I)/(\ln I_0 - \ln I_{m,w})$. { tər'bid·əd·ē ˌfak·tər }
turbonada [METEOROL] A short thundersquall on the north Spanish coast, sometimes accompanied by waterspouts. { tər·bə'näd·ə }
turbopause *See* homopause. { 'tər·bəˌpȯz }
turbosphere [METEOROL] The region of the atmosphere in which turbulence frequently exists. { 'tər·bəˌsfir }
turbulent heat conduction [OCEANOGR] Conduction of heat in water by lateral and vertical eddy diffusion, with currents. { 'tər·byə·lənt 'hēt kənˌdək·shən }
turn of the tide *See* change of tide. { 'tərn əv thə 'tīd }
Turonian [GEOL] A European stage of geologic time: Upper or Middle Cretaceous (above Cenomanian, below Coniacian). { tü'rō·nē·ən }
turret ice *See* ropak. { 'tə·rət ˌīs }
turtle stone *See* septarium. { 'tərd·əl ˌstōn }
twilight arch *See* bright segment. { 'twīˌlīt ˌärch }
twister [METEOROL] In the United States, a colloquial term for tornado. { 'twis·tər }
two-layer ocean [OCEANOGR] An idealized ocean in which a layer of uniform density near the surface overlays a deep layer of uniform but distinctly higher-density water. { 'tü ¦lā·ər 'ō·shən }
two-year ice *See* second-year ice. { 'tü ¦yir 'īs }
tying bar *See* tombolo. { 'tī·iŋ ˌbär }
Tyndall flowers [HYD] Small water-filled cavities, often of basically hexagonal shape, which appear in the interior of ice masses upon which light is falling. { 'tind·əl ˌflaü·erz }
type-α leader [GEOPHYS] A stepped leader of lightning which exhibits very little branching and whose individual steps are short and so weakly luminous as to be difficult to discern. { 'tīp 'al·fə ˌlēd·ər }
type-β leader [GEOPHYS] A stepped leader of lightning in which the upper portion of the channel is characterized by longer and brighter steps than those found in the lower portion of the channel, a consequence of excessive branching in the upper parts under the influence of strong fields set up by heavy space charges near and around the upper end of the channel. { ¦tīp 'bād·ə ˌlēd·ər }
type C1 carbonaceous chondrite [GEOL] A type of carbonaceous chondrite that is strongly magnetic, has a lower density than the other two types, contains sulfates, and has a carbon content of about 3.5%. { 'tīp ¦sē¦wən ˌkär·bə'nā·shəs 'känˌdrīt }

type C2 carbonaceous chondrite [GEOL] A type of carbonaceous chondrite that is weakly magnetic or nonmagnetic, has most of its sulfur present as free sulfur, and contains about 2.5% carbon. { 'tīp ¦sē¦tü ,kär·bə'nā·shəs 'kän,drīt }

type C3 carbonaceous chondrite [GEOL] A type of carbonaceous chondrite that has a lower percentage of water and a higher density than the other two types, and usually consists largely of olivine. { 'tīp ¦sē¦thrē ,kär·bə'nā·shəs 'kän,drīt }

type locality [GEOL] **1.** The place at which a stratigraphic unit is typically displayed and from which it derives its name. **2.** The place where a geologic feature was first recognized and described. { 'tīp lō,kal·əd·ē }

type section [GEOL] That sequence of strata identified as the original sequence for a location or area; the standard against which other stratigraphy of parts of the area are compared. Also known as section. { 'tīp ,sek·shən }

typhoon [METEOROL] A severe tropical cyclone in the western Pacific. { tī'fün }

typhoon wind *See* hurricane wind. { tī'fün ,wind }

typography point *See* point. { tī'päg·rə·fē ,point }

U

ubac [METEOROL] The shady (usually north) side of an Alpine mountain, characterized by a lower timberline and snow line than the sunny side. { 'ü,bäk }

Udalf [GEOL] A suborder of the soil order Alfisol; brown soil formed in a udic moisture regime and in a mesic or warmer temperature regime. { 'ü,dälf }

Udert [GEOL] A suborder of the soil order Vertisol; formed in a humid region so that surface cracks remain open only for 2–3 months. { 'üd,ərt }

Udoll [GEOL] A suborder of the Mollisol soil order; found in humid, temperate, and warm regions where maximum rainfall comes during growing season; has thick, very dark A horizons, brown B horizons, and paler C horizons. { 'üd,ól }

Udult [GEOL] A suborder of the soil order Ultisol; organic-carbon content is low, argillic horizons are reddish or yellowish; formed in a udic moisture regime. { 'üd,əlt }

U figure *See* U index. { 'yü ,fig·yər }

U index [GEOPHYS] The difference between consecutive daily mean values of the horizontal component of the geomagnetic field. Also known as U figure. { 'yü ,in,deks }

Ulatisian [GEOL] A mammalian age in a local stage classification of the Eocene in use on the Pacific Coast based on foraminifers. { ,yü·lə'tē·zhən }

Ulloa's ring *See* Bouguer's halo. { ü'yō·əz ,riŋ }

ulmic acid *See* ulmin. { 'əl·mik 'as·əd }

ulmin [GEOL] Alkali-soluble organic substances derived from decaying vegetable matter; occurs as amorphous brown to black gel material. Also known as carbohumin; fundamental jelly; fundamental substance; gelose; humin; humogelite; jelly; ulmic acid; vegetable jelly. { 'əl·mən }

Ultisol [GEOL] A soil order characterized by typically moist soils, with horizons of clay accumulation and a low supply of bases. { 'əl·tə,sól }

ultralow-velocity zone [GEOPHYS] Thin, mushy layer detected in some places along the earth's core-mantle boundary where seismic waves slow down. { ,əl·trə·lō və'läs·əd·ē ,zōn }

ultravulcanian [GEOL] A type of volcanic eruption characterized by periodic violent gaseous explosions of lithic dust and solid blocks, with little if any fiery scoria. { ¦əl·trə·vəl'kā·nē·ən }

Umbrept [GEOL] A suborder of the Inceptisol soil order; has dark A horizon more than 10 inches (25 centimeters) thick, brown B horizons, and slightly paler C horizons; soil is strongly acid, and clay minerals are crystalline; occurs in cool or temperate climates. { 'əm,brept }

unaka [GEOL] A large residual mass rising above a peneplain that is less well developed than one having a monadnock. { ü'näk·ə }

unconcentrated wash *See* sheet erosion. { ¦ən'käns·ən,trād·əd 'wäsh }

unconformable [GEOL] Pertaining to strata that do not conform in position, dip, or strike to the older underlying rocks. { ¦ən·kən'fór·mə·bəl }

unconformity [GEOL] The relation between adjacent rock strata whose time of deposition was separated by a period of nondeposition or of erosion; a break in a stratigraphic sequence. { ¦ən·kən'fór·məd·ē }

unconformity iceberg [OCEANOGR] An iceberg consisting of more than one kind of ice, such as blue water-formed ice and névé; such an iceberg often contains many crevasses and silt bands. { ¦ən·kən'fór·məd·ē 'īs,bərg }

unconsolidated material [GEOL] Loosely arranged or unstratified sediment whose particles are not cemented together. { ¦ən·kən'säl·ə,dād·əd mə'tir·ē·əl }

undercast [METEOROL] A cloud layer of ten-tenths (1.0) coverage as viewed from an observation point above the layer; the term is used in pilot reporting of in-flight weather conditions. { 'ən·dər,kast }

underclay [GEOL] A layer of clay or other fine-grained detrital material underlying a coal bed or comprising the floor of a coal seam. Also known as coal clay; root clay; seat clay; seat earth; thill; underearth; warrant. { 'ən·dər,klā }

underclay limestone [GEOL] A thin, fresh-water limestone that is relatively free of fossils and is dense and nodular; found in underlying coal deposits. { 'ən·dər,klā 'līm,stōn }

undercliff [GEOL] A subordinate cliff or terrace formed by material which has fallen or slid from above. { 'ən·dər,klif }

underconsolidation [GEOL] Less than normal consolidation of sedimentary material for the existing overburden. { ¦ən·dər·kən,säl·ə'dā·shən }

undercurrent [OCEANOGR] A water current flowing beneath a surface current at a different speed or in a different direction. { 'ən·dər,kə·rənt }

undercutting [GEOL] Erosion of material at the base of a steep slope, cliff, or other exposed rock. { ¦ən·dər¦kəd·iŋ }

underearth *See* underclay. { 'ən·dər,ərth }

underfit stream [HYD] A misfit stream that appears to be too small to have eroded the valley in which it flows. { 'ən·dər,fit 'strēm }

underflow conduit [GEOL] A permeable deposit underlying a surface stream channel. { 'ən·dər,flō 'kän,dü·ət }

underground geology *See* subsurface geology. { ¦ən·dər¦graủnd jē'äl·ə·jē }

underground ice *See* ground ice. { ¦ən·dər¦graủnd 'īs }

underground stream [HYD] A subsurface body of water flowing in a definite current in a distinct channel. { ¦ən·dər¦graủnd 'strēm }

underlie [GEOL] To lie or be situated under; to occupy a lower position, or to pass beneath. { 'ən·dər,lī }

undermelting [HYD] The melting from below of any floating ice. { ¦ən·dər¦melt·iŋ }

undermining *See* sapping. { 'ən·dər,mīn·iŋ }

underthrust [GEOL] A thrust fault in which the lower, active rock mass has been moved under the upper, passive rock mass. { ¦ən·dər¦thrəst }

undertow [OCEANOGR] A subsurface seaward movement by gravity flow of water carried up on a sloping beach by waves or breakers. { 'ən·dər,tō }

underwater vehicle [OCEANOGR] A submersible work platform designed to be operated either remotely or directly. { ,ən·dər,wȯd·ər 'vē·ə·kəl }

undulatus *See* billow cloud. { ən·jə'läd·əs }

unfreezing [GEOL] The upward movement of stones to the surface as a result of repeated freezing and thawing of the containing soil. { ¦ən'frēz·iŋ }

uniformitarianism [GEOL] Classically, the concept that the present is the key to the past; the principle that contemporary geologic processes have occurred in the same regular manner and with essentially the same intensity throughout geologic time, and that events of the geologic past can be explained by phenomena observable today. Also known as actualism; principle of uniformity. { ,yü·nə,fȯr·mə'ter·ē·ə,niz·əm }

United States airways code [METEOROL] A synoptic code for communicating aviation weather observations. Also known as airways code. { yə'nīd·əd 'stāts 'er,wāz ,kōd }

United States Survey foot [GEOD] The foot used by the U.S. Coast and Geodetic Survey in which 1 inch is equal to 2.540005 centimeters. { ¦yü¦es 'sər,vā 'fůt }

universal transmission function [GEOPHYS] A mathematical relationship that attempts to describe quantitatively the complex infrared propagation (including absorption and reradiation) in the atmosphere. { ¦yü·nə¦vər·səl tranz'mish·ən ,fəŋk·shən }

unlimited ceiling [METEOROL] A ceiling that exists when the total sky cover is less than 0.6%, or when the total transparent sky cover is 0.5% or more, or when surface-based obscuring phenomena are classed as partial obscuration (that is, they obscure 0.9%

or less of the sky) and no layer aloft is reported as broken or overcast. { ¦ən'lim·əd·əd 'sē·liŋ }

unrestricted visibility [METEOROL] The visibility when no obstruction to vision exists in sufficient quantity to reduce the visibility to less than 7 miles (11.3 kilometers). { ¦ən·ri'strik·təd ˌviz·ə'bil·əd·ē }

unsaturated zone *See* zone of aeration. { ¦ən'sach·əˌrād·əd 'zōn }

unsettled [METEOROL] Pertaining to fair weather which may at any time become rainy, cloudy, or stormy. { ¦ən'sed·əld }

updrift [OCEANOGR] The direction which is opposite that of the prevailing movement of littoral material. { 'əpˌdrift }

uphole time [GEOPHYS] The time that a seismic pulse requires to travel from an explosion at some depth in a shot hole to the surface of the earth. { 'əpˌhōl ˌtīm }

upland [GEOGR] **1.** An extensive region of high land. **2.** The higher ground of a region, in contrast to a valley, plain, or other low-lying land. **3.** The elevated land above the low areas along a stream or between hills. { 'əp·lənd }

upper [GEOL] Pertaining to rocks or strata that normally overlie those of earlier formations of the same subdivision of rocks. { 'əp·ər }

upper air [METEOROL] The region of the atmosphere which is above the lower troposphere; although no distinct lower limit is set, the term is generally applied to levels above that at which the pressure is 850 millibars. { 'əp·ər 'er }

upper-air chart *See* upper-level chart. { ¦əp·ər ¦er 'chärt }

upper-air disturbance [METEOROL] A disturbance of the flow pattern in the upper air, particularly one which is more strongly developed aloft than near the ground. Also known as upper-level disturbance. { ¦əp·ər ¦er diˌstər·bəns }

upper-air observation [METEOROL] A measurement of atmospheric conditions aloft, above the effective range of a surface weather observation. Also known as sounding; upper-air sounding. { ¦əp·ər ¦er ˌäb·zərˌvā·shən }

upper-air sounding *See* upper-air observation. { ¦əp·ər ¦er ˌsau̇nd·iŋ }

upper anticyclone *See* upper-level anticyclone. { 'əp·ər ¦ant·i'sī·klōn }

upper atmosphere [METEOROL] The general term applied to the atmosphere above the troposphere. { 'əp·ər 'at·məˌsfir }

upper-atmosphere dynamics [METEOROL] Motion of the atmosphere above 300 miles (500 kilometers); predominant dynamical phenomena are internal gravity waves, tides, sound waves, turbulence, and large-scale circulation. { ¦əp·ər ¦at·məˌsfir dī'nam·iks }

upper band *See* upper bright band. { 'əp·ər 'band }

upper branch [GEOD] That half of a meridian or celestial meridian from pole to pole which passes through a place or its zenith. { 'əp·ər 'branch }

upper bright band [METEOROL] A level of enhanced radar echo occasionally observed at a higher altitude than the bright band of the melting level; it is attributable to the growth of a layer of ice crystals in a supercooled cloud into snow pellets. Also known as radar upper band; upper band. { 'əp·ər 'brīt ˌband }

Upper Cambrian [GEOL] The latest epoch of the Cambrian period of geologic time, beginning approximately 510 million years ago. { 'əp·ər 'kam·brē·ən }

Upper Carboniferous [GEOL] The European epoch of geologic time equivalent to the Pennsylvanian of North America. { 'əp·ər ˌkär·bə'nif·ə·rəs }

Upper Cretaceous [GEOL] The late epoch of the Cretaceous period of geologic time, beginning about 90 million years ago. { 'əp·ər kri'tā·shəs }

upper cyclone *See* upper-level cyclone. { 'əp·ər 'sīˌklōn }

Upper Devonian [GEOL] The latest epoch of the Devonian period of geologic time, beginning about 365 million years ago. { 'əp·ər də'vō·nē·ən }

upper front [METEOROL] A front which is present in the upper air but does not extend to the ground. { 'əp·ər 'frənt }

upper high *See* upper-level anticyclone. { 'əp·ər 'hī }

Upper Huronian *See* Animikean. { 'əp·ər hyu̇'rō·nē·ən }

upper inversion *See* tropopause inversion. { 'əp·ər in¦vər·zhən }

Upper Jurassic [GEOL] The latest epoch of the Jurassic period of geologic time, beginning approximately 155 million years ago. { 'əp·ər jü'ras·ik }

upper-level anticyclone [METEOROL] An anticyclonic circulation existing in the upper air; this often refers to such anticyclones only when they are much more pronounced at upper levels than at and near the earth's surface. Also known as high aloft; high-level anticyclone; upper anticyclone; upper high; upper-level high. { ¦əp·ər ¦lev·əl 'ant·i'sī,klōn }

upper-level chart [METEOROL] A synoptic chart of meteorological conditions in the upper air, almost invariably referring to a standard constant-pressure chart. Also known as upper-air chart. { ¦əp·ər ¦lev·əl 'chärt }

upper-level cyclone [METEOROL] A cyclonic circulation existing in the upper air, and specifically, as seen on an upper-level constant-pressure chart; often restricted to describe cyclones associated with relatively little cyclonic circulation in the lower atmosphere. Also known as high-level cyclone; low aloft; upper cyclone; upper-level low; upper low. { ¦əp·ər ¦lev·əl 'sī,klōn }

upper-level disturbance *See* upper-air disturbance. { ¦əp·ər ¦lev·əl di'stər·bəns }

upper-level high *See* upper-level anticyclone. { ¦əp·ər ¦lev·əl 'hī }

upper-level low *See* upper-level cyclone. { ¦əp·ər ¦lev·əl 'lō }

upper-level ridge [METEOROL] A pressure ridge existing in the upper air, especially one that is stronger aloft than near the earth's surface. Also known as high-level ridge; ridge aloft; upper ridge. { ¦əp·ər ¦lev·əl 'rij }

upper-level trough [METEOROL] A pressure trough existing in the upper air, but sometimes restricted to the troughs that are much more pronounced aloft than near the earth's surface. Also known as high-level trough; trough aloft; upper trough. { ¦əp·ər ¦lev·əl 'tröf }

upper-level winds *See* winds aloft. { ¦əp·ər ¦lev·əl 'winz }

upper low *See* upper-level cyclone.

upper mantle [GEOL] The portion of the mantle lying above a depth of about 600 miles (1000 kilometers). Also known as outer mantle; peridotite shell. { 'əp·ər 'mant·əl }

Upper Mississippian [GEOL] The latest epoch of the Mississippian period of geologic time. { 'əp·ər ,mis·ə'sip·ē·ən }

upper mixing layer [METEOROL] The region of the upper mesophere between about 30 and 50 miles (50 and 80 kilometers; that is, immediately above the mesopeak) through which there is a rapid decrease of temperature with height and where there appears to be considerable turbulence. { 'əp·ər 'miks·iŋ ,lā·ər }

Upper Ordovician [GEOL] The latest epoch of the Ordovician period of geologic time, beginning approximately 440 million years ago. { 'əp·ər ,ȯr·də'vish·ən }

Upper Pennsylvanian [GEOL] The latest epoch of the Pennsylvanian period of geologic time. { 'əp·ər ,pen·səl'vā·nyən }

Upper Permian [GEOL] The latest epoch of the Permian period of geologic time, beginning about 245 million years ago. { 'əp·ər 'pər·mē·ən }

upper ridge *See* upper-level ridge. { 'əp·ər 'rij }

Upper Silurian [GEOL] The latest epoch of the Silurian period of geologic time. { 'əp·ər sə'lür·ē·ən }

Upper Triassic [GEOL] The latest epoch of the Triassic period of geologic time, beginning about 200 million years ago. { 'əp·ər trī'as·ik }

upper trough *See* upper-level trough. { 'əp·ər 'tröf }

upper winds *See* winds aloft. { 'əp·ər 'winz }

uprush [METEOROL] The strong upward-flow air current in cumulus clouds during their stage of rapid development, often preceding a thunderstorm. Also known as vertical jet. [OCEANOGR] *See* swash. { 'əp,rəsh }

upsetted moraine *See* push moraine. { ¦əp¦sed·əd mə'rān }

upslope fog [METEOROL] A type of fog formed when air flows upward over rising terrain and, consequently, is adiabatically cooled to or below its dew point. { 'əp,slōp 'fäg }

upstream [HYD] Toward the source of a stream. { 'əp¦strēm }

upthrow [GEOL] **1.** The fault side that has been thrown upward. **2.** The amount of vertical fault displacement. { 'əp,thrō }

upwarp [GEOL] A broad anticline with gently sloping limbs formed as a result of differential uplift. { 'əp,wȯrp }

upwelling [OCEANOGR] The process by which water rises from a deeper to a shallower depth, usually as a result of divergence of offshore currents. { ¦əp¦wel·iŋ }

upwind [METEOROL] In the direction from which the wind is flowing. { 'əp¦wind }

upwind effect [METEOROL] The effect of an orographic barrier in producing orographic precipitation windward of the base of the barrier, because the airflow is forced upward before the barrier slope is actually reached. { 'əp¦wind i,fekt }

Uralean [GEOL] A stage of geologic time in Russia: uppermost Carboniferous (above Gzhelian, below Sakmarian of Permian). { yu̇'rāl·ē·ən }

uralitization [GEOL] **1.** A process of replacement whereby pyroxene undergoes alteration resulting in uralite. **2.** Development of amphibole from pyroxene. { yə,ral·əd·ə'zā·shən }

uranium age [GEOL] The age of a mineral as calculated from the numbers of ionium atoms present originally, now, and when equilibrium is established with uranium. { yə'rā·nē·əm ,āj }

uranium-lead dating [GEOL] A method for calculating the geologic age of a material in years based on the radioactive decay rate of uranium-238 to lead-206 and of uranium-235 to lead-207. { yə'rā·nē·əm 'led 'dād·iŋ }

urban geography [GEOGR] The study of the site, evolution, morphology, spatial patterns, and classification of densely populated areas. { ¦ər·bən jē'äg·rə·fē }

urban geology [GEOL] The study of geological aspects of planning and managing high-density population centers and their surroundings. { ¦ər·bən jē'äl·ə·jē }

urban heat island [METEOROL] Increased urban temperatures of 1–2°C higher for daily maxima and 1–9°C for daily minima compared to rural environs resulting from changes in moisture balance due to impermeable surfaces, decreased humidity, or alteration in heat balance. { 'ər·bən 'hēt ,ī·lənd }

ureilite [GEOL] An achondritic stony meteorite consisting principally of olivine and clinobronzite, with some nickel-iron, troilite, diamond, and graphite. { yə'rē·ə,līt }

urstromthal [GEOL] A large channel cut by a stream of water from melting ice, flowing along the edge of an ice sheet. { 'u̇r,strōm,täl }

U-shaped valley [GEOL] A type of valley with a broad floor and steep walls produced by glacial erosion. Also known as trough valley; U valley. { 'yü ¦shāpt 'val·ē }

Ustalf [GEOL] A suborder of the soil order Alfisol; red or brown soil formed in a ustic moisture regime and in a mesic or warmer temperature regime. { 'üst,älf }

Ustert [GEOL] A suborder of the Vertisol soil order; has a faint horizon and is dry for an appreciable period or more than one period of the year. { 'üst,ərt }

Ustoll [GEOL] A suborder of the soil order Mollisol; formed in a ustic moisture regime and in a mesic or warmer temperature regime; may have a calcic, petrocalcic, or gypsic horizon. { 'üst,ȯl }

Ustox [GEOL] A suborder of the soil order Oxisol that is low to moderate in organic matter, well drained, and dry for at least 90 cumulative days each year. { 'üst,äks }

Ustult [GEOL] A suborder of the soil order Ultisol; brownish or reddish, with low to moderate organic-carbon content; a well-drained soil of warm-temperate and tropical climates with moderate or low rainfall. { 'üst,əlt }

UV-A [METEOROL] Ultraviolet radiation produced by the sun, ranging in wavelength from 320 to 400 nanometers, biologically; it is the least damaging of the sun's rays.

UV-B [METEOROL] Ultraviolet radiation produced by the sun, ranging in wavelength from 280 to 320 nanometers; it is biologically damaging. Stratospheric ozone absorbs much of it.

UV-C [METEOROL] Ultraviolet radiation produced by the sun, ranging in wavelength from 200 to 280 nanometers, biologically, it is the most damaging of the sun's rays. Stratospheric ozone strongly absorbs it and, as a result, the solar spectrum at the earth's surface contains only the UV-A and UV-B radiation.

uvala [GEOGR] Broad-bottomed lowlands. { 'ü·və·lə }

U valley *See* U-shaped valley. { 'yü ,val·ē }

V

vacuole *See* vesicle. { 'vak·yə,wōl }

vadose water [HYD] Water in the zone of aeration. Also known as kremastic water; suspended water; wandering water. { 'vā,dōs ,wȯd·ər }

vadose zone *See* zone of aeration. { 'vā,dōs ,zōn }

valais wind [METEOROL] The notable valley wind that blows along the Rhone Valley from the upper end of Lake Geneva (Valais Canton); it is sufficiently strong and regular to distort the growth of trees. { va'lā ,wind }

valley [GEOGR] A generally broad area of flat, low-lying land bordered by higher ground. [GEOL] A relatively shallow, wide depression of the sea floor with gentle slopes. Also known as submarine valley. { 'val·ē }

valley bottom *See* valley floor. { 'val·ē ,bäd·əm }

valley breeze [METEOROL] A gentle wind blowing up a valley or mountain slope in the absence of cyclonic or anticyclonic winds, caused by the warming of the mountainside and valley floor by the sun. { 'val·ē ,brēz }

valley fill [GEOL] Unconsolidated sedimentary deposit which fills or partly fills a valley. { 'val·ē ,fil }

valley flat [GEOL] The small plain at the bottom of a narrow valley with steep sides. { 'val·ē ,flat }

valley floor [GEOL] The broad, flat bottom of a valley. Also known as valley bottom; valley plain. { 'val·ē ,flȯr }

valley glacier [HYD] A glacier that flows down the walls of a mountain valley. { 'val·ē ,glā·shər }

valley iceberg [OCEANOGR] An iceberg weathered in such a manner that a large U-shaped slot extends through the iceberg. Also known as drydock iceberg. { 'val·ē 'īs,bərg }

valley line *See* thalweg. { 'val·ē ,līn }

valley plain *See* valley floor. { 'val·ē ,plān }

valley train [GEOL] A long, narrow body of outwash, deposited by meltwater far beyond the margin of an active glacier and extending along the floor of a valley. Also known as outwash train. { 'val·ē ,trān }

valley wind [METEOROL] A wind which ascends a mountain valley (up-valley wind) during the day; the daytime component of a mountain and valley wind system. { 'val·ē ,wind }

Van Allen radiation belt [GEOPHYS] One of the belts of intense ionizing radiation in space about the earth formed by high-energy charged particles which are trapped by the geomagnetic field. { va'nal·ən ,rād·ē'a·shən ,belt }

vanishing tide [OCEANOGR] When a high water and low water "melt" together into a period of several hours with a nearly constant water level. { 'van·ish·iŋ 'tīd }

vapor-dominated hydrothermal reservoir [GEOL] Any geothermal system mainly producing dry steam; the Geysers area of northern California and the Larderelle region of Italy are two examples. { 'vā·pər ¦dom·ə,nād·əd ¦hī·drə¦thər·məl 'rez·əv,wär }

vapor pressure [METEOROL] The partial pressure of water vapor in the atmosphere. { 'vā·pər ,presh·ər }

vapor-pressure deficit *See* saturation deficit. { 'vā·pər ¦presh·ər 'def·ə·sət }

vapor trail *See* condensation trail. { 'vā·pər ,trāl }

vardar [METEOROL] A cold fall wind blowing from the northwest down the Vardar valley in Greece to the Gulf of Salonica; it occurs when atmospheric pressure over eastern Europe is higher than over the Aegean Sea, as is often the case in winter. Also known as vardarac. { 'vär,där }

vardarac *See* vardar. { 'vär·də,rak }

variable ceiling [METEOROL] After United States weather-observing practice, a condition in which the ceiling rapidly increases and decreases while the ceiling observation is being made; the average of the observed values is used as the reported ceiling, and it is reported only for ceilings of less than 3000 feet (914 meters). { 'ver·ē·ə·bəl 'sēl·iŋ }

variable visibility [METEOROL] After United States weather observing practice, a condition in which the prevailing visibility fluctuates rapidly while the observation is being made; the average of the observed values is used as the reported visibility, and it is reported only for visibilities of less than 3 miles (4.8 kilometers). { 'ver·ē·ə·bəl ,viz·ə'bil·əd·ē }

variation *See* declination. { ,ver·ē'ā·shən }

variation of latitude [GEOPHYS] Change of the latitude of a place on earth because of the irregular movement of the north and south poles; the movement is caused by the earth's shifting on its axis. { ,ver·ē'ā·shən əv 'lad·ə,tüd }

variation per day [GEOPHYS] The change in the value of any geophysical quantity during 1 day. { ,ver·ē'ā·shən pər 'dā }

variation per hour [GEOPHYS] The change in the value of any geophysical quantity during 1 hour. { ,ver·ē'ā·shən pər 'aùr }

variation per minute [GEOPHYS] The change in the value of any geophysical quantity during 1 minute. { ,ver·ē'ā·shən pər 'min·ət }

variole [GEOL] A spherule the size of a pea, usually consisting of radiating plagioclase or pyroxene crystals. { 'ver·ē,ōl }

Variscan orogeny [GEOL] The late Paleozoic orogenic era in Europe, extending through the Carboniferous and Permian. Also known as Hercynian orogeny. { va'ris·kən ȯ'räj·ə·nē }

varve [GEOL] A sedimentary bed, layer, or sequence of layers deposited in a body of still water within a year's time, and usually during a season. Also known as glacial varve. { 'värv }

varve clay *See* varved clay. { 'värv ,klā }

varved clay [GEOL] A lacustrine sediment of distinct layers consisting of varves. Also known as varve clay. { 'värvd ,klā }

vaudaire [METEOROL] A violent south wind; a foehn of Lake Geneva in Switzerland. Also known as vauderon. { vō'der }

vauderon *See* vaudaire. { vō·də'rōn }

Vectian *See* Aptian. { 'vek·chən }

vectorial structure *See* directional structure. { vek'tȯr·ē·əl 'strək·chər }

veering [METEOROL] **1.** In international usage, a change in wind direction in a clockwise sense (for example, south to southwest to west) in either hemisphere of the earth. **2.** According to widespread usage among United States meteorologists, a change in wind direction in a clockwise sense in the Northern Hemisphere, counterclockwise in the Southern Hemisphere. { 'vir·iŋ }

vegetable jelly *See* ulmin. { 'vej·tə·bəl ,jel·ē }

veil [METEOROL] A very thin cloud through which objects are visible. { vāl }

vein [GEOL] A mineral deposit in tabular or shell-like form filling a fracture in a host rock. { vān }

veinite [GEOL] A genetic type of veined gneiss in which the vein material was secreted from the rock itself. { 'vā,nīt }

velocity discontinuity *See* seismic discontinuity. { və'läs·əd·ē dis,känt·ən'ü·əd·ē }

velocity gradient *See* seismic gradient. { və'läs·əd·ē ,grād·ē·ənt }

velocity ratio [OCEANOGR] The ratio of the speed of tidal current at a subordinate station to the speed of the corresponding current at the reference station. { və'läs·əd·ē ,rā·shō }

velum [METEOROL] An accessory cloud veil of great horizontal extent draped over or penetrated by cumuliform clouds; velum occurs with cumulus and cumulonimbus. { 'vē·ləm }
vendaval [METEOROL] A stormy southwest wind on the southern Mediterranean coast of Spain and in the Straits of Gibraltar; it occurs with a low advancing from the west in late autumn, winter, or early spring, and is often accompanied by thunderstorms and violent squalls. { ˌven·də'väl }
vent [GEOL] The opening of a volcano on the surface of the earth. { vent }
vent da Mùt [METEOROL] A strong, wet wind of Lake Garda in Italy. { ˌvent dä 'müt }
vent des dames [METEOROL] A daily sea breeze of about 15 miles (24 kilometers) per hour from the southwest in summer on the Mediterranean coast east of the Rhone delta, extending some 20 miles (32 kilometers) inland. { vòn de 'däm }
vent du midi [METEOROL] A south wind in the center of the Massif Central and the southern Cevennes (France); it is warm, moist, and generally followed by a southwest wind with heavy rain. { vòn dyů mē'dē }
ventifact [GEOL] A stone or pebble whose shape, wear, faceting, cut, or polish is the result of sandblasting. Also known as glyptolith; rillstone; wind-cut stone; wind-grooved stone; wind-polished stone; wind-scoured stone; wind-shaped stone. { 'ven·təˌfakt }
ventilation [METEOROL] The process of causing representative air to be in contact with the sensing elements of observing instruments; especially applied to producing a flow of air past the bulb of a wet-bulb thermometer. { ˌvent·əl'ā·shən }
vento di sotto [METEOROL] Breezes blowing up-lake on Lake Garda in Italy. { ˌven·tō dī 'sò·tō }
Venturian [GEOL] A North American stage of middle Pliocene geologic time, above Repettian and below Wheelerian. { ven'chůr·ē·ən }
veranillo [CLIMATOL] The lesser dry season, made up of a few weeks of hot dry weather, that breaks up the summer rainy season on the Pacific coast of Mexico and Central America. { ver·ə'nēl·yō }
verano [CLIMATOL] In Mexico and Central America, the main dry season, generally occurring from November through April. { ve'rä·nō }
verdant zone *See* frostless zone. { 'vərd·ənt ˌzōn }
vergence [GEOL] The direction of overturning or of inclination of a fold. { 'vər·jəns }
vernal [GEOPHYS] Pertaining to spring. { 'vərn·əl }
verrou *See* riegel. { və'rü }
vertebratus [METEOROL] A cloud variety (applied mainly to the genus cirrus), the elements of which are arranged in a manner suggestive of vertebrae, ribs, or a fish skeleton. { ˌvərd·ə'bräd·əs }
vertical anemometer [METEOROL] An instrument which records the vertical component of the wind speed. { 'vərd·ə·kəl ˌan·ə'mäm·əd·ər }
vertical differential chart [METEOROL] A synoptic chart showing the difference in value of a meteorological element between two levels in the atmosphere; a common example is the thickness chart. { 'vərd·ə·kəl ˌdif·ə'ren·chəl ˌchärt }
vertical dip slip *See* vertical slip. { 'vərd·ə·kəl 'dip ˌslip }
vertical intensity [GEOPHYS] The magnetic intensity of the vertical component of the earth's magnetic field, reckoned positive if downward, negative if upward. { 'vərd·ə·kəl in'ten·səd·ē }
vertical jet *See* uprush. { 'vərd·ə·kəl 'jet }
vertical separation [GEOL] The vertical component of the dip slip in a fault. { 'vərd·ə·kəl ˌsep·ə'rā·shən }
vertical slip [GEOL] The vertical component of the net slip in a fault. Also known as vertical dip slip. { 'vərd·ə·kəl 'slip }
vertical stability *See* static stability. { 'vərd·ə·kəl stə'bil·əd·ē }
vertical stretching [METEOROL] A process in which ascending vertical motion of air increases with altitude, or descending motion decreases with (increasing) altitude. { 'vərd·ə·kəl 'strech·iŋ }
vertical visibility [METEOROL] According to United States weather observing practice,

the distance that an observer can see vertically into a surface-based obscuring phenomenon, such as fog, rain, or snow. { 'vərd·ə·kəl ˌvis·ə'bil·əd·ē }

Vertisol [GEOL] A soil order formed in regoliths high in clay; subject to marked shrinking and swelling with changes in water content; low in organic content and high in bases. { 'vərd·əˌsȯl }

very close pack ice [OCEANOGR] Sea ice so concentrated that there is little if any open water. { ¦ver·ē ¦klōs 'pak ˌīs }

very open pack ice [OCEANOGR] Sea ice whose concentration ranges between one-tenth and three-tenths of the sea surface. { ¦ver·ē ¦ō·pən 'pak ˌīs }

vesicle [GEOL] A cavity in lava formed by entrapment of a gas bubble during solidification. Also known as wing. { 'ves·ə·kəl }

Vesuvian eruption *See* Vulcanian eruption. { və'sü·vē·ən i'rəp·shən }

VFR weather [METEOROL] In aviation terminology, route or terminal weather conditions which allow operation of aircraft under visual flight rules. { ˌvēˌef'är 'weth·ər }

Vindobonian [GEOL] A European stage of geologic time, middle Miocene. { ˌvin·də'bō·nē·ən }

virazon [METEOROL] **1.** The very strong southwesterly sea breeze experienced where the coastal chains of the Andes Mountains descend steeply to the sea; it sets in about 10 a.m. and reaches its greatest strength at about 3 p.m. **2.** A westerly sea breeze of Spain and Portugal. { vir·ə'zȯn }

virga [METEOROL] Wisps or streaks of water or ice particles falling out of a cloud but evaporating before reaching the earth's surface as precipitation. Also known as fall streaks; Fallstreifen; precipitation trails. { 'vər·gə }

virtual gravity [METEOROL] The force of gravity on a parcel of air, reduced by centrifugal force due to the motion of the parcel relative to the earth. { 'vər·chə·wəl 'grav·əd·ē }

virtual height [GEOPHYS] The apparent height of a layer in the ionosphere, determined from the time required for a radio pulse to travel to the layer and return, assuming that the pulse propagates at the speed of light. Also known as equivalent height. { 'vər·chə·wəl 'hīt }

virtual pressure [METEOROL] The pressure of a parcel of moist air when it has the same density as a parcel of dry air at the same temperature. { 'vər·chə·wəl 'presh·ər }

virtual temperature [METEOROL] In a system of moist air, the temperature of dry air having the same density and pressure as the moist air. { 'vər·chə·wəl 'tem·prə·chər }

viscous magnetization *See* viscous remanent magnetization. { 'vis·kəs ˌmag·nəd·ə'zā·shən }

viscous remanent magnetization [GEOPHYS] A process in which grains of magnetic minerals, which are either too small or too finely divided by undergrowths of different chemical composition to retain a permanent magnetization indefinitely, acquire a new direction of magnetization when the direction of the earth's magnetic field changes. Abbreviated VRM. Also known as viscous magnetization. { 'vis·kəs 'rem·ə·nənt ˌmag·nəd·ə'zā·shən }

Viséan [GEOL] A European stage of lower Carboniferous geologic time forming the lowermost Upper Mississippian, above Tournaisian and below lower Namurian. { vi'sā·ən }

visibility [METEOROL] In weather observing practice, the greatest distance in a given direction at which it is just possible to see and identify with the unaided eye, in the daytime, a prominent dark object against the sky at the horizon and, at nighttime, a known, preferably unfocused, moderately intense light source. { ˌviz·ə'bil·əd·ē }

visual range [METEOROL] The distance, under daylight conditions, at which the apparent contrast between a specified type of target and its background becomes just equal to the threshold contrast of an observer. { 'vizh·ə·wəl 'rānj }

vitavite *See* moldavite. { 'vīd·əˌvīt }

vitrain [GEOL] A brilliant black coal lithotype with vitreous luster and cubical cleavage. Also known as pure coal. { 'viˌtrān }

vitric [GEOL] Referring to a pyroclastic material which is characteristically glassy, that is, contains more than 75% glass. { 'vi·trik }

vitric tuff [GEOL] Tuff composed principally of volcanic glass fragments. { 'vi·trik 'təf }

vitrification [GEOL] Formation of a glassy or noncrystalline material. { ˌvi·trə·fə'kā·shən }
vitrinite [GEOL] A maceral group that is rich in oxygen and composed of humic material associated with peat formation; characteristic of vitrain. { 'vi·trəˌnīt }
vitrinoid [GEOL] Vitrinite occurring in bituminous coking coals; characterized by a reflectance of 0.5–2.0%. { 'vi·trəˌnȯid }
viuga [METEOROL] A cold north or northeast storm of the Russian steppes, lasting about 3 days. { 'vyü·gə }
volatile component [GEOL] A component of magma whose vapor pressures are high enough to allow them to be concentrated in any gaseous phase. Also known as volatile flux. { 'väl·əd·əl kəm'pō·nənt }
volatile flux *See* volatile component. { 'väl·əd·əl 'fləks }
volcanello *See* spatter cone. { ˌväl·kə'nel·ō }
volcanic arc *See* island arc. { välˌkan·ik 'ärk }
volcanic ash [GEOL] Fine pyroclastic material; particle diameter is less than 4 millimeters. { väl'kan·ik 'ash }
volcanic bombs [GEOL] Pyroclastic ejecta; the lava fragments, liquid or plastic at the time of ejection, acquire rounded forms, markings, or internal structure during flight or upon landing. { väl'kan·ik 'bämz }
volcanic foam *See* pumice. { väl'kan·ik 'fōm }
volcanic gases [GEOL] Volatile matter composed principally of about 90% water vapor, and carbon dioxide, sulfur dioxide, hydrogen, carbon monoxide, and nitrogen, released during an eruption of a volcano. { väl'kan·ik 'gas·əz }
volcanic glass [GEOL] Natural glass formed by the cooling of molten lava, or one of its liquid fractions, too rapidly to allow crystallization. { väl'kan·ik 'glas }
volcanicity *See* volcanism. { ˌväl·kə'nis·əd·ē }
volcanic mud [GEOL] Sediment containing large quantities of ash from a volcanic eruption, mixed with water. { väl'kan·ik 'məd }
volcanic mudflow [GEOL] The flow of volcanic mud down the slope of a volcano. { väl'kan·ik 'mədˌflō }
volcanic neck [GEOL] A residual remnant of the pipe or throat of a volcano that was filled with solidified lava after its final eruption. { väl'kan·ik 'nek }
volcanic rift zone [GEOL] A zone comprising volcanic fissures with underlying dike assemblages; occurs in Hawaii. { väl'kan·ik 'rift ˌzōn }
volcanic rock [GEOL] Finely crystalline or glassy igneous rock resulting from volcanic activity at or near the surface of the earth. Also known as extrusive rock. { väl'kan·ik 'räk }
volcanic vent [GEOL] The channelway or opening of a volcano through which magma ascends to the surface; two general types are fissure and pipelike vents. { väl'kan·ik 'vent }
volcanism [GEOL] The movement of magma and its associated gases from the interior into the crust and to the surface of the earth. Also known as volcanicity. { 'väl·kəˌniz·əm }
volcano [GEOL] **1.** A mountain or hill, generally with steep sides, formed by the accumulation of magma erupted through openings or volcanic vents. **2.** The vent itself. { väl'kā·nō }
volcanology [GEOL] The branch of geology that deals with volcanism. { ˌväl·kə'näl·ə·jē }
volume phase *See* surface phase. { 'väl·yəm ˌfāz }
volume transport [OCEANOGR] The volume of moving water measured between two points of reference and expressed in cubic meters per second. { 'väl·yəm ˌtranzˌpȯrt }
vriajem *See* friagem. { 'frē·əˌjem }
VRM *See* viscous remanent magnetization.
V-shaped depression [METEOROL] On a surface chart, a low or trough about which the isobars display a pronounced V shape, with the point of the V usually extending equatorward from the parent low. { 'vē ¦shāpt di'presh·ən }

V-shaped valley [GEOL] A valley having a cross-sectional profile in the form of the letter V, commonly produced by stream erosion. Also known as V valley. { 'vē ¦shāpt 'val·ē }

Vulcanian eruption [GEOL] A volcanic eruption characterized by periodic explosive events. Also known as paroxysmal eruption; Plinian eruption; Vesuvian eruption. { ¦vəl¦kā·nē·ən i'rəp·shən }

vulgar establishment *See* high-water full and change. { 'vəl·gər i'stab·lish·mənt }

vuthan [METEOROL] In southern South America, an intense storm. { ¦vü¦tän }

V valley *See* V-shaped valley. { 'vē ˌval·ē }

Wadati-Benioff zone *See* Benioff zone. { ¦wä¦dä·tē 'ben·ē,ȯf ,zōn }

wadi [GEOL] In the desert regions of southwestern Asia and northern Africa, a stream bed or channel, or a steep-sided ravine, gulley, or valley, which carries water only during the rainy season. Also spelled wady. { 'wäd·ē }

wady *See* wadi. { 'wäd·ē }

wake stream theory [OCEANOGR] The theory that, in a stratified ocean, a compensation current must develop on the right side of a wake stream, flowing in the same direction, and a countercurrent in the opposite direction must appear to the left. { 'wāk ¦strēm ,thē·ə·rē }

wall [GEOL] The side of a cave passage. { wȯl }

wall reef [GEOL] A linear, steep-sided coral reef constructed on a reef wall. { 'wȯl ¦rēf }

wall rock [GEOL] Rock that encloses a vein. { 'wȯl ¦räk }

wall-rock alteration [GEOL] Alteration of wall rock adjacent to hydrothermal veins by the fluid responsible for formation of the mineral deposit. { 'wȯl ¦räk ,ȯl·tə'rā·shən }

wall-sided glacier [HYD] A glacier unconfined by a marked ravine or valley. { 'wȯl ¦sīd·əd 'glā·shər }

wander *See* apparent wander. { 'wän·dər }

wandering dune [GEOL] A sand dune that has moved as a unit in the leeward direction of the prevailing winds, and that is characterized by the lack of vegetation to anchor it. Also known as migratory dune; traveling dune. { 'wän·də·riŋ 'dün }

wandering water *See* vadose water. { 'wän·də·riŋ 'wȯd·ər }

want *See* nip. { wänt }

warm-air drop *See* warm pool. { 'wȯrm ¦er 'dräp }

warm air mass [METEOROL] An air mass that is warmer than the surrounding air; an implication that the air mass is warmer than the surface over which it is moving. { 'wȯrm ¦er 'mas }

warm anticyclone *See* warm high. { 'wȯrm ¦ant·i'sī,klōn }

warm braw [METEOROL] A warm, dry, foehn wind which persists for up to 8 days during the east monsoon in the Schouten Islands off the north coast of New Guinea. { 'wȯrm 'brȯ }

warm-core anticyclone *See* warm high. { 'wȯrm ¦kȯr ¦ant·i'sī,klōn }

warm-core cyclone *See* warm low. { 'wȯrm ¦kȯr 'sī,klōn }

warm-core high *See* warm high. { 'wȯrm ¦kȯr 'hī }

warm-core low *See* warm low. { 'wȯrm ¦kȯr 'lō }

warm cyclone *See* warm low. { 'wȯrm 'sī,klōn }

warm drop *See* warm pool. { 'wȯrm 'dräp }

warm front [METEOROL] Any nonoccluded front, or portion thereof, which moves in such a way that warmer air replaces colder air. { 'wȯrm ,frənt }

warm high [METEOROL] At a given level in the atmosphere, any high that is warmer at its center than at its periphery. Also known as warm anticyclone; warm-core anticyclone; warm-core high. { 'wȯrm 'hī }

warm low [METEOROL] At a given level in the atmosphere, any low that is warmer at its center than at its periphery; the opposite of a cold low. Also known as warm-core cyclone; warm-core low; warm cyclone. { 'wȯrm 'lō }

warm pool [METEOROL] A region, or pool, of relatively warm air surrounded by colder air; the opposite of a cold pool; commonly applied to warm air of appreciable vertical extent isolated in high latitudes when a cutoff high is formed. Also known as warm-air drop; warm drop. { 'wȯrm 'pül }

warm sector [METEOROL] The area of warm air, within the circulation of a wave cyclone, which lies between the cold front and warm front of a storm. { 'wȯrm ˌsek·tər }

warm tongue [METEOROL] A pronounced poleward extension or protrusion of warm air. { 'wȯrm 'təŋ }

warm-tongue steering [METEOROL] The steering influence apparently exerted upon a tropical cyclone by an upper-level warm tongue which often extends a considerable distance into regions adjacent to the cyclone. { 'wȯrm ¦təŋ 'stir·iŋ }

warm wave *See* heat wave. { 'wȯrm ˌwāv }

warning stage [HYD] The stage, on a fixed river gage, at which it is necessary to begin issuing warnings or river forecasts if adequate precautionary measures are to be taken before flood stage is reached. { 'wȯrn·iŋ ˌstāj }

warp [GEOL] **1.** An upward or downward flexure of the earth's crust. **2.** A layer of sediment deposited by water. { wȯrp }

warrant *See* underclay. { 'wär·ənt }

Wasatch winds [METEOROL] Strong, easterly, jet-effect winds blowing out of the mouths of the canyons of the Wasatch Mountains onto the plains of Utah. { 'wäˌsach 'winz }

wash [GEOL] **1.** An alluvial placer. **2.** A piece of land washed by a sea or river. **3.** *See* alluvial cone. { wäsh }

wash-and-strain ice foot [OCEANOGR] An ice foot formed from ice casts and slush and attached to a shelving beach, between the high and low waterlines; high waves and spray may cause it to build up above the high waterline. { ¦wäsh ən ¦strān 'īs ˌfu̇t }

wash-built terrace *See* alluvial terrace. { 'wäsh ¦bilt 'ter·əs }

wash load [GEOL] The finer part of the total sediment load of a stream which is supplied from bank erosion or an external upstream source, and which can be carried in large quantities. { 'wäsh ˌlōd }

Washoe zephyr [METEOROL] The chinook on the Nevada side of the Sierra Nevada Mountains of northern California. { 'wäˌshō 'zef·ər }

washover [GEOL] Material deposited by overwash, especially a small delta produced by storm waves and built on the landward side of a bar or barrier. Also known as storm delta; wave delta. { 'wäshˌō·vər }

wash plain *See* outwash plain. { 'wäsh ˌplān }

wash slope [GEOL] The gentle slope on a hillside occurring below the gravity slope and lying at the foot of an escarpment or steep rock face; usually covered by an accumulation of talus. Also known as haldenhang. { 'wäsh ˌslōp }

waste plain *See* alluvial plain. { 'wāst ˌplān }

water atmosphere [METEOROL] The concept of a separate atmosphere composed only of water vapor. { 'wȯd·ər 'at·məˌsfir }

water-bearing strata [GEOL] Ground layers below the standing water level. { 'wȯd·ər ¦ber·iŋ 'strad·ə }

water budget *See* hydrologic accounting. { 'wȯd·ər ˌbəj·ət }

water cloud [METEOROL] Any cloud composed entirely of liquid water drops; to be distinguished from an ice-crystal cloud and from a mixed cloud. { 'wȯd·ər ˌklau̇d }

water content [HYD] The liquid water present within a sample of snow (or soil) usually expressed in percent by weight; the water content in percent of water equivalent is 100 minus the quality of snow. Also known as free-water content; liquid-water content. { 'wȯd·ər ˌkänˌtent }

watercourse [HYD] **1.** A stream of water. **2.** A natural channel through which water may run or does run. { 'wȯd·ərˌkȯrs }

water cycle *See* hydrologic cycle. { 'wȯd·ər ˌsī·kəl }

water equivalent [METEOROL] The depth of water that would result from the melting of the snowpack or of a snow sample; thus, the water equivalent of a new snowfall is the same as the amount of precipitation represented by that snowfall. { 'wȯd·ər i'kwiv·ə·lənt }

water exchange [OCEANOGR] The volume and rate of water exchange between air and a body of water in a specific location, or between several bodies of water, controlled by such factors as tides, winds, river discharge, and currents. { 'wȯd·ər iks,chānj }

waterfall [HYD] A perpendicular or nearly perpendicular descent of water in a stream. { 'wȯd·ər,fȯl }

waterfall lake *See* plunge pool. { 'wȯd·ər,fȯl ,lāk }

water front [GEOGR] An area partly bounded by water. { 'wȯd·ər ,frənt }

water gap [GEOL] A deep and narrow pass that cuts to the base of a mountain ridge, and through which a stream flows; the Delaware Water Gap is an example. { 'wȯd·ər ,gap }

waterless zone [HYD] The lowest hydrologic zone, generally beginning several miles beneath the land surface and characterized by the absence of water in the pore spaces due to the great pressure and density of the rock. { 'wȯd·ər·ləs 'zōn }

water level *See* water table. { 'wȯd·ər ,lev·əl }

waterline *See* shoreline; water table. { 'wȯd·ər,līn }

water loss *See* evapotranspiration. { 'wȯd·ər ,lȯs }

water mass [OCEANOGR] A body of water identified by its temperature-salinity curve or chemical composition, and normally consisting of a mixture of two or more water types. { 'wȯd·ər ,mas }

water opening [OCEANOGR] A break in sea ice, revealing the sea surface. { 'wȯd·ər ,ō·pə·niŋ }

water requirement [HYD] The total quantity of water required to mature a specified crop under field conditions; includes applied irrigation, water precipitation, and groundwater available to the crop. { 'wȯd·ər ri,kwīr·mənt }

watershed [HYD] The drainage area of a stream. { 'wȯd·ər,shed }

water sky [METEOROL] The dark appearance of the underside of a cloud layer when it is over a surface of open water. { 'wȯd·ər ,skī }

water smoke *See* steam fog. { 'wȯd·ər ,smōk }

water snow [HYD] Snow that, when melted, yields a more than average amount of water; thus, any snow with a high water content. { 'wȯd·ər ,snō }

waterspout [METEOROL] A tornado occurring over water; rarely, a lesser whirlwind over water, comparable in intensity to a dust devil over land. { 'wȯd·ər,spau̇t }

water table [HYD] The planar surface between the zone of saturation and the zone of aeration. Also known as free-water elevation; free-water surface; groundwater level; groundwater surface; groundwater table; level of saturation; phreatic surface; plane of saturation; saturated surface; water level; waterline. { 'wȯd·ər ,tā·bəl }

water trap [GEOL] A chamber or part of a cave system that is filled with water, due to the dipping of the roof or ceiling below the water level. { 'wȯd·ər ,trap }

water type [OCEANOGR] Ocean water of a specified temperature and salinity. { 'wȯd·ər ,tīp }

water-vapor absorption [METEOROL] The absorption of certain wavelengths of infrared radiation by atmospheric water vapor; a process of fundamental importance in the energy budget of the earth's atmosphere. { 'wȯd·ər ¦vā·pər əb,sȯrp·shən }

water year [HYD] Any 12-month period, usually selected to begin and end during a relatively dry season, used as a basis for processing streamflow and other hydrologic data; the period from October 1 to September 30 is most widely used in the United States. { 'wȯd·ər 'yir }

wave base [HYD] The depth at which sediments are not stirred by wave action, usually about 33 feet (10 meters). Also known as wave depth. { 'wāv ,bās }

wave basin [GEOGR] A basin close to the inner entrance of a harbor in which the waves from the outer entrance are absorbed, thus reducing the size of the waves entering the inner harbor. { 'wāv ,bās·ən }

wave-built platform *See* alluvial terrace. { 'wāv ¦bilt 'plat,fȯrm }

wave-built terrace *See* alluvial terrace. { 'wāv ¦bilt 'ter·əs }

wave-cut bench [GEOL] A level or nearly level narrow platform produced by wave erosion and extending outward from the base of a wave-cut cliff. Also known as beach platform; high-water platform. { 'wāv ¦kət 'bench }

wave-cut cliff [GEOL] A cliff formed by the erosive action of waves on rock. { 'wāv ¦kət 'klif }
wave-cut notch [GEOL] An indentation cut into a sea cliff at water level by wave action. { 'wāv ¦kət 'näch }
wave-cut plain *See* wave-cut platform. { 'wāv ¦kət 'plān }
wave-cut platform [GEOL] A gently sloping surface which is produced by wave erosion and which extends into the sea for a considerable distance from the base of the wave-cut cliff. Also known as cut platform; erosion platform; strand flat; wave-cut plain; wave-cut terrace; wave platform. { 'wāv ¦kət 'plat,fȯrm }
wave-cut terrace *See* wave-cut platform. { 'wāv ¦kət 'ter·əs }
wave cyclone [METEOROL] A cyclone which forms and moves along a front; the circulation about the cyclone center tends to produce a wavelike deformation of the front. Also known as wave depression. { 'wāv ,sī,klōn }
wave delta *See* washover. { 'wāv ,del·tə }
wave depression *See* wave cyclone. { 'wāv di,presh·ən }
wave depth *See* wave base. { 'wāv ,depth }
wave disturbance [METEOROL] In synoptic meteorology, the same as wave cyclone, but usually denoting an early state in the development of a wave cyclone, or a poorly developed one. { 'wāv di,stər·bəns }
wave erosion *See* marine abrasion. { 'wāv i,rō·zhən }
wave forecasting [OCEANOGR] The theoretical determination of future wave characteristics based on observed or forecasted meteorological phenomena. { 'wāv 'fȯr ,kast·iŋ }
wave height [OCEANOGR] The height of a water-surface wave is generally taken as the height difference between the wave crest and the preceding trough. { 'wāv ,hīt }
wave line *See* swash mark. { 'wāv ,līn }
wavemark *See* swash mark. { 'wāv,märk }
wave platform *See* wave-cut platform. { 'wāv 'plat,fȯrm }
wave ripple mark *See* oscillation ripple mark. { 'wāv 'rip·əl ,märk }
wave setdown [OCEANOGR] A decrease in the mean water level in the region in which breakers form near the seashore, caused by the presence of a pressure field. { 'wāv 'set,dau̇n }
wave setup [OCEANOGR] An increase in the mean water level shoreward of the region in which breakers form at the seashore, caused by the onshore flux of momentum against the beach. { 'wāv 'sed,əp }
wave system [OCEANOGR] In ocean wave studies, a group of waves which have the same height, length, and direction of movement. { 'wāv ,sis·təm }
wave theory of cyclones [METEOROL] A theory of cyclone development based upon the principle of wave formation on an interface between two fluids; in the atmosphere, a front is taken as such an interface. { 'wāv 'thē·ə·rē əv 'sī,klōnz }
weather [METEOROL] **1.** The state of the atmosphere, mainly with respect to its effects upon life and human activities; as distinguished from climate, weather consists of the short-term (minutes to months) variations of the atmosphere. **2.** As used in the making of surface weather observations, a category of individual and combined atmospheric phenomena which must be drawn upon to describe the local atmospheric activity at the time of observation. { 'weth·ər }
weather central [METEOROL] An organization which collects, collates, evaluates, and disseminates meteorological information in such a manner that it becomes a principal source of such information for a given area. { 'weth·ər 'sen·trəl }
weathered iceberg [OCEANOGR] An iceberg which is irregular in shape, due to an advanced stage of ablation; it may have overturned. { 'weth·ərd 'īs,bərg }
weathered layer [GEOPHYS] The zone of the earth which lies immediately below the surface and is characterized by low wave velocities. { 'weth·ərd 'lā·ər }
weather forecast [METEOROL] A forecast of the future state of the atmosphere with specific reference to one or more associated weather elements. { 'weth·ər ,fȯr,kast }
weathering [GEOL] Physical disintegration and chemical decomposition of earthy and

rocky materials on exposure to atmospheric agents, producing an in-place mantle of waste. Also known as clastation; demorphism. { 'weth·ə‚riŋ }

weathering correction [GEOPHYS] A velocity correction which is applied to seismic data, necessitated by the diminished velocity of seismic wave propagation in weathered rock. { 'weth·ə‚riŋ kə‚rek·shən }

weathering-potential index [GEOL] A measure of the susceptibility of a rock or mineral to weathering. { 'weth·ə‚riŋ pə¦ten·chəl ‚in‚deks }

weathering rind [GEOL] The outer layer of a pebble, boulder, or other rock fragment that has formed as a result of chemical weathering. { 'weth·ər·iŋ ‚rīnd }

weathering velocity [GEOPHYS] The velocity of propagation of seismic waves through weathered rock. { 'weth·ə‚riŋ və‚läs·əd·ē }

weather map [METEOROL] A chart portraying the state of the atmospheric circulation and weather at a particular time over a wide area; it is derived from a careful analysis of simultaneous weather observations made at many observing points in the area. { 'weth·ər ‚map }

weather-map type *See* weather type. { 'weth·ər ‚map ‚tīp }

weather minimum [METEOROL] The worst weather conditions under which aviation operations may be conducted under either visual or instrument flight rules; usually prescribed by directives and standing operating procedures in terms of minimum ceiling, visibility, or specific hazards to flight. { 'weth·ər ‚min·ə·məm }

weather modification [METEOROL] The changing of natural weather phenomena by technical means; so far, only on the microscale of condensation and freezing nuclei has it been possible to exert modifying influences. { 'weth·ər ‚mäd·ə·fə'kā·shən }

weather observation [METEOROL] An evaluation of one or more meteorological elements that describe the state of the atmosphere either at the earth's surface or aloft. { 'weth·ər ‚äb·zər‚vā·shən }

weather pit [GEOL] A shallow depression (depth up to 6 inches or 15 centimeters) on the flat or gently sloping summit of large exposures of granite or granitic rocks, attributed to strongly localized solvent action of impounded water. { 'weth·ər ‚pit }

weather shore [METEOROL] As observed from a vessel, the shore lying in the direction from which the wind is blowing. { 'weth·ər ‚shȯr }

weather side [METEOROL] The side of a ship exposed to the wind or weather. { 'weth·ər ‚sīd }

weather signal [METEOROL] A visual signal displayed to indicate a weather forecast. { 'weth·ər ‚sig·nəl }

weather station [METEOROL] A place and facility for the observation, measurement, and recording and transmission of data of the variable elements of weather; one of the most effective network facilities is that of the U.S. Weather Bureau. { 'weth·ər ‚stā·shən }

weather type [METEOROL] A series of generalized synoptic situations, usually presented in chart form; weather types are selected to represent typical pressure patterns, and were originally devised as a method for lengthening the effective time-range of forecasts. Also known as weather-map type. { 'weth·ər ‚tīp }

Weddell Current [OCEANOGR] A surface current which flows in an easterly direction from the Weddell Sea outside the limit of the West Wind Drift. { we'del 'kə·rənt }

Wedener-Bergeron process *See* Bergeron-Findeisen theory. { 'vād·ən·ər ¦ber·zhə¦rōn ‚prä·səs }

wedge *See* ridge. { wej }

weeping spring *See* spring seepage. { 'wēp·iŋ 'spriŋ }

welding [GEOL] Consolidation of sediments by pressure; water is squeezed out and cohering particles are brought within the limits of mutual molecular attraction. { 'weld·iŋ }

wellhead [HYD] The place where a stream emerges from the ground. { 'wel‚hed }

well-sorted [GEOL] Referring to a sorted sediment that consists of particles of approximately the same size and has a sorting coefficient of less than 2.5. { 'wel ¦sȯrd·əd }

Wenlockian [GEOL] A European stage of geologic time: Middle Silurian (above Tarannon, below Ludlovian). { wen'läk·ē·ən }

Wentworth classification [GEOL] A logarithmic grade for size classification of sediment particles starting at 1 millimeter and using the ratio of 1/2 in one direction (and 2 in the other), providing diameter limits to the size classes of 1, 1/2, 1/4, etc. and 1, 2, 4, etc. { 'went,wərth ,klas·ə·fə'kā·shən }

Wentworth scale [GEOL] A geometric grade scale for sedimentary particles ranging from clay particles (diameter less than 1/250 millimeter) to boulders (diameters greater than 256 millimeters), in which the size classes are related to one another by a constant ratio of 1/2 (4, 2, 1, 1/2, etc.). { 'went,wərth ,skāl }

Werfenian stage *See* Scythian stage. { ver'fē·nē·ən ,stāj }

west [GEOGR] The direction 90° to the left or 270° to the right of north. { west }

West Australia Current [OCEANOGR] The complex current flowing northward along the west coast of Australia; it is strongest from November to January, and weakest and variable from May to July; it curves toward the west to join the South Equatorial Current. { 'west ȯ'strāl·yə 'kə·rənt }

westerlies [METEOROL] The dominant west-to-east motion of the atmosphere, centered over the middle latitudes of both hemispheres; at the earth's surface, the westerly belt (or west-wind belt) extends, on the average, from about 35 to 65° latitude. Also known as circumpolar westerlies; middle-latitude westerlies; mid-latitude westerlies; polar westerlies; subpolar westerlies; subtropical westerlies; temperate westerlies; zonal westerlies; zonal winds. { 'wes·tər·lēz }

westerly wave [METEOROL] An atmospheric wave disturbance embedded in the midlatitude westerlies. { 'wes·tər·lē 'wāv }

Western Equatorial Countercurrent [OCEANOGR] Weak, narrow bands of eastward-flowing water observed in some winter months in the western Atlantic near the equator. { 'wes·tərn ,ek·wə'tȯr·ē·əl kaunt·ər,kə·rənt }

West Greenland Current [OCEANOGR] The current flowing northward along the west coast of Greenland into the Davis Strait; part of this current joins the Labrador Current, while the other part continues into Baffin Bay. { 'west ¦grēn·lənd 'kə·rənt }

Westphalian [GEOL] A European stage of Upper Carboniferous geologic time, forming the Middle Pennsylvanian, above upper Namurian and below Stephanian. { west 'fāl·yən }

westward intensification [OCEANOGR] The intensification of ocean currents to the west, derived from a mathematical model that includes the effects of zonal wind stress at the sea surface and internal friction. { 'west·wərd in,ten·sə·fə'kā·shən }

West Wind Drift *See* Antarctic Circumpolar Current. { 'west ¦wind 'drift }

wet adiabat *See* saturation adiabat. { 'wet 'ad·ē·ə,bat }

wet-bulb depression [METEOROL] The difference in degrees between the dry-bulb temperature and the wet-bulb temperature. { 'wet ¦bəlb di'presh·ən }

wet-bulb temperature [METEOROL] **1.** Isobaric wet-bulb temperature, that is, the temperature an air parcel would have if cooled adiabatically to saturation at constant pressure by evaporation of water into it, all latent heat being supplied by the parcel. **2.** The temperature read from the wet-bulb thermometer; for practical purposes, the temperature so obtained is identified with the isobaric wet-bulb temperature. { 'wet ¦bəlb 'tem·prə·chər }

wet climate [CLIMATOL] A climate whose vegetation is of the rainforest type. Also known as rainforest climate. { 'wet 'klī·mət }

wet season *See* rainy season. { 'wet ,sēz·ən }

wet snow [METEOROL] Deposited snow that contains a great deal of liquid water. { 'wet 'snō }

wetted perimeter [GEOL] The portion of the perimeter of a steam channel cross section which is in contact with the water. { 'wed·əd pə'rim·əd·ər }

whaleback dune [GEOL] A smooth, elongated mound or hill of desert sand shaped generally like a whale's back; formed by passage of a succession of longitudinal dunes along the same path. Also known as sand levee. { 'wāl,bak ,dün }

Wheelerian [GEOL] A North American stage of upper Pliocene geologic time, above the Venturian and below the Hallian. { wē'lir·ē·ən }

whippoorwill storm *See* frog storm. { ¦wip·ər¦wil ,stȯrm }

whirlpool [OCEANOGR] Water in rapid rotary motion. { 'wərl,pül }

whirly [METEOROL] A small violent storm, a few yards (or meters) to 100 yards (91 meters) or more in diameter, frequent in Antarctica near the time of the equinoxes. { 'wər·lē }

whistler [GEOPHYS] An effect that occurs when a plasma disturbance, caused by a lightning discharge, travels out along lines of magnetic force of the earth's field and is reflected back to its origin from a magnetically conjugate point on the earth's surface; the disturbance may be picked up electromagnetically and converted directly to sound; the characteristic drawn-out descending pitch of the whistler is a dispersion effect due to the greater velocity of the higher-frequency components of the disturbance. { 'wis·lər }

whitecap [OCEANOGR] A cloud of bubbles at the sea surface caused by a breaking wave. { 'wīt,kap }

white coal *See* tasmanite. { 'wīt 'kōl }

white frost *See* hoarfrost. { 'wīt 'fro̊st }

whiteout [METEOROL] An atmospheric optical phenomenon of the polar regions in which the observer appears to be engulfed in a uniformly white glow: shadows, horizon, and clouds are not discernible; sense of depth and orientation are lost; dark objects in the field of view appear to float at an indeterminable distance. Also known as milky weather. { 'wīd,au̇t }

Whiterock [GEOL] A North American stage of lowermost Middle Ordovician geologic time, above lower Ordovician and below Marmor. { 'wīt,räk }

white squall [METEOROL] A sudden squall in tropical or subtropical waters, which lacks the usual squall cloud and whose approach is signaled only by the whiteness of a line of broken water or whitecaps. { 'wīt 'skwo̊l }

white water [OCEANOGR] Frothy water, as in whitecaps or breakers. { 'wīt ,wo̊d·ər }

whiting [OCEANOGR] A patch of seawater that contains a substantial amount of calcium carbonate and therefore appears white relative to surrounding water. { 'wīd·iŋ }

whitleyite [GEOL] An achondritic stony meteorite consisting essentially of enstatite with fragments of black chondrite. { 'wit·lē,īt }

whole gale [METEOROL] **1.** In storm-warning terminology, a wind of 48 to 63 knots (55 to 72 miles, or 89 to 133 kilometers, per hour). **2.** In the Beaufort wind scale, a wind whose speed is from 48 to 55 knots (55 to 63 miles, or 89 to 102 kilometers, per hour). { 'hōl 'gāl }

wichtisite *See* tachylite. { 'wik·tə,sīt }

Widmanstatten patterns [GEOL] Characteristic figures that appear on the surface of an iron meteorite when the meteorite is cut, polished, and etched with acid. { 'vit·mən,shtät·ən ,pad·ərnz }

Wiik classification [GEOL] A classification of carbonaceous chondrites into three types, C_1, C_2, and C_3. { 'wik ,klas·ə·fə'kā·shən }

Wilderness [GEOL] A North American stage of Middle Ordovician geologic time, above Porterfield and below Trentonian. { 'wil·dər·nəs }

wildflysch [GEOL] A type of flysch facies that represents a stratigraphic unit with irregularly sorted boulders resulting from fragmentation, and twisted, confused beds resulting from slumping or sliding due to the influence of gravity. { 'vilt,flish }

wild snow [METEOROL] Newly deposited snow which is very fluffy and unstable; in general, it falls only during a dead calm at very low air temperatures. { 'wīld 'snō }

williwaw [METEOROL] A very violent squall in the Straits of Magellan; it may occur in any month but occurs most frequently in winter. { 'wil·ē,wo̊ }

willy-willy [METEOROL] In Australia, a severe tropical cyclone. { 'wil·ē'wil·ē }

wind [METEOROL] The motion of air relative to the earth's surface; usually means horizontal air motion, as distinguished from vertical motion, and air motion averaged over the response period of the particular anemometer. { wind }

wind chill [METEOROL] That part of the total cooling of a body caused by air motion. { 'win ,chil }

wind-chill index [METEOROL] The cooling effect of any combination of temperature and wind, expressed as the loss of body heat in kilogram calories per hour per square meter of skin surface; it is only an approximation because of individual body variations in shape, size, and metabolic rate. { 'win ¦chil ˌinˌdeks }

wind crust [HYD] A type of snow crust, formed by the packing action of wind on previously deposited snow; wind crust may break locally but, unlike wind slab, does not constitute an avalanche hazard. { 'win 'krəst }

wind current [METEOROL] Generally, any of the quasi-permanent, large-scale wind systems of the atmosphere, for example, the westerlies, trade winds, equatorial easterlies, or polar easterlies. { 'win 'kəˌrənt }

wind-cut stone *See* ventifact. { 'win ¦kət 'stōn }

wind direction [METEOROL] The direction from which wind blows. { 'win dəˌrek·shən }

wind-direction shaft [METEOROL] A representational mark for wind direction on a synoptic chart, it is a straight line drawn directly upwind from the station circle; the wind arrow is completed by adding the wind-speed barbs and pennants to the outer end of the shaft. { 'win də¦rek·shən ˌshaft }

wind divide [METEOROL] A semipermanent feature of the atmospheric circulation (usually a high-pressure ridge) on opposite sides of which the prevailing wind directions differ greatly. { 'win dəˌvīd }

wind drift *See* drift current. { 'win ˌdrift }

wind-driven current *See* drift current. { 'win ¦driv·ən 'kə·rənt }

wind erosion [GEOL] Detachment, transportation, and deposition of loose topsoil or sand by the action of wind. { 'wind iˌrō·zhən }

wind gap [GEOL] A shallow, relatively high-level notch in the upper part of a mountain ridge, usually an abandoned water gap. Also known as air gap; wind valley. { 'win ˌgap }

wind-grooved stone *See* ventifact. { 'win ¦grüvd 'stōn }

wind measurement [METEOROL] The determination of three parameters: the size of an air sample, its speed, and its direction of motion. { 'win ˌmezh·ər·mənt }

window [GEOL] A break caused by erosion of a thrust sheet or a large recumbent anticline that exposes the rocks beneath the thrust sheet. Also known as fenster. [GEOPHYS] Any range of wavelengths in the electromagnetic spectrum to which the atmosphere is transparent. [HYD] The unfrozen part of a river surrounded by river ice during the winter. { 'win·dō }

window frost [HYD] A thin deposit of hoarfrost often found on interior surfaces of windows in winter, and frequently exhibiting beautiful fernlike patterns. { 'win·dō ˌfrȯst }

window ice [HYD] A thin deposit of ice which forms by the freezing of many tiny drops of water that have condensed on the indoors side of a cold window surface. { 'win·dō ˌīs }

wind-polished stone *See* ventifact. { 'win ¦päl·əsht 'stōn }

wind ripple [METEOROL] One of a series of wavelike formations on a snow surface, an inch or so in height, at right angles to the direction of wind. Also known as snow ripple. { 'win ˌdrip·əl }

wind rose [METEOROL] A diagram in which statistical information concerning direction and speed of the wind at a location may be summarized; a line segment is drawn in each of perhaps eight compass directions from a common origin; the length of a particular segment is proportional to the frequency with which winds blow from that direction; thicknesses of a segment indicate frequencies of occurrence of various classes of wind speed. { 'win ˌdrōz }

windrow [GEOL] Any accumulation of material formed by wind or tide action. { 'winˌdrō }

winds aloft [METEOROL] Generally, the wind speeds and directions at various levels in the atmosphere above the domain of surface weather observations, as determined by any method of winds-aloft observation. Also known as upper-level winds; upper winds. { 'winz ə'lȯft }

winds-aloft observation [METEOROL] The measurement and computation of wind speeds and directions at various levels above the surface of the earth. { 'winz ə'lȯft ˌäb·zər'vā·shən }

wind scoop [METEOROL] A saucerlike depression in the snow near obstructions such as trees, houses, and rocks, caused by the eddying action of the deflected wind. { 'win ˌsküp }

wind-scoured stone *See* ventifact. { 'win ¦skau̇rd 'stōn }

wind-shaped stone *See* ventifact. { 'win ¦shāpt 'stōn }

wind shear [METEOROL] The local variation of the wind vector or any of its components in a given direction. { 'win ˌshir }

wind-shift line [METEOROL] A line or narrow zone along which there is an abrupt change of wind direction. { 'win ¦shift ˌlīn }

wind slab [HYD] A type of snow crust; a patch of hard-packed snow, which is packed as it is deposited in favored spots by the wind, in contrast to wind crust. { 'wind ˌslab }

wind speed [METEOROL] The rate of motion of air. { 'win ˌspēd }

windstorm [METEOROL] A storm in which strong wind is the most prominent characteristic. { 'winˌstȯrm }

wind stress [METEOROL] The drag or tangential force per unit area exerted on the surface of the earth by the adjacent layer of moving air. { 'win ˌstres }

wind tide [OCEANOGR] **1.** The vertical rise in the still-water level on the leeward side of a body of water, particularly the ocean or other large body, caused by wind stresses on the surface of the water. **2.** The difference in still-water level between the windward and leeward sides of such a body caused by wind stresses. { 'win ˌtīd }

wind valley *See* wind gap. { 'win ˌval·ē }

wind velocity [METEOROL] The speed and direction of wind. { 'win vəˌläs·əd·ē }

windward [METEOROL] In the general direction from which the wind blows. { 'win·wərd }

wind wave [OCEANOGR] A wave resulting from the action of wind on a water surface. { 'win ˌwāv }

wing *See* vesicle. { wiŋ }

winged headland [GEOGR] A seacliff with two bays or spits, one on either side. { 'wiŋd 'hed·lənd }

winter ice [OCEANOGR] Level sea ice more than 8 inches (20 centimeters) thick, and less than 1 year old; the stage which follows young ice. { 'win·tər 'īs }

winter-talus ridge [GEOL] A wall-like arcuate ridge on the floor of a cirque formed by freezing activity that dislodged boulders from a cirque wall covered with a snowbank. Also known as nivation ridge. { 'win·tər 'tā·ləs ˌrij }

Wisconsin [GEOL] Pertaining to the fourth, and last, glacial stage of the Pleistocene epoch in North America; followed the Sangamon interglacial, beginning about 85,000 ± 15,000 years ago and ending 7000 years ago. { wi'skän·sən }

wisper wind [METEOROL] A cold night wind, blowing out of the valley of the Wisper River in Germany during clear weather. { 'wis·pər ˌwind }

Witte-Margules equation [OCEANOGR] A formula expressing the slope of the boundary layer between two water masses of different densities and velocities, taking into account the rotation of the earth. Also known as Margules equation. { 'vid·ə 'mär·gyə·lēz iˌkwā·zhən }

Wolfcampian [GEOL] A North American provincial series of geologic time; lowermost Permian (below Leonardian, above Virgilian of Pennsylvania). { wu̇lf'kam·pē·ən }

wood coal *See* bituminous wood. { 'wu̇d 'kōl }

woodstone *See* silicified wood. { 'wu̇dˌstōn }

woody lignite *See* bituminous wood. { 'wu̇d·ē 'ligˌnīt }

Workman-Reynolds effect [GEOPHYS] A mechanism for electric charge separation during freezing of slightly impure water; when a very dilute solution of certain salts freezes rapidly, a strong potential difference is established between the solid and liquid phases; for some salts, the ice attains negative charge, for others, positive; this mechanism has been suggested as one possible mode of thunderstorm charge separation in those portions of a thunderstorm downdraft where snow-pellet or hail particles sweep out supercooled waterdrops. { 'wərk·mən 'ren·əlz iˌfekt }

world rift system [GEOL] The system of interconnected midocean ridges which is the locus of tensional splitting and magma upwelling believed responsible for sea-floor spreading. { 'wərld 'rift ˌsis·təm }

wrench fault [GEOL] A lateral fault with a more or less vertical fault surface. Also known as basculating fault; torsion fault. { 'rench ˌfȯlt }

Würm [GEOL] **1.** A European stage of geologic time: uppermost Pleistocene (above Riss, below Holocene). **2.** Pertaining to the fourth glaciation of the Pleistocene epoch in the Alps, equivalent to the Wisconsin glaciation in North America, following the Riss-Würm interglacial. { vu̇rm }

wurtzilite [GEOL] A black, massive, sectile, infusible, asphaltic pyrobitumen derived from the metamorphosis of petroleum. { 'wərt·səˌlīt }

Xeralf [GEOL] A suborder of the soil order Alfisol, having good drainage, and found in regions with rainy winters and dry summers in mediterranean climates; the surface horizons tend to become massive and hard during the dry seasons, with some soils having duripans that interfere with root growth. { 'zir,älf }

Xerert [GEOL] A suborder of the soil order Vertisol, formed in a Mediterranean climate; wide surface cracks open and close once a year. { 'zir,ərt }

Xeroll [GEOL] A suborder of the soil order Mollisol, formed in a xeric moisture regime; may have a calcic, petrocalcic, or gypsic horizon, or a duripan. { 'zir,ȯl }

xerothermal period *See* xerothermic period. { ¦zir·ə¦thər·məl 'pir·ē·əd }

xerothermic [CLIMATOL] Characterized by dryness and heat. { ¦zir·ə¦thər·mik }

xerothermic period [GEOL] A postglacial interval of a warmer, drier climate. Also known as xerothermal period. { ¦zir·ə¦thər·mik 'pir·ē·əd }

Xerult [GEOL] A suborder of the soil order Ultisol, formed in a xeric moisture regime; brownish or reddish soil with a low to moderate organic-carbon content. { 'zir,əlt }

X wave *See* extraordinary wave. { 'eks ,wāv }

xylinite [GEOL] A variety of provitrinite consisting of xylem or lignified tissue. { 'zī·lə,nīt }

xyloid coal *See* bituminous wood. { 'zī,lȯid 'kōl }

xyloid lignite *See* bituminous wood. { 'zī,lȯid 'lig,nīt }

Y

yalca [METEOROL] A local name for a severe snowstorm with a strong squally wind which occurs in the Andes Mountain passes of northern Peru. { 'yäl·kə }

yamase [METEOROL] A cool, onshore, easterly wind in the Senriku district of Japan in summer. { yä'mä·sē }

yardang [GEOL] A long, irregular ridge with a sharp crest sited between two round-bottomed troughs that have been carved by wind erosion in a desert region. { 'yär,daŋ }

yardang trough [GEOL] A long, shallow, round-bottomed groove, furrow, or trough cut into a desert floor by wind erosion and separated by a yardang from the neighboring trough. { ¦yär,daŋ 'trȯf }

Yarmouth interglacial [GEOL] The second interglacial stage of the Pleistocene epoch in North America, following the Kansan glacial stage and before the Illinoian. { 'yär·məth ¦in·tər'glā·shəl }

yellow coal *See* tasmanite. { 'yel·ō 'kōl }

yellow mud [GEOL] Mud containing sediment having a characteristic yellow color, resulting from certain iron compounds. { 'yel·ō 'məd }

Yellow Sea [GEOGR] An inlet of the Pacific Ocean between northeastern China and Korea. { 'yel·ō 'sē }

yellow snow [HYD] Snow with a golden or yellow appearance because of the presence of pine or cypress pollen. { 'yel·ō 'snō }

yoked basin *See* zeugogeosyncline. { 'yōkt 'bās·ən }

Yorkian [GEOL] A European stage of geologic time forming part of the lower Upper Carboniferous, above Lanarkian and below Staffordian, equivalent to part of the lower Westphalian. { 'yȯr·kē·ən }

youg [METEOROL] A hot wind during unsettled summer weather in the Mediterranean. { yȯg }

young ice [HYD] Newly formed ice in the transitional stage of development from ice crust to winter ice. { 'yəŋ ¦īs }

youth [GEOL] The first stage of the cycle of erosion in which the original surface or structure is the dominant topographic feature; characterized by broad, flat-topped interstream divides, numerous swamps and shallow lakes, and progressive increase of local relief. Also known as topographic youth. { yüth }

Yucatán Current [OCEANOGR] A rapid northward flowing current along the western side of the Yucatán Strait; generally loops to the north and exits as the Florida Current. { ¦yü·kə¦tän 'kə·rənt }

Y

Z

zastruga *See* sastruga. { 'zas·trə·gə }

Zechstein [GEOL] A European series of geologic time, especially in Germany: Upper Permian (above Rothliegende). { 'zek,shtīn }

Zemorrian [GEOL] A North American stage of Oligocene and Miocene geologic time, above Refugian and below Saucesian. { zə'mȯr·ē·ən }

zenithal rain [METEOROL] In the tropics or subtropics, the rainy season which recurs annually or semiannually at about the time that the sun is most nearly overhead (at zenith). { 'zē·nə·thəl 'rān }

zeolitization [GEOL] Introduction of or replacement by a zeolite mineral. { zē,äl·əd·ə'zā·shən }

zephyr [METEOROL] Any soft, gentle breeze. { 'zef·ər }

zero curtain [GEOL] The layer of ground between the active layer and permafrost where the temperature remains nearly constant at 0°C. { 'zir·ō ,kərt·ən }

zero layer [OCEANOGR] A reference level in the ocean, at which horizontal motion is at a minimum. { 'zir·ō ,lā·ər }

zeugogeosyncline [GEOL] A geosyncline in a craton or stable area, within which is also an uplifted area, receiving clastic sediments. Also known as yoked basin. { ¦zü·gō¦jē·ō'sin,klīn }

zigzag lightning [GEOPHYS] Ordinary lightning of a cloud-to-ground discharge that appears to have a single, but very irregular, lightning channel; when viewed from a suitable angle, this may be observed as beaded lightning. { 'zig,zag ,līt·niŋ }

zobaa [METEOROL] In Egypt, a lofty whirlwind of sand resembling a pillar, moving with great velocity. { zō'bä }

zodiacal cone *See* zodiacal pyramid. { zō'dī·ə·kəl 'kōn }

zodiacal light [GEOPHYS] A diffuse band of luminosity occasionally visible on the ecliptic; it is sunlight diffracted and reflected by dust particles in the solar system within and beyond the orbit of the earth. { zō'dī·ə·kəl 'līt }

zodiacal pyramid [GEOPHYS] The pattern formed by the zodiacal light. Also known as zodiacal cone. { zō'dī·ə·kəl 'pir·ə·mid }

zonal [METEOROL] Latitudinal, easterly or westerly, opposed to meridional. { 'zōn·əl }

zonal circulation *See* zonal flow. { 'zōn·əl ,sər·kyə'lā·shən }

zonal flow [METEOROL] The flow of air along a latitude circle; more specifically, the latitudinal (east or west) component of existing flow. Also known as zonal circulation. { 'zōn·əl 'flō }

zonal index [METEOROL] A measure of strength of the middle-latitude westerlies, expressed as the horizontal pressure difference between 35° and 55° latitude, or as the corresponding geostrophic wind. { 'zōn·əl 'in,deks }

zonal kinetic energy [METEOROL] The kinetic energy of the mean zonal wind, obtained by averaging the zonal component of the wind along a fixed latitude circle. { 'zōn·əl ki'ned·ik 'en·ər·jē }

zonal soil [GEOL] In early classification systems in the United States, a soil order including soils with well-developed characteristics that reflect the influence of agents of soil genesis. Also known as mature soil. { 'zōn·əl 'sȯil }

zonal theory [GEOL] A theory of the formation of mineral deposition and sequence patterns, based on the changes in a mineral-bearing fluid as it passes upward from a magmatic source. { 'zōn·əl 'thē·ə·rē }
zonal westerlies *See* westerlies. { 'zōn·əl 'wes·tər·lēz }
zonal wind [METEOROL] The wind, or wind component, along the local parallel of latitude, as distinguished from the meridional wind. { 'zōn·əl 'wind }
zonal winds *See* westerlies. { 'zōn·əl 'winz }
zonal wind-speed profile [METEOROL] A diagram in which the speed of the zonal flow is one coordinate and latitude the other. { 'zōn·əl 'win ˌspēd ˌprōˌfīl }
zonation [GEOL] The condition of being arranged in zones. { zō'nā·shən }
zonda [METEOROL] A hot wind in Argentina. { 'zän·də }
zone [GEOGR] An area or region of latitudinal character. [GEOL] A belt, layer, band, or strip of earth material such as rock or soil. { zōn }
zone of accumulation *See* B horizon. { 'zōn əv əˌkyü·mə'lā·shən }
zone of aeration [GEOL] The subsurface sediment above the water table containing air and water. Also known as unsaturated zone; vadose zone; zone of suspended water. { 'zōn əv e'rā·shən }
zone of cementation [GEOL] The layer of the earth's crust in which unconsolidated deposits are cemented by percolating water containing dissolved minerals from the overlying zone of weathering. Also known as belt of cementation. { 'zōn əv ˌsēˌmen'tā·shən }
zone of illuviation *See* B horizon. { 'zōn əv iˌlü·vē'ā·shən }
zone of maximum precipitation [METEOROL] In a mountain region, the belt of elevation at which the annual precipitation is greatest. { 'zōn əv 'mak·sə·məm priˌsip·ə'tā·shən }
zone of saturation [HYD] A subsurface zone in which water fills the interstices and is under pressure greater than atmospheric pressure. Also known as phreatic zone; saturated zone. { 'zōn əv ˌsach·ə'rā·shən }
zone of soil water *See* belt of soil water. { 'zōn əv 'sȯil ˌwȯd·ər }
zone of suspended water *See* zone of aeration. { 'zōn əv sə¦spen·dəd ˌwȯd·ər }
Zwischengebirge *See* median mass. { 'tsfish·ənˌgə'bir·gə }

Appendix

Appendix

Equivalents of commonly used units for the U.S. Customary System and the metric system

1 inch = 2.5 centimeters (25 millimeters)	1 centimeter = 0.4 inch	1 inch = 0.083 foot
1 foot = 0.3 meter (30 centimeters)	1 meter = 3.3 feet	1 foot = 0.33 yard (12 inches)
1 yard = 0.9 meter	1 meter = 1.1 yards	1 yard = 3 feet (36 inches)
1 mile = 1.6 kilometers	1 kilometer = 0.62 mile	1 mile = 5280 feet (1760 yards)
1 acre = 0.4 hectare	1 hectare = 2.47 acres	
1 acre = 4047 square meters	1 square meter = 0.00025 acre	
1 gallon = 3.8 liters	1 liter = 1.06 quarts = 0.26 gallon	1 quart = 0.25 gallon (32 ounces; 2 pints)
1 fluid ounce = 29.6 milliliters	1 milliliter = 0.034 fluid ounce	1 pint = 0.125 gallon (16 ounces)
32 fluid ounces = 946.4 milliliters		1 gallon = 4 quarts (8 pints)
1 quart = 0.95 liter	1 gram = 0.035 ounce	1 ounce = 0.0625 pound
1 ounce = 28.35 grams	1 kilogram = 2.2 pounds	1 pound = 16 ounces
1 pound = 0.45 kilogram	1 kilogram = 1.1×10^{-3} ton	1 ton = 2000 pounds
1 ton = 907.18 kilograms		
°F = (1.8 × °C) + 32	°C = (°F − 32) ÷ 1.8	

Appendix

Conversion factors for the U.S. Customary System, metric system, and International System

A. Units of length

Units	*cm*	*m*	*in.*	*ft*	*yd*	*mi*
1 cm	= 1	0.01	0.3937008	0.03280840	0.01093613	6.213712×10^{-6}
1 m	= 100.	1	39.37008	3.280840	1.093613	6.213712×10^{-4}
1 in.	= 2.54	0.0254	1	0.08333333...	0.02777777...	1.578283×10^{-5}
1 ft	= 30.48	0.3048	12.	1	0.3333333...	$1.893939... \times 10^{-4}$
1 yd	= 91.44	0.9144	36.	3.	1	$5.681818... \times 10^{-4}$
1 mi	= 1.609344×10^{5}	1.609344×10^{3}	6.336×10^{4}	5280.	1760.	1

B. Units of area

Units	cm^2	m^2	$in.^2$	ft^2	yd^2	mi^2
1 cm^2	= 1	10^{-4}	0.1550003	1.076391×10^{-3}	1.195990×10^{-4}	3.861022×10^{-11}
1 m^2	= 10^{4}	1	1550.003	10.76391	1.195990	3.861022×10^{-7}
1 $in.^2$	= 6.4516	6.4516×10^{-4}	1	$6.944444... \times 10^{-3}$	7.716049×10^{-4}	2.490977×10^{-10}
1 ft^2	= 929.0304	0.09290304	144.	1	0.1111111...	3.587007×10^{-8}
1 yd^2	= 8361.273	0.8361273	1296.	9.	1	3.228306×10^{-7}
1 mi^2	= 2.589988×10^{10}	2.589988×10^{6}	4.014490×10^{9}	2.78784×10^{7}	3.0976×10^{6}	1

C. Units of volume

Units	m^3	cm^3	liter	$in.^3$	ft^3	qt	gal
1 m^3	= 1	10^6	10^3	6.102374×10^4	35.31467×10^{-3}	1.056688	264.1721
1 cm^3	= 10^{-6}	1	10^{-3}	0.06102374	3.531467×10^{-5}	1.056688×10^{-3}	2.641721×10^{-4}
1 liter	= 10^{-3}	1000.	1	61.02374	0.03531467	1.056688	0.2641721
1 $in.^3$	= 1.638706×10^{-5}	16.38706	0.01638706	1	5.787037×10^{-4}	0.01731602	4.329004×10^{-3}
1 ft^3	= 2.831685×10^{-2}	28316.85	28.31685	1728.	1	2.992208	7.480520
1 qt	= 9.463529×10^{-4}	946.3529	0.9463529	57.75	0.03342014	1	0.25
1 gal (U.S.)	= 3.785412×10^{-3}	3785.412	3.785412	231.	0.1336806	4.	1

D. Units of mass

Units	g	kg	oz	lb	metric ton	ton
1 g	= 1	10^{-3}	0.03527396	2.204623×10^{-3}	10^{-6}	1.102311×10^{-6}
1 kg	= 1000.	1	35.27396	2.204623	10^{-3}	1.102311×10^{-3}
1 oz (avdp)	= 28.34952	0.02834952	1	0.0625	2.834952×10^{-5}	3.125×10^{-5}
1 lb (avdp)	= 453.5924	0.4535924	16.	1	4.535924×10^{-4}	$5. \times 10^{-4}$
1 metric ton	= 10^8	1000.	35273.96	2204.623	1	1.102311
1 ton	= 907184.7	907.1847	32000.	2000.	0.9071847	1

Conversion factors for the U.S. Customary System, metric system, and International System (*cont.*)

E. Units of density

Units	$g \cdot cm^{-3}$	$g \cdot L^{-1}$, $kg \cdot m^{-3}$	$oz \cdot in.^{-3}$	$lb \cdot in.^{-3}$	$lb \cdot ft^{-3}$	$lb \cdot gal^{-1}$
1 $g \cdot cm^{-3}$	= 1	1000.	0.5780365	0.03612728	62.42795	8.345403
1 $g \cdot L^{-1}$, $kg \cdot m^{-3}$	= 10^{-3}	1	5.780365×10^{-4}	3.612728×10^{-5}	0.06242795	8.345403×10^{-3}
1 $oz \cdot in.^{-3}$	= 1.729994	1729.994	1	0.0625	108.	14.4375
1 $lb \cdot in.^{-3}$	= 27.67991	27679.91	16.	1	1728.	231.
1 $lb \cdot ft^{-3}$	= 0.01601847	16.01847	9.259259×10^{-3}	5.787037×10^{-4}	1	0.1336806
1 $lb \cdot gal^{-1}$	= 0.1198264	119.8264	4.749536×10^{-3}	4.329004×10^{-3}	7.480519	1

F. Units of pressure

Units	Pa, $N \cdot m^{-2}$	$dyn \cdot cm^{-2}$	*bar*	*atm*	$kgf \cdot cm^{-2}$	mmHg (torr)	in. Hg	$lbf \cdot in.^{-2}$
1 Pa, 1 $N \cdot m^{-2}$	= 1	10	10^{-5}	9.869233×10^{-6}	1.019716×10^{-5}	7.500617×10^{-3}	2.952999×10^{-4}	1.450377×10^{-4}
1 $dyn \cdot cm^{-2}$	= 0.1	1	10^{-6}	9.869233×10^{-7}	1.019716×10^{-6}	7.500617×10^{-4}	2.952999×10^{-5}	1.450377×10^{-5}
1 bar	= 10^{5}	10^{6}	1	0.9869233	1.019716	750.0617	29.52999	14.50377
1 atm	= 101325	1013250	1.01325	1	1.033227	760.	29.92126	14.69595
1 $kgf \cdot cm^{-2}$	= 98066.5	980665	0.980665	0.9678411	1	735.5592	28.95903	14.22334
1 mmHg (torr)	= 133.3224	1333.224	1.333224×10^{3}	1.315789×10^{-3}	1.359510×10^{-3}	1	0.03937008	0.01933678
1 in. Hg	= 3386.388	33863.88	0.03386388	0.03342105	0.03453155	25.4	1	0.4911541
1 $lbf \cdot in.^{-2}$	= 6894.757	68947.57	0.06894757	0.06804596	0.07030696	51.71493	2.036021	1

G. Units of energy

Units	g mass (energy equiv)	J	eV	cal	cal_{IT}	Btu_{IT}	kWh	hp-h	ft-lbf	$ft^3 \cdot lbf \cdot in.^{-2}$	liter-atm
1 g mass (energy equiv)	= 1	8.987552×10^{13}	5.609589×10^{32}	2.148076×10^{3}	2.146640×10^{13}	8.518555×10^{10}	2.496542×10^{7}	3.347918×10^{7}	6.628878×10^{13}	4.603388×10^{11}	8.870024×10^{11}
1 J	$= 1.112650 \times 10^{-14}$	1	6.241510×10^{18}	0.2390057	0.2388459	9.478172×10^{-4}	$2.777777... \times 10^{-7}$	3.725062	0.7375622	5.121960×10^{-3}	9.869233×10^{-3}
1 eV	$= 1.782662 \times 10^{-33}$	1.602176×10^{-19}	1	3.829293×10^{-20}	3.826733×10^{-20}	1.518570×10^{-22}	4.450490×10^{-26}	5.968206×10^{-26}	1.181705×10^{-19}	8.206283×10^{-22}	1.581225×10^{-21}
1 cal	$= 4.655328 \times 10^{-14}$	4.184	2.611448×10^{19}	1	0.9993312	3.965667×10^{-3}	$1.1622222... \times 10^{-6}$	1.558562×10^{-6}	3.085960	2.143028×10^{-2}	0.04129287
1 cal_{IT}	$= 4.658443 \times 10^{-14}$	4.1868	2.613195×10^{19}	1.000669	1	3.968321×10^{-3}	1.163×10^{-6}	1.559609×10^{-6}	3.088025	2.144462×10^{-2}	0.04132050
1 Btu_{IT}	$= 1.173908 \times 10^{-11}$	1055.056	6.585141×10^{21}	252.1644	251.9958	1	2.930711×10^{-4}	3.930148×10^{-4}	778.1693	5.403953	10.41259
1 kWh	$= 4.005540 \times 10^{-8}$	3600000.	2.246944×10^{25}	860420.7	859845.2	3412.142	1	1.341022	2655224.	18349.06	35529.24
1 hp-h	$= 2.986931 \times 10^{-8}$	2384519.	1.675545×10^{25}	641615.6	641186.5	2544.33	0.7456998	1	1980000.	13750.	26494.15
1 ft-lbf	$= 1.508551 \times 10^{-14}$	1.355818	8.462351×10^{18}	0.3240483	0.3238315	1.285067×10^{-3}	3.766161×10^{-7}	$5.050505... \times 10^{-7}$	1	$6.944444... \times 10^{-3}$	0.01338088
1 ft^3 lbf · $in.^{-2}$	$= 2.172313 \times 10^{-12}$	195.2378	1.218579×10^{21}	46.66295.	46.63174	0.1850497	5.423272×10^{-5}	$7.272727... \times 10^{-5}$	144.	1	1.926847
1 liter-atm	$= 1.127393 \times 10^{-12}$	101.325	6.324210×10^{20}	24.21726	24.20106	0.09603757	2.814583×10^{-5}	3.774419×10^{-5}	74.73349	0.5189825	1

Appendix

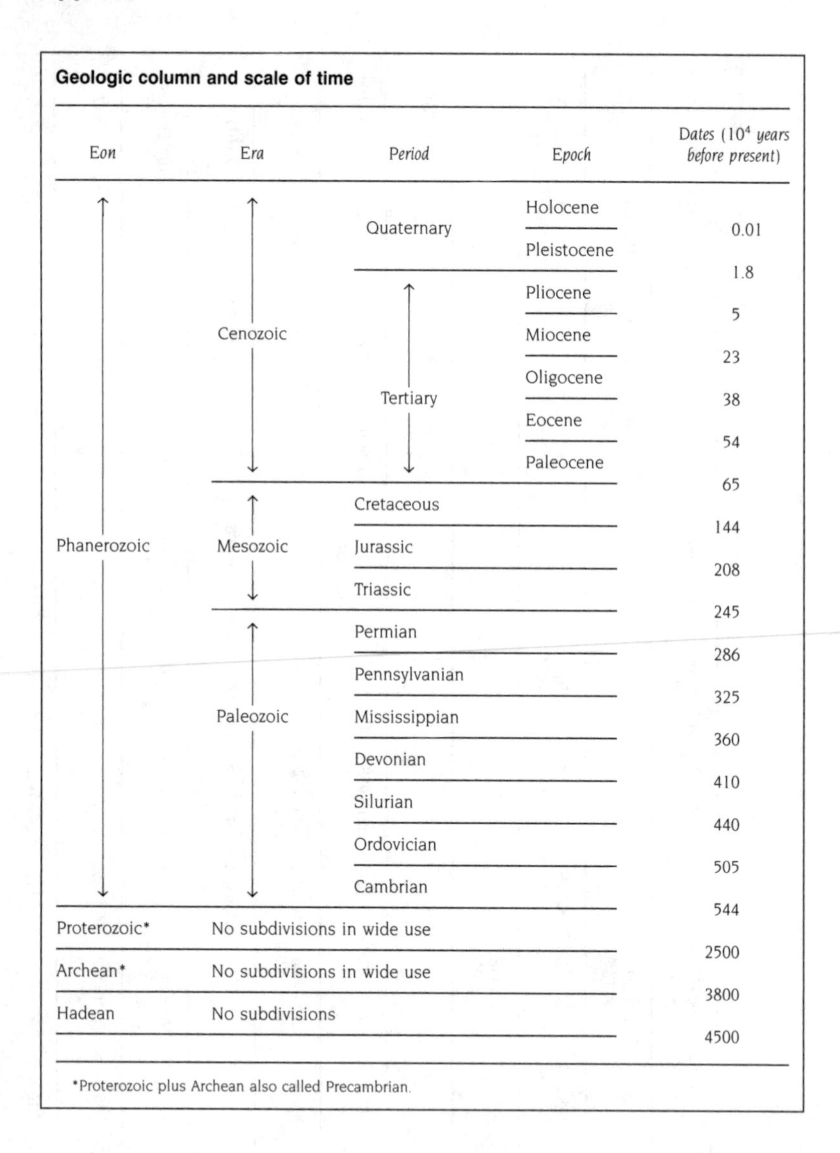

Geologic column and scale of time

Eon	*Era*	*Period*	*Epoch*	*Dates (10^4 years before present)*
Phanerozoic	Cenozoic	Quaternary	Holocene	0.01
			Pleistocene	1.8
		Tertiary	Pliocene	5
			Miocene	23
			Oligocene	38
			Eocene	54
			Paleocene	65
	Mesozoic	Cretaceous		144
		Jurassic		208
		Triassic		245
	Paleozoic	Permian		286
		Pennsylvanian		325
		Mississippian		360
		Devonian		410
		Silurian		440
		Ordovician		505
		Cambrian		544
Proterozoic*	No subdivisions in wide use			2500
Archean*	No subdivisions in wide use			3800
Hadean	No subdivisions			4500

*Proterozoic plus Archean also called Precambrian.

Appendix

Some historical volcanic eruptions

Volcano	*Year*	*Estimated casualties*	*Principal causes of death*
Merapi (Indonesia)	1006	>1000	Explosions
Kelut (Indonesia)	1586	10,000	Lahars (mudflows)
Vesuvius (Italy)	1631	18,000	Lava flows, mudflows
Etna (Italy)	1669	10,000	Lava flows, explosions
Merapi (Indonesia)	1672	>300	Nuées ardentes, lahars
Awu (Indonesia)	1711	3,200	Lahars
Papandayan (Indonesia)	1772	2,957	Explosions
Laki (Iceland)	1783	10,000	Lava flows, volcanic gas, starvation*
Asama (Japan)	1783	1,151	Lava flows, lahars
Unzen (Japan)	1792	15,000	Lahars, tsunami
Mayon (Philippines)	1814	1,200	Nuées ardentes, lava flows
Tambora (Indonesia)	1815	92,000	Starvation*
Galunggung (Indonesia)	1822	4,000	Lahars
Awu (Indonesia)	1856	2,800	Lahars
Krakatau (Indonsia)	1883	36,000	Tsunami
Awu (Indonesia)	1892	1,500	Nuées ardentes, lahars
Mont Pelée, Martinique (West Indies)	1902	36,000	Nuées ardentes
Soufrière, St. Vincent (West Indies)	1902	1,565	Nuées ardentes
Taal (Philippines)	1911	1,332	Explosions
Kelut (Indonesia)	1919	5,000	Lahars
Lamington (Papua New Guinea)	1951	3,000	Nuées ardentes, explosions
Merapi (Indonesia)	1951	1,300	Lahars
Agung (Indonesia)	1963	3,800	Nuées ardentes, lahars
Taal (Philippines)	1965	350	Explosions
Mount St. Helens (United States)	1980	57	Lateral blast, mudflows
El Chichón (Mexico)	1982	>2,000	Explosions, nuées ardentes
Nevado del Ruiz (Colombia)	1985	>25,000	Mudflows
Unzen (Japan)	1991	41	Nuées ardentes
Pinatubo (Philippines)	1991	>300	Nuées ardentes, mudflows, ash fall (roof collapse)
Merapi (Indonesia)	1994	>41	Nuées ardentes from dome collapse
Soufrière Hills, Montserrat (West Indies)	1997	19	Nuées ardentes

*Deaths directly attributable to the destruction or reduction of food crops, livestock, agricultural lands, pasturage, and other disruptions of food chain.

Appendix

Principal regions of a standard earth model

Layer	Approximate depth range, mi (km)
Ocean layer	0–1.8 (0–3)
Upper and lower crust	1.8–15 (3–24)
Lithosphere below the crust	15–50 (24–80)
Asthenosphere	50–140 (80–220)
Upper mantle above phase or compositional changes near 240 and 416 mi (400 and 670 km)	240–416 (220–400)
Transition region between phase or compositional changes near 240 and 416 (400 and 670 km)	240–416 (400–670)
Lower mantle above core-mantle boundary layer	416–1703 (670–2741)
Core-mantle boundary layer	1703–1796 (2741–2891)
Outer core	1796–3200 (2891–5150)
Inner core	3200–3959 (5150–6371)

Physical properties of some common rocks

Rock	Specific gravity	Porosity, %	Compressive strength, lb/in.^{2}	Tensile strength, lb/in.^{2}
Igneous				
Granite	2.67	1	30,000–50,000	500–1000
Basalt	2.75	1	25,000–30,000	
Sedimentary				
Sandstone	2.1–2.5	5–30	5,000–15,000	100–200
Shale	1.9–2.4	7–25	5,000–10,000	
Limestone	2.2–2.5	2–20	2,000–20,000	400–850
Metamorphic				
Marble	2.5–2.8	0.5–2	10,000–30,000	700–1000
Quartzite	2.5–2.6	1–2	15,000–40,000	
Slate	2.6–2.8	0.5–5	15,000–30,000	

Appendix

Approximate concentration of ore elements in earth's crust and in ores

Element	*In average igneous rocks, %*	*In ores, %*
Iron	5.0	50
Copper	0.007	0.5–5
Zinc	0.013	1.3–13
Lead	0.0016	1.6–16
Tin	0.004	0.01*–1
Silver	0.00001	0.05
Gold	0.0000005	0.0000015*–0.01
Uranium	0.0002	0.2
Tungsten	0.003	0.5
Molybdenum	0.001	0.6

*Placer deposits.

Soil orders

Order	*Formative element in name*	*General nature*
Alfisols	alf	Soils with gray to brown surface horizons, medium to high base supply, with horizons of clay accumulation; usually moist, but may be dry during summer
Aridisols	id	Soils with pedogenic horizons, low in organic matter, and usually dry
Entisols	ent	Soils without pedogenic horizons
Histosols	ist	Organic soils (peats and mucks)
Inceptisols	ept	Soils that are usually moist, with pedogenic horizons of alteration of parent materials but not of illuviation
Mollisols	oll	Soils with nearly black, organic-rich surface horizons and high base supply
Oxisols	ox	Soils with residual accumulations of inactive clays, free oxides, kaolin, and quartz; mostly tropical
Spodosols	od	Soils with accumulations of amorphous materials in subsurface horizons
Ultisols	ult	Soils that are usually moist, with horizons of clay accumulation and a low supply of bases
Vertisols	ert	Soils with high content of swelling clays and wide deep cracks during some season

Appendix

Elemental composition of earth's crust based on igneous and sedimentary rock

Element	*Weight %*	*Atomic %*	*Volume %*
Oxygen	46.71	60.5	94.24
Silicon	27.69	20.5	0.51
Titanium	0.62	0.3	0.03
Aluminum	8.07	6.2	0.44
Iron	5.05	1.9	0.37
Magnesium	2.08	1.8	0.28
Calcium	3.65	1.9	1.04
Sodium	2.75	2.5	1.21
Potassium	2.58	1.4	1.88
Hydrogen	0.14	3.0	

World's estimated water supply

Location	*Surface area, mi² (km²)*	*Water volume, mi³ (km³)*	*Percentage of total water*
Surface water			
Fresh-water lakes	330,000 (855,000)	30,000 (130,0000)	0.009
Saline lakes and inland seas	270,000 (700,000)	25,000 (104,000)	0.008
Average in stream channels	—	300 (1300)	0.0001
Subsurface water			
Vadose water (includes soil moisture)		16,000 (67,000)	0.005
Groundwater within depth of a half mile	50,000,000 (130,000,000)	1,000,000 (4,200,000)	0.31
Groundwater, deep-lying		1,000,000 (4,200,000)	0.31
Other water locations			
Ice caps and glaciers	6,900,000 (18,000,000)	7,000,000 (29,000,000)	2.15
Atmosphere (at sea level)	197,000,000 (510,000,000)	3,100 (12,900)	0.001
World ocean	139,500,000 (361,300,000)	317,000,000 (1,321,000,000)	97.2
TOTALS (rounded)		326,000,000 (1,360,000,000)	100

Cloud classification based on air motion and associated physical characteristics

Kind of motion	*Kind of cloud*	*Name*	*Characteristic precipitation*
Widespread slow ascent, associated with cyclones (stable atmosphere)	Thick layers	Cirrus, later becoming: cirrostratus, altostratus, altocumulus	Snow trails
		nimbostratus	Prolonged moderate rain or snow
Convection, due to passage over warm surface (unstable atmosphere)	Small heap cloud	Cumulus	None
	Shower- and thundercloud	Cumulonimbus	Intense showers of rain or hail
Irregular stirring causing cooling during passage over cold surface (stable atmosphere)	Shallow low layer clouds, fogs	Stratus Stratocumulus	None, or slight drizzle or snow

Appendix

Simplified classification of major igenous rocks on the basis of composition and texture

	SiO_2-rich (acidic) ←		→		SiO_2-poor (basic)
	Light colored ←	Gray	Dark colored	→	Black
Mineral composition:	Quartz, potash feldspar, biotite	Potash feldspar, biotite, or amphibole	Sodic plagioclase, hornblende, or augite	Augite, olivine, hypersthene, calcic plagioclase	Olivine, enstatite, augite
INTRUSIVE					
Medium-grained	Granite	Syenite	Diorite	Gabbro	Peridotite
EXTRUSIVE					
Fine-grained to aphanitic	Rhyolite	Trachyte	Andesite	Basalt	
	←	Felsite	→		
Porphyritic	Rhyolite porphyry	Trachyte porphyry	Andesite porphyry	Basalt porphyry	
Glassy	Obsidian				
Vesicular	Pumice			Scoria	
Fragmental		Tuff and agglomerate of each type			

Appendix

Average chemical compositions of igneous rocks (totals reduced to 100%)

	Plutonic rocks							
Components	*Granite*	*Grano-diorite*	*Quartz diorite*	*Syenite*	*Monzonite*	*Diorite*	*Gabbro*	*Nepheline syenite*
SiO_2	70.18	65.01	61.59	60.19	56.12	56.77	48.24	54.63
TiO_2	0.39	0.57	0.66	0.67	1.10	0.84	0.97	0.86
Al_2O_3	14.47	15.94	16.21	16.28	16.96	16.67	17.88	19.89
Fe_2O_3	1.57	1.74	2.54	2.74	2.93	3.16	3.16	3.37
FeO	1.78	2.65	3.77	3.28	4.01	4.40	5.95	2.20
MnO	0.12	0.07	0.10	0.14	0.16	0.13	0.13	0.35
MgO	0.88	1.91	2.80	2.49	3.27	4.17	7.51	0.87
CaO	1.99	4.42	5.38	4.30	6.50	6.74	10.99	2.51
Na_2O	3.48	3.70	3.37	3.98	3.67	3.39	2.55	8.26
K_2O	4.11	2.75	2.10	4.49	3.76	2.12	0.89	5.46
H_2O	0.84	1.04	1.22	1.16	1.05	1.36	1.45	1.35
P_2O_3	0.19	0.20	0.26	0.28	0.47	0.25	0.28	0.25

	Aphanitic rocks							
Components	*Rhyolite*	*Quartz latite*	*Dacite*	*Trachyte*	*Latite*	*Andesite*	*Basalt*	*Phonolite*
SiO_2	72.80	62.43	65.68	60.68	57.65	59.59	49.06	57.45
TiO_2	0.33	0.85	0.57	0.38	1.00	0.77	1.36	0.41
Al_2O_3	13.49	16.15	16.25	17.74	16.68	17.31	15.70	20.60
Fe_2O_3	1.45	4.04	2.38	2.64	2.29	3.33	5.38	2.35
FeO	0.88	1.20	1.90	2.62	4.07	3.13	6.37	1.03
MnO	0.08	0.09	0.06	0.06	0.10	0.18	0.31	0.13
MgO	0.38	1.74	1.41	1.12	3.22	2.75	6.17	0.30
CaO	1.20	4.24	3.46	3.09	5.74	5.80	8.95	1.50
Na_2O	3.38	3.34	3.97	4.43	3.59	3.58	3.11	8.84
K_2O	4.46	3.75	2.67	5.74	4.39	2.04	1.52	5.23
H_2O	1.47	1.90	1.50	1.26	0.91	1.26	1.62	2.04
P_2O_5	0.08	0.27	0.15	0.24	0.36	0.26	0.45	0.12

Appendix

Dimensions of some major lakes

Lake	Area, mi^2 (km^2)	Volume (approx.), 10^3 acre-ft (10^6 m^3)	Shoreline, mi (km)	Depth	
				Av., ft (m)	Max., ft (m)
Caspian Sea	168,500 (436,412)	71,300 (87,947)	3,730 (6,003)	675 (207)	3,080 (939)
Superior	32,200 (83,398)	9,700 (11,965)	1,860 (2,993)	475 (145)	1,000 (305)
Victoria	26,200 (67,858)	2,180 (2,689)	2,130 (3,428)		
Aral Sea	26,233 (67,943)*	775 (956)			
Huron	23,010 (59,596)	3,720 (4,589)	1,680 (2,704)		
Michigan	22,400 (58,016)	4,660 (5,748)			870 (256)
Baikal	13,300 (34,447)*	18,700 (23,066)		2,300 (701)	5,000 (1,524)
Tanganyika	12,700 (32,893)	8,100 (9,991)			4,700 (1,433)
Great Bear	11,490 (29,759)*		1,300 (2,092)		
Great Slave	11,170 (28,930)*		1,365 (2,197)		
Nyasa	11,000 (28,490)	6,800 (8,388)		900 (274)	2,310 (704)
Erie	9,940 (25,744)	436 (538)			
Winnipeg	9,390 (24,320)*		1,180 (1,899)		
Ontario	7,540 (19,529)	1,390 (1,714)			
Balkash	7,115 (18,428)				
Ladoga	7,000 (18,130)	745 (919)			
Chad	6,500 (16,835)*				
Maracaibo	4,000 (10,360)*				
Eyre	3,700 (9,583)*				
Onega	3,764 (9,749)	264 (326)			
Rudolf	3,475 (9,000)*				
Nicaragua	3,089 (8,000)	87 (107)			
Athabaska	3,085 (7,990)				
Titicaca	3,200 (8,288)	575 (709)			
Reindeer	2,445 (6,332)				

*Area fluctuates.

Appendix

Characteristics of some of the world's major rivers

River	*Average discharge, ft^3/s (m^3/s)*	*Drainage area, 10^3 mi^2 (10^3 km^2)*	*Average annual sediment load, 10^3 tons (10^3 metric tons)*	*Length, mi (km)*
Amazon	6,390,000 (181,000)	2770 (7180)	990,000 (900,000)	3899 (6275)
Congo	1,400,000 (39,620)	1420 (3690)	71,300 (64,680)	2901 (4670)
Orinoco	800,000 (22,640)	571 (1480)	93,130 (86,490)	1600 (2570)
Yangtze	770,000 (21,790)	749 (1940)	610,000 (550,000)	3100 (4990)
Brahmaputra	706,000 (19,980)	361 (935)	880,000 (800,000)	1700 (2700)
Mississippi-Missouri	630,000 (17,830)	1240 (3220)	379,000 (344,000)	3890 (6260)
Yenisei	614,000 (17,380)	1000 (2590)	11,600 (10,520)	3550 (5710)
Lena	546,671 (15,480)	1170 (3030)	—	2900 (4600)
Mekong	530,000 (15,000)	350 (910)	206,850 (187,650)	2600 (4180)
Parana	526,000 (14,890)	1200 (3100)	90,000 (81,650)	2450 (3940)
St. Lawrence	500,000 (14,150)	564 (1460)	4,000 (3,630)	2150 (3460)
Ganges	497,600 (14,090)	451 (1170)	1,800,000 (1,600,000)	1640 (2640)
Irrawaddy	478,900 (13,560)	140 (370)	364,070 (330,280)	1400 (2300)
Ob	440,700 (12,480)	1000 (2590)	15,700 (14,240)	2800 (4500)
Volga	350,000 (9,900)	591 (1530)	20,780 (18,840)	2320 (3740)
Amur	338,000 (9,570)	788 (2040)	—	2900 (4670)

Appendix

The 100 highest mountain peaks

Mountain peak	Range	Location	Height	
			Feet	Meters
Everest	Himalayas	Nepal-China	29,028	8,848
K2 (Godwin Austen)	Karakoram	Kashmir	28,250	8,611
Kanchenjunga	Himalayas	Nepal-India	28,208	8,598
Lhotse I	Himalayas	Nepal-China	27,923	8,511
Makalu I	Himalayas	Nepal-China	27,824	8,481
Lhatose II	Himalayas	Nepal-China	27,560	8,400
Dhaulagri	Himalayas	Nepal	26,810	8,172
Manaslu I	Himalayas	Nepal	26,760	8,156
Cho Oyu	Himalayas	Nepal-China	26,750	8,153
Nanga Parbat	Himalayas	Kashmir	26,660	8,126
Annapurna	Himalayas	Nepal	26,504	8,078
Gasherbrum	Karakoram	Kashmir	26,470	8,068
Broad	Karakoram	Kashmir	26,400	8,047
Gosainthan	Himalayas	China	26,287	8,012
Annapura II	Himalayas	Nepal	26,041	7,937
Gyachung Kang	Himalayas	Nepal-China	25,910	7,897
Disteghil Sar	Karakoram	Kashmir	25,858	7,882
Hinalchuli	Himalayas	Nepal	25,801	7,864
Nuptse	Himalayas	Nepal-China	25,726	7,841
Masherbrum	Karakoram	Kashmir	25,660	7,821
Nandi Devi	Himalayas	India	25,645	7,817
Rakaposhi	Karakoram	Kashmir	25,550	7,788
Kanjut Sar	Karakoram	Kashmir	25,461	7,761
Kamet	Himalayas	India-China	25,447	7,756
Namcha Barwa	Himalayas	China	25,445	7,756
Gurla Manghata	Himalayas	China	25,355	7,728
Ulugh Mustagh	Kunlun	China	25,340	7,724
Kungur	Mustagh Ata	China	25,325	7,719
Tirich Mir	Hindu Kush	Pakistan	25,230	7,690
Saser Kangri	Karakoram	Kashmir	25,172	7,672

Makalu II	Himalayas	Nepal-China	25,120	7,657
Conggashan	Daxue Shan	China	24,900	7,590
Kula Kangri	Himalayas	Bhuran-China	24,784	7,554
Chang-tzu	Himalayas	Nepal-China	24,780	7,533
Muztagh Ata	Muztagh Ata	China	24,757	7,546
Skyang Kangri	Himalayas	Kashmir	24,750	7,544
Ismail Samani Peak	Pamirs	Tajikistan	24,590	7,495
Jongsong Peak	Himalayas	Nepal-India	24,472	7,459
Pobeda Peak	Tian Shan	Kyrgyzstan	24,406	7,439
Sia Kangri	Himalayas	Kashmir	24,350	7,422
Haramosh Peak	Karakoram	Kashmir	24,270	7,397
Istoro Nal	Hindu Kush	Pakistan	24,240	7,388
Tent Peak	Himalayas	Nepal-India	24,165	7,365
Chomo Lhari	Himalayas	Bhutan-China	24,040	7,327
Chamlang	Himalayas	Nepal	24,012	7,319
Kabru	Himalayas	Nepal-India	24,002	7,316
Alung Gangri	Himalayas	China	24,000	7,315
Baltoro Knagri	Himalayas	Kashmir	23,990	7,312
Muztagh Ata	Kunlun	China	23,890	7,282
Mana	Himalayas	India	23,860	7,273
Baruntse	Himalayas	Nepal	23,688	7,220
Nepal Peak	Himalayas	Nepal-India	23,500	7,163
Amne Machin	Kunlun	China	23,490	7,160
Gauri Sankar	Himalayas	Nepal-China	23,440	7,145
Badrinath	Himalayas	India	23,420	7,138
Nunkun	Himalayas	Kashmir	23,410	7,135
Lenin Peak	Pamirs	Tajikistan/Kyrgyzstan	23,405	7,134
Pyramid	Himalayas	Nepal-India	23,400	7,132
Api	Himalayas	Nepal	23,399	7,132
Pauhunri	Himalayas	India-China	23,385	7,128
Trisul	Himalayas	India	23,360	7,120
Korzhenevski Peak	Pamirs	Tajikistan	23,310	7,105
Kangto	Himalayas	India-China	23,260	7,090
Nyainqentanglha	Nyainqentanglha Shan	China	23,255	7,088
Trisuli	Himalayas	India	23,210	7,074
Dunagiri	Himalayas	India	23,184	7,066
Revolution Peak	Pamirs	Tajikistan	22,880	6,974

The 100 highest mountain peaks (cont.)

Mountain peak	Range	Location	Height	
			Feet	Meters
Aconcagua	Andes	Argentina	22,834	6,960
Ojos del Salado	Andes	Argentina-Chile	22,572	6,880
Bonete	Andes	Argentina	22,546	6,872
Tupungato	Andes	Argentina-Chile	22,310	6,800
Moscow Peak	Pamirs	Tajikistan	22,260	6,785
Pissis	Andes	Argentina	22,241	6,779
Mercedario	Andes	Argentina	22,211	6,770
Huascaran	Andes	Peru	22,205	6,768
Llullaillaco	Andes	Argentina-Chile	22,057	6,723
El Libertador	Andes	Argentina	22,047	6,720
Cachi	Andes	Argentina	22,047	6,720
Kailas	Himalayas	China	22,027	6,714
Incahusai	Andes	Argentina-Chile	21,720	6,620
Yerupaja	Andes	Peru	21,709	6,617
Kurumda	Pamirs	Tajikistan	21,686	6,610
Galan	Andes	Argentina	21,654	6,600
El Muerto	Andes	Argentina-Chile	21,457	6,540
Sajama	Andes	Bolivia	21,391	6,520
Nacimiento	Andes	Argentina	21,302	6,493
Illimani	Andes	Bolivia	21,201	6,462
Coropuna	Andes	Peru	21,083	6,426
Laudo	Andes	Argentina	20,997	6,400
Ancohuma	Andes	Bolivia	20,958	6,388
Ausangate	Andes	Peru	20,945	6,384
Toro	Andes	Argentina-Chile	20,932	6,380
Illampu	Andes	Bolivia	20,873	6,362
Tres Cruces	Andes	Argentina-Chile	20,853	6,356
Huandoy	Andes	Peru	20,852	6,356
Parinacota	Andes	Bolivia-Chile	20,768	6,330
Tortolas	Andes	Argentina-Chile	20,745	6,323
Ampato	Andes	Peru	20,702	6,310
El Condor	Andes	Argentina	20,669	6,300
Salcantay	Andes	Peru	20,574	6,271